T0207316

Universitext

Universitext

Series Editors:

Sheldon Axler
San Francisco State University

Vincenzo Capasso
Università degli Studi di Milano

Carles Casacuberta
Universitat de Barcelona

Angus MacIntyre
Queen Mary University of London

Kenneth Ribet
University of California, Berkeley

Claude Sabbah
CNRS, École Polytechnique, Paris

Endre Süli
University of Oxford

Wojbor A. Woyczynski
Case Western Reserve University, Cleveland, OH

Universitext is a series of textbooks that presents material from a wide variety of mathematical disciplines at master's level and beyond. The books, often well class-tested by their author, may have an informal, personal even experimental approach to their subject matter. Some of the most successful and established books in the series have evolved through several editions, always following the evolution of teaching curricula, to very polished texts.

Thus as research topics trickle down into graduate-level teaching, first textbooks written for new, cutting-edge courses may make their way into *Universitext*.

More information about this series at http://www.springer.com/series/223

Paul A. Fuhrmann • Uwe Helmke

The Mathematics of
Networks of Linear Systems

Springer

Paul A. Fuhrmann
Department of Mathematics
Ben-Gurion University of the Negev
Beer-Sheva, Southern, Israel

Uwe Helmke
Universität Würzburg
Institute of Mathematics
Würzburg, Germany

ISSN 0172-5939   ISSN 2191-6675   (electronic)
Universitext
ISBN 978-3-319-16645-2   ISBN 978-3-319-16646-9   (eBook)
DOI 10.1007/978-3-319-16646-9

Library of Congress Control Number: 2015935740

Mathematics Subject Classification (2010): 14D21, 53C05, 55R10, 58A05, 78A25, 81T13, 93-XX, 11Cxx, 15Axx, 57M15, 11T71, 47–XX, 37–XX, 94–XX

Springer Cham Heidelberg New York Dordrecht London
© Springer International Publishing Switzerland 2015
This work is subject to copyright. All rights are reserved by the Publisher, whether the whole or part of the material is concerned, specifically the rights of translation, reprinting, reuse of illustrations, recitation, broadcasting, reproduction on microfilms or in any other physical way, and transmission or information storage and retrieval, electronic adaptation, computer software, or by similar or dissimilar methodology now known or hereafter developed.
The use of general descriptive names, registered names, trademarks, service marks, etc. in this publication does not imply, even in the absence of a specific statement, that such names are exempt from the relevant protective laws and regulations and therefore free for general use.
The publisher, the authors and the editors are safe to assume that the advice and information in this book are believed to be true and accurate at the date of publication. Neither the publisher nor the authors or the editors give a warranty, express or implied, with respect to the material contained herein or for any errors or omissions that may have been made.

Printed on acid-free paper

Springer International Publishing AG Switzerland is part of Springer Science+Business Media (www. springer.com)

To Julia and Nilly,
Amir, Galit and Oded

Preface

The task of controlling large-scale networks of interconnected dynamic systems poses fascinating challenges, in terms of both the mathematical foundations and future technological implementations. Dynamic networks of systems is a very active area of current research that lies at the crossroads of diverse disciplines such as biology, physics, computer science, and systems engineering. Examples of areas in which research is being conducted can be found everywhere in science and technology, including synchronization phenomena in laser physics, neural and genetic networks, Internet research, social networks, and cooperating sensor-actor networks in robotics. This line of research fits very well into current trends in systems engineering of analyzing heterogeneous networks of cyber-physical systems by combining methods from mathematics, control theory, and information theory.

How does one stably form a flock of birds or a school of fish? How does one control ensembles of quantum mechanical systems using open-loop control? How can one identify nodes with the strongest control potentials in a huge social network, and what are the fastest means to achieve consensus among network participants? How does one control macroscopic phenomena such as synchronization and clustering? Would the information flow in a big company or a stock market be better organized tall or flat? These are some of the immediate questions that come to mind when dealing with large-scale dynamic systems. Of course, one first needs to come up with precise mathematical models for the node systems as well as for the interconnection devices. Where better to start than with networks of linear systems?

The study of interconnected systems is not new. An early instance is the classic network synthesis problem of realizing electrical circuits using resistors, capacitors, and inductive elements, with pioneering contributions due to, for example, Foster (1924), Brune (1931), and Bott and Duffin (1949). The circuit synthesis problem remains unsolved to this day, but the subject has attracted some attention recently; see, for example, Kalman (2010) and Jiang and Smith (2011). The simplest coupling structures of linear systems are the series, parallel, and feedback interconnections. They have been studied since the very beginning of mathematical systems theory in the 1960s. Early work includes that by Callier and Nahum (1975) and

Fuhrmann (1975), who found frequency domain characterizations of controllability and observability for the three interconnection types. This early focus on simple interconnection schemes seemed a natural starting point for the analysis of more complicated interconnection structures, as is exemplified by the work of Rosenbrock and Pugh (1974) on the control of hierarchical systems. Though it is clear that the control theory for these special coupling structures can be deduced from a more general analysis of networks of interconnected systems, it may come as a surprise that the study of dynamic networks is in fact very closely related to feedback connection.

When analyzing large-scale interconnected systems one is faced with at least two different approaches that have been developed over the decades. A standard one is that of decentralized control, as covered in the book by Siljak (1991). Here one considers a multivariable plant that is to be controlled by a structured controller. The design of such decentralized controllers then becomes the main issue, whereas the interconnection between controllers and subsystems is often thought of as being fixed and does not constitute the central purpose of design. The situation is quite different than in the analysis of, for example, switching circuits, as in Leonard and Krishnaprasad (1994). Here the interconnections are regarded as active control elements that are tuned to achieve better performance of the overall network. This leads to a scenario that is almost dual to decentralized control, i.e., to the study of feedback connections, where the plant model has a decoupled structure, whereas the controller represents the network parameters that are to be tuned. Interesting cases in point are the control of the parallel interconnection of finitely many nonlinear systems using broadcast input signals, as in Brockett (2010); the control of spatially invariant systems and platoons of systems, for example Bamieh, Paganini and Dahleh (2002); the identification of driver nodes in very large-scale networks of linear systems, as in Liu, Slotine and Barabasi (2011), or the ensemble control of quantum dynamical systems, for example Li and Khaneja (2006). In all such applications, the plant consists of a finite or infinite number of decoupled node systems that may or may not be interconnected. External control signals are designed to be broadcasted to all node systems, and one is asked to characterize the controllability and observability properties of the network.

In practical implementations of network control systems, models for the node systems may be represented by first-order or higher-order difference and differential equations. This concerns a classical question from control theory, first discussed by Rosenbrock (1970), of the equivalence between first-order and higher-order system representations. Of course, by writing down first-order representations of both the node systems and the interconnection dynamics, one can apply standard controllability and observability tests for closed-loop systems. However, such an approach using the Hautus criterion or the Kalman rank test is problematic because it requires rank tests for large matrices. Moreover, the linear interconnection structure of the network is lost through the repeated appearance of matrix powers in the Kalman reachability or observability matrices. Thus, this approach quickly becomes inefficient with respect to, for example, medium-scale networks of high-dimensional subsystems. In contrast, we show that using polynomial matrix fraction representations leads to very efficient coprimeness conditions for the reachability and observability of networks of systems. To this end we develop a permanence principle for the strict system equivalence of networks. Although this permanence principle is already known from earlier work by Rosenbrock and Pugh (1974), it has apparently been forgotten in the network control community over the past few years. We present a reformulation and extension of this principle to dynamic interconnections within the unifying framework of polynomial models. From this principle one can easily deduce full characterizations on the controllability and observability of interconnected systems. Applied to classical parallel, series, and feedback interconnections, this leads to elegant new proofs for controllability and observability.

Our approach is based on polynomial model representations of linear systems constructed on the basis of coprime factorizations of the transfer function of the system. Using the shift realization, as in Fuhrmann (1976), such factorizations lead to powerful techniques for analyzing controllability properties of higher-order systems. Somewhat less attention has been paid to the problem of computing the controller, or controllers, that steer the system to a required state. In principle, this can be achieved by computing a right inverse to the reachability map. If the McMillan degree of the system under consideration is large, then this may be a daunting task. This problem may arise when the system is the result of interconnecting a large number of smaller systems. Our intention in this book is, in order to reduce the complexity of computation, to efficiently utilize controllability information on individual nodes in order to compute a minimal controller that steers the system to a given state. In its full generality, this goal remains elusive. For the inversion of a reachability map, which is a module isomorphism over a ring of polynomials, the main instruments are doubly coprime factorizations, i.e., the embedding of an intertwining relation, which is a consequence of coprime factorizations, in two unimodular polynomial matrices that are related by inversion. This presents us with the problem of constructing a doubly coprime factorization for a large-scale interconnected system from the doubly coprime factorizations related to the individual nodes. We treat this construction in full detail for the case of systems connected in parallel. This analysis brings to light the complexity of the problem. A side benefit is the clarification of the role of interpolation theory in the construction of the required control sequences.

This book aims at providing the mathematical foundations of networks of linear control systems, developed from an algebraic systems theory perspective. This includes a thorough treatment of questions of controllability, observability, and realization theory, as well as feedback control and observer theory. The material covered is quite extensive and is not intended as a textbook for a single course. Part I can be used as the basis for a first course in algebraic systems theory. The approach presented has the advantage that, with suitable modifications, it can be used in connection with the study of some infinite-dimensional systems, for example, H^∞-control. It also has proved itself to be instrumental in bringing together state space, polynomial system matrices, and geometric control methods. Part II could be used for a second, advanced, course on linear systems. Finally, Part III, which is largely independent of the previous parts, is ideally suited for advanced research seminars aimed at preparing graduate students for independent research. It addresses some of the basic research questions and approaches pertinent to analyzing networks of linear systems, such as, for example, synchronization, clustering, ensemble control, and controlling infinite platoons.

This book, which was written over the course of the past 2 years, is the culmination of our long-term joint research project that we carried out together over the past few decades. Our interest in networks started with the realization that functional models provide a very convenient setting for studying system interconnections. This led to a joint paper on the controllability of linear networks, and new insights for tackling other problems arose. Very quickly we became fascinated by the extremely rich facets, insights, and enormous intellectual challenges that the field of networks of systems offers. The book is a partial summary of what we have learned so far. Several short mutual visits in Tel Aviv and Würzburg helped to keep the stream of ideas flowing. The research for this book was supported by several people and grant organizations. We thank the Marina Hotel at Tel Aviv, and in particular the staff of the business lounge, for providing such convenient surroundings and a pleasant atmosphere for our research. Partial support was provided by Grant HE 1805/12-2 from the German Research Foundation within Priority Program SPP 1305, "Control Theory for Digitally Networked Dynamical Systems." We thank our colleagues, collaborators, and (former) students Gunther Dirr, Anna von Heusinger, Knut Hüper, Jens Jordan, Christian Lageman, Julia Lieb, Frederike Rüppel, and Michael Schönlein from Würzburg for helpful discussions, corrections, and proofreading of the manuscript. Rudolf Sailer and Frederike Rüppel helped as well with the figures and matrices. Thank you all! Finally, we would like to thank our families. Without their continual support and love this project would not have been possible. To them this book is dedicated.

Beer-Sheva, Israel   Paul A. Fuhrmann
Würzburg, Germany   Uwe Helmke

Contents

Chapter 1
Introduction

In this introductory chapter, the subject of the book is explained in more detail by focusing on two central issues of the control of networks: open-loop control of interconnected systems and synchronization. While the first topic emphasizes the computational aspects involved in controlling interconnected systems, the second theme of synchronization discusses the impact of couplings on the shaping of macroscopic properties that are pertinent in self-organizing systems. Instead of aiming for a more complete analysis, which is postponed to later chapters of the book, we confine ourselves to the discussion of some illustrative examples. At the end of the chapter the main new achievements of this book will be discussed and a brief overview of the contents presented.

1.1   Control of Parallel Connections

Parallel coupling is certainly one of the easiest ways to interconnect linear systems. However, controlling such a network is not that easy. Thus, for a finite number of linear control systems

$$\dot{x}_k(t) = A_k x_k(t) + B_k u_k(t), \quad k = 1, \dots, N, \tag{1.1}$$

with system matrices $(A_k, B_k) \in \mathbb{R}^{n_k \times (n_k + m)}$, consider the **parallel interconnected system**

$$\dot{x}_1(t) = A_1 x_1(t) + B_1 u(t)$$

$$\vdots \tag{1.2}$$

$$\dot{x}_N(t) = A_N x_N(t) + B_N u(t).$$

© Springer International Publishing Switzerland 2015
P.A. Fuhrmann, U. Helmke, *The Mathematics of Networks of Linear Systems*, Universitext, DOI 10.1007/978-3-319-16646-9_1

We refer to the subsystems (A_k, B_k) as the **node systems** of system (1.2). Of course, the difference between the two systems (1.1) and (1.2) is that the first one is completely decoupled, having independent input functions, while in the second one all subsystems share the same input. One could say that the input $u(t) \in \mathbb{R}^m$ is **broadcast** to all systems. When is such a system reachable, and how does one control such a system? The first question is, of course, a classical one whose answer goes back to the very beginnings of state-space theory in the 1970s. The reachability of a linear system

$$\dot{x}(t) = Ax(t) + Bu(t)$$

refers to the ability to choose a piecewise continuous input function $u : [0, T] \longrightarrow \mathbb{R}^m$ that steers the initial state $x(0) = 0$ to an arbitrarily prescribed terminal state vector $x_* = x(T)$ in finite time $T > 0$. Already in one of the first contributions to linear systems theory, Gilbert (1963) characterized the reachability properties of (1.2) for scalar first-order single input node systems. The general case turned out to be more difficult and was first solved by Fuhrmann (1975) for $N = 2$. It was subsequently extended by Fuhrmann (1976a) to certain classes of infinite-dimensional systems. The solution depends on polynomial algebra methods and coprime factorizations. Thus, let

$$(zI_{n_1} - A_1)^{-1}B_1 = N_1(z)D_1(z)^{-1}, \quad (zI_{n_2} - A_2)^{-1}B_2 = N_2(z)D_2(z)^{-1}$$

be two factorizations by polynomial matrices $N_k(z) \in \mathbb{R}[z]^{n_k \times m}, D_k(z) \in \mathbb{R}[z]^{m \times m}$ such that $D_1(z), D_2(z)$ are nonsingular polynomial matrices and the pairs $N_1(z), D_1(z)$ and $N_2(z), D_2(z)$ are right coprime for the respective matrices (for the algebraic terminology see Chapter 2).

Theorem 1.1 (Fuhrmann (1975)). *The parallel connection (1.2) of two linear systems* $(A_1, B_1), (A_2, B_2)$ *is reachable if and only if*

1. *Both systems* (A_1, B_1) *and* (A_2, B_2) *are reachable;*
2. *The* $m \times 2m$ *polynomial matrix* $(D_1(z), D_2(z))$ *has full row rank* m *for all complex numbers* $z \in \mathbb{C}$.

The proof by Fuhrmann, although concise, is not trivial and depends crucially on the theory of polynomial models that was developed by him around the same time (1975); see Fuhrmann (1976). It should be emphasized that, while equivalent state-space characterizations of the reachability of (1.2) are possible, these depend on more complicated formulations using the Jordan canonical forms of $(A_1, B_1), (A_2, B_2)$. Of course, in the single-input case where $m = 1$, everything is easy. In fact, then $D_i(z) = \det(zI - A_i)$, and condition 2 is equivalent to the coprimeness of the characteristic polynomials $\det(zI - A_i), i = 1, 2$. Thus, Theorem 1.1 implies the following corollary.

Corollary 1.2. *Let* $m = 1$. *The parallel connection (1.2) of two linear single-input systems* $(A_1, b_1), (A_2, b_2)$ *is reachable if and only if*

1. *The systems* $(A_1, b_1), (A_2, b_2)$ *are reachable;*
2. *The spectra of* A_1 *and* A_2 *are disjoint.*

In Chapter 9 we will prove a general reachability result for parallel connections of a finite number of systems that contains Theorem 1.1 and Corollary 1.2 as special cases. While Theorem 1.1 effectively solves the reachability problem for the parallel interconnection of two systems, the question remains as to how one can compute a control function $u : [0, T] \longrightarrow \mathbb{R}^m$ for the parallel connection (1.2) that steers the zero state $x = 0$ in finite time $T > 0$ to a desired vector $x^* = \mathrm{col}\,(x_1^*, \dots, x_N^*) \in \mathbb{R}^{n_1 + \dots + n_N}$ of terminal states. Moreover, one would like to be able to compute such a control u for (1.2) from a knowledge of local controls $u_k^* : [0, T] \longrightarrow \mathbb{R}^m$ that steer node systems (1.1) from the zero state to the local terminal states x_k^*, $k = 1, \dots, N$. Although such a problem looks daunting at first glance, we will develop in Chapter 10.1 an approach that will enable us to tackle such problems. To explain the basic idea of the open-loop control of parallel interconnection, we now focus on single-input/single-output (SISO) systems, defined in a discrete-time case, because these are slightly easier to handle than continuous-time systems. We first establish some useful notation.

Consider an nth-order, discrete-time, single-input system

$$x(t+1) = Ax(t) + bu(t), \quad t = 0, 1, 2, \dots .$$

Then a finite input sequence $u_0, u_1, \dots, u_{T-1}$ steers the zero state to a desired state $x(T) = x^* \in \mathbb{R}^n$ if and only if the scalar polynomial

$$\mathbf{u}(z) = \sum_{k=0}^{T-1} u_{T-1-k} z^k$$

satisfies

$$\mathbf{u}(A)b = x^*.$$

We refer to $\mathbf{u}(z)$ as the **input polynomial** for x^*. The reachability of (A, b) is equivalent to the existence of an input polynomial for every $x^* \in \mathbb{R}^n$. Moreover, assuming the reachability of (A, b), the unique input polynomial of minimal degree $n - 1$ is given by

$$\mathbf{u}^*(z) = (1, \dots, z^{n-1})(b, Ab, \dots, A^{n-1}b)^{-1} x^*.$$

Now consider the interconnected single-input systems

$$
\begin{aligned}
x_1(t+1) &= A_1 x_1(t) + b_1 u(t), \\
x_2(t+1) &= A_2 x_2(t) + b_2 u(t),
\end{aligned}
\tag{1.3}
$$

where $A_i \in \mathbb{R}^{n_i \times n_i}$ and $b_i \in \mathbb{R}^{n_i}$ for $i = 1,2$. Assume that system (1.3) is reachable, i.e., that (A_1,b_1) and (A_2,b_2) are both reachable and the characteristic polynomials

$$q_1(z) = \det(zI - A_1), \quad q_2(z) = \det(zI - A_2)$$

are coprime. For local state vectors $x_1^* \in \mathbb{R}^{n_1}, x_2^* \in \mathbb{R}^{n_2}$ there exist unique input polynomials $\mathbf{u}_1^*(z)$ and $\mathbf{u}_2^*(z)$ of degrees $n_1 - 1$ and $n_2 - 1$, respectively, with

$$u_1^*(A_1)b_1 = x_1^*, \quad u_2^*(A_2)b_2 = x_2^*.$$

By the coprimeness of the characteristic polynomials $q_1(z)$ and $q_2(z)$, there exist unique polynomials $c(z), d(z) \in \mathbb{R}[z]$ of degrees $\deg c(z) < n_2, \deg d(z) < n_1$ that satisfy the Bezout identity

$$c(z)q_1(z) + d(z)q_2(z) = 1.$$

By the Cayley–Hamilton theorem, $q_i(A_i) = 0$ for $i = 1,2$. Thus, by substituting A_1 and A_2 into the Bezout identity, we obtain $d(A_1)q_2(A_1) = I$ and $c(A_2)q_1(A_2) = I$. Consider the polynomial

$$\mathbf{u}(z) = d(z)q_2(z)\mathbf{u}_1^*(z) + c(z)q_1(z)\mathbf{u}_2^*(z).$$

Again, by substituting the matrices A_i into both sides of this equation, we obtain

$$\mathbf{u}(A_1)b_1 = d(A_1)q_2(A_1)\mathbf{u}_1^*(A_1)b_1 = x_1^*,$$
$$\mathbf{u}(A_2)b_2 = c(A_2)q_1(A_2)\mathbf{u}_2^*(A_2)b_2 = x_2^*.$$

Thus the (reverse) coefficients of the polynomial $u(z)$ yield an input sequence that steers (1.3) to the desired states.

The preceding approach can be extended in several directions. First, note that it is very closely related to the Chinese remainder theorem; see Chapter 2, Theorem 2.11, for a statement and proof of this central algebraic result. Second, the degree of $\mathbf{u}(z)$ can be upper bounded by $n_1 + n_2$ by replacing the factors $d(z)\mathbf{u}_1^*(z)$ and $c(z)\mathbf{u}_2^*(z)$ with their remainders modulo $q_1(z)$ and $q_2(z)$, respectively. Third, the construction of the control can be easily carried out for an arbitrary number of parallel connected reachable SISO systems:

$$x_1(t+1) = A_1 x_1(t) + b_1 u(t)$$

$$\vdots \qquad\qquad (1.4)$$

$$x_N(t+1) = A_N x_N(t) + b_N u(t).$$

Assume that for each of the N local subsystems (A_j, b_j) local control sequences $\mathbf{u}_j$ are known that steer the zero state to a desired terminal state x_j^*. How can one compute from such local controls a single global input sequence $\mathbf{u}$ that steers all subsystems simultaneously to the desired terminal states? Ideally, one would like to obtain a formula, $\mathbf{u} = \sum_{j=1}^{N} \mathscr{F}_j \mathbf{u}_j$, that expresses the desired control as the weighted sum of local controls, where $\mathscr{F}_j$ are suitable filter operators that act on the local inputs. To answer this question, we need just a little bit more notation.

Let $q_j(z) = \det(zI - A_j)$ denote the characteristic polynomial of A_j, and define $\hat{q}_j(z) = \prod_{i \neq j} q_i(z)$. Assume that the pairs (A_j, b_j) are reachable for each $j = 1, \ldots, N$. The reachability of the parallel connection (1.4) is then equivalent to the coprimeness of $q_j, \hat{q}_j$ for each $j = 1, \ldots, N$. Let the polynomials $c_j(z)$ and $d_j(z)$ denote the unique solutions of the Bezout equation

$$c_j(z)q_j(z) + d_j(z)\hat{q}_j(z) = 1, \quad j = 1, \ldots, N,$$

with degrees $\deg d_j < n_j$. Proceeding as before, we store the input sequences $u_0, \ldots, u_{M-1}$ for controlling (1.4) as coefficients of the associated input polynomial $\mathbf{u}(z) = \sum_{j=0}^{M-1} u_{M-j-1} z^j$. Our basic control result for (1.4) is stated as follows.

Theorem 1.3. *Assume that (1.4) is reachable. For local state vectors $x_1^*, \ldots, x_N^*$ let $\mathbf{u}_1(z), \ldots, \mathbf{u}_N(z) \in \mathbb{R}[z]$ be polynomials with*

$$\mathbf{u}_1(A_1)b_1 = x_1^*, \ldots, \mathbf{u}_N(A_N)b_N = x_N^*.$$

Then

$$\mathbf{u}(z) = \sum_{j=1}^{N} d_j(z)\hat{q}_j(z)\mathbf{u}_j(z) \tag{1.5}$$

satisfies $\mathbf{u}(A_j)b_j = x_j^*$ for all j.

Proof. From the Bezout equation we obtain $d_k(A_k)\hat{q}_k(A_k) = I$ and $d_j(A_k)\hat{q}_j (A_k) = 0$ for $j \neq k$. This implies

$$\mathbf{u}(A_k)b_k = \sum_{j=1}^{N} d_j(A_k)\hat{q}_j(A_k)\mathbf{u}_j(A_k)b_k = d_k(A_k)\hat{q}_k(A_k)\mathbf{u}_k(A_k)b_k = \mathbf{u}_k(A_k)b_k = x_k^*.$$

∎

The preceding result exhibits a very simple formula (1.5) for the global control of (1.4) in terms of a weighted sum of the given local controls for the decoupled systems (A_j, b_j). The polynomial weights $d_j(z)\hat{q}_j(z)$ are independent of the desired local states x_j^* and can be computed beforehand by solving N Bezout equations. Since multiplication by the polynomials $p(z) \mapsto a(z)p(z)$ acts as a Toeplitz operator on the space of coefficients of polynomials $p(z)$ of bounded degrees, formula (1.5) represents the desired filter operator $\mathbf{u} = \sum_{j=1}^{N} \mathscr{F}_j \mathbf{u}_j$ for computing global controls from local ones. Moreover, by replacing $\mathbf{u}(z)$ with the remainder polynomial $\mathbf{u}^*(z)$ obtained by division with the remainders $\mathbf{u}(z) = m(z)q_1(z)\cdots q_N(z) + \mathbf{u}^*(z)$, one obtains an input polynomial of minimal degree bounded by $n_1 + \cdots + n_N$. All this requires only elementary operations from polynomial algebra (Euclidean algorithm, division with remainders) for which efficient implementations are known.

At this stage the reader may wonder why we did not simply invert the reachability matrix of the parallel interconnection scheme (1.3) to compute the minimal length input sequence. The reason for this is that we wanted to find controls that are computed from local control sequences, and the inversion of the global reachability matrix does not allow one to do so. In fact, our solution, although of higher degree than necessary, has an advantage in comparison to the global inversion method. The solutions $c_j(z), d_j(z)$ to the N Bezout equations must be computed only once. Choosing solutions of the Bezout equations in an appropriate way, for example, by using a Newton interpolation, may even open the way to recursive computations of $\mathbf{u}(z)$. This scheme is therefore favorable if the number of parallel connected systems changes or, more generally, if the effects of changing the coupling parameters in an interconnected system are of interest. For a general analysis of open-loop control for linear systems we refer the reader to Chapter 4. The multivariable case poses more difficulties and will be treated in full generality in Chapter 10.1.

1.2 Synchronization of Coupled Harmonic Oscillators

We next turn to a different topic, that of synchronization in the self-organization of coupled systems. This subject has attracted a lot of attention because synchronization is a fundamental process in nature that is observed in many interconnected systems. In the computer science literature, the synchronization property is often referred to as consensus, with obvious appeal to areas such as social networks and opinion dynamics. Being in a synchronized state is a truly macroscopic property that is often due to the coupling effects inherent in a network. We refer the reader to Chapter 11 for a detailed analysis of synchronization phenomena and clustering effects. Here we restrict ourselves to a short discussion of a simple model for synchronization: the case of two coupled linear oscillators. The subsequent elementary discussion of synchronization is based on Fuhrmann, Priel, Sussmann and Tsoi (1987), an unpublished manuscript dating back to the mid-1980s. We are grateful to M. Pavon for providing us with fragments of it.

The coupling of two damped harmonic oscillators is related to resonance phenomena. Consider a damped harmonic oscillator ($r > 0, \omega > 0$) acted on by a periodic force of the form $F\cos\Omega t$, i.e.,

$$\ddot{x} + r\dot{x} + \omega^2 x = F\cos\Omega t.$$

Resonance in system (1.2) is probably the simplest example of **synchronization**. It is well known that the solutions of (1.2) converge from all initial conditions $x(0), \dot{x}(0)$ to the periodic trajectory $MF\cos(\Omega t - \phi)$. Here the magnification factor M is

$$M = \frac{1}{\sqrt{(\Omega^2 - \omega^2)^2 + r^2 \Omega^2}},$$

and the phase delay $0 \le \phi \le \pi$ is determined through the equations

$$\sin \phi = \frac{r\Omega}{\sqrt{(\Omega^2 - \omega^2)^2 + r^2 \Omega^2}}, \quad \cos \phi = \frac{\omega^2 - \Omega^2}{\sqrt{(\Omega^2 - \omega^2)^2 + r^2 \Omega^2}}.$$

Let us next study in some detail the interconnection of two linear oscillators. The model for a single, damped oscillator is

$$\ddot{x}(t) + r\dot{x}(t) + \omega^2 x(t) = v(t),$$
$$w(t) = \dot{x}(t). \tag{1.6}$$

Here $v(t)$ is the input acting on the oscillator while the velocity term $w = \dot{x}$ is regarded as the output of the system. In contrast to standard approaches, we do not want to transform the system into first-order state-space form but rather try to retain the second-order structure by associating to (1.6) the **polynomial system matrix**

$$\begin{pmatrix} z^2 + rz + \omega^2 & -1 \\ z & 0 \end{pmatrix}.$$

On the other hand, using the standard procedure for reducing (1.6) to a first-order system and defining $X = \begin{pmatrix} x \\ \dot{x} \end{pmatrix}$, we obtain

$$\dot{X} = \alpha X + \beta v,$$
$$w = \gamma X, \tag{1.7}$$

where α, β, γ are defined by

$$\alpha = \begin{pmatrix} 0 & 1 \\ -\omega^2 & -r \end{pmatrix}, \qquad \beta = \begin{pmatrix} 0 \\ 1 \end{pmatrix}, \qquad \gamma = \begin{pmatrix} 0 & 1 \end{pmatrix}. \tag{1.8}$$

The transfer function of system (1.7) is

$$g(z) = \gamma(zI - \alpha)^{-1}\beta = \frac{z}{z^2 + rz + \omega^2}.$$

Before proceeding to analyze the case of two identical, symmetrically coupled (damped) oscillators, consider first the decoupled system of uncontrolled oscillators

$$\ddot{x}_i(t) + r\dot{x}_i(t) + \omega^2 x_i(t) = 0,$$
$$w_i(t) = \dot{x}_i(t), \qquad\qquad i = 1, 2. \tag{1.9}$$

The state space of the coupled system is four-dimensional and can be coordinated by $x = (x_1, \dot{x}_1, x_2, \dot{x}_2)^\top$. The state system representation of (1.9) becomes

$$\frac{d}{dt} \begin{pmatrix} x_1 \\ \dot{x}_1 \\ x_2 \\ \dot{x}_2 \end{pmatrix} = \begin{pmatrix} 0 & 1 & 0 & 0 \\ -\omega^2 & -r & 0 & 0 \\ 0 & 0 & 0 & 1 \\ 0 & 0 & -\omega^2 & -r \end{pmatrix} \begin{pmatrix} x_1 \\ \dot{x}_1 \\ x_2 \\ \dot{x}_2 \end{pmatrix}.$$

We observe for $r < \omega^2$ that the state transition matrix of this autonomous system has two identical pairs of complex conjugate eigenvalues ($i := \sqrt{-1}$):

$$\lambda_\pm = \frac{-r \pm i\sqrt{4\omega^2 - r^2}}{2}.$$

When considering couplings there are various possibilities. In the sequel we assume that the coupling mechanism transmits the effect of the motion of the oscillators from one to the other. Furthermore, let us assume the absence of an external driving force, i.e., we consider an autonomous system. Thus, in our model, the equations of motion for the two coupled oscillators are

$$\ddot{x}_1 + r\dot{x}_1 + \omega^2 x_1 = \kappa \dot{x}_2,$$
$$\ddot{x}_2 + r\dot{x}_2 + \omega^2 x_2 = \kappa \dot{x}_1. \tag{1.10}$$

Again, the state space of the coupled system is four-dimensional. Hence, equation (1.10) can be rewritten in first-order form $\dot{X} = \mathscr{A}_\kappa X$ as

$$\frac{d}{dt} \begin{pmatrix} x_1 \\ \dot{x}_1 \\ x_2 \\ \dot{x}_2 \end{pmatrix} = \begin{pmatrix} 0 & 1 & 0 & 0 \\ -\omega^2 & -r & 0 & \kappa \\ 0 & 0 & 0 & 1 \\ 0 & \kappa & -\omega^2 & -r \end{pmatrix} \begin{pmatrix} x_1 \\ \dot{x}_1 \\ x_2 \\ \dot{x}_2 \end{pmatrix}.$$

Computing the characteristic polynomial of $\mathscr{A}_\kappa$ yields

$$\det(zI - \mathscr{A}_\kappa) = (z^2 + rz + \omega^2)^2 - \kappa^2 z^2$$
$$= (z^2 + (r - \kappa)z + \omega^2)(z^2 + (r + \kappa)z + \omega^2). \tag{1.11}$$

We conclude, under the assumption that $r + |\kappa| \leq 2\omega$, that the eigenvalues of $\mathscr{A}_\kappa$ are

$$\frac{-(r \pm \kappa) \pm i\sqrt{4\omega^2 - (r \pm \kappa)^2}}{2}.$$

Our next step is to compute the corresponding eigenvectors. The eigenvector equations are

$$\begin{pmatrix} \lambda & -1 & 0 & 0 \\ \omega^2 & \lambda + r & 0 & -\kappa \\ 0 & 0 & \lambda & -1 \\ 0 & -\kappa & \omega^2 & \lambda + r \end{pmatrix} \begin{pmatrix} a \\ b \\ c \\ d \end{pmatrix} = \begin{pmatrix} 0 \\ 0 \\ 0 \\ 0 \end{pmatrix}.$$

If the two oscillators were synchronized, then $x_2(t) = x_1(t)$ and $\dot{x}_2(t) = \dot{x}_1(t)$. Therefore, we look for an eigenvector of the form $X = (\xi, \eta, \xi, \eta)^\top$. The characteristic equation $\mathscr{A}_\kappa X = \lambda X$ is given by

$$\begin{pmatrix} 0 & 1 & 0 & 0 \\ -\omega^2 & -r & 0 & \kappa \\ 0 & 0 & 0 & 1 \\ 0 & \kappa & -\omega^2 & -r \end{pmatrix} \begin{pmatrix} \xi \\ \eta \\ \xi \\ \eta \end{pmatrix} = \lambda \begin{pmatrix} \xi \\ \eta \\ \xi \\ \eta \end{pmatrix},$$

which reduces to the pair of equations

$$\eta = \lambda \xi,$$
$$-\omega^2 \xi - r\eta + \kappa\eta = \lambda \eta.$$

By substitution, we obtain $(\lambda^2 + (r - \kappa)\lambda + \omega^2)\xi = 0$. Since $\xi \neq 0$ for a nontrivial eigenvector, $\lambda^2 + (r - \kappa)\lambda + \omega^2 = 0$. This implies

$$\lambda = \frac{-(r - \kappa) \pm i\sqrt{4\omega^2 - (r - \kappa)^2}}{2}.$$

Thus, the synchronized eigenvalues are one pair of complex conjugate eigenvalues of $\mathscr{A}_\kappa$. In fact, assuming $\kappa < r$, they correspond to the pair with the larger, though negative, real part. In a completely analogous fashion, we consider the antisynchronized solution $x = (\xi, \eta, -\xi, -\eta)^\top$. In this case, the characteristic equation is given by

$$\begin{pmatrix} 0 & 1 & 0 & 0 \\ -\omega^2 & -r & 0 & \kappa \\ 0 & 0 & 0 & 1 \\ 0 & \kappa & -\omega^2 & -r \end{pmatrix} \begin{pmatrix} \xi \\ \eta \\ -\xi \\ -\eta \end{pmatrix} = \lambda \begin{pmatrix} \xi \\ \eta \\ -\xi \\ -\eta \end{pmatrix}.$$

This reduces to the pair of equations

$$\eta = \lambda \xi,$$
$$-\omega^2 \xi - r\eta - \kappa \eta = \lambda \eta,$$

and thus leads to $(\lambda^2 + (r + \kappa)\lambda + \omega^2) = 0$. This shows that the eigenvalues corresponding to the antisynchronized solution are

$$\lambda_{\pm} = \frac{-(r + \kappa) \pm i\sqrt{4\omega^2 - (r + \kappa)^2}}{2}.$$

So the antisynchronized modes correspond to a pair of complex conjugate eigenvalues of $\mathscr{A}_{\kappa}$. We note that if the coupling term, measured by κ, is too large, i.e., $\kappa > r$, then the system becomes unstable.

The moral of the spectral analysis, in the case of a symmetrically coupled pair of identical damped harmonic oscillators, is that there are two basic eigenmodes, the synchronized and the antisynchronized. Moreover, because the real part of the antisynchronized mode is to the left of the real part of the synchronized mode, this mode decays faster with time. Thus, asymptotically, the dominant mode would be the synchronized one. Of course, in the absence of a forcing term that provides energy to the system to offset the energy dissipated by friction, the solution would tend asymptotically to zero. We see from this example that synchronization is the way the coupled system organizes itself to minimize energy consumption. This leads one to conjecture the existence of a variational principle that determines the dynamics of the system. Moreover, a comparison of the case of coupled versus uncoupled pairs of oscillators suggests strongly that synchronization is related to the problem of pole placement by output feedback.

A natural question arises now: Is it necessary for the spectral analysis to go over to a first-order, or state, representation? Synchronization phenomena are usually defined and studied in the state-space context. However, our example of coupled oscillators shows that it might be preferable to perform the study in a functional context and use the concept of state maps to connect with the state-space analysis. To do this, we go back to the system equations (1.10) and rewrite them into differential operator form as

$$\begin{pmatrix} \frac{d^2}{dt^2} + r\frac{d}{dt} + \omega^2 \, \mathrm{id} & -\kappa\frac{d}{dt} \\ -\kappa\frac{d}{dt} & \frac{d^2}{dt^2} + r\frac{d}{dt} + \omega^2 \, \mathrm{id} \end{pmatrix} \begin{pmatrix} x_1 \\ x_2 \end{pmatrix} = \begin{pmatrix} 0 \\ 0 \end{pmatrix}. \qquad (1.12)$$

However, if we apply the Laplace transform to equation (1.12), we can identify, with

$$Q(z) = \begin{pmatrix} z^2 + rz + \omega^2 & -\kappa z \\ -\kappa z & z^2 + rz + \omega^2 \end{pmatrix},$$

the solution space with the space of tuples of strictly proper rational functions,

$$X^Q = \{h = \begin{pmatrix} h_1 \\ h_2 \end{pmatrix} \mid h_i \text{ strictly proper rational, } Q(z)h(z) \text{ is a polynomial}\},$$

referred to as the **rational model** of (1.12). The space X^Q has a natural structure as a module over the ring $\mathbb{R}[z]$ of polynomials, and we develop in Chapter 3 a comprehensive algebraic theory of such modules. In fact, differentiation leaves invariant the space of solutions of (1.12). Via the Laplace transform this corresponds exactly to the **shift operator** $\sigma : X^Q \longrightarrow X^Q$

$$\sigma(h) = \pi_-(zh(z)),$$

where π_- denotes projection onto the strictly proper part. Clearly,

$$\det \begin{pmatrix} z^2 + rz + \omega^2 & -\kappa z \\ -\kappa z & z^2 + rz + \omega^2 \end{pmatrix} = (z^2 + (r - \kappa)z + \omega^2)(z^2 + (r + \kappa)z + \omega^2).$$
(1.13)

A comparison of (1.13) with (1.11) shows that it might be possible to derive state representations directly from polynomial data. Indeed, this is achieved by using polynomial and rational models as well as shift realization, a topic that will be discussed extensively in Chapter 4. Note that, defining $\omega_{\pm} = \sqrt{\omega^2 - \frac{(r \pm \kappa)^2}{4}}$, the factors in (1.13) can be further factored as

$$z^2 + (r - \kappa)z + \omega^2 = (z + \frac{r - \kappa}{2} + i\omega_-)(z + \frac{r - \kappa}{2} - i\omega_-),$$

$$z^2 + (r + \kappa)z + \omega^2 = (z + \frac{r + \kappa}{2} + i\omega_+)(z + \frac{r + \kappa}{2} - i\omega_+).$$

Thus, the zeros of $Q(z)$, i.e., the points at which $Q(z)$ drops rank, are $\frac{-(r \pm \kappa)}{2} \pm i\omega_+$. This is relevant to the computation of the eigenvectors of the shift in X^Q. An element $h \in X^Q$ is an eigenvector of the backward shift σ if and only if there exists a $\lambda \in \mathbb{C}$ for which $\sigma h(z) = \pi_-(zh) = \lambda h(z)$, which is the case if and only if $h(z) = \frac{\xi}{z - \lambda}$ for some constant vector ξ. Furthermore, $h(z) \in X^Q$ if and only if

$$\pi_- \frac{Q(z)\xi}{z - \lambda} = \frac{Q(\lambda)\xi}{z - \lambda} = 0,$$

which is the case if and only if $Q(\lambda)\xi = 0$. Taking note of factorizations (1.13), we clearly have $z^2 + rz + \omega^2|_{z=-\frac{r-\kappa}{2}+i\omega_-} = \kappa z|_{z=-\frac{r-\kappa}{2} \pm i\omega_-}$. This allows us to evaluate $Q(z)$ at $z = -\frac{r-\kappa}{2} \pm i\omega_-$ to obtain

$$Q(-\frac{r - \kappa}{2} \pm i\omega_-) = \kappa \begin{pmatrix} -\frac{r-\kappa}{2} \pm i\omega_- & \frac{r-\kappa}{2} \mp i\omega_- \\ \frac{r-\kappa}{2} \mp i\omega_- & -\frac{r-\kappa}{2} \pm i\omega_- \end{pmatrix}$$

This implies that $\xi = \text{col}\,(\xi_1, \xi_2) \in \text{Ker}\,Q(-\frac{r-\kappa}{2} \pm i\omega_-)$ if and only if $\xi_2 = \xi_1$. In an analogous fashion, $\xi = \text{col}\,(\xi_1, \xi_2) \in \text{Ker}\,Q(-\frac{r+\kappa}{2} \pm i\omega_+)$ if and only if $\xi_2 = -\xi_1$. The implications, as far as synchronization is concerned, are now clear.

So far, we have studied a pair of coupled oscillators independently in state space as well as in polynomial terms. It is time to elaborate on the connection between the two methods. Let us start with the autonomous system

$$\ddot{x} + r\dot{x} + \omega^2 x = 0,$$
$$y = x.$$

Defining α and γ as in (1.8), the state to output transfer function is easily computed:

$$(1\ \ 0) \begin{pmatrix} z & -1 \\ \omega^2 & z+r \end{pmatrix}^{-1} = (z^2 + rz + \omega^2)^{-1} (z+r\ \ 1).$$

The intertwining relation

$$(z+r\ \ 1) \begin{pmatrix} z & -1 \\ \omega^2 & z+r \end{pmatrix} = (z^2 + rz + \omega^2)(1\ \ 0) \tag{1.14}$$

is embeddable in a doubly coprime factorization as

$$\left(\begin{array}{cc|c} z & -1 & 0 \\ \omega^2 & z+r & 1 \\ \hline 1 & 0 & 0 \end{array} \right) \left(\begin{array}{cc|c} 0 & 0 & 1 \\ -1 & 0 & z \\ \hline -(z+r) & -1 & z^2 + rz + \omega^2 \end{array} \right) = \begin{pmatrix} 1 & 0 & 0 \\ 0 & 1 & 0 \\ 0 & 0 & 1 \end{pmatrix}.$$

This embedding allows one to construct a state map $X(\sigma)$ that reconstructs a state from the output y via

$$X = X(\sigma)y = \begin{pmatrix} 1 \\ \sigma \end{pmatrix} x = \begin{pmatrix} x \\ \dot{x} \end{pmatrix}.$$

In the case of two uncoupled oscillators, the intertwining relation (1.14) is replaced by

$$\begin{pmatrix} z+r & 1 & 0 & 0 \\ 0 & 0 & z+r & 1 \end{pmatrix} \begin{pmatrix} z & -1 & 0 & 0 \\ \omega^2 & z+r & 0 & 0 \\ 0 & 0 & z & -1 \\ 0 & 0 & \omega^2 & z+r \end{pmatrix}$$

$$= \begin{pmatrix} z^2 + rz + \omega^2 & 0 \\ 0 & z^2 + rz + \omega^2 \end{pmatrix} \begin{pmatrix} 1 & 0 & 0 & 0 \\ 0 & 0 & 1 & 0 \end{pmatrix}. \tag{1.15}$$

Proceeding to the case of two identical coupled harmonic oscillators

$$\ddot{x}_1 + r\dot{x}_1 + \omega^2 x_1 = \kappa v_1,$$
$$\ddot{x}_2 + r\dot{x}_2 + \omega^2 x_2 = \kappa v_2,$$
(1.16)

$$v_1 = \left(0 \; \tfrac{d}{dt}\right)\begin{pmatrix} x_1 \\ x_2 \end{pmatrix}, \quad v_2 = \left(\tfrac{d}{dt} \; 0\right)\begin{pmatrix} x_1 \\ x_2 \end{pmatrix},$$

$$y_1 = \left(1 \; 0\right)\begin{pmatrix} x_1 \\ x_2 \end{pmatrix}, \quad y_2 = \left(0 \; 1\right)\begin{pmatrix} x_1 \\ x_2 \end{pmatrix}.$$

Note that here there are two distinct kinds of output. The first, specified by v_i, is used to describe the interconnection, whereas the second, specified by y_i, is the observed output. We encounter a similar situation in observer theory; see Fuhrmann (2008). We say that system (1.16) **synchronizes** if $y_2(t) - y_1(t)$ tends asymptotically to zero. Of course, for a meaningful definition this must be expressed in terms of precise stability requirements, a topic we shall discuss in Chapter 11.

In analogy to (1.15), we consider the intertwining relation

$$\begin{pmatrix} z+r & 1 & 0 & 0 \\ 0 & 0 & z+r & 1 \end{pmatrix}\begin{pmatrix} z & -1 & 0 & 0 \\ \omega^2 & z+r & -\kappa z & 0 \\ 0 & 0 & z & -1 \\ -\kappa z & 0 & \omega^2 & z+r \end{pmatrix}$$
$$= \begin{pmatrix} z^2 + rz + \omega^2 & -\kappa z \\ -\kappa z & z^2 + rz + \omega^2 \end{pmatrix}\begin{pmatrix} 1 & 0 & 0 & 0 \\ 0 & 0 & 1 & 0 \end{pmatrix}.$$
(1.17)

This connects frequency-domain and state-space descriptions of the interconnected system. In fact, a state map that reconstructs the state from the output x_1, x_2 can be obtained from the embedding of (1.17) in a doubly coprime factorization. We will elaborate on this in Chapters 4 and 11.

1.3 Outline of the Book

We aim at developing a coherent approach to the system-theoretic analysis of inter-connected linear dynamic control systems. While previous research on networks of systems primarily focused on state-space models, our approach is motivated by the early work of Rosenbrock, who emphasized as early as the 1970s the need for a systematic investigation of coupled higher-order system representations. Historically, the early investigation of higher-order system representations led naturally to the development of methods from algebraic systems theory and operator theory, such as polynomial and rational models. We are therefore convinced that such an algebraic approach might serve to properly lay the foundations for a mathematical theory of networks of linear systems. Indeed, our very motivation for writing this book was to provide such foundations.

Our main achievements are as follows.

- We develop a general polynomial model approach to the study of the reachability and observability properties of networks of linear systems. The models for the node subsystems in a network can be defined either in the state space or using higher-order polynomial matrix representations. There is a considerable advantage in allowing for different system representations; it opens the way to studying the distinct effects caused by the coupling parameters and the dynamics of node systems. Including discrete-time linear system representations defined over, for example, finite fields, could lead to new extensions of the theory to other research areas, such as, for example, network coding.
- We develop a comprehensive theory of open-loop controls, for both single systems and networks of systems. While there are numerous ways to analyze the reachability of linear systems, the computational aspect of determining open-loop controls has been somehow neglected in the literature. This is in contrast to feedback control, where quite a bit of research has been done on numerical computations of stabilizing state feedback or output feedback controllers. Similar work on open-loop control is lacking.
- We present a unified approach to synchronization and clustering for both first-order and higher-order systems. Previous work on synchronization mainly addressed the problem in the state space, whereas we consider synchronization for partial states as well.
- We present a number of new ideas for controlling the macroscopic properties of networks. Here we only scratch the surface. Our main focus is on ensemble control for parameter-dependent families of linear systems. Such systems constitute the simplest kinds of infinite networks, i.e., networks possessing an infinite number of node systems. Our tools here are mainly function theoretic. We show the connection with controlling partial differential equations, infinite platoons, and systems defined by the action of composition operators on Hardy spaces of analytic functions.

During the process of writing this book we became increasingly aware of how vast the subject of networks of linear systems really is. Therefore, the table of contents not only reflects our own choices of topics but also indicates what material has been left out. Thus, one should regard this book more as a first step rather than as a definitive treatment of the subject. We list a few of the most desirable items that are missing from this book. The reasons for their omission are, of course, varied, one of the most important being simply our lack of sufficient energy and time.

- The first issue to mention is the use of behaviors as general models for dynamics and interconnections. Clearly, it is very desirable to consider interconnections of behaviors rather than of input–output models and thus develop a theory of networks of behaviors. However, a problem that we encountered in such a behavioral approach is that the interconnection structure becomes more exposed in an input–output framework and less so in a behavioral context. For example, it is very easy to define a directed interconnection graph for networks of input–output systems. For these reasons we decided to focus on the simpler problem of studying the interconnections of systems with inputs and outputs.

- Another topic to mention here is that of infinite-dimensional systems. We briefly address such systems when discussing infinite platoons and spatially invariant systems. Including systematically infinite-dimensional systems would require expanding all chapters by including material on Hardy spaces of analytic functions and operator theory. While this would have been possible, it would have required us to expand our exposition in much vaster generality.
- Somehow related to the preceding point is our focus on modules over the ring of polynomials $\mathbb{F}[z]$ in a single variable z over the field $\mathbb{F}$. Of course, one can try to replace this ring by localizations of $\mathbb{F}[z]$, because this would, for example, include the principal ideal domain of stable proper rational functions. Thus proceeding in this direction would have allowed us to include rational Hardy spaces in our framework. Again, we feel that the level of generality that would have been gained might not have been in perfect balance with the level of abstractions required.
- The field of networked control systems is currently in a very rapid development, with major results being derived by various methods, analytic, algebraic, and graph-theoretic. Thus, we obviously had to leave out several other research directions from the text in order to achieve a somewhat coherent description. Such topics include, for example, distributed optimization (Nedic, Ozdaglar and Parrilo (2010)), structural reachability (Dion, Commault and van der Woude (2013), Meshbahi and Egerstedt (2010)), and formation control (Anderson et. al. (2007), Jadbabaie, Lin and Morse (2003)).

We now turn to a more detailed description of the contents of this book. The book makes heavy use of the results and techniques of algebraic systems theory and, in particular, the theory of functional models. For an elementary introduction to scalar functional models and background material on linear algebra we refer to the book on linear algebra by Fuhrmann (2012), where the focus is on SISO systems. The functional model approach to multivariable linear systems has been treated so far only in various journal publications; it has never received systematic treatment in a textbook. This book serves to fill that gap and presents a general account of those aspects of the theory of polynomial and rational models that we regard as important for analyzing networks of systems. Other textbooks, for example, those by Rosenbrock (1970), Wolovich (1974), Kailath (1980), Vardulakis (1991), and Antsaklis and Michel (2005), offer complementary material and viewpoints on polynomial approaches to systems theory.

The book is divided into three parts. Part I provides the necessary background on linear algebra over rings of polynomials, functional model spaces, and algebraic systems theory. Part II discusses more advanced topics, such as stability theory, feedback control, and observer theory, and provides a brief introduction to conditioned and controlled invariant subspaces. The material in these two parts can be used as a self-contained course in linear systems, with a special focus on the structure theory of linear systems. The later chapters provide additional material for seminars and advanced courses in linear systems theory.

Chapter 2 summarizes the basic tools from linear algebra over rings and modules that are necessary for our subsequent analysis. Our focus is on modules over the ring of polynomials $\mathbb{F}[z]$. Although most of the results in this chapter can be extended to principal ideal domains, we refused the temptation of trying to develop linear systems theory over a general principal ideal domain and preferred to stick to the ring of polynomials $\mathbb{F}[z]$. We define the coprimeness and divisibility of polynomials, for both scalar and matrix polynomials, and introduce basic equivalence notions for polynomial matrices. Proofs of the Hermite and Smith canonical forms are provided. We then study quotient modules of the free module $\mathbb{F}[z]^m$ by a full submodule. They serve as abstract versions of the polynomial models introduced in Chapter 3. Theorem 2.54 is an algebraic version of the celebrated commutant lifting theorem and allows us to characterize module homomorphisms on quotient modules and polynomial models via intertwining relations. We introduce doubly coprime factorizations and characterize them via unimodular extensions and the solvability of associated matrix Bezout equations. An important notion in a mathematical theory is that of an isomorphism. For quotient modules this leads to two different notions of the equivalence of polynomial matrices: by unimodular equivalence or using polynomial equivalence via left and right coprime intertwining relations. Theorem 2.46 shows that these two definitions are essentially equivalent. In our subsequent discussion of isomorphism notions for linear systems representations, this theorem basically proves the equivalence of the two notions on strict system equivalence proposed by Rosenbrock (1970) and Fuhrmann (1977).

Chapter 3 introduces the two main players in our algebraic approach to linear systems, i.e., polynomial models X_D and their dual counterpart, rational models X^D. While rational models X^D correspond to solution spaces of systems of higher-order differential or difference equations, polynomial models serve as a concretization of quotient modules. The basic idea behind studying the two spaces comes from operator theory; the connection to systems theory was first expounded by Fuhrmann (1976). Both spaces X_D and X^D are $\mathbb{F}[z]$-modules that are associated with a nonsingular polynomial matrix $D(z)$. The module structure endows both spaces with the action of a shift operator S_D and S^D, respectively. Understanding the algebraic properties of these functional model spaces, together with the action of the shift operators, is key to all subsequent analysis. The invariant subspaces of the shift are naturally identified with polynomial matrix factorizations and direct sum decompositions into shift-invariant subspaces correspond to mutually coprime or skew-prime factorizations. Theorem 3.14 establishes a bijective correspondence between the Boolean lattices of invariant subspaces and polynomial matrix factorizations. The module homomorphisms of polynomial models are characterized via the intertwining relations among polynomial matrices. The standard properties of such module homomorphisms are injective, surjective, or bijective and are characterized in terms of left and right coprimeness. Dual results are obtained for rational model spaces and their associated module homomorphisms. Finally, we prove a version of the Chinese remainder theorem for polynomial matrices.

Chapter 4 contains a concise treatment of linear systems theory, with a special focus on structural issues such as system representations, realization theory, reachability, and observability. We focus on discrete-time systems, which enables us to work over an arbitrary field. Various system models are introduced, such as input/output representations, state-space models, and higher-order polynomial matrix descriptions, such as Rosenbrock models. We briefly discuss behaviors, as introduced by J.C. Willems, and identify them with rational models associated with rectangular polynomial matrices. Although behaviors define a very elegant and powerful approach to modeling linear systems, we will not treat them systematically in the book and restrict ourselves to input–output or state-space descriptions.

Following a self-contained discussion of realization theory, we introduce the shift realization for polynomial matrix representations. The reachability and observability of the shift realization are characterized in terms of the left and right coprimeness of polynomial matrix representations. We construct a reachability map and compute its kernel. Polynomial system matrices are shown to be polynomially equivalent if and only if the respective shift realizations are similar. Thus the full counterpart to Kalman's realization theorem is developed for the more general Rosenbrock representations. Poles and zeros of linear systems are defined and characterized for polynomial system representations.

We then turn to the problem of open-loop control. Surprisingly, at least to us, this problem has not been solved in textbooks on linear systems, at least to the best of our knowledge. Thus, Section 4.8 is devoted to the solution of this problem. We do so by inverting the restricted reachability map, using the technique of embedding polynomial intertwining relations into suitable doubly coprime factorizations. An explicit inversion formula for the restricted reachability map is given in terms of appropriate doubly coprime factorizations. Thus, without being overly explicit at this point, we realize flatness as a special case of our construction. Explicit formulas for open-loop control of higher-order Rosenbrock models are derived. Characterizations of minimal-degree solutions via reachability indices are given. Open-loop controls for steering into a subspace are discussed, as is the dual problem of state reconstruction.

Chapter 5 introduces a new tensor product approach to the classical stability criteria of Routh–Hurwitz and Hermite–Fujiwara. We consider tensor products of polynomial and rational models, both for the underlying vector space structures and for the module structures of functional spaces. Identifying the space of module homomorphisms between polynomial models with a suitable tensor product representation leads us to a coordinate-free approach to Bezoutian maps and associated polynomial Sylvester equations. We are convinced that the potential of this tensor product approach to Bezoutians has not been exhausted. A polynomial extension of the classical Ostrowski–Schneider inertia theorem is proven, from which we then deduce the Hermite–Fujiwara stability criterion.

Chapter 6 provides a study of state feedback control, and contains complete proofs of basic pole placement results, such as Wonham's theorem and Rosenbrock's theorem. A study of reachability indices and the Brunovsky canonical form is developed from a polynomial point of view. We then consider the stabilization task by dynamic feedback controllers. Thus, we introduce coprime factorizations over the ring RH_∞ of stable proper rational functions and derive the Youla–Kucera parameterization of stabilizing controllers. We include an introduction to the standard notions and results from geometric control theory, such as conditioned and controlled invariant subspaces. These subspaces generalize the familiar notion of an invariant subspace for a linear operator. Using functional models, polynomial characterizations of controlled and conditioned invariant subspaces are developed. These concepts will appear again in Chapter 7 on observer theory and in Chapter 11 on the synchronization of networks.

The main focus of Chapter 7 is observer theory. In contrast to many textbooks, here the theory is developed for general functional observers, i.e., for the task of estimating a linear function of states. Following a discussion of the central properties of the observation process, such as trackability, detectability, or reconstructibility, we then turn to a study of various classes of functional observers and characterize them in purely polynomial terms. Asymptotic tracking observers are characterized either in terms of the solvability of an associated Sylvester equation, using conditioned invariant subspaces, or via the solvability of certain linear equations of rational functions. Theorem 7.30 seems to be new and allows for a constructive approach to functional observers via shift realizations.

The third part of the book is devoted to an analysis of networks of systems. The analysis of interconnected systems, in particular with respect to stability properties and synchronization phenomena, often requires the use of methods from graph theory. Thus, we summarize in Chapter 8 several results and concepts from graph theory. This includes a discussion of the Perron–Frobenius theorem and weighted adjacency and Laplacian matrices and their spectral properties. Our proof of the Perron–Frobenius theorem depends on the properties of the Hilbert metric on convex cones, together with a contraction mapping theorem on pointed convex cones. We believe that this approach, due to Birkhoff (1957), is of independent interest. From the Perron–Frobenius theorem we obtain a simple finite-dimensional version of the ergodic theorem, which suffices for a study of the elementary stochastic properties of Markov chains. We characterize connectivity properties both for directed and undirected graphs and introduce weighted adjacency matrices and Laplacians for directed graphs. Although we do not discuss aspects of formation control in detail, we present a brief introduction to Euclidean distance geometry and relevant notions from graph theory, such as graph rigidity. We then study the spectral properties of the Laplacian of a strongly connected weighted digraph. Explicit formulas for the eigenvalues and eigenvectors of Laplacians of simple graphs such as paths and cycles are given. Following Doerfler and Bullo (2013), we introduce the Kron reduction of a graph and relate it to the Schur complement of the graph Laplacian. This may be of interest when studying the model reduction of networks.

Chapter 9 goes beyond the previous theory and starts with a reachability and observability analysis of networks of linear systems. The basic network model that we adopt consists of a finite number of dynamic node systems that are interconnected, either statically or dynamically, by coupling parameters. We assume that one knows how to control such node systems individually and that one knows the interconnection structure between the node systems. Using these data, we address the problem of how to construct a control for the entire network. This calls for finding reachability criteria for interconnected systems that reveal the impact of the interconnection structure as clearly as possible. To obtain such informative conditions for the reachability and observability of networks is in fact the very purpose of this chapter.

Following a discussion of simple examples, we introduce the basic models for representations of interconnected systems. Both state-space formulations and higher-order polynomial system representations are considered. We prove an equivalence principle that basically asserts that the system-theoretic properties of such networks are independent of the various representations of the node systems. This principle was first established by Rosenbrock and Pugh (1974) and is here extended to a general tool analyzing the strict system equivalence for interconnected systems. As a consequence, we derive explicit Hautus-type characterizations of reachability and observability for interconnected systems, possibly including dynamic couplings. Next, we study the influence of coupling structure on a network. The easiest networks are homogeneous ones, namely, those where the node systems are all identical SISO systems. A simple reachability characterization of homogeneous networks appears in Theorem 9.15, thereby reproving a result by Hara, Hayakawa and Sugata (2009) and, earlier, by Sontag (1979). Subsequent sections deal with the goal of characterizing the reachability of networks with coupling patterns defined by trees or circular or periodic couplings.

Chapter 10 discusses in detail reachability and observability properties for standard interconnection structures, such as parallel, series, and feedback connections, leading to new, elegant proofs for the classical characterizations of reachability and observability by Callier and Nahum (1975) and Fuhrmann (1975). The remaining parts of the chapter then focus on the problem of computing open-loop controls for standard interconnections of systems. This is already considered in the preceding Section 1.1 for the special case of parallel connected SISO systems. Here we are interested in the general multivariable situation described as follows. Suppose that "local" controls are known that steer N linear systems from the zero state individually to given states. If the systems are coupled, how does one compute from the local controls a single global input function that steers the zero state of the network to the terminal states of the node systems? More precisely, can one derive a formula that inputs the local controls and the coupling parameters and outputs the desired global control? In Section 10.2, we solve this problem for parallel interconnections. In between, we discuss the relation of computing open-loop controls to the problem of tangential interpolation. Our solution minimizes the complexity of computations by efficiently using local information on the interconnected nodes.

The last two chapters shift the focus by turning to an analysis of the control of macroscopic states, such as synchronization and consensus, clustering, and ensemble control. Chapter 11 presents a brief introduction to the tasks of consensus and synchronization for networks of linear multiagent systems. Following a brief review of established linear models for consensus, we describe two nonlinear models, those by Vicsek and Krause–Hegselmann. We illustrate how consensus algorithms can be used to design distributed solution methods for linear systems of equations. The remaining parts of Chapter 11 then focus on the important issue of synchronization in networks of linear systems. We first define synchronizability in the state space for first-order representations of nodes and derive necessary and sufficient conditions for the interconnection parameters to achieve synchronization and clustering. We then develop an independent and equivalent approach to the synchronization of higher-order node systems, where synchronizability is defined for the partial states. Specific conditions for synchronization are obtained for various arrays of second-order oscillators.

Chapter 12 deals with the task of controlling ensembles of states. To get an idea of what is meant by that, consider a task where a probability distribution on the state space is to be steered to a terminal distribution using a linear control system. This is an infinite-dimensional control problem for the induced transport equation on the space of probability density functions. In simplified form, this problem arises in the control of parameter-dependent linear systems, where a single, parameter-dependent input function is sought that steers the zero state arbitrarily close to a desired family of terminal states. This property is termed *ensemble reachability* and has been studied in the context of quantum control and nuclear magnetic resonance spectroscopy; see Li and Khaneja (2006). Ensemble control is the open-loop version of what used to be called the blending problem in robust feedback control. We derive necessary as well as sufficient conditions for ensemble reachability for single-input systems. One can view a parameter-dependent family of systems as an associated system in a Hilbert space and, thus, apply tools from infinite-dimensional systems theory. Using the Fourier transform, this implies, for example, a weak version of reachability for certain partial differential equations, such as the one-dimensional heat equation. By sampling a parameter-dependent family of systems at an infinite set of parameter values, an infinite network of linear systems is obtained. This connects to recent work on controlling infinite platoons or spatially invariant systems. From an operator-theoretic viewpoint, this amounts to studying control problems defined by the action of composition operators on Hardy spaces. Since reachability is equivalent to the cyclicity of a state-transition operator, there are connections with the work by Beurling on cyclic shifts. A sufficient condition for the reachability of such composition operators is shown in a very special situation. It should be interesting to extend these results to linear systems defined on vectorial Hardy spaces.

Part I
Algebraic Systems Theory: Foundations

Chapter 2
Rings and Modules of Polynomials

Since concepts from algebra such as rings and modules play a dominant role in our polynomial approach to linear systems, it seems useful to collect in this chapter some of the most relevant concepts and results from linear algebra. We assume that the reader is already familiar with the basic definitions and facts of linear algebra over a field. Starting from such a background, our aim is to generalize the well-known theory of vector spaces and linear operators over a field to an analogous theory of modules, defined over the ring of polynomials in one variable. From a system-theoretic point of view this corresponds to passing from a state-space point of view to a frequency-domain description of linear systems, and even further on to a module-theoretic approach. This algebraic approach to linear systems theory was first expounded in the very influential book by Kalman, Falb and Arbib (1969) and has led to a new area in applied mathematics called **algebraic systems theory**. It underpins our work on the functional model approach to linear systems, a theory that forms the mathematical basis of this book. In doing so it proves important to be able to extend the classical arithmetic properties of scalar polynomials, such as divisibility, coprimeness, and the Chinese remainder theorem, to rectangular polynomial matrices. This is done in the subsequent sections of this chapter. Key facts are the Hermite and Smith canonical forms for polynomial matrices and rational matrices, respectively, Wiener–Hopf factorizations, and the basic existence and uniqueness properties for left and right coprime factorizations of rational matrix-valued functions. We study doubly coprime factorizations that are associated with intertwining relations between pairs of left and right coprime polynomial matrices. The existence of such doubly coprime factorizations will play a crucial role later on in our approach to open-loop control.

Notation. Throughout this book we denote by $\mathbb{N} = \{0,1,2,\ldots\}$ and $\mathbb{N}_+ = \{1,2,3,\ldots\}$ the sets of nonnegative and positive integers, respectively, while $\mathbb{Z} = \{\ldots,-2,-1,0,1,2,\ldots\}$ denotes the ring of integers. The fields of rational, real, and complex numbers are denoted by $\mathbb{Q}$, $\mathbb{R}$, and $\mathbb{C}$, respectively. A general field is denoted by $\mathbb{F}$, while $\overline{\mathbb{F}}$ denotes its algebraic closure.

© Springer International Publishing Switzerland 2015
P.A. Fuhrmann, U. Helmke, *The Mathematics of Networks of Linear Systems*, Universitext, DOI 10.1007/978-3-319-16646-9_2

2.1   Rings and Ideals

We assume that the reader knows the definitions of groups, rings, and fields from basic courses on abstract linear algebra; the elementary textbook by Fuhrmann (2012) provides an introduction to linear algebra that is very much in the spirit of this book. Recall that a **ring** R is a set with two operations, "*addition*" $+$ and "*multiplication*" $\cdot$, such that the following rules apply:

1. $(R,+)$ is an abelian group with additive identity element 0;
2. Multiplication is associative and distributive, i.e., $x \cdot (y+z) = x \cdot y + x \cdot z$ for all $x, y, z \in R$. There exists an identity element $1 \in R$ such that $1 \cdot x = x$ for all $x \in R$.

In the sequel the shorter notation xy will be used to denote the product $x \cdot y$ of two ring elements. Throughout this book a ring is always assumed to be commutative, i.e., R satisfies the following additional property:

3. $xy = yx$ for all $x, y \in R$.

The identity element 1 for multiplication is allowed to be equal to 0. In this case, one obtains the zero ring $R = \{0\}$.

An element $b \in R$ is called a **divisor** of $a \in R$ if there exists $c \in R$, with $a = bc$. A **zero divisor** in R is an element $a \in R$ for which there exists a nonzero element $x \in R$, with $ax = 0$. A ring R (with $1 \neq 0$) is called an **integral domain** if R possesses no zero divisor $a \neq 0$. Standard examples of integral domains are the ring $\mathbb{Z}$ of integers and the ring $\mathbb{F}[z]$ of polynomials in a single variable z over a field $\mathbb{F}$.

The most interesting subsets in a ring are the ideals. An **ideal** I of R is a nonempty subset $I \subset R$ with the properties

$$a, b \in I \implies a + b \in I,$$

$$a \in I \text{ and } r \in R \implies ra \in I.$$

Obviously, $I = R$ is an ideal; every other ideal is called **proper**. The zero ideal $\{0\}$ is another trivial example of an ideal; it is contained in every ideal. The set of zero divisors in a ring $R \neq \{0\}$ is an ideal, too. A very important construction of ideals is as follows. Let $a_1, \ldots, a_n$ be ring elements. Then the set

$$(a_1, \ldots, a_n) := a_1 R + \cdots + a_n R = \left\{ \sum_{i=1}^{n} a_i r_i \mid r_1, \ldots, r_n \in R \right\}$$

defines an ideal in R. It is called the **ideal generated** by $a_1, \ldots, a_n$. Ideals

$$(a) = aR := \{ar \mid r \in R\}$$

that are generated by single ring elements a are called **principal ideals**. A **principal ideal domain** (**PID**) is an integral domain such that every ideal of the ring is a principal ideal. Standard examples of PIDs are the ring of integers and the ring of polynomials in a single variable over a field.

Since the additive group $(R, +)$ of a ring is abelian, every ideal I of R is a normal subgroup of $(R, +)$. Therefore, the **factor group** of coset elements

$$R/I := \{x + I \mid x \in R\}$$

is a group. Here we use the notation

$$x + I := \{x + a \mid a \in I\}.$$

The elements $x + I$ of R/I are called **residue classes** (or **cosets**) of x modulo I. The equality of residue classes defines an equivalence relation

$$x \equiv a \bmod I \iff x - a \in I.$$

The sum and product of two elements in R/I are defined by

$$(x + I) + (y + I) := (x + y) + I \tag{2.1}$$

and

$$(x + I) \cdot (y + I) := xy + I, \tag{2.2}$$

respectively. Note that (2.2) makes sense only because of the second closure property that defines the notion of an ideal. A map $f : R \longrightarrow S$ between rings is called a **ring homomorphism** if it satisfies $f(x + y) = f(x) + f(y)$, $f(xy) = f(x)f(y)$, and $f(1) = 1$ for all $x, y \in R$. This implies $f(0) = 0$ and $f(-x) = -f(x)$. A **ring isomorphism** is a bijective ring homomorphism. The inverse of a ring isomorphism is then an isomorphism, too. The **kernel** of a ring homomorphism f is defined as

$$\mathrm{Ker}\ f = \{x \subset R \mid f(x) = 0\}.$$

Thus the kernel of a ring homomorphism on R is an ideal. It is the zero ideal if and only if f is injective. The next result shows that every ideal of R is the kernel of a suitable ring homomorphism $f : R \longrightarrow S$.

Theorem 2.1. *Let I be an ideal of R. Then the operations (2.1), (2.2) define a ring structure on R/I such that the quotient map*

$$\pi : R \to R/I, \quad \pi(x) := x + I$$

becomes a surjective ring homomorphism with kernel $\text{Ker } \pi = I$. *The ring* R/I *is called the* **quotient ring** *of* R *by* I. *The zero element of* R/I *is the coset* $0 + I = I$, *and the identity element of* R/I *is* $1 + I$.

Consider a ring homomorphism $f : R \longrightarrow S$ and an ideal I of R, with $I \subset \text{Ker } f$. Then $f(a + x) = f(a)$ for $x \in I$. This shows that the map $\bar{f} : R/I \longrightarrow S, \bar{f}(a + I) := f(a)$, is well defined and is in fact a ring homomorphism. We obtain the following commutative diagram of homomorphisms:

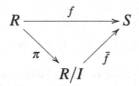

The kernel of the induced map $\bar{f}$ on the quotient ring R/I is equal to

$$\text{Ker } \bar{f} = \{a + I \mid a \in \text{Ker } f\}$$

and, hence, is isomorphic to $\text{Ker } f/I$. In particular, for a surjective homomorphism $f : R \longrightarrow S$ the induced map $\bar{f} : R/\text{Ker } f \longrightarrow S$ defines a ring isomorphism.

The ideals in a quotient ring are characterized as follows.

Proposition 2.2. *Let* I *be an ideal of a ring* R *and* $\pi : R \longrightarrow R/I$ *denote the quotient map. Then* $J \mapsto J' := \pi^{-1}(J)$ *defines a bijective correspondence between ideals* J *of* R/I *and ideals* J' *of* R *that contain* I.

Proof. For ideals J of R/I one verifies that $\pi^{-1}(J)$ is an ideal of R that contains I. Conversely, if J' is an ideal of R with $I \subset J'$, then $J = \pi(J')$ is an ideal of R/I with $J' = \pi^{-1}(J)$.   ∎

The **units** of a ring are defined as the invertible elements of R, i.e., the elements $u \in R$ for which there exists $b \in R$, with $ub = 1$. The set of units of R forms an abelian group $R^{\times}$ under multiplication. A ring $R \neq \{0\}$ is a field if and only if the set of units is equal to $R - \{0\}$. The group of units of $\mathbb{Z}$ is $\{-1, 1\}$, while the units of the ring of polynomials $\mathbb{F}[z]$ are the nonzero constant polynomials. A **prime** of an integral domain R is a nonzero element $p \in R \setminus R^{\times}$ with the property that whenever p divides a product ab of ring elements $a, b \in R$, then p divides a or p divides b. In the ring of integers, the primes are $\pm p$ for p a prime number. A generalization of the notion of primes is that of irreducibility. An element $a \in R \setminus \{0\}$ that is not a unit is called **irreducible** if it cannot be written as a product $a = bc$ of elements $b, c \in R$ that are not units. It is easily seen that the primes in a ring are irreducible. In a PID, the converse is true, too. We next determine the units of a quotient ring.

Lemma 2.3. *Let* I *be an ideal of a ring* R. *A coset* $a + I$ *is a unit of* R/I *if and only if*

$$aR + I = R.$$

Proof. $a + I$ is a unit of R/I if and only if there exists $b \in R$, with $ab + I = 1 + I$. But this is equivalent to $1 \in aR + I$, i.e., to $aR + I = R$. ∎

If an ideal I contains a unit u of R, then $1 = uu^{-1} \in I$, which is equivalent to $I = R$. A proper ideal I of R is called **prime** if

$$ab \in I \implies a \in I \text{ or } b \in I.$$

Thus the zero ideal $I = \{0\}$ is prime if and only if R is an integral domain. A nonzero element p of an integral domain R is prime if and only if pR is a prime ideal. An ideal $I \neq R$ is called **maximal** if I is maximal with respect to the inclusion of ideals, i.e., if I is not strictly contained in a proper ideal J of R. By Zorn's lemma, or equivalently by the axiom of choice, every ring $R \neq \{0\}$ has a maximal ideal; more precisely, every proper ideal I is contained in a maximal ideal.

There is a simple characterization of maximal and prime ideals in a ring.

Proposition 2.4. *Let I denote an ideal in a ring R.*

1. *I is maximal if and only if R/I is a field.*
2. *I is prime if and only if R/I is an integral domain.*

Proof. I is a maximal ideal if and only if $aR + I = R$ for all $a \in R \setminus I$. By Lemma 2.3, this is equivalent to $a + I$ being a unit in R/I for all $a + I \neq I$. Thus I is maximal if and only if all nonzero elements in R/I are invertible, i.e., if and only if R/I is a field. For the second claim, note that a ring element a satisfies $a \notin I$ if and only if $a + I$ is nonzero in R/I. Therefore, the condition $ab \in I \implies a \in I$ or $b \in I$ is equivalent to $(a + I)(b + I) = I \implies a + I = I$ or $b + I = I$, i.e., to the zero divisor property. The result follows. ∎

Proposition 2.4 implies that every maximal ideal is prime, but the converse is not true in general. For the ring of integers $\mathbb{Z}$, the maximal ideals coincide with the nonzero prime ideals and are of the form $p\mathbb{Z}$ for $p \geq 1$ a prime number. The reason is that $\mathbb{Z}$ is a PID. In fact, in a PID the nonzero prime ideals coincide with the maximal ideals.

2.2 Divisibility and Coprimeness of Polynomials

The divisibility of elements in a ring is an algebraic notion that is defined in exactly the same way as it is for integers. Here we focus on explaining this concept for the ring of polynomials $\mathbb{F}[z]$ in a single variable z. Thus the elements of $\mathbb{F}[z]$ are polynomials $p(z) = p_0 + \cdots + p_n z^n$ with coefficients $p_0, \ldots, p_n$ in the field $\mathbb{F}$. In the sequel we will often denote a polynomial $p(z)$ simply by p whenever this notation is more convenient. The sums and products of two polynomials $p(z), q(z) \in \mathbb{F}[z]$ are defined in the usual way. The zero polynomial is the constant polynomial $p(z) = 0$, while the constant polynomial $p(z) = 1$ is the multiplicative unit of $\mathbb{F}[z]$. Thus $\mathbb{F}[z]$

is a ring that has no nonzero zero divisors. If the leading coefficient $p_n \in \mathbb{F}$ of the polynomial $p(z)$ is nonzero, then n is called the **degree** of $p(z)$ and is denoted by $\deg p = n$. The degree of the zero polynomial is defined as $\deg 0 = -\infty$, with the convention that $-\infty < n$ for each nonnegative integer $n \in \mathbb{N}$. A polynomial $d(z) \in \mathbb{F}[z]$ is called a **divisor** of $a(z) \in \mathbb{F}[z]$ (symbolically: $d|a$) if there exists $b(z) \in \mathbb{F}[z]$, with $a(z) = d(z)b(z)$. Using the notion of ideals, one sees that the divisibility of ring elements is equivalent to the inclusion of principal ideals, i.e.,

$$d|a \iff a\mathbb{F}[z] \subset d\mathbb{F}[z].$$

Let $a_1, \ldots, a_n \in \mathbb{F}[z]$ be polynomials that are not all equal to zero. A polynomial $d = \gcd(a_1, \ldots, a_n) \in \mathbb{F}[z]$ is called the **greatest common divisor** of $a_1, \ldots, a_n$ if

1. $d|a_1, d|a_2, \ldots, d|a_n$;
2. If $d' \in \mathbb{F}[z]$ divides $a_1, a_2, \ldots, a_n$, then d' is a divisor of d.

Similarly, the **least common multiple** of $a_1, \ldots, a_r \in \mathbb{F}[z] \setminus \{0\}$ is defined as a polynomial $c = \operatorname{lcm}(a_1, \ldots, a_n) \in \mathbb{F}[z] \setminus \{0\}$ satisfying

1. $a_1|c, a_2|c, \ldots, a_n|c$;
2. If $c' \in \mathbb{F}[z]$ satisfies $a_1|c', a_2|c', \ldots, a_n|c'$, then c divides c'.

It follows directly from the definition that two polynomials d, d' are greatest common divisors of $a_1, \ldots, a_n \in \mathbb{F}[z]$ (not all zero) if and only if there exists a nonzero constant $u \in \mathbb{F}$ with $d' = ud$. Similarly, two least common multiples c and c' of $a_1, \ldots, a_n \in \mathbb{F}[z] \setminus \{0\}$ are unique up to multiplication by nonzero constants, i.e., $c' = uc$ for $u \in \mathbb{F}$ nonzero. Polynomials $a_1, \ldots, a_n \in \mathbb{F}[z]$ that are not all zero are called **coprime** if the greatest common divisor is a unit of $\mathbb{F}[z]$. They are called **mutually coprime** or **pairwise coprime** if a_k and a_l are coprime for all $k \neq l$. The mutual coprimeness of finitely many polynomials implies coprimeness, but not the other way around. In fact, the polynomials $(z-1)(z-2), (z-1)(z-3), (z-2)(z-3)$ are clearly coprime but not mutually coprime.

We next prove that $\mathbb{F}[z]$ is a PID. It is easy to check that the degree satisfies the following two properties:

1. $\deg (pq) = \deg (p) + \deg (q)$;
2. $\deg (p+q) \leq \max\{\deg (p), \deg (q)\}$.

This leads to a proof of a very important property of a ring of polynomials, i.e., uniqueness by division with remainders.

Lemma 2.5 (Division with Remainders). *Let $p(z), q(z) \in \mathbb{F}[z]$ be polynomials, with p not the zero polynomial. Then there exists a unique representation*

$$q(z) = a(z)p(z) + r(z), \tag{2.3}$$

with polynomials $a(z), r(z) \in \mathbb{F}[z]$ satisfying $\deg r < \deg p$.

Proof. Let $p(z) = \sum_{j=0}^{m} p_j z^j$ and $q(z) = \sum_{j=0}^{n} q_j z^j$. For $\deg q < \deg p$ the unique choice in (2.3) is $a = 0, r = q$. Assume $n := \deg q \geq \deg p = m$. To show the existence of the representation, we proceed by induction on n. The polynomial $q_1(z) := q(z) - q_n p_m^{-1} z^{n-m} p(z)$ has degree $\leq n - 1$. Therefore, by induction, $q_1 = a_1 p + r_1$, with $\deg r_1 < \deg p$. This implies $q = ap + r_1$, with $a = a_1 + q_n p_m^{-1} z^{n-m}$. For the uniqueness, assume that $q = a_1 p + r_1 = a_2 p + r_2$. Thus $(a_1 - a_2)p = r_1 - r_2$. But the right-hand side has degree smaller than that of p. Thus $a_1 = a_2$ and, therefore, also $r_1 = r_2$.   ∎

Theorem 2.6. *The ring of polynomials $\mathbb{F}[z]$ is a PID. Moreover, for coprime polynomials $p, q \in \mathbb{F}[z]$ there exist unique polynomials $a, b \in \mathbb{F}[z]$, with*

$$p(z)a(z) + q(z)b(z) = 1, \quad \deg(b) < \deg(p).$$

Proof. The trivial ideal $I = \{0\}$ is a principal ideal. If $I \neq \{0\}$ is a nonzero ideal in $\mathbb{F}[z]$, then there exist $b \in I \setminus \{0\}$ such that $\deg(b) \geq 0$ has the smallest possible value. Obviously, $b\mathbb{F}[z] \subset I$. Let $a \in I \setminus \{0\}$. Using division with remainders we obtain $a = qb + r$ and $r = 0$ or $\deg(r) < \deg(b)$. Clearly, $\deg(r) < \deg(b)$ is impossible because otherwise $r = a - qb \in I \setminus \{0\}$, in contradiction to the choice of b. Hence, $r = 0$, and therefore $a = qb \in b\mathbb{F}[z]$. Since a was chosen arbitrarily, we conclude $I \subset bR$. Thus $b\mathbb{F}[z] \subset I$ implies that $I = b\mathbb{F}[z]$ is a principal ideal.

Since $\mathbb{F}[z]$ is a PID, there exist solutions $\bar{u}, \bar{v} \in \mathbb{F}[z]$ of $a(z)\bar{u}(z) + b(z)\bar{v}(z) = 1$. By division with remainder, $\bar{v} = \alpha a + r$ for elements $\alpha, r \in \mathbb{F}[z]$, with $\deg(r) < \deg(a)$. Thus $u := \bar{u} + b\alpha$ and $v := r$ satisfy $au + bv = 1$, with $\deg(v) < \deg(a)$. For uniqueness, consider two solutions of $au_i + bv_i = 1, \deg(v_i) < \deg(a), i = 1, 2$. Then $u := u_1 - u_2$ and $v := v_2 - v_1$ satisfy $au = bv$. Since a and b are coprime, a must divide v, and therefore $\deg(a) \leq \deg(v)$. Assume $y \neq 0$. Using $\deg(-v_1) = \deg(v_1)$, we obtain $\deg(v) = \deg(v_2 - v_1) \leq \max\{\deg(v_2), \deg(v_1)\} < \deg(a)$. But this is a contradiction. Thus $v = 0$, and therefore also $u = 0$.   ∎

The next result characterizes the main divisibility properties of the ring of polynomials.

Proposition 2.7. *1. The greatest common divisor and the least common multiple of a finite number of elements $a_1, \ldots, a_n \in \mathbb{F}[z] \setminus \{0\}$ always exist.*
2. An element $d \in \mathbb{F}[z]$ is the greatest common divisor of $a_1, \ldots, a_n \in \mathbb{F}[z] \setminus \{0\}$ if and only if

$$a_1 \mathbb{F}[z] + \cdots + a_n \mathbb{F}[z] = d\mathbb{F}[z].$$

In particular, if a polynomial $d(z)$ is a greatest common divisor of $a_1, \ldots, a_n \in \mathbb{F}[z] \setminus \{0\}$, then there exist polynomials $f_1, \ldots, f_n \in \mathbb{F}[z]$ such that

$$a_1(z)f_1(z) + \cdots + a_n(z)f_n(z) = d(z). \tag{2.4}$$

3. *Polynomials* $a_1, \ldots, a_n \in \mathbb{F}[z]$, *not all zero, are coprime if and only if the principal ideals* $a_1\mathbb{F}[z], \ldots, a_n\mathbb{F}[z]$ *satisfy*

$$a_1\mathbb{F}[z] + \cdots + a_n\mathbb{F}[z] = \mathbb{F}[z]. \tag{2.5}$$

Equivalently, there exist polynomial solutions $f_1, \ldots, f_n \in \mathbb{F}[z]$ *to the* **Bezout equation**

$$a_1(z)f_1(z) + \cdots + a_n(z)f_n(z) = 1. \tag{2.6}$$

4. $d(z) \in \mathbb{F}[z]$ *is a least common multiple of* $a_1, \ldots, a_n \in \mathbb{F}[z] \setminus \{0\}$ *if and only if*

$$a_1\mathbb{F}[z] \cap \cdots \cap a_n\mathbb{F}[z] = d\mathbb{F}[z].$$

Proof. An element d' of $\mathbb{F}[z]$ is a divisor of $a_1, \ldots, a_n \in \mathbb{F}[z] \setminus \{0\}$ if and only if d' divides every element a_i. But this is equivalent to the condition that $a_1\mathbb{F}[z] + \cdots + a_n\mathbb{F}[z] \subset d'\mathbb{F}[z]$. Since $\mathbb{F}[z]$ is a PID, the ideal $a_1\mathbb{F}[z] + \cdots + a_n\mathbb{F}[z]$ is a principal ideal, i.e., there exists $\delta \in \mathbb{F}[z]$, with $a_1\mathbb{F}[z] + \cdots + a_n\mathbb{F}[z] = \delta\mathbb{F}[z]$. The preceding argument implies that each common divisor d' satisfies $\delta\mathbb{F}[z] \subset d'\mathbb{F}[z]$, i.e., d' divides δ. Moreover, δ is a common divisor and therefore a greatest common divisor. Conversely, if $d = \gcd(a_1, \ldots, a_n)$, then $a_1\mathbb{F}[z] + \cdots + a_n\mathbb{F}[z] = \delta\mathbb{F}[z] \subset d\mathbb{F}[z]$, and therefore d divides δ. Since d is the greatest common divisor, we conclude $d = c\delta$, for a nonzero constant $c \in \mathbb{F} \setminus \{0\}$. This proves both the first and second claims. It also proves that a greatest common divisor d satisfies (2.4). Equation (2.6) implies that 1 is a greatest common divisor of $a_1, \ldots, a_n$, and therefore $a_1, \ldots, a_n$ are coprime. Conversely, if they are coprime, then 1 is a greatest common divisor and, therefore, by (2.5), satisfies the Bezout equation. The last claim concerning the least common multiple is proven analogously. ∎

We briefly address some computational issues concerning greatest common divisors and the Bezout identity. The classical Euclidean algorithm enables us to determine the greatest common divisor of a pair of polynomials.

Theorem 2.8 (Euclidean Algorithm). *For nonzero polynomials* $a(z), b(z) \in \mathbb{F}[z] \setminus \{0\}$ *there exist uniquely determined polynomials* $r_i(z) \in \mathbb{F}[z]$ *and* $q_i(z) \in \mathbb{F}[z]$, *with* $0 \le \deg r_{i+1} < \deg r_i$, $r_0 = a, r_1 := b, r_{n-1} := 0$, *and*

$$r_0(z) = q_1(z)r_1(z) + r_2(z),$$
$$r_1(z) = q_2(z)r_2(z) + r_3(z),$$
$$\vdots$$
$$r_{n-1}(z) = q_n(z)r_n(z),$$

where $r_n(z) \ne 0$. *The polynomial* $r_n(z)$ *is a greatest common divisor of* $a(z)$ *and* $b(z)$.

Proof. $r_n(z)$ is a divisor of $r_{n-1}(z)$, and $r_{n-2}(z) = q_{n-1}(z)r_{n-1}(z) + r_n(z)$ implies that the polynomial $r_n(z)$ divides $r_{n-2}(z)$. Proceeding inductively backward in n, we obtain that $r_n(z)$ is a common divisor of $r_0(z), \ldots, r_{n-1}(z)$ and, in particular, is a common divisor of $a(z), r_1(z) = b(z)$. Suppose that $d(z)$ is a common divisor of $a(z), b(z)$. Then $d(z)$ is a common divisor of $r_2(z) = a(z) - q_1(z)r_1(z) = a(z) - q_1(z)b(z)$. We conclude that $d(z)$ is a common divisor of $r_0(z), r_1(z), r_2(z)$. Thus, proceeding by induction, we see that $d(z)$ is a common divisor of $r_0(z), \ldots, r_n(z)$. Every common divisor of $a(z), b(z)$ therefore divides $r_n(z)$. Thus $r_n(z)$ is a greatest common divisor. The uniqueness of r_i, q_i follows from the uniqueness property of division with remainders. ∎

The Euclidean algorithm can be used to compute a solution to the Bezout identity. Thus, for polynomials $a(z), b(z) \in \mathbb{F}[z]$, assume that $r_k(z), q_k(z) \in \mathbb{F}[z]$ have been computed using the Euclidean algorithm, $k = 0, \ldots, n$. Consider for $k = 0, \ldots, n-2$ the pair of second-order recursions in $\mathbb{F}[z]$:

$$u_{k+2}(z) = u_k(z) - q_{k+1}(z)u_{k+1}(z),$$
$$v_{k+2}(z) = v_k(z) - q_{k+1}(z)v_{k+1}(z),$$

with initial conditions $u_0(z) = 1, u_1(z) = 0, v_0(z) = 0, v_1(z) = 1$. By the Euclidean algorithm, $r_{k+2} = r_k - q_{k+1}r_{k+1}$ for $k = 0, \ldots n-1$. Thus $s_k := u_k a + v_k b$ satisfies the same recursion $s_{k+2} = s_k - q_{k+1}s_{k+1}$ with $s_0 = a = r_0, s_1 = b = r_1$. This implies $r_k = s_k$ for $k = 0, \ldots, n-1$. We conclude that $u_n(z), v_n(z)$ is the solution to the Diophantine equation $r_n = u_n a + v_n b$, where r_n is a greatest common divisor of a, b. If $a(z)$ and $b(z)$ are coprime, then $r_n \in \mathbb{F}$ is a nonzero constant, and therefore $u(z) = r_n^{-1} u_n(z), v(z) = r_n^{-1} v_n(z)$ solves the Bezout identity $u(z)a(z) + v(z)b(z) = 1$.

Another application of the Euclidean algorithm is to compute continued fraction expansions of rational functions. A **rational function** is expressed uniquely as a quotient of polynomials,

$$g(z) = \frac{p(z)}{q(z)}, \tag{2.7}$$

with $q(z)$ monic and $p(z), q(z) \in \mathbb{F}[z]$ coprime. Addition and multiplication of rational functions are defined in the usual way as

$$\frac{p_1(z)}{q_1(z)} + \frac{p_2(z)}{q_2(z)} = \frac{p_1(z)q_2(z) + p_2(z)q_1(z)}{q_1(z)q_2(z)}, \quad \frac{p_1(z)}{q_1(z)} \cdot \frac{p_2(z)}{q_2(z)} = \frac{p_1(z)p_2(z)}{q_1(z)q_2(z)},$$

and with such operations the set of all rational functions $\mathbb{F}(z)$ becomes a field, the rational function field. The **McMillan degree** $\delta(g)$ of a rational function $g(z)$ is defined as

$$\delta(g) = \max\{\deg p, \deg q\}.$$

A rational function $g(z)$ is called **proper** or **strictly proper** whenever deg $p \leq$ deg q or deg $p <$ deg q, respectively. For a proper rational function $g(z) \in \mathbb{F}(z)$ with coprime factorization (2.7), the Euclidean algorithm yields the unique representation

$$q(z) = a_1(z)p(z) + r_1(z),$$
$$p(z) = a_2(z)r_1(z) + r_2(z),$$

$$\vdots$$

$$r_{N-1}(z) = a_{N+1}(z)r_N,$$

with $r_N \in \mathbb{F}$ a nonzero constant. From the first equation we obtain

$$\frac{p(z)}{q(z)} = \frac{1}{a_1(z) + \frac{r_1(z)}{p(z)}}.$$

Proceeding recursively yields the **continued fraction expansion** as

$$\frac{p(z)}{q(z)} = \cfrac{1}{a_1(z) + \cfrac{1}{a_2(z) + \cfrac{1}{a_3(z) + \ldots + \cfrac{1}{a_{N+1}(z)}}}}.$$

Note that deg $q = \sum_{j=1}^{N+1}$ deg a_j, as is easily verified by induction on N.

Simple conditions for the coprimeness of polynomials are available in terms of rank tests for associated resultant matrices. For simplicity we focus on the case of two polynomials. We associate with scalar polynomials $p(z) = \sum_{j=0}^{m} p_j z^j, q(z) = \sum_{j=0}^{n} q_j z^j \in \mathbb{F}[z]$ the **Sylvester resultant matrix** (plotted here for $m \leq n$):

$$\mathrm{Res}(p,q) = \begin{pmatrix} p_0 & & & & q_0 & & & \\ p_1 & p_0 & & & q_1 & q_0 & & \\ \vdots & p_1 & \ddots & & \vdots & q_1 & \ddots & \\ \vdots & \vdots & \ddots & \ddots & \vdots & \vdots & \ddots & q_0 \\ p_m & \vdots & & \ddots & p_0 & \vdots & \vdots & & q_1 \\ & p_m & \cdots & \cdots & p_1 & q_n & \vdots & & \vdots \\ & & \ddots & & \vdots & & q_n & & \vdots \\ & & & \ddots & \vdots & & & \ddots & \vdots \\ & & & & p_m & & & & q_n \end{pmatrix} \in \mathbb{F}^{(m+n) \times (m+n)}.$$

Notice that the first block of columns of $\mathrm{Res}(p,q)$ has n columns while the second block has m columns.

Theorem 2.9. *Two polynomials* $p(z), q(z) \in \mathbb{F}[z]$ *are coprime if and only if the Sylvester resultant* $\mathrm{Res}(p,q)$ *is invertible.*

Proof. Let $\mathbb{F}[z]_{<k}$ denote the k-dimensional $\mathbb{F}$-vector space consisting of all polynomials of degree $< k$. The Sylvester resultant then is the matrix representation of the $\mathbb{F}$-linear map

$$\mathscr{R}_{pq} : \mathbb{F}[z]_{<n} \oplus \mathbb{F}[z]_{<m} \longrightarrow \mathbb{F}[z]_{<n+m}, \quad (a,b) \mapsto a(z)p(z) + b(z)q(z)$$

with respect to the basis $\{(1,0), \ldots, (z^{n-1},0), (0,1), \ldots, (0,z^m)\}$ and $\{1, \ldots, z^{n+m-1}\}$ of $\mathbb{F}[z]_{<n} \oplus \mathbb{F}[z]_{<m}$ and $\mathbb{F}[z]_{<n+m}$, respectively. Both vector spaces $\mathbb{F}[z]_{<n} \oplus \mathbb{F}[z]_{<m}$ and $\mathbb{F}[z]_{<n+m}$ have the same dimension $m + n$. Thus injectivity of $\mathscr{R}_{pq}$ is equivalent to surjectivity and is, hence, equivalent to bijectivity. If $\mathscr{R}_{pq}$ is bijective, then there exist polynomials $a(z) \in \mathbb{F}[z]_{<n}, b(z) \in \mathbb{F}[z]_{<m}$, with $a(z)p(z) + b(z)q(z) = 1$. This implies the coprimeness of $p(z)$ and $q(z)$. Conversely, assume that $p(z)$ and $q(z)$ are coprime. Consider the polynomials $a(z) \in \mathbb{F}[z]_{<n}, b(z) \in \mathbb{F}[z]_{<m}$, with $a(z)p(z) + b(z)q(z) = 0$. Then $p(z)$ must divide $b(z)$ and $q(z)$ must divide $a(z)$. Therefore, $\deg p \leq \deg b < m$ and $\deg q \leq \deg a < n$. This implies $a(z) = 0, b(z) = 0$ and, therefore, the injectivity of $\mathscr{R}_{pq}$. Thus the coprimeness of $p(z)$ and $q(z)$ is equivalent to the invertibility of $\mathscr{R}_{pq}$. This completes the proof. ∎

An important consequence of the Bezout identity is the ability to glue local objects to global ones. We illustrate this property using the Chinese remainder theorem. It is closely related to the interpolation theory of polynomials. Consider the nonzero polynomials $a_1, \ldots, a_n \in \mathbb{F}[z]$ and so-called remainder polynomials $r_1, \ldots, r_n \in \mathbb{F}[z]$. The Chinese remainder theorem in its most basic form asks for the existence of a polynomial $f \in \mathbb{F}[z]$ that satisfies the divisibility relations

$$a_1 | (f - r_1), \ \ldots, \ a_n | (f - r_n).$$

Equivalently, f must satisfy the relations

$$f \in r_1 + a_1 \mathbb{F}[z], \ \ldots, \ f \in r_n + a_n \mathbb{F}[z],$$

or

$$f \equiv r_1 \bmod a_1, \ \ldots, \ f \equiv r_1 \bmod a_n$$

must be satisfied. Here and in the sequel we write $f \equiv a \bmod b$ for polynomials a, b, f if and only if b divides $f - a$. Using this congruence notation we prove the following existence result.

Theorem 2.10. *Let* $a_1,\ldots,a_n \in \mathbb{F}[z] \setminus \{0\}$ *and* $r_1,\ldots,r_n \in \mathbb{F}[z]$. *A solution* $f \in \mathbb{F}[z]$ *of*

$$f(z) \equiv r_1(z) \bmod a_1(z)$$

$$\vdots \tag{2.8}$$

$$f(z) \equiv r_n(z) \bmod a_n(z)$$

exists if and only if, for all $i \neq j$,

$$r_i - r_j \in a_i\mathbb{F}[z] + a_j\mathbb{F}[z].$$

Equivalently, $\gcd(a_i, a_j)$ *divides* $r_i - r_j$ *for* $1 \leq i \neq j \leq n$.

Proof. For simplicity we give the argument only for $n = 2$; the general case is treated by induction on n. Suppose a solution f exists with $f - r_i \in a_i\mathbb{F}[z]$. Then clearly for $i \neq j$ one has $r_j - r_i = (f - r_i) - (f - r_j) \in a_i\mathbb{F}[z] + a_j\mathbb{F}[z]$. Conversely, consider an element $f := r_1 + ca_1 \in r_1 + a_1\mathbb{F}[z]$ for a to-be-determined $c \in \mathbb{F}[z]$. Then $f \in r_2 + a_2\mathbb{F}[z]$ if and only if $f - r_2 = ca_1 + r_1 - r_2 \in a_2\mathbb{F}[z]$. By assumption, $r_1 - r_2 = a_1r + a_2s$ for suitable $r, s \in \mathbb{F}[z]$. Thus $f - r_2 = (c + r)a_1 + sa_2 \in a_2\mathbb{F}[z]$, provided $c = -r$. Thus $f = r_1 - ra_1$ does the job. ∎

Note that the pairwise coprimeness of $a_1,\ldots,a_n \in \mathbb{F}[z]$ implies $a_i\mathbb{F}[z] + a_j\mathbb{F}[z] = \mathbb{F}[z]$ for all $i \neq j$. Thus pairwise coprimeness implies the sufficient condition $r_i - r_j \in a_i\mathbb{F}[z] + a_j\mathbb{F}[z]$ for the existence of solutions to (2.8).

Theorem 2.11 (Chinese Remainder Theorem). *Let* $q_1,\ldots,q_n \in \mathbb{F}[z]$ *be pairwise coprime, and let* $r_1,\ldots,r_n \in \mathbb{F}[z]$ *satisfy* $\deg r_i < \deg q_i, i = 1,\ldots,n$. *Define* $q := q_1\cdots q_n$ *and* $\hat{q}_i = \prod_{j \neq i} q_j, i = 1,\ldots,n$. *For* $j = 1,\ldots,n$ *choose unique solutions* $a_j, b_j \in \mathbb{F}[z], \deg a_j < \deg \hat{q}_j$, *to the Bezout equation*

$$q_j(z)a_j(z) + \hat{q}_j(z)b_j(z) = 1.$$

Let $u_j(z), v_j(z) \in \mathbb{F}[z], \deg v_j < \deg q_j$, *be the unique polynomials obtained by division with remainder*

$$r_j(z)b_j(z) = u_j(z)q_j(z) + v_j(z).$$

Then

$$r(z) = \sum_{j=1}^{n} \hat{q}_j(z)v_j(z)$$

is the unique solution, with $\deg r < \deg q$, *such that*

$$q_1|(r - r_1), \ldots, q_n|(r - r_n).$$

Proof. Each summand of r has degree $\deg \hat{q}_j v_j = \deg \hat{q}_j + \deg v_j < \deg \hat{q}_j + \deg q_j = \deg q$. Thus $\deg r < \deg q$. Moreover, for each $i = 1, \ldots, n$,

$$r = \sum_{j=1}^{n} \hat{q}_j v_j = r_i b_i \hat{q}_i - u_i q + \sum_{j \neq i} \hat{q}_j v_j.$$

The first term satisfies

$$r_i b_i \hat{q}_i - u_i q + q_i \mathbb{F}[z] = r_i b_i \hat{q}_i + q_i \mathbb{F}[z] = r_i (1 - q_i a_i) + q_i \mathbb{F}[z] = r_i + q_i \mathbb{F}[z].$$

For $j \neq i$, the fraction $q_i^{-1} \hat{q}_j = \prod_{k \neq i, j} q_k \in \mathbb{F}[z]$ is a polynomial, and therefore each summand of the second term satisfies $\hat{q}_j v_j \in q_i \mathbb{F}[z]$. Thus

$$r + q_i \mathbb{F}[z] = r_i + q_i \mathbb{F}[z]$$

for all $i = 1, \ldots, n$. This completes the proof. ∎

The following abstract version of the Chinese remainder theorem has the advantage of being valid in every ring R.

Theorem 2.12 (Abstract Chinese Remainder Theorem). *Let R be a ring and $I_1, \ldots, I_n$ ideals of R satisfying*

$$I_k + I_l = R \quad for\ all \quad k \neq l.$$

Then

$$\rho : R \longrightarrow (R/I_1) \times \cdots \times (R/I_n), \quad \rho(x) = (x + I_1, \ldots, x + I_n)$$

is a surjective ring homomorphism, with kernel equal to the ideal

$$I_1 \cdots I_n = \left\{ \sum_{j=1}^{m} x_{j1} \cdots x_{jn} \mid x_{jk} \in I_k, \ k = 1, \ldots, n \right\}$$

generated by the products of elements of $I_1, \ldots, I_n$. In particular, ρ induces a ring isomorphism

$$R/I_1 \cdots I_n \simeq (R/I_1) \times \cdots \times (R/I_n)$$

that maps the group of units $(R/I_1 \cdots I_n)^{\times}$ isomorphically to the direct product $(R/I_1)^{\times} \times \cdots \times (R/I_n)^{\times}$ of the groups of units of R/I_k.

Proof. Obviously, ρ is a ring homomorphism with kernel equal to the intersection of ideals $I_1 \cap \cdots \cap I_n \supset I_1 \cdots I_n$. We proceed by induction on n, the case $n = 1$ being trivially satisfied. Assume that the theorem is true for ideals $I_1, \ldots I_{n-1}$. Then $J_1 :=$

$I_1 \cap \cdots \cap I_{n-1} = I_1 \cdots I_{n-1}$. Trivially, $J_1 I_n \subset J_1 \cap I_n$ and $(J_1 + I_n)(J_1 \cap I_n) \subset J_1(J_1 \cap I_n) + I_n(J_1 \cap I_n) \subset J_1 I_n$. By mutual coprimeness, there exist elements $x_k \in I_k, z_k \in I_n, k = 1, \ldots, n-1$, with $x_k + z_k = 1$. This implies $x_1 \cdots x_{n-1} - 1 = (1 - z_1) \cdots (1 - z_{n-1}) - 1 \in I_n$ and, therefore, $J_1 + I_n = R$. Hence, $I_1 \cdots I_n = J_1 I_n = (I_1 \cap \cdots \cap I_{n-1}) \cap I_n$. This shows the induction step for the kernel formula. Since $J_1 + I_n = R$, there exist $a \in I_1 \cdots I_{n-1}, b \in I_n$, with $a + b = 1$. Now consider elements $r_1, \ldots, r_n$. By induction, there exists $y \in R$ satisfying $y - r_k \in I_k$ for $k = 1, \ldots, n-1$. Set

$$x := by + ar_n.$$

For $k = 1, \ldots, n-1$, we have $x - r_k = by - r_k + ar_n = y - r_k + a(r_n - y) \in I_k$ since $a(r_n - y) \in J_1 \subset I_k$. For $k = n$, thus, $x - r_n = by + ar_n - r_n = b(y - r_n) \in I_n$. This shows that $\rho(x) = (r_1 + I_1, \ldots, r_n + I_n)$, and we are done. ∎

2.3 Modules

A (left) **module** M over a ring R is an abelian group, endowed with an operation $R \times M \longrightarrow M, (r,x) \mapsto rx$, of R on M (called the scalar multiplication on M) that satisfies the following conditions for all $x, y \in M$ and $r, s \in R$:

$$r(x+y) = rx + ry,$$
$$(r+s)x = rx + sx, (rs)x = r(sx),$$
$$1x = x.$$

Thus modules over a field are simply the vector spaces. Moreover, if R is a ring, then multiplication in R turns it into a module. An abelian group can be regarded in a natural way as a $\mathbb{Z}$-module by defining the scalar multiplation $n \cdot g$ for each $n \in \mathbb{N}$ as the n-fold sum $g + \cdots + g$ and by setting $(-n) \cdot g := n \cdot (-g)$. Thus the theory of modules comprises in a very natural way both the theory of vector spaces and that of abelian groups. An additive subgroup $N \subset M$ that is closed under scalar multiplication, i.e., satisfies $RN \subset N$, is called a **submodule** of M. The R-submodules of a ring R are therefore simply the ideals. A **module homomorphism**, or an R-linear map, between R-modules M, N is a map $f : M \longrightarrow N$ that satisfies

$$f(x+y) = f(x) + f(y),$$
$$f(rx) = rf(x)$$

for all $x, y \in M$ and $r \in R$. For R-linear maps the **kernel** and **image** spaces

$$\mathrm{Ker}\, f = \{x \in M \mid f(x) = 0\},$$
$$\mathrm{Im}\, f = \{f(x) \in N \mid x \in M\}$$

are submodules of M and N, respectively. A **module isomorphism** is a bijective R-linear map $f : M \longrightarrow N$. Note that the inverse of a module isomorphism is automatically R-linear. The set of all R-linear maps between modules M, N is an R-module denoted by $\mathrm{Hom}(M, N)$. Scalar multiplication on $\mathrm{Hom}(M, N)$ is defined pointwise via $(r \cdot f)(m) = rf(m)$. The algebraic dual space of M is the R-module $M' := \mathrm{Hom}(M, R)$. A sequence of R-module homomorphisms

$$M \xrightarrow{g} N \xrightarrow{f} P$$

is called **exact** at N if $\mathrm{Ker}\, f = \mathrm{Im}\, g$.

Direct products and quotient modules are defined in the same way as for vector spaces. Thus, for R-modules M, N scalar multiplication $r \cdot (m, n) := (rm, rn)$ defines the unique R-module structure on $M \times N$ such that the projections $M \times N \longrightarrow M, (x, y) \mapsto x$ and $M \times N \longrightarrow N, (x, y) \mapsto y$ are module homomorphisms. The R-module $M \times N$ is called the **direct product** of M with N. In particular, the n-fold direct product R^n is an R-module. If $N \subset M$ is a submodule, then scalar multiplication $r \cdot (x + N) := rx + N$ defines the unique R-module structure on the coset space

$$M/N := \{x + N \mid x \in M\}$$

such that $\pi : M \longrightarrow M/N, x \mapsto x + N$, becomes a module homomorphism. We refer to M/N as the **quotient module** of M by N. One obtains an exact sequence of module homomorphisms as

$$0 \longrightarrow N \xrightarrow{i} M \xrightarrow{\pi} M/N \longrightarrow 0.$$

The following isomorphism result is standard.

Theorem 2.13 (Isomorphism Theorem). *Let $N \subset M$ be a submodule with canonical projection map $\pi : M \longrightarrow M/N$. For each module homomorphism $f : M \longrightarrow P$, with $N \subset \mathrm{Ker}\, f$, there exists a unique module homomorphism $\bar{f} : M/N \longrightarrow P$ such that the diagram commutes*

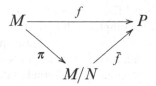

In particular, every surjective module homomorphism $f : M \longrightarrow P$ induces an isomorphism of modules

$$\bar{f} : M/\mathrm{Ker}\, f \longrightarrow P.$$

A family $(x_i|i \in I)$ of elements of an R-module M is said to **generate** M if every element x of M can be written as a finite linear combination,

$$x = r_{i_1} x_{i_1} + \cdots + r_{i_n} x_{i_n},$$

for suitable indices $i_j \in I$ and coefficients $r_{i_j} \in R$. A module M is called **finitely generated** if it has a finite system of generators. A family $(x_i|i \in I)$ of elements in M is called **linearly independent** if every finite subsequence $(x_{i_1}, \ldots, x_{i_n})$ satisfies

$$r_{i_1} x_{i_1} + \cdots + r_{i_n} x_{i_n} = 0 \implies r_{i_1} = \cdots = r_{i_n} = 0.$$

A linearly independent system of generators is called a **basis** of M. A module M is called **free** if a basis of M exists. The free modules M with a finite basis are exactly those that are isomorphic to R^n for some $n \in \mathbb{N}$. One main difference between vector spaces and modules is that a module may not necessarily have a basis. Moreover, a minimal system of generators for a module is not necessarily linearly independent. The general theory of modules is full of pathologies. The following example exhibits such phenomena very clearly.

Example 2.14. Let $R = \mathbb{C}[x,y,z]$ denote the ring of complex polynomials in three variables. Then the R-module

$$M = \{(u,v,w) \in \mathbb{C}[x,y,z]^3 \mid ux + vy + wz = 0\}$$

does not have a basis. The three vectors $(y,-x,0), (z,0,-x), (0,z,y)$ are linearly dependent and form a minimal system of generators for M.

We proceed with a statement of a general fact concerning the existence of a basis for modules over the ring of polynomials $\mathbb{F}[z]$. The subsequent result (and its proof) extends to modules over a PID; see, for example, Hungerford (1974), Theorem IV. 6.1.

Theorem 2.15. *Each submodule M of $\mathbb{F}[z]^n$ is free with a basis of $k \leq n$ elements. Two bases of M have the same number of elements.*

Proof. The proof is by induction on n. For $n = 1$, the submodules M of $\mathbb{F}[z]$ coincide with the ideals. Since $\mathbb{F}[z]$ is a PID, every submodule $M \subset \mathbb{F}[z]$ is generated by a single element $x_1 \in M$. This completes the proof for $n = 1$. Now assume $n \geq 2$. Consider the projection $\pi : \mathbb{F}[z]^n \longrightarrow \mathbb{F}[z]^{n-1}$ onto the first $n-1$ components. Then $N := \pi(M)$ is a submodule of $\mathbb{F}[z]^{n-1}$ and thus, by induction, has a basis $y_1, \ldots, y_k$ of $k \leq n-1$ elements. Moreover, the kernel of the restriction $\pi_M : M \longrightarrow N$ coincides with $M \cap \mathbb{F}[z]e_n$, where e_n denotes the last standard basis vector of $\mathbb{F}[z]^n$. Thus $M \cap \mathbb{F}[z]e_n$ is isomorphic to the ideal

$$\{f(z) \in \mathbb{F}[z] \mid f(z)e_n \in M\}$$

of $\mathbb{F}[z]$. Since $\mathbb{F}[z]$ is a PID, there exists $d(z) \in \mathbb{F}[z]$, with $M \cap \mathbb{F}[z]e_n = d(z)\mathbb{F}[z]e_n$. Consider the preimages $x_1, \ldots, x_k \in M$ with $\pi(x_i) = y_i, i = 1, \ldots, k$. Obviously, $x_1, \ldots, x_k$ are linearly independent. Let $x \in M$. Then

$$y = \pi(x) = a_1 y_1 + \ldots + a_k y_k$$

for suitable $a_1, \ldots, a_k \in \mathbb{F}[z]$. Thus

$$x = a_1 x_1 + \ldots + a_k x_k + bd(z)e_n$$

for a suitable polynomial b. Therefore, $\{x_1, \ldots, x_k, de_n\}$ generates M. If $d = 0$, i.e., if $M \cap \mathbb{F}[z]e_n = \{0\}$, then $\{x_1, \ldots, x_k\}$ is a linearly independent system of generators, i.e., a basis of M. If $d(z) \neq 0$, then every linear combination $a_1 x_1 + \ldots a_k x_k + bd(z)e_n = 0$ implies $0 = \pi(a_1 x_1 + \ldots + a_k x_k + bd(z)e_n) = a_1 y_1 + \ldots + a_k y_k$. Thus $a_1 = \cdots = a_k = 0$, and therefore also $b = 0$. Thus $\{x_1, \ldots, x_k, de_n\}$ is a basis of M. This shows that M is free with at most $k \leq n$ basis elements. To see that two bases of M have the same cardinality, we refer the reader to the proof of Corollary IV. 2.12 in Hungerford (1974). ∎

The second statement in Theorem 2.15 is valid in greater generality. In fact, it is known that two bases of a free module M over a commutative ring have the same cardinalities; see Hungerford (1974), Corollary IV. 2.12. This common cardinality is called the **rank** of M and is denoted by rk M.

In the sequel we continue to denote polynomial matrices $A(z) \in \mathbb{F}[z]^{n \times m}$ also simply by A. This should cause no serious confusion but rather will help to avoid otherwise excessive notation. The **rank** of a polynomial matrix $A \in \mathbb{F}[z]^{n \times m}$ is defined as the rank of the free submodule $A\mathbb{F}[z]^m \subset \mathbb{F}[z]^n$. This coincides with the vector space rank when $A(z)$ is considered as a matrix $A(z) \in \mathbb{F}(z)^{n \times m}$ over the field of rational functions $\mathbb{F}(z)$. This in turn is equal to the **normal rank**, i.e., for $\overline{\mathbb{F}}$ denoting the algebraic closure,

$$\mathrm{rk}\, A = \max_{z \in \overline{\mathbb{F}}} \mathrm{rk}\, A(z).$$

In particular, rk $A = r \leq \min\{m, n\}$. Note that we can always choose a full column rank matrix $B(z) \in \mathbb{F}[z]^{n \times r}$, with Im A = Im B. Recognizing a basis matrix for a submodule of $\mathbb{F}^n$ is easy. In fact, a matrix $A \in \mathbb{F}^{n \times k}$ is a basis matrix of $A\mathbb{F}^k$ if and only if $\mathrm{rk} A = k$. A submodule $M \subset \mathbb{F}[z]^n$ is called **full** if M has a basis of n elements. This is equivalent to the existence of a nonsingular polynomial matrix $D(z) \in \mathbb{F}[z]^{n \times n}$ such that

$$M = D(z)\mathbb{F}[z]^n.$$

Note that a polynomial matrix $D(z) \in \mathbb{F}[z]^{n \times k}$ is called **nonsingular** if $k = n$ and the determinant det $D(z)$ is not the zero polynomial. Thus the full submodules of $\mathbb{F}[z]^n$

correspond to nonsingular polynomial matrices. The equivalent characterizations of the rank of a polynomial matrix are summarized as follows (the trivial proof is omitted).

Proposition 2.16. *For a polynomial matrix $A(z) \in \mathbb{F}[z]^{n \times m}$ the following properties are equivalent:*

1. *The rank of $A(z)$ is equal to r;*
2. *The normal rank is $\max_{z \in \bar{\mathbb{F}}} \text{rk } A(z) = r$;*
3. *The rank of the $\mathbb{F}(z)$-linear map $A(z) : \mathbb{F}(z)^m \longrightarrow \mathbb{F}(z)^n$ is equal to r.*

A square matrix $S(z) \in \mathbb{F}[z]^{n \times n}$ is called **unimodular** if there exists $T(z) \in \mathbb{F}[z]^{n \times n}$ such that $S(z)T(z) = I_n$. It follows then that also $T(z)S(z) = I_n$. Thus the unimodular matrices are exactly the matrices that are invertible over the ring $\mathbb{F}[z]$. Using the multiplicative property of the determinant, this implies the well-known fact that a matrix $S(z) \in \mathbb{F}[z]^{n \times n}$ is unimodular if and only if the determinant $\det S \in \mathbb{F}$ is a nonzero constant. Let $GL_n(\mathbb{F}[z])$ denote the group of unimodular $n \times n$ matrices over $\mathbb{F}[z]$. The following result is often used.

Proposition 2.17. *Every submodule $M \subset \mathbb{F}[z]^n$ is of the form*

$$M = A\mathbb{F}[z]^k$$

for a polynomial matrix $A(z)$ of full column rank k. Suppose that at least one of the matrices $A(z), B(z) \in \mathbb{F}[z]^{n \times k}$ has full column rank. Then $A\mathbb{F}[z]^k = B\mathbb{F}[z]^k$ if and only if there exists a unimodular matrix $U(z) \in GL_k(\mathbb{F}[z])$, with $A(z) = B(z)U(z)$.

Proof. Theorem 2.15 implies that M has a basis consisting of $k \leq n$ elements $a_1(z), \ldots, a_k(z) \in \mathbb{F}[z]^n$. Let $A(z)$ denote the $n \times k$ polynomial matrix with columns $a_1(z), \ldots, a_k(z)$. Then $M = A\mathbb{F}[z]^k$ and $A(z)$ has rank k. $A\mathbb{F}[z]^k = B\mathbb{F}[z]^k$ implies that $B(z) = A(z)U(z)$ and $A(z) = B(z)V(z)$ for matrices $U(z), V(z) \in \mathbb{F}[z]^{k \times k}$. Thus $A(z) = A(z)U(z)V(z)$ and $B(z) = B(z)V(z)U(z)$. Since $A(z)$ has full column rank, this implies $U(z)V(z) = I_k$, and therefore $U(z)$ and $V(z)$ are unimodular. ∎

2.4 Minimal Basis of Modules of Polynomials

In this section we show the existence of certain special basis matrices in modules of polynomial matrices. The existence of such basis matrices for $\mathbb{F}[z]$-modules of rational function spaces goes back to the early work by Dedekind and Weber (1882), where they are called *normal bases*. More recent contributions have been made by Forney (1975) and Münzner and Prätzel-Wolters (1979).

We introduce basis representations of a submodule $M = H(z)\mathbb{F}[z]^k$ that are minimal with respect to their column degrees. Thus, starting with a submodule of $\mathbb{F}[z]^n$, we construct a specific set of generators. The basis thus constructed is closely associated with, for example, state feedback, reachability indices, and Wiener–Hopf factorizations. These connections will be clarified later. We define the **degree of a polynomial vector** $f(z) = (f_1(z), \ldots, f_n(z))^\top \in \mathbb{F}[z]^n$ by

$$\deg f = \max_{1 \le i \le n} \deg f_i.$$

If $h_1(z), \ldots, h_k(z)$ are the columns of an $n \times k$ polynomial matrix $H(z)$, then the numbers $\gamma_i = \deg h_i$, $i = 1, \ldots, k$, are called the **column degrees** of $H(z)$, and the sum $\gamma = \sum_{i=1}^k \gamma_i$ is called the **order** of $H(z)$. A polynomial basis matrix $H(z) \in \mathbb{F}[z]^{n \times k}$ is called a **minimal basis** if its order is minimal among all basis matrices of $M = H(z)\mathbb{F}[z]^k$. We will always assume, by rearranging the basis elements, that they are ordered so that $\gamma_1 \ge \cdots \ge \gamma_k$. Since the order of a polynomial matrix is a nonnegative integer, minimal bases of a submodule of $\mathbb{F}[z]^n$ always exist.

Proposition 2.18. *Every submodule of $\mathbb{F}[z]^n$ has a minimal basis.*

For a polynomial vector $f(z) = (f_1(z), \ldots, f_n(z))^\top \in \mathbb{F}[z]^n$, with $\deg f = \gamma$, let $[f]$ denote the coefficient of z^γ in the expansion of $f(z)$ as a vector polynomial. If $H(z) \in \mathbb{F}[z]^{n \times k}$ denotes a matrix polynomial with columns $h_i(z)$, then define

$$[H]_{hc} = ([h_1], \ldots, [h_k]) \in \mathbb{F}^{n \times k}$$

and call it the **highest column degree coefficient matrix**. A polynomial matrix $H(z)$ is called **column proper** if $[H]_{hc}$ has full column rank. Clearly, $[H]_{hc}$ has full column rank if and only if at least one of its $k \times k$ minors is nonzero.

A basis matrix for a submodule $M \subset \mathbb{F}[z]^n$ is unique up to a right unimodular factor. By the Smith form, every unimodular polynomial matrix is the product of elementary unimodular polynomial matrices, which in turn represent elementary column operations.

Proposition 2.19. *Every full column rank polynomial matrix $H(z) \in \mathbb{F}[z]^{n \times k}$ can be reduced to column proper form by elementary column operations.*

Proof. Assume the column degrees $\gamma_i = \deg p_i$ satisfy $\gamma_1 \ge \cdots \ge \gamma_k$. If $[H]_{hc}$ has full column rank, then we are done. Otherwise, there exists a nontrivial linear combination satisfying

$$\sum_{j=1}^k c_j[h_j] = 0. \tag{2.9}$$

Let $i = \min\{j \,|\, c_j \ne 0\}$. By dividing (2.9) by c_i, we may assume without loss of generality that $c_i = 1$. Clearly, $h_i' = h_i + \sum_{j=i+1}^k c_j z^{\gamma_i - \gamma_j} h_j$ satisfies $\deg h_i' < \deg h_i$. Thus $H' = (h_1, \ldots, h_i', \ldots, h_k)$ satisfies order $H' < \sum_{i=1}^k \gamma_i$. The proof is completed by induction. Note that $H'(z) = H(z)U(z)$, where

$$U(z) = \begin{pmatrix} 1 & & & & \\ & \ddots & & & \\ & & 1 & & \\ & & c_{i+1}z^{\gamma_i - \gamma_{i+1}} & & \\ & & \vdots & \ddots & \\ & & c_k z^{\gamma_i - \gamma_k} & & 1 \end{pmatrix}$$

is unimodular.   ∎

Theorem 2.20. *Let $H(z)$ be an $n \times k$ polynomial matrix with full column rank and column degrees $\gamma_1 \geq \cdots \geq \gamma_k$; $\gamma = \gamma_1 + \cdots + \gamma_k$. The following statements are equivalent.*

1. *$H(z)$ is a minimal basis for the submodule $H(z)\mathbb{F}[z]^k$ of $\mathbb{F}[z]^n$.*
2. *$H(z)$ is column proper, i.e., $\mathrm{rk}[H]_{hc} = k$.*
3. *The maximal degree of the $k \times k$ minors of $H(z)$ is γ.*
4. *For all nonzero polynomials $p = (p_1, \ldots, p_k)^\top \in \mathbb{F}[z]^k$,*

$$\deg H(z)p(z) = \max_i (\deg p_i + \gamma_i).$$

5. *For all $d \in \mathbb{N}$,*

$$\dim M_d = \sum_{i:\gamma_i \leq d} (d - \gamma_i),$$

where

$$M_d = \{ f(z) \in \mathbb{F}[z]^k \mid \deg H(z)f(z) < d \}.$$

Proof. That statement 1 implies statement 2 is a straightforward consequence of Proposition 2.19. To prove the equivalence of statements 2 and 3, let $E(z)$ be a $k \times k$ submatrix of $H(z)$, and let $[E]^*$ denote the corresponding $k \times k$ submatrix of $[H]_{hc}$. Note that $[E]^*$ is not necessarily equal to the highest coefficient matrix $[E]_{hc}$ of $E(z)$. It is easily verified that

$$\det E(z) = \det[E]^* z^\gamma + p(z), \tag{2.10}$$

with $\deg p(z) < \gamma$. Thus the degree of $\det E(z)$ for each $k \times k$ submatrix $E(z)$ of $H(z)$ is at most γ. If $[H]_{hc}$ has full rank, then there exists a nonsingular $k \times k$ submatrix $[E]^*$, and the degree of $\det E(z)$ is equal to γ. Therefore, the maximal degree is γ. Conversely, assuming the maximal degree is γ, then there exists a $k \times k$ submatrix $E(z)$ whose determinant $\det E(z)$ has degree γ. By (2.10), therefore, $\det[E]^* \neq 0$. Thus $[H]_{hc}$ has full rank. This shows that statements 2 and 3 are equivalent.

Next let $p(z) = (p_1(z), \ldots, p_k(z))^\top \in \mathbb{F}[z]^k$; then

$$H(z)p(z) = \sum_{i=1}^{k} p_i(z)h_i(z).$$

Thus

$$\deg(Hp) = \deg \sum_{i=1}^{k} p_i h_i \le \max_i \deg(p_i h_i) = \max_i(\deg p_i + \gamma_i) = \delta.$$

Hence, $H(z)p(z)$ has degree bounded by δ. We show now that the upper bound δ is achieved. Let $I = \{i \mid p_i \ne 0, \deg p_i + \gamma_i = \delta\}$, and let $[p_i] \in \mathbb{F}$ denote the leading coefficients of the scalar polynomials $p_i(z)$, $i \in I$. Then

$$H(z)p(z) = z^\delta \sum_{i \in I} [p_i][h_i] + f(z),$$

with $\deg f < \delta$. If we assume $[H]_{hc}$ to be of full rank, then the column vectors $[h_i], i \in I$ are linearly independent, and therefore

$$\sum_{i=1}^{k} [p_i][h_i] \ne 0,$$

and so statement 2 implies statement 4.

Next assume that statement 4 is satisfied. Suppose there exist α_i such that

$$\sum_{i=1}^{k} \alpha_i[h_i] = 0;$$

then

$$\sum_{i=1}^{k} \alpha_i z^{\gamma_1 - \gamma_i} h_i(z) = z^{\gamma_1} \sum_{i=1}^{k} \alpha_i[h_i] + \text{lower degree terms.}$$

However, by (4), then

$$\deg \sum_{i=1}^{k} \alpha_i z^{\gamma_1 - \gamma_i} h_i(z) = \max[\deg(\alpha_i z^{\gamma_1 - \gamma_i}) + \gamma_i].$$

But the last term is equal to γ_1 if not all α_i are zero. Thus statement 4 implies statement 2.

For every polynomial vector $p(z) \in \mathbb{F}[z]^k$ and $d \in \mathbb{N}$, the bound $\deg H(z)p(z) < d$ if and only if $\max_{p_i \neq 0}(\deg p_i + \gamma_i) < d$ or, equivalently, if and only if $\deg p_i < d - \gamma_i$ for all $i = 1, \ldots, k$. The dimension of the space of such polynomial vectors $(p_1(z), \ldots, p_k(z))^\top$ is equal to $\sum_{i:\gamma_i \leq d}(d - \gamma_i)$, and hence

$$\dim M_d = \sum_{i:\gamma_i \leq d} (d - \gamma_i).$$

This shows that statement 4 implies statement 5. Moreover, the same argument shows that every full column rank matrix $H(z) \in \mathbb{F}[z]^{n \times k}$ satisfies $\deg H(z)p(z) \leq \max_{p_i \neq 0}[\deg p_i + \gamma_i]$. This implies the inequality

$$\dim M_d \geq \sum_{i:\gamma_i \leq d} (d - \gamma_i)$$

for all $d \in \mathbb{N}_0$ and all full column rank matrices $H(z)$.

To prove that statement 5 implies statement 1, let $\sigma(d)$ denote the number of indices j of $H(z)$, with $\gamma_j = d$. Thus, for each $d \in \mathbb{N}_0$, statement 5 implies that

$$\dim M_{d+1} = \sum_{i:\gamma_i \leq d+1} (d+1-\gamma_i) = \sum_{i:\gamma_i \leq d} (d+1-\gamma_i)$$

$$= \sum_{i:\gamma_i \leq d} (d-\gamma_i) + \sum_{i:\gamma_i \leq d} 1$$

$$= \dim M_d + \sum_{r=0}^{d} \sigma(r).$$

By adding and subtracting terms we conclude that

$$\sigma(d) = \dim M_{d+1} + \dim M_{d-1} - 2\dim M_d$$

for all d. By definition of $\sigma(d)$, $\gamma = \sum_{i=1}^{k} \gamma_i = \sum_{d=0}^{\infty} d\sigma(d)$, and therefore γ assumes the same value for all full column rank matrices $H(z)$ that satisfy statement 5. By the preceding arguments, statement 5 is satisfied for each minimal basis $H(z)$. Thus γ is the same for all minimal basis matrices of $H(z)\mathbb{F}[z]^k$. This shows that statement 5 implies statement 1, and the proof is complete.   ∎

Note that knowledge of the numbers $\gamma_1 \geq \ldots \geq \gamma_k$ is equivalent to knowledge of the function $d \mapsto \sigma(d)$. In particular, the column degrees of a minimal basis matrix $H(z)$ depend only on the module $H(z)\mathbb{F}[z]^k$.

Corollary 2.21. *The column degrees of a minimal basis for submodule M are an invariant of the module.*

We therefore define, for a submodule $M \subset \mathbb{F}[z]^n$, the **order indices**, or **minimal indices**, of M as the column degrees of a minimal basis of M. The **order** of a submodule M is defined as the order of a minimal basis of M.

One can characterize all unimodular transformations $U(z) = (u_{ij}(z)) \in GL_k(\mathbb{F}[z])$ that transform a minimal basis matrix $H(z) \in \mathbb{F}[z]^{n \times k}$ into a minimal basis matrix $H(z)U(z)$. The class of all such transformations forms a subgroup of the group of unimodular transformations.

Theorem 2.22. *Let $H(z) \in \mathbb{F}[z]^{n \times k}$ be a minimal basis with column indices $\gamma_1 \geq \dots \geq \gamma_k$, and let $U(z) \in GL_k(\mathbb{F}[z])$ be unimodular. Then $H(z)U(z)$ is a minimal basis if and only if*

1. $\deg u_{ij}(z) \leq \gamma_j - \gamma_i$ for $\gamma_i \leq \gamma_j$.
2. $u_{ij}(z) = 0$ for $\gamma_i > \gamma_j$.

Proof. If $G(z) := H(z)U(z)$ is minimal, then the $j - th$ column $g_j(z)$ of G satisfies

$$\deg g_j = \deg \sum_{i=1}^{k} h_i(z) u_{ij}(z) = \max_i (\deg u_{ij} + \gamma_i) = \gamma_j.$$

This implies $\deg u_{ij} \leq \gamma_j - \gamma_i$ for $\gamma_i \leq \gamma_j$, and $u_{ij} = 0$ otherwise. Conversely, suppose that the unimodular matrix $U(z)$ satisfies statements 1 and 2. Let $U^* = (u_{ij}^*) \in \mathbb{F}^{k \times k}$ denote the constant matrix whose ij-entry u_{ij}^* is defined by the coefficient of $z^{\gamma_j - \gamma_i}$ in $u_{ij}(z)$. Then $U(z)$ and U^* are lower block-triangular invertible matrices, and the leading coefficient matrices $[G]_{hc}$ and $[H]_{hc}$ satisfy

$$[G]_{hc} = [H]_{hc} U^*.$$

Thus $[G]_{hc}$ is full column rank, and therefore $G(z)$ is a minimal basis. ∎

2.5 Divisibility and Coprimeness of Polynomial Matrices

There is a very close connection between the geometry of the lattice of submodules of $\mathbb{F}[z]^m$ and the arithmetic of polynomial matrices. This connection will be a common thread throughout this book. We will return to this topic in Theorem 3.14 when we examine the lattice of invariant subspaces of polynomial, or rational, models. We begin by defining the appropriate notions of divisibility and coprimeness for rectangular polynomial matrices. The noncommutativity of matrices forces one to distinguish between left and right divisors.

Definition 2.23. A polynomial matrix $D(z) \in \mathbb{F}[z]^{n \times q}$ is called a **common left divisor** of matrices $A_i(z) \in \mathbb{F}[z]^{n \times m_i}, i = 1, \dots, r$, if there exist matrices $X_i(z) \in \mathbb{F}[z]^{q \times m_i}, i = 1, \dots, r$, with

$$A_i(z) = D(z)X_i(z), \quad i = 1, \dots, r.$$

A common left divisor $D(z)$ is called a **greatest common left divisor**, denoted by $D(z) = \mathrm{gcld}(A_1(z),\ldots,A_r(z))$, if for every other common left divisor $D'(z) \in \mathbb{F}[z]^{n \times q'}$ there exists $Q(z) \in \mathbb{F}[z]^{q' \times q}$ with $D(z) = D'(z)Q(z)$. A matrix $E(z) \in \mathbb{F}[z]^{n \times q}$ is called a **common right multiple** if there exist matrices $X_i(z) \in \mathbb{F}[z]^{m_i \times q}, i = 1,\ldots,r$, with

$$A_i(z)X_i(z) = E(z), \quad i = 1,\ldots,r.$$

A common right multiple $E(z)$ is called a **least common right multiple**, and is denoted by $E(z) = \mathrm{lcrm}(A_1(z),\ldots,A_r(z))$, if for every other common right multiple $E'(z) \in \mathbb{F}[z]^{n \times q'}$ there exists $Q(z) \in \mathbb{F}[z]^{q' \times q}$, with $E(z)Q(z) = E'(z)$. A matrix $D(z) \in \mathbb{F}[z]^{q \times m}$ is a **common right divisor** of matrices $A_i(z) \in \mathbb{F}[z]^{n_i \times m}, i = 1,\ldots,r$, if there exist matrices $X_i(z) \in \mathbb{F}[z]^{n_i \times q}, i = 1,\ldots,r$, with

$$A_i(z) = X_i(z)D(z), \quad i = 1,\ldots,r.$$

A **greatest common right divisor** $D(z) := \mathrm{gcrd}(A_1(z),\ldots,A_r(z)) \in \mathbb{F}[z]^{q \times m}$ is a common right divisor such that for every other common right divisor $D'(z) \in \mathbb{F}[z]^{q' \times m}$ there exists $Q(z) \in \mathbb{F}[z]^{q' \times q}$, with $D(z) = Q(z)D'(z)$. A matrix $E(z) \in \mathbb{F}[z]^{q \times m}$ is called a **common left multiple** if there exist matrices $X_i(z) \in \mathbb{F}[z]^{q \times n_i}, i = 1,\ldots,r$, with

$$X_i(z)A_i(z) = E(z), \quad i = 1,\ldots,r.$$

A common left multiple $E(z)$ is called a **least common left multiple**, and is denoted by $E(z) = \mathrm{lclm}(A_1(z),\ldots,A_r(z))$, if for every other common left multiple $E'(z) \in \mathbb{F}[z]^{q \times m}$ there exists $Q(z) \in \mathbb{F}[z]^{q' \times q}$, with $Q(z)E(z) = E'(z)$.

Observe that the matrices $A(z) \in \mathbb{F}[z]^{n \times q}$, $B(z) \in \mathbb{F}[z]^{n \times q'}$, $C(z) \in \mathbb{F}[z]^{q \times m}$, and $D(z) \in \mathbb{F}[z]^{q' \times m}$ satisfy the following conditions:

$$A\mathbb{F}[z]^q \subset B\mathbb{F}[z]^{q'} \iff B(z) \quad \text{is a left divisor of} \quad A(z);$$

$$\mathbb{F}[z]^{1 \times q}C \subset \mathbb{F}[z]^{1 \times q'}D \iff C(z) \quad \text{is a right divisor of} \quad D(z). \tag{2.11}$$

This basic property leads to the following polynomial module characterizations of left and right divisors.

Theorem 2.24. *Greatest common left/right divisors and least common left/right multiples of rectangular polynomial matrices $A_1,\ldots,A_r$ exist. They are characterized as follows.*

1. Let $A_i(z) \in \mathbb{F}[z]^{n \times m_i}, i = 1,\ldots,r$. A matrix $D(z) \in \mathbb{F}[z]^{n \times q}$ is a common left divisor if and only if

$$A_1\mathbb{F}[z]^{m_1} + \cdots + A_r\mathbb{F}[z]^{m_r} \subset D\mathbb{F}[z]^q. \tag{2.12}$$

$D(z) \in \mathbb{F}[z]^{n \times q}$ *is a greatest common left divisor if and only if*

$$A_1 \mathbb{F}[z]^{m_1} + \cdots + A_r \mathbb{F}[z]^{m_r} = D\mathbb{F}[z]^q. \tag{2.13}$$

$E(z) \in \mathbb{F}[z]^{n \times q}$ *is a common right multiple if and only if*

$$E\mathbb{F}[z]^q \subset A_1 \mathbb{F}[z]^{m_1} \cap \cdots \cap A_r \mathbb{F}[z]^{m_r}. \tag{2.14}$$

$E(z) \in \mathbb{F}[z]^{n \times q}$ *is a least common right multiple if and only if*

$$E\mathbb{F}[z]^q = A_1 \mathbb{F}[z]^{m_1} \cap \cdots \cap A_r \mathbb{F}[z]^{m_r}. \tag{2.15}$$

2. *Let* $A_i(z) \in \mathbb{F}[z]^{n_i \times m}, i = 1, \ldots, r.$ *A matrix* $D(z) \in \mathbb{F}[z]^{q \times m}$ *is a common right divisor if and only if*

$$\mathbb{F}[z]^{1 \times n_1} A_1 + \cdots + \mathbb{F}[z]^{1 \times n_r} A_r \subset \mathbb{F}[z]^{1 \times q} D. \tag{2.16}$$

$D(z) \in \mathbb{F}[z]^{q \times m}$ *is a greatest common right divisor if and only if*

$$\mathbb{F}[z]^{1 \times n_1} A_1 + \cdots + \mathbb{F}[z]^{1 \times n_r} A_r = \mathbb{F}[z]^{1 \times q} D. \tag{2.17}$$

$E(z) \in \mathbb{F}[z]^{q \times m}$ *is a common left multiple if and only if*

$$\mathbb{F}[z]^{1 \times q} E \subset \mathbb{F}[z]^{1 \times n_1} A_1 \cap \cdots \cap \mathbb{F}[z]^{1 \times n_r} A_r. \tag{2.18}$$

$E(z) \in \mathbb{F}[z]^{q \times m}$ *is a least common left multiple if and only if*

$$\mathbb{F}[z]^{1 \times q} E = \mathbb{F}[z]^{1 \times n_1} A_1 \cap \cdots \cap \mathbb{F}[z]^{1 \times n_r} A_r. \tag{2.19}$$

Proof. Define $A = (A_1, \ldots, A_r) \in \mathbb{F}[z]^{n \times m}, m = \sum_{i=1}^{r} m_i.$ Then D is a greatest common left divisor of $A_1, \ldots, A_r$ if and only if D is a left divisor of A. Using (2.11) we see that this is equivalent to $A_1 \mathbb{F}[z]^{m_1} + \cdots + A_r \mathbb{F}[z]^{m_r} = A\mathbb{F}[z]^m \subset D\mathbb{F}[z]^q$. This proves (2.12). Similarly, characterizations (2.14), (2.16), and (2.18) are shown. We prove (2.13). If $D = \gcd(A_1, \ldots, A_r)$, then (2.12) implies the inclusion $A_1 \mathbb{F}[z]^{m_1} + \cdots + A_r \mathbb{F}[z]^{m_r} \subset D\mathbb{F}[z]^q$. Since $A_1 \mathbb{F}[z]^{m_1} + \cdots + A_r \mathbb{F}[z]^{m_r}$ is a submodule of $\mathbb{F}[z]^n$, Theorem 2.15 implies that $A_1 \mathbb{F}[z]^{m_1} + \cdots + A_r \mathbb{F}[z]^{m_r}$ has a basis. Thus there exists $D' \in \mathbb{F}[z]^{n \times q'}$, with

$$A_1 \mathbb{F}[z]^{m_1} + \cdots + A_r \mathbb{F}[z]^{m_r} = D'\mathbb{F}[z]^{q'}. \tag{2.20}$$

Thus D' is a common left divisor. Since D is the greatest common left divisor, $D = D'Q$ for some $Q \in \mathbb{F}[z]^{q' \times q}$. This implies $D\mathbb{F}[z]^q \subset D'\mathbb{F}[z]^{q'}$. Therefore, $D\mathbb{F}[z]^q = D'\mathbb{F}[z]^{q'} = A_1 \mathbb{F}[z]^{m_1} + \cdots + A_r \mathbb{F}[z]^{m_r}$. Conversely, from $A_1 \mathbb{F}[z]^{m_1} + \cdots + A_r \mathbb{F}[z]^{m_r} = D\mathbb{F}[z]^q$ we conclude that D is a common left divisor. Moreover, by (2.12), each common left divisor D' satisfies $D\mathbb{F}[z]^q = A_1\mathbb{F}[z]^{m_1} + \cdots + A_r\mathbb{F}[z]^{m_r} \subset D'\mathbb{F}[z]^q$, and therefore $D = D'Q$ for some $Q \in \mathbb{F}[z]^{q' \times q}$. Thus D is a greatest common left divisor. This completes the proof of (2.13). One proves (2.17) similarly. For (2.15) we proceed as follows. If E is the least common right multiple, then $E\mathbb{F}[z]^q \subset A_1\mathbb{F}[z]^{m_1} \cap \cdots \cap A_r\mathbb{F}[z]^{m_r}$. The intersection $A_1\mathbb{F}[z]^{m_1} \cap \cdots \cap A_r\mathbb{F}[z]^{m_r}$ is a free submodule of $\mathbb{F}[z]^n$, and thus there exists a matrix $E' \in \mathbb{F}[z]^{n \times q'}$, with $E\mathbb{F}[z]^q \subset A_1\mathbb{F}[z]^{m_1} \cap \cdots \cap A_r\mathbb{F}[z]^{m_r} = E'\mathbb{F}[z]^{q'}$. Thus $E = E'X$ for some $X \in \mathbb{F}[z]^{q' \times q}$. Since E is the least common right multiple, there exists $Q \in \mathbb{F}[z]^{q \times q'}$ with $EQ = E'$. But then $E'\mathbb{F}[z]^{q'} = EQ\mathbb{F}[z]^{q'} \subset E\mathbb{F}[z]^q$, and therefore $E\mathbb{F}[z]^q = A_1\mathbb{F}[z]^{m_1} \cap \cdots \cap A_r\mathbb{F}[z]^{m_r}$. The converse is proven similarly. The case (2.19) runs similarly and is omitted. ∎

We now establish the existence and uniqueness properties of greatest common divisors of matrices; we omit the obvious statements for least common multiples.

Theorem 2.25. *Let $A_i \in \mathbb{F}[z]^{n \times m_i}, i = 1, \ldots, r$, with rank $q := \mathrm{rk}(A_1, \ldots, A_r)$.*

1. *There exists a greatest common left divisor $D \in \mathbb{F}[z]^{n \times q}$ of $A_1, \ldots, A_r$, with $\mathrm{rk}\, D = q$. If D and $D' \in \mathbb{F}[z]^{n \times q'}$ are greatest common left divisors with full column ranks, then $q = q'$, and there exists a unique unimodular transformation $U \in GL_q(\mathbb{F}[z])$, with $D' = DU$. Similarly for greatest common right divisors.*
2. *A square greatest common left divisor $D \in \mathbb{F}[z]^{n \times n}$, with $\det D \neq 0$, exists if and only if $A_1\mathbb{F}[z]^{m_1} + \cdots + A_r\mathbb{F}[z]^{m_r}$ is a full submodule of $\mathbb{F}[z]^n$, i.e., if and only if $\mathrm{rk}(A_1, \ldots, A_r) = n$. Similarly for square greatest common right divisors.*

Proof. The submodule $A_1\mathbb{F}[z]^{m_1} + \cdots + A_r\mathbb{F}[z]^{m_r} \subset \mathbb{F}[z]^n$ is free and has rank $q \leq n$. Thus there exists a full column rank matrix $D \in \mathbb{F}[z]^{n \times q}$, with

$$A_1\mathbb{F}[z]^{m_1} + \cdots + A_r\mathbb{F}[z]^{m_r} = D\mathbb{F}[z]^q.$$

Thus $D(z)$ is a greatest common left divisor. Let $D' \in \mathbb{F}[z]^{n \times q'}$ denote another greatest common left divisor of full column rank. Then $D'\mathbb{F}[z]^{q'} = A_1\mathbb{F}[z]^{m_1} + \cdots + A_r\mathbb{F}[z]^{m_r} = D\mathbb{F}[z]^q$, which implies $\mathrm{rk}\, D' = q$, as well as $D' = DX$ and $D = D'Y$ for suitable matrices $X \in \mathbb{F}[z]^{q \times q'}, Y \in \mathbb{F}[z]^{q' \times q}$. Thus $D' = D'YX$ and $D = DXY$. Since D has full column rank, we obtain $XY = I_q$. Similarly, $YX = I_{q'}$. Thus $q = q'$ and X, Y are unimodular. This proves the first claim. The second claim follows from the first one. ∎

We next define coprimeness for polynomial matrices.

Definition 2.26. 1. Polynomial matrices $A_i \in \mathbb{F}[z]^{n \times m_i}, i = 1, \ldots, r$, are called **left coprime** if there exists a matrix $X \in \mathbb{F}[z]^{q \times n}$ such that $D = \mathrm{gcld}(A_1, \ldots, A_r)$ satisfies $DX = I_n$ or, equivalently, if and only if

$$D\mathbb{F}[z]^q = \mathbb{F}[z]^n.$$

2. The polynomial matrices $A_i \in \mathbb{F}[z]^{n_i \times m}, i = 1,\ldots,r$, are called **right coprime** if there exists a matrix $X \in \mathbb{F}[z]^{m \times q}$ such that $D = \gcd(A_1,\ldots,A_r)$ satisfies $XD = I_m$ or, equivalently, if and only if

$$\mathbb{F}[z]^{1 \times q} D = \mathbb{F}[z]^{1 \times m}.$$

The preceding definition applies to single matrices as well. Thus a polynomial matrix $A(z) \in \mathbb{F}[z]^{m \times n}$ is called **left prime** (or **right prime**) if there exists a polynomial matrix $X(z) \in \mathbb{F}[z]^{m \times n}$ [or $X(z) \in \mathbb{F}[z]^{n \times m}$], with $A(z)X(z) = I_n$ [or $X(z)A(z) = I_m$], respectively. The coprimeness of polynomial matrices is characterized as follows. Let $\overline{\mathbb{F}}$ denote the algebraic closure of the field $\mathbb{F}$.

Theorem 2.27. *1. The matrices $A_i \in \mathbb{F}[z]^{n \times m_i}, i = 1,\ldots,r$, are left coprime if and only if there exist solutions $X_i \in \mathbb{F}[z]^{n_i \times n}, i = 1,\ldots,r$, to the matrix Bezout equation*

$$A_1(z)X_1(z) + \cdots + A_r(z)X_r(z) = I_n \tag{2.21}$$

or, equivalently,

$$\mathrm{rk}\,(A_1(z),\ldots,A_r(z)) = n \quad \forall z \in \overline{\mathbb{F}}.$$

2. The matrices $A_i \in \mathbb{F}[z]^{n_i \times m}, i = 1,\ldots,r$, are right coprime if and only if there exist solutions $Y_i \in \mathbb{F}[z]^{m \times n_i}, i = 1,\ldots,r$, to the matrix Bezout equation

$$Y_1(z)A_1(z) + \cdots + Y_r(z)A_r(z) = I_m$$

or, equivalently,

$$\mathrm{rk}\begin{pmatrix} A_1(z) \\ \vdots \\ A_r(z) \end{pmatrix} = m \quad \forall z \in \overline{\mathbb{F}}.$$

Proof. We only prove the claims concerning left coprimeness. The proof for right coprimeness proceeds by similar arguments. Polynomial matrices $A_1,\ldots,A_r$ are left coprime if and only if $A_1\mathbb{F}[z]^{m_1} + \cdots + A_r\mathbb{F}[z]^{m_r} = \mathbb{F}[z]^n$. But this is equivalent to the solvability of the Bezout identity (2.21). The same argument shows that $A = (A_1,\ldots,A_r)$ is left prime if and only if there exists a polynomial matrix X, with $AX = I_n$. But then clearly $A(z)X(z) = I_n$ for all $z \in \overline{\mathbb{F}}$. Thus rk $A(z) = n$ for all $z \in \overline{\mathbb{F}}$. Conversely, assume the rank conditions are satisfied. For this direction of the proof we refer to a result that is developed later in this chapter, i.e., to the Smith canonical form of a matrix; see Theorem 2.42. Without loss of generality we can assume that A is in Smith normal form; let $d_1,\ldots,d_n \in \mathbb{F}[z]$ be the invariant factors. Since $A(z)$ has full row rank for all $z \in \overline{\mathbb{F}}$, it follows that the invariant factors cannot have a zero in the algebraic closure $\overline{\mathbb{F}}$. But then $d_1,\dots,d_n$ must be nonzero constants, which implies that the Smith normal form is equal to $[I_n,0]$. But then $A(z)$ is left coprime, and we are done. ∎

2.6 Coprime Factorizations of Rational Matrix Functions

Left and right coprime factorizations of polynomial matrices enable one to construct unique factorizations of rational matrix functions. We begin with the following useful lemma.

Lemma 2.28. *Let* $N(z),D(z),M(z),P(z)$ *be polynomial matrices and* $D(z)$ *be nonsingular. Assume that* $P(z) = N(z)D(z)^{-1}M(z)$. *If* N *and* D *are right coprime, then* $M(z) = D(z)M_1(z)$ *for some polynomial matrix* M_1.

Proof. From the Bezout identity $X(z)N(z)+Y(z)D(z) = I$ we conclude that

$$X(z)N(z)D(z)^{-1} + Y(z) = D(z)^{-1}$$

and therefore $X(z)P(z)+Y(z)M(z) = D(z)^{-1}M(z)$. Thus

$$M(z) = D(z)(X(z)P(z)+Y(z)M(z)) = D(z)M_1(z).$$

∎

We now examine coprime factorizations of matrices of rational functions.

Theorem 2.29. *Let* $G(z) \in \mathbb{F}(z)^{p\times m}$.

1. *There exist right coprime polynomial matrices* $N(z) \in \mathbb{F}[z]^{p\times m}, D(z) \in \mathbb{F}[z]^{m\times m}$, *with* $\det D \neq 0$, *such that*

$$G(z) = N(z)D(z)^{-1}. \tag{2.22}$$

If $N_1(z) \in \mathbb{F}[z]^{p\times m}, D_1(z) \in \mathbb{F}[z]^{m\times m}$ *are right coprime with* $\det D_1 \neq 0$ *and*

$$N_1(z)D_1(z)^{-1} = N(z)D(z)^{-1} = G(z),$$

then there exists a unique unimodular matrix $U \in GL_m(\mathbb{F}[z])$ *with* $(N_1(z),D_1(z)) = (N(z)U(z),D(z)U(z))$.
2. *There exist left coprime matrices* $N_\ell(z) \in \mathbb{F}[z]^{p\times m}, D_\ell(z) \in \mathbb{F}[z]^{p\times p}$, *with* $\det D_\ell \neq 0$, *such that*

$$G(z) = D_\ell(z)^{-1}N_\ell(z).$$

If $N_{\ell,1}(z) \in \mathbb{F}[z]^{p\times m}, D_{\ell,1}(z) \in \mathbb{F}[z]^{p\times p}$ *are left coprime with* $\det D_{\ell,1}(z) \neq 0$ *and*

$$D_{\ell,1}(z)^{-1}N_{\ell,1}(z) = D_\ell(z)^{-1}N_\ell(z) = G(z),$$

then there exists a unique unimodular matrix $U \in GL_m(\mathbb{F}[z])$ with $(N_{\ell,1}(z),$ $D_{\ell,1}(z)) = (U(z)N_\ell(z), U(z)D_\ell(z))$.

Proof. We focus on the first claim; the proof of the second claim runs similarly. Let $G = (g_{ij})$, with $g_{ij} = \frac{p_{ij}}{q_{ij}}$, and let $q(z) \in \mathbb{F}[z] \setminus \{0\}$ denote the least common multiple of the denominators q_{ij}. Then $N(z) := q(z)G(z) \in \mathbb{F}[z]^{p \times m}$, and thus the factorization $G(z) = N(z)(q(z)I_m)^{-1}$ exists. The submodule

$$\mathcal{M} := \mathbb{F}[z]^{1 \times p}N + \mathbb{F}[z]^{1 \times m}(qI_m) \subset \mathbb{F}[z]^{1 \times m}$$

contains the linearly independent elements $qe_i^\top, i = 1, \ldots, m$, and therefore has rank $\geq m$. Since $\mathcal{M}$ is a submodule of $\mathbb{F}[z]^{1 \times m}$, one also has rk $\mathcal{M} \leq m$. Thus $\mathcal{M} \subset \mathbb{F}[z]^{1 \times m}$ is a full submodule and, due to Theorem 2.25, there exists a greatest common right divisor $\Delta \in \mathbb{F}[z]^{m \times m}$, det $\Delta \neq 0$, of N, qI_m. Therefore, there exist $\tilde{N} \in \mathbb{F}[z]^{p \times m}$ and $\tilde{D} \in \mathbb{F}[z]^{m \times m}$, with

$$N = \tilde{N}\Delta, \quad qI_m = \tilde{D}\Delta.$$

From

$$\mathbb{F}[z]^{1 \times p}\tilde{N}\Delta + \mathbb{F}[z]^{1 \times m}\tilde{D}\Delta = \mathbb{F}[z]^{1 \times m}\Delta,$$

and observing that Δ is nonsingular, we conclude

$$\mathbb{F}[z]^{1 \times p}\tilde{N} + \mathbb{F}[z]^{1 \times m}\tilde{D} = \mathbb{F}[z]^{1 \times m}.$$

Thus $\tilde{N}(z)$ and $\tilde{D}(z)$ are right coprime. Moreover,

$$G(z) = N(qI_m)^{-1} = (\tilde{N}\Delta)(\tilde{D}\Delta)^{-1} = \tilde{N}\tilde{D}^{-1}.$$

This proves the existence of right coprime factorizations (2.22). If $(N(z), D(z))$ and $(N_1(z), D_1(z))$ are right coprime factorizations of $G(z)$, then $N(z)D(z)^{-1} = G(z) = N_1(z)D_1(z)^{-1}$. Thus $N_1(z) = N(z)D(z)^{-1}D_1(z)$, and Lemma 2.28 implies the existence of $U(z) \in \mathbb{F}[z]^{m \times m}$, with $D_1(z) = D(z)U(z)$. Similarly, $D(z) = D_1(z)V(z)$ for a suitable matrix $V(z) \in \mathbb{F}[z]^{m \times m}$. This implies $D(z) = D(z)U(z)V(z)$. Since det $D \neq 0$, the matrix $D(z)$ is invertible in $\mathbb{F}(z)^{m \times m}$, and therefore $U(z)V(z) = I_m$. Thus $U(z)$ and $V(z)$ are unimodular, with

$$N_1(z) = N(z)D(z)^{-1}D_1(z) = N(z)U(z), \quad D_1(z) = D(z)U(z).$$

This completes the proof.   ■

Unlike the scalar case, it is in general not possible to bound the degrees of the entries of matrix polynomials $N(z), D(z)$ in a coprime factorization $G(z) = N(z)D(z)^{-1}$. In fact, this requires special coprime factorizations where, for example, $D(z)$ is in column proper form. Recall that for rectangular polynomial matrices $P(z) \in \mathbb{F}[z]^{p \times m}$ we define the jth column degree $\deg_j P(z)$ as the column degree of the jth column of $P(z)$. Generalizing from the scalar case, a rational matrix-valued function $G(z) \in \mathbb{F}(z)^{p \times m}$ is called **proper** if $G(z)$ has a Laurent expansion of the form

$$G(z) = \sum_{j=0}^{\infty} G_j z^{-j}.$$

Rational functions of the form

$$G(z) = \sum_{j=1}^{\infty} G_j z^{-j}$$

are called **strictly proper**. $G(z)$ is called **biproper** provided $m = p$ and both $G(z)$ and $G(z)^{-1}$ are proper. Equivalently, the Laurent expansion is of the form $G(z) = \sum_{j=0}^{\infty} G_j z^{-j}$, with G_0 invertible.

Proposition 2.30. *Let* $G(z) = N(z)D(z)^{-1} \in \mathbb{F}(z)^{p \times m}$, *with* $N(z) \in \mathbb{F}[z]^{p \times m}$ *and* $D(z) \in \mathbb{F}[z]^{m \times m}$ *column proper. Then* $G(z)$ *is proper (or strictly proper) if and only if, for all* $j = 1, \ldots, m$,

$$\deg_j N(z) \le \deg_j D(z) \quad (or \ \deg_j N(z) < \deg_j D(z)). \tag{2.23}$$

In particular, this shows that $D(z)^{-1}$ *is proper whenever* $D(z)$ *is row proper or column proper.*

Proof. Let $\gamma_1 \ge \ldots \ge \gamma_m$ denote the column degrees of $D(z)$, and let $\Delta(z) = \mathrm{diag}\,(z^{\gamma_1}, \ldots, z^{\gamma_m})$. Then $D(z)\Delta(z)^{-1}$ is biproper, i.e., $D(z)\Delta(z)^{-1}$ and $\Delta(z)D(z)^{-1}$ are proper. Therefore, $G(z) = N(z)\Delta(z)^{-1}(\Delta(z)^{-1}D(z))^{-1}$ is proper or strictly proper if and only if $N(z)\Delta(z)^{-1}$ is proper or strictly proper, respectively. But this is equivalent to condition (2.23). ∎

We can now prove a version of division with remainders for matrix polynomials.

Theorem 2.31. *Let* $D(z) \in \mathbb{F}[z]^{m \times m}$ *be nonsingular. For every* $N(z) \in \mathbb{F}[z]^{p \times m}$ *there exist unique polynomial matrices* $Q(z) \in \mathbb{F}[z]^{p \times m}, R(z) \in \mathbb{F}[z]^{p \times m}$, *with*

$$N(z) = Q(z)D(z) + R(z),$$

such that $R(z)D(z)^{-1}$ *is strictly proper. If* $D(z)$ *is column proper with column degrees* $\gamma_1 \ge \cdots \ge \gamma_m$, *then such a decomposition exists, with* $\deg_j R(z) < \gamma_j, j = 1, \ldots, m$.

Proof. A rational matrix function has a unique decomposition $N(z)D(z)^{-1} = Q(z) + F(z)$ into a polynomial $Q(z)$ and strictly proper part $F(z)$. Thus $R(z) := F(z)D(z) = N(z) - Q(z)D(z)$ is a polynomial matrix, with $R(z)D(z)^{-1} = F(z)$ strictly proper. This completes the proof of the first part. The second part follows from Proposition 2.30. ∎

A similar version of Theorem 2.31 is true for the solutions of the dual equation

$$N_\ell(z) = D_\ell(z)Q(z) + R(z),$$

provided $D_\ell(z)$ is row proper with minimal row indices $v_1, \ldots, v_m$. This implies the following result.

Proposition 2.32. *Let $D_\ell(z) \in \mathbb{F}[z]^{m \times m}$ be nonsingular and row proper with row degrees $v_1, \ldots, v_m$. Let $f \in \mathbb{F}[z]^m$. Then $f(z)$ can be uniquely represented in the form*

$$f(z) = D_\ell(z)g(z) + r(z), \tag{2.24}$$

with $g(z) \in \mathbb{F}[z]^m$, $r(z) = (r_1(z), \ldots, r_m(z))^\top \in \mathbb{F}[z]^m$, and $\deg r_i < v_i$. Moreover, if $\gamma = \max_{1 \le i \le m}(\deg f_i - v_i) \ge 0$, then $\deg g = \gamma$. Otherwise $g = 0$.

We next relate coprimeness to the very important concept of doubly coprime factorizations. Let

$$G(z) = N_r(z)D_r(z)^{-1} = D_\ell(z)^{-1}N_\ell(z)$$

be a right and left coprime factorization of $G(z) \in \mathbb{F}(z)^{p \times m}$, respectively. This implies the **intertwining relation**

$$N_\ell(z)D_r(z) = D_\ell(z)N_r(z). \tag{2.25}$$

By the assumed coprimeness, one deduces the existence of polynomial matrices $X(z) \in \mathbb{F}[z]^{m \times p}, \bar{X}(z) \in \mathbb{F}[z]^{p \times m}, Y(z) \in \mathbb{F}[z]^{m \times m}, \bar{Y}(z) \in \mathbb{F}[z]^{p \times p}$, with

$$X(z)N_r(z) + Y(z)D_r(z) = I_m$$
$$N_\ell(z)\bar{X}(z) + D_\ell(z)\bar{Y}(z) = I_p. \tag{2.26}$$

Equations (2.25) and (2.26) are equivalent to the matrix identity

$$\begin{pmatrix} Y(z) & X(z) \\ -N_\ell(z) & D_\ell(z) \end{pmatrix} \begin{pmatrix} D_r(z) & -\bar{X}(z) \\ N_r(z) & \bar{Y}(z) \end{pmatrix} = \begin{pmatrix} I & Z \\ 0 & I \end{pmatrix}, \tag{2.27}$$

where $Z := X\overline{Y} - Y\overline{X}$. Therefore, the polynomial matrix $\begin{pmatrix} D_r(z) & -\overline{X}(z) \\ N_r(z) & \overline{Y}(z) \end{pmatrix}$ is

unimodular. Multiplying (2.27) on the right by $\begin{pmatrix} I & -Z \\ 0 & I \end{pmatrix}$, and modifying $\overline{X}(z)$ and

$\overline{Y}(z)$ accordingly, leads directly to the following central result.

Theorem 2.33 (Doubly Coprime Factorization). *Let $N_\ell(z) \in \mathbb{F}[z]^{p \times m}$ and $D_\ell(z) \in \mathbb{F}[z]^{p \times p}$ be right coprime and $N_r(z) \in \mathbb{F}[z]^{p \times m}$ and $D_r(z) \in \mathbb{F}[z]^{m \times m}$ be left coprime, with*

$$D_\ell(z)N_r(z) = N_\ell(z)D_r(z).$$

Then there exist unique polynomial matrices $X(z) \in \mathbb{F}[z]^{m \times p}, \overline{X}(z) \in \mathbb{F}[z]^{m \times p}, Y(z) \in \mathbb{F}[z]^{m \times m}, \overline{Y}(z) \in \mathbb{F}[z]^{p \times p}$, with

$$\begin{pmatrix} Y(z) & X(z) \\ -N_\ell(z) & D_\ell(z) \end{pmatrix} \begin{pmatrix} D_r(z) & -\overline{X}(z) \\ N_r(z) & \overline{Y}(z) \end{pmatrix} = \begin{pmatrix} I_m & 0 \\ 0 & I_p \end{pmatrix},$$

$$\begin{pmatrix} D_r(z) & -\overline{X}(z) \\ N_r(z) & \overline{Y}(z) \end{pmatrix} \begin{pmatrix} Y(z) & X(z) \\ -N_\ell(z) & D_\ell(z) \end{pmatrix} = \begin{pmatrix} I_p & 0 \\ 0 & I_m \end{pmatrix},$$

such that $X(z)D_\ell(z)^{-1}$ and $D_r(z)^{-1}\overline{X}(z)$ are strictly proper.

Proof. The existence of doubly coprime factorizations has been shown already. By Theorem 2.31, there exist unique polynomial matrices $L(z)$ and $X_1(z)$ such that $X(z) = L(z)D_\ell(z) + X_1(z)$ and $X_1(z)D_\ell(z)^{-1}$ is strictly proper. Define $Y_1(z) = Y(z) + L(z)N_\ell(z)$, with $(Y,X) = L(-N_\ell, D_\ell) + (Y_1, X_1)$. Then X_1, Y_1 and $\overline{X}_1 := \overline{X} + D_r L, \overline{Y}_1 := \overline{Y}$ is a doubly coprime factorization, with $X_1 D_\ell^{-1}$ strictly proper. The equality $D_r X_1 = \overline{X}_1 D_\ell$ shows that

$$X_1(z)D_\ell(z)^{-1} = D_r(z)^{-1}\overline{X}(z).$$

Thus the strict properness of $X_1 D_\ell^{-1}$ implies that of $D_r^{-1}\overline{X}_1$. To prove uniqueness observe that the difference $X = X_1 - X_2, Y = Y_1 - Y_2, \overline{X} = \overline{X}_1 - \overline{X}_2, \overline{Y} = \overline{Y}_1 - \overline{Y}_2$ of solutions $X_i, Y_i, \overline{X}_i, \overline{Y}_i$ for the doubly coprime factorizations solves the Sylvester equation

$$D_\ell(z)\overline{Y}(z) + N_\ell(z)\overline{X}(z) = 0.$$

Thus

$$D_\ell(z)\overline{Y}(z) + N_\ell(z)D_r(z)D_r(z)^{-1}\overline{X}(z) = 0.$$

Since $N_\ell(z)D_r(z) = D_\ell(z)N_r(z)$, we obtain

$$D_\ell(z)(\overline{Y}(z) + N_r(z)D_r(z)^{-1}\overline{X}(z)) = 0.$$

Therefore, $N_r(z)D_r(z)^{-1}\overline{X}(z) = -\overline{Y}(z)$ is a polynomial. Lemma 2.28 implies, together with the right coprimeness of $N_r(z), D_r(z)$, that there exists a polynomial matrix $P(z)$, with $\overline{X}(z) = D_r(z)P(z)$. But then $P(z)$ is strictly proper and a polynomial, hence $P(z) = 0$. Therefore, $\overline{X}(z) = 0$ and $\overline{Y}(z) = 0$. Similarly, we conclude that $X(z) = 0$ and $Y(z) = 0$, and the uniqueness part is shown. ∎

Theorem 2.33 easily implies the following result.

Corollary 2.34. *Let $N_r(z) \in \mathbb{F}[z]^{p\times m}$, $D_r(z) \in \mathbb{F}[z]^{m\times m}$ be right coprime, with $\det D_r \neq 0$. Then there exists a unique extension to a unimodular matrix*

$$\begin{pmatrix} D_r(z) & -X_r(z) \\ N_r(z) & Y_r(z) \end{pmatrix} \in GL_{m+p}(\mathbb{F}[z]) \tag{2.28}$$

such that $D_r(z)^{-1}X_r(z)$ is strictly proper.

Proof. Let $D_\ell(z)^{-1}N_\ell(z) = G(z)$ be a left coprime factorization of the rational matrix function $G(z) := N_r(z)D_r(z)^{-1}$. Choose the unique solution $X_r(z), Y_r(z)$ of the Bezout equation $D_\ell(z)Y_r(z) + N_\ell(z)X_r(z) = I_p$ such that $D_r(z)^{-1}X_r(z)$ is strictly proper. The preceding calculation shows that, with this choice of $X_r(z)$ and $Y_r(z)$, matrix (2.28) is unimodular. ∎

Doubly coprime factorizations will play a fundamental role in our subsequent approach to open-loop control.

2.7 Wiener–Hopf Factorizations

Wiener–Hopf factorizations, and the corresponding factorization indices, play a central role in algebraic systems theory, in particular in the study of state feedback. They are also of independent interest in the study of the invertibility of Toeplitz operators and systems of singular integral operators. A basic existence and uniqueness result for Wiener–Hopf factorizations is due to Dedekind and Weber (1882), with subsequent studies by, for example, Gohberg and Krein (1960), Gohberg, Lerer and Rodman (1978), and Fuhrmann and Willems (1979).

Definition 2.35. Let $G(z) \in \mathbb{F}(z)^{p\times m}$ be rational. A **left Wiener–Hopf factorization at infinity** is a factorization of the form

$$G(z) = G_-(z)D(z)G_+(z), \tag{2.29}$$

with $G_+(z) \in \mathbb{F}[z]^{m\times m}$ unimodular, $G_-(z) \in \mathbb{F}[[z^{-1}]]^{p\times p}$ biproper rational, and

$$D(z) = \begin{pmatrix} \Delta(z) & 0 \\ 0 & 0 \end{pmatrix},$$

where $\Delta(z) = \mathrm{diag}\,(z^{\kappa_1},\dots,z^{\kappa_r})$. The integers $\kappa_i \in \mathbb{Z}$, assumed decreasingly ordered, are called the **left factorization indices** at infinity, or the **left Wiener–Hopf indices**. **Right Wiener–Hopf factorization** and **right factorization indices** are defined analogously, with the roles of G_+ and G_- in (2.29) interchanged.

Theorem 2.36. *Let $G(z) \in \mathbb{F}(z)^{p \times m}$ be rational.*

1. *There exist left and right Wiener–Hopf factorizations of $G(z)$.*
2. *The left and right Wiener–Hopf factorization indices are uniquely determined, up to permutations.*
3. *If $G(z)$ is polynomial, then all its right (left) Wiener–Hopf factorization indices are nonnegative. Similarly, if $G(z)$ is a proper rational function, then all its right (left) Wiener–Hopf factorization indices are nonpositive.*

Proof. We prove statement 1. Let us assume first that $G(z) \in \mathbb{F}[z]^{p \times m}$. By Proposition 2.19, there exists a unimodular matrix $V(z)$ such that $G(z)V(z) = (G_1(z)\ 0)$, where $G_1(z)$ is a $p \times r$ column proper matrix with column degrees $\kappa_1 \geq \cdots \geq \kappa_r$. The column properness of $G_1(z)$ implies the left invertibility of the highest coefficient matrix $[G_1]_{hc}$, and we denote by E_0 a left inverse of $[G_1]_{hc}$. Let E be an invertible $p \times p$ matrix whose first r rows coincide with E_0. Then $E\left(G_1(z)\ 0\right)$ is of the form

$$E\left(G_1(z)\ 0\right) = \begin{pmatrix} \Omega_{11}(z) & 0 \\ \Omega_{21}(z) & 0 \end{pmatrix} \begin{pmatrix} \Delta(z) & 0 \\ 0 & I_{m-r} \end{pmatrix} = \begin{pmatrix} \Omega_{11}(z) & 0 \\ \Omega_{21}(z) & I_{p-r} \end{pmatrix} \begin{pmatrix} \Delta(z) & 0 \\ 0 & 0 \end{pmatrix},$$

where

$$\Omega_{11}(z) = I + \Omega'_{11}(z),$$

and $\Omega'_{11}(z) \in z^{-1}\mathbb{F}[[z^{-1}]]^{r \times r}$, $\Omega_{21}(z) \in \mathbb{F}[[z^{-1}]]^{(p-r) \times r}$,

$$\Delta(z) = \mathrm{diag}\,(z^{\kappa_1},\dots,z^{\kappa_r}) \in \mathbb{F}[z]^{r \times r}.$$

Since $\Omega_{11}(z)$ is biproper, it has a proper inverse $\Gamma_{11}(z)$. Define $\Gamma_0(z) \in \mathbb{F}[[z^{-1}]]^{p \times p}$ by

$$\Gamma_0(z) = \begin{pmatrix} \Gamma_{11}(z) & 0 \\ -\Omega_{21}(z)\Gamma_{11}(z) & I \end{pmatrix};$$

then

$$\Gamma_0(z) \begin{pmatrix} \Omega_{11}(z) & 0 \\ \Omega_{21}(z) & I \end{pmatrix} \begin{pmatrix} \Delta(z) & 0 \\ 0 & 0 \end{pmatrix} = \begin{pmatrix} \Delta(z) & 0 \\ 0 & 0 \end{pmatrix}.$$

Altogether, $(\Gamma_0(z)E)G(z)V(z) = \begin{pmatrix} \Delta(z) & 0 \\ 0 & 0 \end{pmatrix}$, implying the factorization $G(z) = G_-(z)D(z)G_+(z)$, with

$$G_-(z) = (\Gamma_0(z)E)^{-1}, \quad G_+(z) = V(z)^{-1}, \quad D(z) = \begin{pmatrix} \Delta(z) & 0 \\ 0 & 0 \end{pmatrix}.$$

We note that the proof also shows that the polynomial matrix $G(z)$ has nonnegative left Wiener–Hopf indices, which are the column degrees of G.

If $G(z)$ is rational, then there exists a nonzero scalar polynomial $g(z) = z^n + g_{n-1}z^{n-1} + \cdots + g_0$ such that $g(z)G(z)$ is a polynomial matrix. Let $g(z)G(z) = H_-(z)D(z)H_+(z)$ be a left factorization. Since $g(z) = z^n\gamma(z)$, with

$$\gamma(z) = 1 + g_{n-1}z^{-1} + \cdots + g_0 z^{-n},$$

it follows that a left factorization of $G(z)$ exists, with $G_-(z) = \gamma(z)^{-1}H_-(z)$, $G_+(z) = H_+(z)$, and $D_1(z) = z^{-n}D(z)$. This proves the existence of left factorizations. To obtain a right factorization of $G(z)$, we transpose a left factorization of $G(z)^\top$. This completes the proof of statement 1.

To prove the uniqueness of Wiener–Hopf indices, we can assume without loss of generality that $p = m$. Thus, assume that $G(z) = \Gamma_1(z)\Delta_1(z)U_1(z)$ and $G(z) = \Gamma_2(z)\Delta_2(z)U_2(z)$ are two left Wiener–Hopf factorizations, with $\Gamma_1(z), \Gamma_2(z)$ biproper, $U_1(z), U_2(z)$ unimodular, and $\Delta_1 = \mathrm{diag}(z^{\kappa_1}, \ldots, z^{\kappa_m})$, $\Delta_2 = \mathrm{diag}(z^{\lambda_1}, \ldots, z^{\lambda_m})$. Define $|\kappa| = \sum_{i=1}^m \kappa_i$ and $|\lambda| = \sum_{i=1}^m \lambda_i$. Taking the determinant of $G(z)$ we obtain

$$\det \Gamma_1(z)z^{|\kappa|} = c \det \Gamma_2(z)z^{|\lambda|} \tag{2.30}$$

for a nonzero constant $c \in \mathbb{F}$. Since $\Gamma_1(z)$ is biproper, so is the determinant $\det \Gamma_1(z)$. Thus $\det \Gamma_1(z) = \sum_{j=0}^\infty \gamma_j z^{-j}, \gamma_0 \neq 0$, and similarly for $\det \Gamma'(z)$. Comparing coefficients in (2.30) we conclude that $|\kappa| = |\lambda|$. From $\Gamma_1(z)\Delta_1(z)U_1(z) = \Gamma_2(z)\Delta_2(z)U_2(z)$, we have

$$\Gamma(z) = \Delta_2(z)U(z)\Delta_1(z)^{-1}$$

for $\Gamma(z) = \Gamma_2(z)^{-1}\Gamma_1(z)$ biproper and $U(z) = U_2(z)U_1(z)^{-1}$ unimodular. Thus the entries $u_{ij}(z)$ of the unimodular polynomial matrix $U(z)$ satisfy

$$u_{ij}(z)z^{\lambda_i - \kappa_j} = \gamma_{ij}(z)$$

for the biproper rational functions $\gamma_{ij}(z)$. Thus either $u_{ij} = 0$ or $\lambda_i \leq \kappa_j$. Since $U(z)$ is nonsingular, there exists a permutation π such that $u_{i\pi(i)} \neq 0$ for all $i = 1, \ldots, m$. This implies that $\lambda_i \leq \kappa_{\pi(i)}$ for $i = 1, \ldots, m$. Since $|\lambda| = |\kappa|$, we conclude that $\lambda_i = \kappa_{\pi(i)}$ for all i. Thus the Wiener–Hopf indices coincide up to a permutation π. This proves the second part.

By the preceding argument for the first part, every polynomial matrix has nonnegative left (or right) Wiener–Hopf indices. Now assume that $G(z)$ is a strictly proper rational function. We set $G(z) = K(z)/t(z)$, with $K(z)$ a polynomial matrix and $t(z)$ the lowest common multiple of the denominators of all entries of $G(z)$. Clearly, $t(z) = z^\tau \gamma(z)$ for some nonnegative τ and a scalar, rational biproper $\gamma(z)$. Let $K(z) = \Gamma(z)\Delta(z)U(z)$ be a left Wiener–Hopf factorization and $\Delta(z) = \operatorname{diag}(z^{v_1}, \ldots, z^{v_p})$. Thus

$$G(z) = \frac{\Gamma(z)}{\gamma(z)} \operatorname{diag}(z^{v_1 - \tau}, \ldots, z^{v_p - \tau}) U(z).$$

This shows that

$$\operatorname{diag}(z^{v_1 - \tau}, \ldots, z^{v_p - \tau}) U(z) = \gamma(z) \Gamma(z)^{-1} G(z)$$

is strictly proper. Since $U(z)$ is a polynomial matrix, we must have $v_i - \tau < 0$, and these are the factorization indices of $G(z)$.   ∎

It is clear that if $G(z)$ is singular, then the right and left factorizations of $G(z)$ are not unique, although the Wiener–Hopf indices are. However, even in the nonsingular case we do not have uniqueness of the factorization. The next result is due to Gohberg and Krein (1960).

Theorem 2.37. *Let $G(z) \in \mathbb{F}(z)^{m \times m}$ be rational and nonsingular, and let*

$$G(z) = G_-(z)\Delta(z)G_+(z) = G'_-(z)\Delta'(z)G'_+(z)$$

be two left factorizations. Then $\Delta(z) = \Delta'(z)$, and there exists a unimodular matrix $U(z) \in \mathbb{F}[z]^{m \times m}$ satisfying

$$\begin{aligned} u_{ij} &= 0 && \text{if } \kappa_i > \kappa_j, \\ \deg(u_{ij}) &\leq \kappa_j - \kappa_i && \text{if } \kappa_j \geq \kappa_i, \end{aligned} \qquad (2.31)$$

for which $G'_+(z) = U(z)G_+(z)$ and $G'_-(z) = G_-(z)\Delta(z)U(z)^{-1}\Delta(z)^{-1}$.

Proof. The equality $\Delta(z) = \Delta'(z)$ follows from Part 2 of Theorem 2.36. Thus $U(z) := G'_+(z)G_+(z)^{-1}$ is unimodular and satisfies $G'_+(z) = U(z)G_+(z)$ and $G'_-(z) = G_-(z)\Delta(z)U(z)^{-1}\Delta(z)^{-1}$. Moreover, each entry $u_{ij}(z)$ of $U(z)$ is a polynomial such that $u_{ij}(z)z^{\kappa_i - \kappa_j}$ is biproper. Therefore,

$$\deg u_{ij} + \kappa_i - \kappa_j \leq 0$$

for all i, j. The result follows.   ∎

Clearly, the set of all unimodular matrices $U(z) \in \mathbb{F}[z]^{m \times m}$ satisfying condition (2.31) forms a multiplicative group, which we will call the **left factorization group**. An analogous result is true for right factorizations.

2.8 Hermite and Smith Normal Forms

In this section we explain one of the main structural tools for analyzing modules of polynomials, the Smith normal form of matrices.

Definition 2.38. Let $A(z)$ and $B(z)$ be polynomial matrices in $\mathbb{F}[z]^{m \times n}$. We say $A(z)$ and $B(z)$ are **unimodularly equivalent** if there exist unimodular polynomial matrices $U(z) \in \mathbb{F}[z]^{m \times m}$ and $V(z) \in \mathbb{F}[z]^{n \times n}$ such that $B(z) = U(z)A(z)V(z)$. Similarly, $A(z)$ and $B(z)$ are called **unimodular left equivalent** (or **unimodular right equivalent**) if there exists a unimodular matrix $U(z) \in GL_m(\mathbb{F}[z])$ [or $V(z) \in GL_n(\mathbb{F}[z])$], with $B(z) = U(z)A(z)$ [or $B(z) = A(z)V(z)$].

Clearly, unimodular equivalence defines an equivalence relation on the set of rectangular polynomial matrices, i.e., it is a reflexive, symmetric, and transitive relation. From standard linear algebra it is known that two matrices A and B over a field are unimodularly equivalent if and only if they have the same rank. In fact, every matrix A over a field is unimodularly equivalent to $\text{diag}(I_r, 0)$, with $r = \text{rk} A$. For matrices over rings the situation is not so easy, and deciding unimodular equivalence of matrices can become difficult. However, for PIDs there is a simple answer. We begin with the somewhat simpler problem of finding normal forms for unimodular left equivalence. The next theorem is stated only for full column rank polynomial matrices. A similar result is true for full row rank polynomial matrices $A(z) \in \mathbb{F}[z]^{m \times n}$, with $m \leq n$.

Theorem 2.39 (Hermite Normal Form). *Let $A(z) \in \mathbb{F}[z]^{m \times n}$ be full column rank. Then there exists a unimodular matrix $U(z) \in GL_m(\mathbb{F}[z])$ such that $B(z) = U(z)A(z)$ has the form*

$$
B(z) = \left(
\begin{array}{ccc|ccc}
b_{11} & b_{12} & \ldots & b_{1n} \\
0 & b_{22} & \ldots & b_{2n} \\
\vdots & \vdots & \ddots & \vdots \\
0 & 0 & \ldots & b_{nn} \\
\hline
0 & 0 & \ldots & 0 \\
\vdots & \vdots & \ddots & \vdots \\
0 & 0 & \ldots & 0
\end{array}
\right).
$$

The polynomials $b_{ij}(z)$ are uniquely determined, with $b_{ii}(z)$ monic and $\deg b_{ij}(z) < \deg b_{jj}(z)$, for $1 \leq i < j \leq n$.

Proof. Using an elementary row operation one can bring the entry in the first column of $A(z)$ with the smallest degree to position 11. By division with remainders, one then has $a_{i1}(z) = q_i(z)a_{11}(z) + r_i(z)$ for $i = 2, \ldots, m$. Applying elementary row operations we can thus achieve $b_{11} = a_{11}, b_{i1} = r_i$, where $\deg b_{i1} < \deg b_{11}$. Repeating the process of moving the entry of smallest degree to the 11-position and reducing degrees in the entries one obtains after finitely many steps a unimodular left equivalent matrix of the form

$$B(z) = \begin{pmatrix} b_{11}(z) & B_{12}(z) \\ 0 & B_{22}(z) \end{pmatrix},$$

with $b_{i1} = 0$ for $i = 2, \ldots, m$ and $B_{12}(z) \in \mathbb{F}[z]^{1 \times (n-1)}$. Here $B_{22}(z) \in \mathbb{F}[z]^{(m-1) \times (n-1)}$ has full column rank. Proceeding with $B_{22}(z)$, one sees that it is left unimodularly equivalent to a matrix $B'_{22}(z)$ with entries $b_{i2} = 0$ for $i > 2$. Suppose that $\deg b_{12} \geq \deg b_{22}$. Using division with remainders and adding a suitable multiple of the second row to the first one, one obtains $\deg b_{12} < \deg b_{22}$. Proceeding inductively with the other columns the Hermite form is obtained. The uniqueness property of the polynomials $b_{ij}(z)$ in the Hermite normal form is easily established. ∎

Theorem 2.20 implies that every full column rank polynomial matrix can be reduced to column proper form by right multiplication with a unimodular matrix. Because the column proper form is not unique, one can use the extra freedom to reduce it to a canonical form. This can be done by the Kronecker–Hermite form, which is stated in the following theorem. For proofs we refer the reader to Hinrichsen and Prätzel-Wolters (1983) and Fuhrmann and Helmke (2001), for example.

Theorem 2.40. *Let $A(z) \in \mathbb{F}[z]^{m \times n}$ be full column rank. Let $a_1, \ldots, a_n$ denote the columns of $A(z)$, and let a_{ij} denote the ith element of a_j. Then there exists a unimodular transformation $V(z)$ such that $A(z)V(z)$ is in **Kronecker–Hermite canonical form**, i.e., there exists a uniquely determined set of indices $1 \leq \mu_1 < \cdots < \mu_n \leq m$ such that*

(a) $a_{\mu_j j}$ is monic with degree $\delta_j := \deg a_{\mu_j j} = \deg a_j$;
(b) $\deg a_{\mu_j k} < \delta_j$ for $1 \leq k \leq n, k \neq j$.
(c) If $i > \mu_j$, then $\deg a_{ij} < \delta_j$ for $j = 1, \ldots, n$.

When Theorems 2.39 and 2.40 are specialized to nonsingular polynomial matrices, the following canonical representations are obtained.

Corollary 2.41. *Let $A(z) \in \mathbb{F}[z]^{m \times m}$ denote a nonsingular polynomial matrix. Then there exists a unimodular matrix $W(z) \in GL_m(\mathbb{F}[z])$ such that*

$$A(z)W(z) = \begin{pmatrix} b_{11} & 0 & \ldots & 0 \\ b_{21} & b_{22} & \ldots & 0 \\ \vdots & \vdots & \ddots & \vdots \\ b_{m1} & b_{m2} & \ldots & b_{mm}. \end{pmatrix}$$

The polynomials $b_{ij}(z)$ are uniquely determined, with $b_{ii}(z)$ monic and $\deg b_{ij}(z) < \deg b_{ii}(z)$, for $1 \leq j < i \leq m$.

Moreover, there exists a unimodular transformation $V(z) \in GL_m(\mathbb{F}[z])$ and polynomials $b_{ij}(z)$ such that

$$A(z)V(z) = \begin{pmatrix} b_{11} & b_{12} & \dots & b_{1m} \\ b_{21} & b_{22} & \dots & b_{2m} \\ \vdots & \vdots & \ddots & \vdots \\ b_{m1} & b_{m2} & \dots & b_{mm} \end{pmatrix}$$

satisfies the following conditions:

(a) $b_{jj}(z)$ *is monic with degree* $\delta_j = \max_i \deg b_{ij}$;
(b) $\deg b_{jk} < \delta_j$ *for all* $k \neq j$;
(c) *If* $i > j$, *then* $\deg b_{ij} < \delta_j$.

Equivalently, $A(z)V(z)$ is in both row proper and column proper form.

It is in general not possible, using elementary row operations, only to reduce the structure of a rectangular polynomial matrix to diagonal form. This is achieved by the Smith canonical form, which employs both elementary row and column operations.

Theorem 2.42 (Smith Normal Form). *Let $A(z) \in \mathbb{F}[z]^{m \times n}$. Then there exist nonzero polynomials $d_1(z), \dots, d_r(z) \in \mathbb{F}[z] \setminus \{0\}$, with $d_1(z)|d_2(z)|\cdots|d_r(z)$ and unimodular matrices $U(z) \in GL_m(\mathbb{F}[z])$ and $V(z) \in GL_n(\mathbb{F}[z])$, such that*

$$U(z)A(z)V(z) = \mathrm{diag}\,(d_1(z), \dots, d_r(z), 0, \dots, 0).$$

Proof. Without loss of generality, assume that $A \neq 0$. The ring of polynomials is a unique factorization domain, and therefore every nonzero element $f(z) \in \mathbb{F}[z]$ has a unique prime factor decomposition as $f(z) = p_1(z)^{n_1} \cdots p_r(z)^{n_r}$, with $p_i(z)$ irreducible polynomials in $\mathbb{F}[z]$. Define the height of f as $h(f) := \sum_{i=1}^{r} n_i$. Then h satisfies the following obvious properties (for $x, y \in \mathbb{F}[z]$):

- $h(xy) = h(x) + h(y)$;
- $x|y \implies h(x) \leq h(y)$;
- $x|y$ and $h(x) = h(y)$ are valid if and only if $y = ux$ for a nonzero constant $u \in \mathbb{F}$.

Let $\delta \in \mathbb{N}_0$ denote the smallest value among all heights $h(b_{ij})$ of nonzero entries $b_{ij}, i = 1, \dots m, j = 1, \dots, n$, of polynomial matrices $B = UAV$, where U and V vary independently over all unimodular matrices in $GL_m(\mathbb{F}[z])$ and $GL_n(\mathbb{F}[z])$, respectively. Without loss of generality, one can assume that $U = I_m, V = I_n$, and $\delta = h(a_{11})$. We now prove the following claims:

1. For all $j = 1, \dots, n$, either $a_{1j} = 0$ or $a_{11}|a_{1j}$.
2. For all $i = 1, \dots, m$, either $a_{i1} = 0$ or $a_{11}|a_{i1}$.

To prove the first claim, let $j = 2$, and assume that $a_{12} \neq 0$ and a_{11} does not divide a_{12}. Let d be the greatest common divisor of a_{11}, a_{12}. Thus $a_{11} = da, a_{12} = db$ for suitable coprime elements $a, b \in \mathbb{F}[z]$. By coprimeness, there exists a solution $x, y \in \mathbb{F}[z]$ to the Bezout equation $ax + by = 1$. This implies that the 2×2 matrix

$$U = \begin{pmatrix} x & -b \\ y & a \end{pmatrix}$$

is unimodular and, therefore, $Q := \mathrm{diag}\,(U, I_{n-2})$ is unimodular, too. We conclude that the 11-entry of AQ is d and the 12-entry of AQ is 0. Since a_{11} does not divide a_{12}, we conclude $h(d) < h(a_{11})$. But this is a contradiction to the choice of a_{11}. Thus $a_{11}|a_{12}$. Similarly for $j > 2$ and claim 2. Similarly to the preceding procedure in constructing U, we obtain after finitely many steps the unimodular matrices P and Q such that PAQ is block-diagonal of the form $\mathrm{diag}\,(a_{11}, A_{22})$, where $A_{22} \in \mathbb{F}[z]^{(m-1)\times(n-1)}$. Proceeding by induction, one obtains the existence of unimodular matrices P and Q and nonzero ring elements $d_1, \ldots, d_r$ such that PAQ has the form of a generalized diagonal matrix $\mathrm{diag}\,(d_1, \ldots, d_r, 0, \ldots, 0)$, possibly augmented by zero columns or rows. Moreover, by construction, $h(d_1) \leq \ldots \leq h(d_r)$. It remains to show that $d_1 | \cdots | d_r$. Suppose that d_1 does not divide d_2. Certainly the 2×2 matrices

$$A := \begin{pmatrix} d_1 & 0 \\ 0 & d_2 \end{pmatrix}, \quad B := \begin{pmatrix} d_1 & d_2 \\ 0 & d_2 \end{pmatrix}$$

are equivalent. Arguing as before, one can reduce the height of d_1 and eliminate the 12-entry in the second matrix. Therefore, A is equivalent to a diagonal matrix $\mathrm{diag}\,(d_1', d_2)$, with $h(d_1') < h(d_1)$. But this contradicts the minimality of $h(d_1)$. Therefore, $d_1|d_2$, and similarly $d_i|d_j$ for $i < j$. ∎

The polynomials $d_1, \ldots, d_r$ are called the **invariant factors** of A. They are uniquely determined up to multiplication by nonzero constants. In fact, it is known that for each $i = 1, \ldots, r$ the product $d_1(z) \cdots d_i(z)$ defines the principal ideal $d_1(z) \cdots d_i(z)\mathbb{F}[z]$, which is generated by the $i \times i$-minors of $A(z)$. We conclude with the following corollary.

Corollary 2.43. *Two matrices $A, B \in \mathbb{F}[z]^{m\times n}$ are unimodularly equivalent if and only if they have the same invariant factors.*

The Smith form extends as follows to matrices of rational functions.

Theorem 2.44 (Smith–McMillan Form). *Let $A(z) \in \mathbb{F}(z)^{m\times n}$ be a matrix of rational functions. Then there exist unimodular matrices $U(z) \in GL_m(\mathbb{F}[z])$ and $V(z) \in GL_n(\mathbb{F}[z])$ and pairwise coprime polynomials $a_i, b_i \in \mathbb{F}[z]$, $i = 1, \ldots, r$, with $a_1 | \cdots | a_r,\ b_r | \cdots | b_1$, such that*

$$U(z)A(z)V(z) = \mathrm{diag}\,\left(\frac{a_1(z)}{b_1(z)}, \ldots, \frac{a_r(z)}{b_r(z)}, 0, \ldots, 0 \right).$$

The rational functions $\frac{a_1}{b_1}, \ldots, \frac{a_r}{b_r} \in \mathbb{F}(z)$ are uniquely determined by A and $r = \mathrm{rk}_{\mathbb{F}(z)}\, A$.

Proof. Let $A = (\frac{a_{ij}}{b_{ij}})$ and $d \in \mathbb{F}[z]$ be a least common multiple of the b_{ij}. Then $dA \in \mathbb{F}[z]^{m\times n}$. By the Smith form, there exist unimodular matrices U and V and uniquely determined elements $d_1, \ldots d_r \in \mathbb{F}[z], d_1 | \cdots | d_r$ such that

$$dUAV = \text{diag}(d_1, \ldots, d_r, 0, \ldots, 0).$$

Choose coprime elements $a_i, b_i \in \mathbb{F}[z]$, with

$$\frac{a_i}{b_i} = \frac{d_i}{d}, \quad i = 1, \ldots, r.$$

Then $a_1 | \cdots | a_r$ and $b_r | \cdots | b_1$. The result follows. ∎

2.9 Equivalence of Polynomial Matrices

It is possible to generalize the concept of unimodular equivalence to the case of nonsingular polynomial matrices of different sizes. We formalize this as follows.

Definition 2.45. Let $D_1(z)$ and $D_2(z)$ be nonsingular polynomial matrices in $\mathbb{F}[z]^{m \times m}$ and $\mathbb{F}[z]^{p \times p}$, respectively. We say $D_1(z)$ and $D_2(z)$ are **polynomially equivalent** if there exist polynomial matrices $N_2(z)$ and $N_1(z)$ such that

$$N_2(z)D_1(z) = D_2(z)N_1(z)$$

and

1. $N_2(z)$ and $D_2(z)$ are left coprime,
2. $D_1(z)$ and $N_1(z)$ are right coprime.

It is easily seen from the coprimeness conditions and use of Bezout equations that polynomial equivalence is a bona fide equivalence relation, i.e., it is reflexive, symmetric, and transitive. Clearly, the unimodular equivalence of $D_1(z)$ and $D_2(z)$ implies polynomial equivalence.

One situation we have in mind for applying this concept is as follows. Consider a rational matrix function $G(z) \in \mathbb{F}(z)^{p \times m}$ with left and right coprime factorizations by polynomial matrices

$$D_\ell(z)^{-1}N_\ell(z) = G(z) = N_r(z)D_r(z)^{-1}.$$

Then $D_r(z) \in \mathbb{F}[z]^{m \times m}$ and $D_\ell(z) \in \mathbb{F}[z]^{p \times p}$ are polynomially equivalent, and we are interested in deducing consequences of this fact. For instance, the next result implies that the determinants $\det D_\ell(z)$ and $\det D_r(z)$ differ by a constant and therefore have the same zeros.

Theorem 2.46. *For nonsingular polynomial matrices $D_1(z) \in \mathbb{F}[z]^{m \times m}$ and $D_2(z) \in \mathbb{F}[z]^{p \times p}$ the following conditions are equivalent:*

1. $D_1(z)$ and $D_2(z)$ are polynomially equivalent;
2. $D_1(z)$ and $D_2(z)$ have the same nonconstant invariant factors;
3. For integers $q \geq \max\{m, p\}$ the extended polynomial matrices

$$\begin{pmatrix} D_1(z) & 0 \\ 0 & I_{q-m} \end{pmatrix}, \quad \begin{pmatrix} D_2(z) & 0 \\ 0 & I_{q-p} \end{pmatrix} \tag{2.32}$$

are unimodularly equivalent.
4. There exists an integer $q \geq \max\{m, p\}$ such that the extended polynomial matrices (2.32) are unimodularly equivalent.

Proof. Let $D_1(z)$ and $D_2(z)$ be polynomially equivalent. By Theorem 2.33, there exist polynomial matrices $X(z)$ and $Y(z)$ such that

$$U(z) = \begin{pmatrix} -N_2(z) & D_2(z) \\ Y(z) & X(z) \end{pmatrix} \in \mathbb{F}[z]^{(m+p) \times (m+p)}$$

is unimodular, with

$$\begin{pmatrix} -N_2(z) & D_2(z) \\ Y(z) & X(z) \end{pmatrix} \begin{pmatrix} D_1(z) \\ N_1(z) \end{pmatrix} = \begin{pmatrix} 0 \\ I_m \end{pmatrix}.$$

Therefore,

$$U(z) \begin{pmatrix} D_1(z) & 0 \\ 0 & I_p \end{pmatrix} = \begin{pmatrix} -N_2(z) & D_2(z) \\ Y(z) & X(z) \end{pmatrix} \begin{pmatrix} D_1(z) & 0 \\ 0 & I_p \end{pmatrix} = \begin{pmatrix} -N_2(z)D_1(z) & D_2(z) \\ Y(z)D_1(z) & X(z) \end{pmatrix}$$

$$= \begin{pmatrix} -D_2(z)N_1(z) & D_2(z) \\ I_m - X(z)N_1(z) & X(z) \end{pmatrix} = \begin{pmatrix} D_2(z) & 0 \\ 0 & I_m \end{pmatrix} V(z),$$

where

$$V(z) := \begin{pmatrix} -N_1(z) & I_p \\ I - X(z)N_1(z) & X(z) \end{pmatrix} = \begin{pmatrix} 0 & I \\ I & X \end{pmatrix} \begin{pmatrix} I & 0 \\ -N_1 & I \end{pmatrix}$$

is a product of unimodular matrices and, therefore, is unimodular. It follows that

$$\begin{pmatrix} D_1(z) & 0 \\ 0 & I_p \end{pmatrix}, \quad \begin{pmatrix} D_2(z) & 0 \\ 0 & I_m \end{pmatrix} \tag{2.33}$$

are unimodularly equivalent; hence they have the same invariant factors. By the uniqueness of the Smith form, the matrices in (2.33) have the same nonconstant invariant factors as $D_1(z)$ and $D_2(z)$, respectively. Thus the nonconstant invariant factors of $D_1(z)$ and $D_2(z)$ coincide. This shows the implication condition 1 $\Longrightarrow$ condition 2.

Suppose that $D_1(z)$ and $D_2(z)$ have the same nonconstant invariant factors. The uniqueness of the invariant factors thus implies that the extended polynomial matrices (2.32) have the same invariant factors and therefore are unimodularly equivalent. This proves that condition 2 $\implies$ condition 3. Obviously, condition 3 $\implies$ condition 4. Finally, assume that condition 4 is satisfied. Without loss of generality, we can assume that $m \le p$ and $D_1(z) = \mathrm{diag}\,(d_1,\ldots,d_m)$ and $D_2(z) = \mathrm{diag}\,(\delta_1,\ldots,\delta_p)$ are in Smith canonical form. By condition 4, the nonconstant invariant polynomials of $D_1(z)$ and $D_2(z)$ coincide, and therefore $d_1 = \delta_1,\ldots,d_m = \delta_m$, $\delta_{m+1} = \ldots = \delta_p \ne 0$. Thus $N_2(z)D_1(z) = D_2(z)N_1(z)$ for the constant matrices

$$N_1 = N_2 = \begin{pmatrix} I_m \\ 0 \end{pmatrix}.$$

Since $N_1(z)$ and $D_1(z)$ are right coprime and $N_2(z)$ and $D_2(z)$ are left coprime, we conclude that $D_1(z)$ and $D_2(z)$ are polynomially equivalent. This completes the proof. ∎

2.10   Structure Theorem and Quotient Modules

The reduction to Smith form in Theorem 2.42 leads directly to the structure theory of finitely generated modules over $\mathbb{F}[z]$. The subsequent proof of the structure theorem is based on the following lemma.

Lemma 2.47. *Let M be a free module of rank s over $\mathbb{F}[z]$. Let $N \subset M$ be a submodule. Then there exists a basis $\{m_1,\ldots,m_s\}$ of M and polynomials $d_1,\ldots,d_s \in \mathbb{F}[z]$ satisfying $d_1 | \cdots | d_s$ such that*

$$\{d_1 m_1,\ldots,d_s m_s\}$$

is a basis of N.

Proof. If $N = \{0\}$, then we are done. Thus, assume $N \ne \{0\}$. Choose a basis $m_1',\ldots,m_s'$ of M. Since N is free, there exists a basis $\{n_1',\ldots,n_t'\}$ of N. Representing the basis vectors n_j' in terms of the m_i' as

$$n_j' = \sum_{i=1}^{t} a_{ij} m_i', \quad j = 1,\ldots,s$$

yields a matrix $A \in \mathbb{F}[z]^{t \times s}$. Equivalently, we can write these equations in a condensed matrix form as

$$(n_1',\ldots,n_s') = (m_1',\ldots,m_t')A.$$

Bringing A into Smith form as

$$UAV = \mathrm{diag}(d_1,\ldots,d_r,0,\ldots,0)$$

via unimodular matrices $U \in GL_t(\mathbb{F}[z]), V \in GL_s(\mathbb{F}[z])$ produces new bases

$$(m_1,\ldots,m_t) = (m_1',\ldots,m_t')P^{-1}, \quad (n_1,\ldots,n_s) = (n_1',\ldots,n_s')Q$$

of M and N, respectively, with $n_i = d_i m_i$ for $i = 1,\ldots,s$. ∎

We can now state and prove the main result on finitely generated $\mathbb{F}[z]$-modules. The result actually is true for finitely generated R-modules over a PID R.

Theorem 2.48 (Structure Theorem). *Every finitely generated $\mathbb{F}[z]$-module M is isomorphic to a finite direct sum*

$$M \simeq \mathbb{F}[z]^r \oplus \mathbb{F}[z]/d_1\mathbb{F}[z] \oplus \cdots \oplus \mathbb{F}[z]/d_t\mathbb{F}[z],$$

*with nonzero polynomials $d_i(z)$ satisfying $d_1(z)|\cdots|d_t(z)$. The integers $r,t \in \mathbb{N}_0$ and $d_1,\ldots,d_t$ are uniquely determined, up to possible factors by nonzero constants. The $d_i(z)$ are called the **invariant factors of the module M**.*

Proof. Since M is finitely generated, there exists a surjective module homomorphism $f : \mathbb{F}[z]^k \longrightarrow M$. The kernel Ker f is a submodule of the free module $\mathbb{F}[z]^k$. By Lemma 2.47, there exists a basis $\{e_1,\ldots,e_k\}$ of $\mathbb{F}[z]^k$ and nonzero elements $d_1,\ldots,d_t \in \mathbb{F}[z], d_1|\cdots|d_t, t \le k$ such that $\{d_1e_1,\ldots,d_te_t\}$ form a basis of Ker f. Set $r = k-t$. Then $\bigoplus_{i=t+1}^{k} \mathbb{F}[z]e_i \simeq \mathbb{F}[z]^r$ is free and Ker $f = \bigoplus_{i=1}^{t} \mathbb{F}[z]d_ie_i$. We conclude

$$M \simeq \mathbb{F}[z]^r \oplus \mathbb{F}[z]^t/\mathrm{Ker}\ f.$$

The result follows since

$$\mathbb{F}[z]^t/\mathrm{Ker}\ f \simeq \mathbb{F}[z]/d_1\mathbb{F}[z] \oplus \cdots \oplus \mathbb{F}[z]/d_t\mathbb{F}[z].$$ ∎

Using the fact that every PID is a unique factorization domain, one can decompose the invariant factors uniquely (i.e., up to units and permutations) into prime factors. Taken together with the Chinese remainder theorem 2.11, this yields the following variant of the structure theorem.

Theorem 2.49. *Every finitely generated $\mathbb{F}[z]$-module M is isomorphic to*

$$M \simeq \mathbb{F}[z]^r \oplus \mathbb{F}[z]/p_1(z)^{n_1}\mathbb{F}[z] \oplus \cdots \oplus \mathbb{F}[z]/p_k(z)^{n_k}\mathbb{F}[z],$$

*where the polynomials $p_1(z),\ldots,p_k(z)$ are primes of $\mathbb{F}[z]$. The polynomials $p_1,\ldots,p_k$ are called the **elementary divisors** of M.*

Recall that a submodule $M \subset \mathbb{F}[z]^n$ is called **full** if M has a basis of exactly n elements. Thus M is full if and only if there exists a nonsingular polynomial matrix $D(z) \in \mathbb{F}[z]^{n \times n}$, with $M = D(z)\mathbb{F}[z]^n$. Two nonsingular polynomial matrices $D_1(z)$ and $D_2(z)$, with $D_1(z)\mathbb{F}[z]^n = M = D_2(z)\mathbb{F}[z]^n$, are related as $D_2(z) = D_1(z)U(z)$ by a uniquely unimodular $n \times n$ matrix $U(z)$.

Since $\mathbb{F}[z]$ has no nontrivial zero divisors, the **torsion submodule** of an $\mathbb{F}[z]$-module M is

$$\mathrm{Tor}(M) = \{x \in M \mid \exists r \in \mathbb{F}[z] \setminus \{0\} \text{ with } rx = 0\}.$$

The elements of $\mathrm{Tor}(M)$ are called torsion elements. M is called **torsion free** if $\mathrm{Tor}(M) = \{0\}$; M is a **torsion module** whenever $\mathrm{Tor}(M) = M$. Thus the quotient $M/\mathrm{Tor}(M)$ is torsion free and $M/\mathrm{Tor}(M) \simeq \mathbb{F}[z]^r$.

For a nonzero ring element $d \in \mathbb{F}[z]$, the quotient $\mathbb{F}[z]/d\mathbb{F}[z]$ is a torsion module, with $\mathrm{Tor}(\mathbb{F}[z]/d\mathbb{F}[z]) = d\mathbb{F}[z]$. Therefore, the structure theorem for finitely generated $\mathbb{F}[z]$-modules yields

$$M \simeq \mathbb{F}[z]^r \oplus \mathrm{Tor}(M),$$

and the torsion submodule is $\mathrm{Tor}(M) = \bigoplus_{j=1}^{s} \mathbb{F}[z]/d_j\mathbb{F}[z]$. This implies the following corollary.

Corollary 2.50. *Finitely generated $\mathbb{F}[z]$-modules M are free if and only if they are torsion free.*

The preceding result implies the following characterization of torsion modules.

Theorem 2.51. *Let $D(z)$ be an $n \times n$ polynomial matrix. For a quotient module $M = \mathbb{F}[z]^n/D(z)\mathbb{F}[z]^n$ the following statements are equivalent:*

(a) M is a torsion module.
(b) The polynomial matrix $D(z)$ is nonsingular.
(c) M is a finite-dimensional $\mathbb{F}$-vector space.

In each of these cases the dimension of M as an $\mathbb{F}$-vector space is equal to

$$\dim \mathbb{F}[z]^n/D(z)\mathbb{F}[z]^n = \deg \det D(z).$$

Proof. Without loss of generality, we can assume that

$$D(z) = \mathrm{diag}(d_1(z), \ldots, d_r(z), 0, \ldots, 0), \quad 0 \le r \le n,$$

is in Smith canonical form, with d_i nonzero polynomials. Then M is isomorphic to the direct sum of $\mathbb{F}[z]$-modules $\mathbb{F}[z]/d_1\mathbb{F}[z] \oplus \cdots \oplus \mathbb{F}[z]/d_r\mathbb{F}[z] \oplus \mathbb{F}[z]^{n-r}$. Thus M is a torsion module if and only if $r = n$, i.e., if and only if $D(z)$ is nonsingular. Moreover, for a polynomial $q(z) \in \mathbb{F}[z]$ of degree n, the coset classes

$$[z]^i := z^i + q(z)\mathbb{F}[z], \quad i = 0, \ldots, n-1$$

form a basis of the quotient module $\mathbb{F}[z]/q(z)\mathbb{F}[z]$. Thus $\mathbb{F}[z]/q(z)\mathbb{F}[z]$ is a finite-dimensional $\mathbb{F}$-vector space of dimension equal to $\deg q$. Therefore, M is a finite-dimensional $\mathbb{F}$-vector space if and only if $D(z)$ is nonsingular. Moreover, in each of these cases the torsion module M has $\mathbb{F}$-dimension equal to $\sum_{i=1}^{n} \deg d_i = \deg \det D(z)$. $\blacksquare$

We characterize the case where a submodule has a kernel representation.

Theorem 2.52. *Let $A(z) \in \mathbb{F}[z]^{n \times r}$ be full column rank r. The following conditions are equivalent:*

(a) The image space Im $A \subset \mathbb{F}[z]^n$ has a **kernel representation,** *i.e., there exists $B \in \mathbb{F}[z]^{(n-r) \times n}$, with Im $A = \text{Ker } B$.*
(b) The cokernel $\mathbb{F}[z]^n / \text{Im } A$ is **torsion-free.**
(c) A is **right prime.**

Proof. Without loss of generality, we can assume that A is in Smith normal form, i.e.,

$$A(z) = \begin{pmatrix} D(z) \\ 0 \end{pmatrix},$$

with $D = \text{diag}(a_1, \ldots, a_r)$ and $a_1 | \cdots | a_r$, $a_i \neq 0$. Thus the cokernel is isomorphic to

$$\mathbb{F}[z]^n / \text{Im } A \simeq \mathbb{F}[z]/a_1\mathbb{F}[z] \oplus \cdots \oplus \mathbb{F}[z]/a_r\mathbb{F}[z]$$

and, therefore, is torsion free if and only if $a_1, \ldots, a_r$ are units of $\mathbb{F}[z]$. Equivalently, the Smith form of A is equal to

$$\begin{pmatrix} I_r \\ 0 \end{pmatrix}.$$

The result follows. $\blacksquare$

We next characterize the submodules of the quotient modules

$$M = \mathbb{F}[z]^m / D(z)\mathbb{F}[z]^m.$$

Theorem 2.53. *Let $D(z) \in \mathbb{F}[z]^{m \times m}$ be a nonsingular polynomial matrix. A subset $X \subset \mathbb{F}[z]^m / D(z)\mathbb{F}[z]^m$ is a submodule if and only if there exist nonsingular polynomial matrices $E(z), F(z)\mathbb{F}[z]^{m \times m}$, with $D(z) = E(z)F(z)$ and*

$$X = E(z)\mathbb{F}[z]^m / D(z)\mathbb{F}[z]^m.$$

Proof. Let $\pi : \mathbb{F}[z]^n \longrightarrow \mathbb{F}[z]^n / D\mathbb{F}[z]^n, x \mapsto x + D\mathbb{F}[z]^n$, denote the canonical projection map. Then X is a submodule of $\mathbb{F}[z]^n / D\mathbb{F}[z]^n$ if and only if $\pi^{-1}(X)$ is a submodule of $\mathbb{F}[z]^n$ that contains $D\mathbb{F}[z]^n$. Since $\mathbb{F}[z]$ is a PID, a submodule of $\mathbb{F}[z]^n$ is free and $\pi^{-1}(X)$ is finitely generated. Since $D\mathbb{F}[z]^n \subset \pi^{-1}(X)$ is a full submodule of $\mathbb{F}[z]^n$, this implies that $\pi^{-1}(X)$ is a full submodule, too. Thus there exists a nonsingular polynomial matrix $E \in \mathbb{F}[z]^{n \times n}$ with $\pi^{-1}(X) = E\mathbb{F}[z]^n$. The condition $D\mathbb{F}[z]^n \subset E\mathbb{F}[z]^n$ is equivalent to the existence of a polynomial matrix $F \in \mathbb{F}[z]^{n \times n}$, with $D(z) = E(z)F(z)$. That D is nonsingular implies that F is nonsingular. The result follows since $X = \pi(E\mathbb{F}[z]^n) = E\mathbb{F}[z]^n/D\mathbb{F}[z]^n$. ∎

The module homomorphisms of quotient modules $M = \mathbb{F}[z]^n/D(z)\mathbb{F}[z]^n$ of full submodules $D(z)\mathbb{F}[z]^n$ are easily characterized. We will study such quotient modules more concretely in Chapter 3 on functional model spaces.

The next result is an algebraic version of the celebrated **commutant lifting theorem** from operator theory. It states that every $\mathbb{F}[z]-$ module homomorphism on quotient modules $Z : \mathbb{F}[z]^m/D_1\mathbb{F}[z]^m \longrightarrow \mathbb{F}[z]^p/D_2\mathbb{F}[z]^p$ has a lift to a module homomorphism $\hat{Z} : \mathbb{F}[z]^m \longrightarrow \mathbb{F}[z]^p$, defined by multiplication with a polynomial matrix $N_2(z)$. Let $\pi_D : \mathbb{F}[z]^n \longrightarrow \mathbb{F}[z]^n/D(z)\mathbb{F}[z]^n$ denote the quotient map.

Theorem 2.54 (Algebraic Commutant Lifting Theorem). *Let $D_1(z) \in \mathbb{F}[z]^{m \times m}$ and $D_2(z) \in \mathbb{F}[z]^{p \times p}$ be nonsingular polynomial matrices. A map*

$$Z : \mathbb{F}[z]^m/D_1(z)\mathbb{F}[z]^m \longrightarrow \mathbb{F}[z]^p/D_2(z)\mathbb{F}[z]^p$$

is an $\mathbb{F}[z]$-homomorphism if and only if there exist $N_1(z), N_2(z) \in \mathbb{F}[z]^{p \times m}$ such that

$$N_2(z)D_1(z) = D_2(z)N_1(z) \tag{2.34}$$

and

$$Z(\pi_{D_1} f) = \pi_{D_2}(N_2 f) \quad \text{for all} \quad f \in \mathbb{F}[z]^m. \tag{2.35}$$

Proof. Z is a module homomorphism if and only if $\overline{Z} := Z \circ \pi_{D_1} : \mathbb{F}[z]^m \longrightarrow \mathbb{F}[z]^p/D_2(z)\mathbb{F}[z]^p$ is a module homomorphism. Let $N_2(z) \in \mathbb{F}[z]^{p \times m}$ be a polynomial matrix whose ith column $n_i(z) \in \mathbb{F}[z]^m$ satisfies

$$\overline{Z}(e_i) = n_i(z) + D_2(z)\mathbb{F}[z]^p.$$

Since $\overline{Z}$ is $\mathbb{F}[z]$-linear, we obtain for all $i = 1, \dots, m$ and $j \geq 0$ that $\overline{Z}(z^j e_i) = z^j n_i(z) + D_2(z)\mathbb{F}[z]^p$. This shows that

$$\overline{Z}(f) = N_2(z)f(z) + D_2(z)\mathbb{F}[z]^p = \pi_{D_2}(N_2 f)$$

for all $f(z) \in \mathbb{F}[z]^m$. Since the kernel of $\overline{Z} := Z \circ \pi_{D_1}$ contains $D_1(z)\mathbb{F}[z]^m$, then $N_2(z)D_1(z)g(z) \in D_2(z)\mathbb{F}[z]^p$ for all $g(z) \in \mathbb{F}[z]^m$. Thus there exists a polynomial matrix $N_1(z) \in \mathbb{F}[z]^{p \times m}$, with $N_2(z)D_1(z) = D_2(z)N_1(z)$. This proves (2.34) and (2.35). Conversely, assume that (2.34) and (2.35) are satisfied. Then the map $\hat{Z} : \mathbb{F}[z]^m \longrightarrow \mathbb{F}[z]^p, \hat{Z}f = N_2(z)f(z)$ is $\mathbb{F}[z]$-linear and satisfies

$$\hat{Z}(D_1(z)\mathbb{F}[z]^m) \subset D_2(z)\mathbb{F}[z]^p.$$

Thus $\hat{Z}$ induces a well-defined map $Z : \mathbb{F}[z]^m/D_1(z)\mathbb{F}[z]^m \longrightarrow \mathbb{F}[z]^p/D_2(z)\mathbb{F}[z]^p$ and, indeed, an $\mathbb{F}[z]$-homomorphism, with $\pi_{D_2} \circ \hat{Z} = Z \circ \pi_{D_1}$. This completes the proof. ∎

2.11 Rings of Rational Functions

In this last section we mention an important construction from algebra, the localization of rings and modules. This leads to a straightforward construction of PIDs of rational functions. Although the material in this section will not be of much use in this book (except for the Youla–Kucera parameterization of stabilizing controllers in Chapter 6), it is of considerable interest in applications to, for example, rational H^∞-theory, model reduction, and spectral factorizations.

1. Localization of Rings and Modules.

The construction of the fraction field of a ring very much resembles that of the field of rational numbers from the ring of integers. A subset $S \subset R$ of an integral domain R is called **multiplicatively closed**, provided the following conditions are satisfied:

- $0 \notin S$;
- $1 \in S$;
- $a, b \in S \implies ab \in S$.

Define an equivalence relation on $R \times S$ via

$$(a,s) \sim (b,t) \iff at - bs = 0.$$

It is easily seen that this indeed defines an equivalence relation on $R \times S$; the equivalence classes are denoted by $\frac{a}{s}$. The set of all such equivalence classes is denoted by

$$R_S = S^{-1}R := \{\frac{a}{s} \mid s \in S\}$$

and is called the **localization** or the **ring of fractions** of R by S. Two fraction elements are added and multiplied according to the usual rules:

$$\frac{a}{s} + \frac{b}{t} = \frac{at+bs}{st},$$

$$\frac{a}{s} \cdot \frac{b}{t} = \frac{ab}{st}.$$

Endowed with these operations, one easily verifies that $S^{-1}R$ becomes a ring with identity element $\frac{1}{1}$. By a simple modification of the preceding equivalence relation one can extend this construction of the ring of fractions to more general rings than integral domains. In this more general context one then checks that the ring of fractions $S^{-1}R$ is an integral domain if and only if R is an integral domain.

The well-known field of fractions is obtained by choosing S to be $R \setminus \{0\}$. Since R is assumed to be integral, $R \setminus \{0\}$ is multiplicatively closed, and we obtain the **field of fraction** of R as

$$\mathbb{K} = (R \setminus \{0\})^{-1}R := \{\frac{r}{s} \mid r, s \in R, s \neq 0\}.$$

In fact, $\mathbb{K}$ is a field that contains R as a subring obtained by identifying ring elements $r \in R$ with fractions $\frac{r}{1}$. Notice that for general rings R the map $r \mapsto \frac{r}{1}$ from R to $S^{-1}R$ need not be injective; however, for integral domains R this is true.

Many desirable properties of rings are preserved by localization; a sample is summarized as follows.

Proposition 2.55. *Let S be a multiplicatively closed subset of an integral domain R.*

1. I is an ideal of R if and only if

$$S^{-1}I := \{\frac{r}{s} \mid r \in I, s \in S\}$$

is an ideal of $S^{-1}R$. Ideals I and J of R satisfy the identities

$$S^{-1}(I+J) = S^{-1}I + S^{-1}J,$$
$$S^{-1}(I \cap J) = S^{-1}I \cap S^{-1}J,$$
$$S^{-1}(R/I) = S^{-1}R/S^{-1}I.$$

2. If R is a PID, then $S^{-1}R$ is a PID.
3. If R is a factorial ring, then $S^{-1}R$ is factorial, too. The prime elements of R_S are those primes p of R with $pR \cap S = \emptyset$.

Proof. We only prove the second claim and leave the others as an exercise to the reader. Let R be a PID and $i : R \longrightarrow R_S, i(a) = \frac{a}{1}$ the canonical inclusion map. If $I \subset R_S$ is an ideal, then $J := i^{-1}(I) = i(R) \cap I$ is an ideal in R and is therefore of the form $J = dR, d \in R$. We show that

$$dR_S = \{\frac{a}{s} \mid a \in dR, s \in S\} = I,$$

which then completes the proof. Obviously, $a \in dR$ implies $\frac{a}{1} \in I$. Since I is an ideal, we conclude that $\frac{a}{s} = \frac{1}{s}\frac{a}{1} \in R_S I \subset I$ for all $s \in S$. Thus $dR_S \subset I$. Conversely, if $\frac{a}{s}$ denotes an element of $I \subset R_S$, then $\frac{s}{1} \in R_S$, and therefore $\frac{a}{1} = \frac{s}{1}\frac{a}{s} \in I$. Thus $I \subset dR_S$, and we are done. ∎

A simple example of the process of localization is the ring of Laurent polynomials $\mathbb{F}[x,x^{-1}]$. This ring is the localization

$$\mathbb{F}[x,x^{-1}] = \mathbb{F}[x]_S$$

of the polynomial ring $\mathbb{F}[x]$ with respect to the multiplicatively closed subset $S = \{1,x,x^2,\ldots\}$. Proposition 2.55 thus implies that $\mathbb{F}[x,x^{-1}]$ is a PID.

A very important case arises when S is the **complement of a prime ideal** $\mathfrak{p}$ of R. It is easy to show that the complement $S := R \setminus I$ of an ideal I of R is multiplicatively closed if and only if I is a prime ideal.

Definition 2.56 (Localization by a Prime Ideal). The localization of a ring R by a prime ideal $\mathfrak{p}$ is

$$R_{\mathfrak{p}} := (R \setminus \mathfrak{p})^{-1}R = \{\frac{r}{s} \mid r \in R,\ s \notin \mathfrak{p}\}.$$

If $p \in R$ is a prime element and $\mathfrak{p} = pR$ the associated prime ideal, then we write

$$R_{(p)} := R_{\mathfrak{p}} = \{\frac{r}{s} \mid r \in R,\ p \nmid s\}.$$

Thus $R_{\mathfrak{p}}$ is a ring. In fact, it is a local ring, as the next result shows. A **local ring** R is defined by the property that R has a unique maximal ideal $\mathfrak{m}$. The quotient field $K = R/\mathfrak{m}$ is called the residue field.

Theorem 2.57. *Let $\mathfrak{p} \neq \{0\}$ be a prime ideal of a ring R. Then the localization $R_{\mathfrak{p}}$ satisfies the following claims:*

1. $I \subset R_{\mathfrak{p}}$ is a prime ideal if and only if

$$I = \mathfrak{q}R_{\mathfrak{p}} = \{\frac{r}{s} \mid r \in \mathfrak{q},\ s \notin \mathfrak{p}\},$$

for a (unique) prime ideal $\mathfrak{q}$ contained in $\mathfrak{p}$;

2. $R_{\mathfrak{p}}$ is a local ring, i.e., it has a unique maximal ideal $\mathfrak{p}R_{\mathfrak{p}} = \{\frac{r}{s} \mid r \in \mathfrak{p},\ s \notin \mathfrak{p}\}$; the units of $R_{\mathfrak{p}}$ are the elements rs^{-1}, with $r \notin \mathfrak{p}$;

3. For every maximal ideal $\mathfrak{m}$ of R, the quotient field $R/\mathfrak{m}R$ is isomorphic to $R_{\mathfrak{m}}/\mathfrak{m}R_{\mathfrak{m}}$.

Proof. The first claim is an immediate consequence of Proposition 2.55. Thus the prime ideals of $R_{\mathfrak{p}}$ are of the form $\mathfrak{q}R_{\mathfrak{p}}$, where $\mathfrak{q} \subset \mathfrak{p}$ is a prime ideal of R. It follows that the only maximal ideal of $R_{\mathfrak{p}}$ is $\mathfrak{p}R_{\mathfrak{p}}$. Thus $R_{\mathfrak{p}}$ is a local ring. For the third statement consider the map

$$\tau : R \longrightarrow R_{\mathfrak{m}}/\mathfrak{m}R_{\mathfrak{m}}, \quad r \mapsto \frac{r}{1} + \mathfrak{m}R_{\mathfrak{m}}.$$

Since $\mathfrak{m}$ is a maximal ideal, $\mathfrak{m}R + sR = R$ for $s \notin \mathfrak{m}$. Therefore, every element $r \in R$ can be written as $r = sr_1 + mr_2$ for suitable $r_1, r_2 \in R$. Thus, the elements $\frac{r}{s} \in R_{\mathfrak{m}}$ are equal to

$$\frac{r}{s} = \frac{r_1}{1} + \frac{mr_2}{s},$$

and therefore $\tau(r_1) = \frac{r}{s} + \mathfrak{m}R_{\mathfrak{m}}$. This shows that τ is surjective. An element $r \in R$ is in the kernel of τ if and only if $\frac{r}{1} \in \mathfrak{m}R_{\mathfrak{m}}$. Equivalently, there exists $m \in \mathfrak{m}, a \in R, s \notin \mathfrak{m}$, with $rs = ma \in \mathfrak{m}R$. Since $s \notin \mathfrak{m}$, this is equivalent to $r \in \mathfrak{m}$. Thus the kernel of τ is equal to $\mathfrak{m}$, and we are done.   ■

The localizations of the ring of integers are the subsets of rational numbers

$$\mathbb{Z}_{(p)} = \{\frac{m}{n} \mid m \in \mathbb{Z},\, n \in \mathbb{N},\, p \nmid n\}$$

for p a prime number. The only nontrivial ideal of $\mathbb{Z}_{(p)}$ is the maximal ideal

$$p\mathbb{Z}_{(p)} = \{\frac{m}{n} \in \mathbb{Q} \mid p \mid m,\, p \nmid n\},$$

whose quotient field is the finite field $\mathbb{Z}/p\mathbb{Z}$ of integers modulo p. We note in passing that the p-adic completion of $\mathbb{Z}_{(p)}$ yields the local ring of p-adic integers $\mathbb{Z}_p$.

Localization of Modules. We begin with an explanation of how the technique of ring localization can be extended to modules. Let M be a module over a ring R, and let S be a multiplicative subset of R. We then define the S-localization of M as the module M_S, whose elements are the fractions $\frac{m}{s}$ of module elements $m \in M$ by ring elements $s \in S$. The usual rules for addition and scalar multiplication fraction elements in M_S are imposed. By defining

$$\frac{r}{s}\frac{m}{s'} = \frac{rm}{ss'} \quad \forall r \in R, s, s' \in S, m \in M,$$

we see that M_S becomes a module over the rings of fractions R_S. The R_S-module M_S is called the S-**localization** of the R-module M. It is easily seen that an R-linear map $f : M \longrightarrow N$ defines an R_S-linear map

$$f_S : M_S \longrightarrow N_S, \quad f(\frac{m}{s}) := \frac{f(m)}{s},$$

called the **localization of** f. The operation of localization satisfies the usual functorial properties, i.e., $(f \circ g)_S = f_S \circ g_S$ and $(id_M)_S = id_{M_S}$. Moreover, it is easily seen that $f_S : M_S \longrightarrow N_S$ is injective, surjective, or bijective if and only if $f : M \longrightarrow N$ is injective, surjective, or bijective. A localized module M_S can be trivial without M being trivial. More precisely, define the **torsion submodule** of M as

$$Tor(M) = \{m \in M \mid \exists s \in R_0 \quad \text{with} \quad sm = 0\}.$$

Here $R_0 \subset R$ denotes the subset of all ring elements of R that are not zero divisors. The elements of $Tor(M)$ are called torsion elements. M is called **torsion free** if $Tor(M) = \{0\}$; M is a **torsion module** whenever $Tor(M) = M$. It follows that $M_S = \{0\}$ if and only if for every $m \in M$ there exists $s \in S$, with $sm = 0$. If R is an integral domain and $S = R \setminus \{0\}$, then $M_S = \{0\}$ if and only if M is a torsion module. The reader is referred to the book by Atiyah-Macdonald (1969) for proofs of the next three results.

Proposition 2.58. *1. If M is finitely generated, then so is M_S. If every submodule of M is finitely generated, then every submodule of M_S is finitely generated.*

2. Let N be a submodule of M. Then N_S is a submodule of M_S with the isomorphism of R_S-modules

$$f : (M/N)_S \longrightarrow M_S/N_S, \qquad \frac{m+N}{s} \mapsto \frac{m}{s} + N_S.$$

3. The R_S-submodules of M_S are exactly of the form N_S, where N denotes a R-submodule of M.

4. If $M \xrightarrow{f} N \xrightarrow{g} P$ is an exact sequence of R-modules, then

$$M_S \xrightarrow{fs} N_S \xrightarrow{gs} P_S$$

is an exact sequence of R_S-modules.

The local-global principle asserts the following.

Theorem 2.59 (Local-Global Principle). *Let M be an R-module.*

1. Then $M = \{0\}$ if and only if $M_{\mathfrak{m}} = 0$ for all maximal ideals $\mathfrak{m}$ of R.

2. An element $x \in M$ is an element of a submodule $N \subset M$ if and only if $\frac{m}{1} \in N_{\mathfrak{m}}$ for all maximal ideals $\mathfrak{m}$ of R.

3. An R-linear map $f : M \longrightarrow N$ between R-modules is surjective (resp. injective, bijective) if and only if the localizations

$$f_{\mathfrak{m}} : M_{\mathfrak{m}} \longrightarrow N_{\mathfrak{m}}$$

are surjective (resp. injective, bijective) for all maximal ideal $\mathfrak{m}$ of R.

The preceding result can be restated in a more applicable form to check the surjectivity of a module homomorphism. Note that the factor ring $R_{\mathfrak{m}}$ is a local ring with unique maximal ideal $\mathfrak{m}R_{\mathfrak{m}} = \{\frac{p}{s} \mid p \in \mathfrak{m}, s \in R \setminus \mathfrak{m}\}$. Thus in part 3 of the local-global principle it suffices to check surjectivity (resp. injectivity, bijectivity) for linear maps $f_{\mathfrak{m}} : M_{\mathfrak{m}} \longrightarrow N_{\mathfrak{m}}$ over local rings. Alternatively, one can consider the family of induced linear maps

$$\bar{f} : M/\mathfrak{m}M \longrightarrow N/\mathfrak{m}N \qquad (2.36)$$

over the fraction field $R/\mathfrak{m}$. We emphasize that if M is finitely generated, then $M/\mathfrak{m}M$ is a finite-dimensional vector space over $R/\mathfrak{m}$. Checking the surjectivity of (2.36) thus becomes a standard matter of linear algebra over a field. A central result for modules over local rings is the following well-known lemma.

Lemma 2.60 (Nakayama Lemma). *Let R be a local ring with unique maximal ideal $\mathfrak{m}$. Let M be a finitely generated R-module with canonical projection map $\pi : M \longrightarrow M/\mathfrak{m}M, \pi(x) = [x] := x + \mathfrak{m}M$. Then $\{x_1, \ldots, x_n\}$ generates M if and only if $\{[x_1], \ldots, [x_n]\}$ generates $M/\mathfrak{m}M$.*

Using this lemma it is easy to verify the following useful criterion for surjectivity and injectivity. We recall that the **Jacobson radical** of a ring is defined as the intersection of all maximal ideals $\mathfrak{m}$ of R, i.e., rad $R = \bigcap_{\mathfrak{m}} \mathfrak{m}$.

Theorem 2.61. *Let M and N be finitely generated R-modules and $f : M \longrightarrow N$ be R-linear.*

1. *f is surjective if and only if the induced $R/\mathfrak{m}$-linear maps*

$$\bar{f} : M/\mathfrak{m}M \longrightarrow N/\mathfrak{m}N$$

   are surjective for all maximal ideals $\mathfrak{m}$ of R.
2. *Assume that the Jacobson radical of R satisfies rad $(R) = \{0\}$. Then f is injective, provided the induced maps (1) are injective for all maximal ideals $\mathfrak{m}$ of R. The converse is false.*

Proof. It is trivial to check that the surjectivity of $f : M \longrightarrow N$ implies the surjectivity of $\bar{f} : M/\mathfrak{m}M \longrightarrow N/\mathfrak{m}N$ for all maximal ideals $\mathfrak{m}$. Conversely, assume that $\bar{f} : M/\mathfrak{m}M \longrightarrow N/\mathfrak{m}N$ is surjective for a maximal ideal $\mathfrak{m}$. Let $\pi_M : M \longrightarrow M/\mathfrak{m}M$ denote the canonical projection map. Then $g = \bar{f} \circ \pi_M : M \longrightarrow N/\mathfrak{m}N$ is surjective and defines an R-linear map. This implies that the induced $R_{\mathfrak{m}}$-linear map $g_{\mathfrak{m}} : M_{\mathfrak{m}} \longrightarrow (N/\mathfrak{m}N)_{\mathfrak{m}}$ is surjective. There is the canonical $R_{\mathfrak{m}}$-module isomorphism $(N/\mathfrak{m}N)_{\mathfrak{m}} \simeq N_{\mathfrak{m}}/\mathfrak{m}N_{\mathfrak{m}}$ via $\frac{n+\mathfrak{m}N}{s} \mapsto \frac{n}{s} + \mathfrak{m}N_{\mathfrak{m}}$. Note that g factorizes as $g = \pi_N \circ f$, with $\pi_N : N \longrightarrow N/\mathfrak{m}N$ the canonical quotient map. Therefore, the localized map $g_{\mathfrak{m}} : M_{\mathfrak{m}} \longrightarrow N_{\mathfrak{m}}/\mathfrak{m}N_{\mathfrak{m}}$ is surjective and coincides with the composition of the $R_{\mathfrak{m}}$-linear map

$$f_{\mathfrak{m}} : M_{\mathfrak{m}} \longrightarrow N_{\mathfrak{m}} \qquad (2.37)$$

and the canonical quotient map $N_{\mathfrak{m}} \longrightarrow (N/\mathfrak{m}N)_{\mathfrak{m}} = N_{\mathfrak{m}}/\mathfrak{m}N_{\mathfrak{m}}$. Thus we are in the situation of the Nakayama lemma and conclude that $f_{\mathfrak{m}} : M_{\mathfrak{m}} \longrightarrow N_{\mathfrak{m}}$ is surjective. Thus the result follows from the local-global principle.

Note that the preceding characterization does not apply for injectivity. In fact, the map $f : \mathbb{Z} \longrightarrow \mathbb{Z}, x \mapsto 2x$, is injective, while the induced map $\bar{f} : \mathbb{Z}/2\mathbb{Z} \longrightarrow \mathbb{Z}/2\mathbb{Z}$ is a zero map. Now suppose that an R-linear map $f : M \longrightarrow N$ is such that the quotient map $\bar{f} : M/\mathfrak{m}M \longrightarrow N/\mathfrak{m}N$ is injective for every maximal ideal $\mathfrak{m}$ of R. Let $x, y \in M$, with $f(x) = f(y)$. Then $\bar{f}([x-y]_{\mathfrak{m}} = 0$, and therefore the injectivity of $\bar{f}$ implies that $x - y \in \mathfrak{m}M$ for all maximal ideals $\mathfrak{m}$. Thus $x - y \in \mathrm{rad}\,(R)M$. Since rad $(R) = \{0\}$, we conclude that $x = y$. The result follows. ■

The elements $x \in \mathrm{rad}\,(R)$ of the Jacobson radical are characterized by the property that $1 - rx$ is a unit of R for all elements $r \in R$. In particular, rad $(R) = \{0\}$ for $R = \mathbb{Z}$ and $R = \mathbb{F}[x]$. More generally, rad $(R) = \{0\}$ is satisfied if R is a semisimple ring. For a local ring R, the Jacobson radical rad (R) coincides with the unique maximal ideal. In particular, for local rings the condition for injectivity is not satisfied.

Corollary 2.62. *A system of linear equations over a ring R*

$$a_{11}x_1 + \cdots + a_{1n}x_n = b_1$$

$$\vdots$$

$$a_{m1}x_1 + \cdots + a_{mn}x_n = b_m$$

is solvable in R if and only if the system of congruences

$$a_{11}x_1 + \cdots + a_{1n}x_n \equiv b_1 \quad \mathrm{mod}\ \mathfrak{m}$$

$$\vdots$$

$$a_{m1}x_1 + \cdots + a_{mn}x_n \equiv b_m \quad \mathrm{mod}\ \mathfrak{m}$$

is solvable in the fraction field $R/\mathfrak{m}$ for all maximal ideals $\mathfrak{m}$ of R.

2. Rings of Rational Functions

We now illustrate the technique of localization by constructing several rings of rational functions and prove that they are PIDs. For simplicity we focus on the field $\mathbb{R}$ of real numbers with algebraic closure $\mathbb{C}$. Let $\Omega \subset \mathbb{C} \cup \{\infty\}$ be a nonempty subset of complex numbers. Define

$$\mathbb{R}(z)_\Omega := \{f \in \mathbb{R}(z) \mid f(z) \neq \infty \quad \forall z \in \Omega\}.$$

Thus $\mathbb{R}(z)_\Omega$ consists of all real rational functions with no pole in Ω.

Theorem 2.63. $\mathbb{R}(z)_\Omega$ *is a PID. The units of* $\mathbb{R}(z)_\Omega$ *are exactly those real rational functions $f(z)$ that have no poles or zeros in Ω.*

Proof. The set

$$S = S_\Omega := \{f \in \mathbb{R}[z] \mid f(z) \neq 0 \quad \forall z \in \Omega\}$$

is multiplicatively closed. Since $\mathbb{R}[z]$ is a PID, the localization

$$\mathbb{R}[z]_S = \mathbb{R}(z)_\Omega$$

is a PID. The units are characterized by f and $1/f$ being elements of $\mathbb{R}(z)_\Omega$. The result follows.   ∎

Theorem 2.63 implies the following result for rings of proper rational functions.

Corollary 2.64. *Let Ω denote a nonempty subset of $\mathbb{C}$. Then*

$$\mathbb{R}(z)_\Omega^{\mathrm{pr}} := \{\frac{p}{q} \in \mathbb{R}(z) \mid \deg p \leq \deg q, \quad q(z) \neq 0 \quad \forall z \in \Omega\}$$

is a PID.

Proof. Applying Theorem 2.63 for $\Omega \cup \{\infty\}$ yields the result.   ∎

One can even show that a stronger property is shared by the rings $\mathbb{R}(z)_\Omega^{\mathrm{pr}}$.

Definition 2.65. An integral domain R is called a **Euclidean ring** if there exists a function $\delta : R \setminus \{0\} \to \mathbb{N}_0$, $\delta(0) := -\infty$, with the following properties:

1. **Monotonicity:** $a|b \implies \delta(a) \leq \delta(b)$ for $a,b \in R \setminus \{0\}$;
2. **Division with remainder:**   For $a \in R, b \in R \setminus \{0\}$ there exist $q, r \in R$ such that

$$a = qb + r,$$

with $r = 0$ or $0 \leq \delta(r) < \delta(b)$.

The function δ is called a **Euclidean valuation**. A Euclidean ring (R, δ) is called **strict** if the second condition is satisfied for **unique** elements q, r.

The standard example for a strict Euclidean ring is the ring of polynomials, with the degree as the Euclidean valuation function $\delta(f) = \deg f$. This valuation δ satisfies the ultrametric inequality

$$\deg (f + g) \leq \max\{\deg (f), \deg (g)\}.$$

Unlike the ring of polynomials $\mathbb{F}[z]$, the uniqueness property in division with remainders gets lost in the localizations $\mathbb{R}(z)_\Omega$. To show that they are nevertheless Euclidean rings, we first introduce the valuation function.

Definition 2.66. For $f \in \mathbb{R}(z)_\Omega$ let $\delta(f)$ denote the number of zeros of f in Ω, counted with multiplicities. Then δ is called the valuation function on $\mathbb{R}(z)_\Omega$.

One has $\delta(0) = \infty$, $\delta(1) = 0$ and $\delta(fg) = \delta(f)\delta(g)$ for all $f,g \in \mathbb{R}(z)_\Omega$. However, $\delta(f + g) \leq \max\{\delta(f), \delta(g)\}$ is not always satisfied, as the example $f(z) = \frac{4-z}{z-1}, g(z) = \frac{2z-3}{z-1} \in \mathbb{R}(z)_{\mathbb{C}_-}$ for the open left half-plane $\mathbb{C}_-$ shows. In fact, $\delta(f) = \delta(g) = 0$ and $\delta(f+g) = 1$. This shows that $(\mathbb{R}(z)_\Omega, \delta_\Omega)$ is not a strict Euclidean ring and the uniqueness of division with remainders in $\mathbb{R}(z)_\Omega$ does not apply.

One can even show that $\mathbb{R}(z)_\Omega$ and $\mathbb{R}(z)_\Omega^{\mathrm{pr}}$ are Euclidean rings.

Theorem 2.67. $\mathbb{R}(z)_\Omega$ *is a Euclidean ring for every nonempty set Ω with $\mathbb{R} \not\subset \Omega \subset \mathbb{C}$.*

Proof. One must establish the existence of division with remainders for δ. Consider elements f and g, with $g \neq 0$. If $\delta(f) < \delta(g)$, then we are done because $f = 0g + r$, with $\delta(r) < \delta(g)$. Next assume that $\delta(f) \geq \delta(g)$. If $\delta(g) = 0$, then g is a unit, and we are done. Thus suppose $\delta(g) \geq 1$. Write

$$f = \frac{p_\Omega p_{\Omega'}}{q},$$

$$g = \frac{u_\Omega u_{\Omega'}}{v},$$

where the polynomials p_Ω, u_Ω and $p_{\Omega'}, u_{\Omega'}, q, v$ are assumed to have all their roots in Ω and $\Omega' := \mathbb{C} \setminus \Omega$, respectively. Choose $\alpha \in \Omega' \cap \mathbb{R}$ and $t = \delta(g)$. Then

$$e = \frac{u_{\Omega'}(z - \alpha)^t}{v} \in \mathbb{R}(z)_\Omega$$

is a unit and, therefore, invertible in $\mathbb{R}(z)_\Omega$. Since u_Ω and q are coprime polynomials, there exist real polynomials $\phi(z)$ and $\psi(z)$, with $\deg \psi < \deg u_\Omega$, that satisfy the Bezout identity

$$u_\Omega \phi + q\psi = p(z - \alpha)^{t-1}.$$

Dividing both sides of this equation by $p(z - \alpha)^{t-1}$ we obtain $f = ag + r$, with

$$a = \frac{1}{e} \frac{\phi(z - \alpha)}{q}, r = \frac{\psi}{(z - \alpha)^{t-1}}.$$

Since e is a unit, a is an element of $\mathbb{R}(z)_\Omega$. Also, $r \in \mathbb{R}(z)_\Omega$:

$$\delta(r) \leq \deg \psi < \deg u_\Omega \leq t.$$

This completes the proof.   ∎

Finally, let us consider one example that will be important in Chapter 6 when we discuss the feedback stabilization of linear systems. Let $\Omega = \overline{\mathbb{C}_+} := \{z \in \mathbb{C} \mid \operatorname{Re}(z) \geq 0\} \cup \{\infty\}$ denote the extended right half-plane. Let

$$RH_\infty := \mathbb{R}(z)_\Omega$$

denote the set of stable proper rational functions.

Theorem 2.68. *1. RH_∞ is a PID and even a Euclidean ring.*
2. $f_1, \ldots, f_r \in RH_\infty$ are coprime if and only if there exists a solution $a_1, \ldots, a_r \in RH_\infty$ to the Bezout equation

$$a_1 f_1 + \cdots + a_r f_r = 1.$$

3. The units of RH_∞ are all rational functions $\frac{p}{q}$, with $\deg p = \deg q$, such that p and q have only zeros in the open left half-plane $\mathbb{C}_-$.

4. $f \in RH_\infty$ *divides* $g \in RH_\infty$ *if and only if the following conditions are satisfied:*

   (a) *The relative degree of* f *is less than or equal to the relative degree of* g;
   (b) *Every finite zero of* f *with multiplicity* m *in* $\overline{\mathbb{C}}_+$ *is a finite zero of* g *with multiplicity* $\geq m$.

5. *The maximal ideals of* RH_∞ *coincide with the prime ideals and are equal to* fRH_∞, *where* f *is one of the following rational functions:*

$$f(z) = \frac{1}{z - \alpha}, \quad \alpha < 0;$$

$$f(z) = \frac{z - \alpha}{z - \beta}, \quad \alpha \geq 0, \beta < 0;$$

$$f(z) = \frac{(z - \lambda)(z - \bar{\lambda})}{z^2 + \alpha z + \beta}, \quad \text{Re } \lambda \geq 0, \text{Im } \lambda > 0, \alpha > 0, \beta > 0.$$

Proof. Theorem 2.63 implies that RH_∞ is a PID; Theorem 2.67 shows that it is a Euclidean ring. The characterization of coprimeness via the Bezout equation is valid in an arbitrary PID, and hence also here. A rational function p/q is a unit if and only if $p/q, q/p \in RH_\infty$, i.e., if and only if deg $p =$ deg q and p and q are Hurwitz polynomials. Since RH_∞ is a PID, the maximal ideals coincide with the prime ideals. $RH_\infty = A_S$ is the localization of the PID $A = \mathbb{R}(z)_{z^{-1}} := \{ \frac{p(z)}{q(z)} \mid$ deg $p \leq$ deg $q\}$ at the multiplicatively closed set $S_+ = \{f \in A \mid f(z) \neq 0 \quad \forall z \in \overline{\mathbb{C}}_+\}$. By Proposition 2.63, the prime ideals are of the form fRH_∞, where f is a prime in A satisfying $fA \cap S_+ = \emptyset$. These in turn are characterized by all primes in A, with $f(z) = 0$ for some $z \in \overline{\mathbb{C}}_+$. Since the rational function f is prime, it has degree either one or two. If $z = \infty$, then f is strictly proper, and therefore f is of the first form. Let z be finite, with Re $z \geq 0$. If $z = \alpha$ is real, then we obtain the second form. If $z = \lambda$ is complex, then f must be of the third form. The result follows. ∎

2.12 Exercises

1. Find solutions $x, y \in \mathbb{Z}$ of $49x + 13y = 1$.
2. Describe all integer solutions of the following system of congruences:

$$2x + 4y \equiv 7 \mod 3,$$

$$3x - y + 2z \equiv 6 \mod 3,$$

$$-8x + 2y - z + 2 \equiv 4 \mod 3.$$

3. Let $m_1, \ldots, m_s \in \mathbb{N}$ be mutually prime integers. Prove that for integers $a_1, \ldots, a_s$ the system of congruences

$$x \equiv a_1 \bmod m_1$$

$$\vdots$$

$$x \equiv a_s \bmod m_s$$

has an integer solution. It is unique up to multiples of $m_1 \cdots m_s$.

4. For $n \in \mathbb{N}, n \geq 2$, define the additive factor group $\mathbb{Z}_n := \mathbb{Z}/n\mathbb{Z}$. Prove:

   (a) $\mathbb{Z}_n$ is a field if and only if n is a prime number;
   (b) The additive subgroups of $\mathbb{Z}_n$ are exactly of the form

   $$H_t = t\mathbb{Z}_{n/t} := \{[0], [t], [2t], \ldots, [(a-1)t]\}$$

   for a divisor t of n and $a := n/t$. For divisors s, t of n show that

   $$H_t \subset H_s \iff s|t.$$

5. Determine the greatest common divisor of $z^3 - 1$ and $z^3 - 2z + 1$ in $\mathbb{R}[z]$.
6. Prove that $z^4 + z + 1$ is irreducible in $\mathbb{Z}_2[z]$ and reducible in $\mathbb{Z}_3[z]$.
7. Calculate the generator polynomial of the ideal

$$< z^3 + z^2 - 4z - 4, z^3 - z^2 - 4z + 4, z^3 - 2z^2 - z + 2 >,$$

   and decide whether or not the polynomial $z^2 - 4$ belongs to the ideal.
8. For the polynomials $p(z) = 1 + z^2$ and $q(z) = 1 + z + z^3$ find a unimodular matrix $U(z) \in \mathbb{F}^{2 \times 2}[z]$ such that $(p(z), q(z))$ is the first row vector of $U(z)$.
9. Let $\mathbb{F}[z]_{<m}$ denote the $\mathbb{F}$-vector space of polynomials of degree $< m$, $m \in \mathbb{N}$. Consider the polynomials $p(z), q(z) \in \mathbb{F}[z]$, with $\deg p = m, \deg q = n$. Prove that there exist nonzero polynomials $a(z) \in \mathbb{F}[z]_{<n}$ and $b(z) \in \mathbb{F}[z]_{<m}$, with $a(z)p(z) + b(z)q(z) = 0$, if and only if p and q are not coprime.
10. Show that the polynomials $a_1(z), \ldots, a_N(z) \in \mathbb{F}[z]$ are coprime if and only if there exists a unimodular matrix $U(z) \in \mathbb{F}^{N \times N}[z]$ such that $(a_1(z), \ldots, a_N(z))$ is the first row of $U(z)$.
11. Let $q_1(z), \ldots, q_N(z) \in \mathbb{F}[z]$ have degrees $d_1, \ldots, d_N$. Define the polynomials $\hat{q}_i(z) = \prod_{j \neq i} q_j(z), i = 1, \ldots, N$. Let $\mathbb{F}[z]_{<k}$ denote the set of polynomials with degree $< k$. Show that the following conditions are equivalent:

   (a) $q_1(z), \ldots, q_N(z)$ are coprime.
   (b) $\hat{q}_1(z), \ldots, \hat{q}_N(z)$ are coprime.
   (c) The $\mathbb{F}$-linear map

$$\mathscr{R} : \mathbb{F}[z]_{<d_1} \oplus \ldots \oplus \mathbb{F}[z]_{<d_N} \longrightarrow \mathbb{F}[z]_{<d_1 + \ldots + d_N}, (a_1, \ldots, a_N) \mapsto \sum_{j=1}^{N} a_j \hat{q}_j$$

   is invertible.

Deduce a matrix resultant test for the coprimeness of $q_1(z), \ldots, q_N(z)$.

12. Prove that an element of a PID is irreducible if and only if it is prime.

13. Prove that the prime ideals in a PID coincide with the maximal ideals.

14. Determine whether the following pairs of polynomial matrices are right coprime. If they are not, compute a greatest common right divisor.

(a)

$$N_r(z) = \begin{pmatrix} 0 & z^2 \\ -z & z^2 \end{pmatrix}, \quad D_r(z) = \begin{pmatrix} 0 & (z+1)^2(z+3) \\ (z+3)^2 & z+3 \end{pmatrix};$$

(b)

$$N_r(z) = \begin{pmatrix} z & z \\ 0 & z(z+1)^2 - z \end{pmatrix}, \quad D_r(z) = \begin{pmatrix} (z+1)^2(z+2)^2 & 0 \\ 0 & (z+2)^2 \end{pmatrix}.$$

15. Obtain a coprime column-reduced right polynomial fraction description for

$$G(z) = \begin{pmatrix} z & z+2 \\ 1 & z+1 \end{pmatrix} \begin{pmatrix} z^2+2 & (z+1)^2 \\ z+1 & z \end{pmatrix}^{-1}.$$

16. Show that the polynomial matrix

$$D(z) = \begin{pmatrix} z^2+2 & z^3 & z^5+z+1 \\ 2z+1 & z^4+3z+1 & 2z^2 \\ z+2 & 2z^2+1 & z^6-2z^4 \end{pmatrix}$$

is nonsingular and column proper with column indices $2, 4, 6$.

17. Let $A, B \in \mathbb{F}^{n \times n}$ be square matrices that satisfy $U(z)(zI - A) = (zI - B)V(z)$ for suitable unimodular polynomial matrices $U(z), V(z) \in \mathbb{F}[z]^{n \times n}$. Show that there exists a constant invertible matrix $S \in GL_n(\mathbb{F})$, with $B = SAS^{-1}$.

18. Prove that a polynomial matrix $A(z) \in \mathbb{F}[z]^{p \times m}$ is left and right coprime if and only if $p = m$ and $A(z)$ is unimodular.

19. Show that a regular matrix pencil $zE - F \in \mathbb{F}[z]^{n \times n}$ is unimodular if and only if there exists a nilpotent matrix $N \in \mathbb{F}^{n \times n}$ such that $zE - F$ is unimodularly equivalent to $I - zN$.

20. Let $A(z) \in \mathbb{R}[z]^{m \times n}$ be a full column rank polynomial matrix. Show that the linear equation $A(z)x(z) = b(z)$ has for each $b(z) \in \mathbb{R}[z]^m \cap A(z)\mathbb{R}(z)^n$ a polynomial solution $x(z) \in \mathbb{R}[z]^n$ if and only if the matrix $A(z)$ has full column rank n for all $z \in \mathbb{C}$.

21. Let $A(z) \in \mathbb{R}[z]^{m \times n}$ be a polynomial matrix and $b(z) \in \mathbb{R}[z]^m$. Prove that $A(z)X(z) = b(z)$ has a polynomial solution $X(z) \in \mathbb{R}[z]^n$ if and only if, for all $z \in \mathbb{C}$,

$$\text{rk}\,(A(z), b(z)) = \text{rk}\,A(z).$$

22. Prove the following abstract version of the Chinese remainder theorem for modules. Let M be a module over a ring R, and let $I_1, \ldots, I_n$ be ideals of R satisfying $I_i + I_j = R$ for all $i \neq j$. Let $I_1 \cdots I_n$ denote the ideal generated by the products of elements in $I_1, \ldots, I_n$. Then $I_1 \cdots I_n M = I_1 M \cap \cdots \cap I_n M$, and the map

$$f : M/(I_1 \cdots I_n M) \longrightarrow \prod_{j=1}^{n} (M/I_j M)$$

$$x + I_1 \cdots I_n M \;\mapsto\; (x + I_1 M, \ldots, x + I_n M)$$

is a module isomorphism.

23. Let $A \in \mathbb{F}[z]^{m \times n}$. Show that the following conditions are equivalent:

    (a) $A : \mathbb{F}[z]^n \longrightarrow \mathbb{F}[z]^m$ is surjective.
    (b) A has a right inverse, i.e., there exists $B \in \mathbb{F}[z]^{n \times m}$, with $AB = I_m$.
    (c) The $m \times m$ subminors of A generate $\mathbb{F}[z]$.

24. Let $P(z) \in \mathbb{F}[z]^{p \times m}$ be full column rank. Show that there exist a right prime polynomial matrix $M(z) \in \mathbb{F}[z]^{p \times m}$ and a nonsingular polynomial matrix $\Delta(z) \in \mathbb{F}[z]^{m \times m}$ such that

$$P(z) = M(z)\Delta(z).$$

25. Let $A(z) \in \mathbb{R}[z]^{m \times n}$ be a real polynomial matrix. Prove that the quotient module $\mathbb{R}[z]^m / A(z)\mathbb{R}[z]^n$ is free of rank r if and only if $\operatorname{rk} A(z) = m - r$ for all $z \in \mathbb{C}$.

2.13 Notes and References

The use of polynomial matrices for a study of linear algebra is not new and goes back to E. Noether; see MacDuffee (1933).

Jodeit (1967) has shown that $\mathbb{F}[z]$ is the only Euclidean ring for which division with remainders is unique. All results in this chapter, as well as the proofs, remain true if one replaces the ring of polynomials $\mathbb{F}[z]$ by a PID. The only exceptions are the results in Section 2.4 (Minimal Basis) and 2.7 (Wiener–Hopf Factorizations), which would require appropriate modifications. In particular, the existence of doubly coprime factorizations is valid for the PID RH^∞ of proper rational stable transfer functions. A discussion of doubly coprime factorizations can be found in Kailath (1980) or Vidyasagar (1987). Doubly coprime factorizations are key to many duality considerations as well as to different applications; see, for example, Fuhrmann (1994b) and Fuhrmann and Ober (1993). Skew primeness was considered in Wolovich (1978), and its geometric interpretation is due to Khargonekar, Georgiou and Özgüler (1983); see also Fuhrmann (2005) for additional work.

Minimal bases were first introduced in the famous work by Dedekind and Weber (1882). That classical paper introduced valuation theory for the first time and proved an equivalent form of the Wiener–Hopf factorization. An elegant approach to minimal bases via filtrations of subspaces is due to Mackey (2012). Wiener–Hopf factorizations in analysis are a central tool in the solution of several classes of integral equations; see Gohberg and Feldman (1971). Our focus is on the purely algebraic theory of Wiener–Hopf factorizations at infinity. This is based on Fuhrmann and Willems (1979). These factorizations are closely related to the reduction of polynomial matrices to row or column proper form. In the systems theory context, proper bases were rediscovered by Forney (1975), with subsequent contributions by Münzner and Prätzel-Wolters (1979).

The classical Bezout equation $ax + by = 1$ can be viewed as the problem of extending the 1×2 matrix (a,b) over a ring to a 2×2 matrix of determinant one. More general extension problems of this kind have been studied in algebraic K-theory; see, for example, Newman (1985) and Vaserstein (1986) on matrix completion problems over a PID.

Localizations of rings and modules are an important tool from algebra and appear in several textbooks on commutative algebra; see, for example, Atiyah-Macdonald (1969) and Kunz (1980). An interesting application for solving parameter-dependent linear equations can be found in Adkins (1985) and Mather (1973). For applications to families of linear control systems, see, for example, Hazewinkel and Perdon (1981). Rings of stable rational functions were studied in a systems theory context using valuation theory by Vidyasagar (1987); see also Morse (1975).

Chapter 3
Functional Models and Shift Spaces

Some mathematical problems can be treated efficiently in one context but may remain intractable in a different context. An example, relevant to our interests, is the characterization and parameterization of invariant subspaces of shift operators. If the problem is set in the Hilbert space $l^2(\mathbb{N})$ of one-sided square summable sequences, it looks intractable; however, after a Fourier transformation into a suitable Hardy space, one has the full machinery of complex analytic functions at one's disposal. This enabled Beurling (1949) to solve the problem. This example shows clearly the advantage of using functional spaces rather than spaces of scalar sequences, an advantage that is the result of having extra structure. It is the same basic idea that underpins our algebraic approach to the analysis of linear systems. Basically, this is the idea, traceable to Emmy Noether, of studying a linear operator in terms of the module structure induced by it.

Our approach, however, is different in that we emphasize computational aspects. The underlying idea is that a linear transformation is replaced by a functional model of it, that is, by an equivalent transformation that acts on an appropriate function space, which in our case is going to be a space of polynomials or rational functions. This idea goes back to the pioneering work of Livsic, Sz.-Nagy and Foias, and De Branges, and it gave great impetus to research in the theory of nonselfadjoint operators. These techniques were quickly applied to system theory and culminated in the widely used H^∞-control theory; see, for example, Vidyasagar (1987). Operator-theoretic methods were adapted to the finite-dimensional, algebraic context in Fuhrmann (1976, 1977) and proved themselves a very useful tool for the study of linear systems. In later chapters, these methods will be applied to the investigation of problems arising in the study of interconnected linear systems, especially to problems not very amenable to state-space techniques.

© Springer International Publishing Switzerland 2015
P.A. Fuhrmann, U. Helmke, *The Mathematics of Networks of Linear Systems*, Universitext, DOI 10.1007/978-3-319-16646-9_3

3.1 Polynomial Models and the Shift Operator

In Chapter 2, the study of nonsingular polynomial matrices and associated quotient modules was approached from a rather abstract algebraic point of view. This study depended on the fact that the ring of polynomials $\mathbb{F}[z]$ is a principal ideal domain. The familiar process of division with remainders that utilizes the Euclidean ring structure of $\mathbb{F}[z]$ allows for a much more concrete approach by identifying coset classes in a quotient module with polynomials obtained by division with remainders. This rather concrete theory of polynomial and rational models has been developed in various papers and books during the past 40 years; see Fuhrmann (1976, 2012). It leads to a very efficient computational approach to modules of polynomial matrices and constitutes our basic approach to the analysis of networks of systems. In the following sections, we present a review of the basic results on polynomial and rational models and introduce concrete representations of homomorphisms between such spaces.

We start by recalling some general facts about **Laurent series** and **formal power series**. Let $\mathbb{F}((z^{-1}))^m$ denote the vector space of truncated Laurent series in z^{-1}, i.e., the elements of $\mathbb{F}((z^{-1}))^m$ are $f(z) = \sum_{j=-\infty}^{n_f} f_j z^j$, with coefficients $f_j \in \mathbb{F}^m$. The **residue** of $f(z)$ is the coefficient of z^{-1}, i.e., the coefficient vector f_{-1}. If $m = 1$, then the sum, product, and quotient of two scalar Laurent series are again a Laurent series. This shows that $\mathbb{F}((z^{-1}))$ is a field that is in fact the quotient field of $z^{-1}\mathbb{F}[z^{-1}]$. By construction, $\mathbb{F}((z^{-1}))^m$ contains the sets of polynomials $\mathbb{F}[z]^m$ and of **strictly proper power series** $z^{-1}\mathbb{F}[z^{-1}]^m$ as subspaces, leading to the following direct sum decomposition of $\mathbb{F}$-vector spaces:

$$\mathbb{F}((z^{-1}))^m = \mathbb{F}[z]^m \oplus z^{-1}\mathbb{F}[[z^{-1}]]^m. \tag{3.1}$$

Denote by

$$\pi_- : \mathbb{F}((z^{-1}))^m \longrightarrow z^{-1}\mathbb{F}[[z^{-1}]]^m \quad \text{and} \quad \pi_+ : \mathbb{F}((z))^m \longrightarrow \mathbb{F}[z]^m$$

the canonical projections onto the strictly proper and polynomial parts, respectively.

With these projection operators at hand, it is easy to derive an explicit formula for division with remainders. In fact, by applying division with remainders to scalar polynomials $f(z), d(z) \in \mathbb{F}[z]$, there is a unique representation $f(z) = a(z)d(z) + r(z)$ by polynomials $a(z), r(z) \in \mathbb{F}[z]$, once we require $\deg r(z) < \deg d(z)$. Thus the rational function $r(z)/d(z)$ is strictly proper, and therefore $r(z)/d(z)$ and $a(z)$ are the strictly proper and the polynomial parts of the rational function $f(z)d(z)^{-1}$, respectively. This implies $r(z) = d(z)\pi_-(d(z)^{-1}f(z))$. Note that

$$\pi_d : \mathbb{F}[z] \longrightarrow \mathbb{F}[z], \quad \pi_d(f) = d(z)\pi_-(d(z)^{-1}f(z))$$

defines a projection operator onto the $\mathbb{F}$-linear subspace of $\mathbb{F}[z]$,

$$X_d := \{f(z) \in \mathbb{F}[z] \mid \deg f < \deg d\},$$

of all polynomials $f(z)$ that have degree $< \deg d$. Note that although X_d does define a finite-dimensional subspace of $\mathbb{F}[z]$, it is not a submodule of $\mathbb{F}[z]$.

Polynomial models are a natural extension of this construction for nonsingular matrix polynomials $D(z) \in \mathbb{F}[z]^{m \times m}$. Thus, polynomial models define concrete representations of quotient modules $\mathbb{F}[z]^m/M$, where $M \subset \mathbb{F}[z]^m$ is a full submodule, i.e., $\mathbb{F}[z]^m/M$ is required to be a torsion module. By Theorem 2.51, this is equivalent to a representation $M = D(z)\mathbb{F}[z]^m$, with $D(z) \in \mathbb{F}[z]^{m \times m}$ nonsingular.

Definition 3.1. The **polynomial model** associated with $D(z)$ is defined as

$$X_D = \{f(z) \in \mathbb{F}[z]^m \mid D(z)^{-1}f(z) \text{ strictly proper}\}. \tag{3.2}$$

By definition, a polynomial vector $f(z) \in \mathbb{F}[z]^m$ satisfies $f(z) \in X_D$ if and only if $D(z)^{-1}f(z)$ is strictly proper. It follows that polynomial models X_D are $\mathbb{F}$-vector subspaces of $\mathbb{F}[z]^m$. To clarify the connection of this construction to modules, and in particular to quotient modules, consider the $\mathbb{F}$-linear map

$$\pi_D : \mathbb{F}[z]^m \longrightarrow \mathbb{F}[z]^m$$

defined by

$$\pi_D f = D(z)\pi_-(D(z)^{-1}f(z)), \qquad f(z) \in \mathbb{F}[z]^m. \tag{3.3}$$

The basic properties of π_D are summarized in the next result.

Lemma 3.2. *Let $D(z) \in \mathbb{F}[z]^{m \times m}$ be a nonsingular polynomial matrix. Then:*

1. *π_D, defined by (3.3), is an $\mathbb{F}$-linear projection operator onto X_D;*
2. *The kernel of π_D is equal to $D(z)\mathbb{F}[z]^m$;*
3. *The following is a direct sum decomposition of $\mathbb{F}$-vector spaces:*

$$X_D \oplus D(z)\mathbb{F}[z]^m = \mathbb{F}[z]^m.$$

Proof. The linearity of π_D is obvious, and the identity

$$D\pi_-(D^{-1}f) = f - D\pi_+(D^{-1}f)$$

shows that, for polynomials $f(z) \in \mathbb{F}[z]^m$, $\pi_D(f)$ is a polynomial. For $f(z) \in \mathbb{F}[z]^m$,

$$\pi_D(\pi_D(f)) = D\pi_-(\pi_-(D^{-1}f)) = D\pi_-(D^{-1}f) = \pi_D(f),$$

which shows that π_D is a projection operator. Every element $f = \pi_D(g)$ in the image space of π_D is a polynomial such that $D^{-1}f = \pi_-(D^{-1}g)$ is strictly proper. Conversely, if $f \in \mathbb{F}[z]^m$ is a polynomial such that $D^{-1}f$ is strictly proper, then $\pi_-(D^{-1}f) = D^{-1}f$. Therefore, $\pi_D(f) = D\pi_-(D^{-1}f) = DD^{-1}f = f$ is in the image of π_D. This proves statement 1. Let $f(z) \in \mathbb{F}[z]^m$. Then $\pi_D(f) = 0$ if and only if $\pi_-(D^{-1}f) = 0$, i.e., if and only if $D(z)^{-1}f(z)$ is a polynomial. Therefore, $f(z) \in$ Ker π_D if and only if $f(z) \in D(z)\mathbb{F}[z]^m$. Finally, statement 3 is a trivial consequence of statements 1 and 2. This completes the proof. $\blacksquare$

In the sequel, the notation

$$[f]_D = f(z) + D(z)\mathbb{F}[z]^m$$

will be frequently used for the associated coset class of a polynomial $f(z)$ in the quotient space $\mathbb{F}[z]^m/D(z)\mathbb{F}[z]^m$. Lemma 3.2 implies that the projection operator π_D induces an $\mathbb{F}$-vector space isomorphism

$$\bar{\pi}_D : \mathbb{F}[z]^m/D(z)\mathbb{F}[z]^m \longrightarrow X_D, \quad \bar{\pi}_D([f]_D) = \pi_D(f). \tag{3.4}$$

Next, using the shift operator, a module structure on X_D is defined so that $\bar{\pi}_D$ becomes an $\mathbb{F}[z]$-module isomorphism.

Definition 3.3. The **shift operator** $S_D : X_D \longrightarrow X_D$ is the $\mathbb{F}$-linear map defined by

$$S_D f = \pi_D(zf), \quad f(z) \in X_D.$$

The polynomial model X_D becomes an $\mathbb{F}[z]$-module using the S_D-induced module structure, i.e., by defining

$$p \cdot f = \pi_D(pf), \qquad p(z) \in \mathbb{F}[z], \ f(z) \in X_D. \tag{3.5}$$

The preceding analysis is summarized by the following result.

Theorem 3.4. *Let $D(z)$ be a nonsingular element in $\mathbb{F}[z]^{m \times m}$. Then:*

1. *π_D is a projection in $\mathbb{F}[z]^m$ and Ker $\pi_D = D\mathbb{F}[z]^m$;*
2. *A vector polynomial $f(z) \in \mathbb{F}[z]^m$ belongs to X_D if and only if $D(z)^{-1}f(z) \in z^{-1}\mathbb{F}[[z^{-1}]]^m$, i.e., if and only if $D(z)^{-1}f(z)$ is a strictly proper rational vector function; so*

$$X_D = \{f(z) \in \mathbb{F}[z]^m \mid \pi_+(D(z)^{-1}f(z)) = 0\};$$

an alternative description of the polynomial model X_D is the following:

$$X_D = \left\{ f(z) \in \mathbb{F}[z]^m \mid f(z) = D(z)h(z), \ h(z) \in z^{-1}\mathbb{F}[[z^{-1}]]^m \right\}.$$

3. *The module structure in X_D defined by (3.5) implies the isomorphism*

$$X_D \simeq \mathbb{F}[z]^m / D\mathbb{F}[z]^m, \tag{3.6}$$

with the isomorphism $\overline{\pi}_D : \mathbb{F}[z]^m / D\mathbb{F}[z]^m \longrightarrow X_D$ defined by (3.4).

Proof. 1. This follows from Lemma 3.2.

2. If $f \in X_D$, then $\pi_D f = f$, and so $D\pi_- D^{-1} f = f$ or $\pi_- D^{-1} f = D^{-1} f$, which implies that $\pi_+ D^{-1} f = 0$. Conversely, if $f(z) \in \mathbb{F}[z]^m$ and $\pi_+ D^{-1} f = 0$, then, as $\pi_+ + \pi_- = I$, the equality $\pi_- D^{-1} f = D^{-1} f$ follows, which implies $f = D\pi_- D^{-1} f = \pi_D f$, that is, $f(z) \in X_D$.

3. By definition, the projection π_D is actually an $\mathbb{F}[z]$-homomorphism. It is surjective, and $\operatorname{Ker} \pi_D = D\mathbb{F}[z]^m$. Hence, the isomorphism (3.6) follows. ∎

Note that the shift operator S_D becomes a module homomorphism with respect to this module structure on X_D. In fact, $z \cdot f = \pi_D(zf) = S_D f$. Moreover, the module structure (3.5) is a unique one such that $\overline{\pi}_D$ becomes a module isomorphism.

Theorem 3.5. *The polynomial model X_D is a finite-dimensional $\mathbb{F}$-vector space of dimension*

$$\dim X_D = \deg \det D(z).$$

Moreover, (3.5) defines the unique module structure on X_D such that

$$\overline{\pi}_D : \mathbb{F}[z]^m / D(z)\mathbb{F}[z]^m \longrightarrow X_D$$

becomes an $\mathbb{F}[z]$-module isomorphism. The inverse of $\overline{\pi}_D$ is given by

$$\overline{\pi}_D^{-1}(f) = f(z) + D(z)\mathbb{F}[z]^m.$$

In particular, X_D is a finitely generated $\mathbb{F}[z]$ torsion module.

Proof. By Theorem 2.51, the quotient module $\mathbb{F}[z]^m / D(z)\mathbb{F}[z]^m$ is a finitely generated torsion module and has vector space dimension $\deg \det D(z)$. Endow X_D with the module structure defined by (3.5). Then the vector space isomorphism $\overline{\pi}_D : \mathbb{F}[z]^m / D(z)\mathbb{F}[z]^m \longrightarrow X_D$ becomes an $\mathbb{F}[z]$-module isomorphism. The result follows. ∎

As a simple computational example, one may consider the nonsingular polynomial matrix

$$D(z) = zI - A$$

defined by a constant matrix $A \in \mathbb{F}^{n \times n}$. The elements of the polynomial model X_{zI-A} are then all vector polynomials $f(z) \in \mathbb{F}[z]^n$ such that $(zI - A)^{-1} f(z)$ is strictly proper. This implies that $f(z) = f_0$ is constant, and therefore

$$X_{zI-A} = \mathbb{F}^n$$

consists of all constant polynomial vectors $\xi \in \mathbb{F}^n \subset \mathbb{F}[z]^n$. The action of the shift operator S_{zI-A} on X_{zI-A} is computed as follows. Let $f(z) = \xi$ be a constant polynomial. Then

$$S_{zI-A}\xi = (zI-A)\pi_-(z(zI-A)^{-1}\xi) = (zI-A)\pi_-(\xi) + (zI-A)\pi_-(A(zI-A)^{-1}\xi)$$
$$= A\xi.$$

Therefore, the shift operator on X_{zI-A} is conjugate, i.e., similar to the linear operator A acting in $\mathbb{F}^n$. Conversely, this proves the next result which states that every square matrix A can be regarded as the shift operator of the associated matrix pencil.

Proposition 3.6. *Every matrix $A \in \mathbb{F}^{n \times n}$ is conjugate to the shift operator of $zI - A$, i.e.,*

$$S_{zI-A} \simeq A.$$

For a slightly more advanced example, consider nonsingular matrix pencils

$$D(z) = zE - A,$$

with $E, A \in \mathbb{F}^{n \times n}$ and $\det(zE - A)$ not identical to zero. Then a polynomial $f(z) \in \mathbb{F}[z]^n$ is contained in X_D if and only if there exists a strictly proper formal power series $h(z) = \sum_{j=1}^{\infty} h_j z^{-j}$, with

$$(zE - A)h(z) = f(z).$$

This implies that $f(z) = f_0 \in \text{Im}\,E$ is constant. Therefore, it follows that $X_D \subset \mathbb{F}^n$. Using the Weierstrass decomposition

$$LER^{-1} = \begin{pmatrix} I_r & 0 \\ 0 & N \end{pmatrix}, \quad LAR^{-1} = \begin{pmatrix} A_1 & 0 \\ 0 & I \end{pmatrix}$$

with suitable invertible matrices L, R, and N nilpotent, we can rewrite the linear descriptor system

$$Ex_{t+1} = Ax_t \tag{3.7}$$

equivalently as a decoupled system,

$$x_{1,t+1} = Ax_{1,t},$$
$$Nx_{2,t+1} = x_{2,t}.$$

Here the first dynamics of (3.7) refers to the so-called slow dynamics, whereas the second equation refers to the fast dynamics of the descriptor system. This implies that the elements of the polynomial model X_D correspond exactly to the states of the slow dynamics of (3.7). In particular, this implies $\dim X_D = \mathrm{rank}\, E$. An equivalent characterization of X_D is stated without proof.

Proposition 3.7. *The polynomial model X_{zE-A} is the largest linear subspace $\mathcal{V} \subset \mathbb{F}^n$ satisfying $A\mathcal{V} \subset E\mathcal{V}$.*

This example shows that the theory of polynomial models captures in a very natural way interesting properties of the dynamics of higher-order differential equations.

3.2   The Lattice of Shift-Invariant Subspaces

In this section, the lattice of submodules of polynomial models X_D, i.e., the lattice of shift-invariant subspaces of X_D, will be explored. To begin, some basic results on the geometry of subspaces of a linear space $\mathcal{X}$ are recalled. The natural set operations on subspaces are sums and intersections of which a special case is that of direct sums. This is in line with the strategy of reducing, if possible, the study of a complex object to the study of a bunch of simpler ones. If A is a linear transformation acting in $\mathcal{X}$, then the subspaces of interest are the **invariant subspaces**, namely, subspaces for which $A\mathcal{V} \subset \mathcal{V}$. As is the case for subspaces, the set of invariant subspaces is closed under sums and intersections. In the sequel, the following lemma from linear algebra, whose elementary proof is left to the reader, will be frequently referred to.

Lemma 3.8. *Let $A : \mathcal{X} \longrightarrow \mathcal{X}$ and $W, Z : \mathcal{X} \longrightarrow \mathcal{Y}$ be $\mathbb{F}$-linear transformations of vector spaces $\mathcal{X}, \mathcal{Y}$. The following statements are equivalent:*

(a)

$$A \operatorname{Ker} W \subset \operatorname{Ker} Z.$$

(b)

$$\operatorname{Ker} W \subset \operatorname{Ker} ZA.$$

(c) There exists a linear transformation $F : \mathcal{Y} \longrightarrow \mathcal{Y}$ such that

$$ZA = FW.$$

Recall that, for a linear space $\mathscr{X}$ and a finite number of $\mathbb{F}$-linear subspaces $\mathscr{V}_j$, $j = 1, \ldots, s$, a subspace $\mathscr{V} \subset \mathscr{X}$ is the **direct sum** of the $\mathscr{V}_i$, written as

$$\mathscr{V} = \mathscr{V}_1 \oplus \cdots \oplus \mathscr{V}_s = \bigoplus_{j=1}^{s} \mathscr{V}_j,$$

if every $x \in \mathscr{V}$ has a unique representation of the form $x = \sum_{j=1}^{s} x_j$, with $x_j \in \mathscr{V}_j$. A subspace $\mathscr{V} = \mathscr{V}_1 \cap \cdots \cap \mathscr{V}_s$ is called the **transversal intersection** of the $\mathscr{V}_1, \ldots, \mathscr{V}_s$ whenever

$$\mathscr{X} = \mathscr{V}_i + \bigcap_{j \neq i} \mathscr{V}_j \tag{3.8}$$

for $i = 1, \ldots, s$. Direct sums and transversal intersections are characterized as follows.

Proposition 3.9. *Let* $\mathscr{V}_i, i = 1, \ldots, s$, *be linear subspaces of an* $\mathbb{F}$-*linear space* $\mathscr{X}$.

1. *A linear subspace* $\mathscr{V} \subset \mathscr{X}$ *is the direct sum* $\mathscr{V} = \bigoplus_{j=1}^{s} \mathscr{V}_j$ *if and only if the following conditions are satisfied:*

   (a) *The subspace* $\mathscr{V}$ *is equal to the sum* $\sum_{j=1}^{s} \mathscr{V}_j$;
   (b) $\mathscr{V}_i \cap \sum_{j \neq i}^{s} \mathscr{V}_j = \{0\}$ *for all* $i = 1, \ldots, s$.

2. *Define*

$$\mathscr{V} = \bigcap_{j=1}^{s} \mathscr{V}_j, \quad \mathscr{W}_i = \bigcap_{j \neq i} \mathscr{V}_j.$$

   $\mathscr{V}$ *is the transversal intersection of* $\mathscr{V}_1, \ldots, \mathscr{V}_s$ *if and only if one has the direct sum decomposition*

$$\mathscr{X} / \mathscr{V} = \mathscr{W}_1 / \mathscr{V} \oplus \cdots \oplus \mathscr{W}_s / \mathscr{V}. \tag{3.9}$$

Proof. The proof of the first part is obvious and is omitted. Assume that the direct sum decomposition (3.9) is true. This can be rewritten as $\mathscr{X} / \mathscr{V} = (\oplus_{j \neq i} \mathscr{W}_j / \mathscr{V}) \oplus \mathscr{W}_i / \mathscr{V}$. The inclusions $\mathscr{V}_i \supset \mathscr{W}_j$, valid for $j \neq i$, imply $\mathscr{V}_i \supset \sum_{j \neq i} \mathscr{W}_j$. So

$$\mathscr{X} / \mathscr{V} = \mathscr{V}_i / \mathscr{V} + \mathscr{W}_i / \mathscr{V},$$

which clearly implies $\mathscr{X} = \mathscr{V}_i + \mathscr{W}_i$. Since $\mathscr{V}_i \cap \mathscr{W}_i = \mathscr{V}$ is satisfied, $\mathscr{V}$ is the transversal intersection of the $\mathscr{V}_i$.

To prove sufficiency, one proceeds by induction on s. Assume the statement is true for all integers up to $s - 1$ and that (3.8) is satisfied for all $i = 1, \ldots, s$. One concludes that $\mathscr{X} / \mathscr{V} = \mathscr{W}_s / \mathscr{V} \oplus \mathscr{V}_s / \mathscr{V}$. Clearly, the inclusions $\mathscr{W}_i \subset \mathscr{V}_s$, valid for $i = 1, \ldots, s-1$, imply $\sum_{i=1}^{s-1} \mathscr{W}_i \subset \mathscr{V}_s$. This yields the direct sum $\mathscr{X}/\mathscr{V} = \sum_{i=1}^{s-1} \mathscr{W}_i/\mathscr{V} \oplus \mathscr{W}_s/\mathscr{V}$. The fact that, for $i \neq j$, $\mathscr{W}_i \cap \mathscr{W}_j = \mathscr{V}$, implies, using a dimensionality argument, that $\sum_{i=1}^{s-1} \mathscr{W}_i/\mathscr{V} = \oplus_{i=1}^{s-1} \mathscr{W}_i/\mathscr{V}$, and hence the direct sum representation (3.9) follows. ∎

By identifying the direct sum of free modules $\mathbb{F}[z]^m \oplus \mathbb{F}[z]^n$ with $\mathbb{F}[z]^{m+n}$, one obtains the following useful decomposition result, which will be used later on to characterize the module structure of polynomial models X_D via the Smith normal form.

Proposition 3.10. *Let $D_i(z) \in \mathbb{F}[z]^{m_i \times m_i}, i = 1, 2$, be nonsingular polynomial matrices. This implies the following equality of polynomial models:*

$$X_{\begin{pmatrix} D_1 & 0 \\ 0 & D_2 \end{pmatrix}} = X_{D_1} \oplus X_{D_2}.$$

Proof. For polynomials $f_1(z) \in \mathbb{F}[z]^{m_1}, f_2(z) \in \mathbb{F}[z]^{m_2}$,

$$f = \begin{pmatrix} f_1 \\ f_2 \end{pmatrix} \in X_{\begin{pmatrix} D_1 & 0 \\ 0 & D_2 \end{pmatrix}}$$

is true if and only if $D_i(z)^{-1} f_i(z)$ are strictly proper for $i = 1, 2$. But this is equivalent to $f_i \in X_{D_i}, i = 1, 2$. ∎

Using isomorphism (3.4) between a polynomial model X_D and the associated quotient module, it becomes a trivial exercise to translate results about quotient modules into corresponding results about polynomial models. To illustrate this process, consider the simple but very important case of characterizing submodules.

Theorem 3.11. *Let $D(z)$ be a nonsingular $m \times m$ polynomial matrix. The submodules $\mathscr{V}$ of X_D are exactly the subspaces of the form*

$$\mathscr{V} = EX_F,$$

where $E(z), F(z)$ are nonsingular $m \times m$ polynomial matrices satisfying $D(z) = E(z)F(z)$.

Proof. By Theorem 2.53, the submodules of $\mathbb{F}[z]^m/D\mathbb{F}[z]^m$ are of the form $\mathscr{X} = E\mathbb{F}[z]^m/D\mathbb{F}[z]^m$ for nonsingular polynomial matrices E, F satisfying $D = EF$. For $g \in \mathbb{F}[z]^m$ and $f = Eg$, one computes

$$\pi_D(f) = D\pi_-(D^{-1}Eg) = D\pi_-(F^{-1}g) = E\pi_F(g).$$

Therefore, the isomorphism $\bar{\pi}_D : \mathbb{F}[z]^m/D\mathbb{F}[z]^m \longrightarrow X_D$ in (3.4) maps $E\mathbb{F}[z]^m/D\mathbb{F}[z]^m$ isomorphically onto EX_F. This proves the statement. ∎

With X_D being recognized as a module, its submodules and module isomorphisms are at our disposal. Two polynomial model spaces X_{D_1} and X_{D_2} are called **isomorphic** if there exists an $\mathbb{F}[z]$-module isomorphism $Z : X_{D_1} \longrightarrow X_{D_2}$ between them. Subsequently, isomorphisms of polynomial models will be characterized via coprimeness and intertwining relations. The following lemma is instrumental for studying the direct sum decompositions of the shift operator.

Lemma 3.12. *Let $D(z), R(z) \in \mathbb{F}[z]^{m \times m}$ be nonsingular polynomial matrices. Then*

$$RX_D \subset X_{RD}$$

is a submodule and $Z_R : X_D \longrightarrow X_{RD}, f \mapsto Rf$ is an injective module homomorphism. The subspace RX_D is invariant under the shift S_{RD}, and one obtains the conjugacy of shifts

$$S_{RD}|RX_D \simeq S_D.$$

Proof. Let $f \in \mathbb{F}[z]^m$ and $g = Rf$. Then $(RD)^{-1}g = D^{-1}f$ is strictly proper if and only if $f \in X_D$. Thus $f \in X_D$ if and only if $g \in X_{RD}$. This shows that $RX_D \subset X_{RD}$ and that $f \mapsto Rf$ is an injection from X_D onto X_{RD}. Thus it suffices to show that Z_R is a module homomorphism. For $f \in X_D$, one computes

$$Z_R(z \cdot f) = R\pi_D(zf) = RD\pi_-(D^{-1}zf) = \pi_{RD}(zRf) = z \cdot Z_R f.$$

Moreover, for $g \in X_D$,

$$S_{RD}(Rg) = \pi_{RD}(zRg) = RD\pi_-(zD(z)^{-1}g(z)) = RS_D(g).$$

Thus one obtains the conjugacy of shift operators $Z_R \circ S_D = (S_{RD}|RX_D) \circ Z_R$. The result follows. ∎

To study the decomposition of a polynomial model X_D into a direct sum of S_D-invariant subspaces or, equivalently, submodules, one needs a weaker concept than coprimeness.

Definition 3.13. 1. Let $E_i(z) \in \mathbb{F}[z]^{m \times m}$, $i = 1, \ldots, s$, be nonsingular polynomial matrices. The $E_i(z)$ are called **mutually left coprime** if, for each i, $E_i(z)$ is left coprime with $E_{\mu_i} = l.c.r.m.\{E_j\}_{j \neq i}$, the unique, up to a right unimodular factor, least common right multiple (l.c.r.m.) of all $\{E_j(z)\}_{j \neq i}$.
2. Let $F_i(z) \in \mathbb{F}[z]^{m \times m}$, $i = 1, \ldots, s$, be nonsingular polynomial matrices. One says that the $F_i(z)$ are **mutually right coprime** if, for each i, $F_i(z)$ is right coprime with $F_{\nu_i} = l.c.l.m.\{F_j\}_{j \neq i}$, the unique, up to a left unimodular factor, least common left multiple (l.c.l.m.) of all $\{F_j(z)\}_{j \neq i}$.

The notion of mutual coprimeness is closely related to transversality conditions. In fact, nonsingular polynomial matrices $E_1(z), \ldots, E_s(z) \in \mathbb{F}[z]^{m \times m}$ are mutually left coprime if and only if the submodules $E_1(z)\mathbb{F}[z]^m, \ldots, E_s(z)\mathbb{F}[z]^m$ intersect transversally in $\mathbb{F}[z]^m$. Similarly, $F_1(z),\ldots,F_s(z) \in \mathbb{F}[z]^{m\times m}$ are mutually right coprime if and only if the submodules $\mathbb{F}[z]^{1\times m}F_1(z),\ldots,\mathbb{F}[z]^{1\times m}F_s(z)$ intersect transversally in $\mathbb{F}[z]^{1\times m}$. These observations follow directly from the identities

$$(l.c.r.m._{j\neq i}E_j)\mathbb{F}[z]^m = \bigcap_{j\neq i} E_j(z)\mathbb{F}[z]^m,$$

$$\mathbb{F}[z]^{1\times m}(l.c.l.m.\{F_j \mid j \neq i\}) = \bigcap_{j\neq i} \mathbb{F}[z]^{1\times m}F_j(z).$$

Mutual coprimeness can also be characterized in terms of a full rank condition on Sylvester-type matrices; see Chapter 9. Note that in the matrix case, mutual left coprimeness is a stronger condition than pairwise left coprimeness. The same is true, of course, of mutual right coprimeness. Another thing worth pointing out is that every subset of a mutually left coprime set of polynomials is itself mutually left coprime.

Following the analysis carried out in Theorem 2.24, the next result explores the close relationship between factorizations of polynomial matrices and shift-invariant subspaces. Thus it provides a link between vector space geometry and the arithmetic of polynomial matrices. It is one of the principal results that makes the study of polynomial models so useful.

Theorem 3.14. *Let* $\mathcal{V}_1,\ldots,\mathcal{V}_s i$ *be submodules of* X_D, *i.e.,* S_D-*invariant subspaces* $\mathcal{V}_i = E_iX_{F_i}$ *that correspond to the factorizations*

$$D(z) = E_i(z)F_i(z), \quad i = 1,\ldots,s.$$

Then:

1. $\mathcal{V}_1 \subset \mathcal{V}_2$ *if and only if there exists a nonsingular polynomial matrix* $R(z)$, *with* $E_1(z) = E_2(z)R(z)$, *i.e., if and only if* $E_2(z)$ *is a left factor of* $E_1(z)$;
2. $\bigcap_{i=1}^s \mathcal{V}_i$ *has the representation* $E_\vee X_{F_\vee}$, *with* $E_\vee(z)$ *a l.c.r.m. of the* $E_i(z)$ *and* $F_\vee(z)$ *a greatest common right divisor (g.c.r.d.) of the* $F_i(z)$;
3. *The equality* $\bigcap_{i=1}^s E_iX_{F_i} = 0$ *is valid if and only if the* $F_i(z)$ *are right coprime;*
4. *The sum* $\mathcal{V}_1 + \cdots + \mathcal{V}_s$ *has the representation* $E_\mu X_{F_\mu}$, *with* $E_\mu(z)$ *a greatest common left divisor (g.c.l.d.) of the* $E_i(z)$ *and* $F_\mu(z)$ *a l.c.l.m. of all the* $F_i(z)$;
5. *The factorizations* $D(z) = E_i(z)F_i(z)$, $i = 1,\ldots,s$, *imply*

$$X_D = E_1X_{F_1} + \cdots + E_sX_{F_s}$$

if and only if the $E_i(z)$ *are left coprime;*
6. *The factorizations* $D(z) = E_i(z)F_i(z)$, $i = 1,\ldots,s$, *imply the direct sum*

$$X_D = E_1X_{F_1} \oplus \cdots \oplus E_sX_{F_s}$$

if and only if the $E_i(z)$ *are left coprime and the* $F_i(z)$ *are mutually right coprime.*

Proof. 1. Assume $E_1(z) = E_2(z)R(z)$. Clearly, $D(z) = E_1(z)F_1(z) = E_2(z)$ $R(z)F_1(z) = E_2(z)F_2(z)$, so $R(z)F_1(z) = F_2(z)$. Hence, $E_1X_{F_1} = E_2RX_{F_1} \subset$ $E_2X_{RF_1} = E_2X_{F_2}$.

Conversely, assume

$$E_1X_{F_1} \subset E_2X_{F_2}. \tag{3.10}$$

Both $E_1X_{F_1} + D\mathbb{F}[z]^m$ and $E_2X_{F_2} + D\mathbb{F}[z]^m$ are submodules of $\mathbb{F}[z]^m$ and satisfy

$$E_iX_{F_i} + D\mathbb{F}[z]^m = E_iX_{F_i} + E_iF_i\mathbb{F}[z]^m = E_i[X_{F_i} + F_i\mathbb{F}[z]^m] = E_i\mathbb{F}[z]^m.$$

Thus (3.10) implies the inclusion

$$E_1\mathbb{F}[z]^m \subset E_2\mathbb{F}[z]^m.$$

From this the factorization $E_1(z) = E_2(z)R(z)$ follows.

2. The intersection of submodules is a submodule. Hence, letting $\mathscr{V}_v = \bigcap_{i=1}^s \mathscr{V}_i$, the representation $\mathscr{V}_v = E_vX_{F_v}$ follows for some factorization $D(z) = E_v(z)F_v(z)$. The inclusions $\mathscr{V}_v \subset \mathscr{V}_i$, for $i = 1,\ldots,s$, imply $E_vX_{F_v} \subset E_iX_{F_i}$ and, hence, the factorizations

$$E_v(z) = E_i(z)R_i(z).$$

In turn, these imply

$$F_i(z) = R_i(z)F_v(z).$$

This shows that $E_v(z)$ is a common right multiple of the $E_i(z)$ and $F_v(z)$ a common right divisor of the $F_i(z)$. Clearly, $D(z)$ is a common left multiple of all the $E_i(z)$, and hence, the l.c.r.m. of all the $E_i(z)$ must be a left factor of $D(z)$. Thus, let $E(z)$ be a common right multiple of the $E_i(z)$, which is also a left factor of $D(z)$. Then $E(z) = E_i(z)Q_i(z)$ and $D(z) = E(z)F(z)$. Clearly,

$$EX_F = E_iQ_iX_F \subset E_iX_{Q_iF} = E_iX_{F_i},$$

so

$$EX_F \subset \bigcap_{i=1}^s E_iX_{F_i} = E_vX_{F_v},$$

and this implies $E(z) = E_v(z)G(z)$. The last equality shows that $E_v(z)$ is the l.c.r.m. of the $E_i(z)$.

Similarly, let $F(z)$ be another common right divisor of the $F_i(z)$. This implies the existence of factorizations $D(z) = E(z)F(z) = E_V(z)F_V(z)$, from which the inclusion $E_V X_{F_V} \supset E X_F$ follows. In particular, $F(z)$ is a right divisor of $F_V(z)$, which shows that $F_V(z)$ is the g.c.r.d. of the $F_i(z)$.

3. Follows from Theorem 3.11.

4. Let $\mathcal{V}_\mu = \mathcal{V}_1 + \cdots + \mathcal{V}_s = E_\mu X_{F_\mu}$. Since $\mathcal{V}_i \subset \mathcal{V}_\mu$, there exist factorizations $E_i(z) = E_\mu(z)S_i(z)$ for all i. This means that $E_\mu(z)$ is a common left divisor of all $E_i(z)$. Let $E(z)$ be another common left divisor of the $E_i(z)$. Thus the factorizations

$$E_i(z) = E(z)R_i(z) \tag{3.11}$$

and

$$E_i(z)F_i(z) = E(z)R_i(z)F_i(z) = E(z)F(z)$$

exist, with

$$F(z) = R_i(z)F_i(z), \qquad 1 \le i \le s.$$

Now equations (3.11) imply $E_i X_{F_i} \subset E X_F$, and hence

$$E_\mu X_{F_\mu} = \mathcal{V}_1 + \cdots + \mathcal{V}_s \subset E X_F.$$

But this implies, by part 1, that $E_\mu(z) = E(z)G(z)$ and, hence, that $E_\mu(z)$ is a g.c.l.d. of the $E_i(z)$. Similarly, one can show that $F_\mu(z)$ is the l.c.l.m. of all the $F_i(z)$.

5. Follows from the previous theorem.

6. The left coprimeness condition is equivalent to $X_D = E_1 X_{F_1} + \cdots + E_s X_{F_s}$, whereas the mutual right coprimeness condition is equivalent to $E_i X_{F_i} \cap \sum_{j \ne i} E_j X_{F_j} = \{0\}$.   ∎

One way to obtain direct sum decompositions of a polynomial model is by factoring its characteristic polynomial. This is our first encounter with spectral factorizations and spectral decompositions.

Theorem 3.15. *Let $D(z) \in \mathbb{F}[z]^{n \times n}$ be nonsingular, and let $d(z) = \det D(z)$. Suppose $d(z)$ has a factorization $d(z) = e_1(z) \cdots e_s(z)$, with the $e_i(z)$ pairwise coprime. Then:*

1. $D(z)$ admits factorizations

$$D(z) = D_i(z)E_i(z),$$

with $\det D_i(z) = d_i(z)$, $\det E_i(z) = e_i(z)$, *and such that*

$$X_D = D_1 X_{E_1} \oplus \cdots \oplus D_s X_{E_s}; \tag{3.12}$$

2. *The characteristic polynomial of* $S_D | D_i X_{E_i}$ *is* $e_i(z) = \det E_i(z)$.

Proof. 1. Define subspaces X_i of X_D by

$$X_i = \{ f \in X_D | \pi_D e_i f = e_i(S_D) f = 0 \} = \operatorname{Ker} e_i(S_D).$$

Clearly, the X_i are submodules of X_D, so there exist factorizations $D(z) = D_i(z) E_i(z)$ such that $X_i = D_i X_{E_i}$. Note that $d_i(z) = \Pi_{j \neq i} e_j(z)$. In particular, $d_1(z), \ldots, d_s(z)$ are coprime, and there exist polynomials $g_1(z), \ldots, g_s(z)$ for which $\sum_{i=1}^s d_i(z) g_i(z) = 1$. This implies that every $f \in X_D$ has a representation $f = \sum_{i=1}^s d_i(S_D) g_i(S_D) f$. Now, $d_i(S_D) g_i(S_D) f \in \operatorname{Ker} e_i(S_D)$ as, by the Cayley–Hamilton theorem,

$$e_i(S_D) d_i(S_D) g_i(S_D) f = d(S_D) g_i(S_D) f = 0.$$

This shows that $X_D = \sum_{i=1}^s D_i X_{E_i}$. Furthermore, letting $\overline{E}_i X_{\overline{D}_i} = \sum_{j \neq i} D_j X_{E_j}$, for $i \neq j$, implies the coprimeness of $E_i(z)$ and $\overline{D}_i(z)$. This implies $D_i X_{E_i} \cap \overline{E}_j X_{\overline{D}_i} = \{0\}$, and hence the direct sum representation (3.12) is proved.
2. Follows from the isomorphism of $S_D | D_i X_{E_i}$ and S_{E_i}, as proven in Lemma 3.12, and from the spectral mapping theorem. ∎

The preceding discussion reveals that one feature that accounts for the applicability of polynomial models is the relation between the geometry of polynomial models and the factorizations of polynomial matrices. To see this, note that, for a nonsingular polynomial matrix $D(z) \in \mathbb{F}[z]^{m \times m}$, a subspace $\mathcal{V} \subset X_D$ is a submodule or, equivalently, an S_D-invariant subspace if and only if it has a representation $\mathcal{V} = D_1 X_{D_2}$ for some factorization $D(z) = D_1(z) D_2(z)$. In Theorem 3.14, it was shown that direct sum decompositions of polynomial models X_D are related to the notion of mutual coprimeness. Another closely related notion is that of **skew primeness**, which is introduced next.

Definition 3.16. Let $D_1(z), D_2(z) \in \mathbb{F}[z]^{m \times m}$ be nonsingular polynomial matrices. The ordered pair $(D_1(z), D_2(z))$ is called **skew prime** if there exist polynomial matrices $\overline{D}_1(z)$ and $\overline{D}_2(z)$ such that:

1. The **intertwining relation**

$$D_1(z) D_2(z) = \overline{D}_2(z) \overline{D}_1(z) \tag{3.13}$$

is satisfied;

2. $D_1(z)$ and $\overline{D}_2(z)$ are left coprime;
3. $D_2(z)$ and $\overline{D}_1(z)$ are right coprime.

In this case, one says that the pair $(\overline{D}_2(z), \overline{D}_1(z))$ is a **skew complement** of $(D_1(z), D_2(z))$. Note that a sufficient, but not necessary, condition for a pair $(D_1(z), D_2(z))$ to be skew prime is that $\det D_1(z)$ and $\det D_2(z)$ are coprime.

The following result is due to Fuhrmann (2005); for a geometric interpretation of skew primeness, see Khargonekar, Georgiou and Özgüler (1983).

Proposition 3.17. *Let $D(z) = D_1(z)D_2(z)$, with $D_1(z), D_2(z) \in \mathbb{F}[z]^{m \times m}$ nonsingular polynomial matrices. The following statements are equivalent:*

(a) $D_1(z)$ and $D_2(z)$ are skew prime with skew complement $(\overline{D}_2(z), \overline{D}_1(z))$.
(b) The following is a direct sum decomposition:

$$X_D = D_1 X_{D_2} \oplus \overline{D}_2 X_{\overline{D}_1}.$$

(c) The intertwining relation (3.13) is embeddable in a doubly coprime factorization:

$$\begin{pmatrix} \overline{D}_2(z) & D_1(z) \\ \overline{X}(z) & -Y(z) \end{pmatrix} \begin{pmatrix} Y(z) & \overline{D}_1(z) \\ X(z) & -D_2(z) \end{pmatrix} = \begin{pmatrix} I & 0 \\ 0 & I \end{pmatrix}.$$

(d) The equation

$$X(z)D_1(z) + D_2(z)Y(z) = I \qquad (3.14)$$

has a polynomial solution.

Proof. $(a) \Leftrightarrow (b)$. $D_1(z), D_2(z)$ are skew prime with skew complement $\overline{D}_2(z), \overline{D}_1(z)$ if and only if $D_1(z), \overline{D}_2(z)$ are left coprime and $D_2(z), \overline{D}_1(z)$ are right coprime. Theorem 3.14 thus implies the equivalence.

$(a) \Leftrightarrow (c)$. That (a) implies (c) follows from Theorem 2.33. Conversely, from the doubly coprime factorization (c) one obtains the Bezout identities $D_1(z)X(z) + \overline{D}_2(z)\overline{Y}(z) = I$ and $Y(z)D_2(z) + \overline{X}(z)\overline{D}_1(z) = I$. Thus $D_1(z), \overline{D}_2(z)$ are left coprime and $D_2(z), \overline{D}_1(z)$ are right coprime. Hence, (a) and (c) are equivalent.

$(c) \Leftrightarrow (d)$. Finally, equation (3.14) follows immediately from the doubly coprime factorization (c). Conversely, from (3.14) one concludes that $D_1(z), Y(z)$ are left coprime and $X(z), D_2(z)$ are right coprime. Embed $D_1(z), -Y(z)$ into a unimodular matrix

$$V(z) = \begin{pmatrix} \overline{D}_2(z) & D_1(z) \\ \overline{X}(z) & -Y(z) \end{pmatrix}$$

such that

$$(X(z) - D(z)) \begin{pmatrix} \overline{D}_2(z) & D_1(z) \\ \overline{X}(z) & -Y(z) \end{pmatrix} = (0 \ I).$$

This is clearly possible. Choosing an embedding of $(X, -D_2)$ in a unimodular matrix

$$U(z) = \begin{pmatrix} \overline{V}(z) & \overline{W}(z) \\ X(z) & -D_2(z) \end{pmatrix}$$

yields the unimodularity of

$$L(z) := U(z)V(z) = \begin{pmatrix} M(z) & N(z) \\ 0 & I \end{pmatrix}.$$

In particular, $M(z)$ is unimodular. Define $\overline{Y}(z) := M(z)^{-1}\overline{V}(z) - M(z)^{-1}N(z)X(z)$ and $\overline{D}_1(z) := M(z)^{-1}\overline{W}(z) + M(z)^{-1}N(z)D_2(z)$. Then

$$\begin{pmatrix} \overline{Y}(z) & \overline{D}_1(z) \\ X(z) & -D_2(z) \end{pmatrix} \begin{pmatrix} \overline{D}_2(z) & D_1(z) \\ \overline{X}(z) & -Y(z) \end{pmatrix} = \begin{pmatrix} I & 0 \\ 0 & I \end{pmatrix}.$$

Thus the unimodular embedding in (c) has been obtained.   ∎

The following technical lemma shows the existence of a special solution to (3.14).

Lemma 3.18. *Let $D_1(z)$ and $D_2(z)$ be nonsingular polynomial matrices such that $D_2(z)^{-1}D_1(z)^{-1}$ is strictly proper. If*

$$X(z)D_1(z) + D_2(z)Y(z) = I$$

has a polynomial matrix solution $X(z), Y(z)$, then it has one with $D_2(z)^{-1}X(z)$ and $Y(z)D_1(z)^{-1}$ strictly proper.

Proof. Let $X(z) = X_1(z) + D_2(z)X_2(z)$ and $Y(z) = Y_1(z) + Y_2(z)D_1(z)$, with $D_2(z)^{-1}X(z)$ and $Y(z)D_1(z)^{-1}$ strictly proper. Then (3.14) implies

$$X_1(z)D_1(z) + D_2(z)Y_1(z) + D_2(z)(X_2(z) + Y_2(z))D_1(z) = I.$$

Equivalently,

$$D_2(z)^{-1}X_1(z) + Y_1(z)D_1(z)^{-1} + X_2(z) + Y_2(z) = (D_1(z)D_2(z))^{-1}.$$

This implies $X_2(z) + Y_2(z) = 0$ and $X_1(z)D_1(z) + D_2(z)Y_1(z) = I$.   ∎

Next, a useful characterization of direct sum decompositions of polynomial models is proved.

Theorem 3.19. *Let* $D_i(z) \in \mathbb{F}[z]^{m \times m}$, $i = 1, \ldots, N$, *be nonsingular polynomial matrices. Define*

$$
\begin{aligned}
D(z) &= l.c.l.m.\{D_i(z)|i = 1, \ldots, N\} \\
\bar{L}_i(z) &= l.c.l.m.\{D_j(z)|j \neq i\},
\end{aligned}
\tag{3.15}
$$

and let $L_i(z)$ *and* $\bar{D}_i(z)$ *denote the unique polynomial matrices defined by the factorizations*

$$
D(z) = L_i(z)D_i(z)
\tag{3.16}
$$

and

$$
D(z) = \bar{D}_i(z)\bar{L}_i(z),
\tag{3.17}
$$

respectively. The following statements are equivalent:

(a) *The dimension formula* $\dim X_D = \sum_{i=1}^{N} \dim X_{D_i}$ *is valid.*
(b) *Factorizations (3.16) are skew prime factorizations.*
(c) *The polynomial matrices* $D_i(z)$, $i = 1, \ldots, N$, *are mutually right coprime.*
(d) *The polynomial matrices* $\bar{D}_i(z)$, $i = 1, \ldots, N$, *are mutually left coprime.*
(e) *The following is a direct sum representation:*

$$
X_D = L_1 X_{D_1} \oplus \cdots \oplus L_N X_{D_N}.
\tag{3.18}
$$

Proof. (a) $\Leftrightarrow$ (e). By the construction of $D(z)$, the polynomial matrices $L_1(z), \ldots, L_N(z)$ in (3.16) are left coprime. By the same reasoning, the matrices $\bar{L}_1(z), \ldots, \bar{L}_N(z)$ are right coprime. Therefore, using Theorem 3.14, $X_D = L_1 X_{D_1} + \cdots + L_N X_{D_N}$. This implies $\dim X_D \leq \sum_{i=1}^{N} \dim X_{D_i}$. Thus (a) is valid if and only if $X_D = L_1 X_{D_1} + \cdots + L_N X_{D_N}$ is a direct sum, i.e., if and only if (e) is true.

(c) $\Leftrightarrow$ (e). This equivalence is a direct consequence of Theorem 3.14.

(b) $\Leftrightarrow$ (c). Assume the $D_i(z)$ are mutually right coprime, and let $\bar{L}_i(z)$ be defined by (3.15). Clearly, this implies, for all i, the right coprimeness of $D_i(z), \bar{L}_i(z)$. On the other hand, $L_i(z), \bar{D}_i(z)$ are left coprime as $D(z) = l.c.l.m.\{D_i(z), \bar{L}_i(z)\}$. This shows that the factorizations $D(z) = L_i(z)D_i(z)$ are skew prime. Conversely, assume the factorizations $D(z) = L_i(z)D_i(z)$ are skew prime. Let $D(z) = \bar{D}_i(z)\bar{L}_i(z)$ be complementary factorizations. This implies the direct sum $X_D = L_i X_{D_i} \oplus \bar{D}_i X_{\bar{L}_i}$ and, hence, also the equality $\deg \det D = \deg \det D_i + \deg \det \bar{L}_i$. Now, $\sum_{j \neq i} L_j X_{D_j} \subset \bar{D}_i X_{\bar{L}_i}$, and it follows that

$$
\begin{aligned}
\deg \det D &= \dim X_D = \dim \sum_{j=1}^{N} L_j X_{D_j} \leq \dim L_i X_{D_i} + \dim \bar{D}_i X_{\bar{L}_i} \\
&= \deg \det D_i + \deg \det \bar{L}_i = \deg \det D.
\end{aligned}
$$

Necessarily, the equality holds throughout, which implies that $\sum_{j \neq i} L_j X_{D_j}$ is a direct sum and $\bigoplus_{j \neq i} L_j X_{D_j} = \overline{D}_i X_{\overline{L}_i}$. This implies the direct sum (3.18) and, consequently, the mutual right coprimeness of the $D_i(z)$.

$(c) \Leftrightarrow (d)$. Follows from duality considerations.   ∎

3.3 Module Homomorphisms and Intertwining Maps

In the class of linear spaces, natural homomorphisms are linear transformations. Similarly, for the class of modules, module homomorphisms become a natural object. For close to a century it was apparent that the most efficient way to study linear transformations is via module theory. Specifically, for a linear space $\mathscr{X}$ over the field $\mathbb{F}$, a linear transformation $A : \mathscr{X} \longrightarrow \mathscr{X}$ induces an $\mathbb{F}[z]$-module structure on $\mathscr{X}$ defined by

$$p \cdot x = p(A)x, \qquad p(z) \in \mathbb{F}[z], x \in \mathscr{X}.$$

By $\mathscr{X}_A$ will be denoted the space $\mathscr{X}$ with an A-induced module structure. Let $\mathscr{X}_A$ and $\mathscr{Y}_B$ be vector spaces, with the module structures induced by A and B, respectively. It is easily verified that a linear map $Z : \mathscr{X} \longrightarrow \mathscr{Y}$ is an $\mathbb{F}[z]$-homomorphism if and only if

$$BZ = ZA,$$

i.e., Z is a solution to a homogeneous Sylvester equation. In this case, one says that Z **intertwines** A and B. The set of all linear transformations intertwining A and B is a linear space that we denote by $\mathrm{Intw}\,(A, B)$ or, equivalently, by $\mathrm{Hom}_{\mathbb{F}[z]}(\mathscr{X}_A, \mathscr{Y}_B)$. A special case of intertwining maps is the **commutant** $\mathscr{C}(A) = \mathrm{Hom}_{\mathbb{F}[z]}(\mathscr{X}_A, \mathscr{X}_A)$ of a linear transformation A, namely, the set of all operators Z commuting with A, i.e., satisfying $ZA = AZ$. The transformations B and A are **similar** if there exists an invertible intertwining map. This indicates that one should separate the characterization of intertwining maps from the study of their invertibility properties. If A and B transform by similarity to $A' = PAP^{-1}$ and $B' = RBR^{-1}$, then $ZA = BZ$ transforms into $(RZP^{-1})(PAP^{-1}) = (RBR^{-1})(RZP^{-1})$, which leads to

$$\mathrm{Intw}\,(PAP^{-1}, RBR^{-1}) = R\,\mathrm{Intw}\,(A, B)P^{-1} \qquad (3.19)$$

or, equivalently, to

$$\mathrm{Hom}_{\mathbb{F}[z]}(\mathscr{X}_A, \mathscr{Y}_B) = R\,\mathrm{Hom}_{\mathbb{F}[z]}(\mathscr{X}_{A'}, \mathscr{Y}_{B'})P^{-1}.$$

The isomorphism $A \simeq S_{zI-A}$, that is, A acting in $\mathscr{X}$ is isomorphic to S_{zI-A} acting in the polynomial model X_{zI-A}, indicates that our study should be extended to the more general problem of characterizing the set of $\mathbb{F}[z]$-homomorphisms of polynomial models. This study was first carried out, and invertibility conditions derived, in Fuhrmann (1976). These characterizations are presented in this section.

Having associated a functional model X_D with every nonsingular polynomial matrix $D(z)$, our next step is to study the relation between different models. Of particular interest is the question of the conditions under which the shift operators S_{D_1} and S_{D_2} of two nonsingular polynomial matrices $D_1(z)$ and $D_2(z)$ are similar. Equivalently, the question can be posed as the existence of an invertible map $Z : X_{D_1} \longrightarrow X_{D_2}$ **intertwining** S_{D_1} and S_{D_2}, i.e., satisfying the equation

$$ZS_{D_1} = S_{D_2}Z. \tag{3.20}$$

Now, if Z intertwines S_{D_1} and S_{D_2}, i.e., (3.20) is satisfied, then for each polynomial $p(z)$,

$$Zp(S_{D_1}) = p(S_{D_2})Z,$$

which shows that intertwining maps are just $\mathbb{F}[z]$-module homomorphisms from X_{D_1} into X_{D_2}. To gain some further insight into this problem, consider briefly the scalar case. Thus, assuming $d(z)$ is a nontrivial polynomial, one looks for maps Z satisfying $ZS_d = S_d Z$. Since S_d is cyclic, it follows in this case that there exists a polynomial $p(z) \in \mathbb{F}[z]$ for which $Z = p(S_d)$, that is,

$$Zf = p(S_d)f = \pi_d pf.$$

This leads us to assume that a general characterization of intertwining maps will involve polynomial algebra. This characterization connects with the well-known fact that a map Z commutes with a cyclic map A if and only if $Z = p(A)$ for some polynomial $p(z)$. This is closely related to the theory of Bezoutian matrices. Barnett (1972) showed that a matrix intertwining a companion matrix C_d and its transpose is given by a **Bezoutian matrix** $B(d,n) = (b_{ij})$, defined by

$$\frac{d(z)n(w) - n(z)d(w)}{z - w} = \sum_{i=1}^{n}\sum_{j=1}^{n} b_{ij}z^{i-1}w^{j-1}. \tag{3.21}$$

Later on, in Helmke and Fuhrmann (1989), as well as Fuhrmann (1981), it was shown that the Bezoutian matrix $B(d,n) = (b_{ij})$, defined in (3.21), is the matrix representation of the intertwining map $n(S_d)$ with respect to the control and standard bases of X_d. Now the expansion in (3.21) is only one of several others. This indicates that one should focus on intertwining maps rather than on their matrix representations and that, in generalizing Bezoutian matrices to the polynomial matrix case, one should analyze intertwining maps for polynomial models. We will return to this topic later on, in Chapter 5, when we introduce tensor products.

The following theorem gives a full characterization of intertwining maps.

Theorem 3.20. *Let $D_1(z) \in \mathbb{F}[z]^{m \times m}$ and $D_2(z) \in \mathbb{F}[z]^{p \times p}$ be nonsingular. Then a map $Z : X_{D_1} \longrightarrow X_{D_2}$ is an $\mathbb{F}[z]$-homomorphism if and only if there exist $N_1(z), N_2(z) \in \mathbb{F}[z]^{p \times m}$ such that*

$$N_2(z)D_1(z) = D_2(z)N_1(z) \tag{3.22}$$

and

$$Zf = \pi_{D_2}(N_2 f). \tag{3.23}$$

Proof. By Theorem 2.54, the module homomorphisms between quotient modules $\bar{Z} : \mathbb{F}[z]^m/D_1(z)\mathbb{F}[z]^m \longrightarrow \mathbb{F}[z]^p/D_2(z)\mathbb{F}[z]^p$ are of the form $\bar{Z}(f) = \pi_{D_2}N_2 f$, where N_1 and N_2 are polynomial matrices satisfying $N_2(z)D_1(z) = D_2(z)N_1(z)$. Applying Theorem 3.5, the module homomorphisms $Z : X_{D_1} \longrightarrow X_{D_2}$ are seen to be of the form $Z = \bar{\pi}_{D_2} \circ \bar{Z} \circ \bar{\pi}_{D_1}^{-1}$, and therefore $Zf = \bar{\pi}_{D_2}(\bar{Z}f) = \pi_{D_2}(\pi_{D_2}(N_2 f)) = \pi_{D_2}(N_2 f)$. The result follows. ∎

Proceeding next with the study of the invertibility properties of these homomorphisms, we characterize the kernel and image of the map Z defined by equation (3.23) in terms of the polynomial data.

Theorem 3.21. *Let $Z : X_{D_1} \longrightarrow X_{D_2}$ be the $\mathbb{F}[z]$-module homomorphism defined by equations (3.22)–(3.23). Then:*

1. *$\operatorname{Ker} Z = E_1 X_{F_1}$, where $D_1(z) = E_1(z)F_1(z)$ and $F_1(z)$ is a g.c.r.d. of $D_1(z)$ and $N_1(z)$;*
2. *$\operatorname{Im} Z = E_2 X_{F_2}$, where $D_2(z) = E_2(z)F_2(z)$ and $E_2(z)$ is a g.c.l.d. of $D_2(z)$ and $N_2(z)$;*
3. *The map Z is injective if and only if $D_1(z)$ and $N_1(z)$ are right coprime;*
4. *The map Z is surjective if and only if $D_2(z)$ and $N_2(z)$ are left coprime;*
5. *The map Z is invertible if and only if $D_1(z)$ and $N_1(z)$ are right coprime and $D_2(z)$ and $N_2(z)$ are left coprime.*

Proof. 1. $\operatorname{Ker} Z$ is a submodule of X_{D_1}, and hence, by Theorem 3.11, it is of the form $E_1 X_{F_1}$ for some factorization $D_1(z) = E_1(z)F_1(z)$. This means that $\pi_{D_2}N_2 E_1 g = 0$ for every $g(z) \in X_{F_1}$. On the other hand, if $g(z) \in F_1 \mathbb{F}[z]^m$, i.e., $g(z) = F_1(z)g'(z)$, then, for every $g'(z) \in \mathbb{F}[z]^m$,

$$\pi_{D_2}(N_2 E_1 F_1 g') = \pi_{D_2}(N_2 D_1 g') = \pi_{D_2}(D_2 N_1 g') = D_2 \pi_-(N_1 g') = 0.$$

From the direct sum representation $\mathbb{F}[z]^m = X_{F_1} \oplus F_1 \mathbb{F}[z]^m$ we conclude the identity $\pi_{D_2}N_2 E_1 F_1 g = 0$ for every $g \in \mathbb{F}[z]^m$. This implies that $D_2(z)^{-1}N_2(z)E_1(z) = N_1(z)D_1(z)^{-1}E_1(z) = N_1(z)F_1(z)^{-1} = P(z)$ is a polynomial matrix. It follows that $N_1(z) = P(z)F_1(z)$ and $F_1(z)$ is a common right divisor of $D_1(z)$ and $N_1(z)$.

Conversely, let $F'(z)$ be a common right divisor of $D_1(z)$ and $N_1(z)$. So $D_1(z) = E'(z)F'(z)$ and $N_1(z) = P'(z)F'(z)$. To prove the inclusion $E'X_{F'} \subset \operatorname{Ker} Z$, one computes, for $f \in X_{F'}$,

$$ZE'f = \pi_{D_2} N_2 E'f = D_2 \pi_- D_2^{-1} N_2 E'f = D_2 \pi_- N_1 D_1^{-1} E'f = D_2 \pi_- N_1 (F')^{-1} f$$
$$= D_2 \pi_- P'f = 0.$$

The inclusion $E'X_{F'} \subset E_1 X_{F_1}$ implies $F'(z)$ is a right divisor of $F_1(z)$, so $F_1(z)$ is the g.c.r.d. of $D_1(z), N_1(z)$.

2. $\operatorname{Im} Z$ is a submodule of X_{D_2}, so it has a representation $\operatorname{Im} Z = E_2 X_{F_2}$ for some factorization $D_2(z) = E_2(z)F_2(z)$. This implies $N_2 X_{D_1} \subset E_2 \mathbb{F}[z]^m$. Also, the relation (3.22) implies $N_2 D_1 \mathbb{F}[z]^m \subset D_2 \mathbb{F}[z]^m \subset E_2 \mathbb{F}[z]^m$. The two inclusions imply $N_2 \mathbb{F}[z]^m \subset E_2 \mathbb{F}[z]^m$, and so $N_2(z) = E_2(z)R(z)$ for some polynomial matrix $R(z)$. Thus $E_2(z)$ is a common left divisor of $N_2(z)$ and $D_2(z)$.

Let $E'(z)$ be a common left divisor of $N_2(z)$ and $D_2(z)$, i.e., $N_2(z) = E'(z)R(z)$ and $D_2(z) = E'(z)F'(z)$, and let $f \in X_{D_1}$. Then

$$Zf = \pi_{D_2} N_2 f = D_2 \pi_- D_2^{-1} N_2 f = E'F'\pi_-(F')^{-1}(E')^{-1}E'Rf$$
$$= E'\pi_{F'}Rf \in E'X_{F'}.$$

But this implies that $E_2 X_{F_2} \subset E'X_{F'}$ or, by Theorem 3.14, that $E'(z)$ is a left divisor of $E_2(z)$. Hence $E_2(z)$ is the g.c.l.d. of $D_2(z), N_2(z)$.

3. The other claims follow from the first two statements. ∎

The next result gives an explicit formula for the **inverse** of module isomorphisms in terms of doubly coprime factorizations.

Theorem 3.22. *1. $D_1(z)$ and $N_1(z)$ are right coprime and $D_2(z)$ and $N_2(z)$ are left coprime if and only if there exist polynomial matrices $X_1(z), Y_1(z), X_2(z), Y_2(z)$ that satisfy the following* **doubly coprime factorization:**

$$\begin{pmatrix} Y_2(z) & X_2(z) \\ -N_2(z) & D_2(z) \end{pmatrix} \begin{pmatrix} D_1(z) & -X_1(z) \\ N_1(z) & Y_1(z) \end{pmatrix} = \begin{pmatrix} I & 0 \\ 0 & I \end{pmatrix},$$

$$\begin{pmatrix} D_1(z) & -X_1(z) \\ N_1(z) & Y_1(z) \end{pmatrix} \begin{pmatrix} Y_2(z) & X_2(z) \\ -N_2(z) & D_2(z) \end{pmatrix} = \begin{pmatrix} I & 0 \\ 0 & I \end{pmatrix}. \tag{3.24}$$

2. In terms of the doubly coprime factorizations (3.24), the inverse $Z^{-1} : X_{D_2} \longrightarrow X_{D_1}$ of the module isomorphism $Z : X_{D_1} \longrightarrow X_{D_2}, Zf = \pi_{D_2}(N_2 f)$ is given by

$$Z^{-1}g = \pi_{D_1}(X_1 g), \qquad g \in X_{D_2}.$$

Proof. 1. The existence of polynomial matrices $X_1(z), Y_1(z), X_2(z), Y_2(z)$ satisfying (3.24) implies the Bezout identities

$$Y_2(z)D_1(z) + X_2(z)N_1(z) = I \quad \text{and} \quad D_2(z)Y_1(z) + N_2(z)X_1(z) = I,$$

which in turn imply the two coprimeness conditions. Conversely, assume the coprimeness conditions are satisfied. The coprimeness conditions imply the existence of polynomial matrices $\hat{X}_1(z), \hat{Y}_1(z), X_2(z), Y_2(z)$ that solve the Bezout equations

$$Y_2(z)D_1(z) + X_2(z)N_1(z) = I$$

and

$$D_2(z)\hat{Y}_1(z) + N_2(z)\hat{X}_1(z) = I.$$

Letting $Q(z) = X_2(z)\hat{Y}_1(z) - Y_2(z)\hat{X}_1(z)$, and defining

$$\begin{pmatrix} Y_1(z) \\ X_1(z) \end{pmatrix} = \begin{pmatrix} \hat{Y}_1(z) \\ \hat{X}_1(z) \end{pmatrix} - \begin{pmatrix} D_1(z) \\ N_1(z) \end{pmatrix} Q(z),$$

we obtain matrix relations (3.24).

2. Compute, for $f(z) \in X_{D_1}$,

$$\pi_{D_1} X_1 \pi_{D_2} N_2 f = \pi_{D_1} X_1 N_2 f = \pi_{D_1}(I - D_1 Y_2)f = f$$

and, for $g(z) \in X_{D_2}$,

$$\pi_{D_2} N_2 \pi_{D_1} X_1 g = \pi_{D_2} N_2 X_1 g = \pi_{D_2}(I - D_2 Y_1)g = g. \qquad \blacksquare$$

3.4   Classification of Shift Operators

Closely related to the task of classifying the associated shift operators up to conjugacy is the question of classifying polynomial models up to module isomorphisms. Let $D_1(z) \in \mathbb{F}[z]^{m \times m}$ and $D_2(z) \in \mathbb{F}[z]^{p \times p}$ be nonsingular. One says that the polynomial models X_{D_1} and X_{D_2} are **isomorphic** if there exists a module isomorphism $Z : X_{D_1} \longrightarrow X_{D_2}$.

Theorem 3.23. *Let $D_1(z) \in \mathbb{F}[z]^{m \times m}$ and $D_2(z) \in \mathbb{F}[z]^{p \times p}$ be nonsingular polynomial matrices. The following statements are equivalent:*

(a) *The polynomial models X_{D_1} and X_{D_2} are isomorphic.*
(b) *$D_1(z)$ and $D_2(z)$ are polynomially equivalent.*
(c) *$D_1(z)$ and $D_2(z)$ have the same nontrivial invariant factors.*

Proof. See Definition 2.45 for the notion of polynomial equivalence and Theorem 2.46 for the proof of the equivalence of (b) and (c). $D_1(z)$ and $D_2(z)$ are polynomially equivalent if and only if there exist polynomial matrices $N_1(z)$ and $N_2(z)$, with $N_2(z)D_1(z) = D_2(z)N_1(z)$, such that $(N_1(z), D_1(z))$ and $(N_2(z), D_2(z))$ are right coprime and left coprime, respectively. By Theorem 3.21, this is equivalent to the existence of a module isomorphism $Z : X_{D_1} \longrightarrow X_{D_2}$. The result follows. ∎

According to Theorem 3.23, the polynomial models X_{D_1} and X_{D_2} of two unimodularly equivalent nonsingular polynomial matrices $D_1(z)$ and $D_2(z)$ are $\mathbb{F}[z]$-linear isomorphic. This leads to the following result in the recurring theme of "equivalence and similarity."

Theorem 3.24. *Let $A_1, A_2 \in \mathbb{F}^{n \times n}$ be two linear transformations on $\mathbb{F}^n$. The following conditions are equivalent:*

(a) *The matrices A_1 and A_2 are similar.*
(b) *The pencils $zI - A_1$ and $zI - A_2$ are unimodularly equivalent.*
(c) *The pencils $zI - A_1$ and $zI - A_2$ are polynomially equivalent.*

Proof. If A_1 and A_2 are similar, then there exists an invertible matrix $R \in GL_n(\mathbb{F})$ such that

$$RA_1 = A_2 R.$$

This in turn implies

$$R(zI - A_1) = (zI - A_2)R,$$

and hence, by the invertibility of R, the unimodular equivalence of $zI - A_1$ and $zI - A_2$. Clearly, unimodular equivalence implies polynomial equivalence. Assume that $zI - A_1$ and $zI - A_2$ are polynomially equivalent. By Theorem 3.23, the polynomial models X_{zI-A_1} and X_{zI-A_2} are isomorphic. This implies the similarity of the shift operators S_{zI-A_1} and S_{zI-A_2}. By Proposition 3.6, A_1 is similar to S_{zI-A_1}, and analogously for A_2. Thus, the similarity of A_1 and A_2 follows by transitivity. ∎

An important property a linear operator A acting in a vector space $\mathscr{X}$ may have is **cyclicity**, i.e., the existence of a vector $v \in \mathscr{X}$ such that the sequence of vectors

$$v, Av, A^2 v, \ldots, A^n v, \ldots$$

spans $\mathscr{X}$. The element $v \in \mathscr{X}$ is then called a **cyclic vector** for A. Proceeding to investigate the question of when a shift operator of a polynomial model X_D is cyclic, one begins with the scalar case.

Theorem 3.25. *Let* $d(z) = \sum_{j=0}^{n} a_j z^j \in \mathbb{F}[z]$ *be monic of degree n. The shift operator* $S_d : X_d \longrightarrow X_d$ *is cyclic. An element* $f(z) \in X_d$ *is a cyclic vector for* S_d *if and only if* $d(z)$ *and* $f(z)$ *are coprime.*

Proof. Obviously, the elements $1, z, \ldots, z^{n-1}$ form a basis of X_d and $S_d(z^i) = z^{i+1}$ for $i = 0, \ldots, n-2$, $S_d(z^{n-1}) = -\sum_{j=0}^{n-1} a_j z^j$. This shows that $1 \in X_d$ is a cyclic vector for the shift. For $f(z) \in X_d$, the span $< S_d | f >$ of all elements $S_d^i(f) = S_d(z^i f(z)), i \in \mathbb{N}_0$, is equal to

$$\{\pi_d(a(z)f(z)) \mid a(z) \in \mathbb{F}[z]\}.$$

Thus $f(z)$ is a cyclic vector if and only if $< S_d | f >= X_d$, i.e., if and only if $f(z)\mathbb{F}[z] + d(z)\mathbb{F}[z] = X_d + d(z)\mathbb{F}[z] = \mathbb{F}[z]$, this is equivalent to $f(z)$ and $d(z)$ being coprime. ∎

Let $D(z) \in \mathbb{F}[z]^{m \times m}$ have invariant factors $d_i(z)$, $i = 1, \ldots, m$. Then this implies the following direct sum representation:

$$X_D \simeq \bigoplus_{i=1}^{m} X_{d_i}. \tag{3.25}$$

Continuing our discussion of cyclicity of the shift operator, we note that, by Lemma 3.10, the shift operator S_D is isomorphic to the direct sum of the cyclic shift operators S_{d_i}. It is easily seen that the direct sum of two cyclic, finite-dimensional linear operators A_1 and A_2 is cyclic if and only if the spectra of A_1 and A_2 are disjoint. This proves:

Theorem 3.26. *Let* $D(z) \in \mathbb{F}[z]^{m \times m}$ *be a nonsingular polynomial matrix with monic invariant factors* $d_m(z)|d_{m-1}(z)| \cdots |d_1(z)$. *Then* $S_D : X_D \longrightarrow X_D$ *is cyclic if and only if* $d_2(z) = \ldots = d_m(z) = 1$.

The preceding discussion also leads to a very simple approach to the derivation of canonical forms for matrices, such as the rational canonical form and the Jordan canonical form.

This study begins with the rational canonical form of a matrix $A \in \mathbb{F}^{n \times n}$. Before stating the result, let us briefly digress on the construction of companion matrices and the connection to shift operators. Thus, let $d(z) = \sum_{j=0}^{n} a_j z^j \in \mathbb{F}[z]$ be a monic polynomial of degree n. Then the polynomial model X_d is an n-dimensional $\mathbb{F}$-vector space with basis $B_{\text{st}} = \{1, z, \ldots, z^{n-1}\}$. The shift operator $S_d : X_d \longrightarrow X_d$ acts on this basis via

$$S_d(z^i) = z^{i+1} \quad i = 0, \ldots, n-2,$$

$$S_d(z^{n-1}) = -\sum_{j=0}^{n-1} a_j z^j.$$

Therefore, the basis representation of S_d with respect to B_{st} is the companion matrix

$$C_d = \begin{pmatrix} 0 & & \cdots & & -a_0 \\ 1 & & & & -a_1 \\ & \ddots & \ddots & & \vdots \\ & & \ddots & 0 & -a_{n-2} \\ & & & 1 & -a_{n-1} \end{pmatrix}. \tag{3.26}$$

Since the characteristic polynomial of C_d is $d(z)$, it follows that $d(z)$ coincides with the characteristic polynomial of the shift S_d. Moreover, $1 \in X_d$ is a cyclic vector of S_d, and therefore the minimal polynomial of S_d coincides with the characteristic polynomial $d(z)$. In particular, $d(z)$ divides every polynomial $p(z)$ that annihilates the shift S_d, i.e., for which $p(S_d) = 0$.

Next consider the nonsingular polynomial matrix $D(z) = zI - A$ of a matrix $A \in \mathbb{F}^{n \times n}$ with monic invariant factors $d_1(z), \ldots, d_r(z)$ of the form

$$d_i(z) = \sum_{j=0}^{n_i} d_{ij} z^j, \quad d_{in_i} = 1.$$

From the direct sum decomposition (3.25) the rational canonical form is derived as follows.

Theorem 3.27 (Rational Canonical Form). *For $A \in \mathbb{F}^{n \times n}$, there exists an invertible transformation $S \in \mathbb{F}^{n \times n}$ such that SAS^{-1} has the block-diagonal structure*

$$\begin{pmatrix} C_1 & & \\ & \ddots & \\ & & C_r \end{pmatrix},$$

where

$$C_i = \begin{pmatrix} 0 & 0 & \cdots & & -d_{i0} \\ 1 & 0 & & & -d_{i1} \\ \vdots & \ddots & \ddots & & \vdots \\ \vdots & & \ddots & 0 & -d_{i,n_i-2} \\ 0 & & \cdots & 1 & -d_{i,n_i-1} \end{pmatrix}$$

denotes the companion matrix of the invariant polynomial $d_i(z) = \sum_{j=0}^{n_i} d_{ij} z^j$.

Proof. From the direct sum decomposition

$$X_D \simeq \bigoplus_{i=1}^{r} X_{d_i}$$

we conclude that the shift operator S_D on X_D, which is equal to A, is similar to the direct sum of shift operators S_{d_i}, i.e.,

$$S_D \simeq \mathrm{diag}\,(S_{d_1}, \cdots, S_{d_r}).$$

Thus A is similar to $\mathrm{diag}\,(C_1, \cdots, C_r)$, where C_i are matrix representations of the shift operator S_{d_i} on X_{d_i}. If we choose the standard basis on X_{d_i}, then C_i becomes the companion matrix of $d_i(z)$. This completes the proof.   ∎

Turning to the Jordan canonical form, let

$$d_i(z) = \prod_{j=1}^{n_i} p_{ij}(z)^{v_{ij}}$$

be the **primary decomposition** of the ith invariant factor $d_i(z)$ of a nonsingular polynomial matrix $D(z)$. The polynomials $p_{ij}(z)^{v_{ij}}$ are called the **elementary divisors** of $D(z)$. The diagonal polynomial matrix $\Delta(z)$ with elementary divisors on the diagonal will be called the **polynomial Jordan form**. The same name will be used even if $\Delta(z)$ has a larger size and has extra units on the diagonal. In fact, defining $\pi_{ij}(z) = \Pi_{k \neq j} p_{ik}(z)^{v_{ik}}$, and noting that

$$d_i(z) = \pi_{ij}(z) p_{ij}(z)^{v_{ij}}, \tag{3.27}$$

the following direct sum decomposition into submodules is obtained:

$$X_D = \bigoplus_{i,j} \pi_{ij} X_{p_{ij}^{v_{ij}}},$$

and the isomorphism $X_D \simeq \bigoplus X_{p_{ij}^{v_{ij}}}$ follows.

Proposition 3.28. *Let $D(z) \in \mathbb{F}[z]^{m \times m}$, and let $\Delta(z)$ denote the polynomial Jordan form of $D(z)$. Then $D(z)$ is polynomially equivalent to $\Delta(z)$.*

Proof. Note, using the factorization (3.27), that

$$\begin{pmatrix} 1 \\ \vdots \\ 1 \end{pmatrix} d_i(z) = \begin{pmatrix} p_{i1}(z) & & \\ & \ddots & \\ & & p_{in_i}(z) \end{pmatrix} \begin{pmatrix} \pi_{i1}(z)^{v_{i1}} \\ \vdots \\ \pi_{in_i}(z)^{v_{in_i}} \end{pmatrix}.$$

It is easily checked that the polynomial matrices $\begin{pmatrix} 1 \\ \vdots \\ 1 \end{pmatrix}, \begin{pmatrix} p_{i1}(z) \\ & \ddots \\ & & p_{in_i}(z) \end{pmatrix}$ are left

coprime and $d_i(z), \begin{pmatrix} p_{i1}(z)^{v_{i1}} \\ \vdots \\ p_{in_i}(z)^{v_{in_i}} \end{pmatrix}$ are right coprime. This implies that $d_i(z)$ and

$\begin{pmatrix} p_{i1}(z) \\ & \ddots \\ & & p_{in_i}(z) \end{pmatrix}$ are polynomially equivalent. The result follows.   ∎

By a suitable choice of basis in the polynomial models $X_{p_{ij}^{v_{ij}}}$, one obtains the

Jordan canonical form for a matrix $A \in \mathbb{F}^{n \times n}$.

Theorem 3.29 (Jordan Canonical Form). *Assume that $\mathbb{F}$ is algebraically closed. Let*

$$p_{ij}(z)^{v_{ij}} = (z - a_i)^{v_{ij}}, i = 1, \ldots, m, j = 1, \ldots, n_i,$$

denote the elementary divisors of $zI - A$, with pairwise distinct roots $a_i \in \mathbb{F}$ and $\sum_{ij} v_{ij} = n$. Then A is similar to the block-diagonal matrix

$$J = \mathrm{diag}(J_{11}, \ldots, J_{1n_1}, \ldots, J_{m1}, \ldots, J_{mn_m}),$$

where each Jordan block J_{ij} has the form

$$J_{ij} = \begin{pmatrix} a_i & & & \\ 1 & \ddots & & \\ & \ddots & \ddots & \\ & & 1 & a_i \end{pmatrix}.$$

Proof. Since we assume, for simplicity, that $\mathbb{F}$ is algebraically closed, the irreducible monic polynomials $p(z)$ of $\mathbb{F}[z]$ are linear, i.e., $p(z) = z - a$ for $a \in \mathbb{F}$. A basis for the polynomial model $X_{(z-a_i)^{v_{ij}}}$ is $\{1, (z - a_i), (z - a_i)^2, \ldots, (z - a_i)^{v_{ij}-1}\}$. The shift operator $S_{ij} := S_{(z-a_i)^{v_{ij}}}$ on $X_{(z-a_i)^{v_{ij}}}$ satisfies

$$S_i((z - a_i)^k) = (z - a_i)^{k+1} + a_i(z - a_i)^k, \quad k = 0, \ldots, v_{ij} - 2,$$

$$S_i((z - a_i)^{v_{ij}-1}) = a_i(z - a_i)^{v_{ij}-1}.$$

Thus the matrix representation of the shift operator on $X_{(z-a_i)^{v_{ij}}}$ with respect to this basis has the structure of a Jordan block,

$$J_{ij} = \begin{pmatrix} a_i & & & \\ 1 & \ddots & & \\ & \ddots & \ddots & \\ & & 1 & a_i \end{pmatrix}.$$

Since A coincides with the shift operator S_{zI-A} on X_{zI-A}, applying Proposition 3.28, the shift operator on X_{zI-A} is similar to the direct sum of shift operators $S_{(z-a_i)^{v_{ij}}}$. Thus the Jordan canonical form of A is obtained.   ∎

An implicit part of the preceding discussion is the spectral properties of the shift operator. To this end, the eigenvalues and eigenvectors of the shift operator will be determined.

Theorem 3.30. *Let $D(z) \in \mathbb{F}[z]^{m \times m}$ be nonsingular, and let $S_D : X_D \longrightarrow X_D$ denote the shift operator.*

1. *The characteristic polynomial of S_D is equal to $\det D(z)$. The minimal polynomial of S_D is equal to $d_1(z)$, where $d_m(z)| \cdots |d_1(z)$ are the invariant factors of $D(z)$.*
2. *Let $p(z) \in \mathbb{F}[z]$. Then $p(S_D)$ is invertible if and only if $p(z)$ and $\det D(z)$ are coprime.*
3. *For a polynomial $f(z) \in X_D$, let $\xi_f = (D^{-1}f)_{-1}$. This implies the following representation:*

$$S_D f = zf - D(z)\xi_f.$$

4. *$\alpha \in \mathbb{F}$ is an eigenvalue of S_D if and only if $\det D(\alpha) = 0$. The eigenfunctions in X_D corresponding to α are*

$$f(z) = \frac{D(z)v}{z - \alpha}, \quad v \in \mathrm{Ker}\, D(\alpha). \tag{3.28}$$

Proof. 1. Since $D(z)$ is unimodularly equivalent to the Smith form $\mathrm{diag}\,(d_1, \ldots, d_m)$, $d_m| \cdots |d_1$, the polynomial model X_D is isomorphic to the direct sum $X_{d_1} \oplus \cdots \oplus X_{d_m}$. Thus the shift operator S_D is similar to the direct sum of shift operators $\mathrm{diag}\,(S_{d_1}, \ldots, S_{d_m})$, and the characteristic polynomial of S_D coincides with the product of characteristic polynomials of S_{d_i}. Since the characteristic polynomial of S_{d_i} is equal to $d_i(z)$, it follows that S_D has the characteristic polynomial $d_1(z) \cdots d_m(z) = \det D(z)$. Similarly, let $f(z)$ be a polynomial that annihilates S_D. Then $f(S_{d_i}) = 0$ for $i = 1, \ldots, m$, i.e., $d_i(z)$ divides $f(z)$ for all i. Since $d_m| \cdots |d_1$, this is equivalent to the condition that $d_1(z)$ divides $f(z)$. Therefore, $d_1(z)$ is the minimal polynomial of S_D.

2. The scalar case is considered first. If $Z : X_d \longrightarrow X_d$ is a homomorphism, then $Zf = \pi_d(pf)$ for some polynomial p where the condition $p(z)d(z) = d(z)p(z)$ is trivially satisfied. Now, as $S_d f = \pi_d(zf)$ for all $f \in X_d$, this implies that $Zf = \pi_d pf = p(S_d)f$. Thus $p(S_d)$ is invertible if and only if $p(z)$ and $d(z)$ are coprime.

Thus the result is shown for scalar polynomials. Using the Smith form it follows that $p(S_D)$ is invertible if and only if $p(S_{d_i})$ is invertible for $i = 1, \ldots, m$, i.e., if and only if $p(z)$ and $d_i(z)$ are coprime, for $i = 1, \ldots, m$. This completes the proof.

3. If $f \in X_D$, then $D(z)^{-1} f(z) = \sum_{j=1}^{\infty} g_j z^{-j}$ is strictly proper and $g_1 = (D^{-1}f)_{-1}$. Thus $\pi_+ (D(z)^{-1} z f(z)) = g_1$. The computation

$$S_D(f) = D(z)\pi_-(D(z)^{-1} z f(z)) = z f(z) - D(z)\pi_+(D(z)^{-1} z f(z)) = z f(z) - D(z)g_1$$

completes the proof.

4. The first claim follows trivially from the identity $\det (zI - S_D) = \det D(z)$. Assume that f is an eigenvector of S_D for the eigenvalue α. Then there exists $v \in \mathbb{F}^m$, with

$$S_D(f) = z f(z) - D(z)v = \alpha f(z),$$

and hence $D(\alpha)v = 0$. But this is equivalent to (3.28).   ∎

3.5 Rational Models

The preceding discussion showed that every finite-dimensional linear operator can be considered as the shift operator on a suitable polynomial model X_D. Alternatively, following an ingenious idea of G.-C. Rota, a linear operator can be considered as the compression of the backward shift, acting on the space of strictly proper formal power series, to a backward-shift-invariant subspace. Let A be a linear transformation in $\mathbb{F}^n$, and consider the difference equation in $\mathbb{F}^n$:

$$x_{t+1} = Ax_t, \quad t = 0, 1, 2 \ldots. \tag{3.29}$$

Consider the infinite product space

$$(\mathbb{F}^n)^{\mathbb{N}} = \mathbb{F}^n \times \mathbb{F}^n \times \cdots,$$

which is clearly an infinite-dimensional $\mathbb{F}$-vector space. The linear embedding

$$\Phi_A : \mathbb{F}^n \longrightarrow (\mathbb{F}^n)^{\mathbb{N}}$$
$$\Phi_A(x) = (x, Ax, A^2 x, \ldots)$$

then associates to each initial state $x_0 \in \mathbb{F}^n$ the entire solution trajectory $(x_t)_{t \in \mathbb{N}_0} = (A^t x_0)_{t \in \mathbb{N}_0}$ of (3.29). More generally, one can embed a nonlinear dynamical system

$$x_{t+1} = f(x_t)$$

on $\mathbb{F}^n$ into $(\mathbb{F}^n)^{\mathbb{N}}$ via the map

$$x \mapsto (x, f(x), f(f(x)), \ldots).$$

This idea of embedding a finite-dimensional dynamical system into an infinite-dimensional space via the associated solution trajectories is well known from symbolic dynamics and goes back at least to Poincaré, Birkhoff, and Smale. It also lies at the heart of modern systems theory, in particular in the theory of behaviors as developed by J.C. Willems.

The linear transformation Φ_A has more structure in connection with the backward shift. Let σ denote the **backward shift operator** on $(\mathbb{F}^n)^{\mathbb{N}}$, defined by

$$\sigma(x_0, x_1, x_2, \ldots) \mapsto (x_1, x_2, x_3, \ldots).$$

Clearly, Φ_A intertwines the linear operator A with the shift σ, i.e.,

$$\Phi \circ A = \sigma \circ \Phi.$$

In particular, the image space $\operatorname{Im} \Phi_A$ is shift-invariant and consists of all trajectories of (3.29). Thus one obtains the following commutative diagram:

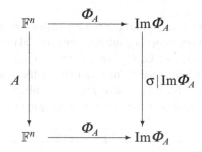

which implies the isomorphism

$$A \simeq \sigma | \operatorname{Im} \Phi_A.$$

Thus, an embedding of a linear operator into the standard backward shift operator was constructed. This far-reaching idea was first expounded in operator theory by Rota (1960). The subspace $\operatorname{Im} \Phi_A$, being finite-dimensional and shift invariant, is a behavior and therefore has a kernel representation. In fact, it is easy to verify that $\operatorname{Im} \Phi_A = \operatorname{Ker}(\sigma I - A)$.

It is possible to reformulate this embedding approach to linear dynamics in a purely algebraic manner. This can be done by identifying the infinite product space $(\mathbb{F}^n)^{\mathbb{N}}$ with the space of strictly proper formal power series $z^{-1}\mathbb{F}[[z^{-1}]]^n$. The embedding Φ_A is then conveniently expressed as $\Phi_A : \mathbb{F}^n \longrightarrow z^{-1}\mathbb{F}[[z^{-1}]]^n$,

$$\Phi_A(\xi) = (zI_n - A)^{-1}\xi.$$

The image space of Φ_A, i.e., the solution space of (3.29), is then identified with the space of rational functions

$$X_{zI-A} = \{(zI_n - A)^{-1}\xi \mid \xi \in \mathbb{F}^n\}.$$

This leads us to rational models whose treatment can be developed in complete duality to that of polynomial models. Recall the splitting of the formal Laurent series

$$\mathbb{F}((z^{-1}))^m = \mathbb{F}[z]^m \oplus z^{-1}\mathbb{F}[[z^{-1}]]^m$$

into polynomial and strictly proper parts. Let

$$\pi_- : \mathbb{F}((z^{-1}))^m \longrightarrow z^{-1}\mathbb{F}[[z^{-1}]]^m$$

denote the canonical projection onto the strictly proper part. The **backward shift** $\sigma : z^{-1}\mathbb{F}[[z^{-1}]]^m \longrightarrow z^{-1}\mathbb{F}[[z^{-1}]]^m$ is the linear operator defined by

$$\sigma(h) = \pi_-(zh(z)).$$

Thus σ acts on strictly proper formal power series via

$$\sigma\left(\sum_{j=1}^{\infty} h_j z^{-j}\right) = \sum_{j=1}^{\infty} h_{j+1} z^{-j},$$

i.e., by shifting the sequence of coefficients backward. Note that $z^{-1}\mathbb{F}[[z^{-1}]]^m$ becomes an $\mathbb{F}[z]$-module by defining, for polynomials $p(z) \in \mathbb{F}[z]$, the module operation as

$$p(z) \cdot h := p(\sigma)h = \pi_-(p(z)h(z)). \tag{3.30}$$

With this module structure at hand, the backward shift σ becomes a module homomorphism on $z^{-1}\mathbb{F}[[z^{-1}]]^m$. More generally, for every nonsingular polynomial matrix $D(z) \in \mathbb{F}[z]^{m \times m}$ and $h(z) \in z^{-1}\mathbb{F}[[z^{-1}]]^m$, the **Toeplitz operator**

$$D(\sigma) : z^{-1}\mathbb{F}[[z^{-1}]]^m \longrightarrow z^{-1}\mathbb{F}[[z^{-1}]]^m,$$

$$D(\sigma)h = \pi_-(D(z)h(z))$$

is a module homomorphism. In particular, the kernel Ker $D(\sigma)$ and the image Im $D(\sigma)$ are σ-invariant and, therefore, are submodules of $z^{-1}\mathbb{F}[[z^{-1}]]^m$.

Definition 3.31. Let $D(z) \in \mathbb{F}[z]^{m \times m}$ be nonsingular. The backward-shift-invariant subspace

$$X^D = \operatorname{Ker} D(\sigma) \subset z^{-1} \mathbb{F}[[z^{-1}]]^m, \qquad (3.31)$$

endowed with the induced $\mathbb{F}[z]$-module structure (3.30), is called a **rational model**. The **restricted shift** is the module homomorphism $S^D : X^D \longrightarrow X^D$ defined by

$$S^D = \sigma|X^D.$$

In the behavioral systems theory literature, for example Polderman and Willems (1997), rational models are also referred to as **autonomous behaviors**. In fact, there is an interesting interpretation of rational models as solution spaces of higher-order matrix difference equations. Consider a **matrix difference equation** on $\mathbb{F}^m$

$$D_r \xi_{t+r} + D_{r-1} \xi_{t+r-1} + \cdots + D_0 \xi_t = 0, \quad t \in \mathbb{N}_0, \qquad (3.32)$$

defined by a nonsingular polynomial matrix

$$D(z) = D_0 + D_1 z + \cdots + D_r z^r \in \mathbb{F}[z]^{m \times m}.$$

By identifying the solutions $(\xi_t)_{t \geq 0}$ in $\mathbb{F}^m$ of (3.32) with the formal power series $h(z) = \sum_{j=0}^{\infty} \xi_j z^{-j-1}$ we see that (3.32) is equivalent to the equation

$$D(\sigma)h = 0$$

in $z^{-1} \mathbb{F}[[z^{-1}]]^m$. Therefore, the rational model X^D can be identified with the solution space of the difference equation (3.32).

Returning to the discussion of the autonomous linear system

$$x_{t+1} = A x_t,$$

it is perhaps useful to emphasize the connection between spectral decompositions of matrices and and factorizations of polynomial matrices. Suppose one wants to decompose this system into a direct sum of subsystems. If $d(z) = \det(zI - A)$ denotes the characteristic polynomial and if $d(z) = d_1(z)d_2(z)$ is a factorization into coprime factors, then $d(z) = d_1(z)d_2(z)$ induces an essentially unique factorization

$$zI - A = \overline{S}_2(z)S_1(z) = \overline{S}_1(z)S_2(z)$$

by nonsingular matrix polynomials $\overline{S}_i(z), S_i(z)$, with $d_i(z) = \det S_i(z) = \det \overline{S}_i(z)$. These factorizations imply the direct sum decomposition of the rational model space

$$X^{zI-A} = \operatorname{Ker} S_1(\sigma) \oplus \operatorname{Ker} S_2(\sigma). \qquad (3.33)$$

One refers to (3.33) as a **spectral decomposition**. Of course, this is just the dynamical systems representation of the spectral decomposition of the state space

$$\mathbb{F}^n = X_1(A) \oplus X_2(A), \tag{3.34}$$

where $X_i(A) = \operatorname{Ker} d_i(A), i = 1, 2$. Note that because $X_i(A)$ are A-invariant subspaces, then with respect to the direct sum decomposition (3.34), one obtains the block-matrix representation

$$A = \begin{pmatrix} A_1 & 0 \\ 0 & A_2 \end{pmatrix}.$$

As a matter of fact, the direct sum representations (3.33) and (3.34) are related via the map $\phi : \mathbb{F}^n \longrightarrow \operatorname{Ker}(\sigma I - A) \subset z^{-1}\mathbb{F}[[z^{-1}]]$, defined by $\phi(x) = (zI - A)^{-1}x$. The map ϕ is an isomorphism satisfying $\phi(Ax) = \sigma\phi(x)$. Therefore, it follows that $\phi(X_i(A)) = \operatorname{Ker} S_i(\sigma)$ for $i = 1, 2$.

Rational models are defined in (3.31) as kernels of special Toeplitz operators. Thus a strictly proper power series $h(z) \in z^{-1}\mathbb{F}[[z^{-1}]]^m$ is an element of X^D if and only if $D(z)h(z)$ is a vector polynomial. Rational models can also be characterized as images of certain projections. For a nonsingular polynomial matrix $D(z) \in \mathbb{F}[z]^{m \times m}$, we define a projection map in $z^{-1}\mathbb{F}[[z^{-1}]]^m$ by

$$\pi^D : z^{-1}\mathbb{F}[[z^{-1}]]^m \longrightarrow X^D$$
$$\pi^D h = \pi_-(D^{-1}\pi_+ Dh), \qquad h \in z^{-1}\mathbb{F}[[z^{-1}]]^m. \tag{3.35}$$

The next theorem examines rational models induced by a nonsingular polynomial matrix $D(z)$ and their relation to the corresponding polynomial model.

Theorem 3.32. *Let $D(z) \in \mathbb{F}[z]^{m \times m}$ be nonsingular. Then:*

1. *The map π^D, defined in (3.35), is a projection operator in $z^{-1}\mathbb{F}[[z^{-1}]]^m$;*
2. *The rational model X^D has the following image representation:*

$$X^D = \operatorname{Im} \pi^D;$$

3. *The map*

$$\tau_D : X_D \longrightarrow X^D, \quad f(z) \mapsto D(z)^{-1} f(z)$$

is an isomorphism of $\mathbb{F}[z]$-modules. In particular, X^D is a finite-dimensional $\mathbb{F}$-vector space of dimension

$$\dim X^D = \deg \det D(z).$$

Proof. 1. Let $h(z)$ be in $z^{-1}\mathbb{F}[[z^{-1}]]^m$; then

$$(\pi^D)^2 h = \pi^D(\pi^D h)$$

$$= \pi_- D^{-1} \pi_+ D\pi_- D^{-1} \pi_+ Dh = \pi_- D^{-1} \pi_+ D(1 - \pi_+) D^{-1} \pi_+ Dh$$

$$= \pi_- D^{-1} \pi_+ DD^{-1} \pi_+ Dh - \pi_- D^{-1} \pi_+ D\pi_+ D^{-1} \pi_+ Dh$$

$$= \pi^D h - \pi_- D^{-1} D\pi_+ D^{-1} \pi_+ Dh = \pi^D h.$$

Here, the fact that $\pi_- \pi_+ = 0$ was used.

2. Assume $h \in X^D = \operatorname{Ker} D(\sigma)$; then Dh is a polynomial, which means $Dh = \pi_+ dh$. Applying the map $\pi_- D^{-1}$ to both sides of the last equality, we obtain $\pi^D h = h$, i.e., $X^D \subset \operatorname{Im} \pi^D$. Conversely, assume $h(z) \in \operatorname{Im} \pi^D$, i.e., $h = \pi^D g$ for some $g(z) \in z^{-1}\mathbb{F}[[z^{-1}]]^m$. Since π^D is a projection, it follows that $\pi^D h = (\pi^D)^2 g = \pi^D g = h$. Next, one computes

$$\pi_- Dh = \pi_- D\pi^D h = \pi_- D\pi_- D^{-1} \pi_+ Dh = \pi_- DD^{-1} \pi_+ Dh = \pi_- \pi_+ Dh = 0.$$

This shows the inclusion $\operatorname{Im} \pi^D \subset X^D$. Here, we used the fact that $\operatorname{Ker} \pi_-$ is invariant under multiplication by $D(z)$ and that $\pi_- \pi_+ = 0$.

3. Clearly, $h \in X^D = \operatorname{Ker} D(\sigma)$ if and only if $D(z)h(z)$ is a polynomial. Since $h(z)$ is assumed to be strictly proper, this shows that $f \in X_D$ if and only if $D(z)^{-1} f(z) \in X^D$. Thus the linear map τ is bijective. From

$$\tau_D(\pi_D(zf(z))) = \pi_-(zD(z)^{-1} f(z)) = \sigma(D(z)^{-1} f(z))$$

we conclude that τ is $\mathbb{F}[z]$-linear and, therefore, a module isomorphism. $\blacksquare$

Let us note the following simple corollary of the characterization of elements of a polynomial model that was given in Theorem 3.4.

Lemma 3.33. *Let $D_1(z), D_2(z) \in \mathbb{F}[z]^{m \times m}$ be nonsingular polynomial matrices. If $D_2(z)^{-1} D_1(z)$ is biproper, i.e., an invertible element of $\mathbb{F}[[z^{-1}]]^{m \times m}$, then X_{D_1} and X_{D_2} are equal as sets, though they may have different module structures.*

Proof. Let $f(z) \in X_{D_1}$. Then $D_2^{-1} f = (D_2^{-1} D_1) D_1^{-1} f$ is strictly proper as a product of the biproper $D_2(z)^{-1} D_1(z)$ and the strictly proper $D_1^{-1} f$, i.e., $f \in X_{D_2}$. The result follows by symmetry, and we note also that $D_1(z)^{-1} D_2(z)$ is proper. $\blacksquare$

In the same way as for polynomial models, there exists an intertwining-map characterization of homomorphisms of rational models. Let $D_1(z) \in \mathbb{F}[z]^{m \times m}$ and $D_2(z) \in \mathbb{F}[z]^{p \times p}$ be nonsingular, and let $\tau_i : X_{D_i} \longrightarrow X^{D_i}$ denote the module isomorphism defined by $f(z) \mapsto D_i(z)^{-1} f(z)$. Then a map $Z : X^{D_1} \longrightarrow X^{D_2}$ is an $\mathbb{F}[z]$-homomorphism if and only if the transformed map $(\tau_2)^{-1} \circ Z \circ \tau_1 : X_{D_1} \longrightarrow X_{D_2}$ is a module homomorphism. Moreover, Z is injective (surjective) if and only if

$(\tau_2)^{-1} \circ Z \circ \tau_1$ is injective (surjective). Therefore,

$$\tau_2^{-1} \circ Z \circ \tau_1(f) = \pi_{D_2}(N_2 f), \quad f \in X_{D_1}$$

for all polynomial matrices $N_1(z), N_2(z)$ satisfying $N_2(z)D_1(z) = D_2(z)N_1(z)$. Hence, for each strictly proper $h(z) = D_1(z)^{-1} f(z) \in X^{D_1}$, we conclude that

$$Z(h) = D_2^{-1} \pi_{D_2}(N_2 f) = \pi_-(D_2^{-1} N_2 f) = \pi_-(N_1 h).$$

This proves the following result.

Theorem 3.34. *Let $D_1(z) \in \mathbb{F}[z]^{m \times m}$ and $D_2(z) \in \mathbb{F}[z]^{p \times p}$ be nonsingular. A map $Z : X^{D_1} \longrightarrow X^{D_2}$ is an $\mathbb{F}[z]$-homomorphism if and only if there exist $N_1(z), N_2(z) \in \mathbb{F}[z]^{p \times m}$ such that*

$$N_2(z)D_1(z) = D_2(z)N_1(z)$$

and

$$Zh = \pi_-(N_1 h) = N_1(\sigma)h, \quad h \in X^{D_1}. \tag{3.36}$$

The homomorphism (3.36) satisfies the following conditions:

1. $\operatorname{Ker} Z = X^{F_1}$, *where $D_1(z) = E_1(z)F_1(z)$ and $F_1(z)$ is a g.c.r.d. of $D_1(z)$ and $N_1(z)$;*
2. $\operatorname{Im} Z = X^{F_2}$, *where $D_2(z) = E_2(z)F_2(z)$ and $E_2(z)$ is a g.c.l.d. of $D_2(z)$ and $N_2(z)$;*
3. *Z is injective if and only if $D_1(z)$ and $N_1(z)$ are right coprime;*
4. *Z is surjective if and only if $D_2(z)$ and $N_2(z)$ are left coprime;*
5. *Z is invertible if and only if $D_1(z)$ and $N_1(z)$ are right coprime and $D_2(z)$ and $N_2(z)$ are left coprime.*

In contrast to the case of polynomials, the module $z^{-1} \mathbb{F}[[z^{-1}]]^m$ is not finitely generated. This makes the analysis of submodules an interesting challenge because such submodules can be either finitely generated or infinitely generated. The algebraic classification of all submodules of $z^{-1}\mathbb{F}[[z^{-1}]]^m$ therefore becomes an extremely difficult, if not impossible, task. The situation is much better if one restricts oneself to rational models. In fact, using the module isomorphism between rational models X^D and polynomial models X_D, results on submodules and homomorphisms of polynomial models in the context of rational model spaces can easily be reinterpreted. The next result shows that the set of submodules of a rational model, ordered by inclusion, forms a lattice.

Theorem 3.35. *(a) Let $D(z)$ be a nonsingular $m \times m$ polynomial matrix. The submodules $\mathscr{V}$ of X^D are exactly the subspaces of the form*

$$\mathscr{V} = X^F,$$

where $E(z)$ and $F(z)$ are nonsingular $m \times m$ polynomial matrices satisfying $D(z) = E(z)F(z)$.

(b) Let X^{F_i}, $i = 1, \ldots, s$ be submodules of X^D corresponding to the factorizations

$$D(z) = E_i(z)F_i(z), \quad i = 1, \ldots, s.$$

Then the following statements are true.

1. $X^{F_1} \subset X^{F_2}$ if and only if there exists a nonsingular polynomial matrix L, with $F_2(z) = L(z)F_1(z)$.
2. The intersection of rational models is a rational model, i.e.,

$$X^{F_1} \cap \cdots \cap X^{F_s} = X^{F_*}, \tag{3.37}$$

where $F_*(z)$ is the g.c.r.d. of the $F_i(z)$.
3. The equality $\bigcap_{i=1}^{s} X^{F_i} = 0$ is valid if and only if the $F_i(z)$ are right coprime.
4. The sum of rational models is a rational model, i.e.,

$$X^{F_1} + \cdots + X^{F_s} = X^{F^*}, \tag{3.38}$$

where $F^*(z)$ is the l.c.l.m. of all the $F_i(z)$.
5. Let $D(z)$ have the factorizations $D(z) = E_i(z)F_i(z)$ for $i = 1, \ldots, s$. Then

$$X^D = X^{F_1} + \cdots + X^{F_s}$$

if and only if $D(z)$ is the l.c.l.m. of $F_1(z), \ldots, F_s(z)$.
6. Let $D(z)$ have the factorizations $D(z) = E_i(z)F_i(z)$ for $i = 1, \ldots, s$. Then

$$X^D = X^{F_1} \oplus \cdots \oplus X^{F_s}$$

if and only if $D(z)$ is the l.c.l.m. of $F_1(z), \ldots, F_s(z)$ and the $F_i(z)$ are mutually right coprime.

Proof. Every submodule of X_D is of the form EX_F, where $D(z) = E(z)F(z)$. Therefore, the submodules of X^D are $D^{-1}EX_F = F^{-1}X_F = X^F$. This proves (a).

1. To prove the first claim in (b), note that $X^{F_1} \subset X^{F_2}$ is equivalent to $E_1X_{F_1} \subset E_2X_{F_2}$. By Theorem 3.14, this is equivalent to the existence of a nonsingular polynomial matrix R, with $E_1 = E_2R$. But then $E_2RF_1 = E_1F_1 = E_2F_2$, which implies $F_2 = RF_1$, as desired.
2. From equation (3.37) one obtains $X^F \subset X^{F_i}$, i.e., $F(z)$ is a common right divisor of $F_i(z), i = 1, \ldots, m$. If $F'(z)$ is another common right divisor, then $X^{F'} \subset \bigcap_{i=1}^{s} X^{F_i} = X^F$. Thus $F'(z)$ is a right divisor of $F(z)$ and this implies that $F(z)$ is the g.c.r.d. of $F_1(z), \ldots, F_s(z)$. This proves statement 2.
3. Statement 2 obviously implies statement 3.
4. Similarly, (3.38) implies $X^{F_i} \subset X^F$, i.e., $F(z) = L_i(z)F_i(z)$ for all i. Thus $F(z)$ is a common left multiple of $F_1(z), \ldots, F_s(z)$. If $F'(z)$ is another common left

multiple, then $X^F = X^{F_1} + \cdots + X^{F_s} \subset X^{F'}$. But this implies that $F'(z)$ is a left multiple of $F(z)$. Therefore, $F(z)$ is a l.c.l.m.

5. $X^D = X^{F_1} + \cdots + X^{F_s}$ is satisfied if and only if $X^D = X^{F^*}$, i.e., if and only if D is a l.c.l.m. of $F_1, \ldots, F_s$.

6. Follows trivially from the previous claims. ∎

The proofs of Theorems 3.35 and 3.14 are independent of each other. In fact, the two results are equivalent, and therefore each version can be deduced from the other. For example, the coprimeness condition for the direct sum decomposition (6), stating that D is the l.c.l.m. of the mutually right coprime polynomials $F_1(z), \ldots, F_s(z)$, is equivalent to left coprimeness of $E_1(z), \ldots, E_s(z)$ and mutual right coprimeness of $F_1(z), \ldots, F_s(z)$.

From these structural properties of rational model spaces one obtains an important abstract characterization of rational models X^D.

Theorem 3.36. *A subset $M \subset z^{-1}\mathbb{F}[[z^{-1}]]^m$ is a finitely generated torsion submodule if and only if there exists a nonsingular $D(z) \in \mathbb{F}[z]^{m \times m}$ such that $M = \operatorname{Ker} D(\sigma)$.*

Proof. The property of being a finitely generated torsion module is invariant under module isomorphisms. Since the polynomial model $X_D \simeq \mathbb{F}[z]^m / D(z)\mathbb{F}[z]^m$ is a finitely generated torsion module, so is X^D. Conversely, assume $M \subset z^{-1}\mathbb{F}[[z^{-1}]]^m$ is a finitely generated torsion submodule. Then there exists a nonzero polynomial $d(z)$ for which $M \subset \operatorname{Ker} d(\sigma) = X^{dI_m}$. Thus M is a submodule of the rational model X^{dI_m}. Theorem 3.35 implies $M = X^F$ for a nonsingular polynomial $m \times m$ matrix $F(z)$. This completes the proof. ∎

Since rational models X^D are simply the solution spaces $\operatorname{Ker} D(\sigma))$ of the higher-order difference equations $D(\sigma)h = 0$, one obtains a Galois correspondence between nonsingular polynomial matrices and solution spaces of difference equations.

Theorem 3.37. *Let $D_i(z) \in \mathbb{F}[z]^{m \times m}, i = 1, \ldots, s$ be nonsingular polynomial matrices. Then:*

1. *$\operatorname{Ker} D_1(\sigma) \subset \operatorname{Ker} D_2(\sigma)$ if and only if $D_2(z) = L(z)D_1(z)$ for a polynomial matrix $L(z)$;*

2. *The intersection of solution spaces is*

$$\operatorname{Ker} D_1(\sigma) \cap \cdots \cap \operatorname{Ker} D_s(\sigma) = \operatorname{Ker} D_*(\sigma),$$

where $D_*(z)$ is the g.c.r.d. of $D_1(z), \ldots, D_s(z)$;

3. *The sum of solution spaces is*

$$\operatorname{Ker} D_1(\sigma) + \cdots + \operatorname{Ker} D_s(\sigma) = \operatorname{Ker} D^*(\sigma),$$

where $D^*(z)$ is the l.c.l.m. of $D_1(z), \ldots, D_s(z)$. ∎

3.6   Duality

Finally, the issue of how to describe the dual vector space of polynomial and rational model spaces will be briefly addressed. This involves polynomial model duality theory as applied to geometric control, an interesting subject that will be treated in Chapter 6. For further details, see Fuhrmann (1981, 2006).

A dual pairing on $\mathbb{F}((z^{-1}))^m \times \mathbb{F}((z^{-1}))^m$ is introduced by defining, for truncated Laurent series $f(z) = \sum_{j=-\infty}^{\infty} f_j z^j$ and $h(z) = \sum_{j=-\infty}^{\infty} h_j z^j \in \mathbb{F}((z^{-1}))^m$,

$$[f,h] := (f(z)^\top h(z))_{-1} = \sum_{j=-\infty}^{\infty} f_j^\top h_{-j-1}. \tag{3.39}$$

It is clear that $[\cdot,\cdot]$ is a well-defined bilinear form on $\mathbb{F}((z^{-1}))^m \times \mathbb{F}((z^{-1}))^m$, because in the defining sum at most a finite number of terms are nonzero. By restriction, one obtains a **bilinear dual pairing** on pairs of polynomials $f(z) = \sum_{j=0}^{N_f} f_j z^j$ and strictly proper Laurent series $h(z) = \sum_{j=1}^{\infty} h_j z^{-j}$ via

$$[\cdot,\cdot] : \mathbb{F}[z]^m \times z^{-1}\mathbb{F}[[z^{-1}]]^m \longrightarrow \mathbb{F},$$

$$[f,h] = \sum_{j=0}^{\infty} f_j^\top h_{j+1}.$$

This bilinear form is nondegenerate in the sense that

(a) $[f,h] = 0$ for all $h(z) \in z^{-1}\mathbb{F}[[z^{-1}]]^m$ if and only if $f(z) = 0$;
(b) $[f,h] = 0$ for all $f(z) \in \mathbb{F}[z]^m$ if and only if $h(z) = 0$.

For subsets $M \subset z^{-1}\mathbb{F}[[z^{-1}]]^m$ and $Q \subset \mathbb{F}[z]^m$, one defines

$$M^\perp = \{f \in \mathbb{F}[z]^m \mid [f,h] = 0 \quad \forall h \in M\},$$
$$^\perp Q = \{h \in z^{-1}\mathbb{F}[[z^{-1}]]^m \mid [f,h] = 0 \quad \forall f \in M\}.$$

A few elementary properties of these spaces are readily established. Let $Q \subset \mathbb{F}[z]^m$ and $M \subset z^{-1}\mathbb{F}[[z^{-1}]]^m$ be submodules of $\mathbb{F}[z]^m$ and $z^{-1}\mathbb{F}[[z^{-1}]]^m$, respectively. Then $^\perp Q \subset z^{-1}\mathbb{F}[[z^{-1}]]^m$ and $M^\perp \subset \mathbb{F}[z]^m$ are submodules, too. To prove this, note that the equality $[zf,h] = [f,\sigma h]$ is valid for all $f \in \mathbb{F}[z]^m$ and $h \in z^{-1}\mathbb{F}[[z^{-1}]]^m$. Let now $f \in Q$ and $h \in {}^\perp Q$; it follows that

$$0 = [zf,h] = [f,\sigma h].$$

This shows that if $Q \subset \mathbb{F}[z]^m$ is a submodule, then $^\perp Q$ is a submodule of $z^{-1}\mathbb{F}[[z^{-1}]]^m$. The other claim follows similarly.

For $\mathbb{F}$-vector spaces $\mathscr{V}$, let $\mathscr{V}^*$ denote the **algebraic dual space**, i.e.,

$$\mathscr{V}^* = \{\lambda : \mathscr{V} \longrightarrow \mathbb{F} \mid \lambda \text{ linear}\}.$$

Note that if $\mathscr{V}$ is an $\mathbb{F}[z]$-module, then $\mathscr{V}^*$ is also an $\mathbb{F}[z]$-module, with the module structure defined by

$$(p \cdot \lambda)(x) := \lambda(p \cdot x), \quad x \in \mathscr{V}, p(z) \in \mathbb{F}[z].$$

The following result is an algebraic version of the Riesz representation theorem. It shows that there is a canonical module isomorphism between the polynomial and rational models X_D and $X^{D^\top}$.

Theorem 3.38. *1. Let $Q \subset \mathbb{F}[z]^m$ be a submodule. The map*

$$L : {}^\perp Q \longrightarrow (\mathbb{F}[z]^m/Q)^*, \quad L(h)f = [f, h] \tag{3.40}$$

is an isomorphism of $\mathbb{F}[z]$-modules. In particular, the following important duality relation is obtained:

$$(\mathbb{F}[z]^m)^* = z^{-1}\mathbb{F}[[z^{-1}]]^m. \tag{3.41}$$

2. For every nonsingular polynomial matrix $D(z) \in \mathbb{F}[z]^{m \times m}$, the space ${}^\perp(D(z)\mathbb{F}[z]^m)$ is equal to the rational model $X^{D^\top}$, i.e.,

$${}^\perp(D(z)\mathbb{F}[z]^m) = X^{D^\top}.$$

3. There is a canonical module isomorphism

$$L : X^{D^\top} \longrightarrow X_D^*, \quad L(h)f = [f, h]. \tag{3.42}$$

4. Assuming $T(z) \in \mathbb{F}[z]^{m \times m}$ is nonsingular with $T(z)^{-1}$ proper rational implies the following direct sum decomposition of $\mathbb{F}$-vector spaces:

$$z^{-1}\mathbb{F}[[z^{-1}]]^m = X^T \oplus \Pi^T, \tag{3.43}$$

where

$$\Pi^T := \{h(z) \in z^{-1}\mathbb{F}[[z^{-1}]]^m \mid T(z)h(z) \in z^{-1}\mathbb{F}[[z^{-1}]]^m\} = \operatorname{Ker}\pi^T. \tag{3.44}$$

Proof. 1. For each $h \in {}^\perp Q$, the map (3.40) is well defined, $\mathbb{F}$-linear, and injective. Consider a linear functional $\lambda : \mathbb{F}[z]^m \longrightarrow \mathbb{F}$ that vanishes on Q. For each $j \geq 0$ define vectors $h_{j+1} \in \mathbb{F}^m$ whose ith component is $\lambda(z^j e_i)$, $i = 1, \ldots, m$. Let $h(z) = \sum_{j=1}^{\infty} h_j z^{-j}$. Then, for a polynomial $f(z) \in \mathbb{F}[z]^m$, the equality $[f, h] = \lambda(f)$

follows. Thus (3.40) is surjective and, therefore, a vector space isomorphism. Since $L(\sigma h)f = [zf(z), h]$, the map (3.40) is $\mathbb{F}[z]$-linear. This proves the claim.

2. A strictly proper power series $h \in z^{-1}\mathbb{F}[[z^{-1}]]^m$ satisfies $h \in^{\perp} (D(z)\mathbb{F}[z]^m)$ if and only if $[D(z)f(z), h(z)] = 0$ for all $f(z) \in \mathbb{F}[z]^m$. Since $[D(z)f(z), h(z)] = [f(z), D(\sigma)h]$, this is equivalent to $h \in X^{D^{\top}}$.

3. For $Q = D(z)\mathbb{F}[z]^m$, the map L in (3.40) yields a module isomorphism

$$X^{D^{\top}} = {}^{\perp}(D(z)\mathbb{F}[z]^m) \simeq (\mathbb{F}[z]^m/D(z)\mathbb{F}[z]^m) \simeq X_D^*.$$

Explicitly, the module isomorphism is given as

$$L : X^{D^{\top}} \longrightarrow (X_D)^*, \qquad L(h)f = [h, f].$$

4. The inclusion $\Pi^T \subset \operatorname{Ker} \pi^T$ is obvious. To prove the converse inclusion, assume $h(z) \in \operatorname{Ker} \pi^T \subset z^{-1}\mathbb{F}[[z^{-1}]]^m$, that is, $\pi_-(T^{-1}\pi_+(Th)) = 0$. Applying T to both sides, this implies $\pi_T\pi_+Th = 0$, and hence, for some $h'(z) \in z^{-1}\mathbb{F}[[z^{-1}]]^m$ and $g(z) \in \mathbb{F}[z]^m$, one obtains $Tg = Th - h'$. From this $g(z) = h(z) - T(z)^{-1}h'(z)$ follows, and hence, under our assumption that $T(z)$ is nonsingular and properly invertible, we conclude that $T(z)^{-1}h'(z)$ and, therefore, $g(z)$ are strictly proper. Thus, $g(z) = 0$, $h(z) = T(z)^{-1}h'(z) \in \Pi^T$, and the equality $\operatorname{Ker} \pi^T = \Pi^T$ follows. The direct sum decomposition (3.43) results from the decomposition $z^{-1}\mathbb{F}[[z^{-1}]]^m = \operatorname{Im} \pi^T \oplus \operatorname{Ker} \pi^T$.   ∎

Using the module isomorphism $X_D \longrightarrow X^D, f \mapsto D^{-1}f$, it follows from Theorem 3.38 that the $\mathbb{F}$-linear maps

$$L : X_D \longrightarrow \mathbb{F}^k$$

are of the form

$$Lf = (P(z)D(z)^{-1}f(z))_{-1}$$

for unique polynomial matrices $P(z) \in \mathbb{F}^{k \times m}$. This proves the following explicit characterizations of the dual spaces X^D and $(X^D)^*$.

Theorem 3.39. *1. Every $\mathbb{F}$-linear function $L : X_D \longrightarrow \mathbb{F}^k$ can be represented as*

$$Lf = (P(z)D(z)^{-1}f(z))_{-1}$$

for a unique polynomial matrix $P(z) \in \mathbb{F}[z]^{k \times m}$.

2. Every $\mathbb{F}$-linear function $L : X^D \longrightarrow \mathbb{F}^k$ can be represented as

$$Lh = (P(z)h(z))_{-1}, \qquad h \in X^D$$

for a unique polynomial matrix $P(z) \in \mathbb{F}[z]^{k \times m}$.   ∎

There is an alternative way to treat duality, solely in the context of polynomial models. The key to this is the canonical isomorphism between polynomial and rational model spaces, defined as

$$\tau_D : X_D \longrightarrow X^D, \quad \tau_D(f) = D(z)^{-1} f(z).$$

Thus, instead of identifying the dual of X_D with the rational model $X^{D^\top}$, as in Theorem 3.38, one can use $X_{D^\top}$ as the dual space. This approach will prove particularly useful in Chapter 5 when we analyze tensor product representations of $\mathbb{F}[z]$-linear homomorphisms. Thus, a new bilinear pairing on $X_D \times X_{D^\top}$ is defined by

$$< f, g > = [D^{-1}f, g] = [f, D^{-\top}g] = \left(f(z)^\top D^\top(z)^{-1} g(z) \right)_{-1} \qquad (3.45)$$

for all $f \in X_D$ and $g \in X_{D^\top}$.

Theorem 3.40. *Let $D(z) \in \mathbb{F}[z]^{m \times m}$ be nonsingular. Then the duality pairing (3.45) on $X_D \times X_{D^\top}$ satisfies the following properties:*

1. The dual space X_D^* of X_D can be identified, via (3.45), with $X_{D^\top}$. Thus

$$X_D^* = X_{D^\top},$$

and moreover the dual S_D^* of the shift S_D is

$$S_D^* = S_{D^\top}; \qquad (3.46)$$

2. Let $M = EX_F \subset X_D$ be a submodule, defined by some factorization $D(z) = E(z)F(z)$ in nonsingular polynomial factors. Then

$$M^\perp := \{ g \in X_{D^\top} \mid < f, g > = 0 \quad \text{for all} \quad f \in X_D \} \qquad (3.47)$$

is a submodule of $X_{D^\top}$ and is equal to

$$(EX_F)^\top = F^\top X_{E^\top}.$$

Proof. By combining the isomorphism $X_{D^\top} \longrightarrow X^{D^\top}, g \mapsto D^{-\top}g$, with the isomorphism $X^{D^\top} \longrightarrow X_D^*$ defined in (3.42), one concludes that $\phi : X_{D^\top} \longrightarrow X_D^*$

$$\phi(g)(f) = [f, D^{-\top}g] = < f, g >$$

defines a module isomorphism. Therefore, $<, >$ is a nondegenerate pairing on $X_D \times X_{D^\top}$. Computing

$$< S_D f, g > = [D^{-1} S_D f, g] = [D^{-1} \pi_D(zf), g] = [\pi_-(zD^{-1}f), g]$$
$$= [zD^{-1}f, g] = [D^{-1}f, zg] = [D^{-1}f, \pi_{D^\top}(zg)]$$
$$= < f, S_{D^\top} g >,$$

(3.46) follows.

That $M^\perp$, defined in (3.47), is a submodule follows from (3.46). Assume now $g \in (EX_F)^\perp$, which implies that $0 = < f, g >= [D^{-1}Ef_1, g] = [F^{-1}f_1, g] = [f_1, F^{-\top}g]$ for all $f = Ef_1 \in EX_F$. Clearly, for all $f_1 = Ff' \in F(z)\mathbb{F}[z]^m$ one obtains $[Ff', F^{-\top}g] = [F^{-1}Ff', g] = 0$. Using the direct sum decomposition $\mathbb{F}[z]^m = X_F \oplus F(z)\mathbb{F}[z]^m$, it follows that $[f', F^{-\top}g] = 0$ for all $f' \in \mathbb{F}[z]^m$, i.e., $g \in F^\top(z)\mathbb{F}[z]^m$. In addition, $g \in X_{D^\top}$ implies that $D^{-\top}g = D^{-\top}F^\top g_1 = E^{-\top}g_1$ is strictly proper, which implies $g_1 \in X_{E^\top}$ and $g \in F^\top X_{E^\top}$. Thus the inclusion $(EX_F)^\perp \subset F^\top X_{E^\top}$ follows.

Conversely, if $f = Ef_1 \in EX_F$ and $g = F^\top g_1 \in F^\top X_{E^\top}$, then

$$< f, g >= [D^{-1}Ef_1, F^\top g_1] = [FD^{-1}Ef_1, g_1] = [f_1, g_1] = 0.$$

This shows that $F^\top X_{E^\top} \subset (EX_F)^\perp$. This completes the proof. ∎

3.7 The Matrix Chinese Remainder Theorem

In Theorems 3.19 and 3.35, direct sum representations of polynomial and rational model spaces were analyzed. Having analyzed module isomorphisms in Section 3.3, we will now explain how a vector of polynomials or rational functions can be decomposed with respect to such a direct sum representation. This is directly related to a matrix version of the Chinese remainder theorem and interpolation theory. Note that in the proof of the subsequent theorem, coprimeness is used in two distinctive ways. One is geometrical, related to direct sum representations, the other is spectral and relates to the inversion of polynomial model homomorphisms.

Proposition 3.41. *Let* $E_1(z), \ldots, E_s(z) \in \mathbb{F}[z]^{m \times m}$ *be mutually left coprime and nonsingular. Define*

$$\begin{aligned} D(z) &= l.c.r.m.\{E_j(z)\}_1^s \\ \overline{F}_i(z) &= l.c.r.m.\{E_j(z)\}_{j \neq i}. \end{aligned} \tag{3.48}$$

Then:

1. There exist right factors $F_i(z) \in \mathbb{F}[z]^{m \times m}$ *of* $D(z) = E_i(z)F_i(z)$;
2. For $i = 1, \ldots, s$, $E_i, \overline{F}_i$ *are left coprime and*

$$D(z) = l.c.r.m.\{E_i(z), \overline{F}_i(z)\}; \tag{3.49}$$

3. *There exist right factors* $\overline{E}_i(z) \in \mathbb{F}[z]^{m \times m}$ *of* $D(z) = \overline{F}_i(z)\overline{E}_i(z)$;
4. *For* $i = 1, \ldots, s$, $F_i(z), \overline{E}_i(z)$ *are right coprime;*
5. *The polynomial model* X_D *has the following direct sum representation:*

$$X_D = \overline{F}_1 X_{\overline{E}_1} \oplus \cdots \oplus \overline{F}_s X_{\overline{E}_s}; \tag{3.50}$$

6. *Let*

$$\begin{pmatrix} \overline{Y}_i(z) & \overline{X}_i(z) \\ -\overline{F}_i(z) & \overline{E}_i(z) \end{pmatrix} \begin{pmatrix} \overline{E}_i(z) & -X_i(z) \\ F_i(z) & Y_i(z) \end{pmatrix} = \begin{pmatrix} I & 0 \\ 0 & I \end{pmatrix},$$

$$\begin{pmatrix} \overline{E}_i(z) & -X_i(z) \\ F_i(z) & Y_i(z) \end{pmatrix} \begin{pmatrix} \overline{Y}_i(z) & \overline{X}_i(z) \\ -\overline{F}_i(z) & \overline{E}_i(z) \end{pmatrix} = \begin{pmatrix} I & 0 \\ 0 & I \end{pmatrix} \tag{3.51}$$

be a doubly coprime embedding of the intertwining relation

$$E_i(z)F_i(z) = \overline{F}_i(z)\overline{E}_i(z), \tag{3.52}$$

with $E_i(z), \overline{F}_i(z)$ *left coprime and* $F_i(z), \overline{E}_i(z)$ *right coprime. Define a map* $Z_i :$ $X_{\overline{F}_i} \longrightarrow X_{E_i}$ *by letting, for* $f_i \in X_{\overline{F}_i}$, $Z_i f_i = \pi_{E_i} \overline{F}_i f_i$. *Then* Z_i *is an invertible map and its inverse is given by* $Z^{-1} : X_{E_i} \longrightarrow X_{\overline{F}_i}$, *defined by* $f_i = Z_i^{-1} g_i = \pi_{\overline{F}_i} \overline{X}_i g_i$.

Proof. Statements 1–4 follow from the definition of $D(z)$.

5. The assumption that the $E_i(z)$ are mutually left coprime implies that the $E_i^\top$ are mutually right coprime. By Theorem 3.35, we have the direct sum representation $X_{D^\top} = \oplus_{i=1}^s X_{E_i^\top}$ or, equivalently, that $X_{D^\top} = \oplus_{i=1}^s F_i^\top X_{E_i^\top}$. By duality considerations, we obtain (3.50).

6. The existence of the unimodular embedding (3.51) follows from the coprimeness conditions. The invertibility of the Z_i follows from Theorem 3.21. ∎

Theorem 3.42 (Chinese Remainder Theorem). *Let* $E_1(z), \ldots, E_s(z) \in \mathbb{F}[z]^{m \times m}$ *be mutually left coprime, nonsingular polynomial matrices, and let* $g_j(z) \in X_{E_j}$ *be polynomial vectors. Then there exists a polynomial vector* $f(z)$ *for which*

$$g_j = \pi_{E_j} f, \qquad j = 1, \ldots, s.$$

The polynomial vector $f(z)$ *is unique if it is assumed that* $f(z) \in X_D$, *where* $D(z)$ *is defined by (3.49).*

Proof. For $f(z) \in X_D$ there is a unique representation $f(z) = \sum_{j=1}^s \overline{F}_j f_j$ with respect to the direct sum (3.50). Applying the projection π_{E_i} and noting that, for $j \neq i$, E_i is a left factor of $\overline{F}_j$, one obtains

$$g_i = \pi_{E_i} f = \pi_{E_i} \sum_{j=1}^s \overline{F}_j f_j = \pi_{E_i} \overline{F}_i f_i = Z_i f_i \in X_{E_i}.$$

By inverting the maps Z_i, as in Proposition 3.41, the existence of $f(z)$ is proved.

To prove uniqueness, assume $f(z), g(z) \in X_D$ and $\pi_{E_j} f = \pi_{E_j} g$ for $j = 1, \ldots, s$. This implies

$$f(z) - g(z) \in \bigcap_{j=1}^{s} \operatorname{Ker} \pi_{E_{\mu_j}} = \bigcap_{j=1}^{s} E_{\mu_j} \mathbb{F}[z]^m = D\mathbb{F}[z]^m.$$

Thus $f(z) - g(z) \in X_D \cap D\mathbb{F}[z]^m = \{0\}$, which proves uniqueness. ∎

As a direct consequence of Theorem 3.42, we can state a polynomial matrix version of the Chinese remainder theorem. It should be emphasized that for $s = 1$ the result is a consequence of Theorem 2.31.

Theorem 3.43. *1. Let $E_i(z) \in \mathbb{F}[z]^{m \times m}$, $i = 1, \ldots, s$, be nonsingular and mutually left coprime, and let $D(z)$ be their l.c.r.m. Then, for polynomial matrices $A_i(z) \in \mathbb{F}[z]^{m \times n}$, $i = 1, \ldots, s$, such that $E_i(z)^{-1} A_i(z)$ is strictly proper, there exist polynomial matrices $A(z)$ and $B_i(z)$ in $\mathbb{F}[z]^{m \times n}$ such that*

$$A(z) = A_i(z) + E_i(z) B_i(z) \qquad i = 1, \ldots, s.$$

$A(z)$ is uniquely determined if one requires $D(z)^{-1} A(z)$ to be strictly proper.
2. Let $F_i(z) \in \mathbb{F}[z]^{m \times m}$, $i = 1, \ldots, s$, be nonsingular and mutually right coprime, and let $D(z)$ be their l.c.l.m. Then, for polynomial matrices $A_i(z) \in \mathbb{F}[z]^{n \times m}$, $i = 1, \ldots, s$, such that $A_i(z) F_i(z)^{-1}$ is strictly proper, there exist polynomial matrices $A(z)$ and $B_i(z) \in \mathbb{F}[z]^{m \times n}$ such that

$$A(z) = A_i(z) + B_i(z) F_i(z) \qquad i = 1, \ldots, s.$$

$A(z)$ is uniquely determined if it is required that $A(z) D(z)^{-1}$ be strictly proper.

Proof. Clearly, it suffices to prove the first claim of the theorem in the special case of $n = 1$. However, this is the content of Theorem 3.42. The second claim is proved in the same way. ∎

3.8 Toeplitz Operators

We present a brief study of the class of Toeplitz operators induced by rational symbols, a class that lends itself to an elegant algebraic treatment. It will become apparent, at a later stage in the study of feedback control, how important this class of operators is for the study of linear systems.

Our point of departure is the infinite system of linear discrete convolution type equations given by

$$\sum_{j=0}^{\infty} A_{i-j}\xi_j = \eta_i \qquad i \in \mathbb{N}. \tag{3.53}$$

We assume that the coefficients are $p \times m$ matrices with entries from a field $\mathbb{F}$ and that $\xi_j \in \mathbb{F}^m$ and $\eta_i \in \mathbb{F}^p$. Furthermore, it is assumed that the matrices A_i are the coefficients of an expansion of a rational function as a truncated Laurent series

$$A(z) = \sum_{i=-\infty}^{n_a} A_i z^i \in \mathbb{F}((z^{-1}))^{p \times m}.$$

This implies that if the sequence (ξ_i) has only a finite number of nonzero terms, then the same is true of the sequence (η_i). This allows us to interpret system (3.53) in functional terms. If the sequences (ξ_j) and (η_j) are identified with the polynomials $x(z) = \sum_j \xi_j z^j \in \mathbb{F}[z]^m$ and $y(z) = \sum_j \eta_j z^j \in \mathbb{F}[z]^p$, respectively, then one can rewrite (3.53) as

$$y = T_A x = \pi_+(Ax), \qquad x \in \mathbb{F}[z]^m,$$

where

$$T_A : \mathbb{F}[z]^m \longrightarrow \mathbb{F}[z]^p, \qquad T_A x = \pi_+(Ax) \tag{3.54}$$

denotes the **Toeplitz operator** with **symbol** $A(z) \in \mathbb{F}((z^{-1}))^{p \times m}$. The reason for this terminology stems from the fact that the matrix representation of (3.54) with respect to the standard basis $B_{\mathrm{st}} = \{z^i e_j\}$ of polynomials is a Toeplitz matrix. For example, if $A(z) = \sum_{j=-\infty}^{0} A_j z^j$, then the matrix representation of T_A is the infinite block upper-triangular Toeplitz matrix

$$\begin{pmatrix} A_0 & A_{-1} & \dots & A_{-N} & 0 & \dots \\ 0 & A_0 & A_{-1} & \dots & A_{-N} & \ddots \\ \vdots & \ddots & \ddots & \ddots & \ddots & \\ 0 & \dots & 0 & A_0 & A_{-1} & \dots \\ & \ddots & & \ddots & \ddots & \ddots \end{pmatrix}$$

with constant entries along the diagonals. Alternatively, a Toeplitz operator $\mathscr{T}_A$ on strictly proper power series is defined by

$$\mathscr{T}_A : z^{-1}\mathbb{F}[[z^{-1}]]^m \longrightarrow z^{-1}\mathbb{F}[[z^{-1}]]^p,$$

$$\mathscr{T}_A h = \pi_-(Ah), \qquad h(z) \in z^{-1}\mathbb{F}[[z^{-1}]]^m. \tag{3.55}$$

In the sequel, we will mainly work with the Toeplitz operator T_A, defined on spaces of polynomials, rather than with the Toeplitz operator $\mathcal{T}_A$, which is defined on spaces of strictly proper Laurent series.

Note that for polynomials $x(z) \in \mathbb{F}[z]^m$, the Toeplitz operator T_A satisfies

$$T_A(zx) = zT_A(x) + (Ax)_{-1}.$$

Due to the presence of the residue term $(Ax)_{-1}$, the Toeplitz operator is not an $\mathbb{F}[z]$-module homomorphism. Probably the main difficulty in the study of Toeplitz maps in our context is the fact that $\mathbb{F}[z]^m$ is not a module over the ring of (matrix) truncated Laurent series; hence, in general, the maps T_{AB} and $T_A T_B$ are different. This means that the set of Toeplitz maps does not have a simple algebraic structure. Still, something can be recovered owing to the fact that $\mathbb{F}[z]^m$ is a module over the rings $\mathbb{F}[z]^{m \times m}$ and $z^{-1}\mathbb{F}[[z^{-1}]]^p$ is a module over the ring $z^{-1}\mathbb{F}[[z^{-1}]]^{p \times p}$.

Recall that a unimodular matrix $U(z) \in \mathbb{F}[z]^{m \times m}$ defines an invertible element in $GL_m(\mathbb{F}[z])$. Similarly, every **biproper function** $\Gamma(z) \in \mathbb{F}[[z^{-1}]]^{m \times m}$, i.e., every proper formal power series in z^{-1} whose inverse is again a proper power series in z^{-1}, defines an invertible element in $GL_m(\mathbb{F}[[z^{-1}]])$. Note that $\Gamma(z) = \sum_{i=0}^{\infty} \Gamma_i z^{-i}$ is invertible if and only if Γ_0 is invertible.

Lemma 3.44. *Let $A(z) \in \mathbb{F}((z^{-1}))^{m \times m}$, $P(z) \in \mathbb{F}[z]^{m \times m}$, and $\Gamma(z) \in \mathbb{F}[[z^{-1}]]^{m \times m}$. Then:*

(a) The Toeplitz operator $T_{\Gamma A P}$ factors as

$$T_{\Gamma A P} = T_\Gamma T_A T_P;$$

(b) Assume $U(z) \in \mathbb{F}[z]^{m \times m}$ is unimodular, and $\Gamma(z) \in \mathbb{F}[[z^{-1}]]^{m \times m}$ is biproper. The Toeplitz operators T_U and T_Γ are invertible with inverses $T_U^{-1} = T_{U^{-1}}$ and $T_\Gamma^{-1} = T_{\Gamma^{-1}}$. Therefore, T_A is invertible if and only if $T_{\Gamma A U}$ is.

Proof. (a). The module $\mathbb{F}[z]^m$ is invariant under multiplication by polynomial matrices, whereas $\operatorname{Ker} \pi_+ = z^{-1}\mathbb{F}[[z^{-1}]]^m$ is invariant under multiplication by $\mathbb{F}[[z^{-1}]]^{m \times m}$ elements. Thus for $f \in \mathbb{F}[z]^m$ one obtains

$$T_\Gamma T_A T_P f = \pi_+ \Gamma \pi_+ A \pi_+ P f = \pi_+ \Gamma \pi_+ A P f = \pi_+ \Gamma A P f = T_{\Gamma A P} f.$$

(b). Part (a) implies that, for unimodular matrices $U(z), V(z) \in \mathbb{F}[z]^{m \times m}$ and biproper matrices $\Gamma(z), \overline{\Gamma}(z) \in \mathbb{F}[[z^{-1}]]^{m \times m}$, respectively, the identities $T_{UV} = T_U T_V$ and $T_{\Gamma\overline{\Gamma}} = T_\Gamma T_{\overline{\Gamma}}$ are satisfied. Since $T_I = \mathrm{Id}$ acts as the identity map, this implies that T_U and T_Γ are invertible with inverses $T_U^{-1} = T_{U^{-1}}$ and $T_\Gamma^{-1} = T_{\Gamma^{-1}}$, and the result follows. ∎

The previous simple, yet important, lemma indicates how Wiener–Hopf factorization theory is related to the problem of Toeplitz operator inversion. As a starting point, one attempts to simplify the problem of inversion through the reduction of $A(z)$ to some simple canonical form using the group of left multiplication

by biproper functions and right multiplication by unimodular functions, a group
operation introduced in Subsection 2.7. Clearly, if a symbol $G(z) \in \mathbb{F}((z^{-1}))^{m \times m}$
is singular, then the Toeplitz operator T_G has an infinite-dimensional kernel or
cokernel. Recall that the **cokernel**, coker f, of a linear map $f : X \longrightarrow Y$ is defined as
the quotient vector space $Y/\operatorname{Im} f$. The case of interest for us is that of a nonsingular
square rational $G(z) \in \mathbb{F}(z)^{m \times m}$. In this case, there exists a coprime factorization
$G(z) = D(z)^{-1} E(z)$, with $D(z)$ and $E(z)$ square, nonsingular polynomial matrices.
The following result is well known.

Proposition 3.45. *Let $G(z) \in \mathbb{F}(z)^{m \times m}$ be a nonsingular, rational matrix function,
and assume*

$$G(z) = G_-(z)\Delta(z)G_+(z)$$

is a left Wiener–Hopf factorization, with $\Delta(z) = \operatorname{diag}(z^{\mu_1}, \ldots, z^{\mu_m})$ and $\mu_1 \geq \cdots \geq \mu_m$. Then:

1. *For the Toeplitz operator T_G one has the following dimension formulas:*

$$\dim \operatorname{Ker} T_G = - \sum_{\mu_i < 0} \mu_i,$$

$$\dim \operatorname{coker} T_G = \operatorname{codim} \operatorname{Im} T_G = \sum_{\mu_i > 0} \mu_i;$$

2. *The Toeplitz operator T_G is invertible if and only if all left Wiener–Hopf
   factorization indices, with $\Delta(z) = \operatorname{diag}(z^{\mu_1}, \ldots, z^{\mu_m})$, are zero, i.e., $\mu_1 = \cdots = \mu_m = 0$. In this case, the factorization $G(z) = G_-(z)G_+(z)$ leads to the inversion
   formula*

$$T_G^{-1} f = G_+^{-1} \pi_+ (G_-^{-1} f), \qquad f \in \mathbb{F}[z]^m.$$

Proof. 1. In view of Lemma 3.44, it suffices to examine the invertibility of Toeplitz
operators with the symbol $\Delta(z) = \operatorname{diag}(z^{\mu_1}, \ldots, z^{\mu_m})$. By the diagonality of $\Delta(z)$,
T_Δ is the direct sum of Toeplitz operators of the form T_{z^κ}, and it suffices to study
those. If $\kappa \geq 0$, then $f \in \mathbb{F}[z]$ implies $T_{z^\kappa} = z^\kappa f(z)$, and so T_{z^κ} is injective with
a κ-dimensional complementary subspace, say, that of all polynomials of degree
$< \kappa$. Thus $\operatorname{codim} T_{z^\kappa} = k$. If $\kappa \leq 0$, then T_{z^κ} is surjective, with $\operatorname{Ker} T_{z^\kappa} = \{f \in \mathbb{F}[z] \mid \deg f < \kappa\}$. So $\dim \operatorname{Ker} T_{z^\kappa} = \kappa$, and one concludes that the invertibility of
T_Δ implies that all κ_j are zero. The converse is trivial.

2. If T_G is invertible, then $\operatorname{Ker} T_G = \{0\}$ and $\operatorname{Im} T_G = \mathbb{F}[z]^m$, which implies that
all left Wiener–Hopf factorization indices are zero. Conversely, assume that all
left Wiener–Hopf factorization indices are zero, which implies the factorization
$G(z) = G_-(z)G_+(z)$. Checking now, for $f \in \mathbb{F}[z]^m$,

$$G_+^{-1} \pi_+ G_-^{-1} \pi_+ G_- G_+ f = G_+^{-1} \pi_+ G_-^{-1} G_- G_+ f = G_+^{-1} \pi_+ G_+ f = f,$$

and, similarly,

$$\pi_+ G_- G_+ G_+^{-1} \pi_+ G_-^{-1} f = \pi_+ G_- \pi_+ G_-^{-1} f = \pi_+ G_- G_-^{-1} f = f.$$

In both computations, the invariance of $\operatorname{Ker} \pi_+$ under multiplication by proper functions was used.

∎

Corollary 3.46. *Let $G(z)$ be an $m \times m$ nonsingular rational matrix function. Then the Toeplitz operator $T_G : \mathbb{F}[z]^m \longrightarrow \mathbb{F}[z]^m$ is injective if and only if the Toeplitz operator $T_{G^{-\top}} : \mathbb{F}[z]^m \longrightarrow \mathbb{F}[z]^m$ is surjective.*

Proof. T_G is injective if and only if all left Wiener–Hopf indices $\mu_1 \geq \ldots \geq \mu_m$ of $G(z)$ are nonnegative. The left Wiener–Hopf indices of $G(z)^{-\top}$ are equal to $-\mu_m \geq \ldots \geq -\mu_1$. By Proposition 3.45, the cokernel of $T_{G^{-\top}}$ has dimension equal to the sum of the positive left Wiener–Hopf indices of $G(z)^{-\top}$, which is zero since $\mu_i \geq 0$ for all i. This completes the proof. ∎

As far as subsequent applications to linear systems theory are concerned, Toeplitz-induced operators, defined on quotient spaces of $\mathbb{F}[z]^m$ and, consequently, on polynomial models, play a most important role, for example, in state feedback and observer theory. Hence, these Toeplitz operators deserve special attention. We focus on the special situation where the symbol $G(z)$ of the Toeplitz operator is a nonsingular rational function, or, equivalently, $G(z)$ is the quotient of two nonsingular polynomial matrices.

Theorem 3.47. *Let $D(z), E(z) \in \mathbb{F}[z]^{m \times m}$ be nonsingular, and let $T_{DE^{-1}} : \mathbb{F}[z]^m \longrightarrow \mathbb{F}[z]^m$ be the Toeplitz operator with symbol $D(z)E(z)^{-1}$. Define a **Toeplitz induced operator** by*

$$\overline{T}_{DE^{-1}} : \mathbb{F}[z]^m / E(z)\mathbb{F}[z]^m \longrightarrow \mathbb{F}[z]^m / D(z)\mathbb{F}[z]^m$$

$$\overline{T}_{DE^{-1}}[f]_E = [T_{DE^{-1}}f]_D. \tag{3.56}$$

Then:

1. *The induced operator $\overline{T}_{DE^{-1}}$ is well defined;*
2. *$\overline{T}_{DE^{-1}}$ is injective if and only if $T_{DE^{-1}}$ is;*
3. *$\overline{T}_{DE^{-1}}$ is surjective if and only if $T_{DE^{-1}}$ is;*
4. *Operator (3.56) and the operator $\pi_D T_{DE^{-1}} : X_E \longrightarrow X_D$ are equivalent. Using the invertible maps $\overline{\pi}_E : \mathbb{F}[z]^m / E(z)\mathbb{F}[z]^m \longrightarrow X_E$, defined by $\overline{\pi}_E[f]_E = \pi_E f$, and $\overline{\pi}_D : \mathbb{F}[z]^m / D(z)\mathbb{F}[z]^m \longrightarrow X_D$, defined by $\overline{\pi}_D[f]_D = \pi_D f$, the equivalence is given by*

$$(\pi_D T_{DE^{-1}})\overline{\pi}_E = \overline{\pi}_D \overline{T}_{DE^{-1}}. \tag{3.57}$$

Proof. 1. Assume $[f_2]_E = [f_1]_E$, i.e., $f_1(z)$ and $f_2(z)$ are two representatives of the same equivalence class in $\mathbb{F}[z]^m/E(z)\mathbb{F}[z]^m$. This assumption implies $f_2(z) - f_1(z) = E(z)g(z)$ for some $g(z) \in \mathbb{F}[z]^m$. Computing

$$\overline{T}_{DE^{-1}}([f_2]_E - [f_1]_E) = [\pi_+ DE^{-1}(f_2 - f_1)]_D = [\pi_+ DE^{-1}Eg]_D = [Dg]_D = 0$$

shows that $\overline{T}_{DE^{-1}}$ is well defined.

2. Assume $T_{DE^{-1}}$ is injective. Let $[f]_E \in \operatorname{Ker} \overline{T}_{DE^{-1}} \subset \mathbb{F}[z]^m/E\mathbb{F}[z]^m$. This means that $T_{DE^{-1}}f \in D\mathbb{F}[z]^m$, or, for some $g(z) \in \mathbb{F}[z]^m$,

$$T_{DE^{-1}}f = Dg = DE^{-1}Eg = \pi_+ DE^{-1}Eg = T_{DE^{-1}}(Eg).$$

Thus $T_{DE^{-1}}(f - Eg) = 0$. Since $T_{DE^{-1}}$ is assumed to be injective, it follows that $f = Eg$ and, hence, $[f]_E = 0$. Conversely, assume $\overline{T}_{DE^{-1}}$ is injective. Let $f \in \operatorname{Ker} T_{DE^{-1}}$, which implies $\overline{T}_{DE^{-1}}[f]_E = [T_{DE^{-1}}f]_D = 0$. Since $\overline{T}_{DE^{-1}}$ is assumed to be injective, it follows that $f = Eg$ and, hence, $T_{DE^{-1}}Eg = Dg = 0$, implying that $g = 0$ and, hence, that $f = 0$.

3. Assume $T_{DE^{-1}}$ is surjective. Let $[g]_D \in \mathbb{F}[z]^m/D(z)\mathbb{F}[z]^m$. Since $T_{DE^{-1}}$ is surjective, there exists an $f \in \mathbb{F}[z]^m$ such that $T_{DE^{-1}}f = g$. This implies $\overline{T}_{DE^{-1}}[f]_E = [T_{DE^{-1}}f]_D = [g]_D$, i.e., $\overline{T}_{DE^{-1}}$ is surjective. Conversely, assume $\overline{T}_{DE^{-1}}$ is surjective. Let $g(z) \in \mathbb{F}[z]^m$; then there exists an $f \in \mathbb{F}[z]^m$ such that $\overline{T}_{DE^{-1}}[f]_E = [T_{DE^{-1}}f]_D = [g]_D$. This implies that $g - \pi_+ DE^{-1}f = Dg'$ for some $g' \in \mathbb{F}[z]^m$. Rewriting this as $g = \pi_+ DE^{-1}f + Dg' = \pi_+ DE^{-1}f + DE^{-1}Eg' = \pi_+ DE^{-1}(f + Eg')$ shows that $T_{DE^{-1}}$ is surjective.

4. The maps $\overline{\pi}_E$ and $\overline{\pi}_D$ are clearly well defined and invertible. The computation

$$\begin{aligned}(\pi_D T_{DE^{-1}})\overline{\pi}_E[f]_E &= \pi_D \pi_+ DE^{-1}\pi_E f = \pi_D \pi_+ DE^{-1}E\pi_- E^{-1}f \\ &= \pi_D \pi_+ D\pi_- E^{-1}f = \pi_D \pi_+ DE^{-1}f = \overline{\pi}_D[\pi_+ DE^{-1}f]_D \\ &= \overline{\pi}_D \overline{T}_{DE^{-1}}[f]_E\end{aligned}$$

proves (3.57).   ∎

The preceding result shows that the Toeplitz operator $T_{DE^{-1}}$, induced by a rational, nonsingular matrix $G(z) = D(z)E(z)^{-1}$, defines the Toeplitz induced operator on polynomial model spaces

$$\pi_D T_{DE^{-1}} : X_E \longrightarrow X_D,$$

$$\pi_D T_{DE^{-1}}(f) = D\pi_-\left(D^{-1}\pi_+(DE^{-1}f)\right) = D\pi^D(E^{-1}f).$$

The next proposition explores the connection between Toeplitz induced operators and projections and how these transform under duality.

Proposition 3.48. *Let $D(z), E(z) \in \mathbb{F}[z]^{m \times m}$ be nonsingular. Then:*

1. Let $\pi^D : z^{-1}\mathbb{F}[[z^{-1}]]^p \longrightarrow z^{-1}\mathbb{F}[[z^{-1}]]^p$ be the rational model projection defined by (3.35), i.e., for $h \in z^{-1}\mathbb{F}[[z^{-1}]]^p$, by $\pi^D h = \pi_- D^{-1}\pi_+ Dh$. Then the following diagram is commutative:

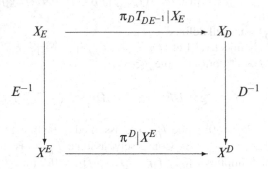

2. The dual map $(\pi^D|X^E)^* : X_{D^\top} \longrightarrow X_{E^\top}$ is $(\pi^D|X^E)^* = \pi_{E^\top}|X_{D^\top}$ and, similarly, $(\pi_D T_{DE^{-1}}|X_E)^* = \mathscr{T}_{E^{-\top}D^\top}|X^{D^\top}$. They satisfy the dual diagram

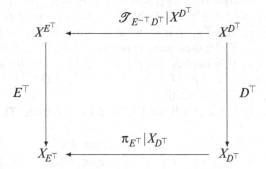

Proof. 1. Note that $f(z) \in X_E$ implies that $E^{-1}f \in X^E$ is strictly proper. Hence, one computes

$$(D^{-1}\pi_D T_{DE^{-1}})f = D^{-1}D\pi_- D^{-1}\pi_+ DE^{-1}f = (\pi^D|X^E)(E^{-1}f).$$

2. Computing, using the duality properties of Theorem 3.38, with $f(z) \in X_E$ and $h(z) \in X^{D^\top}$,

$$[\pi_D \pi_+ DE^{-1}f, h] = [D\pi_- D^{-1}\pi_+ DE^{-1}f, h] = [f, E^{-\top}D^\top \pi_- D^{-\top}\pi_+ D^\top h]$$

$$= [f, \pi_- E^{-\top}D^\top \pi^{D^\top} h] = [f, \mathscr{T}_{E^{-\top}\tilde{D}}\pi^{D^\top} h].$$

Similarly, for $f(z) \in X_{D^\top}$ and $h(z) \in X^E$, one computes

$$[\pi^D h, f] = [h, \pi_{D^\top} f] = [h, f] = [\pi^E h, f] = [h, \pi_{E^\top} f]. \qquad \blacksquare$$

In Theorem 3.47, it was shown that the invertibility properties of the Toeplitz operator $T_{DE^{-1}} : \mathbb{F}[z]^m \longrightarrow \mathbb{F}[z]^m$ are reflected by those of the induced operator $\overline{T}_{DE^{-1}} : \mathbb{F}[z]^m/E\mathbb{F}[z]^m \longrightarrow \mathbb{F}[z]^m/D\mathbb{F}[z]^m$. However, the invertibility properties of $T_{DE^{-1}}$ can be inferred directly from the Wiener–Hopf factorization indices of its symbol.

Theorem 3.49. *Let $D(z), E(z) \in \mathbb{F}[z]^{m \times m}$ be nonsingular. The following statements are equivalent.*

1. *All left Wiener–Hopf factorization indices of $D(z)E(z)^{-1}$ are nonnegative.*
2. *All left Wiener–Hopf factorization indices of $D(z)^{-\top}E(z)^{\top}$ are nonpositive.*
3. *The Toeplitz operator $T_{D^{-\top}E^{\top}} : \mathbb{F}[z]^m \longrightarrow \mathbb{F}[z]^m$ is surjective.*
4. *The Toeplitz operator $T_{DE^{-1}} : \mathbb{F}[z]^m \longrightarrow \mathbb{F}[z]^m$ is injective.*
5. *$\overline{T}_{DE^{-1}} : \mathbb{F}[z]^m/E\mathbb{F}[z]^m \longrightarrow \mathbb{F}[z]^m/D\mathbb{F}[z]^m$ is injective.*
6. *The induced Toeplitz map $\pi_D T_{DE^{-1}} : X_E \longrightarrow X_D$ is injective.*
7. *The reduced projection operator $\pi^D : X^E \longrightarrow X^D$ is injective.*
8. *The reduced projection operator $\pi_{E^{\top}} : X_{D^{\top}} \longrightarrow X_{E^{\top}}$ is surjective.*
9. *The reduced projection operator $\pi_D : X_E \longrightarrow X_D$ is injective.*
10. *The equality $X_E \cap D\mathbb{F}[z]^m = \{0\}$ is true.*
11. *The codimension formula*

$$\mathrm{codim}\,(X_{D^{\top}} \cap E(z)^{\top}\mathbb{F}[z]^m) = \deg \det E(z) \tag{3.58}$$

is satisfied.

Proof. (1)$\Leftrightarrow$(2) Note that, for an invertible $G(z)$, the left factorization indices of $G(z)$ are the right factorization indices of $G^{\top}(z)$, which in turn are the negatives of the left factorization indices of $G^{-T}(z)$.

(2) $\Leftrightarrow$ (3) Follows from Proposition 3.45.

(3) $\Leftrightarrow$ (4) The equivalence follows from Corollary 3.46.

(4) $\Leftrightarrow$ (5) Follows from Theorem 3.47.2.

(5) $\Leftrightarrow$ (6) Follows from Theorem 3.47.4.

(6) $\Leftrightarrow$ (7) Follows from Proposition 3.48.

(7) $\Leftrightarrow$ (8) Follows by duality because the dual of the projection operator $\pi_{E^{\top}}$ reduced to $X_{D^{\top}}$ is the projection π^D reduced to X^E.

(8) $\Leftrightarrow$ (9) Follows from the fact that $\mathrm{Ker}\,(\pi_D|X_E) = X_E \cap D\mathbb{F}[z]^m$.

(9) $\Leftrightarrow$ (10) Follows from the identity $\mathrm{Ker}\,(\pi_D|X_E) = X_E \cap D\mathbb{F}[z]^m$.

(8) $\Leftrightarrow$ (11) Clearly, $\mathrm{Ker}\,(\pi_{E^{\top}}|X_{D^{\top}}) = X_{D^{\top}} \cap E^{\top}\mathbb{F}[z]^m$. The identity $\dim \mathscr{X} = \dim \mathrm{Ker}\,A + \dim \mathrm{Im}\,A$ is valid for all linear transformations A on a finite-dimensional vector space $\mathscr{X}$. Hence, it follows that

$$\dim(X_{D^{\top}} \cap E^{\top}\mathbb{F}[z]^m) = \dim X_{D^{\top}} - \dim \mathrm{Im}\,(\pi_{E^{\top}}|X_{D^{\top}}).$$

Thus, if $\pi_{E^{\top}}|X_{D^{\top}}$ is surjective, then $\dim \mathrm{Im}\,(\pi_{E^{\top}}\|X_{D^{\top}}) = \deg \det E^{\top}(z)$. Hence, $\deg \det E^{\top}(z) = \dim X_{D^{\top}} - \dim X_{D^{\top}} \cap E^{\top}\mathbb{F}[z]^m = \mathrm{codim}\,(X_{D^{\top}} \cap E^{\top}\mathbb{F}[z]^m)$, i.e., the codimension formula (3.58) is satisfied. The argument is clearly reversible. ∎

Our attention turns now to a special, albeit important, case.

Proposition 3.50. *Let* $D(z), E(z) \in \mathbb{F}[z]^{m \times m}$ *be nonsingular, with* $D(z)E(z)^{-1}$ *biproper. Then:*

1. *The polynomial models* $X_{D^\top}$ *and* $X_{E^\top}$ *contain the same elements;*
2. *The Toeplitz operator* $T_{ED^{-1}} : \mathbb{F}[z]^m \longrightarrow \mathbb{F}[z]^m$ *is invertible and induces an invertible map* $\overline{T}_{ED^{-1}} : \mathbb{F}[z]^m/D\mathbb{F}[z]^m \longrightarrow \mathbb{F}[z]^m/E\mathbb{F}[z]^m$ *defined by* $\overline{T}_{ED^{-1}}[f] = [T_{ED^{-1}}f]$;
3. *The adjoint of* $\overline{T}_{DE^{-1}} : X_E \longrightarrow X_D$, *defined in (3.56), is* $\overline{T}^*_{DE^{-1}} : X_{D^\top} \longrightarrow X_{E^\top}$ *given, for* $f(z) \in X_{D^\top}$, *by*

$$\overline{T}^*_{DE^{-1}}f = f. \qquad (3.59)$$

Proof. 1. Clearly, $D(z)E(z)^{-1}$ is biproper if and only if $E(z)^{-\top}D^\top(z)$ is. Let $f(z) \in X_{D^\top}$; then $D^{-\top}f$ is strictly proper and, hence, also $E^{-\top}D^\top D^{-\top}f = E^{-\top}f$ is, i.e., $f(z) \in X_{E^\top}$. This shows that $X_{D^\top} \subset X_{E^\top}$. The opposite inclusion follows by symmetry.

2. That $T_{DE^{-1}}$ is invertible is a special case of Lemma 3.44 or, alternatively, of Proposition 3.45.

3. For $f(z) \in X_E$ and $g(z) \in X_{D^\top}$, one computes

$$< \overline{T}_{DE^{-1}}f, g > = [D^{-1}\pi_D\pi_+DE^{-1}f, g] = [D^{-1}D\pi_-D^{-1}\pi_+DE^{-1}f, g]$$

$$= [\pi_-D^{-1}\pi_+DE^{-1}f, g] = [E^{-1}f, D^\top\pi_-D^{-\top}g] = [E^{-1}f, \pi_{D^\top}g]$$

$$= [E^{-1}f, g] = < f, g >,$$

and (3.59) follows.   ∎

3.9 Exercises

1. Let $D(z) = \begin{pmatrix} 1 & z+z^3 \\ z & 1-z \end{pmatrix}$. Characterize all elements of X_D, find a basis, and obtain a corresponding matrix representation of the shift operator S_D.

2. Let $\mathbb{F}$ be a field of characteristic zero, and assume that $d(z) \in \mathbb{F}[z]$ is a monic polynomial of degree n, which is irreducible. Show

   (a) X_d is a field extension of $\mathbb{F}$ of degree n;
   (b) The trace form $\beta : X_d \times X_d \longrightarrow \mathbb{F}$

$$\beta(f, g) = \operatorname{tr} m(fg)$$

   is a nondegenerate bilinear form, where $m(a) : X_d \longrightarrow X_d$ denotes the $\mathbb{F}$-linear multiplication operator $m(a)b = \pi_d(ab)$, $a, b \in X_d$.

3. Show that a nonzero $\mathbb{F}[z]$-homomorphism $Z : X_{d_1} \longrightarrow X_{d_2}$ exists if and only if the monic polynomials $d_1(z), d_2(z)$ have a nontrivial common factor.

4. Show that $\mathrm{Hom}_{\mathbb{F}[z]}(X_{d_1}, X_{d_2})$ is a finite-dimensional $\mathbb{F}$-vector space, and compute its dimension.

5. Do the same for the space of module homomorphisms $\mathrm{Hom}_{\mathbb{F}[z]}(X_{D_1}, X_{D_2})$ for nonsingular polynomial matrices $D_1(z) \in \mathbb{F}[z]^{m \times m}$ and $D_2(z) \in \mathbb{F}[z]^{p \times p}$.
   [Hint: Use the Smith form!]

6. Prove that $D(z) = A_0 z^2 + A_1 z + A_2 \in \mathbb{F}[z]^{m \times m}$, with A_0 invertible, is polynomially equivalent to

$$\overline{D}(z) = \begin{pmatrix} A_2 + zA_1 & zA_0 \\ zA_0 & -A_0 \end{pmatrix}.$$

Construct an $\mathbb{F}[z]$-module isomorphism $Z : X_D \longrightarrow X_{\overline{D}}$. What happens if A_0 is singular?

7. Let $D(z) = A_d z^d + \cdots + A_0 \in \mathbb{F}[z]^{m \times m}$ be a nonsingular matrix polynomial, with $A_d \neq 0$. Show that every vector of polynomials $f(z) = \mathrm{col}(f_1(z), \cdots, f_m(z)) \in \mathbb{F}[z]^m$, with $D(z)^{-1} f(z)$ strictly proper, satisfies $\deg f_j(z) < d$ for $j = 1, \ldots, m$.

8. Let A_i, $i = 1, 2$, be cyclic matrices with cyclic vectors b_i. Show that $\begin{pmatrix} A_1 & 0 \\ 0 & A_2 \end{pmatrix}$
   is cyclic with cyclic vector $\begin{pmatrix} b_1 \\ b_2 \end{pmatrix}$ if and only if the characteristic polynomials $d_i(z)$ are coprime.
   [Hint: Use the Chinese remainder theorem.]

9. Let $N_2(z) \in \mathbb{F}[z]^{m_2 \times m_1}$. Show that

$$N_2(z)M_1(z)\mathbb{F}[z]^{k_1} \subset M_2(z)\mathbb{F}[z]^{k_2}$$

if and only if there exists an $N_1(z) \in \mathbb{F}[z]^{k_2 \times k_1}$ such that

$$N_2(z)M_1(z) = M_2(z)N_1(z).$$

10. Let $D(z), E(z) \in \mathbb{F}[z]^{m \times m}$ be nonsingular. Prove:

    (a) There exists an injective $\mathbb{F}[z]$-homomorphism $Z : X_D \longrightarrow X_E$ if and only if the invariant factors of $D(z)$ divide those of $E(z)$;

    (b) There exists a surjective $\mathbb{F}[z]$-homomorphism $Z : X_D \longrightarrow X_E$ if and only if the invariant factors of $E(z)$ divide those of $D(z)$.

11. Let $D(z)$ be an $m \times m$ nonsingular polynomial matrix. Show that the polynomial model X_D is isomorphic to the polynomial model X_{UDV} for all $m \times m$ unimodular matrices $U(z), V(z)$.

12. Let $D_1(z), D_2(z) \in \mathbb{F}[z]^{m \times m}$ be nonsingular and $D(z) = D_1(z)D_2(z)$. Assume further that $D_2(z)^{-1}$ is proper. Then:

(a) $D_1(z)D_2(z)$ and

$$D_{12} = \begin{pmatrix} D_1(z) & 0 \\ -I & D_2(z) \end{pmatrix}$$

are polynomially equivalent;

(b) Define a map $Z : X_{D_1 D_2} \longrightarrow X_{D_{12}}$ given by

$$Zf = \pi_{D_{12}} \begin{pmatrix} I \\ 0 \end{pmatrix} f.$$

The map Z is an $\mathbb{F}[z]$-isomorphism that preserves the two direct sum decompositions $X_{D_1 D_2} = X_{D_1} \oplus D_1 X_{D_2}$ and

$$X_{D_{12}} = X_{\begin{pmatrix} D_1(z) & 0 \\ -I & I \end{pmatrix}} \oplus \begin{pmatrix} D_1(z) & 0 \\ -I & I \end{pmatrix} X_{\begin{pmatrix} I & 0 \\ 0 & D_2(z) \end{pmatrix}}.$$

13. Let $D(z)$ be an $m \times m$ nonsingular polynomial matrix that is row proper and has row indices $v_1, \ldots, v_p$. Show that

$$X_D = \{f(z) \in \mathbb{F}[z]^m \mid f(z) = (f_1(z), \ldots, f_m(z))^\top, \deg f_i < v_i\}.$$

Conclude that the sum of the row indices is equal to $\deg \det D(z)$.

14. Let $D_1(z) \in \mathbb{F}[z]^{p \times p}$ and $D_2(z) \in \mathbb{F}[z]^{m \times m}$ be nonsingular polynomial matrices. Define the **tensored polynomial model** $X_{D_1 \otimes D_2}$ as the set of all matrix polynomials $F(z) \in \mathbb{F}[z]^{p \times m}$ such that $D_1(z)F(z)D_2(z)^{-1}$ is strictly proper. Show that $X_{D_1 \otimes D_2}$ is an $\mathbb{F}$-vector space of dimension $(\deg \det D_1) \cdot (\deg \det D_2)$.

15. Prove that the adjoint $\overline{T}^*_{DE^{-1}} : X^{D^\top} \longrightarrow X^{E^\top}$ of the induced Toeplitz operator $\overline{T}_{DE^{-1}}$ satisfies

$$\overline{T}^*_{DE^{-1}} = \mathcal{T}_{E^{-\top}D^\top} | X^{D^\top},$$

where $\mathcal{T}_{E^{-\top}D^\top} : z^{-1}\mathbb{F}[[z^{-1}]]^m \longrightarrow z^{-1}\mathbb{F}[[z^{-1}]]^m$ is the Toeplitz operator defined in (3.55).

16. Prove that the Toeplitz operator $\mathcal{T}_{ED^{-1}} : z^{-1}\mathbb{F}[[z^{-1}]]^m \longrightarrow z^{-1}\mathbb{F}[[z^{-1}]]^m$ is injective if and only if all left Wiener–Hopf indices of $D(z)E(z)^{-1}$ are nonnegative.

17. Let $A : \mathcal{X} \longrightarrow \mathcal{X}$ and $W, Z : \mathcal{X} \longrightarrow \mathcal{Y}$ be $\mathbb{F}$-linear transformations of vector spaces $\mathcal{X}$ and $\mathcal{Y}$. Prove that the following statements are equivalent.

(a)

$$A \operatorname{Ker} W \subset \operatorname{Ker} Z.$$

(b)

$$\operatorname{Ker} W \subset \operatorname{Ker} ZA.$$

(c) There exists a linear transformation $F : \mathcal{Y} \longrightarrow \mathcal{Y}$ such that

$$ZA = FW.$$

3.10 Notes and References

Shift operators were introduced into dynamical systems starting with the classical work by Poincaré and Birkhoff. An early origin arose in the study of dynamics around a periodic orbit by the discrete dynamics of the Poincaré map. In his masterful paper Smale (1967), Smale advocated the use of shift operators as a systematic tool for dynamic systems. Subsequently all these developments led to what is today known as symbolic dynamics. For a beautiful account of symbolic dynamics in connection with coding theory, see Lind and Marcus (1995).

The introduction of polynomial and rational models in an algebraic framework is due to Fuhrmann (1976). The motivation for this arose out of the successful theory of functional models for the study of linear transformations in a Hilbert space. This operator-theoretic work started with the pathbreaking work by Livsic (1966). The classic papers by Beurling (1949) and Lax (1959) on the parameterization of shift- and translation-invariant subspaces in terms of inner functions and Rota (1960) on the universality of shift operators were very influential. Further motivation for this line of research was provided by Masani and Wiener on stochastic processes Helson (1964), the physics of wave scattering as in Lax and Phillips (1967), as well as operator-theoretic contributions by Sz.-Nagy and Foias (1970) and de Branges and Rovnyak (1986). For a comprehensive discussion of the shift operator in the functional analytic context, see Nikolski (1986).

Paraphrasing Wigner (1960), a natural question arises regarding the unreasonable effectiveness of polynomial models in systems theory. The principal reason for that is its role as the lingua franca of systems theory. Before its introduction, the area of systems and control was more or less split into several separate subareas according to the mathematical techniques employed, with very limited communication among the different areas. The schism between state-space and frequency-domain techniques was bridged with polynomial-model-based realization theory, generalized isomorphism theorems, and extensions of strict system equivalence. This was followed by the establishment of a connection with geometric control as well as, somewhat later, with behavioral theory. One factor that was useful in this unification process was the fact that polynomial model theory occupies a middle ground; it uses richer structures than those used by matrix manipulations, yet it is more concrete than abstract module theory. Because its roots lie in operator theory, the language and results obtained by polynomial methods provide an ideal starting point for infinite-dimensional generalizations.

Theorems 3.20 and 3.22 are essential to the usefulness of polynomial models in algebraic systems theory, and they have numerous applications, for example, isomorphism theorems for various system representations, inversion of finite Hankel and Toeplitz matrices, computation of minimal open-loop controls, problems of state reconstruction, the study of generalized Bezoutians, and (tangential) interpolation, to name a few. Both theorems were proved in Fuhrmann (1976). Theorem 3.20 is the algebraic version of the celebrated commutant lifting theorem proved, in the context of operator theory in Hilbert spaces, by Sarason (1967) in the scalar case and by Sz.-Nagy and Foias (1968) in the general case. It should be noted that Sarason's work was motivated by the Nevanlinna-Pick interpolation. In turn, Theorem 3.22, characterizing the invertibility of intertwining maps, is adapted from the papers by Fuhrmann (1968a,b). The usefulness of doubly coprime factorizations is stressed in Vidyasagar (1987). Duality theory for polynomial models was first developed in Fuhrmann (1981) and has been extended to a far-reaching duality theory for multidimensional behaviors by Oberst (1990).

The Chinese remainder theorem has a long history, with modern expositions in, for example, Lang (1965), and Fuhrmann (1983). It bears, by its very construction, a close connection with interpolation theory (Hermite, Lagrange, Newton) if rings of functions are considered. This connection between the Chinese remainder theorem and scalar polynomial interpolation was apparently first observed by M. Riesz Schoenberg (1987). For an algebraic analysis of scalar interpolation, refer to van der Waerden (1949) and Fuhrmann (2012). The matrix version of the Chinese remainder theorem and the multivariable partial fraction decomposition were proved in Fuhrmann (1983).

Note that the idea of weakening the classification of polynomial matrices up to unimodular equivalence by enlarging the matrices via suitably sized identity matrices originates from algebraic K-theory, where it is called stabilization (a concept that has nothing in common with the system-theoretic term of feedback stabilization). In the systems theory context, it was first noticed by Rosenbrock (1970) that introducing this notion of equivalence is a sensible idea.

Chapter 4
Linear Systems

This chapter introduces basic definitions and concepts from linear systems theory. Our focus is on questions of existence, uniqueness, and minimality of state-space realizations and the important issue of system equivalence. We discuss various representations of linear systems, develop state-space realization theory, and provide characterizations of reachability and observability. The approach taken is developed from a coherent point of view that is well known from algebraic systems theory, i.e., the use of functional models and shift operators. This approach will also prove to be very useful in later chapters when we examine interconnected systems.

Linear systems have several different representations. Certain problems may have easier solutions in one representation than in one of the others. Thus, it is best to have a good grasp of the various different representations of linear systems. Therefore, the beginning of this chapter is devoted to several formal definitions of various representations of linear systems. Since most of our emphasis is on algebraic aspects of the theory, it seems best not to restrict the field, and as a consequence we will deal mostly with discrete time systems over a field $\mathbb{F}$. This has the advantage that the achieved results are applicable to closely related research areas such as, for example, convolutional codes and network coding. Our focus will be restricted to finite-dimensional linear systems. From an internal point of view, one encounters first-order state-space representations and higher-order or generalized differential systems. On the other hand, one can take an external point of view, where the interest lies in the interplay between inputs and outputs. This leads to the study of input/output maps and transfer function matrices. Rational matrix transfer functions have a variety of representations, each one carrying its own system-theoretic interpretation. Thus, matrix fractions appear throughout the text and occupy a central part in the analysis of linear systems. The input/output information of the system can be encoded in a corresponding Hankel matrix, which turns out to be particularly well adapted for certain applications. The connecting link between the internal and external points of view is made via the magnificent structure of realization theory. The most general approach to modeling systems

© Springer International Publishing Switzerland 2015
P.A. Fuhrmann, U. Helmke, *The Mathematics of Networks of Linear Systems*, Universitext, DOI 10.1007/978-3-319-16646-9_4

that encompasses all previous ones is that of behavioral representations, where no distinction is made between inputs and outputs. A comprehensive account of the theory of behaviors is given in Polderman and Willems (1997). Although it seems tempting to approach interconnected systems from a behavioral point of view, it is not so obvious how to express in this framework potentially one-sided interconnections in terms of graphs. For the purpose of this book, our preference is to focus on input/output representations because in that case the interconnection structures of the networks can be easily expressed in terms of weighted digraphs. Consequently, behaviors are only occasionally mentioned in passing.

4.1   System Representations

A. Input/Output Representations. From a classical systems theory point of view, a linear system is regarded as a linear operator that transforms input sequences into output sequences. This point of view, in conjunction with the associated state-space formalism, was first pronounced by R. Kalman in the 1960s and was formalized using the language of rings and modules in the very influential textbook by Kalman (1969). Mostly, we shall restrict ourselves to discrete-time systems.

Our discussion begins with some general comments concerning bi-infinite sequences $\xi = (\xi_t)_{t \in \mathbb{Z}}$ of vectors ξ_t in a vector space $\mathcal{V}$. The **past** and **future** of ξ are the subsequence defined by $\xi_+ = (\xi_t)_{t \leq 0}$ and $\xi_- = (\xi_t)_{t > 0}$, respectively. Thus $t = 0$ is regarded as the present. Arbitrarily, for mathematical convenience, the present is adjoined to the past. A sequence ξ is called an **admissible sequence** if there exists an integer N such that $\xi_t = 0$ for all $t \leq N$. Thus, admissible sequences are those that have only a finite past. Applying the Z-transform, one identifies an admissible sequence $\xi = (\xi_t)_{t \in \mathbb{Z}}$ with the Laurent series

$$\sum(z) = \sum_{t=-\infty}^{\infty} \xi_t z^{-t}.$$

Under this identification, the space of all admissible bi-infinite sequences in a vector space $\mathcal{V}$ becomes the space of all Laurent series $\mathcal{V}((z^{-1}))$. Similarly, the set of all past and future sequences is identified with the space of the polynomials $\mathcal{V}[z]$ and strictly proper formal power series $z^{-1}\mathcal{V}[[z^{-1}]]$, respectively. This is expressed by the direct sum decomposition

$$\mathcal{V}((z^{-1})) = \mathcal{V}[z] \oplus z^{-1}\mathcal{V}[[z^{-1}]]$$

together with canonical linear projection maps

$$\pi_+ : \mathcal{V}((z^{-1})) \longrightarrow \mathcal{V}[z], \quad \pi_- : \mathcal{V}((z^{-1})) \longrightarrow z^{-1}\mathcal{V}[[z^{-1}]]$$

onto the past and future subspaces, respectively. The space $\mathcal{V}((z^{-1}))$ comes equipped with a natural $\mathbb{F}[z]$-module structure, defined by the ordinary multiplication of a Laurent series $\xi(z) \in \mathcal{V}((z^{-1}))$ with a polynomial $p(z) \in \mathbb{F}[z]$. In particular, the past subspace $\mathcal{V}[z]$ is a submodule of $\mathcal{V}((z^{-1}))$. In this framework, the **inputs** and **outputs** of a system are defined as Laurent series $u(z) \in \mathbb{F}((z^{-1}))^m$ and $y(z) \in \mathbb{F}((z^{-1}))^p$, respectively.

Definition 4.1. A **linear discrete-time, time-invariant, input/output system** is a map

$$L_G : \mathbb{F}((z^{-1}))^m \longrightarrow \mathbb{F}((z^{-1}))^p, \quad L_G(u) = y(z) := G(z)u(z)$$

defined for a Laurent series $G(z) \in \mathbb{F}((z^{-1}))^{p \times m}$. The system is called **causal** if $G(z) \in \mathbb{F}[[z^{-1}]]^{p \times m}$ and **strictly causal** if $G(z) \in z^{-1}\mathbb{F}[[z^{-1}]]^{p \times m}$. The operator L_G is referred to as the **Laurent operator**, with the symbol G.

Note that a Laurent operator L_G acts as a multiplication operator with the symbol G and therefore satisfies $L_G(zh(z)) = zL_G(h)$. In particular, a Laurent operator L_G is an $\mathbb{F}[z]$-linear homomorphism (however, not every such homomorphism between spaces of Laurent series is a Laurent operator!).

Thus, an input/output system is simply a linear convolution operator between inputs and outputs,

$$y_t = \sum_{k=-\infty}^{\infty} G_k u_{t-k}, \quad t \in \mathbb{Z}. \tag{4.1}$$

Note that the series in (4.1) is indeed a finite sum since both $G(z)$ and $u(z)$ are assumed to be Laurent series. For causal systems, the representation (4.1) simplifies to

$$y_t = \sum_{k=0}^{\infty} G_k u_{t-k}, \quad t \in \mathbb{Z},$$

which implies that the output at time t depends only on input values at times $\leq t$.

In the sequel, the causality of the Laurent operator will always be assumed. In that situation, there is a much cleaner way to describe the action of a system – by expressing the effect of past inputs on future outputs. This leads to the definition of Hankel operators.

Definition 4.2. Let $G(z) \in \mathbb{F}((z^{-1}))^{p \times m}$. Define the **restricted input/output map**, or the associated **Hankel operator** H_G, as the $\mathbb{F}[z]$-module homomorphism

$$H_G : \mathbb{F}[z]^m \longrightarrow z^{-1}\mathbb{F}[[z^{-1}]]^p, \quad H_G(f) = \pi_-(G(z)f(z)).$$

Note that the Hankel operator H_G does not depend on the polynomial part of the symbol $G(z)$. The following result shows that a strictly causal input/output operator is uniquely determined from the Hankel operator. In particular, the action of a strictly causal input/output system is completely encoded in the action of the Hankel operator on past input sequences.

Proposition 4.3. *Two strictly causal transfer functions* $G_1(z), G_2(z) \in z^{-1}$ $\mathbb{F}[[z^{-1}]]^{p \times m}$ *are equal if and only if the Hankel operators* H_{G_1}, H_{G_2} *coincide.*

Proof. Assume $H_{G_1} = H_{G_2}$. This implies that $\pi_-((G_1(z) - G_2(z))f(z)) = 0$ for all polynomials $f(z) \in \mathbb{F}[z]^m$. Because $G_1(z) - G_2(z)$ is strictly proper, $(G_1(z) - G_2(z))\xi = 0$ for all $\xi \in \mathbb{F}^m$. Thus $G_1(z) = G_2(z)$ follows. The converse is clear. ∎

B. State-Space Representations. In practice, a discrete-time linear system is often not given by an input/output operator but rather by a system of difference equations containing input and output variables. The simplest situation is encountered using first-order difference equations. The input and output spaces $\mathscr{U}, \mathscr{Y}$ will be identified with $\mathbb{F}^m, \mathbb{F}^p$, respectively. Since much of our analysis will use functional spaces, a similar identification for the state space will be avoided.

Definition 4.4. A discrete-time, linear, state-space system Σ is a quadruple of $\mathbb{F}$-linear maps (A, B, C, D), with $A : \mathscr{X} \longrightarrow \mathscr{X}$, $B : \mathbb{F}^m \longrightarrow \mathscr{X}$, $C : \mathscr{X} \longrightarrow \mathbb{F}^p$, and $D : \mathbb{F}^m \longrightarrow \mathbb{F}^p$. The quadruple of maps represents the system of equations

$$
\begin{aligned}
x_{t+1} &= Ax_t + Bu_t, \quad t \in \mathbb{Z}, \\
y_t &= Cx_t + Du_t.
\end{aligned}
\tag{4.2}
$$

The **system dimension**, $\dim \Sigma$, is defined by

$$
\dim \Sigma = \dim \mathscr{X}.
$$

The spaces $\mathscr{X}$, $\mathscr{U} = \mathbb{F}^m$, and $\mathscr{Y} = \mathbb{F}^p$ are referred to as the **state space**, **input space**, and **output space**, respectively. In the sequel, we will assume $\mathscr{X}$ to be finite dimensional. In that case we will refer to Σ as a **finite-dimensional linear system**.

An explicit representation of an input/output system in the form (4.2) is called a **state-space description**. Of course, once a state-space description is given and initial conditions specified, the choice of inputs completely determines the choice of outputs. To describe the input/output operator associated with (4.2), it will be found both convenient, as well as reasonable, to assume that the system was at rest in the remote past and that at some point in time a sequence of inputs is applied, bringing the system to, so to speak, life. Thus, one considers u_t, x_t, and y_t to be the input, state, and output, respectively, at time t and assumes that for some $N \in \mathbb{N}$ the state satisfies $x_t = 0$ for $t \leq -N$. To the sequence $\{u_t\}_{t=-N}^{\infty}$ one associates the truncated Laurent series $u(z) = \sum_{t=-N}^{\infty} u_t z^{-t}$. Analogously one defines $x(z)$ and $y(z)$. Thus, powers of z serve as time markers. Clearly, the present corresponds to $t = 0$. A state trajectory $x(z) = \sum_{t=-N}^{\infty} x_t z^{-t}$ decomposes naturally according to the direct sum

$$\mathscr{X}((z^{-1})) = \mathscr{X}[z] \oplus z^{-1}\mathscr{X}[[z^{-1}]]$$

as

$$x(z) = \sum_{t \geq 0} x_{-t}z^t + \sum_{t=1}^{\infty} x_t z^{-t} = (\pi_+ x)(z) + (\pi_- x)(z).$$

To get the input/output relations from the system equations (4.2), one multiplies the equations by appropriate powers of z and sums up, using $x_{-N} = 0$, to obtain

$$z\sum_{t=-N}^{\infty} x_t z^{-t} = \sum_{t=-N}^{\infty} x_{t+1}z^{-t} = A\sum_{t=-N}^{\infty} x_t z^{-t} + B\sum_{t=-N}^{\infty} u_t z^{-t},$$

$$\sum_{t=-N}^{\infty} y_t z^{-t} = C\sum_{t=-N}^{\infty} x_t z^{-t} + D\sum_{t=-N}^{\infty} u_t z^{-t},$$

i.e.,

$$\begin{cases} zx(z) = Ax(z) + Bu(z), \\ \\ y(z) = Cx(z) + Du(z). \end{cases} \tag{4.3}$$

Conversely, equation (4.3) is equivalent to (4.2), with the initial state condition $x_{-N} = 0$ enforced. Under the assumptions of zero initial states, the analysis of the linear system of equations (4.3) on $\mathscr{X}((z^{-1}))$ is equivalent to analyzing the state-space system (4.2).

To obtain the input/output relations, one must solve system (4.2). This leads to

$$x_t = \sum_{j=0}^{\infty} A^j B u_{t-j-1}, \quad t \in \mathbb{N}.$$

Here, the summation is well defined because there is only a finite number of inputs u_j that are nonzero for negative indices, and hence only finitely many nonzero terms occur in the summation. The input/output relations induced by the system can therefore be written in the form of a discrete convolution product,

$$y_t = Du_t + \sum_{j=0}^{\infty} CA^j B u_{t-j-1} = \sum_{j=0}^{\infty} G_j u_{t-j},$$

where

$$G_j = \begin{cases} D & j = 0, \\ CA^{j-1}B & j > 0. \end{cases}$$

This input/output operator is causal because the output at time t depends only on input values u_j, with $j \leq t$. The input/output relation depends on the quadruple (A,B,C,D) only through the maps D and $\{CA^jB\}_0^\infty$, which are called the **Markov parameters** of the system.

Eliminating the state trajectory from equation (4.3) leads to the input/output relation $y(z) = G(z)u(z)$, where

$$G(z) = D + C(zI - A)^{-1}B \in \mathbb{F}[[z^{-1}]]^{p \times m} \tag{4.4}$$

is the **transfer function** of the system. The suggestive notation

$$G(z) = \left[\begin{array}{c|c} A & B \\ \hline C & D \end{array}\right] \tag{4.5}$$

will be used to denote the transfer function defined by (A,B,C,D). Clearly, the map from input functions to output functions is multiplication by $G(z)$, i.e., it is the Laurent operator $L_G : \mathbb{F}((z^{-1}))^m \longrightarrow \mathbb{F}((z^{-1}))^p$ with the symbol $G(z)$.

Every (A,B,C,D) satisfying (4.4) is called a **state-space realization** of $G(z)$. Note that system (4.2) determines a unique transfer function and, hence, the input/output relations. However, the converse is not true. A transfer function $G(z)$ may have several different realizations. This will be elaborated on when we discuss realization theory.

C. Polynomial Matrix Descriptions. In general, when modeling a complex system such as a large electrical network, one first writes down all possible equations, which may be of high order, and proceeds to eliminate the redundant variables. With this in mind, Rosenbrock initiated the **polynomial matrix descriptions** of systems, or what is sometimes also referred to as **generalized descriptor systems**. These representations are in the form of higher-order difference equations,

$$\sum_{j=0}^{M} T_j \xi_{t+j} = \sum_{i=0}^{N} U_i u_{t+i},$$

$$y_t = \sum_{j=0}^{K} V_j \xi_{t+j} + \sum_{i=0}^{L} W_i u_{t+i}, \quad t \in \mathbb{Z}, \tag{4.6}$$

specified by the polynomial matrices $T(z) = \sum_{j=0}^{M} T_j z^j$, $U(z), V(z)$, and $W(z)$. Here again we assume that the system has been at rest in the past, i.e., that $\xi_t = 0$ for sufficiently negative t. In terms of the Z-transforms of the variables ξ, u, y, system (4.6) becomes equivalent to the equations on spaces of Laurent series as described by

$$T(z)\xi(z) = U(z)u(z),$$

$$y(z) = V(z)\xi(z) + W(z)u(z); \tag{4.7}$$

the only assumption being made is that the polynomial matrix $T(z)$ is nonsingular. The interpretation of (4.7) is that of a higher-order difference equation. In such a system interpretation one may look at the variables ξ as **internal**, or **latent variables**, which are sometimes referred to as **partial states**. Here, $\xi(z) \in \mathbb{F}((z^{-1}))^n, u(z) \in \mathbb{F}((z^{-1}))^m$ and $y(z) \in \mathbb{F}((z^{-1}))^p$ are assumed to be Laurent series in z^{-1}. With a polynomial matrix description (4.7), we associate a polynomial matrix $\mathscr{P}(z)$ given by

$$\mathscr{P}(z) = \begin{pmatrix} T(z) & -U(z) \\ V(z) & W(z) \end{pmatrix},$$

which will be referred to as the **polynomial system matrix** of system (4.7). By eliminating the auxiliary variable ξ from equations (4.7), the transfer function of the system is easily computed to be

$$G(z) = V(z)T(z)^{-1}U(z) + W(z).$$

For example, to the state-space system (4.2) corresponds the polynomial system matrix

$$\mathscr{P}(z) = \begin{pmatrix} zI - A & -B \\ C & D \end{pmatrix}.$$

Similarly, the matrix fraction representations of a strictly proper rational function, $G(z) = N_r(z)D_r(z)^{-1} = D_\ell(z)^{-1}N_\ell(z)$, yields the polynomial system matrices

$$\begin{pmatrix} D_r(z) & -I \\ N_r(z) & 0 \end{pmatrix}, \quad \begin{pmatrix} D_\ell(z) & -N_\ell(z) \\ I & 0 \end{pmatrix},$$

respectively.

Of course, the system represented by (4.7) is causal if and only if its transfer function is proper. One should note, however, that a polynomial matrix description of a system transcends the boundaries of input/output systems. To take an extreme case, we consider, for a nonsingular polynomial matrix $T(z)$, the polynomial system matrix $(T(z))$. This describes a nonsingular, autonomous system of higher-order difference equations. The existence of nontrivial solutions depends on the spaces. Considering the Laurent operator $L_T : \mathbb{F}((z^{-1}))^d \longrightarrow \mathbb{F}((z^{-1}))^d$, the assumed nonsingularity of $T(z)$ implies $\operatorname{Ker} L_T = \{0\}$. However, if the action of the Toeplitz operator $\mathscr{T}_T = T(\sigma) : z^{-1}\mathbb{F}[[z^{-1}]]^d \longrightarrow z^{-1}\mathbb{F}[[z^{-1}]]^d$ in $z^{-1}\mathbb{F}[[z^{-1}]]^d$ is considered, one has $\operatorname{Ker} T(\sigma) = X^T$; hence there exists a finite-dimensional solution space.

D. Behaviors. A generalization of the preceding system concepts is the notion of behaviors due to J.C. Willems. We refer the reader to Polderman and Willems (1997) for an account of this theory. The behavioral approach to systems theory allows for a very satisfactory treatment of higher-order polynomial matrix representations, even in the absence of inputs or outputs of the system. Thus, the behavioral approach to systems seems particularly well suited to treating interconnected systems, i.e., to the subject of this book. Although there are thus very clear advantage points for discussing behaviors within the context of interconnected systems, we do refrain in the sequel from such analysis and focus instead on the restricted class of input/output models. The main reason for this lies both in the ubiquity of input/output representations and in our ability to present a very satisfactory functional model approach.

For an $\mathbb{F}$-vector space W, let $W^{\mathbb{N}} = \prod_{\mathbb{N}} W$ denote the infinite-dimensional vector space of all infinite sequences $(w_t)_{t \in \mathbb{N}}$. The space $\mathcal{W}$ will be identified with the space of strictly proper Laurent series $z^{-1}W[[z^{-1}]]$ equipped with the $\mathbb{F}[z]$-module structure induced by the **backward shift operator**

$$\sigma : z^{-1}W[[z^{-1}]] \longrightarrow z^{-1}W[[z^{-1}]], \quad \sigma w = \pi_-(zw(z)). \qquad (4.8)$$

A subspace $\mathcal{B} \subset z^{-1}W[[z^{-1}]]$ is called **shift-invariant** if

$$\sigma\mathcal{B} \subset \mathcal{B}.$$

For $N \in \mathbb{N}$ let $P_N : z^{-1}W[[z^{-1}]] \longrightarrow z^{-1}W[[z^{-1}]]$ denote the linear projection operator defined by

$$P_N\left(\sum_{j=0}^{\infty} \frac{w_j}{z^{j+1}}\right) = \sum_{j=0}^{N} \frac{w_j}{z^{j+1}}.$$

A linear subspace $\mathcal{B}$ is called **complete** if every infinite sequence $w = (w_t)_{t \in \mathbb{N}}$ in $z^{-1}W[[z^{-1}]]$ satisfies

$$w \in \mathcal{B} \iff P_N(w) \in P_N(\mathcal{B}) \quad \forall N \in \mathbb{N}. \qquad (4.9)$$

The general definition of a behavior follows, adapted to the case of discrete-time linear systems on the nonnegative time axis.

Definition 4.5. A **discrete-time linear behavior** with time axis $\mathbb{N}$, or a **linear shift space**, is a pair $\Sigma = (W, \mathcal{B})$, where the signal space W is a finite-dimensional $\mathbb{F}$-vector space and $\mathcal{B}$ is a linear shift-invariant subspace of $z^{-1}W[[z^{-1}]]$ that is complete. A behavior is called **autonomous** whenever $\mathcal{B}$ is finite-dimensional, as an $\mathbb{F}$-vector space.

Thus a behavior with time axis $\mathbb{N}$ is simply an $\mathbb{F}[z]$-submodule of $z^{-1}W[[z^{-1}]]$ that satisfies the completeness condition (4.9). The following result characterizes autonomous behaviors as rational models of nonsingular polynomial models.

Theorem 4.6. *An $\mathbb{F}$-linear subspace $\mathcal{B} \subset z^{-1}\mathbb{F}[[z^{-1}]]^m$ is an autonomous behavior if and only if there exists a nonsingular polynomial matrix $D(z) \in \mathbb{F}[z]^{m \times m}$, with*

$$\mathcal{B} = X^D := \operatorname{Ker} D(\sigma).$$

Proof. A submodule $\mathscr{B}$ of $z^{-1}\mathbb{F}[[z^{-1}]]^m$ is a finite-dimensional $\mathbb{F}$-vector space if and only if $\mathscr{B}$ is a finitely generated torsion submodule of $z^{-1}\mathbb{F}[[z^{-1}]]^m$. By Theorem 3.36, this is equivalent to $\mathscr{B}$ having the representation $\mathscr{B} = \text{Ker } D(\sigma)$ for a nonsingular polynomial matrix $D(z) \in \mathbb{F}[z]^{m \times m}$. This completes the proof. ∎

In greater generality, the main structural result from behavior theory asserts that each discrete-time linear behavior can be identified with the kernel of a suitable Toeplitz operator. For a full row rank polynomial matrix $V(z) \in \mathbb{F}[z]^{m \times (m+p)}$, the rational model is defined by

$$X^V = \text{Ker } V(\sigma) \subset z^{-1}\mathbb{F}[[z^{-1}]]^{m+p}.$$

Thus X^V is the submodule of $z^{-1}\mathbb{F}[[z^{-1}]]^{m+p}$, defined by all strictly proper Laurent series $h(z) \in z^{-1}\mathbb{F}[[z^{-1}]]^{m+p}$ such that $V(z)h(z) \in \mathbb{F}[z]^m$ is a polynomial. Moreover, the elements $h(z) = \sum_{j=0}^{\infty} w_j z^{-j-1}$ of Ker $V(\sigma)$ correspond exactly to the solutions of the system of difference equations

$$\sum_{j=0}^{s} V_j w_{j+t} = 0, \quad t = 0, 1, 2, \ldots.$$

It is easy to see that for a full row rank polynomial matrix $V(z) \in \mathbb{F}[z]^{m \times (m+p)}$ the functional model

$$\mathscr{B} := \text{Ker } V(\sigma) \subset z^{-1}\mathbb{F}[[z^{-1}]]^{m+p} \tag{4.10}$$

defines a behavior. The description (4.10) is called an **autoregressive representation** or, equivalently, an **AR representation** of $\mathscr{B}$. One can show that two full row rank polynomial matrices of the *same size* $V_1(z), V_2(z) \in \mathbb{F}[z]^{m \times (m+p)}$ are AR representations of the same behavior $\mathscr{B} = \text{Ker } V_1(\sigma) = \text{Ker } V_2(\sigma)$ if and only if there exists a unimodular matrix $U(z) \in \mathbb{F}[z]^{m \times m}$ such that $V_2(z) = U(z)V_1(z)$.

The following theorem is the main structural result of behavior theory, as developed by J.C. Willems. The proof, which is a consequence of duality theory, is omitted; see Willems (1986, 1991) and Fuhrmann (2002).

Theorem 4.7 (J.C. Willems, P.A. Fuhrmann).

1. *The discrete-time linear behaviors $\mathscr{B} \subset z^{-1}\mathbb{F}[[z^{-1}]]^{m+p}$ are exactly the submodules of the form Ker $V(\sigma)$, where $V(z) \in \mathbb{F}[z]^{m \times (m+p)}$ is an arbitrary polynomial matrix of full row rank.*
2. *Let $V_i(z) \in \mathbb{F}[z]^{m_i \times (m_i+p_i)}, i = 1, 2$, be full row rank polynomial matrices. The behaviors Ker $V_i(\sigma)$, $i = 1, 2$, are equivalent by a continuous $\mathbb{F}[z]$-isomorphism if and only if $m_1 - p_1 = m_2 - p_2$ and $V_1(z)$ and $V_2(z)$ have the same nontrivial invariant factors.*

This result has some nice consequences, one of which is mentioned here.

Corollary 4.8. *Let $\mathscr{B} \subset z^{-1}\mathbb{F}[[z^{-1}]]^{m+p}$ be a discrete-time linear behavior. The following conditions are equivalent:*

1. *$\mathscr{B}$ is autonomous.*
2. *$\mathscr{B}$ is finitely generated.*
3. *$\mathscr{B}$ is a torsion module.*

In particular, the torsion elements of a behavior $\mathscr{B}$ form a subbehavior $\mathscr{B}_a$, called the autonomous part of $\mathscr{B}$.

Proof. If $\mathscr{B}$ is a behavior, then there exists a full row rank polynomial matrix $V(z) \in \mathbb{F}[z]^{m\times(m+p)}$, with $\mathscr{B} = \operatorname{Ker} V(\sigma)$. Without loss of generality, one can assume that $V(z) = (D(z),0)$ is in Smith normal form, with $D(z) = \operatorname{diag}(d_1(z),\ldots,d_m(z))$. Split the elements $h(z)$ of $z^{-1}\mathbb{F}[[z^{-1}]]^{m+p}$ into components $h = (h_1(z),h_2(z))$, with $h_1(z) \in z^{-1}\mathbb{F}[[z^{-1}]]^m, h_2(z) \in z^{-1}\mathbb{F}[[z^{-1}]]^p$. It follows that $h \in \mathscr{B}$ if and only if $d_1(\sigma)h_1 = 0,\ldots,d_m(\sigma)h_m = 0$. Therefore, $\mathscr{B}$ is autonomous if and only if $p = 0$. This completes the proof. ∎

Why is behavior theory indeed a generalization of linear systems theory? To illustrate this point, consider linear state-space systems of the form

$$x_{t+1} = Ax_t + Bu_t \quad t \geq 0, \quad x_0 = 0. \tag{4.11}$$

Let $\mathscr{B} \subset z^{-1}\mathbb{F}[[z^{-1}]]^{n+m}$ be the set of pairs of strictly proper Laurent series

$$w(z) = \begin{pmatrix} x(z) \\ u(z) \end{pmatrix},$$

with $u(z) = \sum_{j=0}^{\infty} u_j z^{-j-1}, x(z) = \sum_{j=0}^{\infty} x_j z^{-j-1}$, subject to (4.11). Then $\mathscr{B}$ is identical to the behavior described as

$$\mathscr{B} = \operatorname{Ker}(\sigma I - A, B).$$

More generally, consider the polynomial matrices $N_\ell(z) \in \mathbb{F}[z]^{p\times m}, D_\ell(z) \in \mathbb{F}[z]^{p\times p}$, with $D_\ell(z)$ nonsingular. The coefficients of

$$w(z) = \begin{pmatrix} \xi(z) \\ u(z) \end{pmatrix},$$

with $u(z) = \sum_{j=0}^{\infty} u_j z^{-j-1}, \xi(z) = \sum_{j=0}^{\infty} \xi_j z^{-j-1}$, then satisfy the higher-order difference system

$$\sum_{j=0}^{s} D_j \xi_{t+j} = \sum_{i=0}^{r} N_i u_{t+i}, \quad t \geq 0, \tag{4.12}$$

if and only if $D_\ell(z)\xi(z) = N_\ell(z)u(z)$ is satisfied. Thus the solutions to (4.12) define the behavior

$$\mathcal{B} = \mathrm{Ker}\ (D_\ell(\sigma), -N_\ell(\sigma)).$$

One can therefore identify, in a very natural way, the trajectories of a linear input-state system with the elements of a suitably defined behavior.

4.2 Reachability and Observability

The basic notions in the theory of linear control systems are those of reachability and observability. For linear state-space systems they were first introduced by Kalman (1969). The definitions for linear, time-invariant, discrete-time systems follow.

Definition 4.9. 1. A linear system (4.2) is called **reachable** if, for each state vector $x \in \mathcal{X}$, there exists $T \in \mathbb{N}$ and a finite input sequence $u_0, \ldots, u_T$ such that the induced state trajectory (x_t) with initial condition $x_0 = 0$ satisfies $x_T = x$.
2. A linear system (4.2) is called **observable** if, given two state trajectories $(x_t), (\bar{x}_t)$ with the same input sequence (u_t), the condition

$$Cx_t + Du_t = C\bar{x}_t + Du_t \quad \forall t$$

implies $(x_t) = (\bar{x}_t)$.

Intuitively, the observability of a linear system refers to the property whereby the state trajectory of a system can be reconstructed from knowledge of the input and output sequences u_t and y_t, respectively. Using linearity, the observability condition is seen as being equivalent to the condition (with $u_t = 0$)

$$CA^t x_0 = 0 \quad \forall t \geq 0 \implies x_0 = 0.$$

There exist simple equivalent reformulations of these notions via suitable input/output operators. Recall that the polynomial model X_{zI-A} of the linear operator $A : \mathcal{X} \longrightarrow \mathcal{X}$ coincides with $\mathcal{X}$. Moreover, the canonical projection map is $\pi_{zI-A} : \mathbb{F}[z]^m \longrightarrow X_{zI-A} = \mathcal{X}$, with $\pi_{zI-A}(f) = (zI - A)\pi_-((zI - A)^{-1}f)$.

Definition 4.10. Consider a discrete-time linear system (A, B, C, D) with state space $\mathcal{X}$ and input and output space $\mathbb{F}^m$ and $\mathbb{F}^p$, respectively.

(a) The **reachability map** of (A, B, C, D) is the $\mathbb{F}[z]$-linear map

$$\mathcal{R}_{(A,B)} : \mathbb{F}[z]^m \longrightarrow \mathcal{X}, \quad \mathcal{R}_{(A,B)}(u) = \pi_{zI-A}(Bu(z)). \tag{4.13}$$

(b) The **module of zero return** is the kernel of $\mathscr{R}_{(A,B)}$, i.e., the submodule

$$\mathfrak{M}_{(A,B)} = \{u(z) \in \mathbb{F}[z]^m \mid \pi_-((zI-A)^{-1}Bu(z)) = 0\}.$$

(c) The **observability map** of (A,B,C,D) is the $\mathbb{F}[z]$-linear map

$$\mathscr{O}_{(C,A)} : \mathscr{X} \longrightarrow z^{-1}\mathbb{F}[[z^{-1}]]^p, \quad \mathscr{O}_{(C,A)}(\xi) = C(zI-A)^{-1}\xi. \tag{4.14}$$

With this definition, we state the following useful characterizations of reachability and observability.

Theorem 4.11. *1. The image space* $\mathrm{Im}\,\mathscr{R}_{(A,B)} \subset \mathscr{X}$ *of a reachability map consists of all states in* $\mathscr{X}$ *that are reachable from the zero state. System (4.2) is reachable if and only if the reachability map (4.13) is surjective.*
2. Let $N(z) \in \mathbb{F}[z]^{n \times m}$ *and* $D(z) \in \mathbb{F}[z]^{m \times m}$ *be right coprime, with* $D(z)$ *nonsingular and*

$$(zI-A)N(z) = BD(z).$$

The submodule $\mathfrak{M}_{(A,B)}$ *has the representation*

$$\mathfrak{M}_{(A,B)} = D(z)\mathbb{F}[z]^m$$

and therefore is a full submodule. Moreover, the map

$$Z : X_D \longrightarrow \mathrm{Im}\,\mathscr{R}_{(A,B)}, \; Zf = \pi_{zI-A}(Bf) \tag{4.15}$$

defines an $\mathbb{F}[z]$-*linear isomorphism.*
3. System (4.2) is observable if and only if the observability map (4.14) is injective.
4. The Hankel operator of a proper transfer function $G(z) = C(zI-A)^{-1}B+D \in \mathbb{F}[[z^{-1}]]^{p \times m}$ *factorizes as*

$$H_G = \mathscr{O}_{(C,A)} \circ \mathscr{R}_{(A,B)}.$$

Proof. 1. For every polynomial $u(z) = \sum_{j=0}^{s} u_j z^j$, one has $\pi_{zI-A}(Bu(z)) \in X_{zI-A} = \mathscr{X}$. Moreover, $\pi_{zI-A}(z^j f(z)) = A^j \pi_{zI-A}(f)$ for all $j \geq 0$. Therefore,

$$\pi_{zI-A}(Bu(z)) = \sum_{j=0}^{s} \pi_{zI-A}(z^j Bu_j) = \sum_{j=0}^{s} \pi_{zI-A}(z^j Bu_j) = \sum_{j=0}^{s} A^j Bu_j.$$

For the initial condition $x_0 := 0$ and after $T \geq 1$ steps, the iteration $x_{j+1} := Ax_j + Bv_j$ satisfies $x_T = \sum_{j=0}^{T-1} A^j Bv_{T-j-1}$. This implies that for the polynomial $u(z) = \sum_{j=0}^{T-1} v_{T-1-j} z^j$ one obtains $x_T = \pi_{zI-A}(Bu(z))$. Hence, the set of states x that are reachable in finitely many steps from x_0 coincides with the image space of $\mathscr{R}_{(A,B)}$.

In conclusion, an input sequence $u_0, \ldots, u_{T-1}$, with associated input polynomial $u(z) = \sum_{j=0}^{T-1} u_j z^j$, steers x_0 into x if and only if the **reflected polynomial**

$$\mathbf{u}_T(z) := \sum_{j=0}^{T-1} u_{T-1-j} z^j = z^{T-1} u\left(\frac{1}{z}\right)$$

satisfies $x = \pi_{zI-A}(B\mathbf{u}_T(z))$.

2. By definition, the module of zero return is

$$\mathfrak{M}_{(A,B)} = \{u(z) \in \mathbb{F}[z]^m \mid \exists x(z) \in \mathcal{X}[z] \quad \text{with} \quad (zI - A)x(z) + Bu(z) = 0\}.$$

Let $N(z)D(z)^{-1} = (zI - A)^{-1}B$ be a right coprime factorization. Then $u(z) \in \mathfrak{M}_{(A,B)}$ if and only if $N(z)D(z)^{-1}u(z)$ is a polynomial. By Lemma 2.28, this is equivalent to $u(z) \in D(z)\mathbb{F}[z]^m$. This shows that $\mathfrak{M}_{(A,B)} = D(z)\mathbb{F}[z]^m$. Thus the reachability map induces a module isomorphism of $\mathbb{F}[z]^m/D(z)\mathbb{F}[z]^m$ onto the submodule $\operatorname{Im}\mathscr{R}_{(A,B)}$ of X_{zI-A}.

3. Observability is satisfied if and only if the infinite set of linear equations $CA^t x_0 = 0$, $t \in \mathbb{N}, \ldots$, implies $x_0 = 0$. Thus the claim follows from the identity

$$C(zI - A)^{-1}\xi = \sum_{j=0}^{\infty} \frac{CA^j \xi}{z^{j+1}}, \quad \xi \in \mathcal{X}.$$

4. For $u(z) \in \mathbb{F}[z]^m$ one computes

$$\mathcal{O}_{CA}(\mathscr{R}_{AB}(u)) = C(zI - A)^{-1}\pi_{zI-A}(Bu(z)) = \pi_-(C\pi_-((zI - A)^{-1}Bu(z)))$$

$$= \pi_-(C(zI - A)^{-1}Bu(z)) - \pi_-(C\pi_+((zI - A)^{-1}Bu(z)))$$

$$= H_G(u).$$

∎

As a trivial consequence, we obtain the familiar rank conditions by Kalman and Hautus for the reachability and observability of finite-dimensional systems. Let

$$\mathscr{R}_n(A,B) = (B, AB, \cdots, A^{n-1}B)$$

and

$$\mathcal{O}_n(C,A) = \begin{pmatrix} C \\ CA \\ \vdots \\ CA^{n-1} \end{pmatrix}$$

be the length-n reachability and observability matrices, respectively.

Corollary 4.12. *Let (A, B, C, D) be linear, with state space $\mathcal{X} = \mathbb{F}^n$.*

1. *The following conditions are equivalent:*

   (a) *The pair (A, B) is reachable.*
   (b) rank $\mathscr{R}_n(A, B) = n$.
   (c) $zI - A$ *and B are left coprime.*
   (d) rank $[zI - A, B] = n$ *for all $z \in \overline{\mathbb{F}}$.*

2. *The following conditions are equivalent:*

   (a) *The pair (C, A) is observable.*
   (b) rank $\mathscr{O}_n(C, A) = n$.
   (c) C *and $zI - A$ are right coprime.*
   (d) *For all $z \in \overline{\mathbb{F}}$,*

$$\text{rank} \begin{pmatrix} C \\ zI - A \end{pmatrix} = n.$$

Proof. We focus on reachability; the proof for observability runs in exactly the same manner. By the Cayley–Hamilton theorem, the image space of $\mathscr{R}_n(A, B)$ coincides with the span of all vectors of the form $A^k Bu$, $k \in \mathbb{N}, \in \mathbb{F}^m$, i.e., with the set of all states reachable from zero. This proves the equivalence of (a) with (b). Condition (d) is simply the standard rank conditions for left coprimeness. Thus it remains to show the equivalence with (c). Recall the intertwining relation $BD(z) = (zI - A)N(z)$, with $N(z), D(z)$ right coprime. Thus $Z : X_D \longrightarrow X_{zI-A}$ defined in (4.15) is a module isomorphism if and only if $(zI - A), B$ are left coprime. This completes the proof. ∎

Condition (a) in the preceding corollary is called the **Kalman rank condition**, while (c) is referred to as the **Hautus criterion**.

4.3 Abstract Realization Theory

Realization theory was introduced by R.E. Kalman and has become a cornerstone of linear systems theory because it enabled, for the first time, a systematic approach to modeling and system identification. The reader is referred to the book by Kalman (1968) for the first comprehensive exposition on the subject. The fundamental concepts in the theory are those of reachability, observability, and the shift realization of a transfer function. Starting with an abstract approach that connects state-space realizations with canonical factorizations of the Hankel operator, one moves on to general shift realizations via coprime factorizations.

In the preceding section it was shown that every state-space system (A,B,C,D) defines a proper transfer function

$$G(z) = D + C(zI - A)^{-1}B.$$

Realization theory is concerned with the inverse question of associating a state-space realization (A,B,C,D) to a proper transfer function $G(z)$. Here, our focus is on the construction of factorizations of Hankel operators. In later subsections, it will be shown how to construct state-space realizations from such factorizations. Our starting point is a general statement concerning Hankel operators. Let $G(z) \in \mathbb{F}[[z^{-1}]]^{p \times m}$ denote a proper matrix Laurent series, and let $H_G : \mathbb{F}[z]^m \longrightarrow z^{-1}\mathbb{F}[[z^{-1}]]^p$ denote the associated **Hankel operator**

$$H_G(f) = \pi_-(G(z)f(z)),$$

The following result characterizes kernel and image spaces of Hankel operators in terms of the coprime factorizations of its symbol.

Theorem 4.13 (Structure Theorem on Hankel Operators). *Consider a strictly proper formal power series* $G(z) \in z^{-1}\mathbb{F}[[z^{-1}]]^{p \times m}$.

1. *A map* $H : \mathbb{F}[z]^m \longrightarrow z^{-1}\mathbb{F}[[z^{-1}]]^p$ *is an* $\mathbb{F}[z]$-*homomorphism if and only if there exists* $G(z) \in z^{-1}\mathbb{F}[[z^{-1}]]^{p \times m}$, *with* $H = H_G$.
2. *The kernel* $\operatorname{Ker} H_G$ *is a finitely generated submodule of* $\mathbb{F}[z]^m$, *and the image* $\operatorname{Im} H_G$ *is a finitely generated* $\mathbb{F}[z]$-*submodule of* $z^{-1}\mathbb{F}[[z^{-1}]]^p$.
3. *The following statements are equivalent:*

   (a) *The kernel* $\operatorname{Ker} H_G$ *is a full submodule of* $\mathbb{F}[z]^m$.
   (b) *The Hankel operator* H_G *has finite rank, i.e.,* $\operatorname{Im} H_G$ *is a finite-dimensional subspace of* $z^{-1}\mathbb{F}[[z^{-1}]]^p$.
   (c) $G(z)$ *is rational.*

4. *Let* $G(z) \in z^{-1}\mathbb{F}[[z^{-1}]]^{p \times m}$ *be rational. Let* $N_r(z), N_\ell(z) \in \mathbb{F}[z]^{p \times m}$ *and* $D_r(z) \in \mathbb{F}[z]^{m \times m}, D_\ell(z) \in \mathbb{F}[z]^{p \times p}$ *be polynomial matrices that define right and left coprime factorizations*

$$N_r(z)D_r(z)^{-1} = G(z) = D_\ell(z)^{-1}N_\ell(z),$$

respectively. The kernel and image of the Hankel operator have respectively the following representations:

$$\operatorname{Ker} H_G = D_r(z)\mathbb{F}[z]^m$$

and

$$\operatorname{Im} H_G = X^{D_\ell}.$$

Proof. 1. Every Hankel operator H_G is $\mathbb{F}[z]$-linear. Conversely, if $H : \mathbb{F}[z]^m \longrightarrow z^{-1}\mathbb{F}[[z^{-1}]]^p$ is an $\mathbb{F}[z]$-homomorphism, then one constructs a matrix of formal power series $G(z) \in z^{-1}\mathbb{F}[[z^{-1}]]^{p \times m}$ whose ith column is equal to $H(e_i), i = 1, \ldots, m$. Since H is a module homomorphism, $H(e_i z^j) = \pi_-(z^j G(z) e_i)$ for all i and $j \geq 0$. This shows that $H(f) = \pi_-(Gf)$ for all polynomials $f(z) \in \mathbb{F}[z]^m$.

2. This follows easily from the preceding statement by noting that the image of a finitely generated module under a module homomorphism is finitely generated.

3. The Hankel operator H_G induces an isomorphism $\overline{H_G} : \mathbb{F}[z]^m / \mathrm{Ker}\, H_G \longrightarrow \mathrm{Im}\, H_G$ of $\mathbb{F}[z]$-modules and, hence, of vector spaces. The quotient module $\mathbb{F}[z]^m / \mathrm{Ker}\, H_G$ is finite-dimensional if and only if $\mathrm{Ker}\, H_G$ is a full submodule. This shows the equivalence of (a) and (b). Moreover, (b) implies that the finitely generated $\mathbb{F}[z]$-module $\mathrm{Im}\, H_G$ is finite-dimensional and, therefore, a torsion submodule of $z^{-1}\mathbb{F}[[z^{-1}]]^p$. From Theorem 3.36 one concludes that there exists a nonsingular $p \times p$ polynomial matrix $\overline{D}(z)$, with $\mathrm{Im}\, H_G = X^{\overline{D}}$. Thus, for each standard basis vector $e_i \in \mathbb{F}^m$, one obtains

$$G(z)e_i = H_G(e_i) \in X^{\overline{D}}, \quad i = 1, \ldots, m.$$

Since $X^{\overline{D}}$ consists of rational functions, this shows that $G(z)$ is rational. Conversely, if $G(z)$ is rational, then there exist left and right coprime factorizations $G(z) = N(z)D(z)^{-1} = \overline{D}(z)^{-1}\overline{N}(z)$, respectively. For $g \in \mathbb{F}[z]^m$ the polynomial $f = Dg$ satisfies $H_G(f) = \pi_-(ND^{-1}f) = \pi_-(Ng) = 0$, and therefore $D(z)\mathbb{F}[z]^m \subset \mathrm{Ker}\, H_G$. Since $D(z)$ is nonsingular, $D(z)\mathbb{F}[z]^m$ is a full submodule of $\mathbb{F}[z]^m$, and therefore $\mathrm{Ker}\, H_G$ is full, implying that $\mathrm{Im}\, H_G$ is finite-dimensional.

4. It was shown earlier that $D_r(z)\mathbb{F}[z]^m \subset \mathrm{Ker}\, H_G$. Thus, let $f(z) \in \mathrm{Ker}\, H_G$, i.e., $\pi_-(G(z)f(z)) = 0$. Thus there exists a polynomial $g(z)$ with $N_r(z)D_r(z)^{-1}f(z) = g(z)$. By Lemma 2.28, one concludes that $f(z) \in D_r(z)\mathbb{F}[z]^m$. This shows the inclusion $\mathrm{Ker}\, H_G \subset D_r(z)\mathbb{F}[z]^m$ and, therefore, the equality $\mathrm{Ker}\, H_G = D_r(z)\mathbb{F}[z]^m$.

Similarly, using the established relations between polynomial and rational models from Chapter 3, one obtains, for $f(z) \in \mathbb{F}[z]^m$,

$$H_G(f) = \pi_-(Gf) = \pi_-(D_\ell^{-1}N_\ell f) = D_\ell^{-1}\pi_{D_\ell}(N_\ell f) \in D_\ell^{-1}X_{D_\ell} = X^{D_\ell}.$$

This shows the inclusion $\mathrm{Im}\, H_G \subset X^{D_\ell}$. Conversely, for $h \in X^{D_\ell}$, $f(z) = D_\ell(z)h(z)$ is a polynomial. By the left coprimeness of D_ℓ and N_ℓ, there exist polynomial matrix solutions $A(z), B(z)$ to the Bezout equation

$$D_\ell(z)A(z) + N_\ell(z)B(z) = I_p.$$

Thus $h(z) = D_\ell(z)^{-1}f(z) = A(z)f(z) + G(z)B(z)f(z)$, which implies

$$h(z) = \pi_-(h(z)) = \pi_-(A(z)f(z) + G(z)B(z)f(z)) = \pi_-(G(z)B(z)f(z)) \in \mathrm{Im}\, H_G.$$

This completes the proof.   ∎

A **state-space realization** of a Hankel operator H_G is a triple (A, B, C) of $\mathbb{F}$-linear maps

$$A : \mathcal{X} \longrightarrow \mathcal{X}, \quad B : \mathbb{F}^m \longrightarrow \mathcal{X}, \quad C : \mathcal{X} \longrightarrow \mathbb{F}^p$$

such that $G(z) = C(zI - A)^{-1}B$. Each such realization defines a factorization of the Hankel operator H_G

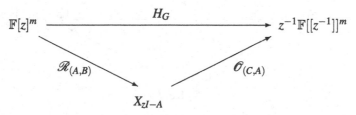

as a product of the reachability and observability operators, respectively. Notice that A acts as the shift operator on $X_{zI-A} = \mathcal{X}$. Therefore, the system (A, B, C) can be reconstructed from the factorization by the formulas

$$Ax = S_{zI-A}x, \quad x \in \mathcal{X},$$

$$Bu = \pi_{zI-A}(Bu) = \mathcal{R}_{(A,B)}(u), \quad u \in \mathbb{F}^m,$$

$$Cx = (C(zI - A)^{-1}x)_{-1} = \mathcal{O}_{(C,A)}x, \quad x \in \mathcal{X}.$$

Conversely, an abstract version of realizations using factorizations of Hankel operators, that is, of $\mathbb{F}[z]$-homomorphisms $f : \mathbb{F}[z]^m \longrightarrow z^{-1}\mathbb{F}[[z^{-1}]]^p$, is introduced.

Definition 4.14. For an $\mathbb{F}[z]$-module homomorphism, $f : \mathbb{F}[z]^m \longrightarrow z^{-1}\mathbb{F}[[z^{-1}]]^p$.

1. An **abstract realization** of f is a factorization of the form $f = h \circ g$ through an $\mathbb{F}[z]$-module $\mathcal{X}$, with $g : \mathbb{F}[z]^m \longrightarrow \mathcal{X}$ and $h : \mathcal{X} \longrightarrow z^{-1}\mathbb{F}[[z^{-1}]]^p$ both being $\mathbb{F}[z]$-homomorphisms. $\mathcal{X}$ is called a **factor module** of the realization.
2. The factorization $f = h \circ g$ is called **minimal** if for every other factorization $f = h' \circ g'$ with state space $\mathcal{X}'$ there exists a unique $\mathbb{F}[z]$-homomorphism $Z : \mathcal{X} \longrightarrow \mathcal{X}'$, with

$$Z \circ g = g', \quad h' \circ Z = h. \tag{4.16}$$

The factor module $\mathcal{X}$ of a minimal realization $f = h \circ g$ is called the **state module** or the **state space**.
3. The factorization $f = h \circ g$ is called **reachable** if g is surjective and **observable** if h is injective. It is called **canonical** if it is both reachable and observable.
4. An abstract realization is called **finite-dimensional** if the factor module $\mathcal{X}$ has finite dimension as an $\mathbb{F}$-linear space.

There are two natural candidates for choosing a state module for a realization of a restricted input/output map f. This is shown by the next result, which also proves the existence of canonical factorizations.

Theorem 4.15. *Let* $f : \mathbb{F}[z]^m \longrightarrow z^{-1}\mathbb{F}[[z^{-1}]]^p$ *be a restricted input/output map. The following commutative diagrams yield canonical and minimal abstract realizations of* f:

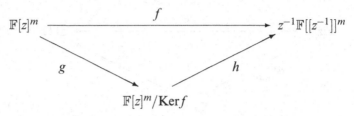

1.

Here g *is the canonical projection and* h *the homomorphism induced by* f *on the quotient module* $\mathbb{F}[z]^m/\mathrm{Ker}\,f$ *endowed with the induced quotient* $\mathbb{F}[z]$-*module structure.*

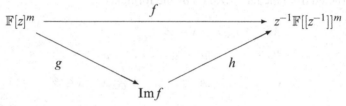

2.

Here g *differs from* f *only in the range module, and* h *is the canonical injection of* $\mathrm{Im}\,f$ *into* $z^{-1}\mathbb{F}[[z^{-1}]]^p$. *Here* $\mathrm{Im}\,f$ *is endowed with the* $\mathbb{F}[z]$-*module structure induced by the module structure on* $z^{-1}\mathbb{F}[[z^{-1}]]^p$.

Proof. Both assertions on canonical realizations are obvious. Concerning minimality, we focus on the first assertion and assume that $f = h' \circ g'$ is a realization. Then $g' : \mathbb{F}[z]^m \longrightarrow \mathscr{X}'$ is a homomorphism with kernel $\mathrm{Ker}\,g' \subset \mathrm{Ker}\,f$. Thus there exists a unique homomorphism $Z : \mathbb{F}[z]^m \longrightarrow \mathbb{F}[z]^m/\mathrm{Ker}\,f$, with $Z \circ g = g'$. This implies $h' \circ Z \circ g = h' \circ g' = f = h \circ g$. By the surjectivity of g, therefore, $h' \circ Z = h$. This completes the proof.   ∎

In the preceding definition, the notion of minimality was introduced via a universal property. It is therefore a trivial consequence that two minimal factorizations $f = h \circ g$ and $f = h' \circ g'$ are related as in (4.16), namely, by a unique module isomorphism $Z : \mathscr{X} \longrightarrow \mathscr{X}'$. That minimal factorizations coincide with canonical ones is shown next.

Proposition 4.16. *A factorization* $f = h \circ g$ *is minimal if and only if it is canonical.*

Proof. Suppose that the factorization $f = h \circ g$ with state space $\mathscr{X}$ is minimal. Consider the first canonical factorization $f = i \circ \pi$ from Theorem 4.15, where $\pi : \mathbb{F}[z]^m \longrightarrow \mathbb{F}[z]^m/\mathrm{Ker}\,f$ is the canonical projection and $i : \mathbb{F}[z]^m/\mathrm{Ker}\,f \longrightarrow z^{-1}\mathbb{F}[[z^{-1}]]^p$ the injective map induced by f. There exists a homomorphism $Z : \mathscr{X} \longrightarrow \mathbb{F}[z]^m/\mathrm{Ker}\,f$, with $Z \circ g = \pi$ and $i \circ Z = h$. Since π is surjective, so is Z. Similarly, the injectivity of i implies the injectivity of Z. Thus Z is an isomorphism, and therefore g is surjective and h is injective. Conversely, assume $f = h \circ g$ is canonical. By the minimality of $f = i \circ \pi$, there exists a homomorphism $Z : \mathbb{F}[z]^m / \mathrm{Ker} f \longrightarrow \mathscr{X}$, with $Z \circ \pi = g$ and $i \circ Z = h$. By the same reasoning as above, it follows that Z is an isomorphism. This implies the minimality of $f = h \circ g$. ∎

This leads immediately to the following version of the state-space isomorphism theorem. Subsequently, this issue will be discussed in greater detail.

Theorem 4.17. *Let $f = h_i \circ g_i$, $i = 1, 2$, be two factorizations of a Hankel operator $f : \mathbb{F}[z]^m \longrightarrow z^{-1}\mathbb{F}[[z^{-1}]]^p$ through the $\mathbb{F}[z]$-factor modules $\mathscr{X}_1$ and $\mathscr{X}_2$.*

1. *Assume that $f = h_1 \circ g_1$ is a canonical factorization. There exists a unique injective $\mathbb{F}[z]$-linear map $Z : \mathscr{X}_1 \longrightarrow \mathscr{X}_2$, with $g_2 = Z \circ g_1$ and $h_1 = h_2 \circ Z$.*
2. *If $f = h_i \circ g_i$, $i = 1, 2$, are both canonical factorizations, then there exists a unique $\mathbb{F}[z]$-isomorphism $Z : \mathscr{X}_1 \longrightarrow \mathscr{X}_2$ for which $g_2 = Z \circ g_1$ and $h_1 = h_2 \circ Z$.*

Proof. The first assertion follows from the universal property of minimality and by Proposition 4.16. Thus there exists a unique homomorphism $Z : \mathscr{X}_1 \longrightarrow \mathscr{X}_2$ with $g_2 = Z \circ g_1$ and $h_1 = h_2 \circ Z$. The injectivity of h_1 implies the injectivity of Z. Similarly, the surjectivity of g_2 implies that Z is surjective. The result follows. ∎

It is very easy to construct a state-space realization of a proper transfer function. Let $G(z) \in \mathbb{F}(z)^{p \times m}$ be proper and having the expansion $G(z) = \sum_{i=0}^{\infty} G_i z^{-i}$. One associates with $G(z)$ the maps

$$A : z^{-1}\mathbb{F}[[z^{-1}]]^p \longrightarrow z^{-1}\mathbb{F}[[z^{-1}]]^p, \quad B : \mathbb{F}^m \longrightarrow z^{-1}\mathbb{F}[[z^{-1}]]^p,$$

$$C : z^{-1}\mathbb{F}[[z^{-1}]]^p \longrightarrow \mathbb{F}^p, \quad D : \mathbb{F}^m \longrightarrow \mathbb{F}^p$$

by

$$A = \sigma, \quad B\xi = \pi_-(G(z)\xi), \quad C\sum_{i=1}^{\infty} h_i z^{-i} = h_1, \quad D\xi = G_0\xi.$$

Here σ denotes the backward shift on $z^{-1}\mathbb{F}[[z^{-1}]]^p$. It is easy to check that (A, B, C, D) is a realization of $G(z)$, as is seen from the following computation, with $\xi \in \mathbb{F}^m$:

$$CA^{j-1}B\xi = CA^{j-1}\sum_{i=1}^{\infty} G_i \xi z^{-i} = C\sigma^{j-1}\sum_{i=1}^{\infty} G_i \xi z^{-i}$$

$$= C\sum_{i=1}^{\infty} G_{j+i-1}\xi z^{-i} = G_j \xi.$$

This is the easiest realization to construct. The downside of this argument is the fact that the state space of this realization, namely $z^{-1}\mathbb{F}[[z^{-1}]]^p$, is infinite-dimensional. In the case where $G(z)$ is rational, it is shown next that finite-dimensional state-space realizations exist.

The basic result connecting concrete state-space realizations (A,B,C,D) with factorizations of the Hankel operator is the following general form of an early result due to Kronecker .

Theorem 4.18 (Kronecker (1890)). *Let* $f : \mathbb{F}[z]^m \longrightarrow z^{-1}\mathbb{F}[[z^{-1}]]^p$ *be an* $\mathbb{F}[z]$-*linear map with the symbol* $G(z) \in z^{-1}\mathbb{F}[[z^{-1}]]^{p \times m}$. f *has a finite-dimensional abstract realization if and only if the state module of each minimal abstract realization is finite-dimensional. The following statements are equivalent:*

(a) $G(z)$ *is rational.*
(b) f *has a finite-dimensional abstract realization.*
(c) f *has a finite-dimensional state-space realization* (A,B,C).

In any of the cases, the minimal dimensions of the respective state spaces coincide.

Proof. Let $f = h' \circ g'$ be an abstract realization with the finite-dimensional factor module $\mathscr{X}$. For a minimal realization $f = h \circ g$ with the state module $\mathscr{X}$ there exists a unique homomorphism $Z : \mathscr{X} \longrightarrow \mathscr{X}'$, with $h' \circ Z = h$ and $Z \circ g = g'$. Since h is injective, Z is also injective, implying that, as vector spaces, the dimension of $\mathscr{X}$ is less than or equal to the dimension of $\mathscr{X}'$. This shows that $\mathscr{X}$ is finite-dimensional. For the remaining parts we proceed as follows.

1. Assume f factors through a finite-dimensional $\mathbb{F}$-vector space $\mathscr{X}$. Clearly, this implies that the image of the Hankel operator $f = H_G$ is finite-dimensional. Thus Theorem 4.13 implies that G is rational. Conversely, if G is rational, then $\operatorname{Im} f$ is finite-dimensional and f factors through the finitely generated torsion module $\operatorname{Im} f$. This shows the equivalence $(a) \Longleftrightarrow (b)$.
2. Let (A,B,C) be a finite-dimensional realization of the transfer function $G(z) = C(zI - A)^{-1}B$. In Theorem 4.11 it was proved that the reachability and observability maps $\mathscr{R}_{(A,B)}, \mathscr{O}_{(C,A)}$, defined in (4.13) and (4.14) respectively, yield a finite-dimensional factorization of the Hankel operator as $H_G = \mathscr{O}_{(C,A)} \circ \mathscr{R}_{(A,B)}$. Thus $(c) \Longrightarrow (b)$. Conversely, assume $f : \mathbb{F}[z]^m \longrightarrow z^{-1}\mathbb{F}[[z^{-1}]]^p$ has a finite-dimensional abstract realization. Let $f = h \circ g$ be a factorization through the finitely generated $\mathbb{F}[z]$-torsion module $\mathscr{X}$. Define the triple of maps (A,B,C) by

$$Ax = z \cdot x,$$
$$Bu = g(u), \qquad u \in \mathbb{F}^m, \tag{4.17}$$
$$Cx = (h(x))_{-1}, \qquad x \in \mathscr{X}.$$

Here $z \cdot x$ denotes the action of the polynomial z on $x \in \mathscr{X}$, and $\mathbb{F}^m \subset \mathbb{F}[z]^m$ is the natural inclusion. It will be shown now that (4.17) defines a realization of the input/output map f. Since $f(\xi) \in z^{-1}\mathbb{F}[[z^{-1}]]^p$, then $f(\xi) = \sum_{j=1}^{\infty}(f(\xi))_{-j}z^{-j}$. Now $f(\xi))_{-j}$, the coefficient of z^{-j} in the expansion, is linear in ξ; hence, there exists $F_i \in \mathbb{F}^{p \times m}$ such that $f(\xi))_{-j} = F_j\xi$. Therefore, it suffices to show that $CA^{j-1}B = F_i$. This follows, for $\xi \in \mathbb{F}^m$, from the computation

$$CA^{j-1}B\xi = (hz^{j-1} \cdot g(\xi))_{-1} = (z^{j-1} \cdot hg(\xi))_{-1}$$

$$= (hg(\xi))_{-j} = (f(\xi))_{-j} = F_j\xi. \qquad \blacksquare$$

The previous result leaves open several important questions. One is the characterization of those input/output maps that lead to finite-dimensional realizations. Then one is interested not in an abstract realization but in one given in terms of linear maps or matrices. Thus, an algorithmic solution is needed in terms of factorizations of the transfer function rather than the Hankel operator. Finally, one is interested in questions of uniqueness.

4.4 Equivalence of Realizations

Any input/output map or, equivalently, transfer function may have several different realizations. Our next aim is to examine the relation between these different realizations. To this end, we introduce the following definition.

Definition 4.19. Let (A_1, B_1, C_1) and (A_2, B_2, C_2) be state-space systems defined in the state spaces $\mathscr{X}_1$ and $\mathscr{X}_2$, respectively. A map $Z : \mathscr{X}_1 \longrightarrow \mathscr{X}_2$ **intertwines** (A_1, B_1, C_1) and (A_2, B_2, C_2) if the following relations are satisfied:

$$ZA_1 = A_2Z, \quad ZB_1 = B_2, \quad C_1 = C_2Z. \tag{4.18}$$

Similarly, if only $ZA_1 = A_2Z$, we say Z intertwines A_1 and A_2, or Z intertwines (A_1, B_1) and (A_2, B_2) if $ZB_1 = B_2$ and $A_2Z = ZA_1$. One says that two systems (A_1, B_1, C_1) and (A_2, B_2, C_2) are **isomorphic**, or **similar**, if there exists an invertible map Z intertwining the two systems.

Equations (4.18) are clearly equivalent to the commutativity of the following basic diagram:

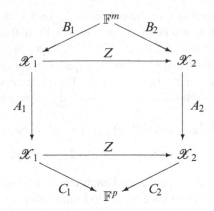

Note that system similarity is an equivalence relation.

Contrary to the trivial case of existence of maps intertwining two matrices A_1 and A_2, say, by the zero map, the existence of an intertwining map for systems yields some nontrivial information.

Lemma 4.20. *Let Z intertwine (A_1,B_1,C_1) and (A_2,B_2,C_2). Then the two systems have the same transfer function.*

Proof. It suffices to show that $C_1 A_1^i B_1 = C_2 A_2^i B_2$ for all $i \geq 0$. This follows from the following computation:

$$C_1 A_1^i B_1 = C_2 Z A_1^i B_1 = C_2 A_2^i Z B_1 = C_2 A_2^i B_2. \qquad \blacksquare$$

While in general two systems with the same transfer function need not be isomorphic, this is the case if they are both canonical systems, as the following important theorem shows.

Theorem 4.21 (State-Space Isomorphism Theorem).   *Two canonical systems $\Sigma_1 = (A_1,B_1,C_1)$ and $\Sigma_2 = (A_2,B_2,C_2)$ are similar if and only if they have the same transfer function. In this case the isomorphism is uniquely determined.*

Proof. The *if* part was proved in Lemma 4.20. To prove the *only if* part, let $\mathscr{R}_1, \mathscr{R}_2$ be the reachability maps of Σ_1 and Σ_2, respectively, and $\mathscr{O}_1$ and $\mathscr{O}_2$ the respective observability maps. Clearly, for the Hankel operator H_G, one has the factorizations $H_G = \mathscr{O}_1 \mathscr{R}_1 = \mathscr{O}_2 \mathscr{R}_2$. By our assumptions, $\mathscr{R}_1$ and $\mathscr{R}_2$ are surjective $\mathbb{F}[z]$-homomorphisms, while $\mathscr{O}_1$ and $\mathscr{O}_2$ are injective; hence, applying Theorem 4.17 one obtains the existence of a unique isomorphism $Z : \mathscr{X}_1 \longrightarrow \mathscr{X}_2$ satisfying $Z\mathscr{R}_1 = \mathscr{R}_2$ and $\mathscr{O}_1 = \mathscr{O}_2 Z$. Since Z is an isomorphism, with $\mathscr{X}_1 = X_{zI-A_1}$ and $\mathscr{X}_2 = X_{zI-A_2}$ taken with the $\mathbb{F}[z]$-module structures induced by A_1 and A_2, respectively, the intertwining relation $ZA_1 = A_2 Z$ necessarily follows. Applying the equality $Z\mathscr{R}_1 = \mathscr{R}_2$ to constant polynomials implies $Z\mathscr{R}_1 u_0 = ZB_1 u_0 = \mathscr{R}_2 u_0 = B_2 u_0$. Since one can choose u_0 freely, the equality $ZB_1 = B_2$ follows. From the equality $\mathscr{O}_1 = \mathscr{O}_2 Z$ it follows that $\sum_{i=0}^{s} C_1 A_1^i z^{-i-1} = \sum_{i=0}^{s} C_2 A_2^i Z z^{-i-1}$. In particular, equating coefficients, one obtains $C_1 = C_2 Z$. The uniqueness of Z, as already mentioned, is a consequence of Theorem 4.17.   ∎

Definition 4.22. The **McMillan degree** of a transfer function $G(z)$, denoted by $\delta(G)$, is defined as the dimension of a minimal state-space realization of $G(z)$. A state-space realization of $G(z)$ of order equal to $\delta(G)$ is called a **minimal realization**.

In view of the state-space isomorphism theorem, the McMillan degree is well defined.

Theorem 4.23. *A realization (A,B,C) of a transfer function $G(z)$ is minimal if and only if it is both reachable and observable, i.e., canonical. The McMillan degree $\delta(G)$ coincides with the dimension of the state space of an arbitrary canonical abstract realization of H_G.*

Proof. This is a simple consequence of Proposition 4.16 and Theorem 4.18. Each realization (A,B,C) of a rational transfer function $G(z)$ yields a factorization of the Hankel operator $H_G = \mathcal{O}_{(C,A)} \circ \mathcal{R}_{(A,B)}$ with a state space $\mathcal{X}$ of dimension $\geq \delta(G)$. Let $H_G = h \circ g$ denote a minimal, i.e., canonical, abstract realization of H_G with state space $\mathcal{X}'$. Then there exists an injection $\mathcal{X}'$ into $\mathcal{X}$, and therefore $\dim \mathcal{X}' \leq \dim \mathcal{X}$. By Theorem 4.18, there exists a realization (A',B',C') of $G(z)$ of order equal to $\dim \mathcal{X}'$. Thus $\dim \mathcal{X}' = \dim \mathcal{X}$ if and only if (A,B,C) is a minimal realization. Thus (A',B',C') is minimal, too. The result follows.   ∎

From now on, the term *minimal* shall be used exclusively. The next result gives an explicit formula for the McMillan degree in terms of coprime factorizations.

Theorem 4.24. *Let $G(z) \in \mathbb{F}[[z^{-1}]]^{p \times m}$ be rational, and assume it has the coprime factorizations*

$$G(z) = N_r(z) D_r(z)^{-1} = D_\ell(z)^{-1} N_\ell(z).$$

Then the McMillan degree $\delta(G)$ has the following representations:

$$\delta(G) = \operatorname{codim} \operatorname{Ker} H_G = \dim(\operatorname{Im} H_G) = \operatorname{rank} H_G,$$
$$\delta(G) = \deg \det D_r = \deg \det D_\ell,$$

or, equivalently, by

$$\delta(G) = \operatorname{rank} \mathcal{H},$$

where $\mathcal{H}$ denotes an infinite matrix with entries $\mathcal{H}_{ij} = G_{i+j-1}$, for $i, j \geq 1$.

Proof. Theorem 4.24 implies that $\delta(G)$ coincides with the dimension of the state space of an arbitrary canonical abstract realization. Thus $\delta(G) = \dim \operatorname{Im} H_G$. By

Theorem 4.13, $\operatorname{codim} \operatorname{Ker} H_G = \dim(\operatorname{Im} H_G) = \operatorname{rank} H_G$. Moreover, $\dim(\operatorname{Im} H_G) = \deg \det D_r$ and $\operatorname{codim} \operatorname{Ker} H_G = \deg \det D_\ell$. Finally, in terms of the powers of z bases in $\mathbb{F}[z]^m$ and $z^{-1}\mathbb{F}[[z^{-1}]]^p$, respectively, the Hankel operator H_G has the foregoing matrix representation. Thus $\dim \operatorname{Im} H_G = \operatorname{rank}(G_{i+j-1})$.   ∎

4.5 The Shift Realization

Theorem 4.18 connects the rationality of a transfer function $G(z)$ with its realizability in a finite-dimensional state space. Rational functions may have several different representations: power series expansions, partial fraction representations, and continued fraction representations, as well as various types of polynomial matrix fraction representations, and this list is not exhaustive. Each class of representations can be taken as the starting point of a realization procedure.

While the factorization approach to realization theory, as manifested in Theorem 4.15, is very elegant; it lacks concreteness. What is required is a concrete description of a system (A, B, C) that lends itself easily to computations. The key to this, in the case of a rational transfer function $G(z)$, is our knowledge, as summarized in Theorem 4.18, of concrete representations of $\operatorname{Ker} H_G$ and $\operatorname{Im} H_G$ in terms of polynomial matrices and the related coprime factorizations of $G(z)$. But rather than starting from a coprime factorization, we prefer to take as our starting point a more general representation of rational functions, introduced by Rosenbrock, and use it directly for constructing realizations. The shift realization, as defined in what follows, serves exactly this purpose.

Definition 4.25. Let $G(z) = G_0 + \sum_{i=1}^{\infty} \dfrac{G_i}{z^i}$ be a $p \times m$ proper rational function, with the representation

$$G(z) = V(z)T(z)^{-1}U(z) + W(z), \tag{4.19}$$

and let $\mathscr{P} = \begin{pmatrix} T(z) & -U(z) \\ V(z) & W(z) \end{pmatrix}$ be the associated polynomial system matrix. Let X_T denote the polynomial model of $T(z)$, with the shift operator S_T. Then the system

$$A : X_T \longrightarrow X_T, \quad B : \mathbb{F}^m \longrightarrow X_T, \quad C : X_T \longrightarrow \mathbb{F}^p, \quad D : \mathbb{F}^m \longrightarrow \mathbb{F}^p,$$

defined by

$$\Sigma_{VT^{-1}U+W} := \begin{cases} Af = S_T f, & f \in X_T, \\ B\xi = \pi_T(U\xi), & \xi \in \mathbb{F}^m, \\ Cf = (VT^{-1}f)_{-1}, & f \in X_T, \\ D = G_0, \end{cases} \tag{4.20}$$

is called the **shift realization** of $G(z)$, given by (4.19).

The fundamental properties of the shift realization are summarized in the following theorem.

Theorem 4.26. *The shift realization (A,B,C,D) has the following properties:*

1. *The transfer function of (A,B,C,D) is $G(z)$.*
2. *The reachability and observability maps of the realization (4.20) are*

$$\mathscr{R}u = \pi_T(Uu), \qquad\qquad u \in \mathbb{F}[z]^m$$

and

$$\mathscr{O}f = \pi_-(VT^{-1}f), \qquad\qquad f(z) \in X_T.$$

3. *The reachable subspace of realization (4.20) is $T_1 X_{T_2}$, where $T_1(z)$ is a g.c.l.d. of $T(z)$ and $U(z)$ and $T(z) = T_1(z)T_2(z)$.*
4. *The unobservable subspace of realization (4.20) is $T_1 X_{T_2}$, where $T_2(z)$ is a g.c.r.d. of $T(z)$ and $V(z)$ and $T(z) = T_1(z)T_2(z)$.*
5. *The realization is observable if and only if $V(z)$ and $T(z)$ are right coprime and reachable if and only if $T(z)$ and $U(z)$ are left coprime.*
6. *If both coprimeness conditions are satisfied, then the McMillan degree is given by*

$$\delta(G) = \deg \det T(z).$$

Proof. 1. To show that (4.20) is a realization of $G(z)$, one computes, with $\xi \in \mathbb{F}^m$,

$$\begin{aligned}
CA^j B\xi &= (VT^{-1}\pi_T z^j \pi_T U\xi)_{-1} = (VT^{-1}\pi_T z^j U\xi)_{-1} \\
&= (VT^{-1}T\pi_- T^{-1}z^j U\xi)_{-1} = (V\pi_- T^{-1}Uz^j\xi)_{-1} \\
&= (VT^{-1}Uz^j\xi)_{-1} = ((VT^{-1}U + W)z^j\xi)_{-1} = (Gz^j\xi)_{-1} \\
&= G_{j+1}\xi.
\end{aligned}$$

2. The reachability and observability maps of this realization are computed next, beginning with the reachability map $\mathscr{R} : \mathbb{F}[z]^m \longrightarrow X_T$. Computing

$$\mathscr{R}\sum_{i=0}^n u_i z^i = \sum_{i=0}^n S_T^i \pi_T U u_i = \sum_{i=0}^n \pi_T z^i \pi_T U u_i$$

$$= \sum_{i=0}^n \pi_T U z^i u_i = \pi_T U \sum_{i=0}^n z^i u_i$$

shows that the reachability map $\mathscr{R} : \mathbb{F}[z]^m \longrightarrow X_T$ is given, for $u \in \mathbb{F}[z]^m$, by

$$\mathscr{R}u = \pi_T U u.$$

Let $f(z) \in X_T$. For the observability map $\mathcal{O}$ one computes

$$\mathcal{O}f = \sum_{i=1}^{\infty} (VT^{-1}\pi_T z^{i-1}f)_{-1}z^{-i} = \sum_{i=1}^{\infty} (VT^{-1}T\pi_- T^{-1}z^{i-1}f)_{-1}z^{-i}$$

$$= \sum_{i=1}^{\infty} (VT^{-1}z^{i-1}f)_{-1}z^{-i} = \pi_-(VT^{-1}f).$$

3. Since $\mathcal{R}$ is an $\mathbb{F}[z]$-homomorphism, its image, the reachable subspace of X_T, is an $\mathbb{F}[z]$-submodule, i.e., an S_T-invariant subspace. By Theorem 3.11, it is given by $\mathrm{Im}\,\mathcal{R} = T_1 X_{T_2}$ for some factorization $T(z) = T_1(z)T_2(z)$ into nonsingular factors. Assuming that T_1 is a nonsingular common left factor of $T(z)$ and $U(z)$ implies the factorizations $T(z) = T_1(z)T_2(z)$ and $U(z) = T_1(z)U_1(z)$. One computes, for $u(z) \in \mathbb{F}[z]^m$, that

$$\pi_T(Uu) = T_1 T_2 \pi_-(T_2^{-1}T_1^{-1}T_1 U_1 u) = T_1 \pi_{T_2}(U_1 u) \in T_1 X_{T_2}.$$

Conversely, assume that for every $u(z) \in \mathbb{F}[z]^m$ one has $\pi_T Uu \in T_1 X_{T_2}$, i.e., that $T_1 T_2 \pi_-(T_2^{-1}T_1^{-1}T_1 U_1 u) \in T_1 X_{T_2}$. This implies $\pi_-(T_2^{-1}T_1^{-1}T_1 U_1 u) \in X^{T_2}$, and therefore $T_2^{-1}T_1^{-1}U_1 u = g + h$, with g a polynomial and $h \in X^{T_2}$. Thus also $T_1^{-1}U_1 u = T_2 g + T_2 h$ is a polynomial for every u. Choosing for u the unit vectors in $\mathbb{F}^m$, it follows that $U_1(z) = T_1(z)^{-1}U(z)$ is necessarily a polynomial, and the factorization $U(z) = T_1(z)U_1(z)$ follows. Clearly, the reachable subspace corresponds to the maximal common left factor of $T(z), U(z)$.

4. Since $\mathcal{O}$ is an $\mathbb{F}[z]$-homomorphism, its kernel, the unobservable subspace of Σ, is a submodule of X_T and, hence, has a representation $\mathrm{Ker}\,\mathcal{O} = T_1 X_{T_2}$ for some factorization $T(z) = T_1(z)T_2(z)$ with nonsingular factors. If $T_2(z)$ is a common right factor of $V(z)$ and $T(z)$, then $V(z) = V_1(z)T_2(z)$. Thus, for $f = T_1 g$ and $g \in X_{T_2}$, one obtains

$$\mathcal{O}f = \pi_-(VT^{-1}f) = \pi_-(V_1 T_2 T_2^{-1}T_1^{-1}f) = \pi_-(V_1 T_1^{-1}T_1 g) = \pi_-(V_1 g) = 0,$$

i.e., $T_1 X_{T_2} \subset \mathrm{Ker}\,\mathcal{O}$. Conversely, if $f \in \mathrm{Ker}\,\mathcal{O} = T_1 X_{T_2}$, then $f = T_1 g$, with $g \in X_{T_2}$. One computes

$$0 = \mathcal{O}f = \pi_- VT^{-1}T_1 g = \pi_- VT_2^{-1}T_1^{-1}T_1 g = \pi_- VT_2^{-1}g.$$

Clearly, for every $g \in T_2\mathbb{F}[z]^q$, the rational vector $VT_2^{-1}g$ is a polynomial. Using the direct sum $\mathbb{F}[z]^q = X_{T_2} \oplus T_2\mathbb{F}[z]^q$, it follows that for every $g \in \mathbb{F}[z]^q$, $VT_2^{-1}g$ is a polynomial vector. Choosing for g the standard basis unit vectors in $\mathbb{F}[z]^q$, we conclude that $V_1(z) = V(z)T_2(z)^{-1}$ is a polynomial, and the factorization $V(z) = V_1(z)T_2(z)$ follows. Clearly, the unobservable subspace corresponds to a greatest common right factor of $V(z), T(z)$.

5. Follows from the previous parts.

6. If both coprimeness conditions are satisfied, then $\Sigma_{VT^{-1}U+W}$ is a minimal realization; hence, the McMillan degree is $\delta(G) = \dim X_T = \deg \det T$. ∎

The importance of the shift realization cannot be overemphasized. Initially, it provided the link between state-space theory, module theory, and the theory of polynomial system matrices. Later, it became the bridge between input/output-based theories of linear systems and the more recent behavioral approach. It serves to extend the theory of strict system equivalence to that of various behavior representations and to elucidate connections between behaviors and geometric control. Thus, it is the ultimate tool for the unification of all existing approaches to linear systems.

Note that, using the standard isomorphism of X_T and X^T, one can transform the realization in (4.20) to a dual realization in the state space X^T defined by

$$\Sigma^{VT^{-1}U+W} := \begin{cases} Ah = S^T h, & h \in X^T, \\ B\xi = \pi_-(T^{-1}U\xi), & \xi \in \mathbb{F}^m, \\ Ch = (Vh)_{-1}, & h \in X^T, \\ D = G_0. \end{cases} \tag{4.21}$$

There exist obvious similar results to Theorem 4.26 for such dual realizations as (4.21); however, the details are omitted.

There are several important special cases of Theorem 4.26. In particular, realizations based on left and right matrix fractions are emphasized. Using the fact that every polynomial model is isomorphic to a rational model, the results can be stated in both settings. This is important in geometric control and in duality analysis. The subsequent results are special cases of Theorem 4.26.

Theorem 4.27. *Let*

$$G(z) = N_r(z)D_r(z)^{-1} = D_\ell(z)^{-1}N_\ell(z) \tag{4.22}$$

be matrix fraction representations of a proper $p \times m$ rational function $G(z)$.

1 In the state space X_{D_r}, the transfer function $G(z)$ is realized by

$$\Sigma_{N_r D_r^{-1}} := \begin{cases} A = S_{D_r}, \\ B\xi = \pi_{D_r}\xi, \\ Cf = (N_r D_r^{-1}f)_{-1}, \\ D = G(\infty). \end{cases} \tag{4.23}$$

Then:

(a) *The unobservable subspace of realization (4.23) is $D_1 X_{D_2}$, where $D_2(z)$ is a g.c.r.d. of $D_r(z)$ and $N_r(z)$ and $D_r(z) = D_1(z)D_2(z)$;*
(b) *This realization is reachable, and it is observable if and only if $N(z)$ and $D(z)$ are right coprime.*

2. In the state space X_{D_ℓ}, the transfer function $G(z)$ is realized by

$$\Sigma_{D_\ell^{-1}N_\ell} := \begin{cases} A = S_{D_\ell}, \\ B\xi = \pi_{D_\ell}(N_\ell\xi), \\ Cf = (D_\ell^{-1}f)_{-1}, \\ D = G(\infty). \end{cases} \tag{4.24}$$

Then:

(a) The reachable subspace of realization (4.24) is $\overline{D}_1 X_{\overline{D}_2}$, where $\overline{D}_1(z)$ is a g.c.l.d. of $D_\ell(z)$ and $N_\ell(z)$ and $D_\ell(z) = \overline{D}_1(z)\overline{D}_2(z)$;
(b) This realization is observable, and it is reachable if and only if $N_\ell(z)$ and $D_\ell(z)$ are left coprime.

The dual formulation of the preceding theorem is as follows.

Theorem 4.28. *Let*

$$G(z) = N_r(z)D_r(z)^{-1} = D_\ell(z)^{-1}N_\ell(z)$$

be matrix fraction representations of a proper, $p \times m$ rational function $G(z)$.

1. In the state space $X^{\overline{D}}$, the transfer function $G(z)$ is realized by

$$\Sigma^{D_\ell^{-1}N_\ell} := \begin{cases} A = S^{D_\ell}, \\ B\xi = \pi_-(D_\ell^{-1}N_\ell\xi), \\ Ch = (h)_{-1}, \\ D = G(\infty). \end{cases} \tag{4.25}$$

Then:

(a) The reachable subspace of realization (4.25) is $X^{\overline{D}_2}$, where $\overline{D}_1(z)$ is the g.c.l.d. of $D_\ell(z)$ and $N_\ell(z)$ and $D_\ell(z) = \overline{D}_1(z)\overline{D}_2(z)$;
(b) This realization is observable, and it is reachable if and only if $N_\ell(z)$ and $D_\ell(z)$ are left coprime.

2. In the state space X^{D_r}, the transfer function $G(z)$ is realized by

$$\Sigma^{N_r D_r^{-1}} := \begin{cases} A = S^{D_r}, \\ B\xi = \pi_-(D_r^{-1}\xi), \\ Ch = (N_r h)_{-1}, \\ D = G(\infty). \end{cases} \tag{4.26}$$

Then:

(a) The unobservable subspace of realization (4.26) is X^{D_2}, where $D_2(z)$ is the g.c.r.d. of $D_r(z)$ and $N_r(z)$ and $D_r(z) = D_1(z)D_2(z)$;

(b) *This realization is reachable, and it is observable if and only if $N_r(z)$ and $D_r(z)$ are right coprime.*

The various realizations (4.23), (4.24), (4.25), and (4.26) will be referred to as the **shift realizations** associated with the corresponding matrix fractions (4.22). The uniqueness properties of such shift realizations are summarized as follows.

Theorem 4.29. *Assuming that $N_r(z)$ and $D_r(z)$ are right coprime and $N_\ell(z)$ and $D_\ell(z)$ are left coprime, the realizations $\Sigma_{N_r D_r^{-1}}$ and $\Sigma_{D_\ell^{-1} N_\ell}$, given in (4.23) and (4.24) respectively, are isomorphic. The isomorphism is given by the map $Z : X_{D_r} \longrightarrow X_{D_\ell}$ defined by*

$$Zf = \pi_{D_\ell} N_\ell f.$$

The isomorphism is equivalent to the commutativity of the following diagram:

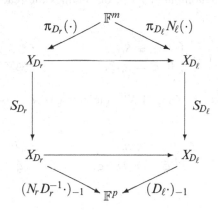

Similarly, under the same coprimeness assumptions, the realizations (4.25) and (4.26) are isomorphic. The isomorphism $\overline{Z} : X^{D_r} \longrightarrow X^{D_\ell}$ is given by

$$\overline{Z}h = \pi_- N_r h = N_r(\sigma)h. \tag{4.27}$$

Proof. Note that the intertwining equality $N_r(z)D_r(z)^{-1} = D_\ell(z)^{-1}N_\ell(z)$ implies the equality $N_\ell(z)D_r(z) = D_\ell(z)N_r(z)$, which is the connecting link to the results about module homomorphisms. Theorems 3.20 and 3.21 are applied to verify that Z is indeed a module isomorphism. Thus $ZS_{D_r} = S_{D_\ell}Z$. To complete the proof, one must show the commutativity of the triangular parts of the diagram. Indeed,

$$Z\pi_{D_r}\xi = \pi_{D_\ell}N_\ell\pi_{D_r}\xi = \pi_{D_\ell}N_\ell\xi.$$

Similarly,

$$(D_\ell^{-1}Zf)_{-1} = (D_\ell^{-1}\pi_{D_\ell}N_\ell f)_{-1} = (D_\ell^{-1}D_\ell\pi_{D_\ell}^{-1}N_\ell f)_{-1}$$
$$= (\pi_{D_\ell}^{-1}N_\ell f)_{-1} = (D_\ell^{-1}N_\ell f)_{-1} = (N_r D_r^{-1}f)_{-1}.$$

To prove (4.27), the isomorphisms $X_{D_r} \simeq X^{D_r}$ and $X_{D_\ell} \simeq X^{D_\ell}$ will be used. For $h \in X^{D_r}$, one computes

$$\bar{Z}h = D_\ell^{-1}Z(D_r^{-1}h) = D_\ell^{-1}\pi_{D_\ell}(N_\ell D_r h) = \pi_-(D_\ell^{-1}N_\ell D_r h) = \pi_-(N_r h) = N_r(\sigma)h.$$

■

If one considers the nonsingular polynomial matrix $T(z)$ to be the left denominator in a left matrix fraction $T(z)^{-1}V(z)$, then $\Sigma_{T^{-1}\bullet} := (C_T, A_T)$ will denote the observable pair defined by

$$\Sigma_{T^{-1}\bullet} := \begin{cases} A_T f = S_T f, \\ C_T f = (T^{-1}f)_{-1}, \end{cases} \qquad f \in X_T.$$

Similarly, for a right denominator $D(z)$, one defines the reachable pair $\Sigma_{\bullet D^{-1}} := (A_D, B_D)$ by

$$\Sigma_{\bullet D^{-1}} := \begin{cases} A_D f = S_D f, & f \in X_D, \\ B_D \xi = \pi_D \xi, & \xi \in \mathbb{F}^m. \end{cases} \qquad (4.28)$$

Using the isomorphism of X_D and X^D, the pair (A_D, B_D) is similar to the pair (A^D, B^D) given by

$$\Sigma^{\bullet D^{-1}} := \begin{cases} A^D h = S^D h, & h \in X^D, \\ B^D \xi = \pi_- D^{-1}\xi, & \xi \in \mathbb{F}^m. \end{cases}$$

4.6   Strict System Equivalence

Having proven the state-space isomorphism theorem, our attention turns to the isomorphism of systems that are described by higher-order equations, or polynomial matrix descriptions of the form (4.7). Recall that associated with such a system, or with its polynomial system matrix, is the shift realization (4.20). This allows us to produce the following definition.

Definition 4.30. Two polynomial system matrices

$$\mathscr{P}_1 = \begin{pmatrix} T_1(z) & -U_1(z) \\ V_1(z) & W_1(z) \end{pmatrix}, \quad \mathscr{P}_1 = \begin{pmatrix} T_2(z) & -U_2(z) \\ V_2(z) & W_2(z) \end{pmatrix}$$

are called **Fuhrmann system equivalent (FSE)** if the shift realizations $\Sigma_{V_i T_i^{-1} U_i + W_i}$ are isomorphic.

Proposition 4.31. *Fuhrmann system equivalence is an equivalence relation.*

Proof. Follows from the fact that system similarity is an equivalence relation. ∎

The next theorem directly characterizes Fuhrmann system equivalence.

Theorem 4.32. *Two polynomial system matrices* $\mathscr{P}_i = \begin{pmatrix} T_i(z) & -U_i(z) \\ V_i(z) & W_i(z) \end{pmatrix}$, $i =$
1,2, are FSE if and only if there exist appropriately sized polynomial matrices $M(z), N(z), X(z), Y(z)$, *with* $M(z)$ *and* $T_2(z)$ *left coprime and* $T_1(z)$ *and* $N(z)$ *right coprime, satisfying*

$$\begin{pmatrix} M(z) & 0 \\ -X(z) & I \end{pmatrix} \begin{pmatrix} T_1(z) & -U_1(z) \\ V_1(z) & W_1(z) \end{pmatrix} = \begin{pmatrix} T_2(z) & -U_2(z) \\ V_2(z) & W_2(z) \end{pmatrix} \begin{pmatrix} N(z) & Y(z) \\ 0 & I \end{pmatrix}. \tag{4.29}$$

Proof. Assume $\mathscr{P}_1$ and $\mathscr{P}_2$ are FSE, i.e., that the shift realizations $\Sigma_{V_iT_i^{-1}U_i+W_i}$ are isomorphic. This means that there exists an invertible map $Z : X_{T_1} \longrightarrow X_{T_2}$ that makes the following diagram commutative:

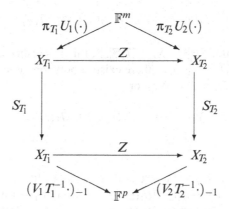

The commutativity of the central part of the diagram means that $Z : X_{T_1} \longrightarrow X_{T_2}$ is a module homomorphism, and for these, a complete characterization has been established. Indeed, by Theorem 3.20, there exist polynomial matrices $M(z)$ and $N(z)$ satisfying

$$M(z)T_1(z) = T_2(z)N(z), \tag{4.30}$$

with $M(z)$ and $T_2(z)$ left coprime and $T_1(z)$ and $N(z)$ right coprime. In these terms, Z is given by

$$Zf = \pi_{T_2}Mf, \qquad f \in X_{T_1}. \tag{4.31}$$

Since Z intertwines the two systems, it follows that, for all $j \geq 0$, $ZA_1^jB_1 = A_2^jB_2$. Equivalently, for each $\xi \in \mathbb{F}^m$,

$$\pi_{T_2}M\pi_{T_1}z^j\pi_{T_1}U_1\xi = \pi_{T_2}z^j\pi_{T_2}U_2\xi.$$

The kernels of π_{T_1} and π_{T_2} are submodules; therefore, $\pi_{T_2}(U_2 - MU_1)z^j\xi = 0$, and thus $\pi_{T_2}(U_2 - MU_1)f = 0$ for all $f \in \mathbb{F}[z]^m$. Equivalently, $\pi_- T_2^{-1}(U_2 - MU_1) = 0$, which means that there exists a polynomial matrix Y for which $T_2^{-1}(U_2 - MU_1) = Y$ or

$$U_2(z) = M(z)U_1(z) + T_2(z)Y(z). \tag{4.32}$$

Next, the relation $C_1 A_1^j = C_2 A_2^j Z$ is used to conclude that, for all $f \in X_{T_1}$,

$$\begin{aligned}
0 &= (V_2 T_2^{-1} \pi_{T_2} z^j \pi_{T_2} M f - V_1 T_1^{-1} \pi_{T_1} z^j f)_{-1} = (V_2 T_2^{-1} \pi_{T_2} M z^j f - V_1 T_1^{-1} \pi_{T_1} z^j f)_{-1} \\
&= (V_2 T_2^{-1} T_2 \pi_- T_2^{-1} M z^j f - V_1 T_1^{-1} T \pi_- T_1^{-1} z^j f)_{-1} = ((V_2 T_2^{-1} M - V_1 T_1^{-1}) z^j f)_{-1} \\
&= ((V_2 N T_1^{-1} - V_1 T_1^{-1}) z^j f)_{-1} = ((V_2 N - V_1) T_1^{-1} z^j f)_{-1}.
\end{aligned}$$

Now clearly, for every $f \in T_1 \mathbb{F}[z]^s$, $f = T_1 g$ and

$$((V_2 N - V_1) T_1^{-1} f)_{-1} = ((V_1 N - V) T_1^{-1} T_1 g)_{-1} = 0,$$

so, using the direct sum $\mathbb{F}[z]^s = X_{T_1} \oplus T_1 \mathbb{F}[z]^s$, this is satisfied for every $f \in \mathbb{F}[z]^s$. Hence, $\pi_-(V_2 M_2 - V_1) T_1^{-1} = 0$, or there exists a polynomial matrix $X(z)$ such that $(V_1(z) - V_2(z)M_2(z))T_1(z)^{-1} = X(z)$, or

$$V_1(z) = V_2(z)N(z) + X(z)T_1(z). \tag{4.33}$$

Finally, we compute

$$\begin{aligned}
X(z)U_1(z) + W_1(z) &= (V_1(z) - V_2(z)M_2(z))T_1(z)^{-1}U_1(z) + W_1(z) \\
&= V_1(z)T_1(z)^{-1}U_1(z) - V_2(z)NT_1(z)^{-1}U_1(z) + W_1(z) \\
&= V_1(z)T_1(z)^{-1}U_1(z) + W_1(z) - V_2(z)T_2(z)^{-1}MU(z)_1(z) \\
&= V_1(z)T_1(z)^{-1}U_1(z) + W_1(z) - V_2(z)T_2(z)^{-1}M(z)U_1(z) \\
&= V_2(z)T_2(z)^{-1}U_2(z) + W_2(z) - V_2(z)T_2(z)^{-1}(U_2(z) - T_2(z)Y(z)) \\
&= V_2(z)Y(z) + W_2(z)
\end{aligned}$$

or

$$X(z)U_1(z) + W_1(z) = V_2(z)Y(z) + W_2(z). \tag{4.34}$$

The four preceding relations, (4.30), (4.32), (4.33), and (4.34), can be written compactly as the one polynomial relation (4.29). Thus $\mathscr{P}_1$ and $\mathscr{P}_2$ are FSE if (4.29) is satisfied, with $M(z)$ and $T_2(z)$ left coprime and $N(z)$ and $T_1(z)$ right coprime.

Conversely, assume (4.29) is satisfied, with $M(z)$ and $T_2(z)$ left coprime and $N(z)$ and $T_1(z)$ right coprime. Define the map $Z : X_{T_1} \longrightarrow X_{T_2}$ as in (4.31). The coprimeness conditions guarantee that Z is an isomorphism. Using the same formulas, it follows that Z intertwines the two realizations; hence, the polynomial system matrices $\mathscr{P}_i$ are FSE. ∎

Note that no assumptions were made in Theorem 4.32 concerning the reachability or observability of the state-space realizations.

It is emphasized that the preceding characterization of strict system equivalence is also equivalent to the characterization proposed initially by Rosenbrock. The proof of the next proposition is left as an exercise to the reader.

Proposition 4.33. *Two polynomial system matrices* $\mathscr{P}_1$ *and* $\mathscr{P}_2$ *of sizes* $(p + r_i) \times (m + r_i)$, $i = 1, 2$, *are FSE if and only if for* $q \geq \max(r_1, r_2)$ *there exist* $q \times q$ *unimodular matrices* $M(z)$ *and* $N(z)$ *and polynomial matrices* $X(z)$ *and* $Y(z)$ *such that*

$$\begin{pmatrix} M(z) & 0 \\ -X(z) & I \end{pmatrix} \begin{pmatrix} I_{q-r_1} & 0 & 0 \\ 0 & T_1(z) & -U_1(z) \\ 0 & V_1(z) & W_1(z) \end{pmatrix} = \begin{pmatrix} I_{q-r_2} & 0 & 0 \\ 0 & T_2(z) & -U_2(z) \\ 0 & V_2(z) & W_2(z) \end{pmatrix} \begin{pmatrix} N(z) & Y(z) \\ 0 & I \end{pmatrix}.$$

Fuhrmann system equivalence has important implications for the understanding of zeros of multivariable systems, as is further explained in Section 4.7. In this regard, the following theorem is of interest.

Theorem 4.34. *Let* $G(z)$ *be a* $p \times m$ *proper rational transfer function, with the coprime representations*

$$G(z) = N_r(z)D_r(z)^{-1} = D_\ell(z)^{-1}N_\ell(z) = V(z)T(z)^{-1}U(z) + W(z). \tag{4.35}$$

Then the polynomial matrices $N_r(z)$, $N_\ell(z)$, *and* $\begin{pmatrix} T(z) & -U(z) \\ V(z) & W(z) \end{pmatrix}$ *have the same nontrivial invariant factors.*

Proof. To representations (4.35) are associated the polynomial system matrices

$$\begin{pmatrix} D_\ell(z) & -N_\ell(z) \\ I & 0 \end{pmatrix}, \quad \begin{pmatrix} D_r(z) & -I \\ N_r(z) & 0 \end{pmatrix}, \quad \begin{pmatrix} T(z) & -U(z) \\ V(z) & W(z) \end{pmatrix},$$

respectively. By Proposition 4.33, and for some nonnegative integers a, b, c, the polynomial matrices

$$\begin{pmatrix} I_a & 0 & 0 \\ 0 & D_\ell(z) & -N_\ell(z) \\ 0 & I & 0 \end{pmatrix}, \quad \begin{pmatrix} I_b & 0 & 0 \\ 0 & D_r(z) & -I \\ 0 & N_r(z) & 0 \end{pmatrix}, \quad \begin{pmatrix} I_c & 0 & 0 \\ 0 & T(z) & -U(z) \\ 0 & V(z) & W(z) \end{pmatrix}$$

are unimodularly equivalent. Clearly, the nontrivial invariant factors of

$$\begin{pmatrix} I_a & 0 & 0 \\ 0 & D_\ell(z) & -N_\ell(z) \\ 0 & I & 0 \end{pmatrix}$$

and $N_\ell(z)$ coincide. Similarly, the nontrivial invariant factors of

$$\begin{pmatrix} I_b & 0 & 0 \\ 0 & D_r(z) & -I \\ 0 & N_r(z) & 0 \end{pmatrix}$$

coincide with those of $N_r(z)$, which proves the theorem.   ∎

This result indicates, as was realized by Rosenbrock, the importance of polynomial system matrices in the study of zeros of multivariable systems. If we remove the coprimeness constraints, the theorem as stated is no longer true. This is related to a more elaborate study of input decoupling and output decoupling zeros Rosenbrock (1970). The study of zeros is also closely connected to geometric control theory, which will be briefly discussed in Chapter 6.

Using these characterizations of strict system equivalence, we proceed to deduce some important invariance properties of that notion.

Theorem 4.35. *1. If two polynomial system matrices* $\mathscr{P}_i = \begin{pmatrix} T_i(z) & -U_i(z) \\ V_i(z) & W_i(z) \end{pmatrix}$, $i =$ *1,2, are FSE, then they have the same transfer function, that is,*

$$V_1(z)T_1(z)^{-1}U_1(z) + W_1(z) = V_2(z)T_2(z)^{-1}U_2(z) + W_2(z).$$

Moreover, there exists a nonzero constant $c \in \mathbb{F}$, *with*

$$\det T_2(z) = c \det T_1(z). \tag{4.36}$$

2. Conversely, if

$$V_1(z)T_1(z)^{-1}U_1(z) + W_1(z) = G(z) = V_2(z)T_2(z)^{-1}U_2(z) + W_2(z)$$

are two right and left coprime factorizations of a proper rational transfer function $G(z)$, *then the associated polynomial system matrices* $\mathscr{P}_1$ *and* $\mathscr{P}_2$ *are FSE.*

Proof. The first part follows from Lemma 4.20 and the similarity of the two shift realizations $\Sigma_{V_1 T_1^{-1} U_1 + W_1}$ and $\Sigma_{V_2 T_2^{-1} U_2 + W_2}$. Alternatively, it follows by a simple computation from equations (4.30), (4.32), (4.33), and (4.34). By Proposition 4.33, there exist unimodular matrices $M(z)$ and $N(z)$ with

$$M(z)\begin{pmatrix} I_{q-r_1} & 0 \\ 0 & T_1(z) \end{pmatrix} = \begin{pmatrix} I_{q-r_2} & 0 \\ 0 & T_2(z) \end{pmatrix} N(z).$$

This implies $\det M(z) \det T_1(z) = \det T_2(z) \det N(z)$ and, therefore, (4.36).

Let (A_i, B_i, C_i, D_i) denote the shift realization associated with the strictly proper transfer function $G(z) = V_i(z)T_i(z)^{-1}U_i(z)$, $i = 1, 2$. By Theorem 4.26, the realizations (A_i, B_i, C_i, D_i) of $G(z)$ are reachable and observable. By applying the state-space isomorphism theorem, (A_1, B_1, C_1, D_1) is similar to (A_2, B_2, C_2, D_2). This completes the proof of the second part. ∎

Fuhrmann system equivalence is now applied to the problem of finding a polynomial characterization of input/state and state/output systems. This leads to an explicit construction of a basis matrix for the polynomial model associated with a minimal realization.

Proposition 4.36. *1. Let $G(z) = T(z)^{-1}U(z)$ be a strictly proper, left coprime matrix fraction. The following statements are equivalent:*

(a) There exists an observable pair (C, A) for which

$$C(zI - A)^{-1} = T(z)^{-1}U(z). \tag{4.37}$$

*(b) The polynomial matrix $U(z)$ is a **basis matrix** for the polynomial model X_T, i.e., the columns of $U(z)$ form a basis.*

2. Let (C, A) be an observable pair, and let $T(z)$ and $U(z)$ be left coprime matrix polynomials that satisfy (4.37). Then $U(z)$ is a basis matrix for the polynomial model X_T and

$$X^T = \{C(zI - A)^{-1}\xi \mid \xi \in \mathbb{F}^n\}.$$

Moreover, if C has full row rank, then $U(z)$ is left prime.
3. Let $G(z) = V(z)T(z)^{-1}$ be a strictly proper matrix fraction. Then the following statements are equivalent:

(a) There exists a reachable pair (A, B) for which

$$(zI - A)^{-1}B = V(z)T(z)^{-1}. \tag{4.38}$$

*(b) The polynomial matrix $V(z)$ is a **dual basis matrix** for the polynomial model X_T, i.e., the columns of $V(z)^\top$ form a basis for $X_{T^\top}$.*

4. Let (A, B) be a reachable pair and $V(z), T(z)$ right coprime matrix polynomials that satisfy (4.38). Then $V(z)$ is a dual basis matrix for the polynomial model X_T and

$$\{\xi^\top(zI - A)^{-1}B \mid \xi \in \mathbb{F}^n\} = \{f(z)^\top \mid f \in X^{T^\top}\}.$$

Moreover, if B is full column rank, then $V(z)$ is right prime.

Proof. 2. The observability of the pair (C, A) is equivalent to the right coprimeness of $C, zI - A$. Condition (4.37) is equivalent to the identity

$$\begin{pmatrix} U(z) & 0 \\ 0 & I \end{pmatrix} \begin{pmatrix} zI - A & -I \\ C & 0 \end{pmatrix} = \begin{pmatrix} T(z) & -U(z) \\ I & 0 \end{pmatrix} \begin{pmatrix} C & 0 \\ 0 & I \end{pmatrix}.$$

Using the left coprimeness of $T(z)$ and $U(z)$, this implies

$$\begin{pmatrix} zI - A & -I \\ C & 0 \end{pmatrix} \simeq_{FSE} \begin{pmatrix} T(z) & -U(z) \\ I & 0 \end{pmatrix}.$$

Applying Theorem 4.32, it follows that (A, I, C) is similar to the shift realization (A_s, B_s, C_s) associated with the matrix fraction $T(z)^{-1}U(z)$. In particular, this implies the invertibility of B_s. Since $B_s \xi = \pi_T U \xi = U \xi$, the injectivity of B_s implies the linear independence of the columns of $U(z)$, whereas the surjectivity of B_s implies that the columns of $U(z)$ span X_T. Taken together, this shows that $U(z)$ is a basis matrix for X_T. Moreover, if C is full row rank and v satisfies $v^\top U(z) = 0$, then, using (4.37), $v^\top T(z)C = 0$ follows. Since C is full row rank, this implies $v^\top T(z) = 0$. Since $T(z)$ is nonsingular, this implies $v = 0$. This completes the proof of 2.

1. Let $U(z)$ denote a basis matrix of X_T. Associate with the matrix fraction $T(z)^{-1}U(z)$ the shift realization (A_s, B_s, C_s), defined by (4.20). Since $U(z)$ is a basis matrix, B_s is invertible and

$$T(z)^{-1}U(z) = C_s(zI - A_s)^{-1}B_s = C_s B_s(zI - B_s^{-1}A_s B_s)^{-1}.$$

Thus, $(C_s B_s, B_s^{-1} A_s B_s)$ is an observable pair satisfying (4.37). Conversely, if (C, A) is observable with (4.37), then, by 2, $U(z)$ is a basis matrix.

3, 4. Follows from the first two parts by duality considerations. The details are omitted.   ∎

As a consequence, the class of all realizations of strictly proper transfer functions $N(z)D(z)^{-1}$ with varying numerator polynomial $N(z)$ can be described.

Theorem 4.37. *Let $G(z)$ be strictly proper with minimal realization (A, B, C).*

1. *If $G(z) = N(z)D(z)^{-1}$ is a right coprime factorization and $M(z)D(z)^{-1}$ is strictly proper, then there exist C_0 such that*

$$M(z)D(z)^{-1} = C_0(zI - A)^{-1}B.$$

2. *If $G(z) = T(z)^{-1}U(z)$ is a left coprime factorization and $T(z)^{-1}L(z)$ is strictly proper, then there exist B_0 such that*

$$T(z)^{-1}L(z) = C(zI - A)^{-1}B_0.$$

Proof. Only the first claim will be proved; the second follows mutatis mutandis. Without loss of generality, it can be assumed that (A, B, C) is the shift realization of $G(z)$. Thus, $A = S_D$, $B = \pi_D$, and $Cf = (N(z)D(z)^{-1}f(z))_{-1}$. Similarly, the shift realization of $M(z)D(z)^{-1}$ is $A = S_D$, $B = \pi_D$, and $C_0 = (M(z)D(z)^{-1}f(z))_{-1}$. This completes the proof. ∎

Connecting continuous- and discrete-time systems.
The next proposition describes the relation between a class of continuous-time behaviors and related discrete-time systems. Since the analysis of continuous-time systems is not the main focus of this book, more general results will not be sought.

For an analytic function $f(z)$ of exponential growth, the Taylor expansion $f(t) = \sum_{i=0}^{\infty} \frac{f^{(i)}(0)}{i!} t^i$ and the Laplace transform are related as

$$F(s) = \mathcal{L}(f) = \sum_{i=0}^{\infty} \frac{f^{(i)}(0)}{s^{i+1}}.$$

Proposition 4.38. *Let $D(z) \in \mathbb{C}[z]^{p \times p}$ be nonsingular. Let $\operatorname{Ker}D(\frac{d}{dt}) = \{f | D(\frac{d}{dt}) f = 0\}$, and let X^D be the rational model associated with D. Then the following assertions are true:*

1. *The Laplace transform is a $\mathbb{C}[z]$-homomorphism.*
2. *$\mathcal{L}$, the restriction of the Laplace transform to $\operatorname{Ker}D(\frac{d}{dt})$, is a bijective map of $\operatorname{Ker}D(\frac{d}{dt})$ onto X^D.*
3. *Let (C^D, A^D) be defined by (4.28). Define the pair $(\hat{C}_D, \hat{A}_D)$ by*

$$\hat{A}_D \phi = \frac{d\phi}{dt}, \quad \phi \in \operatorname{Ker}D(\frac{d}{dt}),$$

$$\hat{C}_D \phi = \phi(0), \quad \phi \in \operatorname{Ker}D(\frac{d}{dt}).$$

Then the following diagram is commutative:

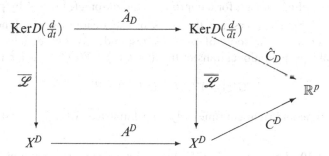

Proof. 1. Follows from

$$\mathcal{L}(\frac{d}{dt}f) = \mathcal{L}(f') = s\mathcal{L}(f) - f(0) = \pi_- sF(s) = \sigma\mathcal{L}(f). \tag{4.39}$$

2. From (4.39), by induction, one obtains

$$\mathcal{L}(\phi^{(j)}) = s^j \mathcal{L}(\phi) - s^{j-1}\phi(0) - \cdots - \phi^{(j-1)}(0) = s^j \mathcal{L}(\phi) - \sum_{v=0}^{j-1} s^{j-v-1}\phi^{(v)}(0).$$

Assuming that $D(z) = \sum_{j=0}^{r} D_j z^j$ and $\phi \in \operatorname{Ker} D(\frac{d}{dt})$, or equivalently that $D(\frac{d}{dt})\phi = 0$, implies

$$\mathcal{L}\sum_{j=0}^{r} D_j \phi^{(j)} = 0 = \sum_{j=0}^{r} D_j [s^j \mathcal{L}(\phi) - \sum_{v=0}^{j-1} s^{j-v-1}\phi^{(v)}(0)].$$

Changing the order of summation, one gets

$$D(s)\mathcal{L}(\phi) = \sum_{j=0}^{r}\sum_{v=0}^{j-1} D_j s^{j-v-1}\phi^{(v)}(0)$$

$$= \sum_{j=0}^{r-1}\sum_{v=j+1}^{r} D_j s^{j-v-1}\phi^{(v)}(0).$$

From this it follows that $\mathcal{L}$ maps $\operatorname{Ker} D(\frac{d}{dt})$ into X^D. It is easy to check, see Hinrichsen and Prätzel-Wolters (1980), that the restricted map $\overline{\mathcal{L}} : \operatorname{Ker} D(\frac{d}{dt}) \longrightarrow X^D$ is actually a bijection.

3. This is a simple verification. ∎

4.7   Poles and Zeros

For scalar rational transfer functions $g(z) \in \mathbb{F}(z)$, the notions of poles and zeros are very easy to define. In fact, for a coprime factorization $g(z) = \frac{p(z)}{q(z)}$ by polynomials $p(z), q(z) \in \mathbb{F}[z]$, the poles are the roots of the denominator polynomial $q(z)$, while the zeros are the roots of $p(z)$. Correspondingly, one defines poles for a multivariable, proper rational transfer function $G(z) \in \mathbb{F}(z)^{p \times m}$ via a representation

$$G(z) = V(z)T^{-1}(z)U(z) + W(z) \tag{4.40}$$

by left coprime and right coprime polynomial matrices $V(z), T(z)$ and $T(z), U(z)$, respectively.

Definition 4.39. Let $G(z) \in \mathbb{F}(z)^{p \times m}$ be a strictly proper rational transfer function with the representation (4.40) and satisfying the previously given coprimeness conditions. The **poles** of $G(z)$ are defined as the roots of the scalar polynomial equation

$$\det T(z) = 0.$$

Equivalently, they are the eigenvalues of the shift operator S_T on X_T.

By Corollary 4.35, any two representations (4.40) of $G(z)$ that satisfy the coprimeness conditions are Fuhrmann strict system equivalent and thus define the same poles. This shows that poles are well defined and independent of the choice of coprime factors. In particular, for reachable and observable linear systems

$$x_{t+1} = Ax_t + Bu_t$$
$$y_t = Cx_t + Du_t,$$

with transfer function $G(z) = C(zI - A)^{-1}B + D$, the poles of $G(z)$ are identical to the zeros of the characteristic polynomial $\det (zI - A)$, i.e., they are equal to the eigenvalues of system matrix A.

So far, poles of transfer functions have been introduced and their relation to the eigenvalues of a minimal first-order realization observed. In contrast, the analysis of zeros of multivariable systems becomes more complicated. Heuristically, zeros correspond to input frequencies that annihilate the output of a system. The scalar case is easy. In fact, if (A, b, c) is a minimal state-space SISO representation of a **scalar**, strictly proper transfer function,

$$g(z) = \frac{p(z)}{q(z)} = c(zI - A)^{-1}b,$$

then $q(z) = \det (zI - A)$ and

$$p(z) = \det \begin{pmatrix} zI - A & -b \\ c & 0 \end{pmatrix}.$$

This shows that zeros of SISO systems correspond to those values z of the algebraic closure $\overline{\mathbb{F}}$ with

$$\mathrm{rank} \begin{pmatrix} zI - A & -b \\ c & 0 \end{pmatrix} < n + 1.$$

This is equivalent to the classical definition of zeros given in, for example, Kailath (1980) for SISO state-space systems.

Our next aim is to extend the classical definition of zeros from SISO first-order representation systems to higher-order systems of the form

$$T(\sigma)\xi = U(\sigma)\xi,$$
$$y = V(\sigma)\xi + W(\sigma)u.$$

Here $V(z) \in \mathbb{F}[z]^{p \times r}, T(z) \in \mathbb{F}[z]^{r \times r}, U(z) \in \mathbb{F}[z]^{r \times m}, W(z) \in \mathbb{F}[z]^{p \times m}$ are polynomial matrices, with $T(z)$ nonsingular and the transfer function $G(z) = V(z)T(z)^{-1}U(z) + W(z)$ assumed to be proper. Note that the **normal rank** grk $G(z)$ of a rational matrix function $G(z) \in \mathbb{F}(z)^{p \times m}$ is defined as the classical rank of $G(z)$ when viewed as a

$p \times m$ matrix over the rational function field $\mathbb{F}(z)$. For an algebraically closed field $\mathbb{F}$, this coincides with the generic rank

$$\text{grk } G(z) = \max\{\text{rk } G(z) \mid z \in \overline{\mathbb{F}} \text{ not a pole}\}.$$

Definition 4.40. Let $U(z), V(z), T(z), W(z)$ be polynomial matrices, with $T(z) \in \mathbb{F}[z]^{r \times r}$ nonsingular, such that the $p \times m$ rational function $V(z)T(z)^{-1}U(z) + W(z)$ is proper. A **finite zero** of a polynomial system matrix

$$\mathscr{P}(z) = \begin{pmatrix} T(z) & -U(z) \\ V(z) & W(z) \end{pmatrix}$$

is a complex number $z_0 \in \overline{\mathbb{F}}$ such that

$$\text{rk } \mathscr{P}(z_0) < \text{grk } \mathscr{P}(z).$$

$\mathscr{P}(z)$ is said to have a **zero at infinity** if

$$\text{rk } \pi_+(V(z)T(z)^{-1}U(z) + W(z)) < \text{grk } (V(z)T(z)^{-1}U(z) + W(z)).$$

Note that, by assuming the properness of $W(z) + V(z)T(z)^{-1}U(z)$, the polynomial part $\pi_+(V(z)T(z)^{-1}U(z) + W(z))$ is a constant matrix. Definition 4.40 applies, in particular, to first-order state-space representations (A, B, C, D):

$$\begin{aligned} x_{t+1} &= Ax_t + Bu_t, \\ y_t &= Cx_t + Du_t, \end{aligned} \tag{4.41}$$

with the associated polynomial system matrix

$$\Sigma(z) = \begin{pmatrix} zI - A & -B \\ C & D \end{pmatrix}.$$

We conclude that system (4.41) has a zero at infinity if and only if

$$\text{rk } D < \text{grk } \left(D + C(zI - A)^{-1}B\right).$$

One complication with the notion of zeros for multivariable systems is that there may be a common pole and zero occurring in a coprime factorization of a transfer function. For example,

$$G(z) = \begin{pmatrix} \frac{z}{(z+1)^2} & 0 \\ 0 & \frac{z+1}{z^2} \end{pmatrix}$$

has both poles and zeros at $z = 0$ and $z = -1$ and a zero at infinity. Moreover, a $p \times m$ proper rational transfer function with $p \geq m$ may not have finite or infinite zeros, as the example

$$G(z) = \begin{pmatrix} \frac{z}{z+1} & 0 \\ 0 & \frac{z+1}{z} \end{pmatrix}$$

shows.

A very useful result, relating the computation of zeros of first-order representations (4.41) to higher-order representations, is proved next.

Theorem 4.41. *Let (A,B,C,D) be reachable and observable with the transfer function*

$$G(z) = D + C(zI - A)^{-1}B.$$

1. For a factorization

$$G(z) = V(z)T(z)^{-1}$$

with right coprime polynomial matrices $V(z)$ and $T(z)$, an element $z_0 \in \mathbb{F}$ is a finite zero of (A,B,C,D) if and only if

$$\text{rank } V(z_0) < \text{grk } V(z).$$

2. For a factorization

$$G(z) = T(z)^{-1}U(z)$$

with left coprime polynomial matrices $T(z)$ and $U(z)$, an element $z_0 \in \mathbb{F}$ is a finite zero of (A,B,C,D) if and only if

$$\text{rank } U(z_0) < \text{grk } U(z).$$

3. (A,B,C,D) has a zero at infinity if and only if

$$\text{rank } D < \text{grk } U(z) \quad or \quad \text{rank } D < \text{grk } V(z).$$

Proof. Since (A,B,C,D) is reachable and observable, and using the right coprimeness of $V(z)$ and $T(z)$, one obtains the following equivalence:

$$\Sigma(z) = \begin{pmatrix} zI - A & -B \\ C & D \end{pmatrix} \simeq_{FSE} \begin{pmatrix} T(z) & -I_m \\ V(z) & 0 \end{pmatrix} = \mathscr{P}(z).$$

Applying a suitable unimodular transformation on the right of $\mathscr{P}(z)$, one sees that $\mathscr{P}(z)$ and

$$\begin{pmatrix} I_m & 0 \\ 0 & V(z) \end{pmatrix}$$

have the same Smith form. By Theorem 4.35, this implies that $\Sigma(z)$ and $V(z)$ have the same nontrivial invariant factors and, thus, the same zeros. This proves assertion 1. The proof is similar for the other assertions.   ∎

There is a straightforward relation of the preceding definition of zeros to the Smith–McMillan form of a rational function.

Proposition 4.42. *Let* $G(z) = V(z)T(z)^{-1}U(z) + W(z) \in \mathbb{F}(z)^{p \times m}$ *be a proper rational function with Smith–McMillan form*

$$\mathrm{diag}\left(\frac{\varepsilon_1(z)}{\psi_1(z)}, \cdots, \frac{\varepsilon_r(z)}{\psi_r(z)}, 0, \cdots, 0\right).$$

If $V(z)$ *and* $T(z)$ *are right coprime and* $T(z)$ *and* $U(z)$ *are left coprime, then the zeros of the polynomials* $\varepsilon_1(z), \ldots, \varepsilon_r(z)$ *coincide with the finite zeros of the polynomial system matrix*

$$\mathscr{P}(z) = \begin{pmatrix} T(z) & -U(z) \\ V(z) & W(z) \end{pmatrix}.$$

In particular, the numerator polynomials $V(z)$ *in a right coprime factorization* $V(z)T(z)^{-1} = G(z)$ *all have the same Smith form, and similarly for right numerators.*

Proof. Let $L(z), R(z)$ be unimodular matrices such that $L(z)G(z)R(z)$ is in Smith–McMillan form. Consider the $p \times m$ and $m \times m$ polynomial matrices

$$N(z) = \mathrm{diag}\left(\varepsilon_1(z), \cdots, \varepsilon_r(z), 0\right), \quad D(z) = \mathrm{diag}\left(\psi_1(z), \cdots, \psi_r(z), I_{m-r}\right).$$

Then $P(z) := L(z)^{-1}N(z)$ and $Q(z) = Z(z)D(z)$ are polynomial matrices, and

$$G(z) = P(z)Q(z)^{-1}$$

is a right coprime factorization. Therefore, the polynomial matrices $\mathscr{P}(z)$ and

$$M(z) = \begin{pmatrix} Q(z) & -I \\ P(z) & 0 \end{pmatrix}$$

are Fuhrmann strict system equivalent, and therefore $\mathcal{P}(z)$ has the same nontrivial invariant factors as $P(z)$ or, equivalently, $N(z)$, i.e., as $\varepsilon_1(z), \ldots, \varepsilon_r(z)$. This completes the proof. ∎

In view of Proposition 4.42, it is tempting to define the finite zeros of a linear system, with the proper, rational transfer function $G(z)$, as the zeros of the polynomial system matrix $\begin{pmatrix} T(z) & -U(z) \\ V(z) & W(z) \end{pmatrix}$. However, this is a representation-dependent definition. To circumvent this difficulty, Wyman and Sain (1981) initiated a module-theoretic approach to zeros.

Definition 4.43. Let $G(z) \in \mathbb{F}(z)^{p \times m}$ be rational, with

$$G(z)^{-1}\mathbb{F}[z]^p := \{f(z) \in \mathbb{F}(z)^m \mid G(z)f(z) \in \mathbb{F}[z]^p\},$$
$$\mathrm{Ker}\, G(z) := \{f(z) \in \mathbb{F}(z)^m \mid G(z)f(z) = 0\}.$$

The **pole module** of $G(z)$ is defined as the $\mathbb{F}[z]$-module

$$X(G) = \frac{\mathbb{F}[z]^m}{G(z)^{-1}\mathbb{F}[z]^p \cap \mathbb{F}[z]^m}. \tag{4.42}$$

The **zero module** of $G(z)$ is defined as the $\mathbb{F}[z]$-module

$$Z(G) = \frac{G(z)^{-1}\mathbb{F}[z]^p + \mathbb{F}[z]^m}{\mathrm{Ker}\, G(z) + \mathbb{F}[z]^m}. \tag{4.43}$$

The meaning of the preceding definition becomes clearer if one assumes that $G(z)$ is given in terms of a coprime factorization. It is worth mentioning, without giving a complete proof, one characterization due to Wyman and Sain (1981). The first part shows that the pole module is simply the state space of the system, viewed as a polynomial model.

Theorem 4.44. *1. Let $G(z) = N_r(z)D_r(z)^{-1}$ be a right coprime factorization of the proper transfer function $G(z)$. Then the pole module $X(G)$ is $\mathbb{F}[z]$-linearly isomorphic to the polynomial model X_{D_r}. In particular, one has the dimension formula*

$$\dim X(G) = \deg \det D_r.$$

2. Let $G(z) = D_\ell(z)^{-1}N_\ell(z)$ be a left coprime factorization of the proper transfer function $G(z)$. Then the zero module $Z(G)$ is $\mathbb{F}[z]$-linearly isomorphic to the torsion submodule of

$$\mathbb{F}[z]^p / N_\ell(z)\mathbb{F}[z]^m.$$

In particular, one has the dimension formula

$$\dim Z(G) = \sum_i \deg \det \varepsilon_i,$$

where the ε_i are the nontrivial invariant factors of $N_\ell(z)$.

Proof. 1. Note that $g(z) \in G(z)^{-1}(\mathbb{F}[z]^p) \cap \mathbb{F}[z]^m$ if and only if $N_r D_r^{-1} g \in \mathbb{F}[z]^p$. By the right coprimeness of $N_r(z), D_r(z)$, there exist polynomial matrices $X_\ell(z), Y_\ell(z)$ for which $X_\ell(z)N_r(z) + Y_\ell(z)D_r(z) = I$. This implies

$$D_r(z)^{-1}g(z) = X_\ell(z)N_r(z)D_r(z)^{-1}g(z) + Y_\ell(z)g(z) \in \mathbb{F}[z]^p,$$

i.e., $g(z) \in D_r\mathbb{F}[z]^m$, and so $G(z)^{-1}(\mathbb{F}[z]^p) \cap \mathbb{F}[z]^m = D_r\mathbb{F}[z]^m$. From this we conclude that

$$X(G) = \frac{\mathbb{F}[z]^m}{G(z)^{-1}\mathbb{F}[z]^p \cap \mathbb{F}[z]^m} = \frac{\mathbb{F}[z]^m}{D_r\mathbb{F}[z]^m} \simeq X_{D_r},$$

which completes the proof of the first assertion.

2. Note that $f(z) \in \mathbb{F}(z)^m$ satisfies $G(z)f(z) \in \mathbb{F}[z]^p$ if and only if $N_\ell(z)f(z) = D_\ell(z)a(z)$ is satisfied for a polynomial $a(z) \in \mathbb{F}[z]^p$. Thus, $N_\ell f$ is a polynomial for all $f(z) \in G(z)^{-1}\mathbb{F}[z]^p$. Let $\pi : \mathbb{F}[z]^p \longrightarrow \mathbb{F}[z]^p/N_\ell(z)\mathbb{F}[z]^m$ denote the canonical quotient map. It is verified next that the $\mathbb{F}[z]$-linear map

$$\tau : G(z)^{-1}\mathbb{F}[z]^p + \mathbb{F}[z]^m \longrightarrow \mathbb{F}[z]^p/N_\ell(z)\mathbb{F}[z]^m, \ f(z) \mapsto \pi(N_\ell f)$$

has kernel $\operatorname{Ker} \tau = \operatorname{Ker} G(z) + \mathbb{F}[z]^m$ and, therefore, induces a module isomorphism of $Z(G)$ onto a submodule of $\mathbb{F}[z]^p/N_\ell(z)\mathbb{F}[z]^m$. In fact, for $f \in G(z)^{-1}\mathbb{F}[z]^p + \mathbb{F}[z]^m$ one has $\pi(N_\ell f) = 0$ if and only if $N_\ell(z)f(z) = N_\ell(z)a(z)$ for a polynomial $a(z) \in \mathbb{F}[z]^m$. This is equivalent to $G(z)(f(z) - a(z)) = 0$, i.e., to $f \in \operatorname{Ker} G + \mathbb{F}[z]^m$. Thus $\operatorname{Ker} \tau = \operatorname{Ker} G + \mathbb{F}[z]^m$. For the proof that the image of τ coincides with the torsion submodule, see Wyman and Sain (1981). ∎

Instead of a matrix fraction representation, one can consider a generalized polynomial matrix description of the form

$$G(z) = V(z)T(z)^{-1}U(z) + W(z), \tag{4.44}$$

with $V(z), T(z), U(z)$, and $W(z)$ polynomial matrices and $T(z)$ nonsingular. Assuming the left coprimeness of $T(z)$ and $U(z)$ and the right coprimeness of $T(z)$ and $V(z)$, the pole information is completely determined by $T(z)$. A natural question that arises is the representation of the zero module in terms of the polynomial data that are represented by the polynomial matrices $V(z), T(z), U(z)$, and $W(z)$. As in (4.1), the polynomial system matrix associated with the representation (4.44) is

$$P(z) := \begin{pmatrix} T(z) & -U(z) \\ V(z) & W(z) \end{pmatrix}. \tag{4.45}$$

We already had, in Proposition 4.33, some indications that the polynomial system matrix (4.45) of a transfer function $G(z)$ behaves like the numerator in a matrix fraction description. Therefore, the following theorem, proved in Hautus and Fuhrmann (1980), is not totally unexpected. However, before stating it, the following notation is introduced. Considering $\mathbb{F}(z)^{s+p}$ as the direct sum $\mathbb{F}(z)^s \oplus \mathbb{F}(z)^p$, we define the projections $pr_1 : \mathbb{F}(z)^{s+p} \longrightarrow \mathbb{F}(z)^s$ and $pr_2 : \mathbb{F}(z)^{s+p} \longrightarrow \mathbb{F}(z)^p$ by

$$pr_1 \begin{pmatrix} f_1 \\ f_2 \end{pmatrix} = f_1, \quad pr_2 \begin{pmatrix} f_1 \\ f_2 \end{pmatrix} = f_2.$$

Theorem 4.45. *Let $G(z) \in \mathbb{F}(z)^{p \times m}$ be proper, having the representation (4.44), and let $P(z)$ be the associated $(s+p) \times (s+m)$ polynomial system matrix. Then:*

1. If $T(z)$ and $U(z)$ are left coprime, then

$$Z(G) = \frac{pr_2(P^{-1}(\mathbb{F}[z]^{s+p}) + \mathbb{F}[z]^m}{pr_2(\mathrm{Ker}\, P) + \mathbb{F}[z]^m};$$

2. If in addition $V(z)$ and $T(z)$ are right coprime, then $Z(P) \simeq Z(G)$, the isomorphism being the one induced by pr_2.

Proof. 1. We note first that if $\begin{pmatrix} u \\ v \end{pmatrix} \in \mathbb{F}(z)^{s+m}$, then $\begin{pmatrix} u \\ v \end{pmatrix} \in \mathrm{Ker}\, P$ if and only if $v \in \mathrm{Ker}\, G$ and $u = T^{-1}Uv$. So $\mathrm{Ker}\, G = pr_2 \mathrm{Ker}\, P$, and hence

$$\mathrm{Ker}\, G + \mathbb{F}[z]^m = pr_2(\mathrm{Ker}\, P) + \mathbb{F}[z]^m.$$

Next, we show the inclusion

$$G^{-1}(\mathbb{F}[z]^p) \subset pr_2(P^{-1}(\mathbb{F}[z]^{s+p}). \tag{4.46}$$

Indeed, if $h_2 \in G^{-1}(\mathbb{F}[z]^p)$, then $Gh_2 = p_2 \in \mathbb{F}[z]^p$. Defining $h_1 := -T^{-1}Uh_2$, then $P \begin{pmatrix} h_1 \\ h_2 \end{pmatrix} = \begin{pmatrix} 0 \\ p_2 \end{pmatrix} \in \mathbb{F}[z]^{s+p}$, which proves (4.46). In particular, we obtain the inclusion

$$G^{-1}(\mathbb{F}[z]^p) + \mathbb{F}[z]^m \subset pr_2 P^{-1}(\mathbb{F}[z]^{s+p}) + \mathbb{F}[z]^m. \tag{4.47}$$

Using the left coprimeness of $T(z)$ and $U(z)$, we next show that

$$pr_2 P^{-1}(\mathbb{F}[z]^{s+p}) \subset G^{-1}(\mathbb{F}[z]^p) + \mathbb{F}[z]^m. \tag{4.48}$$

By the left coprimeness of $T(z)$ and $U(z)$, we have

$$T\mathbb{F}[z]^s + U\mathbb{F}[z]^m = \mathbb{F}[z]^s. \tag{4.49}$$

If $h_2 \in pr_2 P^{-1}(\mathbb{F}[z]^{s+p})$, then there exists h_1 such that

$$P\begin{pmatrix} h_1 \\ h_2 \end{pmatrix} = \begin{pmatrix} y_1 \\ y_2 \end{pmatrix} \in \mathbb{F}[z]^{s+p}.$$

By (4.49), there exist polynomial vectors p_1 and p_2 for which $Tp_1 - Up_2 = y_1$. We obtain the equations

$$Th_1 - Uh_2 = Tp_1 - Up_2$$

and

$$Vh_1 + Wh_2 = y_2.$$

Since $h_1 = p_1 + T^{-1}U(h_2 - p_2)$, we obtain, by substitution, that $h_2 - p_2 \in G^{-1}(\mathbb{F}[z]^p)$ or $h_2 \in G^{-1}(\mathbb{F}[z]^p) + \mathbb{F}[z]^m$. This proves (4.48), and hence

$$pr_2(P^{-1}(\mathbb{F}[z]^{s+p}) + \mathbb{F}[z]^m \subset G^{-1}(\mathbb{F}[z]^p) + \mathbb{F}[z]^m.$$

By (4.47), the equality $pr_2(P^{-1}(\mathbb{F}[z]^{s+p}) + \mathbb{F}[z]^m = G^{-1}(\mathbb{F}[z]^p) + \mathbb{F}[z]^m$ follows. This proves 1.

2. If $h_2 \in pr_2(\operatorname{Ker} P) + \mathbb{F}[z]^m$, then there exist a $k \in \mathbb{F}((z^{-1}))^s$ and a polynomial vector g such that

$$P\begin{pmatrix} k \\ h_2 + g \end{pmatrix} = \begin{pmatrix} 0 \\ 0 \end{pmatrix}$$

or

$$\begin{pmatrix} k \\ h_2 \end{pmatrix} \in \operatorname{Ker} P + \mathbb{F}[z]^{s+m},$$

which implies that $h_2 \in pr_2(\operatorname{Ker} P + \mathbb{F}[z]^{s+m})$.

Conversely, if $h_2 \in pr_2(\operatorname{Ker} P + \mathbb{F}[z]^{s+m})$, then there exist h_1 and polynomials g_1 and g_2 such that

$$P\begin{pmatrix} h_1 + g_1 \\ h_2 + g_2 \end{pmatrix} = \begin{pmatrix} 0 \\ 0 \end{pmatrix},$$

or $h_2 - g_2 \in pr_2 \text{Ker} P$, that is, $h_2 \in pr_2(\text{Ker} P) + \mathbb{F}[z]^m$, and so we have

$$pr_2(\text{Ker} P) + \mathbb{F}[z]^m = pr_2(\text{Ker} P + \mathbb{F}[z]^{s+m}). \qquad (4.50)$$

In an analogous way we prove

$$pr_2(P^{-1}(\mathbb{F}[z]^{s+p}) + \mathbb{F}[z]^m = pr_2(P^{-1}(\mathbb{F}[z]^{s+p}) + \mathbb{F}[z]^{s+m}). \qquad (4.51)$$

Equalities (4.50) and (4.51) and Part 1 imply the surjectivity of the induced map $pr_2 : Z(P) \longrightarrow Z(G)$.

As a final step, we show that if V and T are right coprime, then the map pr_2 is also injective and, hence, is an isomorphism. To this end, assume that

$$\begin{pmatrix} h_1 \\ h_2 \end{pmatrix} \in P^{-1}(\mathbb{F}[z]^{s+p}) + \mathbb{F}[z]^{s+m}$$

and

$$pr_2 \begin{pmatrix} h_1 \\ h_2 \end{pmatrix} \in pr_2(\text{Ker} P) + \mathbb{F}[z]^m.$$

We will show that $\begin{pmatrix} h_1 \\ h_2 \end{pmatrix} \in \text{Ker} P + \mathbb{F}[z]^{s+m}$. Our assumption implies that $P \begin{pmatrix} h_1 \\ h_2 \end{pmatrix} \in \mathbb{F}[z]^{s+p}$. Also, there exist k and a polynomial $g \in \mathbb{F}[z]^m$ such that $P \begin{pmatrix} k \\ h_2 + g \end{pmatrix} = \begin{pmatrix} 0 \\ 0 \end{pmatrix}$, and so, as noted earlier, $G(h_2 + g) = 0$. We choose $g_2 = g$ and will show that $g_1 = h_1 + T^{-1}U(h_2 + g)$ is a polynomial. Using right coprimeness, there exist polynomial matrices S and R such that $ST - RV = I$, from which the equality

$$T^{-1}U = SU - RVT^{-1}U = SU + RW - RG$$

follows. Since $P \begin{pmatrix} h_1 \\ h_2 \end{pmatrix} \in \mathbb{F}[z]^{s+p}$, it follows that

$$(S \ R) P \begin{pmatrix} h_1 \\ h_2 \end{pmatrix} = h_1 + (SU + RW)h_2$$

is a polynomial. Furthermore,

$$\begin{aligned} g_1 &= h_1 + T^{-1}U(h_2 + g) \\ &= h_1 + (SU + RW - RG)(h_2 + g) \\ &= h_1 + (SU + RW)h_2 + (SU + RW)g \end{aligned}$$

is a polynomial. Here, we use the fact that $h_2 + g \in \operatorname{Ker} G$. This implies

$$\begin{pmatrix} h_1 + g_1 \\ h_2 + g_2 \end{pmatrix} \in \operatorname{Ker} P,$$

which proves Part 2.

∎

It is worth pointing out that there is an alternative approach to the study of zeros of linear systems that is based on geometric control. We will return briefly to this circle of ideas in Chapter 6.

4.8   Open-Loop Control

In the preceding sections of this chapter, following the path-breaking work of Kalman, the basic concepts of reachability and observability were introduced. The underlying question concerning reachability was: Can every state be reached from the zero state by a suitable control sequence? The reachability of a pair (A, B) was characterized by the Kalman reachability matrix $(B, AB, \ldots, A^{n-1}B)$, which is of full rank. Equivalently, reachability is characterized in terms of the Hautus condition by the left coprimeness of $zI - A, B$. When dealing with linear systems described by higher-order equations, as in (4.7), there is no a priori given state, and therefore reachability and observability are defined in terms of the associated shift realization (4.20). Whatever the representation of the system, the interesting question remains: For a desired state, how does one compute the control sequence (or sequences) that steers (steer) the system from the resting state to that desired state? There are several variants to this problem, some of which will be discussed in this chapter. These include the problem of controlling to zero, controlling in a quotient space, and reconstructing states from observations. Another important direction is toward the construction of steering controls for systems described by higher-order difference or differential equations.

Because a reachability map maps input functions to states, the natural thing to do is to invert it. However, if one assumes the reachability of a system, that reachability map will be surjective but not necessarily injective, so it is not properly invertible. To obtain a right inverse, one needs to factor out its kernel and restrict the reachability map to the associated quotient space. Moreover, the reachability map is a homomorphism over the ring of polynomials $\mathbb{F}[z]$. Thus the reduced reachability map has a representation of the form studied in Theorem 3.20, and its inversion is achieved by embedding an intertwining relation in a doubly coprime factorization. Because such an embedding is not unique, one can parameterize all controllers that steer to a given state. This opens up the possibility of studying various optimal control problems and leads, as will be shown, to explicit constructions of the inputs that steer the zero state to a desired state vector and will present our main approach to open-loop control.

1. Open-Loop Control for State Systems

An approach to the terminal-state problem from the point of view of inverting the reachability map is now presented. Consider a linear system given in state-space form as

$$x_{t+1} = Ax_t + Bu_t, \quad t \in \mathbb{Z},$$

or, equivalently, via the polynomial vector equation

$$(zI - A)x(z) = Bu(z). \tag{4.52}$$

In principle, there are two different ways to formalize the problem of steering from a zero initial state to a terminal state ξ in finite time τ, depending on whether the desired state is to be realized at time $t = 0$ or in the future. The first is the classical approach and assumes that $(A, B) \in \mathbb{F}^{n \times n} \times \mathbb{F}^{n \times m}$ is a reachable pair and that the system has been at rest till time $t = -\tau < 0$. For state vectors $\xi \in \mathbb{F}^n$, our aim is to compute control sequences that steer the system from the origin to state ξ at time $t = 0$. The simplest approach to the solution is to expand the time trajectories given by the polynomial vectors

$$x(z) = \sum_{i=0}^{\tau-1} x_{-i} z^i, \quad u(z) = \sum_{i=0}^{\tau} u_{-i} z^i$$

and equate the coefficients in (4.52). This leads to the equation $\xi = \sum_{i=0}^{\tau} A^i B u_{-i}$ or, in matrix form,

$$\xi = x_0 = \begin{pmatrix} B & AB & \dots & A^\tau B \end{pmatrix} \begin{pmatrix} u_0 \\ \vdots \\ u_{-\tau} \end{pmatrix}. \tag{4.53}$$

By our assumption on the reachability of the pair (A, B), this equation is solvable for all $\tau \geq n - 1$. The condition $\tau \geq n - 1$ is not necessary, and there may exist shorter control sequences that steer the system from the resting state to state ξ. As will be seen later, the exact minimal-length input sequence depends on the minimal indices of an associated polynomial matrix $D(z)$. In Chapter 6, it will be explained how the minimal indices are related to the fine structure of the pair (A, B) as manifested in its reachability indices. The down side of using (4.56) as a basis for the computation of the set of steering controllers is the lack of a convenient parameterization.

This short heuristic analysis indicates some of the shortcomings of the state-space approach and suggests turning to a functional, or module-theoretic, approach. To this end, $\mathbb{F}^n$, endowed with the $\mathbb{F}[z]$-module structure induced by A, is identified with the polynomial model X_{zI-A}. Moreover, to stay in line with standard traditions, the coefficients of the input polynomial are labelled by nonnegative integers, i.e., as $u(z) = \sum_{j=0}^{s} u_j z^j$. This convention implies that the input sequence applied to the system corresponds to the coefficients of $u(z)$, **listed in reverse order**. Next, the definition of the **reachability map** $\mathscr{R}_{(A,B)} : \mathbb{F}[z]^m \longrightarrow X_{zI-A}$ is recalled, namely,

$$\mathscr{R}_{(A,B)} : \mathbb{F}[z]^m \longrightarrow X_{zI-A} = \mathbb{F}^n,$$

$$\mathscr{R}_{(A,B)} u = \pi_{zI-A} Bu, \qquad u(z) \in \mathbb{F}[z]^m,$$

or, equivalently, by

$$\mathscr{R}_{(A,B)} \sum_{i=0}^{s} u_i z^i = \sum_{i=0}^{s} A^i B u_i, \qquad \sum_{i=0}^{s} u_i z^i \in \mathbb{F}[z]^m. \tag{4.54}$$

Thus, the states at time 0 that can be reached from the zero state in the remote past are exactly the points contained in the image of $\mathscr{R}_{(A,B)}$. The reachability map is an $\mathbb{F}[z]$-module homomorphism. Its image, $\mathrm{Im}\,\mathscr{R}_{(A,B)}$, is the **reachable subspace** of the pair (A,B). Thus, to compute controls that steer to a state ξ, one needs to invert the reachability map $\mathscr{R}_{(A,B)}$ [and by reversing the order of the coefficients in the computed polynomial $u(z)$]. Note that $\mathscr{R}_{(A,B)}$ has a large kernel that is a full submodule of $\mathbb{F}[z]^m$ and, hence, is representable as $D\mathbb{F}[z]^m$ for a nonsingular $D(z) \in \mathbb{F}[z]^{m \times m}$. To obtain an invertible map, the kernel must be factored out. Denote by $\mathscr{R}$ the **reduced reachability map**, namely, the map induced by $\mathscr{R}_{(A,B)}$ on the quotient space $\mathbb{F}[z]^m / D\mathbb{F}[z]^m$, which is identified with the polynomial model X_D. Thus the reduced reachability map is equal to

$$\mathscr{R} : X_D \longrightarrow X_{zI-A}, \qquad \mathscr{R} u = \pi_{zI-A} Bu. \tag{4.55}$$

By the assumption of reachability, $\mathscr{R}$ is surjective, whereas by construction it is injective; hence, $\mathscr{R}$ is an isomorphism.

Definition 4.46. Assume that

$$N(z)D(z)^{-1} = (zI - A)^{-1} B$$

is a right coprime factorization. An input sequence $u_0, \dots, u_T \in \mathbb{F}^m$ with associated input polynomial $u(z) = \sum_{i=0}^{T} u_i z^i$ is called **minimal** for $\xi \in \mathbb{F}^n$, provided that

$$\mathscr{R}_{(A,B)} u = \xi$$

and

$$D(z)^{-1} u(z)$$

is strictly proper. Equivalently, $u(z) = \mathscr{R}^{-1} \xi$.

Thus it turns out that computing the minimal control sequence is achieved by inverting the map $\mathscr{R}$. Applying our results on inverting polynomial model homomorphisms, this is best achieved by embedding intertwining maps in doubly coprime factorizations. The following diagram will help to better understand the various maps:

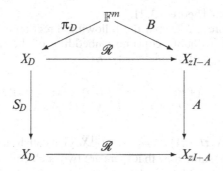

Theorem 4.47. *Let* $(A, B) \in \mathbb{F}^{n \times n} \times \mathbb{F}^{n \times m}$ *be a reachable pair, and let*

$$(zI - A)^{-1} B = N(z) D(z)^{-1} \tag{4.56}$$

be coprime factorizations, with $D(z) \in \mathbb{F}[z]^{m \times m}, N(z) \in \mathbb{F}[z]^{n \times m}$.

1. The intertwining relation

$$BD(z) = (zI - A) N(z) \tag{4.57}$$

can be embedded in the following doubly coprime factorization:

$$\begin{pmatrix} Y(z) & X(z) \\ -B & zI - A \end{pmatrix} \begin{pmatrix} D(z) & -\overline{X}(z) \\ N(z) & \overline{Y}(z) \end{pmatrix} = \begin{pmatrix} I & 0 \\ 0 & I \end{pmatrix}. \tag{4.58}$$

2. There exists a unique embedding (4.58) for which X *is constant. In that case, the pair* (X, A) *is an observable pair,* $D(z)^{-1} \overline{X}(z)$ *is strictly proper, and the columns of* $\overline{X}(z)$ *form a basis for the polynomial model* X_D.

3. The reachability map $\mathscr{R}_{(A,B)}$ *is an* $\mathbb{F}[z]$*-homomorphism, and its kernel is a submodule of* $\mathbb{F}[z]^m$ *with the representation*

$$\operatorname{Ker} \mathscr{R}_{(A,B)} = D\mathbb{F}[z]^m. \tag{4.59}$$

4. For each state $\xi \in \mathbb{F}^n$*, there exists a unique minimal control sequence* $u_{min}(z)$ *that steers the system from the zero state to* ξ *at time* $t = 0$*, and it is given by the coefficients of* $u_{min}(z) = \mathscr{R}^{-1} \xi$*. Specifically, in terms of the doubly coprime factorization (4.58), the minimal input polynomial is*

$$u_{min}(z) = \mathscr{R}^{-1} \xi = \overline{X} \xi. \tag{4.60}$$

5. *Every solution $u_*(z)$ to the steering problem is of the form*

$$u_*(z) = u_{min}(z) + D(z)g(z),$$

with $g(z) \in \mathbb{F}[z]^m$.

Proof. 1. Follows from Theorem 3.21.

2. Follows from Theorem 2.33, but the following direct argument may be useful, too. Let (4.58) be a doubly coprime embedding of the intertwining relation (4.57); then for every polynomial matrix $E(z)$,

$$\begin{pmatrix} Y(z) - E(z)B & X(z) + E(z)(zI - A) \\ -B & zI - A \end{pmatrix} \begin{pmatrix} D(z) & -\overline{X}(z) - D(z)E(z) \\ N(z) & \overline{Y}(z) - N(z)E(z) \end{pmatrix} = \begin{pmatrix} I & 0 \\ 0 & I \end{pmatrix}.$$

Choosing $E = \pi_+ X(zI - A)^{-1} = -\pi_+ D^{-1}\overline{X}$ gives such an embedding, which proves existence. Assume now that there are two such embeddings, with X_1 and X_2 constant matrices, which implies the two Bezout equations $Y_i D - X_i N = I$, $i = 1, 2$. Subtracting, one gets $\left((Y_2 - Y_1), -(X_2 - X_1) \right) \begin{pmatrix} D \\ N \end{pmatrix} = 0$, which implies that there exists a polynomial matrix $F(z)$ for which $\left((Y_2 - Y_1), -(X_2 - X_1) \right) = F(-B, zI - A)$. This implies that $F(z) = (X_2 - X_1)(zI - A)^{-1}$ is both polynomial as well as strictly proper and, hence, necessarily zero. This proves uniqueness. The observability of the pair (X, A) is a consequence of the right coprimeness of X and $zI - A$. That the columns of $\overline{X}$ form a basis for the polynomial model X_D is a consequence of the equality $X(zI - A)^{-1} = D^{-1}\overline{X}$ and Proposition 4.36.

3. Follows from the coprimeness of the factorizations (4.56).

4. The doubly coprime factorization (4.58) implies the intertwining relation

$$D(z)X = \overline{X}(z)(zI - A) \tag{4.61}$$

as well as the necessary coprimeness conditions for applying Theorem 3.21, thus (4.60) follows. Incidentally, the doubly coprime factorization (4.58) shows that the columns of $\overline{X}(z)$ form a basis $\mathcal{B}$ for X_D. In fact, there exists a unique basis $\mathcal{B}$ in X_D that is the inverse image of the standard basis in $\mathbb{F}^n$. This allows us to view the map $\mathcal{R}$ as a change of basis transformation $\mathcal{R}u(z) = [u]^{\mathcal{B}}$. By transposing the doubly coprime factorization (4.58), one infers that the columns of $N(z)^\top$ are the elements of the basis $\mathcal{B}^*$ of $X_{D^\top}$, which is dual to the basis $\mathcal{B}$ of X_D.

5. Follows from (4.59). ∎

An interesting question that is not answered by the preceding theorem is how to estimate the minimal degree of an input polynomial that steers 0 to ξ. Of course, the length of the control sequence $u_0, \ldots, u_T$ required to steer to the state ξ is equal to $\deg u(z) - 1$ of the polynomial $u(z) = \sum_{i=0}^{T} u_i z^i$, and it depends on the state. One might guess that the degree of minimal controllers $u_{min}(z)$ coincides with the minimal length T_{opt} of an input sequence that steers to ξ. In fact, in the single-input case $m = 1$, this is certainly true. Moreover, if $D(z) \in \mathbb{F}[z]^{m \times m}$ is diagonal with column (or row) degrees $\kappa_1, \ldots, \kappa_m$, it is easily seen that for polynomials $u(z) \in X_D$ and $f(z) = \mathrm{col}\,(f_1(z), \ldots, f_m(z)) \in \mathbb{F}[z]^m$ the identity

$$\deg(u + Df) = \max\left(\deg u, \max_{1 \le i \le m}\,(\kappa_i + \deg f_i)\right)$$

is valid. However, the general multivariable case $m > 1$ is not so easy to handle. In particular, it remains an open problem to decide under which conditions $\deg u_{\min}(z) \le \deg u_*(z)$ is satisfied for all open-loop controllers $u_*(z)$. Some of the difficulties that arise are illustrated by the following example.

Example 4.48. Let

$$D(z) = \begin{pmatrix} z^5 & 0 \\ 0 & z^2 \end{pmatrix},$$

and assume that $u_{\min} = \begin{pmatrix} 2z^4 + 3 \\ 3z - 1 \end{pmatrix} \in X_D$ is a "minimal" open-loop controller for a suitable terminal state. Then

$$u_*(z) = \begin{pmatrix} 2z^4 + 3 \\ 3z - 1 \end{pmatrix} + \begin{pmatrix} z^5 & 0 \\ 0 & z^2 \end{pmatrix}\begin{pmatrix} 0 \\ z + 6 \end{pmatrix} = \begin{pmatrix} 2z^4 + 3 \\ z^3 + 6z^2 + 3z - 1 \end{pmatrix}$$

is also an open-loop controller having the same degree. Note that

$$\deg \begin{pmatrix} z^5 & 0 \\ 0 & z^2 \end{pmatrix}\begin{pmatrix} 0 \\ z + 6 \end{pmatrix} = 3 < \deg u_*(z) = 4.$$

In particular, minimal-degree open-loop controllers need not be unique.

Our solution to the problem of computing open-loop controls used polynomial model theory. Because of the isomorphism between polynomial models on the one hand and rational models on the other, one could just as well use the later. From a system-theoretic point of view, this requires that the problem formulation be set in the future. So, for a state $\xi \in \mathbb{F}^n$, one looks for a controller that steers the system from the zero state to the terminal state ξ at time τ. For the time trajectories, we write $x(z) = \sum_{i=0}^{\infty} x_i z^{-i-1}$ and similarly for other trajectories.

2. Open-Loop Control for Higher-Order Systems

In many situations, the system under consideration is represented not in state-space terms, but rather by equations of higher order. Assuming $\xi(z), u(z)$ to be truncated Laurent series, then, over the full time axis $\mathbb{Z}$, the system is given by an equation of the form

$$T(z)\xi(z) = U(z)u(z), \tag{4.62}$$

with $T(z) \in \mathbb{F}[z]^{r \times r}$ nonsingular and $U(z) \in \mathbb{F}[z]^{r \times m}$. In cases with a continuous-time system, the action of z is interpreted as differentiation. Further, it will be assumed that $T(z)^{-1}U(z)$ is strictly proper, i.e., that (4.62) represents a strictly causal system. With the system equation (4.62) one can associate a natural input/state pair (A, B) by defining the polynomial model space X_T as in (3.2) and using the shift realization (4.20). By Theorem 4.26, the pair (A, B) so defined is reachable if and only if $T(z)$ and $U(z)$ are left coprime. Our principal goal is to extend Theorem 4.47 to the present situation, i.e., to compute controls that steer the system from the zero state to a state $f(z)$ in the state space X_T. Of course, one can apply realization theory to represent our system by first-order equations and solve the problem as before. However, our preference is to avoid matrix realization procedures and solve the problem in the context in which it is given, namely, in terms of higher-order equations. The following theorem, the counterpart of Theorem 4.47, summarizes the main results.

Proposition 4.49. *Let* $(T(z), U(z)) \in \mathbb{F}[z]^{r \times r} \times \mathbb{F}[z]^{r \times m}$ *be left coprime polynomial matrices with* $T(z)^{-1}U(z)$ *strictly proper. Then:*

1. *There exist right coprime polynomial matrices* $\overline{T}(z) \in \mathbb{F}[z]^{m \times m}, \overline{U}(z) \in \mathbb{F}[z]^{r \times m}$ *with*

$$T(z)^{-1}U(z) = \overline{U}(z)\overline{T}(z)^{-1} \qquad (4.63)$$

or, equivalently, that satisfy the intertwining relation

$$T(z)\overline{U}(z) = U(z)\overline{T}(z). \qquad (4.64)$$

Moreover, there exists a right coprime factorization $\overline{U}(z)\overline{T}(z)^{-1}$ *of* $T(z)^{-1}U(z)$ *for which the polynomial matrix* $\overline{T}(z) = \left(\overline{T}_1(z) \ldots \overline{T}_m(z) \right)$ *is column proper with column degrees* $\kappa_1 \geq \cdots \geq \kappa_m \geq 0$.

2. *The intertwining relation (4.64) can be embedded in the following doubly coprime factorization:*

$$\begin{pmatrix} Y(z) & X(z) \\ -U(z) & T(z) \end{pmatrix} \begin{pmatrix} \overline{T}(z) & -\overline{X}(z) \\ \overline{U}(z) & \overline{Y}(z) \end{pmatrix} = \begin{pmatrix} I & 0 \\ 0 & I \end{pmatrix}. \qquad (4.65)$$

3. *There exists a unique embedding (4.65) for which* $X(z)T(z)^{-1}$ *and* $\overline{T}(z)^{-1}\overline{X}(z)$ *are strictly proper. In that case, the pair* $(X(z), T(z))$ *induces, by way of the shift realization, an observable pair in the state space* X_T.

Proof. 1. By using the shift realization (4.20), the corresponding reachability map $\mathcal{R}_{(T,U)} : \mathbb{F}[z]^m \longrightarrow X_T$ is given by

$$\mathcal{R}_{(T,U)}u = \pi_T(Uu), \qquad u(z) \in \mathbb{F}[z]^m.$$

Clearly, $\mathcal{R}_{(T,U)}$ is an $\mathbb{F}[z]$-homomorphism that is surjective if and only if $T(z)$ and $U(z)$ are left coprime, which is equivalent to reachability. Assuming that, one notes that $\operatorname{Ker}\mathcal{R}_{(T,U)} \subset \mathbb{F}[z]^m$ is a submodule and, hence, has a representation

$$\operatorname{Ker}\mathcal{R}_{(T,U)} = \overline{T}\mathbb{F}[z]^m. \tag{4.66}$$

Computing, for $g(z) \in \mathbb{F}[z]^m$, $\pi_T U \overline{T} g = T \pi_- T^{-1} U \overline{T} g = 0$ implies that $\overline{U}(z) = T(z)^{-1} U(z) \overline{T}(z)$ is a polynomial matrix. This proves equations (4.63) and (4.64). Equation (4.66) implies the right coprimeness of $\overline{T}(z)$ and $\overline{U}(z)$. Equivalently stated, $\overline{U}(z)\overline{T}(z)^{-1}$ is a right coprime factorization of $T(z)^{-1}U(z)$.

2. Taking note of the intertwining relation (4.64) and the coprimeness assumptions, the statement follows by an application of Theorem 2.33.

3. The existence of such an embedding follows from Theorem 2.33, whereas the observability of the constructed pair follows from Theorem 4.26. ∎

The solution to the minimum-time, open-loop control problem is stated next.

Theorem 4.50. *1. For each state $f(z) \in X_T$, there exists a unique minimal control sequence that steers system (4.62) from the zero state to $f(z)$, and it is given by the (reversed) coefficients of $u_{min}(z) = \mathcal{R}^{-1}f$. Specifically, in terms of the doubly coprime factorization (4.65), one has*

$$u_{min}(z) = \mathcal{R}^{-1}f = \pi_{\overline{T}}\overline{X}f.$$

2. A control $u_*(z)$ steers the system from the zero state to $f(z)$ at time $t = 0$ if and only if there exists a $g(z) \in \mathbb{F}[z]^m$ for which

$$u_*(z) = u_{min}(z) + \overline{T}(z)g(z).$$

3. Assume that in a right coprime factorization $\overline{U}(z)\overline{T}(z)^{-1}$ the polynomial matrix $\overline{T}(z) = \left(\overline{T}_1(z) \ldots \overline{T}_m(z)\right)$ is column proper, with column degrees $\kappa_1 \geq \cdots \geq \kappa_m \geq 0$. Let $\tau > \kappa_1$. Then a controller $u_*(z)$ steers system (4.62) from the zero state at time $t = -\tau$ to a prescribed state $f(z) \in X_T$ at time $t = 0$ if and only if $u_*(z) = u_{min}(z) + \overline{T}(z)g(z)$, with $g(z) = \operatorname{col}(g_1(z),\ldots,g_m(z))$ satisfying $\max_{i=1,\ldots,m}(\deg \overline{T}_i + \deg g_i) = \tau - 1$.

Proof. 1. Since the polynomial matrices $\overline{T}(z)$ and $\overline{U}(z)$ are determined only up to a common right unimodular factor, we can use this freedom to reduce $\overline{T}(z)$ to a column proper form, with column degrees $\kappa_1 \geq \cdots \geq \kappa_m \geq 0$. Applying Theorem 3.20, this implies the existence of an isomorphism $Z : X_{\overline{T}} \longrightarrow X_T$,

$$Zf = \pi_T Uf, \qquad f(z) \in X_{\overline{T}}.$$

The pair $(\overline{T}(z), I)$ induces, by way of the shift realization, a reachable pair $(\overline{A}, \overline{B})$ in the state space $X_{\overline{T}}$. Noting the isomorphism $X_{\overline{T}} \simeq \mathbb{F}[z]^m / \overline{T}\mathbb{F}[z]^m$, we identify the two spaces and define the **reduced reachability map**, $\mathscr{R} : X_{\overline{T}} \longrightarrow X_T$ by $\mathscr{R}u = \pi_T U u$. This is summed up by the following diagram:

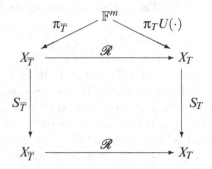

2. In view of the coprimeness assumptions, the intertwining relation (4.64) can be embedded in a doubly coprime factorization (4.65).
3. Assume that $u_*(z)$ steers to $f \in X_T$ and is of degree $\tau - 1 \geq \kappa_1$. Then $u_*(z) = u_{\min}(z) + \overline{T}(z)g(z)$. Since $u_{\min} \in X_{\overline{T}}$, the inequality $\deg u_{\min} < \kappa_1$ follows and therefore $\deg(\overline{T}g) = \max_{1 \leq i \leq m}(\deg \overline{T}_i + \deg g_i) = \deg u_* = \tau - 1$.
   For the converse direction assume that $u_*(z) = u_{\min}(z) + \overline{T}(z)g(z)$ is true for $g(z) = \operatorname{col}(g_1(z), \ldots, g_m(z))$ satisfying $\max_{i=1,\ldots,m}(\deg \overline{T}_i + \deg g_i) = \tau - 1 \geq \kappa_1$. Then u_* steers to f. Since each component of $u_{\min}$ has degree $< \kappa_1 \leq \tau - 1$, we see that $\deg u_* = \tau - 1$. This completes the proof. ∎

Observe that in the proof of Theorem 4.50, a unimodular embedding of the polynomial matrix $\left(T(z) \; -U(z) \right)$ was introduced for the sole purpose of inverting the reduced reachability map $\mathscr{R}$. However, once a unimodular embedding is available, it becomes natural to consider the system described by the equations

$$T(z)\xi = U(z)u,$$
$$y = V(z)\xi + W(z)u.$$

Thus, a **flat output** is added to system (4.62). This suggests strongly that it may be of interest to conduct a deeper study of the use of flat outputs in problems of computing terminally constrained steering controllers. In fact, they are a strong tool in the solution of problems other than that of steering to a prescribed state. Typical examples include steering to constrained partial state trajectories.

3. Reachability in a Quotient Space

There are situations when a subspace of the state space is given, and one need not control the system to a specific state in that subspace. In that case, the aim should be to reach all states of a quotient space, and this is the next subject of discussion. To this end, let (A, B) be a reachable pair and $N(z)D(z)^{-1}$ a right coprime factorization of $(zI - A)^{-1}B$. Assume that the intertwining relation

$$BD(z) = (zI - A)N(z) \tag{4.67}$$

is embedded in the doubly coprime factorization (4.58). Let $\mathcal{V} \subset X_{zI-A}$ be an m-dimensional invariant subspace. Choosing an appropriate basis in X_{zI-A}, we obtain the following representations:

$$A = \begin{pmatrix} A_1 & 0 \\ A_3 & A_2 \end{pmatrix}, \quad B = \begin{pmatrix} B_1 \\ B_2 \end{pmatrix}, \quad N = \begin{pmatrix} N_1 \\ N_2 \end{pmatrix}, \quad \mathcal{V} = \{ \begin{pmatrix} 0 \\ x \end{pmatrix} \mid x \in \mathbb{F}^m \}, \quad (4.68)$$

and we define

$$E(z) = \begin{pmatrix} zI - A_1 & 0 \\ -A_3 & zI - A_2 \end{pmatrix}, \quad E_1(z) = \begin{pmatrix} zI - A_1 & 0 \\ -A_3 & I \end{pmatrix}, \quad E_2(z) = \begin{pmatrix} I & 0 \\ 0 & zI - A_2 \end{pmatrix}.$$

Theorem 4.51. *Let $\mathcal{R} : X_D \longrightarrow X_{zI-A}$ be the reduced reachability map defined by (4.55).*

1. *The invariant subspace $\mathcal{V}$, defined in (4.68), has the representation*

$$\mathcal{V} = X_{\begin{pmatrix} I & 0 \\ 0 & zI - A_2 \end{pmatrix}} = \begin{pmatrix} 0 \\ X_{zI-A_2} \end{pmatrix}$$

which corresponds to the factorization $E(z)) = E_1(z)E_2(z)$, or, explicitly, to

$$\begin{pmatrix} zI - A_1 & 0 \\ -A_3 & zI - A_2 \end{pmatrix} = \begin{pmatrix} zI - A_1 & 0 \\ -A_3 & I \end{pmatrix} \begin{pmatrix} I & 0 \\ 0 & zI - A_2 \end{pmatrix}.$$

2. *Using the isomorphism $X_{E_1} \simeq X_{E_1 E_2}/E_1 X_{E_2}$, we compose $\mathcal{R}$ with the projection π_{E_1} to get a surjective map $\overline{\mathcal{R}} : X_D \longrightarrow X_{E_1}$ given, for $f(z) \in X_D$, by*

$$\overline{\mathcal{R}} f = \pi_{E_1} B f. \quad (4.69)$$

3. *There exists a factorization $D(z) = D_1(z)D_2(z)$, with $D_2 = g.c.r.d.(D, E_2 N)$, for which $\mathrm{Ker}\,\overline{\mathcal{R}} = D_1 X_{D_2}$.*
4. *There exists a factorization*

$$E_2(z)N(z) = \overline{H}(z)D_2(z), \quad (4.70)$$

with a suitable polynomial factor $\overline{H}(z)$ and $D_1(z), \overline{H}(z)$ right coprime. The intertwining relation

$$BD_1(z) = E_1(z)\overline{H}(z) \quad (4.71)$$

can be embedded in the doubly coprime factorization

$$\begin{pmatrix} \Theta_1(z) & \Xi_1(z) \\ -B & E_1(z) \end{pmatrix} \begin{pmatrix} D_1(z) & -\overline{\Xi}_1(z) \\ \overline{H}(z) & \overline{\Theta}_1(z) \end{pmatrix} = \begin{pmatrix} I & 0 \\ 0 & I \end{pmatrix}. \tag{4.72}$$

5. *The inverse of the map $\overline{\mathscr{R}}$, defined in (4.69), is given by*

$$\mathscr{R}^{-1}\xi = \pi_{E_1}\overline{\Xi}_1\xi.$$

Proof. 1. Follows by computing

$$\mathscr{V} = \begin{pmatrix} zI - A_1 & 0 \\ -A_3 & I \end{pmatrix} X \begin{pmatrix} I & 0 \\ 0 & zI - A_2 \end{pmatrix} = \begin{pmatrix} 0 \\ X_{zI-A_2} \end{pmatrix}.$$

2. This follows by computing, for $f(z) \in X_D$,

$$\overline{\mathscr{R}}f = \pi_{E_1}\pi_E Bf = E_1\pi_- E_1^{-1}E_1 E_2\pi_- E_2^{-1}E_1^{-1}Bf = E_1\pi_- E_2\pi_- E_2^{-1}E_1^{-1}Bf$$
$$= E_1\pi_- E_2 E_2^{-1}E_1^{-1}Bf = E_1\pi_- E_1^{-1}Bf = \pi_{E_1}Bf.$$

3. Clearly, $\overline{\mathscr{R}}$ is surjective. However, it has a kernel that is necessarily a submodule of X_D; hence, it has a representation $D_1 X_{D_2}$ for a factorization $D(z) = D_1(z)D_2(z)$.

4. Applying Theorem 3.21 yields $D_2(z) = g.c.r.d.(D(z), E_2(z)N(z))$. This implies the existence of the factorization (4.70). The intertwining relation (4.67), i.e., $B(D_1(z)D_2(z)) = E_1(z)(E_2(z)N(z)) = E_1(z)(\overline{H}(z)D_2(z))$, implies in turn the intertwining relation

$$BD_1(z) = E_1(z)\overline{H}(z) \tag{4.73}$$

as well as the isomorphism $X_{D_1} \simeq X_{E_1}$. Because of the coprimeness conditions, the intertwining relation (4.70) can be embedded in the doubly coprime factorization (4.71). Rather than compute this embedding ab initio, it is related to the doubly coprime factorization (4.58). For this, some preliminary computations are needed. First, note that the intertwining relations (4.67) and (4.71) are compatible. Indeed, if the second relation is multiplied on the right by $D_2(z)$ and (4.73) is used, then

$$BD(z) = BD_1(z)D_2(z) = E_1(z)\overline{H}(z)D_2(z) = E_1(z)E_2(z)N(z) = E(z)N(z).$$

Next, the Bezout equations $-B\overline{\Xi}(z) + E_1(z)E_2(z)\overline{\Theta}(z) = I$ and $-B\overline{\Xi}_1(z) + E_1(z)\overline{\Theta}_1(z) = I$ are compared, which suggests the choice

$$\overline{\Xi}_1(z) = \overline{\Xi}(z)$$
$$\overline{\Theta}_1(z) = E_2(z)\overline{\Theta}(z).$$

By considering the intertwining relations $\overline{\Xi}(z)E(z) = D(z)\Xi(z)$ and $\overline{\Xi}_1(z)E_1(z) = D_1(z)\Xi_1(z)$, one computes

$$\overline{\Xi}(z)E(z) = \overline{\Xi}(z)E_1(z)E_2(z) = D_1(z)\Xi_1(z)E_2(z) = D_1(z)D_2(z)\Xi(z),$$

which implies $\Xi_1(z)E_2(z) = D_2(z)\Xi(z)$ or $\Xi_1(z) = D_2(z)\Xi(z)E_2(z)^{-1}$. Comparing the Bezout equations $\Theta_1(z)D_1(z) - \Xi_1(z)N(z) = I$ and $\Theta(z)D(z) - \Xi(z)N(z) = I$ leads to

$$I = \Theta(z)D_1(z)D_2(z) - \Xi(z)N(z) = \Theta(z)D_1(z)D_2(z) - D_2(z)^{-1}\Xi_1(z)E_2(z)N(z).$$

In turn, this implies $D_2(z)\Theta(z)D_1(z) - \Xi_1(z)E_2(z)N(z)D_2(z)^{-1} = (D_2(z)\Theta(z))$ $D_1(z) - \Xi_1(z)N(z) = I$. Consequently, $\Theta_1(z) = D_2(z)\Theta(z)$ follows.

5. Follows from Theorem 4.47 and the doubly coprime factorization (4.72). ∎

4. Observability and State Reconstruction

We confine ourselves to a very brief exposition of the problem dual to that of computing open-loop controls that steer a system from zero to a desired state. Thus, the problem of interest is the reconstruction of a state from output observations. Assume the (not necessarily observable) pair $(C,A) \in \mathbb{F}^{p \times n} \times \mathbb{F}^{n \times n}$ represents the autonomous system

$$x_{t+1} = Ax_t, \quad x_0 = \xi,$$

$$y_t = Cx_t.$$

The **observability map**

$$\mathcal{O}_{(C,A)} : X_{zI-A} \longrightarrow z^{-1}\mathbb{F}[[z^{-1}]]^p$$

is defined by $\mathcal{O}_{(C,A)}\xi := C(zI - A)^{-1}\xi$. The kernel of the observability map, namely, $\mathcal{O}_* := \operatorname{Ker} \mathcal{O}_{(C,A)}$, is called the **unobservable subspace** of the pair (C,A) and is equal to $\operatorname{Ker} \mathcal{O}_{(C,A)} = \bigcap_{j=0}^{\infty} \operatorname{Ker} CA^j$.

Assume now that the pair (C,A) is observable, an assumption equivalent to the right coprimeness of C and $zI - A$. Let $D_\ell(z)^{-1}N_\ell(z)$ be a left coprime factorization of $C(zI - A)^{-1}$, which leads to the intertwining relation

$$N_\ell(z)(zI - A) = D_\ell(z)C. \tag{4.74}$$

By observability, the observability map $\mathcal{O}_{(C,A)}$ is injective but certainly not surjective. The left coprimeness of $D_\ell(z), N_\ell(z)$ implies

$$\operatorname{Im} \mathcal{O}_{(C,A)} = X^{D_\ell}. \tag{4.75}$$

Therefore, it is natural to consider the **reduced observability map** $\mathscr{O} : X_{zI-A} \longrightarrow X^{D_\ell}$. Taking the shift realization (4.20) associated with the left coprime factorization $D_\ell(z)^{-1}N_\ell(z)$ leads to the following commutative diagram:

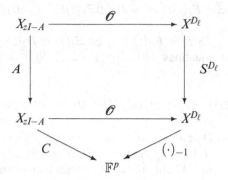

The coprimeness conditions allow us to embed the intertwining relation (4.74) in the following doubly coprime factorization:

$$\begin{pmatrix} Y(z) & X(z) \\ -N_\ell(z) & D_\ell(z) \end{pmatrix} \begin{pmatrix} zI - A & -\overline{X}(z) \\ C & \overline{Y}(z) \end{pmatrix} = \begin{pmatrix} I & 0 \\ 0 & I \end{pmatrix}. \tag{4.76}$$

The embedding is unique if we require $\overline{X}$ to be constant or, equivalently, that $X(z)D_\ell(z)^{-1}$ be strictly proper.

Theorem 4.52. *Let* $(C,A) \in \mathbb{F}^{p \times n} \times \mathbb{F}^{n \times n}$ *be observable, and let* $C(zI - A)^{-1} = D_\ell(z)^{-1}N_\ell(z)$ *be coprime factorizations, with* $D_\ell(z) \in \mathbb{F}[z]^{p \times p}, N_\ell(z) \in \mathbb{F}[z]^{p \times n}$. *Then:*

1. *The intertwining relation (4.74) can be embedded in the doubly coprime factorization (4.76);*
2. *The observability map* $\mathscr{O}_{(C,A)}$ *is an injective* $\mathbb{F}[z]$-*homomorphism onto the torsion submodule of* $z^{-1}\mathbb{F}[[z^{-1}]]^p$, *with representation* $\mathrm{Im}\,\mathscr{O}_{(C,A)} = X^{D_\ell}$;
3. *The reduced observability map* $\mathscr{O} = \mathscr{O}_{(C,A)} : X_{zI-A} \longrightarrow X^{D_\ell}$, *with*

$$\mathscr{O}\xi = D_\ell(z)^{-1}N_\ell(z)\xi,$$

is an isomorphism. For every observation sequence $y \in X^D$ *there is a unique state reconstructed from* y, *and it is given by* $\xi = \mathscr{O}^{-1}y$. *Specifically, in terms of the doubly coprime factorization (4.76),*

$$\xi = \mathscr{O}^{-1}y = \pi_{zI-A}\overline{X}Dy. \tag{4.77}$$

Proof. 1. Follows trivially from Theorem 2.33.
2. Follows from Theorem 4.13.
3. The isomorphism property of $\mathscr{O} : X_{zI-A} \longrightarrow X^{D_\ell}$ was shown in the preceding claim. The rest is clear.   ∎

4.9 Exercises

1. Consider the higher-order system

$$\ddot{x}_1 + x_1 + x_2 = \dot{u}_2 + u_1,$$
$$\dot{x}_1 + \dot{x}_2 + 2x_2 = \dot{u}_2,$$
$$y_1 = x_1,$$
$$y_2 = x_2.$$

(a) Show that the polynomial system matrix is

$$\mathscr{P}(z) = \left(\begin{array}{c|c} T(z) & -U(z) \\ \hline V(z) & W(z) \end{array} \right) = \left(\begin{array}{cc|cc} z^2+1 & 1 & -1 & -z \\ z & z+2 & 0 & -z \\ \hline 1 & 0 & 0 & 0 \\ 0 & 1 & 0 & 0 \end{array} \right),$$

with transfer function

$$G(z) = \frac{1}{z^3 + 2z^2 + 2} \left(\begin{array}{cc} z+2 & z(z+1) \\ -z & z(z^2-z+1) \end{array} \right).$$

(b) Are $T(z)$ and $U(z)$ left coprime and $V(z)$ and $T(z)$ right coprime?
(c) Show that

$$A = \begin{pmatrix} 0 & 0 & -1 \\ 1 & 0 & 0 \\ 0 & 2 & -2 \end{pmatrix}, B = \begin{pmatrix} 1 & -1 \\ 0 & 1 \\ 0 & -2 \end{pmatrix}, C = \begin{pmatrix} 0 & 1 & 0 \\ 0 & -1 & 1 \end{pmatrix}, D = \begin{pmatrix} 0 & 0 \\ 0 & 1 \end{pmatrix}$$

is a realization of $G(z)$. Is (A,B,C) reachable or observable?

2. Let

$$\mathscr{P}(z) = \left(\begin{array}{cc|c} 1 & 0 & 0 \\ 0 & (z+1)^2 & z^3 \\ \hline 0 & -1 & 2-z \end{array} \right)$$

be a polynomial system matrix.

(a) Show that the transfer function is $G(z) = \frac{3z+2}{(z+1)^2}$.

(b) Show that

$$A = \begin{pmatrix} 0 & 1 \\ 1 & -2 \end{pmatrix}, \ B = \begin{pmatrix} 2 \\ 3 \end{pmatrix}, \ C = (0 \ 1)$$

is a matrix representation of the shift realization.

3. Consider the continuous-time system

$$\dot{x}_1 + \dot{x}_2 + x_1 = 0,$$
$$\dot{x}_2 + x_2 = u,$$
$$y = x_1,$$

with zero initial conditions.

(a) Prove that the associated polynomial system matrix is

$$\mathscr{P}(z) = \begin{pmatrix} T(z) & -U(z) \\ V(z) & 0 \end{pmatrix} = \left(\begin{array}{cc|c} z+1 & z^2 & 0 \\ 0 & z+1 & -1 \\ \hline 1 & 0 & 0 \end{array} \right).$$

(b) Prove that the transfer function is $G(z) = -1 + \frac{2z+1}{(z+1)^2}$.

(c) Show that $\left\{ \begin{pmatrix} 1 \\ 0 \end{pmatrix}, \begin{pmatrix} z \\ 1 \end{pmatrix} \right\}$ is a basis for the polynomial model X_T.

(d) Show that

$$A = \begin{pmatrix} -1 & -1 \\ 0 & -1 \end{pmatrix}, \quad B = \begin{pmatrix} 1 \\ 1 \end{pmatrix}, \quad C = (1 \ 1)$$

is a matrix representation of the shift realization. Show that (A,B,C) is reachable and observable.

4. Let (A,B,C,D) denote a continuous-time linear system defined over the field of real numbers $\mathbb{R}$, $v \in \mathbb{R}^m$ a fixed vector, and λ a real number that is not an eigenvalue of A. Let $G(s) = D + C(sI - A)^{-1}B$ denote the transfer function. Then the response of

$$\dot{x} = Ax + Bu,$$
$$y = Cx + Du$$

to the input function $u(t) = ve^{\lambda t}$ is

$$y(t) = Ce^{tA}\left(x_0 - (\lambda I - A)^{-1}Bv \right) + G(\lambda)ve^{\lambda t}.$$

5. Let $A \in \mathbb{F}^{n \times n}$ and $b \in \mathbb{F}^n$, $q(z) = \det(zI - A)$ and $p_1(z), \ldots, p_n(z) \in \mathbb{F}[z]$ be unique polynomials with

$$(zI - A)^{-1} b = \frac{1}{q(z)} \begin{pmatrix} p_1(z) \\ \vdots \\ p_n(z) \end{pmatrix}.$$

Show that (A, b) is reachable if and only if $p_1(z), \ldots, p_n(z)$ are $\mathbb{F}$-linearly independent.

6. Let $(A, b) \in \mathbb{F}^{n \times n} \times \mathbb{F}^n$, with coprime factorization $(zI - A)^{-1} b = \frac{P(z)}{q(z)}$, and let

$$M_P = \begin{pmatrix} p_{1,0} & \cdots & p_{1,n-1} \\ \vdots & \ddots & \vdots \\ p_{n,0} & \cdots & p_{n,n-1} \end{pmatrix} \in \mathbb{F}^{n \times n}$$

denote the coefficient matrix of $P(z)$. That is, $p_i(z) = \sum_{j=0}^{n-1} p_{i,j} z^j$ for $i = 1, \ldots, n$.

a. Prove that the reachability of (A, b) implies that M_P is invertible.
b. Let $T \geq n$, (A, b) be reachable and $x_T \in \mathbb{F}^n$ a terminal state vector. Let $y = (y_T, \ldots, y_{T+n-1})^\top \in \mathbb{F}^n$ be the unique solution of

$$M_P y = x_T. \tag{4.78}$$

Show that the set of all input sequences $u = (u_0, \ldots, u_{T-1})^\top$ that steer $x_0 = 0$ in time T to x_T is the set of all vectors

$$u = \begin{pmatrix} q_n & & \\ \vdots & \ddots & \\ q_0 & \cdots & q_n \\ & \ddots & \ddots \\ & & q_0 & \cdots & q_n \end{pmatrix} \begin{pmatrix} y_n \\ \vdots \\ y_{T+n-1} \end{pmatrix}.$$

Here, the $T - n$ variables $y_n, \ldots, y_{T-1}$ can be arbitrarily chosen, while $y_T, \ldots,$
y_{T+n-1} are uniquely determined by (4.78).

7. Let $N(z), M(z), D(z)$ be polynomial matrices, with $D(z)$ nonsingular, such that $M(z)D(z)^{-1}$ and $N(z)D(z)^{-1}$ are strictly proper. Show that there exists (A, B, C_1, C_2) with

$$G(z) = C_1(zI - A)^{-1} B, \quad M(z)D(z)^{-1} = C_2(zI - A)^{-1} B.$$

8. Let $G(z) = N(z)D(z)^{-1} \in \mathbb{F}(z)^{p \times m}$ be a right coprime factorization. Assume that the strictly proper function $G(z)$ has a minimal realization (A, B, C). Let $M(z)$ be a polynomial matrix such that $M(z)D(z)^{-1}$ is strictly proper. Show that $M(z)D(z)^{-1}$ has a realization of the form (A, B, C_0) for some C_0.

9. Prove that the two polynomial system matrices $\mathscr{P}_1$ and $\mathscr{P}_2$ of sizes $(p + r_i) \times (m + r_i)$, $i = 1, 2$, are FSE if and only if for every $q \geq \max(r_1, r_2)$ there exist $q \times q$-unimodular matrices M and N and polynomial matrices X and Y such that

$$
\begin{pmatrix} M(z) & 0 \\ -X(z) & I \end{pmatrix}
\begin{pmatrix} I_{q-r_1} & 0 & 0 \\ 0 & T_1(z) & -U_1(z) \\ 0 & V_1(z) & W_1(z) \end{pmatrix}
=
\begin{pmatrix} I_{q-r_2} & 0 & 0 \\ 0 & T_2(z) & -U_2(z) \\ 0 & V_2(z) & W_2(z) \end{pmatrix}
\begin{pmatrix} N(z) & Y(z) \\ 0 & I \end{pmatrix}.
$$

10. Let $P(z) \in \mathbb{F}[z]^{p \times m}$ be a full column rank polynomial matrix. Show that $P(z)$ has no finite zeros if and only if $P(z)$ is right prime.

11. A linear parameter-dependent state-space system

$$
x(t+1) = A(\theta)x(t)
$$
$$
y(t) = C(\theta)x(t)
$$

is called globally identifiable, provided that knowledge of the output trajectory $(y(t))_{t \geq 0}$ implies knowledge of both the parameter $\theta \in \mathbf{P}$ and the initial state x_0. Consider a family of left coprime factorizations

$$
\overline{D}_\theta(z)^{-1} \overline{N}_\theta(z) = C(\theta)(zI - A(\theta))^{-1}, \quad \theta \in \mathbf{P},
$$

with polynomial matrices $\overline{D}_\theta(z) \in \mathbb{F}[z]^{p \times p}$, $\overline{N}_\theta(z) \in \mathbb{F}[z]^{p \times n}$. Prove that the system is globally identifiable if and only if the following conditions are true:

(a) $(C(\theta), A(\theta))$ is observable for all $\theta \in \mathbf{P}$;
(b) For all $\theta \neq \theta'$ the polynomial matrices $\overline{D}_\theta(z)$ and $\overline{D}_{\theta'}(z)$ are right coprime.

4.10 Notes and References

The idea that the behavior of dynamical systems can be modeled by differential or difference equations can be traced back to Newton. The notion of state arose in the analysis of uniqueness results in the theory of differential equations. In the engineering context of systems theory, it made its first appearance in automata theory via the notion of Nerode equivalence. This in turn influenced Kalman's abstract module-theoretic approach to linear systems realization theory Kalman, Falb and Arbib (1969). The notions of reachability and observability also originate from there owing to their fundamental role in establishing and characterizing realizations of input/output data with minimal state dimension.

The use of infinite Hankel matrices for the characterization of rationality is due to Kronecker (1890). Kalman's approach to realization theory was initially developed in terms of restricted input/output maps, but concrete algorithms were also available such as the Kalman–Ho algorithm Kalman, Falb and Arbib (1969). In Chapter 10 of the same book, the fundamental insight of Kalman was explained thoroughly to treat finite-dimensional, linear, time-invariant systems as finitely generated $\mathbb{F}[z]$-torsion modules.

The tests for reachability appeared quite early as a technical condition, for example, Gantmacher (1959). However, the concepts of reachability and observability, as well as the realization that a linear system could be treated as a module over the ring of polynomials, are part of Kalman's great contributions to the algebraic theory of linear systems; see Kalman (1968), which contains some interesting historical comments. The Hautus test for reachability is from Hautus (1969) and appears also in the work by Popov (1973). The early work by Gilbert (1963) presents reachability conditions for parallel connected systems that are equivalent to the Hautus–Popov condition in special cases.

Polynomial matrix descriptions of linear systems were introduced by Rosenbrock (1970). It was an extremely influential contribution to the linear system theoretic literature. It prevented the total domination of state-space methods and helped to preserve the diversity of approaches to the study of linear systems. Rosenbrock's proof of the generalized pole placement theorem reinforced the strength of frequency-domain, functional methods.

The idea of using shift operators for the realization of system transfer functions originated in the mid-1970s in papers of Dewilde, Fuhrmann, and Helton, following the functional model results of Sz.-Nagy and Foias. The shift realization, given in Theorem 4.26, was motivated by Rosenbrock's work and introduces a canonical realization associated with every polynomial fraction decomposition of a transfer function. This first appeared in Fuhrmann (1976, 1977) and provides the link that connects the module-theoretic framework of Kalman and the more concrete polynomial matrix description methods of Rosenbrock. The concept of FSE is due to Fuhrmann (1977) and presents a relaxed version of Rosenbrock's strict system equivalence; in this context, see also Kailath (1980) and Özgüler (1994). The equivalence between Rosenbrock's notion and that of Fuhrmann has been shown by Smith (1981).

Polynomial system matrices have their own interpretation as representing the zero structure; see Hautus and Fuhrmann (1980) for details. Thus the study of rectangular polynomial matrices is intimately related to zeros of matrix rational functions. This is in turn the focal point of geometric control, and characterizations of zeros in terms of controlled and conditioned invariant subspaces are well known. An abstract, module-theoretic approach to zeros has been developed in a series of publications; see, for example, Wyman and Sain (1981); Wyman, Sain, Conte and Perdon (1989). Zeros also relate in a natural way to left and right inverses of linear systems and, thus, to factorization theory. Since in a left matrix fraction the numerator polynomial, which relates to zeros, is a rectangular polynomial matrix, one would expect to find strong links between the module-theoretic study of zeros and behavior theory. However, this research direction seems to have been overlooked. To give proper credit to all these developments on zeros of systems by explaining the mutual relations among geometric control, polynomial system matrices, module-theoretic methods, and state-space approaches would require another book.

The problem of open-loop control via inversion of the reachability map leads directly to the problem of unimodular embedding and, hence, indirectly, to the study of flat outputs; see Levine and Nguyen (2003) and Trentelmann (2004) for further information on constructing flat outputs. In fact, the functional model approach can be viewed as a constructive alternative to flat outputs (a concept that was first introduced by Fliess et. al. (1995) in connection with state feedback linearizations) as a tool for the control of higher-order nonlinear systems. Though it might be tempting to elaborate further on the connection to polynomial models and flatness, this has not been done here because the constructive methods we achieved are fully sufficient for our purposes. Theorems 4.50 and 4.52 seem to be new.

Since polynomial models were introduced as an algebraic counterpart to coinvariant, that is backward-shift-invariant, subspaces of vectorial Hardy spaces H^2, it should come as no surprise that results obtained in the algebraic context can be lifted to the analytic context. This concerns in particular the open-loop control of finite-dimensional, continuous-time, antistable systems. In this case, we take the input function space to be $L^2(0, \infty)$. The Fourier–Plancherel transform then allows one to pass from L^2 spaces to the (more structured) Hardy spaces $H^2_\pm$ of the open left and right half-planes. As it turns out, the operator-theoretic results by Fuhrmann (1994b); Fuhrmann and Ober (1993) can be applied to obtain a new approach to optimal, minimum norm control, thereby generalizing Theorem 4.47 in a Hardy space context. This seems to be a promising direction for future research.

Part II
Algebraic Systems Theory: Advanced Topics

Chapter 5
Tensor Products, Bezoutians, and Stability

Mathematical structures often start from simple ones and are extended by various constructions to structures of increasing complexity. This process is to be controlled, and the guiding lines should include, among other things, applicability to problems of interest. The present chapter is devoted to a circle of ideas from abstract linear algebra that covers several topics of interest to us because of their applicability to the study of linear systems. These topics include bilinear forms defined on vector spaces, module homomorphisms over various rings, and the analysis of classes of special structured matrices such as Bezoutian, Hankel, and Toeplitz matrices. The connections to algebraic methods for analyzing 2D systems, i.e., modules over the ring $\mathbb{F}[z, w]$ of polynomials in two variables, are also explored. The unifying tools are tensor products defined for modules and vector spaces. The interaction between the diverse areas of structured matrix analysis, polynomial modules, and quadratic form theory becomes particularly evident within this context.

Tensor products provide a link between multilinear algebra and classical linear algebra by enabling one to represent multilinear maps as linear functions on a tensor product space. This technique applies both to vector spaces over a field and modules over a ring. The scenario becomes interesting and rich in the context of polynomial and rational models, which are modules over the ring $\mathbb{F}[z]$ and, at the same time, finite-dimensional vector spaces over the field $\mathbb{F}$. In this situation, the result of forming the tensor product depends critically on the algebraic context within which the tensor product is formed, i.e., whether one considers the tensor product over the field $\mathbb{F}$ or over the ring $\mathbb{F}[z]$. Taking the tensor product of two polynomial models $\mathscr{X} \otimes_{\mathbb{F}} \mathscr{Y}$ over the field $\mathbb{F}$ yields a space of polynomials in two variables. In contrast, the tensor product $\mathscr{X} \otimes_{\mathbb{F}[z]} \mathscr{Y}$ over the ring $\mathbb{F}[z]$ is a module of polynomials in one variable. In this chapter these constructions will be explained and their algebraic properties worked out in detail. As a useful byproduct, explicit characterizations of intertwining maps and homomorphisms of polynomial

© Springer International Publishing Switzerland 2015
P.A. Fuhrmann, U. Helmke, *The Mathematics of Networks of Linear Systems*, Universitext, DOI 10.1007/978-3-319-16646-9_5

models using tensor products will be derived. Following classical duality theory of polynomial models, the mutual relationships between tensor products of polynomial models and spaces of homomorphisms are expressed via the following commutative diagram:

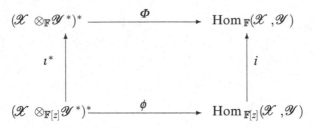

The horizontal maps denote natural isomorphisms and the vertical maps canonical injections. Although these maps are canonically defined, they crucially involve the construction of dual spaces. Therefore, to obtain a more concrete version of this diagram, a suitable duality theory for power series in two variables is first developed. This will enable us to find explicit descriptions for tensor products and duals of polynomial models and lead us, finally, to the concrete form of the preceding commutative diagram as

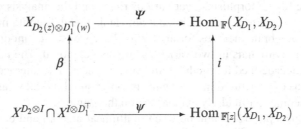

The construction of this commutative diagram is a central purpose of this chapter. It leads to a deeper understanding of the characterization of homomorphisms given in Chapter 3, provides us with a coordinate-free construction of Bezoutian matrices, and yields elegant matrix rank characterizations of coprimeness for matrix polynomials. Historically, the Euclidean algorithm applied to scalar polynomials suggested finding matrix criteria for the coprimeness of two polynomials. This led to the introduction, by Sylvester, of the Bezoutian and resultant matrices. Quadratic (and Hermitian) forms over the field of complex numbers were used efficiently by Hermite (1856) in his work on the root location of polynomials. Generalized Bezoutians, generated by a quadruple of polynomial matrices, originated in the work of Anderson and Jury (1976) in the analysis of coprimeness of polynomial matrices and the solvability of Sylvester-type equations. An interesting advance was the characterization of the Bezoutian matrix as a matrix representation of an intertwining map. Intertwining maps between polynomial models were characterized in Theorem 3.20, providing a powerful tool for the study of multivariable linear systems. Using tensor product representations of module homomorphisms, another proof of this important result will be given. On the way, polynomial versions of classical matrix equations are derived, of which the Sylvester and Lyapunov equations are important special cases. These polynomial Sylvester equations will also prove useful in deriving stability tests for nonsingular matrix polynomials.

5.1   Tensor Products of Modules

Tensor products of modules are at center stage of this chapter, so it is only appropriate to give a working definition. In algebra, the tensor product of two modules is defined abstractly via a universal property as follows.

Definition 5.1. Let R be a commutative ring with identity, and let M, N, L be R-modules. An R-module $M \otimes_R N$ is called a **tensor product** of M and N if there exists an R-bilinear map $\phi : M \times N \longrightarrow M \otimes_R N$ such that for every R-bilinear map $\gamma : M \times N \longrightarrow L$ there exists a unique R-homomorphism $\gamma_* : M \otimes_R N \longrightarrow L$ that makes the following diagram commutative:

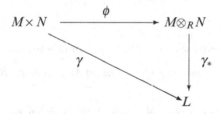

Although the notation $M \otimes_R N$ for the tensor product of two modules seems to suggest this, one should beware of assuming that the elements of $M \otimes_R N$ can be represented as single tensor products $m \otimes_R n$ of the elements $m \in M, n \in N$. In fact, the elements of the tensor product are finite sums $\sum_{i=1}^{k} m_i \otimes_R n_i$ and may not simplify to a decomposable representation of the form $m \otimes_R n$. Note further that, according to Definition 5.1, there may be several, necessarily isomorphic, tensor products.

One can give an abstract construction of a tensor product that is briefly sketched as follows; see, for example, Hungerford (1974). Let $< M \times N >$ denote the free R-module of all finite formal linear combinations $\sum_{i=1}^{k} r_i(m_i, n_i)$, where $r_i \in R, m_i \in M$, and $n_i \in N$. Let I_R denote the R-submodule of $< M \times N >$ that is generated by elements of the form

1. $r(m, n) - (rm, n)$, $r(m, n) - (m, rn)$;
2. $(m_1 + m_2, n) - (m_1, n) - (m_2, n)$;
3. $(m, n_1 + n_2) - (m, n_1) - (m, n_2)$.

Then the quotient R-module

$$M \otimes_R N := < M \times N > / I_R, \tag{5.1}$$

together with the map $\phi : M \times N \longrightarrow M \otimes_R N,\ (m,n) \mapsto m \otimes n := (m,n) + I_R$, satisfies the universal properties of a tensor product and thus serves as a model for the tensor product, unique up to isomorphisms. However, because most of the spaces we use have functional representations, one looks for concrete representations of the various tensor products encountered. As it is, such representations turn out to be amenable to explicit computations.

Listed below are a few basic properties of tensor products. Relative to direct sums, tensor products have the following distributivity and associativity properties:

$$(\oplus_{i=1}^{k} M_i) \otimes_R N \simeq \oplus_{i=1}^{k}(M_i \otimes_R N),$$

$$M \otimes_R (\oplus_{j=1}^{l} N_j) \simeq \oplus_{j=1}^{l}(M \otimes_R N_j), \tag{5.2}$$

$$M \otimes_R (N \otimes_R P) \simeq (M \otimes_R N) \otimes_R P.$$

Another useful isomorphism is

$$M \otimes_R N \simeq N \otimes_R M,$$

via the map that sends $m \otimes n$ to $n \otimes m$.

Concerning ring extensions, the following lemma is useful.

Lemma 5.2. *Let S be a subring of R, and let M and N be R-modules and L an S-module. Then:*

1. *The unique S-linear map $M \otimes_S N \longrightarrow M \otimes_R N$, which maps each element $m \otimes_S n$ to $m \otimes_R n$, is surjective.*
2. *Suppose that $b : M \times N \longrightarrow L$ is an S-bilinear map that satisfies*

$$b(rm,n) = b(m,rn)$$

for $r \in R$, $m \in M$, and $n \in N$. Then there exists a unique S-linear map

$$B : M \otimes_R N \longrightarrow L,$$

with $B(m \otimes n) = b(m,n)$.

Proof. Here it pays off to work with the abstract definition of a tensor product. The map $M \times N \longrightarrow M \otimes_R N,\ (m,n) \mapsto m \otimes_R n$ is S-bilinear and therefore induces a unique S-linear map $M \otimes_S N \longrightarrow M \otimes_R N$ that maps each $m \otimes_S n$ to $m \otimes_R n$. In fact, by inspection of (5.1), one sees that $I_S \subset I_R$ is valid and therefore induces a well-defined surjective map $< M \times N > /I_S \longrightarrow < M \times N > /I_R$, with $(m,n) + I_S \mapsto (m,n) + I_R$. This completes the proof of the first part.

Since each element of $M \otimes_R N$ is of the form $t = m_1 \otimes n_1 + \cdots + m_r \otimes n_r$, the additivity of B implies $B(t) = b(m_1, n_1) + \cdots + b(m_r, n_r)$. This implies the uniqueness of B. To prove the existence of B, define $B(r(m,n)) := b(rm,n)$ for all

$r \in R, m \in M, n \in N$. This extends to a well-defined map $\overline{B}$ on $< M \times N >$. The bilinearity of b implies that $\overline{B}$ is additive. Since $b(rm, n) = b(m, rn)$, it follows that $\overline{B}$ vanishes on I_R and therefore induces a unique S-linear map $B : M \otimes_R N \longrightarrow L$, with $B(m \otimes n) = b(m, n)$. This completes the proof. ∎

The tensor product of linear maps is defined as follows. Let $f : M_1 \longrightarrow N_1$ and $g : M_2 \longrightarrow N_2$ be R-linear maps between R-modules M_1, N_1 and M_2, N_2, respectively. Then $f \times g : M_1 \times M_2 \longrightarrow N_1 \otimes_R N_2, (m_1, m_2) \mapsto f(m_1) \otimes f(m_2)$ is R-bilinear and therefore extends in a natural way to an R-linear map,

$$f \otimes g : M_1 \otimes_R M_2 \longrightarrow N_1 \otimes_R N_2,$$

that maps each element $m_1 \otimes m_2$ to $f(m_1) \otimes f(m_2)$. We refer to $f \otimes g$ as the **tensor product** of f with g. It is clear that the tensor product of two module isomorphisms is a module isomorphism. If concrete matrix representations of f and g are given, the matrix representation of the tensor product $f \otimes g$ is equivalent to the so-called Kronecker product of matrices; we will return to this in Section 5.2 when we examine tensor products of polynomial models. Since polynomial models are concrete representations of polynomial quotient modules, it is only natural to expect that the analysis of tensor products of polynomial quotient modules, and of associated module homomorphisms, should prove useful for linear systems theory, and therefore particular attention will be paid to this topic.

The following result provides a very useful identification of tensor products of quotient modules. Let M_1 and M_2 be R-modules, with R a commutative ring. Let $N_i \subset M_i$ be submodules. The quotient spaces M_i/N_i then have a natural R-module structure.

Proposition 5.3. *Let N be the submodule of $M_1 \otimes_R M_2$ defined as*

$$N := N_1 \otimes_R M_2 + M_1 \otimes_R N_2.$$

The R-linear map

$$f : M_1 \otimes_R M_2 \longrightarrow M_1/N_1 \otimes_R M_2/N_2, \quad m_1 \otimes m_2 \mapsto (m_1 + N_1) \otimes (m_2 + N_2)$$

defines the following isomorphism of R-modules:

$$M_1/N_1 \otimes_R M_2/N_2 \simeq (M_1 \otimes_R M_2)/N.$$

Proof. By the construction of f, the kernel of f is contained in N. Moreover, f is surjective. Thus f induces a surjective homomorphism $\overline{f} : (M_1 \otimes_R M_2)/N \longrightarrow M_1/N_1 \otimes_R M_2/N_2$. The map

$$g : M_1/N_1 \otimes_R M_2/N_2 \longrightarrow (M_1 \otimes_R M_2)/N, \quad (m_1 + N_1) \otimes (m_2 + N_2) \mapsto m_1 \otimes m_2 + N$$

is well defined and is a left inverse to $\overline{f}$. Thus $\overline{f}$ is an isomorphism. ∎

The next result is useful in establishing isomorphisms between tensor product spaces.

Proposition 5.4. *Let S be a subring of R and $k \geq 3$. Let $M_1, \ldots, M_k$ be R-modules and N an S-module. Suppose that $f : M_1 \times \cdots \times M_k \longrightarrow N$ is an S-multilinear function that satisfies*

$$f(m_1, \ldots, m_{k-2}, rm_{k-1}, m_k) = f(m_1, \ldots, m_{k-2}, m_{k-1}, rm_k)$$

for all $r \in R$ and $m_i \in M_i$. Then there exists a unique S-multilinear function

$$F : M_1 \times \cdots \times M_{k-2} \times (M_{k-1} \otimes_R M_k) \longrightarrow N,$$

with

$$F(m_1, \ldots, m_{k-2}, m_{k-1} \otimes m_k) = f(m_1, \ldots, m_k). \tag{5.3}$$

Proof. Since the elements of $M \otimes_R N$ are finite sums of elementary tensors $a \otimes b$, the uniqueness of F follows from (5.3) and the additivity of F in the last argument. To construct F, we fix $m_1, \ldots, m_{k-2}$. By Lemma 5.2, there exists a unique S-linear map $F_{m_1 \cdots m_{k-2}} : M \otimes_R N \longrightarrow L$, with

$$F_{m_1 \ldots m_{k-2}}(m_{k-1} \otimes m_k) = f(m_1, \ldots, m_{k-2}, m_{k-1}, m_k).$$

By the S-multilinearity of f, this yields the desired S-linear map $F : M_1 \times \cdots \times M_{k-2} \times (M_{k-1} \otimes_R M_k) \longrightarrow N$, satisfying

$$F(m_1, \ldots, m_{k-2}, m_{k-1} \otimes m_k) = F_{m_1 \ldots m_{k-2}}(m_{k-1} \otimes m_k).$$

■

Consider a commutative ring R, and let $S \subset R$ be a subring. Let M and N be R-modules, and let L be an S-module. Note that the space $\mathrm{Hom}_S(N, L)$ of S-linear maps becomes an R-module by defining

$$(r \cdot f)(n) = f(rn)$$

for all $r \in R, n \in N$ and $f \in \mathrm{Hom}_S(N, L)$. For greater generality, one denotes by $\mathrm{Bil}_{S,R}(M, N; L)$ the set of all S-bilinear maps $f : M \times N \longrightarrow L$ that satisfy

$$f(rm, n) = f(m, rn) \tag{5.4}$$

for all $(m, n) \in M \times N$ and $r \in R$. In the special case where $S = R$, one uses the simplified notation $\mathrm{Bil}_R(M, N; L)$. It is easily seen that $\mathrm{Bil}_{S,R}(M, N; L)$ is, in a natural way, an R-module with respect to the R-scalar product

$$(r \cdot f)(m, n) = f(rm, n), \quad r \in R.$$

Proposition 5.5. *Let R be a commutative ring, let $S \subset R$ be a subring and let M and N be R-modules and L an S-module. Then the following are R-module isomorphisms:*

$$\mathrm{Hom}_S(M \otimes_R N, L) \simeq \mathrm{Bil}_{S,R}(M,N;L) \simeq \mathrm{Hom}_R(M, \mathrm{Hom}_S(N,L)).$$

Proof. Let $f \in Bil_{S,R}(M,N;L)$, and let $m \in M$. Defining, for $n \in N$, $f_m(n) = f(m,n)$, it follows that $f_m : M \longrightarrow \mathrm{Hom}_S(N,L)$ is S-linear for all $m \in M$. Since f satisfies (5.4), then $(r \cdot f_m)(n) = f(m,rn) = f(rm,n) = f_{rm}(n)$ for all $r \in R$. Thus $f_{rm} = r \cdot f_m$, i.e., the map f_m belongs to $\mathrm{Hom}_R(M, \mathrm{Hom}_S(N,L))$. On the other hand, given $g \in \mathrm{Hom}_R(M, \mathrm{Hom}_S(N,L))$, a map $f : M \times N \longrightarrow L$ is defined by $f(x,y) = g(x)(y)$, which is necessarily S-bilinear and satisfies (5.4) for each $r \in R$. This proves the isomorphism $\mathrm{Bil}_{S,R}(M,N;L) \simeq \mathrm{Hom}_R(M, \mathrm{Hom}_S(N,L))$. To prove the existence of an R-linear isomorphism

$$\mathrm{Hom}_S(M \otimes_R N, L) \simeq \mathrm{Hom}_R(M, \mathrm{Hom}_S(N,L)),$$

consider the map $\Theta : \mathrm{Hom}_S(M \otimes_R N, L) \longrightarrow \mathrm{Hom}_R(M, \mathrm{Hom}_S(N,L)), \Theta f = g$, where

$$\Theta(f)(m)(n) := f(m \otimes_R n), \qquad f \in \mathrm{Hom}_S(M \otimes_R N, L).$$

Clearly, Θ is R-linear. Similarly, the map

$$\Psi : \mathrm{Hom}_R(M, \mathrm{Hom}_S(N,L)) \longrightarrow \mathrm{Hom}_S(M \otimes_R N, L),$$

which is defined by

$$\Psi g(m \otimes_R n) := g(m)(n),$$

is R-linear and $\Psi \circ \Theta$ and $\Theta \circ \Psi$ are identity maps. The result follows. ∎

The duality properties of tensor products are of interest to us. Recall that the **algebraic dual module** M' of an R-module M is defined by

$$M' = \mathrm{Hom}_R(M,R), \tag{5.5}$$

i.e., by the space of all R-linear functionals on M or, equivalently, by the space of all R-homomorphisms of M into R. For the special case $L = R$, Proposition 5.5 implies the module isomorphisms

$$(M \otimes_R N)' = \mathrm{Hom}_R(M \otimes_R N, R) \simeq \mathrm{Hom}_R(M, N'). \tag{5.6}$$

Clearly, if M is an R-torsion module and the ring R has no zero divisors, then $M' = 0$. Unfortunately, the definition of the algebraic dual in the context of modules, namely, by (5.5), is of little use for the applications we have in mind. Of course, given a subring $S \subset R$, one can replace the algebraic dual M' with the S-dual

$$M^* := \mathrm{Hom}_S(M, S).$$

This is still an R-module that has in general better duality properties than M'. The best choice here would be to take S as a subfield of R, and this is what will be done in the sequel. The objects of interest to us are polynomial and rational models. Both have two structures: they are vector spaces over the field $\mathbb{F}$ and modules over the polynomial ring $\mathbb{F}[z]$. As $\mathbb{F}[z]$-modules, they are finitely generated torsion modules, and hence their algebraic dual, defined by (5.5), is the zero module. In much the same way, all objects defined in the subsequent isomorphism (5.6) are trivial. To overcome this problem, two module structures, over $\mathbb{F}$ and $\mathbb{F}[z]$ respectively, will be used and the algebraic dual replaced by the vector space dual.

1. Tensor products of vector spaces

Tensor products of vector spaces over a field $\mathbb{F}$ are studied next in somewhat more detail. This is of course a much simpler situation in which most of the pathologies encountered in studying tensor products over a ring disappear. As indicated previously, one needs to introduce vector space duality. For a finite-dimensional $\mathbb{F}$-vector space $\mathscr{X}$, the **vector space dual** is defined by

$$\mathscr{X}^* = \mathrm{Hom}_{\mathbb{F}}(\mathscr{X}, \mathbb{F}).$$

The **annihilator** of a subspace $\mathscr{V} \subset \mathscr{X}$ is defined as the linear subspace

$$\mathscr{V}^{\perp} := \{\lambda \in \mathscr{X}^* \mid \lambda|_{\mathscr{V}} = 0\}.$$

If $\mathscr{X}$ is finite-dimensional, then so is $\mathscr{V}^{\perp}$ and

$$\dim \mathscr{V}^{\perp} = \dim \mathscr{X} - \dim \mathscr{V}.$$

Finite-dimensional vector spaces are reflexive, i.e., the isomorphism $\mathscr{X}^{**} \simeq \mathscr{X}$ is satisfied. In fact, these two spaces can be identified by letting each vector $x \in \mathscr{X}$ act on an element $x^* \in \mathscr{X}^*$ by $x(x^*) = x^*(x)$. We now take a closer look at the case of tensor products of two finite-dimensional $\mathbb{F}$-linear spaces $\mathscr{X}, \mathscr{Y}$. Let $\mathscr{B}_X = \{f_i\}_{i=1}^n$, $\mathscr{B}_Y = \{g_i\}_{i=1}^m$ be bases of $\mathscr{X}$ and $\mathscr{Y}$, respectively. Let $\mathscr{B}_X^* = \{\phi_i\}_{i=1}^n$ be the basis of $\mathscr{X}^*$, which is dual to $\mathscr{B}_X$, i.e., it satisfies $\phi_i(f_j) = \delta_{ij}$. For a linear transformation $T \in \mathrm{Hom}_{\mathbb{F}}(\mathscr{X}, \mathscr{Y})$, let $t_{ij} \in \mathbb{F}$ be defined by

$$Tf_j = \sum_{i=1}^m t_{ij} g_i, \qquad j = 1, \ldots, n. \tag{5.7}$$

Thus $[T]_{\mathscr{B}_X}^{\mathscr{B}_Y} = (t_{ij})$ is the matrix representation with respect to this pair of bases. On the other hand, we consider the tensor product $\mathscr{Y} \otimes \mathscr{X}^*$, which is generated by the basis elements $g_i \otimes \phi_k$. Associate with $g_i \otimes \phi_k$ the linear map from $\mathscr{X}$ to $\mathscr{Y}$, defined for $x \in \mathscr{X}$ by

$$(g_i \otimes \phi_k)x = \phi_k(x)g_i.$$

We claim that $\{g_i \otimes \phi_k | i = 1, \ldots, m, k = 1, \ldots, n\}$ is a basis for $L(\mathscr{X}, \mathscr{Y})$. Indeed, $T = \sum_{i=1}^{m} \sum_{k=1}^{n} c_{ik} g_i \otimes \phi_k$ implies

$$Tf_j = \sum_{i=1}^{m} \sum_{k=1}^{n} c_{ik}(g_i \otimes \phi_k)f_j = \sum_{i=1}^{m} \sum_{k=1}^{n} c_{ik}\phi_k(f_j)g_i$$

$$= \sum_{i=1}^{m} \sum_{k=1}^{n} c_{ik}\delta_{kj}g_i = \sum_{i=1}^{m} c_{ij}g_i.$$

Comparing this with (5.7), it follows that $c_{ij} = t_{ij}$. Hence,

$$T = \sum_{i=1}^{m} \sum_{j=1}^{n} t_{ij} g_i \otimes \phi_j, \tag{5.8}$$

i.e., $\{g_i \otimes \phi_j\}$ is a basis for $\mathscr{Y} \otimes_{\mathbb{F}} \mathscr{X}^*$. This leads us back to the isomorphism

$$\mathrm{Hom}_{\mathbb{F}}(\mathscr{X}, \mathscr{Y}) \simeq \mathscr{Y} \otimes_{\mathbb{F}} \mathscr{X}^* \tag{5.9}$$

given by (5.8). The representation (5.8) of $T \in \mathrm{Hom}_{\mathbb{F}}(\mathscr{X}, \mathscr{Y})$ can be simplified. If $\mathrm{rank}\, T = k$, then there exists a minimal-length representation

$$T = \sum_{i=1}^{k} \psi_i \otimes \phi_i,$$

where $\{\phi_i\}_{i=1}^{k}$ is a basis of $(\mathrm{Ker}\, T)^{\perp} \subset \mathscr{X}^*$ and $\{\psi_i\}_{i=1}^{k}$ is a basis of $\mathrm{Im}\, T \subset \mathscr{Y}$. For a linear transformation $T \in \mathrm{Hom}_{\mathbb{F}}(\mathscr{X}, \mathscr{Y})$, the **adjoint transformation** $T^* \in \mathrm{Hom}_{\mathbb{F}}(\mathscr{Y}^*, \mathscr{X}^*)$ is defined by

$$(T^*y^*)x = y^*(Tx).$$

The map $T \mapsto T^*$ yields the isomorphism

$$\mathrm{Hom}_{\mathbb{F}}(\mathscr{X}, \mathscr{Y}) \simeq \mathrm{Hom}_{\mathbb{F}}(\mathscr{Y}^*, \mathscr{X}^*).$$

Clearly, (5.9) is a special case of (5.6). Another consequence of (5.9) is the following dimension formula:

$$\dim(\mathscr{X} \otimes_{\mathbb{F}} \mathscr{Y}) = \dim \mathscr{X} \cdot \dim \mathscr{Y}. \tag{5.10}$$

Note that in view of the isomorphism (5.9), for $T \in \operatorname{Hom}_{\mathbb{F}}(\mathscr{X}, \mathscr{Y})$ there are two possible matrix representations that turn out to be equal, namely,

$$[T]_{\mathscr{B}_X}^{\mathscr{B}_Y} = [T]^{\mathscr{B}_Y \otimes \mathscr{B}_X^*}.$$

Here $\mathscr{B}_Y \otimes \mathscr{B}_X^*$ is the tensor product of the basis $\mathscr{B}_Y$ of $\mathscr{Y}$ and the basis $\mathscr{B}_X^*$ of X^*, which is dual to the basis $\mathscr{B}_X$ of X.

2. Tensor products of $\mathbb{F}[z]$-modules

Next, the situation of principal interest to us, namely, the case where the vector spaces $\mathscr{X}$ and $\mathscr{Y}$ are actually $\mathbb{F}[z]$-modules, is addressed. The module structure on $\mathscr{X}$ defines a canonical $\mathbb{F}[z]$-module structure on the dual vector space $\mathscr{X}^*$ via

$$(z \cdot \lambda)(x) = \lambda(z \cdot x)$$

for $x \in \mathscr{X}$ and $\lambda \in \mathscr{X}^*$. As a consequence of the reflexivity property, there exists an isomorphism $\mathscr{X} \simeq \mathscr{X}^{**}$ as $\mathbb{F}[z]$-modules. Of particular interest is the establishment of a relation between the operations of forming the tensor product $\mathscr{X} \otimes \mathscr{Y}$ and the space of homomorphisms $\operatorname{Hom}(\mathscr{X}, \mathscr{Y})$. The dimension formula (5.10) implies that there is a well-defined $\mathbb{F}$-linear isomorphism

$$\mathscr{Y} \otimes_{\mathbb{F}} \mathscr{X}^* \longrightarrow \operatorname{Hom}_{\mathbb{F}}(\mathscr{X}, \mathscr{Y}), \quad (y, \lambda) \mapsto (x \mapsto \lambda(x)y) \qquad (5.11)$$

as long as $\mathscr{X}$ and $\mathscr{Y}$ are finite-dimensional $\mathbb{F}$-vector spaces. It is desirable to establish a similar $\mathbb{F}[z]$-linear isomorphism of modules $\mathscr{Y} \otimes_{\mathbb{F}[z]} \mathscr{X}^* \longrightarrow \operatorname{Hom}_{\mathbb{F}[z]}(\mathscr{X}, \mathscr{Y})$. For a ring R and R-modules M and N, it is well known, see Hilton and Wu (1974), that if M is finitely generated and projective, then the canonical map

$$\phi : N \otimes_R M' \longrightarrow \operatorname{Hom}_R(M, N), \quad (n, \lambda) \mapsto (m \mapsto \lambda(m)n) \qquad (5.12)$$

is a module isomorphism. Since every free module is projective, this covers the vector space case. Thus, it seems to be the appropriate generalization of (5.11) to $\mathbb{F}[z]$-modules. Unfortunately, the isomorphism (5.12) excludes the case of M being a torsion module, which is the case of main interest for us. This shows that the isomorphism (5.12) with the algebraic dual M' instead of the vector space dual M^* is of no further use for us.

Let $D(z) \in \mathbb{F}[z]^{m \times m}$ be a nonsingular polynomial matrix, and let X_D denote the associated polynomial model. Standard duality theory for polynomial models, as summarized in Theorem 3.38, shows that there exists an isomorphism of $\mathbb{F}[z]$-modules

$$X_D \simeq X_{D^\top}^* \simeq X_{D^\top}.$$

In fact, the isomorphism $X_D \simeq X_{D^\top}$ follows from the fact that both polynomial matrices $D(z)$ and $D^\top(z)$ have the same invariant factors. The isomorphism

$X_D \simeq X_{D^\top}^*$ follows from Theorem 3.38. The next result explores in more detail the various isomorphisms between spaces of homomorphisms and tensor products.

Proposition 5.6. *Let* $\mathbb{F}$ *be a field and* $\mathscr{X}$ *and* $\mathscr{Y}$ *be* $\mathbb{F}[z]$*-modules that are finite-dimensional as vector spaces over* $\mathbb{F}$*. The following assertions are true:*

1. There is a natural $\mathbb{F}[z]$*-module isomorphism*

$$\mathscr{X} \simeq \mathscr{X}^{**}.$$

2. There is a natural $\mathbb{F}[z]$*-module isomorphism*

$$\mathrm{Bil}_{\mathbb{F},\mathbb{F}[z]}(\mathscr{X},\mathscr{Y};\mathbb{F}) \simeq (\mathscr{X} \otimes_{\mathbb{F}[z]} \mathscr{Y}^*)^* \simeq \mathrm{Hom}_{\mathbb{F}[z]}(\mathscr{X},\mathscr{Y}). \tag{5.13}$$

3. There are natural $\mathbb{F}$*-vector space isomorphisms*

$$\mathscr{X}^* \otimes_{\mathbb{F}} \mathscr{Y} \simeq \mathrm{Bil}_{\mathbb{F}}(\mathscr{X},\mathscr{Y};\mathbb{F}) \simeq (\mathscr{X} \otimes_{\mathbb{F}} \mathscr{Y}^*)^* \simeq \mathrm{Hom}_{\mathbb{F}}(\mathscr{X},\mathscr{Y}). \tag{5.14}$$

Proof. 1. For $x \in \mathscr{X}$ let $\varepsilon_x : \mathscr{X} \longrightarrow \mathbb{F}$ denote the linear functional defined as $\varepsilon_x(\lambda) = \lambda(x)$. By the finite dimensionality of $\mathscr{X}$, the natural map $\varepsilon : \mathscr{X} \longrightarrow \mathscr{X}^{**}, x \mapsto \varepsilon_x$ is an $\mathbb{F}$-linear isomorphism. The $\mathbb{F}[z]$-module structure on $\mathscr{X}^{**}$ is defined as $(z \cdot \varepsilon_x)(\lambda) := \varepsilon_x(z\lambda) = \lambda(zx)$ for $x \in \mathscr{X}, \lambda \in \mathscr{X}^*$. Therefore, $z \cdot \varepsilon_x = \varepsilon_{zx}$ for all $x \in \mathscr{X}$, which proves that ε is $\mathbb{F}[z]$-linear. The result follows.

2. The isomorphism $(\mathscr{X} \otimes_{\mathbb{F}[z]} \mathscr{Y}^*)^* \simeq \mathrm{Hom}_{\mathbb{F}[z]}(\mathscr{X},\mathscr{Y})$ follows by applying Proposition 5.5 to the case of $S = \mathbb{F}$, $R = \mathbb{F}[z]$, $M = \mathscr{X}$, and $N = \mathscr{Y}^*$, noting that the $\mathbb{F}[z]$-modules $\mathscr{X}^{**}$ and $\mathscr{X}$ are isomorphic.

3. The last two natural isomorphisms in (5.14) follow from Proposition 5.5 by taking $R = S = \mathbb{F}$. The natural isomorphism between $\mathscr{X}^* \otimes_{\mathbb{F}} \mathscr{Y}$ and $\mathrm{Hom}_{\mathbb{F}}(\mathscr{X},\mathscr{Y})$ can be constructed explicitly by mapping the generating elements $x^* \otimes_{\mathbb{F}} y$ of the tensor product $\mathscr{X}^* \otimes_{\mathbb{F}} \mathscr{Y}$ onto the linear transformation from $\mathscr{X}$ to $\mathscr{Y}$, defined by $(x^* \otimes_{\mathbb{F}} y)(x) = (x^*(x))y$. ∎

Proposition 5.6 implies the following commutative diagram of linear maps:

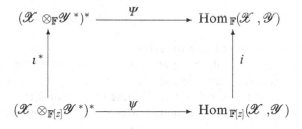

Here ι^* denotes the adjoint transformation of the canonical surjective $\mathbb{F}$-linear map $\iota : \mathscr{X} \otimes_{\mathbb{F}} \mathscr{Y}^* \longrightarrow \mathscr{X} \otimes_{\mathbb{F}[z]} \mathscr{Y}^*$ and therefore is injective. Map i is the canonical inclusion map and ψ and Ψ are the canonical $\mathbb{F}[z]$-linear and $\mathbb{F}$-linear isomorphisms defined by (5.13) and (5.14), respectively. The preceding diagram has the advantage of involving only canonical constructions of polynomial models. Its disadvantage lies in the usage of duality, which makes it hard to write down the maps in concrete form. Therefore, in the sequel, we will aim at a more concrete representation of the spaces appearing in this diagram that does not involve dual spaces. This new diagram will be derived in the following sections and will require concrete representations for tensor product spaces. It will also play an important role in our analysis of Bezoutians.

Next, one extends the duality theory from vector spaces to $\mathbb{F}[z]$-modules.

Definition 5.7. Let M be a module over the ring of polynomials $\mathbb{F}[z]$. The **vector space annihilator** of an $\mathbb{F}[z]$-submodule $N \subset M$ is the submodule of M^* defined as

$$N^\perp = \{\phi \in M^* \mid \phi|_N = 0\}.$$

Using this notation, the dual of a quotient module over $\mathbb{F}[z]$ has a nice representation. In fact, one has the $\mathbb{F}[z]$-linear isomorphism

$$(M/N)^* \simeq N^\perp.$$

The isomorphism (5.6) is extended to the tensor product of two quotient modules.

Proposition 5.8. *Let $N_i \subset M_i$, $i = 1,2$, be $\mathbb{F}[z]$-modules. There is an $\mathbb{F}[z]$-linear isomorphism*

$$(M_1/N_1 \otimes_{\mathbb{F}[z]} M_2/N_2)^* \simeq (M_1 \otimes_{\mathbb{F}[z]} N_2 + N_1 \otimes_{\mathbb{F}[z]} M_2)^\perp$$
$$= (N_1 \otimes_{\mathbb{F}[z]} M_2)^\perp \cap (M_1 \otimes_{\mathbb{F}[z]} N_2)^\perp.$$

Proof. Let N denote the submodule in $M_1 \otimes_{\mathbb{F}[z]} M_2$ that is generated by the spaces $N_1 \otimes_{\mathbb{F}[z]} M_2$ and $M_1 \otimes_{\mathbb{F}[z]} N_2$. Clearly, the equality of annihilators $N^\perp = (N_1 \otimes_{\mathbb{F}[z]} M_2 + M_1 \otimes_{\mathbb{F}[z]} N_2)^\perp$ is true. Thus Proposition 5.3 implies the module isomorphism

$$(M_1/N_1 \otimes_{\mathbb{F}[z]} M_2/N_2)^* \simeq (M_1 \otimes_{\mathbb{F}[z]} M_2)/N \simeq N^\perp = (N_1 \otimes_{\mathbb{F}[z]} M_2 + M_1 \otimes_{\mathbb{F}[z]} N_2)^\perp$$
$$= (N_1 \otimes_{\mathbb{F}[z]} M_2)^\perp \cap (M_1 \otimes_{\mathbb{F}[z]} N_2)^\perp. \qquad \blacksquare$$

3. Tensor product spaces of Laurent series

The ambient space for the algebraic analysis of discrete-time linear systems is $\mathbb{F}((z^{-1}))^m$. Thus as a first step one considers the tensor product of such spaces, both taken over the field $\mathbb{F}$ as well as over the ring of polynomials $\mathbb{F}[z]$. Clearly, the polynomial ring $\mathbb{F}[z]$ is a rank one module over itself but an infinite-dimensional vector space over $\mathbb{F}$.

Proposition 5.9. *The following* $\mathbb{F}$*-linear and* $\mathbb{F}[z]$*-linear isomorphisms are valid:*

$$\mathbb{F}[z]^p \otimes_{\mathbb{F}} \mathbb{F}[z]^m \simeq \mathbb{F}[z,w]^{p \times m}, \quad f \otimes_{\mathbb{F}} g \mapsto f(z)g(w)^{\top},$$

$$(5.15)$$

$$\mathbb{F}[z]^p \otimes_{\mathbb{F}[z]} \mathbb{F}[z]^m \simeq \mathbb{F}[z]^{p \times m}, \quad f \otimes_{\mathbb{F}[z]} g \mapsto f(z)g(z)^{\top}.$$

Proof. To prove (5.15), one notes that the map $\gamma : \mathbb{F}[z]^p \times \mathbb{F}[z]^m \longrightarrow \mathbb{F}[z,w]^{p \times m}$ that maps a pair of polynomials (f,g) to the polynomial matrix in two variables $f(z)g(w)^{\top}$ is $\mathbb{F}$-linear and therefore determines a unique $\mathbb{F}$-linear map $\gamma_* : \mathbb{F}[z]^p \otimes_{\mathbb{F}} \mathbb{F}[z]^m \longrightarrow \mathbb{F}[z,w]^{p \times m}$ for which $\gamma = \gamma_* \phi$. γ_* is surjective because every element $Q(z,w) \in \mathbb{F}[z,w]^{p \times m}$ is a finite sum $Q(z,w) = \sum_{i=1}^q f_i(z)g_i(w)^{\top} = \gamma_* \sum_{i=1}^q f_i \otimes_{\mathbb{F}} g_i$. To prove the injectivity of γ_*, we note that $\mathbb{F}[z]^m$ has a basis $\{z^i e_j | i \in \mathbb{Z}_+, j = 1, \ldots, m\}$. Therefore, each element of $\mathbb{F}[z]^p \otimes_{\mathbb{F}} \mathbb{F}[z]^m$ in the kernel of γ_* has the form $\xi = \sum_{(i,j) \in I} f_{ij} z^i e_j$, with $\gamma_*(\xi) = \sum_{(i,j) \in I} f_{ij}(z) w^i (e_j)^{\top} = 0$. Hence, $\sum_{(i,j) \in I} f_{ij}(z) w^i = 0$ for all j, implying that $f_{ij}(z) = 0$ for all i, j. Thus $\xi = 0$. Mutatis mutandis, using the $\mathbb{F}[z]$-bilinear map $\gamma : \mathbb{F}[z]^p \otimes_{\mathbb{F}} \mathbb{F}[z]^m \longrightarrow \mathbb{F}[z]^{p \times m}$ that maps a pair of polynomials (f,g) to the polynomial matrix $f(z)g(z)^{\top} \in \mathbb{F}[z]^{p \times m}$ exhibits an induced $\mathbb{F}[z]$-isomorphism $\mathbb{F}[z]^p \times \mathbb{F}[z]^m \longrightarrow \mathbb{F}[z]^{p \times m}$, which proves the second isomorphism in (5.15). ∎

The surjectivity of (5.15) can be reformulated as follows.

Proposition 5.10. *Every* $Q(z,w) \in \mathbb{F}[z,w]^{p \times m}$ *has a representation of the form*

$$Q(z,w) = \sum_{i=1}^k R_i(z) P_i^{\top}(w),$$

with $R_i(z) \in \mathbb{F}[z]^p$ *and* $P_i(w) \in \mathbb{F}[w]^m$*. This implies a factorization*

$$Q(z,w) = R(z) P^{\top}(w),$$

with $R(z) \in \mathbb{F}[z]^{p \times k}$ *and* $P(w) \in \mathbb{F}[w]^{m \times k}$*.*

To extend the previous results to Laurent series, several more spaces will be needed. Because the field $\mathbb{F}((z^{-1}))$ of truncated Laurent series has two module structures of interest, namely, with respect to the fields $\mathbb{F}$ and $\mathbb{F}((z^{-1}))$, there are two different tensor products, given by

$$\mathbb{F}((z^{-1}))^p \otimes_{\mathbb{F}} \mathbb{F}((z^{-1}))^m \simeq \mathbb{F}_{\text{sep}}((z^{-1}, w^{-1}))^{p \times m} \qquad (5.16)$$

and

$$\mathbb{F}((z^{-1}))^p \otimes_{\mathbb{F}((z^{-1}))} \mathbb{F}((z^{-1}))^m \simeq \mathbb{F}((z^{-1}))^{p \times m}. \qquad (5.17)$$

These are the analogs of equation (5.15).

Here, $\mathbb{F}_{\mathrm{sep}}((z^{-1},w^{-1}))$ denotes the ring of separable truncated Laurent series in the variables z and w, which are of the form $F(z,w) = \sum_{i=1}^{N} f_i(z)g_i(w)$ for finitely many $f_1, \ldots, f_N \in \mathbb{F}((z^{-1}))$, $g_1, \ldots, g_N \in \mathbb{F}((w^{-1}))$. Thus $\mathbb{F}_{\mathrm{sep}}((z^{-1},w^{-1}))$ is a proper subset of $\mathbb{F}((z^{-1},w^{-1}))$, the field of truncated Laurent series. By $\mathbb{F}_{\mathrm{sep}}((z^{-1},w^{-1}))^{p\times m}$ we denote the module of all $p \times m$ matrices with entries in $\mathbb{F}_{\mathrm{sep}}((z^{-1},w^{-1}))$. Rational elements $H(z,w) \in \mathbb{F}_{\mathrm{sep}}((z^{-1},w^{-1}))^{p\times m}$ have representations of the form $H(z,w) = \sum_{i=1}^{k} f_i(z)g_i(w)^{\top}$, with both $f_i(z)$ and $g_i(z)$ rational. This implies a representation of the form

$$H(z,w) = d(z)^{-1}Q(z,w)e(w)^{-1},$$

with $Q(z,w) \in \mathbb{F}[z,w]^{p\times m}$ and $e(w), d(z)$ nonzero, scalar polynomials.

The isomorphism $\mathbb{F}((z^{-1},w^{-1}))^{p\times m} \simeq \mathbb{F}^{p\times m}((z^{-1},w^{-1}))$ will be routinely used, and we will actually identify the two spaces. The identification $\mathbb{F}^{p\times m}[z,w] = \mathbb{F}[z,w]^{p\times m}$ is a special case. By $\mathbb{F}[z,w]$ we denote the ring of polynomials in the variables z and w and by $\mathbb{F}[[z^{-1},w^{-1}]]$ the ring of formal power series in z^{-1} and w^{-1}. Denote by $\mathbb{F}[z,w]^{p\times m}$ the space of $p \times m$ polynomial matrices. The elements of $z^{-1}\mathbb{F}[[z^{-1},w^{-1}]]^{p\times m}w^{-1}$ are called *strongly strictly proper*. It should be emphasized that not every strictly proper rational function in two variables belongs to $z^{-1}\mathbb{F}[[z^{-1},w^{-1}]]w^{-1}$. For example, $\frac{1}{z-w} \notin z^{-1}\mathbb{F}[[z^{-1},w^{-1}]]w^{-1}$. It will be convenient to use $\mathbb{F}[[z^{-1},w]]^{p\times m}$ to denote the subspace of $\mathbb{F}((z^{-1},w^{-1}))^{p\times m}$ of matrices whose entries are formal power series in z^{-1} and polynomial in w. Thus the elements of $\mathbb{F}[[z^{-1},w]]^{p\times m}$ are of the form $F(z,w) = \sum_{i=0}^{N} F_i(z)w^i$ for suitable $F_i(z) \in \mathbb{F}((z^{-1}))^{p\times m}$. Thus $\mathbb{F}[[z^{-1},w]]^{p\times m} \subset \mathbb{F}_{\mathrm{sep}}((z^{-1},w^{-1}))^{p\times m}$. The space $\mathbb{F}[z,w^{-1}]]^{p\times m}$ is similarly defined. In the same vein, $\mathbb{F}_{\mathrm{sep}}[[z^{-1},w^{-1}]]^{p\times m}$ denotes the space of $p \times m$ matrix functions of separable formal power series $\sum_{i=1}^{N} f_i(z)g_i(w)^{\top}$, with $f_i(z) \in \mathbb{F}[[z^{-1}]]^p$, $g_i(z) \in \mathbb{F}[[w^{-1}]]^m$.

The definition of $\mathbb{F}_{\mathrm{sep}}((z^{-1},w^{-1}))^{p\times m}$ implies the isomorphism

$$\mathbb{F}_{\mathrm{sep}}((z^{-1},w^{-1}))^{p\times m} \simeq \mathbb{F}((z^{-1}))^p \otimes_{\mathbb{F}} \mathbb{F}((z^{-1}))^m.$$

Taking into account the direct sum representations (3.1) as well as (5.2), one computes

$$\mathbb{F}_{\mathrm{sep}}((z^{-1},w^{-1}))^{p\times m} \simeq (\mathbb{F}[z]^p \oplus z^{-1}\mathbb{F}[[z^{-1}]]^p) \otimes_{\mathbb{F}} (\mathbb{F}[z]^m \oplus z^{-1}\mathbb{F}[[z^{-1}]]^m)$$

$$\simeq (\mathbb{F}[z]^p \otimes_{\mathbb{F}} \mathbb{F}[z]^m) \oplus (\mathbb{F}[z]^p \otimes_{\mathbb{F}} z^{-1}\mathbb{F}[[z^{-1}]]^m)$$

$$\oplus (z^{-1}\mathbb{F}[[z^{-1}]]^p \otimes_{\mathbb{F}} \mathbb{F}[z]^m) \oplus (z^{-1}\mathbb{F}[[z^{-1}]]^p \otimes_{\mathbb{F}} z^{-1}\mathbb{F}[[z^{-1}]]^m)$$

$$\simeq \mathbb{F}[z,w]^{p\times m} \oplus \mathbb{F}[z,w^{-1}]]^{p\times m}w^{-1} \oplus z^{-1}\mathbb{F}[[z^{-1},w]]^{p\times m} \oplus z^{-1}\mathbb{F}_{\mathrm{sep}}[[z^{-1},w^{-1}]]^{p\times m}w^{-1}.$$

To these direct sum representations correspond, respectively, the following projection identities:

$$I = \pi_+^z \otimes I + \pi_-^z \otimes I = I \otimes \pi_+^w + I \otimes \pi_-^w$$
$$= \pi_+^z \otimes \pi_+^w + \pi_-^z \otimes \pi_+^w + \pi_+^z \otimes \pi_-^w + \pi_-^z \otimes \pi_-^w.$$

Our next target is the extension of duality theory to the context of polynomial spaces in two variables. To this end, the space of matrix truncated Laurent series in two variables, i.e.,

$$\mathbb{F}((z^{-1}, w^{-1}))^{p \times m} = \{ G(z,w) = \sum_{i=-\infty}^{n_1} \sum_{j=-\infty}^{n_2} G_{ij} z^i w^j \},$$

is introduced. For $G(z,w) \in \mathbb{F}((z^{-1}, w^{-1}))^{p \times m}$, its **residue** is defined as the coefficient of $z^{-1} w^{-1}$, i.e., $G_{-1,-1}$. In analogy with (3.41), for $G(z,w), H(z,w) \in \mathbb{F}((z^{-1}, w^{-1}))^{p \times m}$, we define a bilinear form on $\mathbb{F}((z^{-1}, w^{-1}))^{p \times m}$ by

$$= \text{Trace}(H^\top G)_{-1,-1} = \text{Trace} \sum_{i=-\infty}^{\infty} \sum_{j=-\infty}^{\infty} H^\top_{-i-1,-j-1} G_{ij}$$

$$\tag{5.18}$$

$$= \sum_{i=-\infty}^{\infty} \sum_{j=-\infty}^{\infty} \text{Trace} \, H^\top_{-i-1,-j-1} G_{ij}.$$

Note that the sum defining $[G, H]$ contains only a finite number of nonzero terms. Clearly, the form defined in (5.18) is nondegenerate. If $G(z,w) \in \mathbb{F}((z^{-1}, w^{-1}))^{q \times m}$, $A(z,w) \in \mathbb{F}((z^{-1}, w^{-1}))^{p \times q}$, and $H(z,w) \in \mathbb{F}((z^{-1}, w^{-1}))^{p \times m}$, then $A(z,w) G(z,w) \in \mathbb{F}((z^{-1}, w^{-1}))^{p \times m}$ and

$$[AG, H] = [G, A^\top H].$$

It is easy to see that, with respect to the bilinear form (5.18), one has

$$(\mathbb{F}[z,w]^{p \times m})^\perp = \mathbb{F}[z, w^{-1}]]^{p \times m} + \mathbb{F}[[z^{-1}, w]^{p \times m}. \tag{5.19}$$

The next result gives a concrete representation of the dual space of $\mathbb{F}[z, w]^{p \times m}$ that is an extension of Theorem 3.38.

Proposition 5.11. *The vector space dual of* $\mathbb{F}[z, w]^{p \times m}$ *is* $\mathbb{F}$-*linear isomorphic to the space* $(z^{-1} \mathbb{F}[[z^{-1}, w^{-1}]]w^{-1})^{p \times m}$, *i.e.,*

$$(\mathbb{F}[z, w]^{p \times m})^* \simeq (z^{-1} \mathbb{F}[[z^{-1}, w^{-1}]]w^{-1})^{p \times m}.$$

Proof. Clearly, for $H(z,w) \in (z^{-1} \mathbb{F}[[z^{-1}, w^{-1}]]w^{-1})^{p \times m}$ and $Q(z,w) \in \mathbb{F}[z, w]^{p \times m}$, the map $\Phi : \mathbb{F}[z, w]^{p \times m} \longrightarrow \mathbb{F}$, defined by $\Phi(Q) = [Q, H]$, is a linear functional on $\mathbb{F}[z, w]^{p \times m}$. Conversely, suppose Φ is a linear functional on $\mathbb{F}[z, w]^{p \times m}$. For all

$i, j \geq 0$, Φ induces linear functionals Φ_{ij} on $\mathbb{F}^{p \times m}$ by defining $\Phi_{ij}(A) = \Phi(z^i A w^j)$. Every functional Φ_{ij} on $\mathbb{F}^{p \times m}$ has a representation of the form $\Phi_{ij}(A) = \mathrm{Trace}\,(H_{ij}^\top A)$ for a unique $H_{ij} \in \mathbb{F}^{p \times m}$. Defining

$$H(z,w) = \sum_{i=0}^{\infty} \sum_{j=0}^{\infty} H_{ij} z^{-i-1} w^{-j-1} \in z^{-1} \mathbb{F}[[z^{-1}, w^{-1}]]^{p \times m} w^{-1},$$

it follows that $\Phi(Q) = [Q, H]$.   ∎

5.2   Tensored Polynomial and Rational Models

Turning now to a detailed study of tensor products of polynomial and rational models, taken together with duality theory, will enable us to construct, in concrete terms, an isomorphism

$$\mathscr{X}^* \otimes_{\mathbb{F}[z]} \mathscr{Y} \longrightarrow \mathrm{Hom}_{\mathbb{F}[z]}(\mathscr{X}, \mathscr{Y})$$

and, in the process, develop a coordinate-free approach to Bezoutians.

Spaces like $\mathbb{F}[z]^m$ or, more importantly for our purposes, quotient spaces like $\mathbb{F}[z]^m / D(z) \mathbb{F}[z]^m$ have module structures with respect to both the field $\mathbb{F}$, i.e., vector space structures, and the ring of polynomials $\mathbb{F}[z]$. With respect to the characterization of tensor products, the underlying ring is of utmost importance because the tensor product depends very much on the ring used. These two constructs do not exhaust the possibilities, especially where polynomial models are concerned, and we will also study polynomial models defined by the Kronecker product of polynomial matrices. In analyzing the tensor products of two polynomial models, our first objective will be to find concrete representations of the various tensor products. Furthermore, it will be shown that the class of polynomial models is closed under tensor product operations. The inherent noncommutative situation in the case of nonsingular polynomial matrices makes things more difficult, especially if concrete isomorphisms are to be constructed. Of particular difficulty is the lack of a concrete representation of $X_{D_1} \otimes_{\mathbb{F}[z]} X_{D_2}$. This is the result of the absence of a nice representation of a two-sided greatest common divisor. The tensor products of function spaces given by (5.15), (5.16), and (5.17) just set the stage. In studying tensor products of polynomial or rational models, there are essentially four ways to proceed. For nonsingular polynomial matrices $D_1(z) \in \mathbb{F}[z]^{p \times p}$ and $D_2(z) \in \mathbb{F}[z]^{m \times m}$, one can study the $\mathbb{F}$ – and $\mathbb{F}[z]$ – tensor products, i.e., $X_{D_1} \otimes_{\mathbb{F}} X_{D_2}$ and $X_{D_1} \otimes_{\mathbb{F}[z]} X_{D_2}$, respectively. Additionally, one can study the polynomial, and rational, models defined by the Kronecker products $D_1(z) \otimes D_2^\top(w)$ and $D_1(z) \otimes D_2^\top(z)$, respectively. Later on, it will be shown that the $\mathbb{F}$-tensor product $X_{D_1} \otimes_{\mathbb{F}} X_{D_2}$ and the polynomial model $X_{D_1(z) \otimes D_2^\top(w)}$ are isomorphic, which will reduce the complexity to the study of three distinct spaces.

Before starting the analysis of the four different tensor product representations, some useful notation and terminology must be established. For rectangular matrices $A \in \mathbb{F}^{m \times n}, B \in \mathbb{F}^{k \times \ell}$, and in this chapter only, the **Kronecker product** is defined as the $mk \times n\ell$ matrix

$$A \otimes B = \begin{pmatrix} b_{11}A & \cdots & b_{1\ell}A \\ \vdots & \ddots & \vdots \\ b_{k1}A & \cdots & b_{k\ell}A \end{pmatrix}.$$

Note that this definition is in harmony with the definition of the tensor product $f \otimes g$ of two linear maps $f : \mathbb{F}^n \longrightarrow \mathbb{F}^m$ and $g : \mathbb{F}^\ell \longrightarrow \mathbb{F}^k$. In fact, if A and B denote the matrices of f and g with respect to the standard basis, then $A \otimes B$ is the matrix representation of $f \otimes g$ with respect to the standard basis. Nevertheless, we warn the reader that this definition of the Kronecker product is slightly different from that used by many other authors, in the sense that what we denote as $A \otimes B$ is usually denoted as $B \otimes A$. We will use the preceding definition of the Kronecker product only in this chapter, in order to simplify some of the expressions. Later on, in Part III of this book, we will return to the standard definition of the Kronecker product.

By the definition of the Kronecker product, the Kronecker product of an upper triangular matrix A with an rectangular matrix B is block-upper triangular. In particular, the Kronecker product $B \otimes I_N$ is of the form

$$I_N \otimes B = \begin{pmatrix} b_{11}I_N & \cdots & b_{1\ell}I_N \\ \vdots & \ddots & \vdots \\ b_{k1}I_N & \cdots & b_{k\ell}I_N \end{pmatrix},$$

while

$$A \otimes I_N = \mathrm{diag}(A, \ldots, A) = \begin{pmatrix} A & \cdots & 0 \\ \vdots & \ddots & \vdots \\ 0 & \cdots & A \end{pmatrix}.$$

For invertible, $n \times n$ and $m \times m$, respectively, matrices A and B, the Kronecker product $A \otimes B$ is invertible, and

$$(A \otimes B)^{-1} = A^{-1} \otimes B^{-1}.$$

The following rules for the Kronecker product are easily verified:

$$(A \otimes B) \otimes C = A \otimes (B \otimes C),$$
$$(A \otimes B)(C \otimes D) = AC \otimes BD,$$
$$(A \otimes B)^\top = A^\top \otimes B^\top.$$

Let $\text{vec}(A) \in \mathbb{F}^{mn}$ denote the column vector that is obtained by stacking the second column of A under the first, then the third under the second, and so on. The identity

$$\text{vec}(ABC) = (A \otimes C^\top)\text{vec}(B)$$

is very useful in replacing linear matrix equations in vectorized form. Moreover, there exist permutation matrices P and Q such that

$$P(A \otimes B)Q = B \otimes A$$

is true of all matrices A, B. The eigenvalues of $A \otimes B$ are the products $\lambda_i(A)\lambda_j(B)$ of the eigenvalues $\lambda_i(A)$ and $\lambda_j(B)$ of A and B, respectively. Therefore, the trace and determinant of $A \otimes B$ of matrices A and B are $\text{tr}(A \otimes B) = \text{tr}(A)\text{tr}(B)$ and $\det(A \otimes B) = \det(A)^m \det(B)^n$. Similarly, the eigenvalues of $A \otimes I_m + I_n \otimes B$ are the sums $\lambda_i(A) + \lambda_j(B)$.

1. Kronecker Product Polynomial Models

Our aim is to obtain concrete representations of the tensor products $X_{D_1} \otimes_\mathbb{F} X_{D_2^\top}$ and $X_{D_1} \otimes_{\mathbb{F}[z]} X_{D_2^\top}$. To this end, the theory of polynomial and rational models is extended to the case of models induced by Kronecker products of polynomial matrices in one or two variables. Polynomial models based on Kronecker product representations were first studied by Helmke and Fuhrmann (1998) in order to obtain explicit descriptions for tangent spaces for manifolds of rational transfer functions.

Recalling the identification (5.17) and the fact that $\mathbb{F}((z^{-1}))^p$ is a vector space over the field $\mathbb{F}((z^{-1}))$ allows us to introduce a module structure on the space of truncated matrix Laurent series in two variables, z and w, i.e., on $\mathbb{F}((z^{-1}, w^{-1}))^{p \times m}$, as follows.

Definition 5.12. For Laurent series $A_1(z) \in \mathbb{F}((z^{-1}))^{p \times p}$ and $A_2(w) \in \mathbb{F}((w^{-1}))^{m \times m}$, define their $\mathbb{F}$-**Kronecker product** $A_1(z) \otimes A_2^\top(w)$ as the map

$$(A_1(z) \otimes A_2^\top(w)) : \mathbb{F}((z^{-1}, w^{-1}))^{p \times m} \longrightarrow \mathbb{F}((z^{-1}, w^{-1}))^{p \times m}$$

$$(A_1(z) \otimes A_2^\top(w))F(z, w) = A_1(z)F(z, w)A_2(w).$$

Clearly, $A_1(z) \otimes A_2^\top(w)$ is an $\mathbb{F}((z^{-1}, w^{-1}))$-linear map and, hence, also an $\mathbb{F}$-linear map. Similarly, one defines the $\mathbb{F}[z]$-**Kronecker product** $A_1(z) \otimes A_2^\top(z)$ as the map

$$(A_1(z) \otimes A_2^\top(z)) : \mathbb{F}((z^{-1}))^{p \times m} \longrightarrow \mathbb{F}((z^{-1}))^{p \times m}$$

$$(A_1(z) \otimes A_2^\top(z))F(z) = A_1(z)F(z)A_2(z).$$

There are many derivatives of this definition. In particular, we will look at the restriction to polynomial spaces $\mathbb{F}[z]^{p \times m}$ and $\mathbb{F}[z, w]^{p \times m}$, i.e., to spaces of polynomial matrices in one or two variables. Thus, we define the two projection maps $\pi_{D_1(z) \otimes D_2^\top(w)} : \mathbb{F}[z, w]^{p \times m} \longrightarrow \mathbb{F}[z, w]^{p \times m}$ and $\pi_{D_1(z) \otimes D_2^\top(z)} : \mathbb{F}[z]^{p \times m} \longrightarrow \mathbb{F}[z]^{p \times m}$ by

$$\pi_{D_1(z)\otimes D_2^\top(w)}F(z,w) = (D_1(z)\otimes D_2^\top(w))(\pi_-^z\otimes\pi_-^w)(D_1(z)\otimes D_2^\top(w))^{-1}F(z,w)$$

$$= (\pi_{D_1(z)}\otimes_{\mathbb{F}}\pi_{D_2^\top(w)})F(z,w) \qquad (5.20)$$

and

$$\pi_{D_1(z)\otimes D_2^\top(z)}F(z) = (D_1(z)\otimes D_2^\top(z))\pi_-(D_1(z)\otimes D_2^\top(z))^{-1}F(z)$$

$$= D_1(z)[\pi_-(D_1(z)^{-1}F(z)D_2(z)^{-1})]D_2(z) \qquad (5.21)$$

$$= (\pi_{D_1(z)}\otimes_{\mathbb{F}[z]}\pi_{D_2^\top(z)})F(z),$$

respectively. Clearly, $\pi_-^z\otimes\pi_-^w$ is a projection map in $\mathbb{F}((z^{-1},w^{-1}))^{p\times m}$ and π_- a projection map in $\mathbb{F}((z^{-1}))^{p\times m}$. Hence, $\pi_{D_1(z)\otimes D_2^\top(w)}$ is a projection map in $\mathbb{F}[z,w]^{p\times m}$ and $\pi_{D_1(z)\otimes D_2^\top(z)}$ a projection map in $\mathbb{F}[z]^{p\times m}$. There are two important special cases of these maps, namely,

$$\pi_{D_1(z)\otimes_{\mathbb{F}}I}Q(z,w) = \pi_{D_1(z)}Q(z,w),$$

$$\pi_{I\otimes_{\mathbb{F}}D_2^\top(w)}Q(z,w) = Q(z,w)\pi_{D_2^\top(w)}.$$

To formulate the basic properties of the projection operators, we first prove an elementary result about projections.

Lemma 5.13. *Let $\mathcal{X}$ be a linear space and P_1 and P_2 two commuting linear projections acting in $\mathcal{X}$, i.e., $P_1P_2 = P_2P_1$. Then:*

$$\mathrm{Ker}\,P_1P_2 = \mathrm{Ker}\,P_1 + \mathrm{Ker}\,P_2,$$
$$\mathrm{Im}\,P_1P_2 = \mathrm{Im}\,P_1 \cap \mathrm{Im}\,P_2. \qquad (5.22)$$

Proof. Since $\mathrm{Ker}\,P_1, \mathrm{Ker}\,P_2 \subset \mathrm{Ker}\,P_1P_2 = \mathrm{Ker}\,P_2P_1$, also $\mathrm{Ker}\,P_1 + \mathrm{Ker}\,P_2 \subset \mathrm{Ker}\,P_1P_2$. Conversely, assume $x \in \mathrm{Ker}\,P_1P_2$. This implies $P_2x \in \mathrm{Ker}\,P_1$. The representation $x = (x - P_2x) + P_2x$, with $(x - P_2x) \in \mathrm{Ker}\,P_2$ and $P_2x \in \mathrm{Ker}\,P_1$, shows that $\mathrm{Ker}\,P_1P_2 \subset \mathrm{Ker}\,P_1 + \mathrm{Ker}\,P_2$, and (5.22) follows. By the commutativity assumption, $\mathrm{Im}\,P_1P_2 \subset \mathrm{Im}\,P_1 \cap \mathrm{Im}\,P_2$. Conversely, assuming $x \in \mathrm{Im}\,P_1 \cap \mathrm{Im}\,P_2$, there exist vectors $z, w \in \mathcal{X}$ for which $x = P_1z = P_2w$. Since $P_2^2 = P_2$, this implies $x = P_2w = P_2P_1z \in \mathrm{Im}\,P_1P_2$, i.e., $\mathrm{Im}\,P_1 \cap \mathrm{Im}\,P_2 \subset \mathrm{Im}\,P_1P_2$. The two inclusions imply the equality. ∎

Proposition 5.14. *Let $D_1(z) \in \mathbb{F}[z]^{p\times p}$ and $D_2(w) \in \mathbb{F}[w]^{m\times m}$ be nonsingular polynomial matrices. Then:*

1. *The maps $\pi_{D_1(z)\otimes I}$, $\pi_{I\otimes D_2^\top(w)}$, and $\pi_{D_1(z)}\otimes_{\mathbb{F}}\pi_{D_2^\top(w)}$ are all projections in $\mathbb{F}[z,w]^{p\times m}$;*
2. *The projections $\pi_{D_1(z)\otimes I}$ and $\pi_{I\otimes D_2^\top(w)}$ commute, and*

$$\pi_{D_1(z)\otimes I}\pi_{I\otimes D_2^\top(w)} = \pi_{I\otimes D_2^\top(w)}\pi_{D_1(z)\otimes I} = \pi_{D_1(z)\otimes D_2^\top(w)} = \pi_{D_1(z)}\otimes_{\mathbb{F}}\pi_{D_2^\top(w)};$$

$$(5.23)$$

3. *The following characterizations are valid:*

$$\operatorname{Ker} \pi_{D_1(z) \otimes I} = D_1(z) \mathbb{F}[z, w]^{p \times m},$$

$$\operatorname{Ker} \pi_{I \otimes D_2^\top(w)} = \mathbb{F}[z, w]^{p \times m} D_2(w), \tag{5.24}$$

$$\operatorname{Ker} \pi_{D_1(z) \otimes D_2^\top(w)} = D_1(z) \mathbb{F}[z, w]^{p \times m} + \mathbb{F}[z, w]^{p \times m} D_2(w).$$

Proof. 1. Follows from the fact that π_{D_1} is a projection in $\mathbb{F}[z]^p$ and π_{D_2} a projection in $\mathbb{F}[z]^m$.

2. From the isomorphism (5.15) it follows that elements of the form $f(z) \otimes_\mathbb{F} g(w) = f(z) g^\top(w)$ span $\mathbb{F}[z, w]^{p \times m}$. On elements of this form

$$\begin{aligned}
\pi_{D_1(z) \otimes I} \pi_{I \otimes D_2^\top(w)} f \otimes_\mathbb{F} g &= \pi_{D_1(z) \otimes I} (f \otimes_\mathbb{F} \pi_{D_2^\top(w)} g) \\
&= (\pi_{D_1(z)} f \otimes_\mathbb{F} \pi_{D_2(w)} g) = \pi_{I \otimes D_2^\top(w)} (\pi_{D_1(z)} f \otimes_\mathbb{F} g) \\
&= \pi_{I \otimes D_2^\top(w)} \pi_{D_1(z) \otimes I} (f \otimes_\mathbb{F} g),
\end{aligned}$$

from which (5.23) follows.

3. Clearly, $Q(z, w) \in \operatorname{Ker} \pi_{D_1(z) \otimes I}$ if and only if $\pi_-^z D_1(z)^{-1} Q(z, w) = 0$, i.e., $D_1(z)^{-1} Q(z, w) = P(z, w)$ for some polynomial matrix $P(z, w)$. This is equivalent to $Q(z, w) = D_1(z) P(z, w) \in D_1(z) \mathbb{F}[z, w]^{p \times m}$. The second equality is proved analogously. The third equality follows from Lemma 5.13 and the commutativity of the projections $\pi_{D_1(z) \otimes I}$ and $\pi_{I \otimes D_2^\top(w)}$. ∎ ■

Definition 5.15. Let $D_1(z) \in \mathbb{F}[z]^{p \times p}$ and $D_2(w) \in \mathbb{F}[w]^{m \times m}$ be nonsingular polynomial matrices.

1. The **two-variable Kronecker product polynomial model** is defined by

$$X_{D_1(z) \otimes D_2^\top(w)} = \operatorname{Im} \pi_{D_1(z) \otimes D_2^\top(w)} \subset \mathbb{F}[z, w]^{p \times m}, \tag{5.25}$$

where the projection $\pi_{D_1(z) \otimes D_2^\top(w)}$ is defined by (5.20).

2. The **Kronecker product polynomial model** is defined by

$$X_{D_1(z) \otimes D_2^\top(z)} = \operatorname{Im} \pi_{D_1(z) \otimes D_2^\top(z)} \subset \mathbb{F}[z]^{p \times m}, \tag{5.26}$$

where the projection $\pi_{D_1(z) \otimes D_2^\top(z)}$ is defined by (5.21).

Note that in either of these cases, the spaces $X_{D_1(z) \otimes D_2^\top(w)}$ and $X_{D_1(z) \otimes D_2^\top(z)}$ can be identified with polynomial models for the Kronecker products $D_1(z) \otimes D_2^\top(w)$ and $D_1(z) \otimes D_2^\top(z)$, respectively.

Theorem 5.16. *Let $D_1(z) \in \mathbb{F}[z]^{p \times p}$ and $D_2(w) \in \mathbb{F}[w]^{m \times m}$ be nonsingular polynomial matrices. Then:*

1. *$Q(z,w) \in \mathbb{F}[z,w]^{p \times m}$ satisfies $Q(z,w) \in X_{D_1(z) \otimes D_2^\top(w)}$ if and only if the rational matrix function*

$$D_1(z)^{-1} Q(z,w) D_2(w)^{-1} \in z^{-1} \mathbb{F}[z^{-1}, w^{-1}]^{p \times m} w^{-1}$$

is strictly proper in both variables;

2. *The set $J = D_1(z)\mathbb{F}[z,w]^{p \times m} + \mathbb{F}[z,w]^{p \times m} D_2(w)$ is an $\mathbb{F}[z,w]$-submodule of $\mathbb{F}[z,w]^{p \times m}$. The following isomorphism of $\mathbb{F}[z,w]$-torsion modules is valid:*

$$X_{D_1(z) \otimes D_2^\top(w)} \simeq \mathbb{F}[z,w]^{p \times m} / (D_1(z)\mathbb{F}[z,w]^{p \times m} + \mathbb{F}[z,w]^{p \times m} D_2(w))$$

$$= X_{D_1(z) \otimes I} \cap X_{I \otimes D_2^\top(w)} \tag{5.27}$$

$$\simeq X_{D_1} \otimes_{\mathbb{F}} X_{D_2^\top};$$

3. *The following dimension formula is valid:*

$$\dim X_{D_1(z) \otimes D_2^\top(w)} = \deg(\det D_1) \cdot \deg(\det D_2);$$

4. *A polynomial matrix satisfies $Q(z) \in X_{D_1(z) \otimes D_2^\top(z)}$ if and only if $D_1(z)^{-1} Q(z) D_2(z)^{-1}$ is strictly proper;*

5. *One has*

$$\operatorname{Ker} \pi_{D_1(z) \otimes D_2^\top(z)} = D_1(z)\mathbb{F}[z]^{p \times m} D_2(z), \tag{5.28}$$

and $D_1(z)\mathbb{F}[z]^{p \times m} D_2(z)$ is a full submodule of $\mathbb{F}[z]^{p \times m}$. Hence, there is an isomorphism

$$X_{D_1(z) \otimes D_2^\top(z)} \simeq \mathbb{F}[z]^{p \times m} / (D_1(z)\mathbb{F}[z]^{p \times m} D_2(z)),$$

with both sides being $\mathbb{F}[z]$-torsion modules;

6. *The following dimension formula is valid:*

$$\dim X_{D_1(z) \otimes D_2^\top(z)} = \deg(\det D_1) \cdot \deg(\det D_2).$$

Proof. 1. A $p \times m$ polynomial matrix $Q(z,w)$ is in $X_{D_1(z) \otimes D_2^\top(w)}$ if and only if $Q(z,w) = \pi_{D_1(z) \otimes D_2^\top(w)} Q(z,w)$. In view of (5.20), this is equivalent to

$$(D_1(z) \otimes D_2^\top(w))^{-1} Q(z,w) = (\pi_-^z \otimes_{\mathbb{F}} \pi_-^w)(D_1(z) \otimes D_2^\top(w))^{-1} Q(z,w),$$

i.e., to $D_1(z)^{-1} Q(z,w) D_2(w)^{-1} \in z^{-1} \mathbb{F}[[z^{-1}, w^{-1}]]^{p \times m} w^{-1}$.

2. By Proposition 5.14, the map

$$\pi_{D_1(z)} \otimes_{\mathbb{F}} \pi_{D_2^\top(w)} = \pi_{D_1(z) \otimes D_2^\top(w)} : \mathbb{F}[z,w]^{p \times m} \longrightarrow X_{D_1(z) \otimes D_2^\top(w)}$$

is surjective and $\mathbb{F}[z,w]$-linear, with $\text{Ker} \, \pi_{D_1(z) \otimes D_2^\top(w)} = J$. Thus, the first isomorphism in (5.27) holds. The second equality follows from Lemma 5.13, while the third isomorphism follows directly from Proposition 5.8.

Clearly, J is an $\mathbb{F}[z,w]$-submodule of $\mathbb{F}[z,w]^{p \times m}$. Using, with $d(z) = \det D(z)$, the identity $d(z)I = D(z)\text{adj}\,D(z)$, we get the inclusion $d_1(z)\mathbb{F}[z,w]^{p \times m}d_2(w) \subset D_1(z)\mathbb{F}[z,w]^{p \times m} + \mathbb{F}[z,w]^{p \times m}D_2(w)$. In turn, this implies that

$$\pi_{D_1(z) \otimes D_2^\top(w)} (d_1(z)d_2(w)Q(z,w)) = 0$$

for all $Q(z,w) \in X_{D_1(z) \otimes D_2^\top(w)}$, i.e., the quotient module $\mathbb{F}[z,w]^{p \times m}/J$ is an $\mathbb{F}[z,w]$-torsion module.

3. Follows from (5.27), using the fact that the dimension of the $\mathbb{F}$-tensor product of two $\mathbb{F}$-vector spaces is the product of their dimensions.

4. Follows trivially from (5.21).

5. Clearly, $D_1(z)\mathbb{F}[z]^{p \times m}D_2(z) \subset \text{Ker} \, \pi_{D_1(z) \otimes D_2^\top(z)}$. Conversely, assume that $Q(z) \in \text{Ker} \, \pi_{D_1(z) \otimes D_2^\top(z)}$. By the invertibility of the multiplication operator $D_1(z) \otimes D_2^\top(z)$, this means $\pi_{-}D_1(z)^{-1}Q(z)D_2(z)^{-1} = 0$. Thus there exists a $P(z) \in \mathbb{F}[z]^{p \times m}$ such that $D_1(z)^{-1}Q(z)D_2(z)^{-1} = P(z)$ or $Q(z) = D_1(z)P(z)D_2(z)$, which implies the inclusion $\text{Ker} \, \pi_{D_1 \otimes_{\mathbb{F}[z]} D_2^\top} \subset D_1(z)\mathbb{F}[z]^{p \times m}D_2(z)$, and hence (5.28) follows.

6. Without loss of generality, we can assume that $D_1(z)$ and $D_2(z)$ are in Smith canonical form with invariant factors $d_1^{(1)}(z), \dots, d_p^{(1)}(z)$ and $d_1^{(2)}(z), \dots, d_m^{(2)}(z)$, respectively. Thus, the quotient module $\mathbb{F}[z]^{p \times m}/(D_1(z)\mathbb{F}[z]^{p \times m}D_2(z))$ is a finite-dimensional $\mathbb{F}$-vector space of dimension

$$\sum_{i=1}^{p} \sum_{j=1}^{m} \deg d_i^{(1)} \deg d_j^{(2)} = \deg \det D_1(z) \deg \det D_2(z).$$

This completes the proof.   ∎

2. Tensored Rational Models

In analogy with the introduction of tensored polynomial models, we introduce the tensored rational models. Let $D_1(z) \in \mathbb{F}[z]^{p \times p}$ and $D_2(w) \in \mathbb{F}[w]^{m \times m}$ be nonsingular. Define a projection map

$$\pi^{D_1(z) \otimes D_2^\top(w)} : z^{-1}\mathbb{F}[[z^{-1}, w^{-1}]]^{p \times m}w^{-1} \longrightarrow z^{-1}\mathbb{F}[[z^{-1}, w^{-1}]]^{p \times m}w^{-1}$$

by

$$\pi^{D_1(z)\otimes D_2^\top(w)}H(z,w)$$
$$= (\pi_-^z \otimes_\mathbb{F} \pi_-^w)(D_1(z) \otimes D_2^\top(w))^{-1}(\pi_+^z \otimes_\mathbb{F} \pi_+^w)(D_1(z) \otimes D_2^\top(w))H(z,w).$$

The **two-variable Kronecker product rational model** and the **Kronecker product rational model** are defined as

$$X^{D_1(z)\otimes D_2^\top(w)} := \operatorname{Im}\pi^{D_1(z)\otimes D_2^\top(w)} \subset z^{-1}\mathbb{F}[[z^{-1}, w^{-1}]]^{p\times m}w^{-1},$$
$$X^{D_1(z)\otimes D_2^\top(z)} := \operatorname{Im}\pi^{D_1(z)\otimes D_2^\top(z)} \subset z^{-1}\mathbb{F}[z^{-1}]^{p\times m}, \tag{5.29}$$

respectively.

Equation (5.29) provides an image representation of the rational model $X^{D_1(z)\otimes D_2^\top(w)}$. To derive a kernel representation of rational models, we introduce **two-variable Toeplitz operators** on $z^{-1}\mathbb{F}[[z^{-1}, w^{-1}]]^{p\times m}w^{-1}$. For $P_1(z) \in \mathbb{F}[z]^{p\times p}$ and $P_2(w) \in \mathbb{F}[w]^{m\times m}$, we define the **Toeplitz operator** $P_1(\sigma) \otimes_\mathbb{F} P_2^\top(\tau)$, acting on a truncated Laurent series in two variables $H(z,w) \in z^{-1}\mathbb{F}[[z^{-1}, w^{-1}]]^{p\times m}w^{-1}$, by

$$(P_1(\sigma) \otimes_\mathbb{F} P_2^\top(\tau))H(z,w) = (\pi_-^z \otimes_\mathbb{F} \pi_-^w)(P_1(z)H(z,w)P_2(w)).$$

Special cases are the backward shifts σ and τ in the variables z and w, respectively.

Lemma 5.17. *Let $D_1(z) \in \mathbb{F}[z]^{p\times p}$ and $D_2(w) \in \mathbb{F}[w]^{m\times m}$ be nonsingular polynomial matrices. Let $H(z,w) \in z^{-1}\mathbb{F}[[z^{-1}, w^{-1}]]^{p\times m}w^{-1}$. A necessary and sufficient condition for $H(z,w) \in X^{D_1(z)\otimes D_2^\top(w)}$ to be valid is $D_1(z)H(z,w)D_2(w) \in \mathbb{F}[z,w]^{p\times m}$. This is equivalent to the kernel representation*

$$X^{D_1(z)\otimes D_2^\top(w)} = \operatorname{Ker}(D_1(\sigma) \otimes_\mathbb{F} D_2^\top(\tau)). \tag{5.30}$$

Proof. Clearly, $H(z,w) \in X^{D_1(z)\otimes D_2^\top(w)}$ if and only if $H(z,w) = \pi^{D_1(z)\otimes D_2^\top(w)}H(z,w)$. Assume $H(z,w) \in X^{D_1(z)\otimes D_2^\top(w)}$. Computing

$$(\pi_-^z \otimes_\mathbb{F} \pi_-^w)(D_1(z) \otimes D_2^\top(w))H(z,w)$$
$$= (\pi_-^z \otimes_\mathbb{F} \pi_-^w)(D_1(z) \otimes D_2^\top(w))\pi^{D_1(z)\otimes D_2(w)}H(z,w)$$
$$= (\pi_-^z \otimes_\mathbb{F} \pi_-^w)(\pi_+^z \otimes_\mathbb{F} \pi_+^w)(D_1(z) \otimes D_2^\top(w))H(z,w) = 0$$

implies $X^{D_1(z)\otimes D_2^\top(w)} \subset \operatorname{Ker}(D_1(\sigma) \otimes_\mathbb{F} D_2^\top(\tau))$.

Conversely, assuming $H(z,w) \in \operatorname{Ker}(D_1(\sigma) \otimes_\mathbb{F} D_2^\top(\tau))$ implies $D_1(z)H(z,w)D_2(w) \in \mathbb{F}[z,w]^{p\times m}$. This implies

$$(\pi_-^z \otimes_{\mathbb{F}} \pi_-^w)(D_1(z) \otimes D_2^\top(w))^{-1}(\pi_+^z \otimes_{\mathbb{F}} \pi_+^w)(D_1(z) \otimes D_2^\top(w))H(z,w)$$

$$= (\pi_-^z \otimes_{\mathbb{F}} \pi_-^w)(D_1(z) \otimes D_2^\top(w))^{-1}(D_1(z) \otimes D_2^\top(w))H(z,w) = H(z,w),$$

i.e., $\mathrm{Ker}\,(D_1(\sigma) \otimes_{\mathbb{F}} D_2^\top(\tau)) \subset X^{D_1(z) \otimes D_2^\top(w)}$. The two inclusions imply (5.30). ■

The elements of $X^{D_1(z) \otimes D_2^\top(w)}$ are rational functions of a special type. They are characterized next.

Proposition 5.18. *Let* $D_1(z) \in \mathbb{F}[z]^{p \times p}$ *and* $D_2(w) \in \mathbb{F}[w]^{m \times m}$ *be nonsingular polynomial matrices. Every element* $H(z,w) \in X^{D_1(z) \otimes D_2^\top(w)}$ *is a rational, strictly proper function in two variables that has a representation of the form*

$$H(z,w) = \frac{P(z,w)}{d_1(z)d_2(w)}, \qquad (5.31)$$

with $d_i(z) = \det D_i(z) \in \mathbb{F}[z]$ *nonzero polynomials and* $P(z,w) \in \mathbb{F}[z,w]^{p \times m}$.

Proof. From Lemma 5.17 it follows that $H(z,w) \in X^{D_1(z) \otimes D_2^\top(w)}$ if and only if $Q(z,w) = D_1(z)H(z,w)D_2(w) \in \mathbb{F}[z,w]^{p \times m}$. Letting $d_1(z) = \det D_1(z)$ and $d_2(w) = \det D_1(w)$, we compute

$$H(z,w) = D_1(z)^{-1}Q(z,w)D_2(w)^{-1} = \frac{\mathrm{adj}\,D_1(z)Q(z,w)\mathrm{adj}\,D_2(w)}{d_1(z)d_2(w)} = \frac{P(z,w)}{d_1(z)d_2(w)}. \quad ■$$

Rational functions of the form (5.31) are called **separable** and have the property that the set of poles is a direct product $A \times B$ of two finite subsets of the algebraic closure $\overline{\mathbb{F}}$. We refer to Fliess (1970) for a characterization of rational elements of $\mathbb{F}[[z^{-1}, w^{-1}]]$ in terms of a finite rank condition of an appropriate Hankel matrix.

The study of duality for the tensor product of models is our next topic.

Theorem 5.19. *Let* $D_1(z) \in \mathbb{F}[z]^{p \times p}$ *and* $D_2(z) \in \mathbb{F}[z]^{m \times m}$ *be nonsingular. Then*

$$(X_{D_1(z) \otimes D_2^\top(z)})^* \simeq (\mathbb{F}[z]^{p \times m}/D_1(z)\mathbb{F}[z]^{p \times m}D_2(z))^* \simeq X^{D_1^\top(z) \otimes D_2(z)}$$

and

$$(X_{D_1(z) \otimes D_2^\top(w)})^* \simeq X^{D_1^\top(z) \otimes D_2(w)} \simeq X_{D_1^\top(z) \otimes D_2(w)}$$

are $\mathbb{F}[z]$- *and* $\mathbb{F}[z,w]$-*linear isomorphisms, respectively.*

Proof. Proposition 5.16 implies the following isomorphism of $\mathbb{F}[z]$-modules:

$$(X_{D_1(z) \otimes D_2^\top(z)})^* \simeq (\mathbb{F}[z]^{p \times m}/D_1(z)\mathbb{F}[z]^{p \times m}D_2(z))^*.$$

Using the vec-operator, we can identify the matrix space $\mathbb{F}[z]^{p \times m}$ with the space of vector polynomials $\mathbb{F}[z]^{pm}$. Moreover, the identity

$$(D_1(z) \otimes D_2^\top(z))^\top = D_1^\top(z) \otimes D_2(z)$$

is satisfied. For vectors of polynomials $F(z)$ and strictly proper functions $H(z)$ let $[F, H] = (F^\top(z)H(z))_{-1}$, denote the residue term of $F(z)^\top H(z)$. Using Theorem 3.38, the map

$$X^{D_1^\top(z) \otimes \mathbb{F}[z] D_2(z)} \longrightarrow \left(X_{D_1(z) \otimes D_2^\top(z)}\right)^*$$

$$H(z) \mapsto \left(F(z) \mapsto \text{Trace}(F(z)^\top H(z))_{-1}\right)$$

yields the module isomorphism

$$\left(X_{D_1(z) \otimes D_2^\top(z)}\right)^* \simeq X^{D_1^\top(z) \otimes \mathbb{F}[z] D_2(z)}.$$

This completes the proof for the first isomorphisms.

By Proposition 5.16, the elements of the rational model $X^{D_1^\top(z) \otimes D_2(w)}$ are of the form $H(z, w) = D_1(z)^{-\top} Q(z, w) D_2(w)^{-\top}$. Consider the map

$$L : X^{D_1^\top(z) \otimes D_2(w)} \longrightarrow \left(X_{D_1(z) \otimes D_2^\top(w)}\right)^* \tag{5.32}$$

defined by

$$L(H)F := [F(z, w), H(z, w)] = \text{Trace}(F(z, w)^\top H(z, w))_{-1, -1}.$$

Here, $[\cdot, \cdot]$ denotes the dual pairing on functions of two variables, defined by (5.18). To prove that this pairing is nondegenerate, assume that a polynomial $F(z, w) \in X_{D_1(z) \otimes D_2^\top(w)}$ satisfies $[F, H] = 0$ for all $H \in X^{D_1^\top(z) \otimes D_2(w)}$. Equivalently,

$$[D_1(z)^{-1} F(z, w) D_2(w)^{-1}, Q(z, w)] = 0$$

for all polynomial matrices Q of the appropriate size. By (5.19), this is equivalent to

$$D_1(z)^{-1} F(z, w) D_2(w)^{-1} \in z^{-1} \mathbb{F}[[z^{-1}, w^{-1}]] w^{-1} \cap \left(\mathbb{F}[z, w^{-1}]]^{p \times m} + \mathbb{F}[[z^{-1}, w]^{p \times m}\right)$$
$$= \{0\}.$$

Thus, $F(z, w) = 0$. Similarly, $[F, H] = 0$ for all $F \in X_{D_1(z) \otimes D_2^\top(w)}$ implies $H = 0$. Thus the bilinear form $[\cdot, \cdot]$ on $X_{D_1(z) \otimes D_2^\top(w)} \times X^{D_1^\top(z) \otimes D_2(w)}$ is nondegenerate and therefore induces a vector space isomorphism (5.32). Moreover, the natural $\mathbb{F}[z, w]$-module action on, respectively, $X_{D_1(z) \otimes D_2^\top(w)}$ and $X^{D_1^\top(z) \otimes D_2(w)})$ implies

$$[p(z, w) \cdot F, H] = [p(z, w)F, H] = [F, p(z, w)H] = [F, p(z, w) \cdot H].$$

Therefore, (5.32) defines an $\mathbb{F}[z,w]$-linear isomorphism. The equality

$$X^{D_1^\top(z)\otimes D_2(w)} = D_1(z)^{-\top} \otimes D_2(w)^{-1} X_{D_1^\top(z)\otimes D_2(w)}$$

implies the $\mathbb{F}[z,w]$-linear isomorphism

$$X^{D_1^\top(z)\otimes D_2(w)} \simeq X_{D_1^\top(z)\otimes D_2(w)}.$$

This completes the proof. ∎

4. $\mathbb{F}$-Tensored Polynomial Models

Our attention turns to the study of $\mathbb{F}$-tensor products of vectorial polynomial models. Proposition 5.16 implies that a Kronecker tensored polynomial model, in the sense of (5.25), is isomorphic to the tensor product of polynomial models taken over the field $\mathbb{F}$. This is no longer true if tensored polynomial models in the sense of (5.26) are used, and indeed, the models $X_{D_1(z)\otimes D_2^\top(z)}$ and $X_{D_1(z)} \otimes_\mathbb{F} X_{D_2(z)}$ are generally not isomorphic. The next proposition gives a concrete, functional representation of the $\mathbb{F}$-tensor product of two polynomial models.

Proposition 5.20. *Let $D_1(z) \in \mathbb{F}[z]^{p\times p}$ and $D_2(z) \in \mathbb{F}[z]^{m\times m}$ be nonsingular polynomial matrices. Let $\phi : X_{D_1} \times X_{D_2^\top} \longrightarrow X_{D_1} \otimes_\mathbb{F} X_{D_2^\top}$ be the canonical isomorphism, and let $\gamma : X_{D_1} \times X_{D_2^\top} \longrightarrow X_{D_1(z)\otimes D_2^\top(w)}$ be $\mathbb{F}$-bilinear.*

1. The map $\gamma_* : X_{D_1} \otimes_\mathbb{F} X_{D_2^\top} \longrightarrow X_{D_1(z)\otimes D_2^\top(w)}$ defined by

$$\gamma_*(f_1 \otimes_\mathbb{F} f_2) = f_1(z)f_2^\top(w)$$

is an $\mathbb{F}[z,w]$-linear isomorphism implying

$$X_{D_1} \otimes_\mathbb{F} X_{D_2^\top} \simeq X_{D_1(z)\otimes D_2^\top(w)}. \tag{5.33}$$

In particular, this gives a concrete representation of the tensor product.
2. The following dimension formula is valid:

$$\dim(X_{D_1(z)} \otimes_\mathbb{F} X_{D_2^\top(w)}) = \deg(\det D_1) \cdot \deg(\det D_2).$$

Proof. 1. Noting the isomorphism (5.27), we compute

$$X_{D_1(z)\otimes D_2^\top(w)} \simeq \mathbb{F}[z,w]^{p\times m}/(D_1(z)\mathbb{F}[z,w]^{p\times m} + \mathbb{F}[z,w]^{p\times m}D_2(w))$$

$$\simeq (\mathbb{F}[z]^p/D_1(z)\mathbb{F}[z]^p) \otimes_\mathbb{F} (\mathbb{F}[z]^m/D_2^\top(z)\mathbb{F}[z]^m)$$

$$\simeq X_{D_1(z)} \otimes_\mathbb{F} X_{D_2^\top(z)}.$$

2. Using the dimension formula $\dim X_D = \deg(\det D)$ and the isomorphism (5.33), we obtain for the $\mathbb{F}$-tensor product of two polynomial models the dimension formula

$$\dim \left(\mathbb{F}[z]^p / D_1(z)\mathbb{F}[z] \otimes_{\mathbb{F}} \mathbb{F}[w]^m / D_2^{\top}(w)\mathbb{F}[w]^m \right) = \deg(\det D_1) \cdot \deg(\det D_2). \quad \blacksquare$$

Note that the polynomial models X_{D_1} and X_{D_2} not only have a vector space structure but are actually $\mathbb{F}[z]$-modules. This implies that $X_{D_1(z) \otimes D_2^{\top}(w)}$ and, hence, using the isomorphism (5.33), $X_{D_1} \otimes_{\mathbb{F}} X_{D_2^{\top}}$ have natural $\mathbb{F}[z, w]$-module structures. This is defined by

$$p(z,w) \cdot Q(z,w) = \pi_{D_1(z) \otimes D_2^{\top}(w)} p(z,w) Q(z,w), \qquad Q(z,w) \in X_{D_1(z) \otimes D_2^{\top}(w)},$$

$$(5.34)$$

where $p(z,w) \in \mathbb{F}[z,w]$.

Similarly, we define an $\mathbb{F}[z,w]$-module structure on the tensored rational model $X^{D_1(z) \otimes D_2^{\top}(w)}$ by letting,

$$p(z,w) \cdot H(z,w) = \pi^{D_1(z) \otimes D_2^{\top}(w)} [\textstyle\sum_{i=1}^{k} \sum_{j=1}^{l} p_{ij} z^{i-1} H(z,w) w^{j-1}] \qquad (5.35)$$

for $p(z,w) = \sum_{i=1}^{k} \sum_{j=1}^{l} p_{ij} z^{i-1} w^{j-1} \in \mathbb{F}[z,w]$ and $H(z,w) \in X^{D_1(z) \otimes D_2^{\top}(w)}$.

Proposition 5.21. *Let $D_1(z) \in \mathbb{F}[z]^{p \times p}$ and $D_2(w) \in \mathbb{F}[w]^{m \times m}$ be nonsingular polynomial matrices and $H(z,w) \in X^{D_1(z) \otimes D_2^{\top}(w)}$.*

1. *The $\mathbb{F}[z,w]$-module structure on $X^{D_1(z) \otimes D_2^{\top}(w)}$ defined by (5.35) can be rewritten as*

$$p(z,w) \cdot H(z,w) = (\pi_{-}^{z} \otimes \pi_{-}^{w}) \sum_{i=1}^{k} \sum_{j=1}^{l} p_{ij} z^{i-1} H(z,w) w^{j-1}. \qquad (5.36)$$

2. *With the $\mathbb{F}[z,w]$-module structure on $X_{D_1(z) \otimes D_2^{\top}(w)}$ and $X^{D_1(z) \otimes D_2^{\top}(w)}$, given by (5.36) and (5.35) respectively, the multiplication map*

$$D_1(z) \otimes D_2^{\top}(w) : X^{D_1(z) \otimes D_2^{\top}(w)} \longrightarrow X_{D_1(z) \otimes_{\mathbb{F}} D_2^{\top}(w)}$$

is an $\mathbb{F}[z,w]$-module isomorphism, giving

$$X_{D_1(z) \otimes D_2^{\top}(w)} \simeq X^{D_1(z) \otimes D_2^{\top}(w)}.$$

Proof. 1. Follows from (5.35).

2. Follows, using Lemma 5.17, from the fact that $H(z,w) \in X^{D_1(z) \otimes D_2^{\top}(w)}$ if and only if $H(z,w) = \pi^{D_1(z) \otimes D_2^{\top}(w)} H(z,w)$.

Equivalently, if and only if $\pi_{D_1(z)\otimes D_2^\top(w)} D_1(z) H(z,w) D_2(w) = D_1(z) H(z,w)$
$D_2(w)$, i.e., $D_1(z) H(z,w) D_2(w) \in X_{D_1(z)\otimes D_2^\top(w)}$.   ∎

Special cases of interest are the single-variable shift operators

$$S_z, S_w : X_{D_1(z)\otimes D_2^\top(w)} \longrightarrow X_{D_1(z)\otimes D_2^\top(w)},$$

defined by

$$S_z Q(z,w) = \pi_{D_1(z)\otimes D_2^\top(w)} z Q(z,w) = \pi_{D_1(z)} z Q(z,w),$$

$$S_w Q(z,w) = \pi_{D_1(z)\otimes D_2^\top(w)} Q(z,w) w = \pi_{I\otimes_\mathbb{F} D_2^\top(w)} Q(z,w) w.$$

A concrete representation of the dual space to a tensored polynomial model is given next. For subspaces $\mathscr{U}$ and $\mathscr{V}$ of a linear space $\mathscr{X}$, we shall use the isomorphism $(\mathscr{X}/\mathscr{V})^* \simeq \mathscr{V}^\perp$, as well as the identity $(\mathscr{U}+\mathscr{V})^\perp = \mathscr{U}^\perp \cap \mathscr{V}^\perp$.

Theorem 5.22. *Let $D_1(z) \in \mathbb{F}[z]^{p\times p}$ and $D_2(w) \in \mathbb{F}[w]^{m\times m}$ be nonsingular polynomial matrices. Then the following is an $\mathbb{F}[z,w]$-linear isomorphism:*

$$(X_{D_1} \otimes_\mathbb{F} X_{D_2^\top})^* \simeq X^{D_1(z)\otimes I} \cap X^{I\otimes_\mathbb{F} D_2^\top(w)} \simeq X^{D_1^\top(z)\otimes D_2(w)}.$$

Proof. By Proposition 5.16 and Theorem 5.19, the following are $\mathbb{F}[z,w]$-linear isomorphisms:

$$X_{D_1} \otimes_\mathbb{F} X_{D_2^\top} \simeq X_{D_1(z)\otimes D_2^\top(w)},$$

$$(X_{D_1} \otimes_\mathbb{F} X_{D_2^\top})^* \simeq X^{D_1^\top(z)\otimes D_2(w)}.$$

This implies the $\mathbb{F}[z,w]$-linear isomorphisms

$$(X_{D_1} \otimes_\mathbb{F} X_{D_2^\top})^* \simeq (\mathbb{F}[z,w]^{m\times p}/(D_1(z)\mathbb{F}[z,w]^{p\times m} + \mathbb{F}[z,w]^{p\times m} D_2(w)))^*$$

$$\simeq (D_1(z)\mathbb{F}[z,w]^{p\times m} + \mathbb{F}[z,w]^{p\times m} D_2(w))^\perp$$

$$= (D_1(z)\mathbb{F}[z,w]^{p\times m})^\perp \cap (\mathbb{F}[z,w]^{p\times m} D_2(w))^\perp$$

$$= X^{D_1(z)\otimes I} \cap X^{I\otimes_\mathbb{F} D_2^\top(w)}.$$

Here, the identities

$$\left(D_1^\top(z)\mathbb{F}[z,w]^{p\times m}\right)^\perp = X^{D_1(z)\otimes_\mathbb{F} I},$$

$$\left(\mathbb{F}[z,w]^{p\times m}D_2^\top(w)\right)^\perp = X^{I\otimes_\mathbb{F} D_2^\top(w)}$$

were used. They follow from the duality relation based on the bilinear form (5.18). Indeed, $H(z,w) \in (D_1(z)\mathbb{F}[z,w]^{p\times m})^\perp$ if and only if, for every $Q(z,w) \in \mathbb{F}[z,w]^{p\times m}$,

$$0 = [D_1(z)Q(z,w),H(z,w)] = [Q(z,w),D_1(z)H(z,w)],$$

i.e., if and only if $D_1(z)H(z,w) \in \mathbb{F}[z,w]^{p\times m}$, which implies $H(z,w) \in X^{D_1(z)\otimes I}$. The other formula is proved similarly. ∎

3. $\mathbb{F}[z]$-tensored polynomial models
In the preceding parts, tensor product representations of polynomial models over a field were studied. Things change dramatically when tensor products of polynomial models are taken over the polynomial ring $\mathbb{F}[z]$. This leads directly to the study of intertwining maps, the Sylvester equation, and, in a very natural way, to the study of generalized Bezoutians.

Definition 5.23. Let $D_1(z) \in \mathbb{F}[z]^{p\times p}$ and $D_2(z) \in \mathbb{F}[z]^{m\times m}$ be nonsingular polynomial matrices. The **greatest common left Kronecker divisor** of $D_1(z)$ and $D_2(z)$ is defined as the greatest common left divisor $D_1(z)\wedge D_2(z)$ of the polynomial matrices $D_1(z)\otimes I_m$ and $I_p\otimes D_2(z)^\top$.

Of course, by construction, the greatest common left Kronecker divisor $D_1(z)\wedge D_2(z) \in \mathbb{F}[z]^{pm\times pm}$ is a nonsingular polynomial matrix. Further elementary properties of $D_1(z)\wedge D_2(z)$ are listed subsequently in Corollary 5.25. The $\mathbb{F}[z]$-tensor product of the polynomial models X_{D_1} and X_{D_2} is characterized by the following theorem. It shows in particular that the $\mathbb{F}[z]$-tensor product of two polynomial models is isomorphic to a polynomial model, defined by the greatest common left Kronecker divisor.

Theorem 5.24. *Let $D_1(z) \in \mathbb{F}[z]^{p\times p}$ and $D_2(z) \in \mathbb{F}[z]^{m\times m}$ be nonsingular polynomial matrices. Let J be the submodule of $\mathbb{F}[z]^{p\times m}$ defined by*

$$J = D_1(z)\mathbb{F}[z]^{p\times m} + \mathbb{F}[z]^{p\times m}D_2(z).$$

1. The $\mathbb{F}[z]$-tensor product can be identified by the following isomorphism:

$$X_{D_1}\otimes_{\mathbb{F}[z]}X_{D_2^\top} \simeq \mathbb{F}[z]^{p\times m}/(D_1(z)\mathbb{F}[z]^{p\times m} + \mathbb{F}[z]^{p\times m}D_2(z)).$$

The isomorphism is given by the canonical map $\gamma : X_{D_1} \times X_{D_2} \longrightarrow X_{D_1} \otimes_{\mathbb{F}[z]} X_{D_2}$
defined by

$$\gamma(f_1, f_2) = [f_1 f_2^\top]_J,$$

where $[F]_J$ *denotes the equivalence class of* $F \in \mathbb{F}[z]^{p \times m}$ *with respect to submodule* J.

2. *Let* $D_1 \wedge D_2 \in \mathbb{F}[z]^{pm \times pm}$ *denote the greatest common left divisor of* $D_1(z) \otimes I_m$ *and* $I_p \otimes D_2^\top(z)$. *Then* $X_{D_1} \otimes_{\mathbb{F}[z]} X_{D_2^\top}$ *is* $\mathbb{F}[z]$-*linearly isomorphic to the polynomial model*

$$X_{D_1 \wedge D_2}.$$

Proof. The first claim follows trivially from Proposition 5.3.
By identifying $\mathbb{F}[z]^{p \times m}$ with $\mathbb{F}[z]^{pm}$ we obtain the module isomorphism

$$X_{D_1} \otimes_{\mathbb{F}[z]} X_{D_2^\top} \simeq \mathbb{F}[z]^{pm} / D(z) \mathbb{F}[z]^{2pm},$$

where $D(z) := (D_1(z) \otimes I_m, I_p \otimes D_2^\top(z)) \in \mathbb{F}[z]^{pm \times 2pm}$. Thus

$$D(z) = (D_1(z) \wedge D_2(z)) A(z),$$

where $A(z) \in \mathbb{F}[z]^{pm \times 2pm}$ is left prime. Thus $A(z) \mathbb{F}[z]^{2pm} = \mathbb{F}[z]^{pm}$, and therefore

$$D(z) \mathbb{F}[z]^{2pm} = D_1(z) \wedge D_2(z) \mathbb{F}[z]^{pm}.$$

This implies $X_{D_1} \otimes_{\mathbb{F}[z]} X_{D_2^\top} \simeq \mathbb{F}[z]^{pm} / D_1(z) \wedge D_2(z) \mathbb{F}[z]^{pm} \simeq X_{D_1 \wedge D_2}$. This completes the proof of the second part. ∎

Corollary 5.25. *Let* $D_1(z) \in \mathbb{F}[z]^{p \times p}$ *and* $D_2(z) \in \mathbb{F}[z]^{m \times m}$ *be nonsingular polynomial matrices with invariant factors* $d_1^{(1)}(z), \ldots, d_p^{(1)}(z)$ *and* $d_1^{(2)}(z), \ldots, d_m^{(2)}(z)$, *respectively. Let* $d_i^{(1)} \wedge d_j^{(2)}$ *denote the greatest common divisor of the polynomials* $d_i^{(1)}(z)$ *and* $d_j^{(2)}(z)$.

1. *The tensor product* $X_{D_1} \otimes_{\mathbb{F}[z]} X_{D_2}$ *is isomorphic to the polynomial model* $X_{D_1 \wedge D_2}$.
2. *For scalar polynomials* $d_1(z), d_2(z)$ *there is an isomorphism of* $\mathbb{F}[z]$-*modules*

$$X_{d_1} \otimes_{\mathbb{F}[z]} X_{d_2} \simeq X_{d_1 \wedge d_2}.$$

3. $X_{D_1} \otimes_{\mathbb{F}[z]} X_{D_2}$ *is* $\mathbb{F}[z]$-*linearly isomorphic to the direct sum*

$$\bigoplus_{i=1}^{p} \bigoplus_{j=1}^{m} X_{d_i^{(1)} \wedge d_j^{(2)}}.$$

In particular, the following dimension formula is valid:

$$\dim X_{D_1} \otimes_{\mathbb{F}[z]} X_{D_2} = \sum_{i=1}^{p} \sum_{j=1}^{m} \deg(d_i^{(1)} \wedge d_j^{(2)}).$$

4. *The invariant factors of* $D_1(z) \wedge D_2(z)$ *are* $d_i^{(1)}(z) \wedge d_j^{(2)}(z)$. *In particular,*

$$\det D_1(z) \wedge D_2(z) = \det D_1(z)^m \det D_2(z)^p.$$

Proof. Assertions 1 and 2 follow at once from part 2 of Theorem 5.24.

3. Let $\Delta_1 = \operatorname{diag}(d_1^{(1)}, \ldots, d_p^{(1)})$ and $\Delta_2 = \operatorname{diag}(d_1^{(2)}, \ldots, d_m^{(2)})$ be the respective Smith forms of $D_1(z)$ and $D_2(z)$, and let $U_i(z)$ and $V_i(z)$ be unimodular polynomial matrices satisfying $U_i(z)D_i(z) = \Delta_i(z)V_i(z)$. This implies the $\mathbb{F}[z]$-linear isomorphism

$$X_{\Delta_1} \simeq \bigoplus_{i=1}^{p} X_{d_i^{(1)}}, \quad X_{\Delta_2} \simeq \bigoplus_{j=1}^{m} X_{d_j^{(2)}}.$$

Using 2, the isomorphisms

$$X_{D_1} \otimes_{\mathbb{F}[z]} X_{D_2} \simeq X_{\Delta_1} \otimes_{\mathbb{F}[z]} X_{\Delta_2} \simeq \left(\bigoplus_{i=1}^{p} X_{d_i^{(1)}} \right) \otimes_{\mathbb{F}[z]} \left(\bigoplus_{i=1}^{m} X_{d_j^{(2)}} \right)$$

$$\simeq \bigoplus_{i,j} X_{d_i^{(1)}} \otimes_{\mathbb{F}[z]} X_{d_j^{(2)}} \simeq \bigoplus_{i=1}^{p} \bigoplus_{j=1}^{m} X_{d_i^{(1)} \wedge d_j^{(2)}}$$

follow. This completes the proof of 3.

4. Consider the unimodular polynomial matrices $U_i(z)$ and $V_i(z)$ such that

$$D_1(z) = U_1(z)\Delta_1(z)V_1(z), \quad D_2(z)^\top = U_2(z)\Delta_2(z)^\top V_2(z)$$

are in Smith form. Since $D_1 \wedge D_2$ is the greatest common left factor of $D_1 \otimes I_m$ and $I_p \otimes D_2^\top$ there exist polynomial matrices $M(z)$ and $N(z)$ such that

$$(D_1 \wedge D_2)M = D_1 \otimes I_m = (U_1 \otimes U_2)(\Delta_1 \otimes I)(V_1 \otimes U_2^{-1}),$$

$$(D_1 \wedge D_2)N = (U_1 \otimes U_2)(I_p \otimes \Delta_2^\top)(U_1^{-1} \otimes V_2).$$

Thus there exist unimodular matrices $P(z) = (U_1 \otimes U_2)^{-1}$, $R(z) = (V_1^{-1} \otimes U_2)$, and $S(z) = (U_1 \otimes V_2^{-1})$ such that

$$P(D_1 \wedge D_2)MR = \Delta_1 \otimes I, \quad P(D_1 \wedge D_2)NS = I_p \otimes \Delta_2^\top.$$

This implies that $P(z)(D_1 \wedge D_2)$ is a greatest common left divisor of $\Delta_1 \otimes I$ and $I_p \otimes \Delta_2^\top$ and therefore, up to an irrelevant unimodular factor, coincides with $\Delta_1 \wedge \Delta_2$. Thus $D_1 \wedge D_2$ and $\Delta_1 \wedge \Delta_2$ have the same invariant factors. It is easy to see that $\Delta_1 \wedge \Delta_2$ can be chosen as a diagonal matrix with diagonal entries $d_i^{(1)} \wedge d_j^{(2)}$. This completes the proof.   ∎

Corollary 5.26. *Consider nonsingular polynomial matrices $D(z) \in \mathbb{F}[z]^{p \times p}$ and $\overline{D}(z) \in \mathbb{F}[z]^{m \times m}$ with the same nontrivial invariant factors d_i, ordered so that $d_i | d_{i-1}$. Then the following assertions hold:*

1.

$$\dim \operatorname{Hom}_{\mathbb{F}[z]}(S_{\overline{D}}, S_D) = \sum_i (2i-1) \deg d_i.$$

2. Let $A \in \mathbb{F}^{n \times n}$ have invariant factors $d_1, \ldots, d_n$ ordered such that $d_i | d_{i-1}$. Let $\mathscr{C}(A) = \{X \in \mathbb{F}^{n \times n} \mid AX = XA\}$ denote the centralizer of A. Then:

$$\dim \mathscr{C}(A) = \sum_{i=1}^{n} (2i-1) \deg d_i; \tag{5.37}$$

3. For $A \in \mathbb{F}^{n \times n}$ with invariant factors $d_1, \ldots, d_n$, $\dim \mathscr{C}(A) = n^2$ if and only if there exists an $\alpha \in \mathbb{F}$ such that, for all $i = 1, \ldots, n$, $d_i(z) = z - \alpha$. Equivalently, $A = \alpha I$, i.e., A is a scalar transformation.

The relation of the $\mathbb{F}[z]$-tensor product $X_{D_1} \otimes_{\mathbb{F}[z]} X_{D_2^\top}$ to the tensored Kronecker model $X_{D_1(z) \otimes D_2^\top(z)}$ is examined next.

Proposition 5.27. *Let $D_1(z) \in \mathbb{F}[z]^{p \times p}$ and $D_2(z) \in \mathbb{F}[z]^{m \times m}$ be nonsingular. The following is an $\mathbb{F}[z]$-linear isomorphism:*

$$X_{D_1} \otimes_{\mathbb{F}[z]} X_{D_2^\top} \simeq X_{D_1(z) \otimes D_2^\top(z)} / (D_1 X_{I \otimes D_2^\top} + X_{D_1 \otimes I} D_2). \tag{5.38}$$

Proof. With i the canonical injections, the following diagram is commutative. Here $\pi_{D_1 \otimes D_2^\top} | J$ denotes the restriction of $\pi_{D_1 \otimes D_2^\top}$ to the subspace $J = D_1(z)\mathbb{F}[z]^{p \times m} + \mathbb{F}[z]^{p \times m} D_2(z)$:

$$
\begin{array}{ccc}
D_1(z)\mathbb{F}[z]^{p \times m} + \mathbb{F}[z]^{p \times m} D_2(z) & \xrightarrow{\ \ i\ \ } & \mathbb{F}[z]^{p \times m} \\[2mm]
\Big\downarrow{\scriptstyle \pi_{D_1 \otimes D_2^\top} | J} & & \Big\downarrow{\scriptstyle \pi_{D_1 \otimes D_2^\top}} \\[4mm]
D_1 X_{I \otimes D_2^\top} + X_{D_1 \otimes I} D_2 & \xrightarrow{\ \ i\ \ } & X_{D_1 \otimes D_2^\top}
\end{array}
$$

Moreover, $\pi^{-1}_{D_1 \otimes D_2^\top}(D_1 X_{I \otimes D_2^\top} + X_{D_1 \otimes I} D_2) = D_1(z) \mathbb{F}[z]^{p \times m} + \mathbb{F}[z]^{p \times m} D_2(z)$. Using the surjectivity of $\pi_{D_1 \otimes D_2^\top} : \mathbb{F}[z]^{p \times m} \longrightarrow X_{D_1 \otimes D_2^\top}$ and applying a standard argument, the isomorphism (5.38) follows. ∎

The next theorem yields an explicit description of the module $(X_{D_1} \otimes_{\mathbb{F}[z]} X_{D_2}^*)^*$.

Theorem 5.28. *Let* $D_1(z) \in \mathbb{F}[z]^{p \times p}$ *and* $D_2(z) \in \mathbb{F}[z]^{m \times m}$ *be nonsingular. The following are* $\mathbb{F}[z]$-*linear isomorphisms:*

$$X_{D_1} \otimes_{\mathbb{F}[z]} X_{D_2^\top} \simeq (X_{D_1} \otimes_{\mathbb{F}[z]} X_{D_2}^*)^* \simeq (X_{D_1} \otimes_{\mathbb{F}[z]} X_{D_2^\top})^* \simeq X^{D_1 \otimes I} \cap X^{I \otimes D_2^\top}.$$

Proof. For each submodule $M \subset \mathbb{F}[z]^{p \times m}$, one can identify the annihilator with $M^\perp = \{H \in z^{-1}\mathbb{F}[[z^{-1}]]^{p \times m} \mid \mathrm{Trace}[F,H] = 0 \ \forall F \in M\}$. Here $[F,H] = (F(z)^\top H(z))_{-1}$ denotes the residue. Computing

$$(X_{D_1} \otimes_{\mathbb{F}[z]} X_{D_2^\top})^* \simeq \left(\mathbb{F}[z]^{p \times m} / (D_1(z) \mathbb{F}[z]^{p \times m} + \mathbb{F}[z]^{p \times m} D_2^\top(z)) \right)^*$$

$$= (D_1(z) \mathbb{F}[z]^{p \times m} + \mathbb{F}[z]^{p \times m} D_2^\top(z))^\perp$$

$$= (D_1(z) \mathbb{F}[z]^{p \times m})^\perp \cap (\mathbb{F}[z]^{p \times m} D_2^\top(z))^\perp$$

$$= X^{D_1 \otimes I} \cap X^{I \otimes D_2^\top}$$

proves one isomorphism.

To prove the other isomorphisms, one uses the Smith form. Thus, for each nonsingular polynomial matrix $Q(z)$, the invariant factors of $Q(z)$ and $Q^\top(z)$ are equal, implying the $\mathbb{F}[z]$-linear isomorphism

$$X_Q \simeq X_{Q^\top}.$$

Moreover, Theorem 3.38 implies the isomorphism $X_Q \simeq X_Q^*$. By Theorem 5.24, the tensor product $X_{D_1} \otimes_{\mathbb{F}[z]} X_{D_2^\top}$ is $\mathbb{F}[z]$-linearly isomorphic to the polynomial model $X_{D_1 \wedge D_2^\top}$ and hence also to $X_{D_1 \wedge D_2^\top}^* \simeq (X_{D_1} \otimes_{\mathbb{F}[z]} X_{D_2^\top})^*$. This completes the proof. ∎

5.3 Polynomial Sylvester Equation

Proceeding now to a more detailed study of the Sylvester equation in the tensored polynomial model framework, definition (5.34) is specialized to the polynomial $p(z,w) = z - w$. One obtains, for all $Q(z,w) \in X_{D_1(z) \otimes_{\mathbb{F}} D_2^\top(w)}$, that

$$\mathscr{S}Q(z,w) = (z-w) \cdot Q(z,w) = \pi_{D_1(z) \otimes D_2^\top(w)}(zQ(z,w) - Q(z,w)w). \tag{5.39}$$

The map $\mathscr{S}$ is referred to as the **polynomial Sylvester operator** . In fact, with constant matrices $A_1 \in \mathbb{F}^{p \times p}$ and $A_2 \in \mathbb{F}^{m \times m}$, and defining $D_1(z) = zI - A_1$ and $D_2(w) = wI - A_2$, we obtain $X_{D_1(z) \otimes_\mathbb{F} D_2^\top(w)} = \mathbb{F}^{p \times m}$. Therefore, $Q(z,w) \in X_{D_1(z) \otimes D_2^\top(w)}$ if and only if $Q(z,w) \in \mathbb{F}^{p \times m}$, i.e., $X = Q(z,w)$ is a constant matrix. This implies

$$(z - w) \cdot X = \pi_{(zI-A_1) \otimes (wI-A_2^\top)}(z-w)X = A_1X - XA_2$$

for all $X \in \mathbb{F}^{p \times m}$, i.e., we recover the standard **Sylvester operator**. This computation shows that the **classical Sylvester equation**

$$A_1X - XA_2 = C$$

corresponds to the equation

$$\mathscr{S}X = C,$$

with $X, C \in X_{(zI-A_1) \otimes_\mathbb{F} (wI-A_2^\top)}$ necessarily constant matrices.

Note that every polynomial matrix $T(z,w) \in X_{D_1(z) \otimes D_2^\top(w)}$ has a factorization of the form

$$T(z,w) = R_1(z)R_2^\top(w),$$

with $R_1(z) \in X_{D_1 \otimes I_k} \subset \mathbb{F}[z]^{p \times k}$ and $R_2(w)^\top \in X_{I_k \otimes D_2^\top(w)} \subset \mathbb{F}[w]^{k \times m}$, and both polynomial matrices $R_1(z)$ and $R_2(w)$ have linearly independent columns. The following theorem reduces the analysis of the general Sylvester equation to a polynomial equation of the Bezout type. This extends the method, introduced in Willems and Fuhrmann (1992), for the analysis of the Lyapunov equation. Of course, a special case is the homogeneous Sylvester equation, which has a direct connection to the theory of Bezoutians.

Theorem 5.29. *Let $D_1(z) \in \mathbb{F}[z]^{p \times p}$ and $D_2(w) \in \mathbb{F}[w]^{m \times m}$ be nonsingular, let the Sylvester operator $\mathscr{S} : X_{D_1(z) \otimes D_2^\top(w)} \longrightarrow X_{D_1(z) \otimes D_2^\top(w)}$ be defined by (5.39) and let $R_1(z) \in X_{D_1(z) \otimes_\mathbb{F} I}, R_2^\top(w) \in X_{I \otimes_\mathbb{F} D_2^\top(w)}$. Then:*

1. The Sylvester equation

$$S_{D_1}Q - QS_{D_2} = T(z,w) = R_1(z)R_2^\top(w), \qquad (5.40)$$

   or equivalently

$$\mathscr{S}Q(z,w) = \pi_{D_1(z) \otimes D_2^\top(w)}(z-w)Q(z,w) = R_1(z)R_2^\top(w);$$

is solvable if and only if there exists polynomial matrices $N_1(z) \in X_{D_1(z) \otimes I}$ and $N_2(z) \in X_{I \otimes D_2^\top(z)}$ for which

$$D_1(z)N_2(z) - N_1(z)D_2(z) + R_1(z)R_2^\top(z) = 0. \tag{5.41}$$

*Equation (5.40) will be referred to as the **polynomial Sylvester equation**, or **PSE** for short, and its solutions are given by*

$$Q(z,w) = \frac{D_1(z)N_2(w) - N_1(z)D_2(w) + R_1(z)R_2^\top(w)}{z - w}; \tag{5.42}$$

2. $Q(z,w) \in X_{D_1(z) \otimes D_2^\top(w)}$ *solves the **homogeneous polynomial Sylvester equation**, or **HPSE** for short, if and only if there exist polynomial matrices $N_1(z) \in X_{D_1 \otimes I}$ and $N_2(z) \in X_{I \otimes D_2^\top}$ that satisfy*

$$D_1(z)N_2(z) - N_1(z)D_2(z) = 0, \tag{5.43}$$

in terms of which

$$Q(z,w) = \frac{D_1(z)N_2(w) - N_1(z)D_2(w)}{z - w}.$$

Proof. 1. Assume there exist the polynomial matrices $N_1(z) \in X_{D_1 \otimes I}$ and $N_2(z) \in X_{I \otimes D_2^\top}$, solving equation (5.41), and for which $Q(z,w)$ is defined by (5.42). Note first that, under our assumptions on $R_1(z)$ and $R_2(w)$,

$$D_1(z)^{-1}Q(z,w)D_2(w)^{-1}$$
$$= \frac{N_2(w)D_2(w)^{-1} - D_1(z)^{-1}N_1(z) + D_1(z)^{-1}R_1(z)R_2^\top(w)D_2(w)^{-1}}{z - w}$$

is strictly proper in both variables, i.e., $Q(z,w)$ is in $X_{D_1(z) \otimes D_2^\top(w)}$. Computing

$$\mathscr{S}Q(z,w) = \pi_{D_1(z) \otimes D_2^\top(w)}(z - w)Q(z,w)$$
$$= \pi_{D_1(z) \otimes D_2^\top(w)}(D_1(z)N_2(w) - N_1(z)D_2(w) + R_1(z)R_2(w))$$
$$= R_1(z)R_2(w)^\top,$$

it follows that $Q(z,w)$ is indeed a solution.

To prove the converse, note that, given a nonsingular polynomial matrix $D_1(z) \in \mathbb{F}[z]^{p \times p}$, then, for $f(z) \in X_{D_2}$, $(S_{D_2}f)(z) = zf(z) - D_2(z)\xi_f$, where $\xi_f = (D_2^{-1}f)_{-1}$. This implies that, for $Q(z,w) \in X_{D_1(z) \otimes D_2^\top(w)}$,

$$S_{z\otimes 1}Q(z,w) = zQ(z,w) - D_1(z)N_2(w), \quad S_{1\otimes w}Q(z,w) = Q(z,w)w - N_1(z)D_2(w),$$

with both $N_2 D_2^{-1}$ and $D_1^{-1} N_1$ strictly proper. Assuming $Q(z,w)$ is a solution of the PSE, we compute

$$\begin{aligned} S_{z-w}Q(z,w) &= [zQ(z,w) - D_1(z)N_2(w)] - [Q(z,w)w - N_1(z)D_2(w)] \\ &= R_1(z)R_2(w)^\top, \end{aligned}$$

implying

$$Q(z,w) = \frac{D_1(z)N_2(w) - N_1(z)D_2(w) + R_1(z)R_2(w)^\top}{z-w}.$$

However, because $Q(z,w) \in X_{D_1(z)\otimes D_2^\top(w)}$ is a polynomial matrix, (5.41) necessarily holds.
2. Follows from the previous part. ∎

This leads us to introduce the following object.

Definition 5.30. A polynomial matrix $Q(z,w) \in X_{D_1(z)\otimes D_2^\top(w)}$ is called a **generalized Bezoutian** if it has a representation of the form

$$Q(z,w) = \frac{D_1(z)N_2(w) - N_1(z)D_2(w)}{z-w}, \tag{5.44}$$

with $D_1^{-1}N_1$ and $N_2 D_2^{-1}$ strictly proper and such that the identity

$$D_1(z)N_2(z) = N_1(z)D_2(z) \tag{5.45}$$

is satisfied.

Corollary 5.31. $Q(z,w) \in X_{D_1(z)\otimes D_2^\top(w)}$ is a solution of the HPSE (5.43) if and only if $Q(z,w)$ is a generalized Bezoutian.

Proof. Follows from Theorem 5.29.2. ∎

5.4 Generalized Bezoutians and Intertwining Maps

Proposition 5.6 shows that there is a close connection between tensor products of vector spaces and spaces of $\mathbb{F}$-linear maps between vector spaces. For functional models one can be more specific about the form of such connections, leading to a new interpretation of Bezoutian operators in terms of intertwining maps and module homomorphisms of polynomial models. Denote by $\mathrm{Hom}_{\mathbb{F}}(X_{D_1}, X_{D_2})$ the space of all $\mathbb{F}$-linear maps from X_{D_1} to X_{D_2} and by $\mathrm{Hom}_{\mathbb{F}[z]}(X_{D_1}, X_{D_2})$ the space of all $\mathbb{F}[z]$-

linear maps from X_{D_1} to X_{D_2}, i.e., the space of all $\mathbb{F}$-linear maps Z from X_{D_1} to X_{D_2} that satisfy $ZS_{D_1} = S_{D_2}Z$. The essential information that encodes the mutual interrelations between these spaces is given by the following commutative diagram:

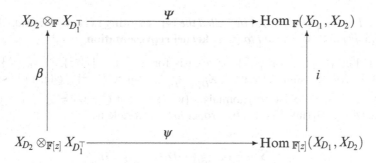

Here i is the natural inclusion of $\mathrm{Hom}_{\mathbb{F}[z]}(X_{D_1},X_{D_2})$ in $\mathrm{Hom}_{\mathbb{F}}(X_{D_1},X_{D_2})$. The map β will be constructed via Bezoutians, and we will establish the $\mathbb{F}$- and $\mathbb{F}[z]$-linear isomorphisms Ψ and ψ, respectively. As is easily seen, this diagram is an equivalent reformulation of the first diagram, mentioned at the beginning of this chapter, insofar as the dual spaces $X_{D_1}^*$ and $(X_{D_2} \otimes X_{D_1}^*)^*$ are replaced by their isomorphic counterparts $X_{D_1^\top}$ and $X_{D_2} \otimes X_{D_1^\top}$, respectively. For the relevant isomorphisms that underpin such reformulations, we refer to Theorem 5.28. It may come as somewhat of a surprise that the study of $\mathbb{F}[z]$-homomorphisms of polynomial models can be based on the study of tensored models. This (see Theorem 5.34) leads to a further clarification of the connection between intertwining maps and Bezoutians.

To achieve an even more concrete form of the previous diagram, we next prove a proposition that establishes a concrete connection between the space of maps Z intertwining the shifts S_{D_2} and S_{D_1} and the $\mathbb{F}[z]$-tensor product of the polynomial models $X_{D_2^\top}$ and X_{D_1}.

Proposition 5.32. *Let $D_1(z) \in \mathbb{F}[z]^{p \times p}$ and $D_2(z) \in \mathbb{F}[w]^{m \times m}$ be nonsingular.*

1. Every $Q(z,w) \in X_{D_2(z) \otimes D_1^\top(w)}$ has a representation of the form

$$Q(z,w) = R_2(z)R_1^\mid(w), \tag{5.46}$$

with $R_2(z) \in X_{D_2(z) \otimes I}$ and $R_1(w)^\top \in X_{I \otimes D_1(w)^\top}$, i.e., both $D_2(z)^{-1}R_2(z)$ and $D_1(w)^{-\top}R_1(w)$ are strictly proper. Without loss of generality, assume that the columns of $R_1(z)$ and $R_2(z)$ are linearly independent.

2. Define a map $\Psi : X_{D_2(z) \otimes_{\mathbb{F}} D_1^\top(w)} \longrightarrow \mathrm{Hom}_{\mathbb{F}}(X_{D_1}, X_{D_2})$, for $f(z) \in X_{D_1}$ and $Q(z,w) \in X_{D_2(z) \otimes D_1^\top(w)}$ having the representation (5.46), by

$$\Psi(Q)f = Z_Q f = \,<f, Q(z,.)^\top> = [D_1^{-1}f, Q(z,.)^\top] = (Q(z,\cdot)D_1^{-1}f)_{-1}. \tag{5.47}$$

Then Ψ *induces the isomorphism*

$$X_{D_2(z) \otimes_{\mathbb{F}} D_1^\top (w)} \simeq \mathrm{Hom}_{\mathbb{F}}(X_{D_1}, X_{D_2}).$$

The polynomial $Q(z, w)$ *will be called the* **representing kernel** *of the map* $\Psi(Q)$, *and* (5.47) *will be referred to as its* **kernel representation**.

Proof. 1. Let $\{u_1(z), \ldots, u_{n_1}(z)\}$ be a basis for X_{D_2} and $\{v_1(z), \ldots, v_{n_2}(z)\}$ be a basis for X_{D_1}. Assume $Q(z, w) \in X_{D_2(z) \otimes D_1^\top(w)}$. Since $D_2(z)^{-1} Q(z, w)$ is strictly proper in z, there exist polynomials $g_i(w)$ such that $Q(z, w) = \sum_{i=0}^{n_2} u_i(z) g_i(w)$. Since $Q(z, w) D_1(w)^{-1}$ is strictly proper in the variable w,

$$\sum_{i=0}^{n_2} u_i(z) \pi_+ (g_i(w) D_1(w)^{-1}) = 0.$$

In turn, this implies $g_i^\top(w) \in X_{D_1^\top}$ and, hence, the existence of $\alpha_{ij} \in \mathbb{F}$ for which $g_i(w) = \sum_{j=0}^{n_1} \alpha_{ij} v_j(w)^\top$. Thus

$$Q(z, w) = \sum_{i=0}^{n_2} \sum_{j=0}^{n_1} \alpha_{ij} u_i(z) v_j(w)^\top = R^{(2)}(z) A R^{(1)}(w)^\top.$$

Here $R^{(2)} = (u_1(z), \ldots, u_{n_1}(z))$, $R^{(1)} = (v_1(z), \ldots, v_{n_2}(z))$ and $A = (\alpha_{ij})$. Next, let $r = \mathrm{rank} A$, which implies the existence of a factorization $A = A_2 A_1^\top$, with $A_i \in \mathbb{F}^{n_i \times r}$ of full column rank. Redefining the $R^{(i)}(z)$, the statement follows.

2. As elements of the form $u_i(z)(v_j(w))^\top$ generate $X_{D_2(z) \otimes_{\mathbb{F}} D_1^\top(w)}$, we compute for such an element $\Psi(u_i(z)(v_j(w))^\top) = u_i(z) < f, v_j >$. This allows us to compute, for $Q(z, w) = \sum_{i=1}^{k} u_i(z) v_i(w)^\top = R_2(z) R_1^\top(w)$,

$$\Psi(\sum_{i=1}^{k} u_i(z) v_i(w)^\top) f = \sum_{i=1}^{k} u_i(z) < f, v_i >$$

$$= (\sum_{i=1}^{k} u_i(z) v_i(w)^\top D_1(w)^{-1} f(w))_{-1} = (\sum_{i=1}^{k} u_i(z) v_i(w)^\top D_1(w)^{-1} f(w))_{-1}$$

$$= < f, \sum_{i=1}^{k} v_i(w) u_i(z)^\top > = < f, Q(z, \cdot)^\top > .$$

Clearly, Ψ defined by (5.47) is $\mathbb{F}$-linear. To show the injectivity of Ψ, assume without loss of generality that the columns of $R_1(z)$ are linearly independent and that $\Psi(Q) = 0$, i.e., that for all $g(z) \in X_{D_1}$,

$$0 = < g, Q(z, .)^\top > = R_2(z) (R_1^\top(w) D_1(w)^{-1} g(w))_{-1}.$$

This implies $(R_1^\top(w)D_1(w)^{-1}g(w))_{-1} = 0$ for all $g(z) \in X_{D_1}$. Since the columns of $R_1(z)$ are in $X_{D_1^\top}$, Theorem 3.38 implies that $R_1 = 0$ and, hence, $Q(z,w) = 0$. That Ψ is an isomorphism follows from the equality of dimension. Indeed,

$$\dim X_{D_2(z)\otimes_\mathbb{F} D_1^\top(w)} = \dim X_{D_2} \otimes_\mathbb{F} X_{D_1^\top}$$
$$= \dim X_{D_2} \cdot \dim X_{D_1^\top} = \deg\det D_1 \cdot \deg\det D_2$$
$$= \dim \mathrm{Hom}_\mathbb{F}(X_{D_1}, X_{D_2}).$$

■

The following lemma will be needed in the sequel.

Lemma 5.33. *Let* $H(z) \in \mathbb{F}((z^{-1}))^{p\times m}$. *Then:*

$$\left(\frac{H(w)}{w-z}\right)_{-1} = \pi_+ H(z).$$

Proof. Let $H(w) = \sum_{k=-\infty}^{n_H} H_{-k} w^k$. One computes

$$\left(\frac{H(w)}{w-z}\right)_{-1} = \left(\sum_{k=-\infty}^{n_H} H_{-k}\frac{w^k}{w-z}\right)_{-1} = \sum_{k=-\infty}^{n_H} H_{-k}\left(\frac{w^k}{w-z}\right)_{-1}$$
$$= \sum_{k=-\infty}^{n_H} H_{-k}\sum_{j=0}^{\infty}\left(w^k\frac{z^j}{w^{j+1}}\right)_{-1} = \sum_{k=0}^{n_H} H_{-k}z^k$$
$$= \pi_+ H(z).$$

Here we used

$$\left(w^k\frac{z^j}{w^{j+1}}\right)_{-1} = \begin{cases} 0 & j \neq k \\ z^k & j = k. \end{cases}$$

■

In Theorem 3.20, a characterization of maps intertwining two polynomial models was derived. In fact, already in equation (3.22) there is a clue to the beautiful link between intertwining maps and the theory of generalized Bezoutians. This connection is now formalized in the following theorem, which plays a central role in our analysis. It allows for a second, independent approach to the characterization of homomorphisms between polynomial models and the commutant lifting theorem (Chapter 3).

Theorem 5.34. *Let* $D_1(z) \in \mathbb{F}[z]^{p\times p}$ *and* $D_2(z) \in \mathbb{F}[w]^{m\times m}$ *be nonsingular. Let* $R_1(z) \in \mathbb{F}[z]^{p\times k}$ *and* $R_2(z) \in \mathbb{F}[z]^{m\times k}$. *Assume that* $D_1(z)^{-\top}R_1(z)$ *and* $D_2(z)^{-1}R_2(z)$ *are strictly proper. Then the following statements are equivalent.*

1. $Q(z,w) = R_2(z)R_1^\top(w)$ *is a solution of the HPSE (5.43).*
2. $Q(z,w) = R_2(z)R_1^\top(w)$ *is a Bezoutian, i.e., it has a representation of the form (5.44) and satisfies (5.45).*
3. *The map* $Z : X_{D_1} \longrightarrow X_{D_2}$ *defined by*

$$
\begin{aligned}
Zg = R_2(z) < g, R_1^\top > &=< g, Q(z,.)^\top > \\
&= [D_1^{-1}g, (R_2(z)R_1^\top(w))^\top] = (R_2(z)R_1^\top(w)D_1(w)^{-1}g(w))_{-1}
\end{aligned}
\tag{5.48}
$$

satisfies

$$
S_{D_2}Z = ZS_{D_1},
\tag{5.49}
$$

i.e., it is an intertwining map or, equivalently, an $\mathbb{F}[z]$*-homomorphism.*
4. *The map* $Z : X_{D_1} \longrightarrow X_{D_2}$ *has the representation*

$$
Zg = \pi_{D_2}N_2g, \qquad g \in X_{D_1},
\tag{5.50}
$$

with $D_1(z), D_2(z)$ *satisfying the intertwining relation*

$$
N_2(z)D_1(z) = D_2(z)N_1(z)
\tag{5.51}
$$

for some $N_1(z), N_2(z) \in \mathbb{F}[z]^{p \times m}$.

Proof. $(1) \Leftrightarrow (2)$
Follows from Theorem 5.29 and Definition 5.30.
$(2) \Rightarrow (3)$
Assume $Q(z,w) = R_2(z)R_1(w)^\top$ is a Bezoutian, i.e., it has a representation of the form (5.44). We compute, for $g(z) \in X_{D_1}$,

$$
\begin{aligned}
(S_{D_2}Z &- ZS_{D_1})g \\
&= S_{D_2}(R_2(z)R_1(w)^\top D_1(w)^{-1}g(w))_{-1} - R_2(z)(R_1(w)^\top D_1(w)^{-1}S_{D_1}g(w))_{-1} \\
&= \pi_{D_2}(zR_2(z)R_1^\top(w)D_1(w)^{-1}g(w))_{-1} - R_2(z)(R_1(w)^\top D_1(w)^{-1}\pi_{D_1}wg(w))_{-1} \\
&= \pi_{D_2}(zR_2(z)R_1^\top(w)D_1(w)^{-1}g(w))_{-1} - R_2(z)(R_1(w)^\top \pi_- D_1^{-1}wg(w))_{-1} \\
&= \pi_{D_2}(zR_2(z)R_1^\top(w)D_1(w)^{-1}g(w))_{-1} - R_2(z)(R_1(w)^\top D_1(w)^{-1}wg(w))_{-1} \\
&= \pi_{D_2}((D_2(z)N_1(w) - N_2(z)D_1(w))D_1(w)^{-1}g(w))_{-1} \\
&= -\pi_{D_2}(N_2(z))(D_2(w)D_2(w)^{-1}g(w))_{-1}) \\
&= -N_2(z)(g(w))_{-1} = 0,
\end{aligned}
$$

using the fact that $g(w)$ is a polynomial. This implies (5.49).

$(2) \Rightarrow (4)$

Assume first that $Q(z,w)$ is a Bezoutian, i.e., has a representation of the form (5.44). We prove now that Z has the alternative representation (5.50). To this end, using Lemma 5.33, one computes

$$
\begin{aligned}
Zg &=< g, Q(z,\cdot)^\top > \\
&= [D_1^{-1}g, Q(z,\cdot)^\top] = \left(Q(z,w)D_1(w)^{-1}g(w)\right)_{-1} \\
&= \left(\frac{D_2(z)N_1(w) - N_2(z)D_1(w)}{z-w}D_1(w)^{-1}g(w)\right)_{-1} \\
&= \left(D_2(z)\frac{N_1(w)D_1(w)^{-1}g(w)}{z-w} - N_2(z)\frac{g(w)}{z-w}\right)_{-1} \\
&= -D_2(z)\pi_+ N_1 D_1^{-1}g + N_2(z)g(z) = N_2(z)g(z) - D_2\pi_+ D_2^{-1}N_2g \\
&= D_2\pi_- D_2^{-1}N_2g = \pi_{D_2}N_2g.
\end{aligned}
$$

$(3) \Rightarrow (2)$

Assume that $Z : X_{D_1} \longrightarrow X_{D_2}$, defined by (5.48), is intertwining. For $g(z) \in X_{D_1}$, one computes, using Lemma 5.33, the fact that $\pi_{D_2}R_2 = R_2$, and that a contribution of a polynomial term to the residue $()_{-1}$ is zero,

$$
\begin{aligned}
0 &= (S_{D_2}Z - ZS_{D_1})g = S_{D_2} < g, Q(z,.)^\top > - < S_{D_1}g, Q(z,.)^\top > \\
&= S_{D_2}\left(R_2(z)R_1(w)^\top D_1(w)^{-1}g(w)\right)_{-1} - \left(R_2(z)R_1(w)^\top D_1(w)^{-1}S_{D_1}g(w)\right)_{-1} \\
&= \pi_{D_2}\left(zR_2(z)R_1(w)^\top D_1(w)^{-1}g(w)\right)_{-1} - \left(R_2(z)R_1(w)^\top D_1(w)^{-1}(\pi_{D_1}wg(w))\right)_{-1} \\
&= \pi_{D_2}\left(zR_2(z)R_1(w)^\top D_1(w)^{-1}g(w)\right)_{-1} - \left(R_2(z)R_1(w)^\top D_1(w)^{-1}wg(w)\right)_{-1} \\
&= \pi_{D_2}\left(R_2(z)(z-w)R_1(w)^\top D_1(w)^{-1}g(w)\right)_{-1}.
\end{aligned}
$$

Since this is true for all $g(z) \in X_{D_1}$, and as it trivially holds for $g(z) \in D_1\mathbb{F}[z]^m$, it is satisfied for all $g(z) \in \mathbb{F}[z]^m$. Hence, $\pi_{D_2}\left(R_2(z)(z-w)R_1(w)^\top D_1(w)^{-1}\right)$ is a polynomial in both variables. It follows that $(\pi_{D_2} \otimes \pi_{D_1})\left(R_2(z)(z-w)R_1(w)^\top\right) = 0$, i.e., $R_2(z)R_1(w)^\top$ is a solution of the HPSE. Applying Theorem 5.29, it follows that $R_2(z)R_1(w)^\top$ is a Bezoutian.

$(4) \Rightarrow (3)$

From representation (5.50) it easily follows that Z is intertwining. Indeed, noting that equality (5.51) implies $N_2 \operatorname{Ker}\pi_{D_2} \subset \operatorname{Ker}\pi_{D_1}$, we compute

$$
S_{D_1}Zg - ZS_{D_2}g = \pi_{D_1}z\pi_{D_1}N_2g - \pi_{D_1}N_2\pi_{D_2}zg = \pi_{D_1}zN_2g - \pi_{D_1}N_2zg = 0.
$$

∎

Proposition 5.35. *Let $D_1(z) \in \mathbb{F}[z]^{m \times m}$ and $D_2(z) \in \mathbb{F}[z]^{p \times p}$ be nonsingular.*

1. Every $H \in X^{D_2 \otimes I} \cap X^{I \otimes D_1^\top}$ has unique representations

$$H(z) = D_2(z)^{-1} N_2(z) = N_1(z) D_1(z)^{-1}, \qquad (5.52)$$

with $N_1(z) \in X^{I \otimes D_1^\top}$ and $N_2(z) \in X_{D_2 \otimes I}$.

2. The map $\psi : X^{D_2 \otimes I} \cap X^{I \otimes D_1^\top} \longrightarrow \mathrm{Hom}_{\mathbb{F}[z]}(X_{D_1}, X_{D_2})$ defined by

$$\psi(H)g = \pi_{D_2} N_2 g, \qquad g(z) \in X_{D_1} \qquad (5.53)$$

induces the isomorphism

$$X^{D_2 \otimes I} \cap X^{I \otimes D_1^\top} \simeq \mathrm{Hom}_{\mathbb{F}[z]}(X_{D_1}, X_{D_2}). \qquad (5.54)$$

Proof. 1. That $H(z) \in X^{D_2 \otimes I} \cap X^{I \otimes D_1^\top}$ has the unique representations (5.52) is obvious from the definitions.

2. Clearly, by Theorem 5.34, $\psi(H) \in \mathrm{Hom}_{\mathbb{F}[z]}(X_{D_1}, X_{D_2})$, i.e., it is an intertwining map. To show the injectivity of the map ψ, assume $\psi(H) = 0$. With the representation (5.52), this implies $\pi_{D_2} N_2 g = 0$ for all $g \in X_{D_1}$. This means that $D_2(z)$ is a left factor of $N_2(z)$. But as $N_2(z) \in X_{D_2 \otimes I}$, necessarily $N_2(z) = 0$. That ψ is surjective follows from Theorem 5.34. ∎

The Bezout Map.

We are now in a position to explain the beautiful connection between tensor products, intertwining maps, and Bezoutians in a very concrete way. Our starting point is the commutative diagram of canonical homomorphisms established in the following diagram:

$$
\begin{array}{ccc}
(X_{D_2} \otimes_{\mathbb{F}} X_{D_1^\top})^* & \xrightarrow{\quad \Phi \quad} & \mathrm{Hom}_{\mathbb{F}}(X_{D_1}, X_{D_2}) \\
\Big\uparrow{\scriptstyle \iota^*} & & \Big\uparrow{\scriptstyle i} \\
(X_{D_2} \otimes_{\mathbb{F}[z]} X_{D_1^\top})^* & \xrightarrow{\quad \phi \quad} & \mathrm{Hom}_{\mathbb{F}[z]}(X_{D_1}, X_{D_2})
\end{array}
$$

In Proposition (5.6), it was shown that $(X_{D_2} \otimes_{\mathbb{F}} X_{D_1^\top})^*$ is naturally isomorphic to $X_{D_2} \otimes_{\mathbb{F}} X_{D_1^\top}$, which in turn is isomorphic to the Kronecker product space $X_{D_2(z) \otimes D_1^\top(w)}$ of polynomials in two variables z, w. Similarly, using duality theory and Theorem 5.28, the module $(X_{D_2} \otimes_{\mathbb{F}[z]} X_{D_1^\top})^*$ can be identified with $X_{D_2} \otimes_{\mathbb{F}[z]} X_{D_1^\top}$, but now at the expense of a less clear interpretation of the maps ι^* and ϕ.

Matters simplify considerably if, instead of using a polynomial model representation $X_{D_2} \otimes_{\mathbb{F}[z]} X_{D_1^\top}$, we pass to a rational model $X^{D_2 \otimes I} \cap X^{I \otimes D_1^\top}$ via the isomorphism (5.54). This leads to a new commutative diagram of concrete maps that is described now in more detail.

Noting that every $H(z) \in X^{D_2 \otimes I} \cap X^{I \otimes D_1^\top}$ has a unique representation of the form

$$H(z) = D_2(z)^{-1} N_2(z) = N_1(z) D_1(z)^{-1},$$

one defines the **Bezout map**

$$\beta : X^{D_2 \otimes I} \cap X^{I \otimes D_1^\top} \longrightarrow X_{D_2(z) \otimes D_1^\top(w)}$$

by associating with the rational function (z) the two-variable polynomial

$$\beta(H) = Q(z, w) = \frac{D_2(z) N_1(w) - N_2(z) D_1(w)}{z - w}. \tag{5.55}$$

Note that β is $\mathbb{F}$-linear and injective and the image space consists of all **Bezoutian forms**. That $Q(z, w) \in X_{D_2(z) \otimes D_1^\top(w)}$ follows from the calculation

$$D_2(z)^{-1} Q(z, w) D_1(w)^{-1} = D_2(z)^{-1} \frac{D_2(z) N_1(w) - N_2(z) D_1(w)}{z - w} D_1(w)^{-1}$$

$$= \frac{H(w) - H(z)}{z - w} = \sum_{k=1}^{\infty} H_k \frac{w^{-k} - z^{-k}}{z - w} = -\frac{1}{zw} \sum_{k=1}^{\infty} H_k \frac{w^{-k} - z^{-k}}{w^{-1} - z^{-1}}$$

$$= -\frac{1}{zw} \sum_{k=1}^{\infty} H_k \left(w^{1-k} + w^{2-k} z^{-1} + \cdots + z^{1-k} \right),$$

which shows that it is indeed strictly proper.

With these definitions, the principal result can be stated.

Theorem 5.36. *Let $D_2(z) \in \mathbb{F}[z]^{p \times p}$ and $D_1(z) \in \mathbb{F}[w]^{m \times m}$ be nonsingular. Let the maps*

$$\Psi : X_{D_2(z) \otimes D_1^\top(w)} \longrightarrow \mathrm{Hom}_{\mathbb{F}}(X_{D_1}, X_{D_2})$$

and

$$\psi : X^{D_2 \otimes I} \cap X^{I \otimes D_1^\top} \longrightarrow \mathrm{Hom}_{\mathbb{F}[z]}(X_{D_1}, X_{D_2})$$

be defined by (5.47) and (5.53), respectively. Let

$$i : \mathrm{Hom}_{\mathbb{F}[z]}(X_{D_1}, X_{D_2}) \longrightarrow \mathrm{Hom}_{\mathbb{F}}(X_{D_1}, X_{D_2})$$

be the canonical embedding, and let

$$\beta : X^{D_2 \otimes I} \cap X^{I \otimes D_1^\top} \longrightarrow X_{D_2(z) \otimes D_1^\top(w)}$$

be the injective Bezout map given in (5.55). Then the following diagram commutes:

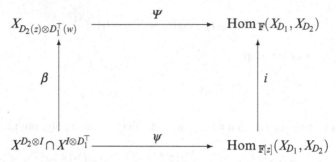

Proof. Note that, by Proposition 5.35, every element $H \in X^{D_2 \otimes I} \cap X^{I \otimes D_1^\top}$ has a unique representation of the form (5.52). With the **Bezout map** defined by (5.55), all that remains is to prove the identity

$$\Psi \circ \beta = i \circ \psi. \tag{5.56}$$

To this end, with $H(z) \in X^{D_2 \otimes I} \cap X^{I \otimes D_1^\top}$, $g \in X_{D_1}$, and using equation (5.50) and Lemma 5.4, we compute

$$(\Psi \circ \beta)(H)g = \Psi(\beta(H))g = \Psi(Q)g$$

$$= <g, Q(z, \cdot)^\top> = \left(\frac{D_2(z)N_1(w) - N_2(z)D_1(w)}{z - w} D_1(w)^{-1} g(w) \right)_{-1}$$

$$= D_2(z) \left(\frac{D_2(w)^{-1} N_2(w) g(w)}{z - w} \right)_{-1} - N_2(z) \left(\frac{g(w)}{z - w} \right)_{-1}$$

$$= -D_2(z)\pi_+ (D_2(z)^{-1} N_2(z) g(z)) + N_2(z)\pi_+(g(z))$$

$$= D_2(z)\pi_- (D_2(z)^{-1} N_2(z) g(z)) - N_2(z)g(z) + N_2(z)\pi_+(g(z))$$

$$= \pi_{D_2} N_2 g = \psi(H)g = (i \circ \psi)(H)g,$$

i.e., (5.56) is proved.   ∎

Theorem 5.36 shows that the $\mathbb{F}$-linear maps $X_{D_1} \longrightarrow X_{D_2}$

$$Zf = \left(Q(z, w) D_1(w)^{-1} f(w) \right)_{-1},$$

whose representing kernel $Q(z, w)$ is a Bezoutian form, correspond exactly to the $\mathbb{F}[z]$-linear homomorphisms $X_{D_1} \longrightarrow X_{D_2}$. Thus the theory of Bezoutians is intimately connected to the structure of module homomorphisms of polynomial models. This is a two-way street because it shows the existence of concrete representations of module homomorphisms by Bezoutian forms. Conversely, the linear maps defined by Bezoutians are seen to be module homomorphisms whose structure is clarified by the commutant lifting theorem, Theorem 2.54.

Generalized Bezoutian Matrices.

Generalized Bezoutian matrices $B(N_1, D_1, N_2, D_2)$, induced by a quadruple of polynomial matrices satisfying an intertwining relation, were introduced in Anderson and Jury (1976) and studied in further detail in Lerer and Tismenetsky (1982). In the sequel we will find it convenient to distinguish between the Bezoutian form as a matrix polynomial in two variables, which is an element of a tensored model, the corresponding intertwining map, and the Bezoutian matrix, which is a specific matrix representation. This is analogous to the distinction between a linear transformation and its matrix representation. There are many choices of bases in polynomial models, and some lead to interesting matrix representations; see, for example, Fuhrmann and Datta (1989) or Mani and Hartwig (1997).

Definition 5.37. Assume that the polynomial matrices $N_1(z)$ and $N_2(z)$ are such that $H(z) = N_1(z)D_1(z)^{-1} = D_2(z)^{-1}N_2(z) \in \mathbb{F}(z)^{p \times m}$ are strictly proper. Let the polynomial matrix $Q(z, w)$ be given as

$$Q(z, w) = \frac{D_2(z)N_1(w) - N_2(z)D_1(w)}{z - w} = \sum_{i,j=1}^{k} Q_{ij} z^{i-1} w^{j-1}.$$

Then the matrix

$$B(N_1, D_1, N_2, D_2) = (Q_{ij}) \in \mathbb{F}^{kp \times km}$$

is called the **generalized Bezoutian matrix**. The linear operator

$$\mathbf{B} : X_{D_1} \longrightarrow X_{D_2}, \ \mathbf{B}f = \left(Q(z, w)D_1(w)^{-1}f(w) \right)_{-1}$$

is called the **Bezout operator** of $H(z) = N_1(z)D_1(z)^{-1} = D_2(z)^{-1}N_2(z)$.

In Chapter 3, coprimeness conditions for the injectivity, surjectivity, and bijectivity of homomorphisms between polynomial models were obtained. Using Theorem 5.36, this result can now be applied to characterize full rank properties of the Bezout operator.

Theorem 5.38. *The Bezout operator* $\mathbf{B}$ *of* $H(z) = N_1(z)D_1(z)^{-1} = D_2(z)^{-1}N_2(z)$ *is*

1. *Full column rank if and only if* $D_1(z)$ *and* $N_1(z)$ *are right coprime,*
2. *Full row rank if and only if* $D_2(z)$ *and* $N_2(z)$ *are left coprime,*
3. *Invertible if and only if* $D_1(z)$ *and* $N_1(z)$ *are right coprime and* $D_2(z)$ *and* $N_2(z)$ *are left coprime.*

Proof. By Theorem 5.36, the Bezout operator **B** coincides with the homomorphism $Z : X_{D_1} \longrightarrow X_{D_2}$ defined by $Zf = \pi_{D_2}(N_2 f)$. The result follows by applying Theorem 3.21.   ∎

The Bezout operator has the advantage of providing a simple rank test for coprimeness of pairs of polynomial matrices. However, to compute a matrix representation, basis vectors in the polynomial model spaces must be chosen. In contrast, the generalized Bezoutian matrix by Anderson and Jury (1976) is directly defined as the matrix $(Q_{k\ell})$ of coefficients of the Bezoutian polynomial

$$Q(z,w) = \sum_{k,\ell} Q_{k\ell} z^{k-1} w^{\ell-1}.$$

This matrix is certainly easier to compute than a matrix representation of the Bezout operator, but it is more difficult to explore the structural properties of this matrix. For instance, using Theorem 5.36, the greatest common left and right divisors of polynomial matrices can be characterized in terms of the kernel of the Bezout operator. For the generalized Bezoutian matrix $(Q_{k\ell})$ of Anderson and Jury (1976), this is much harder to achieve. Thus our preference is to work with the foregoing definition.

The strength of the preceding approach is illustrated by briefly discussing the case of classical Bezoutians. Thus, let

$$q(z) = z^n + q_{n-1} z^{n-1} + \cdots + q_0 \in \mathbb{F}[z]$$

be a scalar monic polynomial, and let

$$p(z) = p_{n-1} z^{n-1} + p_{n-2} z^{n-2} + \cdots + p_0 \in \mathbb{F}[z].$$

The **Bezoutian form** then has the expansion

$$\frac{q(z)p(w) - p(z)q(w)}{z - w} = \sum_{i,j=1}^{n} b_{ij} z^{i-1} w^{j-1},$$

with unique coefficients $b_{ij} \in \mathbb{F}$. The **Bezoutian matrix**, then, is the $n \times n$ matrix

$$B(p,q) = (b_{ij}) \in \mathbb{F}^{n \times n}.$$

The following basic representation theorem for Bezoutian matrices is discussed next. Let $\mathcal{B}_{st} = \{1, z, \ldots, z^{n-1}\}$ denote the **standard basis** of X_q, and let $\mathcal{B}_{co} = \{e_1(z), \ldots, e_n(z)\}$, with

$$e_i(z) = z^{n-i} + q_{n-1} z^{n-i-1} + \cdots + q_i = \pi_+(z^{-i} q(z)),$$

denote the **control basis** of X_q. One checks that $\mathcal{B}_{co}$ is the dual basis to $\mathcal{B}_{st}$ by computing

$$\left(\frac{z^{k-1}e_\ell(z)}{q(z)}\right)_{-1} = \left(\frac{z^{k-1}\pi_+(z^{-\ell}q(z))}{q(z)}\right)_{-1}$$

$$= \left(\frac{z^{k-1}(z^{-\ell}q(z))}{q(z)}\right)_{-1} = \left(z^{k-\ell-1}\right)_{-1} = \delta_{k\ell}.$$

Theorem 5.39. *1. The Bezoutian matrix $B(p,q) = [\mathbf{B}]_{co}^{st}$ is the matrix representation of the Bezout operator $\mathbf{B}$ with respect to the control basis and standard basis on X_q.*
2. The Bezoutian can be expressed, using the shift operator $S_q : X_q \longrightarrow X_q$, as

$$B(p,q) = [p(S_q)]_{co}^{st}.$$

Proof. Computing

$$\mathbf{B}e_j = \left(\frac{q(z)p(w)-p(z)q(w)}{z-w}q(w)^{-1}e_j(w)\right)_{-1}$$

$$= \sum_{r,s=1}^{n} b_{rs}z^{r-1}\left(w^{s-1}q(w)^{-1}\pi_+(w^{-j}q(w))\right)_{-1} = \sum_{r,s=1}^{n} b_{rs}z^{r-1}\left(w^{s-1}q(w)^{-1}(w^{-j}q(w))\right)_{-1}$$

$$= \sum_{r,s=1}^{n} b_{rs}z^{r-1}\left(w^{s-j-1}\right)_{-1} = \sum_{r=1}^{n} b_{rj}z^{r-1}$$

shows that the Bezout matrix $B(p,q) = [\mathbf{B}]_{co}^{st}$ is simply a matrix representation of the Bezout operator $\mathbf{B} : X_q \longrightarrow X_q$. This proves the first claim.

Theorem 5.36 implies that, for $j = 1,\ldots,n$, $\mathbf{B}e_j = \pi_q(p(z)e_j(z)) = p(S_q)e_j$, which completes the proof. $\blacksquare$

Recall from (3.26) that the matrix representation of the shift operator S_q with respect to the standard basis on X_q is the companion matrix

$$[S_q]_{st}^{st} = C_q := \begin{pmatrix} 0 & 0 & \cdots & & -q_0 \\ 1 & 0 & & & -q_1 \\ & \ddots & \ddots & & \vdots \\ & & 1 & 0 & -q_{n-2} \\ & & & 1 & -q_{n-1} \end{pmatrix}.$$

Moreover, the basis change matrix $[I]_{co}^{st}$ coincides with the Bezoutian $B(1,q)$. Thus the explicit description of the Bezoutian matrix as $B(p,q) = p(C_q)B(1,q)$ is deduced. This expression is often referred to as **Barnett's formula**.

As a further simple consequence, one obtains a classical coprimeness test for scalar polynomials.

Theorem 5.40. *The polynomials $p(z)$ and $q(z)$ are coprime if and only if the Bezoutian matrix $B(p,q)$ is invertible.*

Proof. We know already that the Bezout operator $\mathbf{B} : X_q \longrightarrow X_q$ defined by $h(z) = p(z)q(z)^{-1} = q(z)^{-1}p(z)$ is invertible if and only if p and q are coprime. In the preceding theorem it was shown that $B(p,q)$ is a matrix representation of $\mathbf{B}$. Thus, the result follows.   ■

5.5 Stability Characterizations

Characterization of the stability of linear systems is central to systems theory. A discrete-time linear dynamic system

$$x_{t+1} = Ax_t$$

on $\mathbb{F}^n$ is called **asymptotically stable** if the sequences $x_t = A^t x_0$ converge to zero for all initial conditions $x_0 \in \mathbb{F}^n$. Likewise, a continuous-time linear system

$$\dot{x}(t) = Ax(t)$$

is called asymptotically stable whenever

$$\lim_{t \to \infty} e^{tA}x_0 = 0$$

is true for all $x_0 \in \mathbb{F}^n$. Of course, in order for such a notion to make sense, a topology on the field $\mathbb{F}$ must be specified. Throughout this section, we will restrict our discussion to the standard situation where $\mathbb{F}$ denotes either the field of real numbers $\mathbb{R}$ or the field of complex numbers $\mathbb{C}$, both being endowed with their standard Euclidean topology. To streamline the presentation of the subsequent results, we will mainly restrict ourselves to continuous-time systems and mention corresponding results for discrete-time systems only occasionally.

It is easily seen (see Proposition 5.41 below) that a continuous-time system $\dot{x}(t) = Ax(t)$ is asymptotically stable if and only if all eigenvalues of A have a negative real part. Similarly, a discrete-time system is asymptotically stable if and only if all eigenvalues of A are in the open unit disc. Since the eigenvalues of A are the roots of the characteristic polynomial, determination of the stability of finite-dimensional linear systems reduces to the problem of characterizing the location of zeros of real and complex polynomials. This problem has a long history, and there are two basic approaches to it – via the use of quadratic forms on the one hand or the use of special Sylvester equations on the other. The problem of root location was already solved by Hermite (1856) and Hurwitz (1895) using quadratic form approaches. In this connection one should also refer to the work of Routh (1877) because of the efficiency of the computational algorithm proposed by the researcher. In a major contribution to the subject, Lyapunov (1893) offered a completely different approach based on energy considerations. In the linear case, the Lyapunov theory reduces the study of the stability of a system of first-order homogeneous constant coefficient differential equations to the positive definiteness of the solution of the celebrated Lyapunov equation.

In this section, we will first characterize the asymptotic stability of first-order systems using Lyapunov's method and then discuss the generalization to higher-order systems of differential equations. Because it is trivial to reduce a scalar nth-order homogeneous equation

$$\frac{d^n y}{dt^n} + a_{n-1}\frac{d^{n-1}y}{dt^{n-1}} + \cdots + a_0 y = 0$$

to a first-order system $\dot{x} = Ax$, it became possible to derive the classical stability criteria for scalar higher-order systems from Lyapunov theory. This was done surprisingly late, and the paper by Parks (1962) is usually considered the first of such derivations. The various reductions seemed to work also for the case of a higher-order system of matrix differential equations of the form

$$y^{(n)} + P_{n-1}y^{(n-1)} + \cdots + P_0 y = 0,$$

i.e., for which the matrix polynomial $P(z) = I_m z^n + P_{n-1}z^{n-1} + \cdots + P_0$ is monic. Strangely, a gap remained in the theory related to finding an algebraic test for the asymptotic stability of solutions of a system of the form

$$P_n y^{(n)} + P_{n-1}y^{(n-1)} + \cdots + P_0 y = 0,$$

where $P(z) = P_n z^n + P_{n-1}z^{n-1} + \cdots + P_0 \in \mathbb{C}^{m \times m}[z]$ is a nonsingular polynomial matrix. It is our aim in this chapter not only to close this gap but to apply the theory of quadratic forms and tensor products of polynomial and rational models to the derivation of stability criteria for higher-order multivariable systems. This leads to the study of two-variable polynomial matrices. It is worth mentioning that Kalman (1969, 1970) utilized a similar idea of switching from a polynomial equation in one variable to a polynomial in two variables and its associated quadratic form. Both these papers deal solely with the scalar case.

1. Lyapunov Stability

A brief summary of the basic facts from linear algebra on complex Hermitian matrices and adjoint operators is presented. In particular, bilinear forms are replaced by **sesquilinear forms**, i.e., forms $< x, y >$ that are antilinear in x and linear in y. Let $\mathscr{X}$ be a finite-dimensional complex vector space, and let $\mathscr{X}^*$ denote its dual space of complex linear functionals on $\mathscr{X}$. In this context there are two different ways of defining duality. The first is the one already employed in Chapter 3, using the canonical dual pairing $\mathscr{X}^* \times \mathscr{X} \longrightarrow \mathbb{C}$ defined by $(\lambda, x) \mapsto \lambda(x)$. The other definition – and this is the one adopted in this section – is given by the **Hermitian**

dual pairing $< \cdot, \cdot >: \mathscr{X}^* \times \mathscr{X} \longrightarrow \mathbb{C}$ defined as

$$< \lambda, x >_{\mathbb{C}} = \overline{\lambda(\bar{x})}.$$

Note that $< \lambda, x >$ is antilinear in λ and linear in x and therefore defines a sequilinear form. Since finite-dimensional vector spaces are reflexive, we can identify $\mathscr{X}^{**}$ with $\mathscr{X}$. Thus

$$< x, \lambda >_{\mathbb{C}} = \overline{< \lambda, x >_{\mathbb{C}}}.$$

Consider now a linear map $T : \mathscr{X} \longrightarrow \mathscr{Y}$, where $\mathscr{X}$ and $\mathscr{Y}$ are complex vector spaces with duals $\mathscr{Y}^*$ and $\mathscr{X}^*$, respectively. The Hermitian adjoint map $T^* : \mathscr{Y}^* \longrightarrow \mathscr{X}^*$ is determined through the equality

$$< Tx, y^* >_{\mathbb{C}} = < x, T^* y^* >_{\mathbb{C}} .$$

The notion of self-adjointness is now extended to this setting. A map $Z : \mathscr{X} \longrightarrow \mathscr{X}^*$ will be called **self-adjoint** or **Hermitian** if and only if $Z^* = Z$, i.e., if, for all $f, g \in \mathscr{X}$,

$$< Zf, g >_{\mathbb{C}} = < f, Zg >_{\mathbb{C}} .$$

If $\mathscr{B}$ is a basis in $\mathscr{X}$ and $\mathscr{B}^*$ is its dual basis, then the bilinear form $< Zf, g >$ can be evaluated as $([Z]_{\mathscr{B}}^{\mathscr{B}^*} [f]^{\mathscr{B}}, [g]^{\mathscr{B}})$. Here $[Z]_{\mathscr{B}}^{\mathscr{B}^*}$ is the representing matrix of Z and

$$(\xi, \eta) = \xi^* \eta := \overline{\xi}^\top \eta$$

denotes the **standard Hermitian inner product** in $\mathbb{C}^n$. One denotes by $A^* = \overline{A}^\top$ the **Hermitian adjoint** of a complex matrix $A \in \mathbb{C}^{n \times n}$. A^* is a unique matrix such that $(A\xi, \eta) = (\xi, A^* \eta)$ for all $\xi, \eta \in \mathbb{C}^n$. A matrix A is Hermitian if and only if $A^* = A$. Thus the matrix representing the Hermitian adjoint Z^* is the Hermitian adjoint A^* of $A = [Z]_{\mathscr{B}}^{\mathscr{B}^*} \in \mathbb{C}^{n \times n}$. Thus Z is Hermitian if and only if its matrix representation A is Hermitian. The map Z is called **positive definite**, denoted by $Z \succ 0$, if $< Zf, f >_{\mathbb{C}} > 0$ for all nonzero f in $\mathscr{X}$. Similarly, we write $Z \succeq 0$ whenever Z is positive semidefinite. It is easily seen that a Hermitian map Z is positive if and only if $A = [Z]_{\mathscr{B}}^{\mathscr{B}^*}$ is a positive definite Hermitian matrix.

Our starting point for the stability analysis of linear systems, induced by a complex $n \times n$ matrix A, is to derive a characterization linking the stability question with an associated eigenvalue problem.

Proposition 5.41. *1. The continuous-time system $\dot{x} = Ax$ is asymptotically stable if and only if all eigenvalues of A are in the open left half-plane $\mathbb{C}_- = \{z \in \mathbb{C} \mid \operatorname{Re} z < 0\}$.*
2. The discrete-time system $x_{t+1} = Ax_t$ is asymptotically stable if and only if all eigenvalues of A are in the open unit disc $\mathbb{D} = \{z \in \mathbb{C} \mid |z| < 1\}$.

Proof. 1. Assume that $\dot{x} = Ax$ is asymptotically stable, i.e., $\lim_{t \to \infty} e^{tA} x_0 = 0$ is satisfied for all $x_0 \in \mathbb{C}^n$. Let $\lambda \in \mathbb{C}$ be an eigenvalue of A with associated eigenvector $v \in \mathbb{C}^n$. Then $e^{tA} v = e^{t\lambda} v$ is true for all t. Thus

$$\lim_{t \to \infty} e^{tA} v = \lim_{t \to \infty} e^{t\lambda} v = 0,$$

and therefore $\lim_{t \to \infty} e^{t \operatorname{Re} \lambda} = \lim_{t \to \infty} |e^{t\lambda}| = 0$. Thus $\lambda < 0$, and therefore the condition is necessary.

Conversely, assume that all eigenvalues of A have negative real part. Let $\lambda_1, \ldots, \lambda_r$ denote the distinct eigenvalues of A with algebraic multiplicities $n_1, \ldots, n_r$, respectively. Using the Jordan canonical form for A, one obtains the representation

$$e^{tA} = \sum_{j=1}^{r} e^{t\lambda_j} P_j(t), \tag{5.57}$$

where $P_j(t) \in \mathbb{C}^{n \times n}$ are suitable matrix polynomials in t whose entries have degrees $\leq n_j$. Thus, each entry of $e^{t\lambda_j} P_j(t)$ is of the form $e^{t\lambda_j} \pi(t)$ for a certain polynomial $\pi(t)$. But

$$\lim_{t \to \infty} |e^{t\lambda_j} \pi(t)| = \lim_{t \to \infty} e^{t \operatorname{Re} \lambda_j} |\pi(t)| = 0$$

since $\operatorname{Re} \lambda_j < 0$ and the exponential function grows faster than polynomials. This implies $\lim_{t \to \infty} e^{tA} = 0$, i.e., the asymptotic stability of $\dot{x} = Ax$.

2. The proof of the discrete-time case runs completely similar to the preceding case and is therefore omitted. ∎

Because of the preceding characterization, we call $\mathbb{C}_-$ and $\mathbb{D}$ the **stability domain** of $\dot{x} = Ax$ and $x_{t+1} = Ax_t$, respectively.

Definition 5.42. A scalar complex polynomial $p(z) = \sum_{i=0}^{n} a_i z^i \in \mathbb{C}[z]$ is called a **Hurwitz polynomial** (or **Schur polynomial**) if $p(z)$ has roots only in $\mathbb{C}_-$ (or only in $\mathbb{D}$).

Of course, it is by no means obvious how one can recognize whether a particular polynomial $p(z)$ is Hurwitz or Schur. For real polynomials of degree 2 it is an easy exercise to see that $z^2 + az + b$ is a Hurwitz polynomial if and only if $a > 0$ and $b > 0$. The characterization of degree 3 real Hurwitz polynomials is due to Maxwell (1868), who showed that $z^3 + az^2 + bz + c$ is Hurwitz if and only if

$$a > 0, \quad c > 0, \quad ab > c.$$

For complex polynomials, these expressions are more complicated. A full characterization of Hurwitz polynomials, expressed in terms of the positivity of a certain quadratic form whose coefficients are quadratic polynomials in the coefficients of $p(z)$, will be given later on.

Next, the classic characterization by Lyapunov (1893) of asymptotic stability in terms of linear matrix inequalities is presented.

Theorem 5.43 (Lyapunov). *Let $\mathscr{X}$ be a complex n-dimensional vector space and $A : \mathscr{X} \longrightarrow \mathscr{X}$ a linear operator. The following statements are equivalent:*

1. *The system $\dot{x} = Ax$ on $\mathscr{X}$ is asymptotically stable.*
2. *There exists a positive definite Hermitian linear operator $Q : \mathscr{X} \longrightarrow \mathscr{X}^*$ that satisfies the **Lyapunov inequality***

$$A^*Q + QA \prec 0.$$

Proof. Without loss of generality, one can assume that $\mathscr{X} = \mathbb{C}^n$ and $A \in \mathbb{C}^{n \times n}$. Consequently, $0 \prec Q = Q^* \in \mathbb{C}^{n \times n}$. Suppose that Q is a positive definite solution of the Lyapunov inequality. Let λ be an eigenvalue of A with associated eigenvector $v \in \mathbb{C}^n$. Then $Av = \lambda v$ and $v^*A^* = \bar{\lambda}v^*$. Thus

$$0 > v^*A^*Qv + v^*QA^*v = (\bar{\lambda} + \lambda)v^*Qv.$$

By the positive definiteness of Q, we obtain $v^*Qv > 0$, which implies $2\,\mathrm{Re}\,\lambda = \bar{\lambda} + \lambda < 0$. Thus A is asymptotically stable.

For the converse assume that $\dot{x} = Ax$ is asymptotically stable, i.e., A has only eigenvalues with negative real part. From the decomposition (5.57), it can be seen that each entry of e^{tA} is in $L^2([0, \infty), \mathbb{C})$, with $\lim_{t \to \infty} e^{tA} = 0$, and, in particular, the integral

$$Q := \int_0^\infty e^{tA^*} e^{tA} dt$$

exists. Obviously, $Q = Q^*$ is Hermitian and satisfies

$$v^*Qv = \int_0^\infty \|e^{tA}v\|^2 dt \geq 0$$

for all $v \in \mathbb{C}^n$. Thus $v^*Qv = 0$ if and only if $e^{tA}v = 0$ for all $t \geq 0$, i.e., if and only if $v = 0$. This shows that Q is positive definite. Moreover,

$$A^*Q + QA = \int_0^\infty \frac{d}{dt}(e^{tA^*} e^{tA}) dt = -I_n \prec 0.$$

This completes the proof. $\blacksquare$

A useful strengthening of the preceding theorem of Lyapunov is due to Snyders and M. Zakai (1970) and Wimmer (1974a).

Theorem 5.44. *Let $\mathscr{X}$ and $\mathscr{Y}$ be two complex, finite-dimensional vector spaces. Let $A : \mathscr{X} \longrightarrow \mathscr{X}$ be a linear transformation. The following assertions are equivalent:*

1. *The system $\dot{x} = Ax$ on $\mathscr{X}$ is asymptotically stable.*
2. *For a linear transformation $C : \mathscr{X} \longrightarrow \mathscr{Y}$, with (C,A) observable, the **Lyapunov equation***

$$A^*Q + QA = -C^*C \tag{5.58}$$

   has a unique positive definite Hermitian solution $Q : \mathscr{X} \longrightarrow \mathscr{X}^*$.
3. *There exists a linear transformation $C : \mathscr{X} \longrightarrow \mathscr{Y}$, with (C,A) observable, such that the Lyapunov equation (5.58) has a positive definite Hermitian solution $Q : \mathscr{X} \longrightarrow \mathscr{X}^*$.*

Proof. Again, and without loss of generality, assume $\mathscr{X} = \mathbb{C}^n$, $\mathscr{Y} = \mathbb{C}^p$, $A \in \mathbb{C}^{n \times n}$, $Q = Q^* \succ 0$. Obviously, 2 implies 3.

Assume that 3 is satisfied. Let λ be an eigenvalue of A with associated eigenvector $v \in \mathbb{C}^n$. From the Lyapunov equation we get

$$2\operatorname{Re}\lambda v^*Qv = v^*A^*Qv + v^*QA^*v = -\|Cv\|^2 \le 0.$$

Since $v^*Qv > 0$, $\operatorname{Re}\lambda \le 0$. Suppose $\operatorname{Re}\lambda = 0$. Then $Cv = 0$, i.e., v is an eigenvector of A that is contained in the kernel of C. But, by the Hautus criterion, this contradicts observability. Therefore, each eigenvalue of A has negative real part and assertion 1 is proved.

To prove that 1 implies 2, we proceed similarly to the proof of the Lyapunov theorem. Thus, assume that A has only eigenvalues with negative real part. For $C \in \mathbb{C}^{p \times n}$, the integral

$$Q := \int_0^\infty e^{tA^*} C^* C e^{tA} \, dt$$

exists and defines a Hermitian matrix. For each complex vector v this implies

$$v^*Qv = \int_0^\infty \|Ce^{tA}v\|^2 \, dt \ge 0,$$

and therefore $v^*Qv = 0$ if and only if $Ce^{tA}v = 0$ for all $t \ge 0$. Equivalently, $CA^kv = 0$ for all $k \in \mathbb{N}_0$. By the observability of (C,A), $v = 0$. This shows that Q is positive definite. Moreover,

$$A^*Q + QA = \int_0^\infty \frac{d}{dt} e^{tA^*} C^* C e^{tA} \, dt = -C^*C.$$

This completes the proof.   ■

The preceding results by Lyapunov, Snyders-Zakai, and Wimmer have been generalized by Ostrowski and Schneider (1962) into an inertia theorem that relates the number of stable and unstable eigenvalues with the signature of Q. Their result will be proved in a slightly more general form. For a matrix $A \in \mathbb{C}^{n \times n}$, let $n_0(A)$ denote the number of eigenvalues (counting multiplicities) of A with real part zero. Similarly, let $n_{\pm}(A)$ denote the number of eigenvalues (counting multiplicities) of A with positive and negative real parts, respectively. The triple $\mathrm{ind}(A) = (n_0(A), n_+(A), n_-(A))$ is called the **inertia index** of A.

Theorem 5.45 (Inertia Theorem). *Let $A : \mathscr{X} \longrightarrow \mathscr{X}$ and $C : \mathscr{X} \longrightarrow \mathscr{Y}$ be linear transformations between finite-dimensional complex vector spaces. Assume that (C,A) is an observable pair. Then every Hermitian solution $Q : \mathscr{X} \longrightarrow \mathscr{X}^*$ of the* **Lyapunov inequality**

$$A^*Q + QA + C^*C \preceq 0$$

satisfies

$$n_0(Q) = n_0(A) = 0, \quad n_+(Q) = n_-(A), \quad n_-(Q) = n_+(A). \tag{5.59}$$

Proof. First, one proves $n_0(A) = n_0(Q) = 0$. As before, one assumes, without loss of generality, that $\mathscr{X} = \mathbb{C}^n$, and so forth. Suppose that $\lambda = i\omega$ is a purely imaginary eigenvalue of A with eigenvector $v \in \mathbb{C}^n$. Multiplying v^* and v on both sides of the Lyapunov inequality, we obtain

$$0 \geq v^*(A^*Q + QA + C^*C)v = \|Cv\|^2,$$

which implies $Cv = 0$. By the observability of (C,A), $v = 0$. Thus $n_0(A) = 0$. Next, consider $v \in \mathbb{C}^n$, with $Qv = 0$. Then, by the same reasoning, we obtain $Cv = 0$. After applying a suitable unitary state-space similarity transformation, one can assume that

$$A^*Q + QA + C^*C = -\mathrm{diag}\,(I_r, 0).$$

Partition v accordingly as $v = \mathrm{col}\,(x,y)$, with $x \in \mathbb{C}^r$. Thus

$$-\|x\|^2 = v^*(A^*Q + QA + C^*C)v = 0,$$

and therefore $x = 0$. Moreover, $QAv = (A^*Q + QA + C^*C)v = -\mathrm{diag}\,(I_r, 0)v = 0$ for $v = \mathrm{col}\,(0,y)$. This implies $A(\mathrm{Ker}\,Q) \subset \mathrm{Ker}\,Q$. Since $\mathrm{Ker}\,Q \subset \mathrm{Ker}\,C$, the observability of (C,A) implies $\mathrm{Ker}\,Q = \{0\}$. Thus $n_0(Q) = 0$.

For the proof of the remaining equalities (5.59), we proceed by a simple continuity argument. Consider the convex cone of Hermitian $n \times n$ matrices

$$\mathscr{C} = \{Q = Q^* \mid A^*Q + QA + C^*C \preceq 0\}.$$

Using the observability of the pair (C,A), it was just shown that each element of $\mathscr{C}$ is invertible. Thus $\mathscr{C} \subset GL_n(\mathbb{C})$ is a convex and hence connected subset of $GL_n(\mathbb{C})$. Since the inertia index, $\mathrm{ind}(P)$, depends continuously on $P \in GL_n(\mathbb{C})$, it suffices to establish the inertia equalities (5.59) for a single element $Q_0 \in \mathscr{C}$. Since A has no purely imaginary eigenvalues, we can assume without loss of generality that (C,A) are partitioned as

$$A = \begin{pmatrix} A_+ & 0 \\ 0 & A_- \end{pmatrix}, \quad C = \begin{pmatrix} C_+ & C_- \end{pmatrix}.$$

Here the eigenvalues of A_+ are assumed to have positive real part, while the eigenvalues of A_- have negative real part. Using the Hautus criterion one sees that the observability of (C,A) is equivalent to the observability of (C_+,A_+) and (C_-,A_-). Define for real numbers $r > 0$

$$Q_- = r \int_0^\infty e^{tA_-^*} e^{tA_-}\, dt,$$

$$Q_+ = r \int_0^\infty e^{-tA_+^*} e^{-tA_+}\, dt.$$

Then $Q_\pm = Q_\pm^* \succ 0$ are positive definite and Hermitian and the Hermitian matrix

$$Q := \mathrm{diag}\,(-Q_+,Q_-)$$

satisfies

$$A^*Q + QA + C^*C = C^*C - rI.$$

Moreover, $n_+(Q) = n_-(A)$ and $n_-(Q) = n_+(A)$. Choose $r > 0$ such that $rI \succeq C^*C$. Then Q satisfies the Lyapunov inequality. ∎

As an example, we examine the stability of second-order systems of the form

$$\ddot{x}(t) + (M + \Delta)\dot{x}(t) + Nx(t) = 0 \tag{5.60}$$

for $x \in \mathbb{R}^n$. The following assumptions will be made:

$$M = M^\top \succeq 0, \quad N = N^\top \succ 0,$$

$$\Delta^\top = -\Delta.$$

Consider the matrices

$$A = \begin{pmatrix} 0 & I_n \\ -N & -M - \Delta \end{pmatrix}, \quad C = \begin{pmatrix} 0 & M^{\frac{1}{2}} \end{pmatrix}.$$

Using the skew symmetry of Δ, one verifies that the positive definite symmetric matrix

$$Q = \frac{1}{2}\begin{pmatrix} N & 0 \\ 0 & I \end{pmatrix}$$

satisfies the Lyapunov equation

$$A^\top Q + QA = -C^\top C.$$

Moreover, the pair (C,A) is easily seen to be observable. Now Theorem 5.44 implies the asymptotic stability of (5.60).

2. Complex Polynomials and Hermitian Operators.

Sesquilinear forms and adjoint operators are then extended to the context of polynomial and rational models. We will discuss briefly two-variable polynomial matrices and examine naturally induced linear maps and Hermitian forms in complex polynomial models. In the preceding parts of this chapter, the connection between tensor products, linear transformations, and bilinear forms was established. When the field is taken to be the field $\mathbb{C}$ of complex numbers, this must be slightly modified in order for it to be applicable to stability questions. Using sesquilinear forms, the duality theory of polynomial models developed in Chapter 3 can be extended in a rather straightforward way.

Our analysis starts with functions in one variable. Define, for $A(z) = \sum_{j=-\infty}^{n_A} A_j z^j \in \mathbb{C}((z^{-1}))^{p \times m}$, the conjugate power series $A^* \in \mathbb{C}((z^{-1}))^{p \times m}$ by

$$A^*(z) = \overline{A(\bar{z})}^\top = \sum_{j=-\infty}^{n_A} A_j^* z^j \in \mathbb{C}((z^{-1}))^{m \times p}.$$

In $\mathbb{C}((z^{-1}))^m \times \mathbb{C}((z^{-1}))^m$, a sesquilinear form $[g,f]_{\mathbb{C}}$ is defined by

$$[g,f]_{\mathbb{C}} = \sum_{j=-\infty}^{\infty} g_{-j-1}^* f_i = (g^*(z)f(z))_{-1}, \qquad (5.61)$$

where $f(z) = \sum_{j=-\infty}^{\infty} f_j z^j$, $g(z) = \sum_{j=-\infty}^{\infty} g_j z^j$, and $g^*(z) = \sum_{j=-\infty}^{\infty} g_j^* z^j$. Thus $\overline{[g,f]_{\mathbb{C}}} = [f,g]_{\mathbb{C}}$. It is clear that, because both $f(z)$ and $g(z)$ are truncated Laurent series, the sum in (5.61) is well defined, containing only a finite number of nonzero terms. Let $A(z) \in \mathbb{C}((z^{-1}))^{m \times m}$. Then $[g, Af]_{\mathbb{C}} = [A^*g, f]_{\mathbb{C}}$ for all $f(z), g(z) \in \mathbb{C}((z^{-1}))^m$. This global form is used to obtain a concrete representation of X_D^*, the dual space of the polynomial model X_D.

Proposition 5.46. *Let $D(z) \in \mathbb{C}[z]^{m \times m}$ be nonsingular. Then $D^*(z) := \overline{D(\bar{z})}^\top \in \mathbb{C}^{m \times m}[z]$ is nonsingular, and the following assertions are in force.*

1. *The dual space X_D^* of X_D can be identified with X_{D^*} under the nondegenerate pairing*

$$< g, f >_\mathbb{C} = [g, D^{-1}f]_\mathbb{C} = (g^*(z)D(z)^{-1}f(z))_{-1},$$

   for $f(z) \in X_D$ and $g(z) \in X_{D^*}$.

2. *The form $< g, f >_\mathbb{C}$ is sesquilinear, i.e.,*

$$< g, f >_\mathbb{C} = \overline{< f, g >_\mathbb{C}}.$$

3. *The module structures of X_D and X_{D^*} are related through*

$$S_D^* = S_{D^*}.$$

Proof. The nondegeneracy of $< \cdot, \cdot >_\mathbb{C}$ follows from Theorem 3.38. Computing

$$\overline{< f, g >_\mathbb{C}} = \overline{(f^*(z)D^*(z)^{-1}g(z))_{-1}} = (g^*(z)D(z)^{-1}f(z))_{-1}$$

proves the second assertion. The last assertion is proved by a trivial calculation, which is omitted. ∎

Let $\mathbb{C}[z, w]^{n_1 \times n_2}$ denote the $n_1 \times n_2$ complex polynomial matrices in the complex variables z and w. For $M(z, w) \in \mathbb{C}[z, w]^{n_1 \times n_2}$, one defines the conjugate polynomial $M^*(z, w) \in \mathbb{C}[z, w]^{n_2 \times n_1}$ by

$$M^*(z, w) := \overline{M(\overline{w}, \overline{z})}^\top.$$

A polynomial matrix $M(z, w) \in \mathbb{C}[z, w]^{n \times n}$, with

$$M(z, w) = \sum_{i,j=1}^{d} M_{ij} z^{i-1} w^{j-1},$$

will be called **Hermitian** if

$$M^*(z, w) = M(z, w).$$

It is easy to see that this is equivalent to the condition $M_{ij} = M_{ji}^*$. A Hermitian polynomial matrix $M(z, w) \in \mathbb{C}[z, w]^{n \times n}$ is called **nonnegative**, and denoted by $M \geq 0$, if and only if $\sum_{i,j} \xi_i^* M_{ij} \xi_j \geq 0$ for all $\xi_i \in \mathbb{C}^n$. This is equivalent to the matrix $(M_{ij}) \in \mathbb{C}^{dn \times dn}$ being Hermitian and positive semidefinite. Thus there exists a full column rank matrix $C \in \mathbb{C}^{dn \times k}$ with $(M_{ij}) = CC^*$. This implies that the matrix polynomial M is nonnegative if and only if there exists some $C(z) \in \mathbb{C}[z]^{n \times k}$ such that

$$M(z,w) = C(z)C^*(w).$$

In fact, the relation between C and $C(z)$ is simply $C(z) = (I_n, \ldots, z^{d-1}I_n)C$.

Next, the corresponding objects in the context of polynomial models are introduced. Let $D_1(z) \in \mathbb{C}[z]^{m \times m}$ and $D_2(z) \in \mathbb{C}[z]^{p \times p}$ be two nonsingular polynomial matrices, and let X_{D_1} and X_{D_2} be the associated polynomial models. The Kronecker product model $X_{D_2(z) \otimes D_1^\top(w)}$ is defined as the set of all complex $p \times m$ polynomial matrices $M(z,w)$ for which $D_2(z)^{-1}M(z,w)D_1(w)^{-1}$ is strictly proper in both variables. The isomorphism $X_{D_2(z) \otimes D_1^\top(w)} \simeq X_{D_2(z)} \otimes_{\mathbb{C}} X_{D_1^\top(w)}$ implies that every $M(z,w) \in X_{D_2(z) \otimes_{\mathbb{C}} D_1^\top(w)}$ has a representation, in terms of generating elements, of the form

$$M(z,w) = C_2(z)C_1^*(w),$$

with each column of $D_2(z)^{-1}C_2(z)$ and $D_1^*(w)^{-1}C_1(w)$ strictly proper. Note that

$$M^*(z,w) = C_1(z)C_2^*(w).$$

For polynomials $M(z,w) \in X_{D_2(z) \otimes D_1^\top(w)}$, we define a map $\mathbf{M} : X_{D_1} \longrightarrow X_{D_2}$, induced by $M(z,w)$, by

$$\mathbf{M}f(z) := (M(z,w)D_1^{-1}(w)f(w))_{-1} = <M^*(\cdot,\bar{z}),f>_{\mathbb{C}}, \quad f(z) \in X_{D_1}. \quad (5.62)$$

Note that this construction parallels that in Theorem 5.34. Clearly, $\mathbf{M}$ is a linear operator that maps X_{D_1} to X_{D_2}.

Definition 5.47. A Hermitian polynomial $M(z,w) \in X_{D(z) \otimes \bar{D}(w)}$ is D-**positive**, denoted by $M >_D 0$, if the induced Hermitian map $\mathbf{M} : X_{D^*} \longrightarrow X_D$ is positive, that is, if the **quadratic form** $< \mathbf{M}f, f >$ is positive definite, i.e., $< \mathbf{M}f, f > > 0$ for all nonzero $f(z) \in X_{D^*}$.

Theorem 5.48. *Let* $M(z,w) \in X_{D_2(z) \otimes D_1^\top(w)}$. *Let the map* $\mathbf{M} : X_{D_1} \longrightarrow X_{D_2}$ *be defined by* (5.62). *Then:*

1. The Hermitian adjoint map $\mathbf{M}^* : X_{D_2^*} \longrightarrow X_{D_1^*}$ *is given, for* $g(z) \in X_{D_2^*}$, *by*

$$(\mathbf{M}^*g)(w) = <M^*(w,\cdot),g>_{\mathbb{C}}; \quad (5.63)$$

2. $M(z,w) \in X_{D(z) \otimes \bar{D}(w)}$ *is Hermitian if and only if* $M(z,w)$ *has a representation of the form*

$$M(z,w) = \sum_{i=1}^{k} \lambda_i g_i(z)g_i^*(w), \quad (5.64)$$

with $g_i(z) \in X_D$ and $\lambda_i \in \mathbb{R}$. $M(z,w)$ is Hermitian if and only if the induced map $\mathbf{M}: X_{D^*} \longrightarrow X_D$ is Hermitian;

3. Let $D(z) \in \mathbb{C}[z]^{m \times m}$ be nonsingular, and let $M(z,w) \in X_{D(z) \otimes \overline{D}(w)}$ be Hermitian. The following conditions are equivalent:

   a. $M(z,w)$ is D-positive.
   b. There exists a basis $\{g_i(z)\}$ in X_D for which

$$M(z,w) = \sum_{i=1}^{n} g_i(z)g_i^*(w).$$

Proof. From the identity $M(z,w) = C_2(z)C_1^*(w)$ one obtains

$$\mathbf{M}f(z) = (C_2(z)C_1^*(w)D_1(w)^{-1}f(w))_{-1},$$

and thus, for all $f \in X_{D_1}, g \in X_{D_2^*}$,

$$\begin{aligned}
< g, \mathbf{M}f >_{\mathbb{C}} &= \left(g^*(z)D_2(z)^{-1}(C_2(z)C_1^*(w)D_1(w)^{-1}f(w))_{-1}\right)_{-1} \\
&= \left((g^*(z)D_2(z)^{-1}C_2(z))_{-1}C_1^*(w)D_1(w)^{-1}f(w)\right)_{-1} \\
&= < \mathbf{M}^*g, f >_{\mathbb{C}} \\
&= \left((\mathbf{M}^*g)^*(w)D_1(w)^{-1}f(w)\right)_{-1}.
\end{aligned}$$

Thus

$$\begin{aligned}
\mathbf{M}^*g(w) &= C_1(w)\overline{(g^*(z)D_2(z)^{-1}C_2(z))}_{-1} \\
&= \left(C_1(w)C_2^*(z)D_2^*(z)^{-1}g(z)\right)_{-1} \\
&= \left(M^*(w,z)D_2^*(z)^{-1}g(z)\right)_{-1} \\
&= < M^*(w,\cdot), g >_{\mathbb{C}}.
\end{aligned}$$

This proves (5.63).

A complex matrix Q is Hermitian if and only if $Q = C\Lambda C^*$ is satisfied for a real diagonal matrix Λ and a complex matrix C. This shows that a complex polynomial $M(z,w) \in \mathbb{C}[z,w]^{m \times m}$ is Hermitian if and only if it is of the form

$$M(z,w) = C(z)\Lambda C^*(w) = \sum_{i=1}^{k} \lambda_i g_i(z)g_i^*(w)$$

for suitable real numbers $\lambda_1, \ldots, \lambda_k$ and complex polynomial vectors $g_i(z) \in \mathbb{C}[z]^m$. Here $g_i(z)$ denotes the ith column of $C(z)$. Also, $M(z,w) \in X_{D(z) \otimes \overline{D}(w)}$ if and only if $g_i \in X_D$ for all i. This proves (5.64). For $f \in X_{D^*}$, one has

$$\mathbf{M}f = \left(M(z,w)D^*(w)^{-1}f(w) \right)_{-1}$$

$$\mathbf{M}^*f = \left(M^*(w,z)D^*(z)^{-1}f(z) \right)_{-1}.$$

Thus $\mathbf{M} = \mathbf{M}^*$ if and only if $M^*(z,w) = M(z,w)$. This completes the proof of the second assertion.

To prove the last assertion, one computes the inner product $< \mathbf{M}f, f >_{\mathbb{C}}$ for each $f \in X_{D^*}$. Since $M(z,w)$ is Hermitian, it has the representation $M(z,w) = C(z)\Lambda C^*(w)$ for a real diagonal matrix $\Lambda = \mathrm{diag}(\lambda_1, \cdots, \lambda_n)$ and $D(z)^{-1}C(z)$ strictly proper. Without loss of generality, one can assume that the columns $c_1(z), \ldots, c_n(z)$ of $C(z)$ form a basis of X_D. Thus the quadratic form

$$\begin{aligned}
< \mathbf{M}f, f >_{\mathbb{C}} &= \left(f^*(z)D(z)^{-1}(C(z)\Lambda C^*(w)D^*(w)^{-1}f(w))_{-1} \right)_{-1} \\
&= \left(f^*(z)D(z)^{-1}C(z) \right)_{-1} \Lambda \left(C^*(w)D^*(w)^{-1}f(w) \right)_{-1} \\
&= \sum_{i=1}^{n} \lambda_i | < f, c_i >_{\mathbb{C}} |^2
\end{aligned}$$

is positive definite on X_{D^*} if and only if $\lambda_1 > 0, \ldots, \lambda_n > 0$. Thus the elements $g_i(z) = \sqrt{\lambda_i} c_i(z)$, $i = 1, \ldots, n$, define a basis of X_D, with

$$M(z,w) = \sum_{i=1}^{n} g_i(z)g_i^*(w)$$

and

$$< \mathbf{M}f, f >_{\mathbb{C}} = \sum_{i=1}^{n} | < f, g_i >_{\mathbb{C}} |^2. \qquad \blacksquare$$

In Theorem 5.48, the map induced by a two-variable polynomial matrix $M(z,w) \in X_{D_2(z) \otimes D_1^\top(w)}$ was examined. This restriction is unnecessary because one can use the projection $\pi_{D_2(z) \otimes D_1^\top(w)}$ in $\mathbb{C}[z,w]^{p \times m}$, as defined in (5.20). Thus, assume that $M(z,w) \in \mathbb{C}[z,w]^{p \times m}$. Let $D_1(z) \in \mathbb{C}[z]^{m \times m}$ and $D_2(z) \in \mathbb{C}[z]^{p \times p}$ be two nonsingular polynomial matrices, and let X_{D_1} and X_{D_2} be the associated polynomial models. Since $\mathrm{Im}\,\pi_{D_2(z) \otimes D_1^\top(w)} = X_{D_2(z) \otimes_{\mathbb{C}} D_1^\top(w)}$, one can define the induced map $\mathbf{M} : X_{D_1} \longrightarrow X_{D_2}$ by

$$(\mathbf{M}f)(z) = \left(\pi_{D_2(z) \otimes D_1^\top(w)} M(z,w)D_1(w)^{-1}f(w) \right)_{-1}. \qquad (5.65)$$

In view of (5.24), the induced map satisfies $\mathbf{M} = 0$ if and only if there exist polynomial matrices $M_i(z,w)$ such that $M(z,w) = D_2(z)M_1(z,w) + M_2(z,w)D_1(w)$.

3. Stability of Higher-Order Equations

We proceed now to establish stability criteria for complex polynomial matrices. This is done by reduction to the Lyapunov theorem. For simplicity we focus on continuous-time systems; discrete-time systems can be handled similarly. Recall that a complex matrix A is called **stable** if all its eigenvalues lie in the open left half-plane. Equivalently, a matrix A is stable if and only if its characteristic polynomial is stable. One extends this definition to nonsingular polynomial matrices by saying that λ is an **eigenvalue of a nonsingular polynomial matrix** $D(z)$ if there exists a nonzero vector ξ in $\text{Ker}\,D(\lambda)$. In view of Theorem 3.30, λ is an eigenvalue of $D(z)$ if and only if it is an eigenvalue of S_D. A nonsingular polynomial matrix $D(z)$ is a **stable polynomial matrix** if all its eigenvalues lie in the open left half-plane. Note that a matrix A is stable if and only if the polynomial matrix $zI - A$ is stable. With the analysis of stability our present goal, it is important to introduce symmetry with respect to the imaginary axis. For a polynomial matrix $P(z) \in \mathbb{C}[z]^{m \times m}$, one defines its **parahermitian adjoint** $P_*(z)$ by

$$P_*(z) := \overline{P(-\bar{z})}^\top = P^*(-z).$$

One says that $P(z)$ is **parahermitian** if $P_*(z) = P(z)$. Clearly, $P(z)$ is stable if and only if $P_*(z)$ is antistable, i.e., it has all its eigenvalues in the open right half-plane.

Our principal theorem is stated next.

Theorem 5.49. *Let $P(z) \in \mathbb{C}[z]^{m \times m}$ be a nonsingular polynomial matrix. Then $P(z)$ is stable if and only if, for polynomial matrices $R(z)$ with $P(z)$ and $R(z)$ right coprime and $R(z)P(z)^{-1}$ proper, there exists a solution $Q(z)$ of the polynomial Sylvester equation*

$$P_*(z)Q(z) + Q_*(z)P(z) = R_*(z)R(z) \tag{5.66}$$

such that the quadratic form, induced in X_P by

$$V(z,w) = \frac{P^*(z)Q(w) + Q^*(z)P(w) - R^*(z)R(w)}{z + w}, \tag{5.67}$$

is positive definite.

Proof. Assume $P(z)$ is stable and $R(z)$ is right coprime with $P(z)$. The coprimeness condition implies (Theorem 4.28) that the pair (C, A) defined, in the state space X_P, by the shift realization (4.23) is observable. Explicitly, the shift realization of the proper transfer function $R(z)P(z)^{-1}$, with $R(z)P(z)^{-1} - D$ being strictly proper, is given as $A : X_P \longrightarrow X_P$, $C : X_P \longrightarrow \mathbb{C}^p$,

$$A = S_P,$$

$$Cf = \left(R(z)P(z)^{-1}f(z)\right)_{-1}, \quad f \in X_P. \tag{5.68}$$

Note that $Cf = \left((R(z)P(z)^{-1} - D)f(z)\right)_{-1}$, which implies that C does not depend on the constant term D. A straightforward computation shows that the Hermitian adjoint of C is the linear operator $C^* : \mathbb{C}^p \longrightarrow X_{P^*}$, defined as $C^*v = R^*(z) - P^*(z)D^*v$ for $v \in \mathbb{C}^p$. In fact, for all $v \in \mathbb{C}^p, f \in X_P$,

$$< C^*v, f >_{\mathbb{C}} = \left((C^*v)^*(z)P(z)^{-1}f(z)\right)_{-1} = < v, Cf >_{\mathbb{C}}$$

$$= v^* \left((R(z) - DP(z))P(z)^{-1}f(z)\right)_{-1},$$

which is equivalent to $C^* = R^*(z) - P^*(z)D^*$.

The stability of $P(z)$ implies that there exists a solution to equation (5.66). This is shown next. Because $P(z)$ is stable, so is $p(z) = \det P(z)$, and the scalar polynomials $p(z)$ and $p_*(z) := \overline{p}(-z)$ are coprime. Therefore, the polynomial Sylvester equation

$$K(z)p(z) + p_*(z)L(z) = R^*(-z)R(z)$$

is solvable. Moreover, the solution is unique if one assumes $L(z)$ is reduced modulo $p(z)$ and $K(z)$ modulo $p_*(z)$. By a symmetry argument, $K(z) = L_*(z)$, and hence

$$L_*(z)p(z) + p_*(z)L(z) = R_*(z)R(z).$$

Using Cramer's rule, i.e., $p(z)I = \text{adj}\,P(z)P(z)$, and defining $Q(z) := \text{adj}\,P_*(z)L(z)$, with $Q_*(z) = L_*(z)\text{adj}\,P(z)$, the equality $Q_*(z)P(z) + P_*(z)Q(z) = R_*(z)R(z)$ follows.

Thus $Q(z)$ solves the polynomial Sylvester equation (5.66). As a consequence of (5.66), $V(z,w)$ defined by (5.67) is a polynomial matrix in two variables. Let

$$M(z,w) = (z+w)V(z,w) = P^*(z)Q(w) + Q^*(z)P(w) - R^*(z)R(w).$$

Clearly, both $V(z,w)$ and $M(z,w)$ are Hermitian. Moreover,

$$\pi_{P^*(z)\otimes P^\top(w)}M(z,w) = -(R^*(z) - P^*(z)D^*)R(w).$$

The polynomial matrix $M(z,w)$ induces a Hermitian linear operator

$$\mathbf{M} : X_P \longrightarrow X_{P^*}, \quad \mathbf{M}f = \left(\pi_{P^*(z)\otimes P^\top(w)}M(z,w)P(w)^{-1}f(w)\right)_{-1}$$

by equation (5.65). Using (5.68) we obtain

$$\mathbf{M}f = -(R^*(z) - P^*(z)D^*)\left(R(w)P(w)^{-1}f(w)\right)_{-1} = -(R^*(z) - P^*(z)D^*)Cf = -C^*Cf.$$

Thus

$$\mathbf{M} = -C^*C. \tag{5.69}$$

Similarly, using Proposition 5.32, the polynomial $V(z,w)$ induces a Hermitian linear operator $\mathbf{V} : X_P \longrightarrow X_{P*}$ defined by

$$\mathbf{V}f = \left(\pi_{P*(z) \otimes P^\top(w)} V(z,w) P(w)^{-1} f(w) \right)_{-1}.$$

Since the adjoint S_P^* coincides with the shift S_{P*} on X_{P*}, one obtains, for all $f \in X_P$,

$$(S_{P*} \mathbf{V} + \mathbf{V} S_P)f = \left(\pi_{P*(z) \otimes P^\top(w)} (zV(z,w) + V(z,w)w) P(w)^{-1} f(w) \right)_{-1}$$

$$= \left(\pi_{P*(z) \otimes P^\top(w)} M(z,w) P(w)^{-1} f(w) \right)_{-1} = \mathbf{M}f.$$

This proves that

$$S_{P*} \mathbf{V} + \mathbf{V} S_P = \mathbf{M}.$$

Since $A = S_P$, and using (5.69), one concludes that $\mathbf{V}$ satisfies the Lyapunov equation

$$A^* \mathbf{V} + \mathbf{V}A = -C^*C.$$

Since P is stable, the shift operator $A = S_P$ has all its eigenvalues in the open left half-plane. By Theorem 5.44, the quadratic form $< \mathbf{V}f, f >$ is positive definite, or, equivalently, $\mathbf{V} >_P 0$.

Conversely, assume $Q(z)$ is a solution of the polynomial Sylvester equation (5.66) and the quadratic form $< \mathbf{V}f, f >$ induced in X_P by $V(z,w)$, as defined in (5.67), is positive definite. Since $S_{P_*} = S_P^*$, as before, the following Lyapunov equation is satisfied:

$$S_P^* \mathbf{V} + \mathbf{V} S_P = -C^*C.$$

Applying Theorem 5.44, one concludes that S_P is stable and, hence, by Theorem 3.30, that $P(z)$ is, too. ∎

Using the same technique, one can derive a higher-order analog of the Ostrowski and Schneider (1962) inertia theorem.

Theorem 5.50 (Polynomial Inertia Theorem). *Let $G(z) = R(z)P(z)^{-1}$ be a proper complex rational matrix function that is right coprime. Suppose that the polynomial matrix $P(z)$ has n_0 eigenvalues with real part zero, n_+ eigenvalues with positive real part, and n_- eigenvalues with negative real part, all counted with* *multiplicities. Let $Q(z)$ be a solution of the polynomial Sylvester equation*

$$P_*(z)Q(z) + Q_*(z) = R_*(z)R(z),$$

and let

$$V(z,w) = \frac{P^*(z)Q(w) + Q^*(z)P(w) - R^*(z)R(w)}{z+w}$$

denote the associated Hermitian polynomial. Then the Hermitian operator **V** *in* X_P *is invertible with exactly* n_+ *negative and* n_- *positive eigenvalues. Moreover,* $n_0 = 0$.

Proof. Following the notation in the proof of Theorem 5.49, one obtains

$$A^* V + VA = -C^* C$$

for the shift realization (A,C) of $R(z)P(z)^{-1}$. By the right coprimeness of $P(z), R(z)$, the pair (C,A) is observable. Thus the result follows from the inertia theorem 5.45. ∎ ∎

4. Classical Stability Criteria

The results of the previous section can be used for an easy derivation of some of the classical stability criteria for real and complex scalar polynomials. To do this, it will be useful to compute the matrix representation of the Hermitian operator **V** induced by a scalar two-variable polynomial $V(z,w)$. Our derivation of the classical stability criteria is nonstandard insofar as they are deduced from the Lyapunov stability criteria rather than from using winding number arguments and the Cauchy index.

Let

$$V(z,w) = \sum_{i,j=1}^{n} a_{ij} z^{i-1} w^{j-1}$$

denote a Hermitian scalar polynomial, and let $p(z) = \sum_{j=0}^{n} p_j z^j, p_n = 1$, denote a monic complex polynomial. Note that

$$X_{\overline{p}(z) \otimes p(w)} = \{ \sum_{i,j=1}^{n} a_{ij} z^{i-1} w^{j-1} \mid a_{ij} \in \mathbb{C} \}$$

and therefore each polynomial in z,w of degree $< n$ is contained in $X_{\overline{p}(z) \otimes p(w)}$. In particular, $V(z,w) \in X_{\overline{p}(z) \otimes p(w)}$. Recall next the construction of the standard basis and control basis on scalar polynomial models. Thus, let

$$\mathscr{B}_{\text{st}} = \{1, z, \dots, z^{n-1}\}$$

denote the **standard basis** of $X_{\bar{p}}$ and

$$\mathscr{B}_{co} = \{e_1(z), \ldots, e_n(z)\}$$

with

$$e_i(z) = z^{n-i} + p_{n-1}z^{n-i-1} + \cdots + p_i = \pi_+(z^{-i}p(z))$$

denote the **control basis** of X_p. The computation

$$\left(\frac{z^{k-1}e_\ell(z)}{p(z)}\right)_{-1} = \left(\frac{z^{k-1}\pi_+(z^{-\ell}p(z))}{p(z)}\right)_{-1}$$

$$= \left(\frac{z^{k-1}(z^{-\ell}p(z))}{p(z)}\right)_{-1} = \left(z^{k-\ell-1}\right)_{-1} = \delta_{k\ell}$$

shows that $\mathscr{B}_{co}$ is the dual basis to $\mathscr{B}_{st} = \{1, z, \ldots, z^{n-1}\}$.

To obtain a matrix representation of the operator $\mathbf{V} : X_p \longrightarrow X_{\bar{p}}$ with respect to the control basis and standard basis on X_p and X_{p^*}, respectively, one computes, using the fact that the product of two strictly proper functions has zero residue,

$$\left(V(z,w)p(w)^{-1}e_j(w)\right)_{-1} = \sum_{r,s=1}^{n} a_{rs}z^{r-1}\left(w^{s-1}p(w)^{-1}\pi_+(w^{-j}p(w))\right)_{-1}$$

$$= \sum_{r,s=1}^{n} a_{rs}z^{r-1}\left(w^{s-1}p(w)^{-1}(w^{-j}p(w))\right)_{-1}$$

$$= \sum_{r,s=1}^{n} a_{rs}z^{r-1}\left(w^{s-j-1}\right)_{-1}$$

$$= \sum_{r=1}^{n} a_{rj}z^{r-1}.$$

Thus the matrix representation of $\mathbf{V}$ is given by the coefficients of $V(z,w)$, that is, $[\mathbf{V}]_{co}^{st} = (a_{rs})$.

Definition 5.51. Let $p(z) = z^n + p_{n-1}z^{n-1} + \cdots + p_0 \in \mathbb{C}[z]$ be a complex polynomial with Hermitian adjoint $\bar{p}(z) = z^n + \bar{p}_{n-1}z^{n-1} + \cdots + \bar{p}_0 \in \mathbb{C}[z]$. The **Hermite–Fujiwara form** is defined as a Hermitian form with generating function

$$\frac{\bar{p}(z)p(w) - p(-z)\bar{p}(-w)}{z+w} = \sum_{i,j=1}^{n} h_{ij}z^{i-1}w^{j-1},$$

and the **Hermite–Fujiwara matrix** is the Hermitian matrix

$$\mathbf{H}_n(p) = (h_{ij}) \in \mathbb{C}^{n \times n}.$$

As an example, the Hermite–Fujiwara matrix for a complex polynomial $z^2 + p_1 z + p_0$ of degree two is computed. Thus

$$\mathbf{H}_2(p) = 2 \begin{pmatrix} \mathrm{Re}\,(\overline{p}_0 p_1) & -i\mathrm{Im}\,p_0 \\ i\mathrm{Im}\,p_0 & \mathrm{Re}\,p_1 \end{pmatrix},$$

which is positive definite if and only if

$$\mathrm{Re}(p_1) > 0, \quad \mathrm{Re}(p_1)\,\mathrm{Re}(\overline{p}_0 p_1) - \mathrm{Im}^2(p_0) > 0.$$

From the polynomial inertia theorem the following root location result can be deduced.

Theorem 5.52 (Hermite). *Let $p(z) \in \mathbb{C}[z]$ be a monic complex polynomial of degree n that is coprime with $\overline{p}(-z)$. Let n_+ and n_- denote the number of roots of $p(z)$ that are located in the open right half-plane and open left half-plane, respectively. Let $\mathbf{H}_n(p)$ denote the Hermite–Fujiwara matrix of $p(z)$. Then $\mathbf{H}_n(p)$ is invertible with exactly n_- positive eigenvalues and n_+ negative eigenvalues.*

Proof. Defining $r(z) = \overline{p}(-z)$ and $q(z) = \frac{1}{2}p(z)$, the polynomials $r(z)$ and $p(z)$ are coprime, and $q(z)$ solves the polynomial Sylvester equation

$$p(z)q_*(z) + q(z)p_*(z) = r_*(z)r(z).$$

Thus

$$V(z,w) := \frac{p^*(z)q(w) + q^*(z)p(w) - r^*(z)r(w)}{z+w} = \frac{\overline{p}(z)p(w) - p(-z)\overline{p}(-w)}{z+w}$$

coincides with the Hermite–Fujiwara form. Since $\{1, z, \dots, z^{n-1}\}$ is a basis of the polynomial model X_p, the matrix representation of $\mathbf{V}$ on X_p is given by $\mathbf{H}_n(p)$. The result follows from Theorem 5.50. ■

Theorem 5.52 can be generalized by omitting the coprimeness assumption of $p(z)$ and $\overline{p}(-z)$. The result is strong enough for the characterization of asymptotic stability.

Theorem 5.53. *A necessary and sufficient condition for a complex polynomial $p(z)$ to be a Hurwitz polynomial is that the Hermite–Fujiwara form*

$$\frac{\overline{p}(z)p(w) - p(-z)\overline{p}(-w)}{z+w}$$

must be positive definite on X_p *or, equivalently, that the Hermite–Fujiwara matrix* $\mathbf{H}_n(p)$ *must be positive definite.*

Proof. Assuming that $p(z)$ is a Hurwitz polynomial implies that $p(z)$ and $\bar{p}(-z)$ do not have common roots and, hence, are coprime. By Theorem 5.52, the Hermite–Fujiwara matrix $\mathbf{H}_n(p)$ is positive definite. Conversely, assume that $\mathbf{H}_n(p)$ is positive definite or, equivalently, that the Hermite–Fujiwara form

$$\frac{\bar{p}(z)p(w) - p(-z)\bar{p}(-w)}{z+w} \tag{5.70}$$

is positive definite on X_p. The change of variable $w = -\zeta$ transforms (5.70) into the **Bezoutian form**

$$\frac{\bar{p}(z)p(-\zeta) - p(-z)\bar{p}(\zeta)}{z-\zeta}$$

of $\bar{p}(z)$ and $p_*(z) = p(-z)$. This shows that the **Bezoutian matrix** $B(\bar{p}, p_*)$ of $\bar{p}$ and p_* is equal to the product

$$B(\bar{p}, p_*) = \mathbf{H}_n(p)S$$

of the Hermite–Fujiwara matrix with the invertible matrix

$$S = \mathrm{diag}\,(1, -1, \cdots, (-1)^{n-1}).$$

Thus $B(\bar{p}, p_*)$ is invertible and Theorem 5.40 implies that $\bar{p}$ and p_* are coprime. Equivalently, p and $\bar{p}_*$ are coprime. Theorem 5.52 is now applied to infer the stability of $p(z)$. ∎

In the case of real polynomials, the Hermite–Fujiwara form admits a further reduction. To this end, the even and odd parts $p_+(z)$ and $p_-(z)$ of a real polynomial $p(z) = \sum_{j\geq 0} p_j z^j$ are introduced. These are defined as the polynomials

$$p_+(z) = \sum_{j\geq 0} p_{2j} z^j, \quad p_-(z) = \sum_{j\geq 0} p_{2j+1} z^j.$$

Thus

$$p(z) = p_+(z^2) + z p_-(z^2), \quad p_*(z) = p_+(z^2) - z p_-(z^2). \tag{5.71}$$

In the next proposition, it will be shown that both the Bezoutian $B(p, p_*)$ and the Hermite–Fujiwara forms have direct sum decompositions that are useful for reducing the computational complexity of stability analysis.

Proposition 5.54. *Let $p(z)$ be a real polynomial.*

1. *The following isomorphisms of quadratic forms are valid: For the Hermite–Fujiwara form $H(p)$ one has*

$$H(p) \simeq 2B(zp_-, p_+) \oplus 2B(p_+, p_-),$$

whereas for the Bezoutian $B(p, p_*)$,

$$B(p, p_*) \simeq 2B(zp_-, p_+) \oplus 2B(p_-, p_+).$$

2. *The Hermite–Fujiwara form is positive definite if and only if the two Bezoutians $B(q_+, q_-)$ and $B(zq_-, q_+)$ are positive definite.*

Proof. 1. The polynomial $p(z)$ being real implies $p(z) = \bar{p}(z)$. From (5.71) it follows that $p(-z) = p_+(z^2) - zp_-(z^2)$. We compute

$$\frac{p(z)p(w) - p(-z)p(-w)}{(z+w)} = 2\frac{zp_-(z^2)p_+(w^2) + p_+(z^2)wp_-(w^2)}{z+w}$$

$$= 2\frac{z^2 p_-(z^2)p_+(w^2) - p_+(z^2)w^2 p_-(w^2)}{z^2 - w^2} - 2zw\frac{p_-(z^2)p_+(w^2) - p_+(z^2)p_-(w^2)}{z^2 - w^2}.$$

The first summand contains only even terms, while the second contains only odd ones. This proves the first statement. By a change of variable $w = -\zeta$, the Hermite–Fujiwara form transforms into the Bezoutian of $\bar{p}(z)$ and $p(-z)$. However, this change of variable affects only the terms in

$$2zw\frac{p_-(z^2)p_+(w^2) - p_+(z^2)p_-(w^2)}{z^2 - w^2},$$

and this by a change of sign.

2. Follows from the direct sum representation of the Hermite–Fujiwara form. ∎

The following classical result is obtained as a direct corollary of this.

Theorem 5.55. *Let $p(z)$ be a monic real polynomial of degree n. The following statements are equivalent:*

(i)   *$p(z)$ is a Hurwitz polynomial.*
(ii) *The Hermite–Fujiwara matrix $\mathbf{H}_n(p)$ is positive definite.*
(iii) *The two Bezoutian matrices $B(p_+, p_-)$ and $B(zp_-, p_+)$ are positive definite.*

5.6 Exercises

1. Compute the tensor product $\mathbb{Z}_2 \otimes_{\mathbb{Z}} \mathbb{Z}_3$. For which pairs of integers $m, n \in \mathbb{N}$ is $\mathbb{Z}_m \otimes_{\mathbb{Z}} \mathbb{Z}_n = \{0\}$? What is $\mathbb{Z} \otimes_{\mathbb{Z}} \mathbb{Q}$?
2. Prove that $m \otimes n = 0$ is valid in the tensor product $M \otimes_R N$ if and only if every bilinear map $B : M \times N \longrightarrow P$ vanishes at (m, n).

3. For ideals I and J in a commutative ring R there exists a unique R-linear map

$$R/I \otimes_R R/J \simeq R/(I+J)$$

   satisfying $(x+I) \otimes (y+J) \mapsto xy + (I+J)$.
4. Let $M = R^n$ be a free module with $n \geq 2$ and $\{e_1,\ldots,e_n\}$ the standard basis. Check that $e_1 \otimes e_1 + e_2 \otimes e_2$ is not an elementary tensor in $M \otimes_R M$, i.e., that there exists no $v \in M$ with $e_1 \otimes e_1 + e_2 \otimes e_2 = v \otimes v$.
5. Prove that the tensor product $M \otimes_R N$ of torsion modules M and N over a commutative ring is a torsion module. Prove that the algebraic dual $M' = \{0\}$ for all torsion modules M. What happens if R has zero divisors?
6. Prove the R-module isomorphism $M \otimes_R N \simeq N \otimes_R M$.
7. Prove that the tensor product $f \otimes g$ of two surjective R-module homomorphisms $f : M_1 \longrightarrow N_1$ and $g : M_2 \longrightarrow N_2$ is surjective. Is this also true if surjectivity is replaced by injectivity?
8. Explain why the submodule $\{ \begin{pmatrix} a & b \\ b & c \end{pmatrix} | a,b,c \in \mathbb{F}[z] \} \subset \mathbb{F}[z]^{2 \times 2}$ does not have a representation of the form $D_1(z)\mathbb{F}[z]^{2 \times 2} + \mathbb{F}[z]^{2 \times 2}D_2(z)$.
9. Let $D_1(z) \in \mathbb{F}[z]^{m \times m}$ and $D_2(z) \in \mathbb{F}[z]^{p \times p}$ be nonsingular polynomial matrices with invariant factors $d_1,\ldots,d_m$ and $e_1,\ldots,e_p$, respectively. Let $d_i \wedge e_j$ denote the greatest common divisor of the polynomials $d_i(z)$ and $e_j(z)$. Prove the dimension formula

$$\dim_\mathbb{F} \text{Hom}_{\mathbb{F}[z]}(X_{D_1},X_{D_2}) = \sum_{i,j} \deg(d_i \wedge e_j).$$

   Deduce

$$\dim_\mathbb{F} \text{Hom}_{\mathbb{F}[z]}(X_{D_1},X_{D_1}) = \sum_{i=1}^m (2i-1)\deg d_i.$$

10. Let $D_1(z) \in \mathbb{F}[z]^{m \times m}$ and $D_2(z) \in \mathbb{F}[z]^{p \times p}$ be nonsingular. Show that $\text{Hom}_{\mathbb{F}[z]}(X_{D_1},X_{D_2}) = \{0\}$ if and only if $\det D_1(z)$ and $\det D_2(z)$ are coprime.
11. Let $D_1(z) \in \mathbb{F}[z]^{m \times m}$ and $D_2(z) \in \mathbb{F}[z]^{m \times m}$ be nonsingular polynomial matrices with determinants $d_i(z) = \det D_i(z)$. Show that

$$X_{D_1(z) \otimes D_2(z)} = (D_1(z) \otimes I)X_{I \otimes D_2(z)} \oplus X_{D_1(z) \otimes I}(I \otimes D_2(z))$$

   is true if and only if $d_1(z)$ and $d_2(z)$ are coprime.
12. Let A be a linear transformation in $\mathbb{F}^n$ with invariant factors $d_1,\ldots,d_n$. Let $\mathscr{C}(A)$ denote the centralizer of A, i.e., the set of all $Z \in \mathbb{F}^{n \times n}$, with $ZA = AZ$. Prove the following:

   a. Show $\dim_\mathbb{F} \mathscr{C}(A) = n$ if and only if A is cyclic.
   b. Show $\dim_\mathbb{F} \mathscr{C}(A) = n^2$ if and only if A is scalar, i.e., $A = \alpha I$ for some $\alpha \in \mathbb{F}$.

c. Let A and B be linear transformations. Show that there exist no nontrivial maps intertwining them if and only if the minimal polynomials or, equivalently, the characteristic polynomials of A and B are coprime.

13. Show that the discrete-time system $x_{t+1} = Ax_t$ on $\mathbb{C}^n$ is asymptotically stable if and only if for each positive definite Hermitian matrix $Q = Q^*$ there exists a unique positive definite Hermitian solution $P = P^*$ of the Stein equation

$$A^*PA - P = -Q.$$

14. Prove that a real monic polynomial $p(z)$ of degree n is a Schur polynomial if and only if

$$(z-1)^n p\left(\frac{z+1}{z-1}\right)$$

is a Hurwitz polynomial.

15. Prove that every real Hurwitz polynomial $p(z) = z^n + p_{n-1}z^{n-1} + \cdots + p_0$ satisfies $p_0 > 0, p_1 > 0, \ldots, p_{n-1} > 0$.

16. Prove that $\dot{x} = Ax$ is asymptotically stable for the tridiagonal matrix

$$A = \begin{pmatrix} -\frac{1}{2}a_1^2 & -a_2 & \cdots & 0 \\ a_2 & 0 & \ddots & \vdots \\ \vdots & \ddots & \ddots & -a_n \\ 0 & \cdots & a_n & 0 \end{pmatrix}$$

if $a_i \neq 0$ for $i = 1, \ldots, n$.

17. Assume that M and N are real symmetric $n \times n$ matrices that are positive definite. Prove that the second-order system

$$\ddot{x}(t) + M\dot{x}(t) + Nx(t) = 0$$

is asymptotically stable, i.e., $\lim_{t \to \infty} x(t) = 0$ and $\lim_{t \to \infty} \dot{x}(t) = 0$ are true for all solutions.

5.7 Notes and References

Our exposition of the basic theory of tensor products of modules and quotient modules follows Lang (1965) and Hungerford (1974). In the paper by Helmke and Fuhrmann (1998), tensored polynomial and rational models were introduced to describe tangent spaces of manifolds of rational functions. The systematic study of tensor products of functional models is due to Fuhrmann and Helmke (2010) and is continued in Fuhrmann (2010a). The tensor products of polynomial models lead to a polynomial approach to the Sylvester and Stein equations and clarifies the role of Bezoutians in representing solutions. The polynomial approach to Lyapunov equations is due to Willems and Fuhrmann (1992). For matrix versions of these equations see also de Souza and Bhattacharyya (1981) and Heinig and Rost (1984). The study of Bezoutians is old and dates back to the nineteenth century, with important contributions by Cayley, Jacobi, and Sylvester.

The problem of the stability of a linear (control) system was one of the first problems of the area of control theory. The interest in stability analysis is usually traced to J.C. Maxwell's theory of governors Maxwell (1868). However, the problem of root location of polynomials has a longer history. Since, with the work of Galois and Abel, exact determination of zeros of polynomials was proved to be impossible, interest shifted to the problem of localizing the zeros in some region of the complex plane. The unit disc and the major half-planes were the regions of greatest interest. The problem of root location was already solved by Hermite (1856). But in this connection the work of Routh (1877) turned out to be important because of the efficiency of the computational algorithm. In the same way, the work of Hurwitz (1895) was significant for its connection to topological problems. For a derivation of algebraic stability criteria that is close to the spirit of the chapter we refer the reader to Fuhrmann (1982).

In a major contribution to the subject, Lyapunov (1893) offered a completely different approach based on energy considerations. In the linear case, the Lyapunov theory reduces the study of the stability of a system of first-order homogeneous constant coefficient differential equations to the positive definiteness of the solution of the celebrated Lyapunov equation. This reduction is generally attributed to Gantmacher (1959). Our approach to the stability problem of higher-order systems of differential equations is based on a strengthened form of the Lyapunov equation, given in Theorem 5.44, replacing positive definiteness by a reachability, or observability, assumption. Considering Lyapunov equations in this form is due to Snyders and M. Zakai (1970) and Wimmer (1974a). Our approach to this reduction is achieved via the use of polynomial model theory and tensor algebra. The polynomial matrix analog of the Lyapunov equation is identified, and, with a solution to this equation, a two-variable polynomial matrix is constructed. In turn, this polynomial matrix induces an operator between two polynomial models. In the special case of symmetry, this map induces a quadratic form on a polynomial model, which leads to the required reduction.

The classic paper by Krein and Naimark (1936), is an excellent source for much of the older work on root location; however, strangely, no mention of Bezoutians is made there. The study of scalar Bezoutians goes back to Cayley. Multivariable Bezoutians were introduced by Anderson and Jury (1976) and used to derive rank formulas for the McMillan degree and stability test for multivariable linear systems. Their connection to tensor products and homomorphisms of polynomial models is central to our approach. Theorem 5.49 and its application to the stability analysis of higher-order equations are due to Willems and Fuhrmann (1992). The dimension formula (5.37) appears in Gantmacher (1959) and is attributed to Shoda (1929).

Chapter 6
State Feedback and Output Injection

Our attention turns now to the study of the fundamental question: How does one use input variables in the actual control of a system? Naturally, the use of control functions depends on the desired objectives of performance. Moreover, there are various ways in which the control can be applied. One way is to determine, a priori, a control sequence in the discrete-time case, or a control function in the continuous-time case, and apply it. Thus the control is applied without regard to its lasting effects or to the actual system performance, except insofar as the design goals have been taken into account. This is referred to as **open-loop control**. Obviously, this kind of control is often far from being satisfactory. The reasons for this may be manifold, in that there is no exact modeling of systems or that there are no errorless determinations of state vectors or precisely known control mechanisms. Thus open-loop control does not take into account noise in the system or random variations occurring from external influences. However, one advantage of open-loop control lies in the computational ease of determining such controls, for instance using optimization techniques. This aspect becomes particularly important when dealing with the control of large networks of systems.

An alternative to open-loop control is **feedback control**. Our standing assumption is that some knowledge of the state is available to the control mechanism, or to the controller, and the control device takes this information into account. There is a wide range of possibilities in designing feedback laws. If at each moment the controller has access to the full state of the system, then one refers to it as **state feedback**. If only a function, linear in the case of interest to us, of the state variables is available, then this will be referred to as **output feedback**. A controller can be memoryless, that is, the control is determined only by the currently available information. In this case, it is called **constant gain feedback**. Alternatively, the controller itself may be a dynamic system, in which case one speaks of **dynamic feedback control**. Whatever the specific control strategy may be, feedback control has the well-known advantage of ensuring robustness, while its disadvantages lie in the computational burden that comes with either computing the feedback gains

© Springer International Publishing Switzerland 2015
P.A. Fuhrmann, U. Helmke, *The Mathematics of Networks
of Linear Systems*, Universitext, DOI 10.1007/978-3-319-16646-9_6

or estimating the full or partial state variables of the system. Thus feedback control can become an increasingly complex task for large-scale networks. The tasks of controlling large-scale interconnected systems, such as swarms or ensembles of systems, therefore rather ask for a hybrid approach where both open-loop and closed-loop control strategies are employed.

Schematically one considers the following feedback configuration:

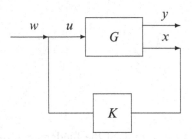

Here G is the plant, or the original system, and K denotes the control device. Thus, in the linear, discrete-time case, the equations of the plant are

$$x_{t+1} = Ax_t + Bu_t,$$

$$y_t = Cx_t + Du_t,$$

and if the controller is memoryless, we consider static output feedback as

$$u_t = -Ky_t + w_t.$$

State feedback is a special case of this, and we set $p = n$ and $C = I_n$. A dynamic situation occurs when K is itself a linear system and hence is given internally by

$$z_{t+1} = Fz_t + Gy_t,$$

$$v_t = Hz_t + Jy_t,$$

while the coupling equation, which allows for another control apart from the feedback already incorporated, is

$$u_t = w_t + v_t.$$

In this chapter we focus on analyzing the effects of state feedback control, while the design of open-loop controls was addressed already in Chapter 4.

For a deeper analysis of state feedback and output injection problems for linear systems (A, B, C), it turns out to be useful to study special classes of linear subspaces in the state space that capture the dynamics of the subsystems of (A, B, C). Such subspaces are the controlled and conditioned invariant subspaces and were first introduced and extensively studied in the early 1970s by Basile and Marro, as well as Wonham and Morse in the context of geometric control theory. The textbooks by Wonham (1979) and Basile and Marro (1992) give comprehensive accounts of the geometric theory. This chapter will be confined to the derivation of basic characterizations of controlled and conditioned invariant subspaces, both in the state space and using functional model representations. In the subsequent Chapter 7, characterizations of functional observers will be given using conditioned invariant subspaces.

6.1 State Feedback Equivalence

This section is devoted to the study of the effects of state feedback transformations that act on a fixed input-to-state pair (A, B). Let $\mathscr{U}$ and $\mathscr{X}$ be finite-dimensional vector spaces over the field $\mathbb{F}$, and let (A, B) be a reachable pair with $A : \mathscr{X} \longrightarrow \mathscr{X}$ and $B : \mathscr{U} \longrightarrow \mathscr{X}$ linear transformations. Our assumption will be that $\dim \mathscr{U} = m$ and $\dim \mathscr{X} = n$. Through a choice of bases one can identify $\mathscr{U}$ and $\mathscr{X}$ with $\mathbb{F}^m$ and $\mathbb{F}^n$, respectively. In that case, A and B are represented by $n \times n$ and $n \times m$ matrices, respectively. The pair (A, B) stands for the linear system

$$x_{t+1} = Ax_t + Bu_t. \tag{6.1}$$

If the system is augmented by the identity readout map

$$y_t = x_t,$$

then the transfer function of the combined system is

$$G(z) = (zI - A)^{-1}B.$$

A **state feedback law** is given by

$$u_t = -Kx_t + w_t, \tag{6.2}$$

where w_t denotes the external input applied at time t. Substituting (6.2) back into (6.1) amounts to transforming the pair (A, B) into the pair $(A - BK, B)$. In this case, one says that $(A - BK, B)$ has been obtained from (A, B) by state feedback. Clearly, the applications of state feedback transformations form a commutative group. If the group is enlarged to the one generated by invertible transformations in the input space $\mathscr{U}$, state-space similarity transformations in $\mathscr{X}$, and state feedback transformations, then the full **state feedback group** $\mathscr{F}$ is obtained. Thus an element of $\mathscr{F}$ is a triple of linear maps (S, K, R), with $S : \mathscr{X} \longrightarrow \mathscr{X}$ and $R : \mathscr{U} \longrightarrow \mathscr{U}$ nonsingular and $K : \mathscr{X} \longrightarrow \mathscr{U}$. The feedback group acts on a pair (A, B) by the following rule:

$$(A, B) \overset{(S,K,R)}{\mapsto} \left(S(A - BR^{-1}K)S^{-1}, SBR^{-1} \right).$$

This implies the group composition law

$$(S,K,R) \circ (S_1,K_1,R_1) = (SS_1, RK_1 + KS_1, RR_1).$$

This composition law is clearly associative since it can be expressed in terms of matrix multiplication as follows:

$$\begin{pmatrix} S & 0 \\ K & R \end{pmatrix} \begin{pmatrix} S_1 & 0 \\ K_1 & R_1 \end{pmatrix} = \begin{pmatrix} SS_1 & 0 \\ KS_1 + RK_1 & RR_1 \end{pmatrix} \in GL_{n+m}(\mathbb{F}).$$

The **state feedback action** can be identified with

$$(A,B) \mapsto S(A,B) \begin{pmatrix} S & 0 \\ K & R \end{pmatrix}^{-1} = \left(S(A - BR^{-1}K)S^{-1}, SBR^{-1} \right).$$

This clearly shows that

$$(S,K,R)^{-1} = (S^{-1}, -R^{-1}KS^{-1}, R^{-1})$$

and, hence, that $\mathscr{F}$ is a bona fide group. It is clear from the matrix representation of the feedback group that every element of $\mathscr{F}$ is the product of three elementary types:

1. Similarity or change of basis in the state space, i.e., elements of the form $(S,0,I)$, with S invertible;
2. Similarity or change of basis in the input space, i.e., elements of the form $(I,0,R)$, with R invertible;
3. Pure feedbacks, i.e., elements of the form (I,K,I).

Indeed, one has the composition law

$$(S,K,R) = (S,0,I)(I,K,I)(I,0,R).$$

The feedback group $\mathscr{F}$ induces a natural equivalence relation in the set of reachable pairs (A,B) with state space $\mathscr{X}$ and input space $\mathscr{U}$. Let (A_i,B_i), $i = 1,2$, be input pairs with state spaces $\mathscr{X}_i$ and input spaces $\mathscr{U}_i$, respectively. The pair (A_2,B_2) is said to be **state feedback equivalent** to (A_1,B_1) if there exist invertible maps $Z : \mathscr{X}_1 \longrightarrow \mathscr{X}_2$ and $R : \mathscr{U}_1 \longrightarrow \mathscr{U}_2$ and a map $K : \mathscr{X}_1 \longrightarrow \mathscr{U}_2$ that satisfy

$$ZA_1 - A_2Z = B_2K,$$

$$ZB_1 = B_2R.$$

It is trivial to check that this is indeed an equivalence relation. The equivalence classes are called the **orbits** of the feedback group, and one would like to obtain the orbit invariants as well as to isolate a single element in each orbit, a canonical form, that exhibits these invariants.

The dual concept the state feedback transformations is that of output injection. Here it is formulated for matrix representations of the system maps $A : \mathcal{X} \longrightarrow \mathcal{X}$, $B : \mathcal{U} \longrightarrow \mathcal{X}$, and $C : \mathcal{X} \longrightarrow \mathcal{Y}$, although a coordinate-free description is of course also possible. Thus two output pairs $(C_i, A_i) \in \mathbb{F}^{p \times n} \times \mathbb{F}^{n \times n}$, $i = 1, 2$, are called **output injection equivalent** if there exist invertible matrices $R \in GL_n(\mathbb{F})$ and $S \in GL_p(\mathbb{F})$ and a matrix $L \in \mathbb{F}^{n \times p}$ such that

$$\begin{pmatrix} A_2 \\ C_2 \end{pmatrix} = \begin{pmatrix} R & L \\ 0 & S \end{pmatrix} \begin{pmatrix} A_1 \\ C_1 \end{pmatrix} R^{-1} = \begin{pmatrix} (RA_1 - LC_1)R^{-1} \\ SC_1R^{-1} \end{pmatrix}. \tag{6.3}$$

The relation of output injection equivalence defines a bona fide equivalence relation on the matrix space $\mathbb{F}^{p \times n} \times \mathbb{F}^{n \times n}$. The equivalence classes are given by the orbits

$$\left\{ (RCS^{-1}, S(A - LC)S^{-1}) \mid S \in GL_n(\mathbb{F}), R \in GL_p(\mathbb{F}), L \in \mathbb{F}^{n \times p} \right\}$$

of the **output injection group** $\mathcal{G}$, where

$$\mathcal{G} = \left\{ \begin{pmatrix} R & L \\ 0 & S \end{pmatrix} \mid R \in GL_n(\mathbb{F}), S \in GL_p(\mathbb{F}), L \in \mathbb{F}^{n \times p} \right\} \subset GL_{n+p}(\mathbb{F}). \tag{6.4}$$

Conceptually, output injection seems much harder to grasp than state feedback. A clarification of its importance comes from a deeper study of observer theory, and this will be taken up in Chapter 7. There is therefore a measure of poetic justice in the fact that the analysis of the output injection case is, from a technical point of view, often significantly easier than that of the feedback case. The notion of output injection bears a natural duality with state feedback. In fact, a pair (C_1, A_1) is output injection equivalent to a pair (C_2, A_2) if and only if the dual pair $(A_1^\top, C_1^\top)$ is state feedback equivalent to $(A_2^\top, C_2^\top)$. This simple fact allows us to translate results for state feedback into corresponding results for output injection, and vice versa. However, while a natural approach is to dualize the feedback result, one can often develop an independent analysis of the output injection case, with the option of deriving results on the feedback group by duality considerations. Sometimes the proofs obtained along such lines become easier than those derived from state feedback analysis using duality arguments.

6.2 Polynomial Characterizations

The feedback group is introduced through a state-space formalism. However, as is seen in several instances, various aspects of linear systems theory are easier to handle if one operates with polynomial data, and this approach is our choice. Henceforth, by a choice of bases, $\mathcal{X}$ will be identified with $\mathbb{F}^n$ and $\mathcal{U}$ with $\mathbb{F}^m$.

Thus, for a reachable pair (A, B), the polynomial matrices $zI - A$ and B are left coprime. Since each factorization of a rational matrix function is associated with a right coprime factorization, one can write

$$(zI - A)^{-1} B = N(z)D(z)^{-1}, \tag{6.5}$$

with $N(z)$ and $D(z)$ right coprime. Furthermore, $N(z)$ and $D(z)$ are uniquely determined up to a common right unimodular factor. Thus, each reachable pair (A, B) is associated with the unique shift realization, defined in (4.23) as

$$S_D : X_D \longrightarrow X_D, \quad \pi_D : \mathbb{F}^m \longrightarrow X_D.$$

By Theorem 4.21, the pairs (A, B) and (S_D, π_D) are similar. Moreover, (6.5) defines a bijective correspondence between the similarity equivalence classes

$$\{(SAS^{-1}, SB) \mid S \in GL_n(\mathbb{F})\}$$

of reachable pairs and the equivalence classes

$$\{(D(z)U(z)) \mid U(z) \in GL_n(\mathbb{F}[z])\},$$

with respect to right multiplication by unimodular polynomial matrices, of nonsingular polynomial matrices $D(z)$. The next theorem characterizes feedback equivalence in terms of the factorizations (6.5).

Theorem 6.1. *Let (A, B) be a reachable pair, with $A \in \mathbb{F}^{n \times n}$ and $B \in \mathbb{F}^{n \times m}$. Let $N(z)D(z)^{-1}$ be a right coprime factorization of $(zI - A)^{-1}B$. Then a necessary and sufficient condition for a reachable pair $(\overline{A}, \overline{B})$ to be feedback equivalent to (A, B) is the existence of $R \in GL_m(\mathbb{F})$, $S \in GL_n(\mathbb{F})$, and $Q(z) \in \mathbb{F}[z]^{m \times m}$, for which $Q(z)D(z)^{-1}$ is strictly proper, such that*

$$(zI - \overline{A})^{-1}\overline{B} = SN(z)(D(z) + Q(z))^{-1}R^{-1}. \tag{6.6}$$

Proof. Assume $G(z) = (zI - A)^{-1}B = N(z)D(z)^{-1}$ are coprime factorizations, and let $(\overline{A}, \overline{B})$ be feedback equivalent to (A, B). Thus, there exist invertible maps S and R such that $\overline{A} = S(A - BK)S^{-1}$ and $\overline{B} = SBR^{-1}$. Hence,

$$(zI - \overline{A})^{-1}\overline{B} = (S(zI - A + BK)^{-1}S^{-1})^{-1}SBR = S(zI - A + BK)^{-1}BR^{-1}.$$

Now, computing

$$(zI - A + BK)^{-1}B = [(zI - A)(I + (zI - A)^{-1}BK)]^{-1}B$$
$$= (I + (zI - A)^{-1}BK)^{-1}(zI - A)^{-1}B$$
$$= (I + G(z)K)^{-1}G(z),$$

and using the equality $G(z)(I + KG(z)) = (I + G(z)K)G(z)$, it follows that $(I + G(z)K)^{-1}G(z) = G(z)(I + KG(z))^{-1}$. Consequently,

$$G_f(z) := (zI - \overline{A})^{-1}\overline{B} = SG(z)(I + KG(z))^{-1}R^{-1}$$
$$= SN(z)D(z)^{-1}(I + KN(z)D(z)^{-1})^{-1}R^{-1}$$
$$= SN(z)(D(z) + KN(z))^{-1}R^{-1}.$$

If one defines $Q(z) = KN(z)$, then clearly $G_f(z) = SN(z)(D(z) + Q(z))^{-1}R^{-1}$, and $Q(z)D(z)^{-1} = KN(z)D(z)^{-1}$ is strictly proper. This proves the necessity part of the theorem.

Conversely, assume that (6.6) is satisfied. Without loss of generality, it suffices to show that if, with $\overline{D}(z) = D(z) + Q(z)$ and $Q(z)D(z)^{-1}$ being strictly proper, the equality $N(z)\overline{D}(z)^{-1} = (zI - \overline{A})^{-1}\overline{B}$ is satisfied for a reachable pair $(\overline{A},\overline{B})$, then $(\overline{A},\overline{B})$ is feedback equivalent to (A,B). Thus it suffices to show that the pairs (S_D, π_D) and $(S_{\overline{D}}, \pi_{\overline{D}})$ are feedback equivalent. Alternatively, it must be shown that, for some invertible $\mathbb{F}$-linear map $Y : X_{\overline{D}} \longrightarrow X_D$ and a linear map $K : X_D \longrightarrow \mathbb{F}^n$, the equality $S_D - YS_{\overline{D}}Y^{-1} = BK$ is valid, where $B : \mathbb{F}^n \longrightarrow X_D$ is defined by $Bu = \pi_{Du}$ for $u \in \mathbb{F}^n$. Clearly, the previous equation is equivalent to $S_DY - YS_{\overline{D}} = BKY = BK_1$. Hence, it suffices to show that

$$\mathrm{Im}\,(S_DY - YS_{\overline{D}}) \subset \mathrm{Im}\,\pi_D,$$

and this we proceed to do. We define the Toeplitz induced map $Y : X_{\overline{D}} \longrightarrow X_D$ by

$$Yf = \pi_D\pi_+D\overline{D}^{-1}f, \qquad f \in X_{\overline{D}}.$$

Showing that Y is an invertible map follows, using Theorem 3.49, from the fact that $D(z)\overline{D}(z)^{-1}$ is biproper. For $f \in X_{\overline{D}}$ one computes

$$(YS_{\overline{D}} - S_{DY})f = \pi_D\pi_+D\overline{D}^{-1}\pi_{\overline{D}}zf - \pi_Dz\pi_D\pi_+D\overline{D}^{-1}f$$
$$= \pi_D\pi_+D\overline{D}^{-1}\overline{D}\pi_-\overline{D}^{-1}zf - \pi_Dz\pi_+D\overline{D}^{-1}f$$
$$= \pi_D\pi_+D\pi_-\overline{D}^{-1}zf - \pi_Dz\pi_+D\overline{D}^{-1}f$$
$$= \pi_D\pi_+D\overline{D}^{-1}zf - \pi_Dz\pi_+D\overline{D}^{-1}f$$
$$= \pi_D(D\overline{D}^{-1}f)_{-1} \in \mathrm{Im}\,\pi_D.$$

This completes the proof. ∎

The following theorem allows us to characterize state feedback equivalence in purely module-theoretic terms.

Theorem 6.2. *Let* $D(z), \overline{D}(z) \in \mathbb{F}[z]^{m \times m}$ *be nonsingular. Then the reachable pairs* (S_D, π_D) *and* $(S_{\overline{D}}, \pi_{\overline{D}})$ *obtained by the shift realizations* $\Sigma_{\bullet D^{-1}}$ *and* $\Sigma_{\bullet \overline{D}^{-1}}$, *respectively, are state feedback equivalent if and only if there exist a unimodular matrix* $U(z) \in GL_m(\mathbb{F}[z])$ *and a biproper rational matrix* $\Gamma(z) \in \mathbb{F}[[z^{-1}]]^{m \times m}$, *with*

$$D(z) = \Gamma(z)\overline{D}(z)U(z).$$

Equivalently, this is satisfied if and only if the left Wiener–Hopf indices of $D(z)$ *and* $\overline{D}(z)$ *are equal.*

Proof. To prove the sufficiency direction, assume that $D(z)$ is of the form $D(z) = \Gamma(z)\overline{D}(z)U(z)$ for a unimodular matrix $U(z)$ and $\Gamma(z)$ is biproper. Since the shift realizations of $\overline{D}(z)$ and $\overline{D}(z)U(z)$ are similar, we can assume without loss of generality that $U(z) = I_m$. Then the assumption is equivalent to $D(z)\overline{D}(z)^{-1}$ being biproper, i.e., the left Wiener–Hopf indices of $D(z)\overline{D}(z)^{-1}$ being zero. Then, by Theorem 3.47, the Toeplitz operator $T_{D\overline{D}^{-1}} : \mathbb{F}[z]^m \longrightarrow \mathbb{F}[z]^m$ being invertible, and so is the induced Toeplitz operator $\pi_D T_{D\overline{D}^{-1}} : X_{\overline{D}} \longrightarrow X_D$. Computing now

$$(\pi_D T_{D\overline{D}^{-1}} S_{\overline{D}} - S_D \pi_D T_{D\overline{D}^{-1}})f = \pi_D \pi_+ D\overline{D}^{-1} \pi_{\overline{D}} zf - \pi_D z \pi_D \pi_+ D\overline{D}^{-1} f$$

$$= \pi_D \pi_+ D\overline{D}^{-1} \overline{D} \pi_- \overline{D}^{-1} zf - \pi_D z \pi_+ D\overline{D}^{-1} f$$

$$= \pi_D \pi_+ D\pi_- \overline{D}^{-1} zf - \pi_D z \pi_+ D\overline{D}^{-1} f$$

$$= \pi_D \pi_+ D\overline{D}^{-1} zf - \pi_D z \pi_+ D\overline{D}^{-1} f$$

$$= \pi_D (D\overline{D}^{-1} f)_{-1} \in \operatorname{Im} \pi_D$$

proves the state feedback equivalence of $\Sigma_{\bullet D^{-1}}$ and $\Sigma_{\bullet \overline{D}^{-1}}$.

Conversely, assume that the pairs (S_D, π_D) and $(S_{\overline{D}}, \pi_{\overline{D}})$ are state feedback equivalent. Choose basis matrices $N(z), \overline{N}(z)$ for X_D and $X_{\overline{D}}$, and let (A, B) and $(\overline{A}, \overline{B})$ be the uniquely determined reachable pairs satisfying $(zI - A)^{-1}B = N(z)D(z)^{-1}$, $(zI - \overline{A})^{-1}\overline{B} = \overline{N}(z)\overline{D}(z)^{-1}$. By the transitivity of state feedback equivalence, the pairs (A, B) and $(\overline{A}, \overline{B})$ are state feedback equivalent. By Theorem 6.1, we obtain

$$\overline{N}(z)\overline{D}(z)^{-1} = (zI - \overline{A})^{-1}\overline{B} = SN(z)(D(z) + Q(z))^{-1}R^{-1},$$

with $Q(z)D(z)^{-1}$ strictly proper and $Q(z) = KN(z)$. Since $N(z)$ and $D(z)$ are assumed to be right coprime, $SN(z), D(z) + KN(z)$ are right coprime, too. Therefore, both pairs $(\overline{N}(z), \overline{D}(z))$ and $SN(z), D(z) + KN(z)$ are right coprime. Thus, there exists a unimodular polynomial matrix $U(z)$ with $\overline{D}(z) = R(D(z) + Q(z))U(z)$. Since QD^{-1} is strictly proper, the matrix

$$\Gamma(z) = D(z)(D(z) + Q(z))^{-1}R^{-1} = \left(I_m + Q(z)D(z)^{-1}\right)^{-1}R^{-1}$$

is biproper, and hence

$$\Gamma(z)\overline{D}(z) = D(z)U(z).$$

This completes the proof. ∎

The next theorem summarizes the preceding results.

Theorem 6.3. *For* $i = 1,2$, *let* $(A_i, B_i) \in \mathbb{F}^{(n \times (n+m))}$, *rank* $B_i = m$, *be reachable pairs, with input-to-state transfer functions* $G_i(z)$, *having the coprime factorizations*

$$G_i(z) = (zI - A_i)^{-1}B_i = N_i(z)D_i(z)^{-1}.$$

The following assertions are equivalent:

(a) (A_2, B_2) *is state feedback equivalent to* (A_1, B_1).
(b) *There exist state feedback transformation matrices* $L \in GL_m(\mathbb{F}), K \in \mathbb{F}^{m \times n}$, $S \in GL_n(\mathbb{F})$ *with*

$$G_2(z) = SG_1(z)(I + KG_1(z))^{-1}L^{-1}.$$

(c) $D_1(z)$ *and* $D_2(z)$ *have the same left Wiener–Hopf indices.*
(d) $G_1(z)$ *and* $G_2(z)$ *have the same left Wiener–Hopf indices.*

Proof. The equivalence $(a) \Longleftrightarrow (b) \Longleftrightarrow (c)$ was shown already in Theorem 6.2. The implication $(b) \Longrightarrow (d)$ is trivial. We prove $(d) \Longrightarrow (c)$. Thus, there exist a biproper rational function $\Gamma(z) \in \mathbb{F}[[z^{-1}]]^{m \times m}$ and a unimodular polynomial matrix $U(z) \in GL_n(\mathbb{F}[z])$ such that

$$N_2(z)D_2(z)^{-1} = U(z)N_1(z)D_1(z)^{-1}\Gamma(z). \tag{6.7}$$

By the reachability of (A_1, B_1), the matrix $N_1(z)$ is right prime, and therefore $U(z)N_1(z)$ is right prime, too. Thus there exists a polynomial matrix $M(z)$ that satisfies $M(z)U(z)N_1(z) = I$. Multiplying the identity (6.7) by $M(z)$ on both sides, it follows that

$$M(z)N_2(z) - D_1(z)^{-1}\Gamma(z)D_2(z).$$

Taking determinants, one obtains

$$\det(M(z)N_2(z)) = \frac{\det D_2(z)}{\det D_1(z)}\det\Gamma(z).$$

Since $\Gamma(z)$ is biproper, the determinant $\det\Gamma(z)$ is biproper, too. Moreover, both $\det D_1(z) = \det(zI - A_1)$ and $\det D_2(z) = \det(zI - A_2)$ have degree n. Thus

$$\frac{\det D_2(z)}{\det D_1(z)}\det\Gamma(z)$$

is biproper. Hence, the polynomial $\det(M(z)N_2(z))$ is also biproper, which implies that $M(z)N_2(z)$ is unimodular. Thus

$$\Gamma(z)^{-1}D_1(z)M(z)N_2(z) = D_2(z)$$

implies that $D_1(z)$ and $D_2(z)$ have the same left Wiener–Hopf indices. This shows $(d) \implies (c)$, and the proof is complete. ∎

The importance of the preceding result lies in showing that the classification of reachable pairs $(A,B) \in \mathbb{F}^{n \times (n+m)}$, up to state feedback equivalence, is equivalent to the classification of nonsingular $m \times m$ polynomial matrices $D(z)$ with identical left Wiener–Hopf indices. The bijective correspondence between orbits of the state feedback group and polynomial matrices with fixed Wiener–Hopf indices will be taken up in the next section.

Duality is used to derive the following two counterparts to Theorems 6.1 and 6.2.

Theorem 6.4. *Let $D(z), \overline{D}(z) \in \mathbb{F}[z]^{p \times p}$ be nonsingular. Then the observable pairs $((D\cdot)_{-1}, S_D)$ and $((\overline{D}\cdot)_{-1}, S_{\overline{D}})$, obtained by the shift realization in the state spaces $X_D, X_{\overline{D}}$, are output injection equivalent if and only if there exist a unimodular matrix $U(z) \in GL_p(\mathbb{F}[z])$ and a biproper rational matrix $\Gamma(z) \in \mathbb{F}[[z^{-1}]]^{p \times p}$ with*

$$D(z) = U(z)\overline{D}(z)\Gamma(z).$$

Equivalently, this is true if and only if the right Wiener–Hopf indices of $D(z)$ and $\overline{D}(z)$ are equal.

Proof. Using Proposition 3.48, this follows from Theorem 6.2 by duality considerations. ∎

Theorem 6.5. *Let $(C_1, A_1), (C_2, A_2) \in \mathbb{F}^{(p+n) \times n}, \operatorname{rk} C_i = p$, be observable pairs with state-to-output transfer functions*

$$G_1(z) = C_1(zI - A_1)^{-1} = D_{\ell,1}(z)^{-1}N_{\ell,1}(z),$$
$$G_2(z) = C_2(zI - A_2)^{-1} = D_{\ell,2}(z)^{-1}N_{\ell,2}(z)$$

and left coprime factorizations $D_{\ell,1}(z), N_{\ell,1}(z)$ and $D_{\ell,2}(z), N_{\ell,2}(z)$, respectively. The following statements are equivalent:

(a) *(C_1, A_1) is output injection equivalent to (C_2, A_2).*
(b) *There exists an output injection transformation matrix $P \in GL_p(\mathbb{F})$, $L \in \mathbb{F}^{n \times p}, S \in GL_n(\mathbb{F})$ with*

$$G_2(z) = P(I_p + G_1(z)L)^{-1}G_1(z)S^{-1}.$$

(c) *$D_{\ell,1}(z)$ and $D_{\ell,2}(z)$ have the same right Wiener–Hopf indices.*
(d) *$G_1(z)$ and $G_2(z)$ have the same right Wiener–Hopf indices.*

Proof. A change of basis in the output space changes the transfer function by a left nonsingular factor. Similarly, a similarity transformation in the state space can be easily coped with. Thus, without loss of generality, one can assume that $A_2 = A_1 - LC_1$ and $C_2 = C_1$. Rewriting the coprime factorization as $N_{\ell,1}(z)(zI - A_1) = D_{\ell,1}(z)C_1$ and adding $N_{\ell,1}(z)LC_1$ to both sides, one obtains the intertwining relation $N_{\ell,1}(z)(zI - A_1 + LC_1) = (D_{\ell,1}(z) + N_{\ell,1}(z)L)C_1$, which can be written as

$$C_1(zI - A_1 + LC_1)^{-1} = (D_{\ell,1}(z) + N_{\ell,1}(z)L)^{-1} N_{\ell,1}(z) = D_{\ell,2}(z)^{-1} N_{\ell,2}(z).$$

It is easily checked that the factorization $\overline{G}_2(z) = (D_{\ell,1}(z) + N_{\ell,1}(z)L)^{-1} N_{\ell,1}(z)$ is left coprime. Thus, there exists a unimodular polynomial matrix $M(z)$ such that $D_{\ell,2}(z) = M(z)(D_{\ell,1}(z) + N_{\ell,1}(z)L)$ and $N_{\ell,2}(z) = M(z)N_{\ell,1}(z)$ are fulfilled, thereby obtaining the right Wiener–Hopf factorization

$$D_{\ell,2}(z) = M(z)D_{\ell,1}(z)\Gamma(z),$$

with $\Gamma(z) = D_{\ell,1}(z)^{-1}(D_{\ell,1}(z) + N_{\ell,1}(z)L) = I_p + G_1(z)L$ biproper. In particular, $D_{\ell,1}(z)$ and $D_{\ell,2}(z)$ have the same right Wiener–Hopf indices. This shows the implications $(a) \implies (b) \implies (c)$. The reverse directions follow as for the proofs of Theorems 6.1 and 6.2. The implication $(b) \implies (d)$ is trivial. The proof that $(d) \implies (c)$ runs parallel to the proof in Theorem 6.3 and is thus omitted.   ∎

6.3   Reachability Indices and the Brunovsky Form

For discrete-time systems

$$x_{t+1} = Ax_t + Bu_t$$
$$y_t = Cx_t,$$

with state space $\mathscr{X}$, input space $\mathscr{U}$, and output space $\mathscr{Y}$, there exists a fine structure in the state space according to how fast the various states are reached. Dually, one can ask how fast one can observe the state from the output. It turns out that this structure, manifested through the so-called reachability and observability indices, is all important for the study of the fundamental problems of systems theory, namely, a description of the inherent limitations of controllers to change the dynamics of the system and for state estimation purposes.

Consider the sequence of subspaces $\mathscr{V}_i(A,B) \subset \mathscr{X}$, defined by

$$\mathscr{V}_i(A,B) = \mathscr{B} + A\mathscr{B} + \cdots + A^i\mathscr{B}, \tag{6.8}$$

where $\mathscr{B} = \operatorname{Im} B$. Thus, in discrete time, the linear subspace $\mathscr{V}_i$ consists of all states that can be reached from zero in at most $i + 1$ steps. Obviously, $\mathscr{V}_i \subset \mathscr{V}_{i+1}$. Applying the Cayley–Hamilton theorem and the assumption of reachability, $\mathcal{V}_{n-1+j} = \mathcal{R}$ is also valid for $j \geq 0$, where $\mathcal{R}$ denotes the reachable space of (A,B). Define a sequence of indices by

$$v_i(A,B) = \begin{cases} \dim \mathcal{B} & i = 0, \\ \dim \mathcal{V}_i - \dim \mathcal{V}_{i-1} & i \geq 1. \end{cases} \tag{6.9}$$

Thus

$$m \geq v_0(A,B) \geq v_1(A,B) \geq \ldots \geq v_n(A,B) = 0.$$

Define the dual set of indices by

$$\kappa_i(A,B) = \#\{v_j(A,B) \mid v_j(A,B) \geq i\}. \tag{6.10}$$

Thus $\kappa_1 \geq \cdots \geq \kappa_m$ and $\sum_{i=1}^{m} \kappa_i = \sum_{j=0}^{n} v_j$. Thus $\kappa = (\kappa_1, \ldots, \kappa_m)$ and $v = (v_0, \ldots, v_n)$ form dual partitions of $r = \dim \mathcal{V}_n(A,B)$. The indices $\kappa_1 \geq \cdots \geq \kappa_m$ are usually called the **controllability indices** of the pair (A,B). In the discrete-time case, it is more appropriate to call them, as we shall, the **reachability indices**. If the pair (A,B) is fixed, then one writes κ_i for $\kappa_i(A,B)$, and so forth. It follows trivially from (6.9) that $\kappa_1 + \cdots + \kappa_m = v_0 + \cdots + v_n = \dim \mathcal{R}$. Therefore, the reachability indices of a reachable pair on an n-dimensional state space form a partition of n, that is, a representation $\kappa_1 + \cdots + \kappa_m = n$. It is easily seen, by examples, that in fact all partitions of n into at most m parts arise as reachability indices of a suitable reachable pair (A,B).

Similarly, the observability indices of a pair $(C,A) \in \mathbb{F}^{p \times n} \times \mathbb{F}^{n \times n}$ will be introduced. To this end, define, for each $i = 1, \ldots, n$, the ranks of the ith partial observability matrix as

$$r_i(C,A) = \text{rank} \begin{pmatrix} C \\ \vdots \\ CA^{i-1} \end{pmatrix}.$$

Thus the differences ($s_0 := 0$)

$$s_i = r_i - r_{i-1}, \quad i = 1, \ldots, n$$

measure the increase in the ranks of the partial observability matrices.

Definition 6.6. The **observability indices** of $(C,A) \in \mathbb{F}^{p \times n} \times \mathbb{F}^{n \times n}$ are the nonnegative integers $\lambda_1(C,A) \geq \ldots \geq \lambda_p(C,A)$ defined by

$$\lambda_i(C,A) = \#\{s_j(C,A) \mid s_j(C,A) \geq i\}.$$

In complete analogy with the reachability indices, the equality

$$\lambda_1 + \cdots + \lambda_p = n$$

is true if and only if (C,A) is observable. The following lemma will be needed.

Lemma 6.7. *The reachability indices are state feedback invariants, i.e.,*

$$\kappa_i(S(A+BK)S^{-1}, SBR^{-1}) = \kappa_i(A,B), \ i = 1,\ldots,m,$$

is true for all matrices $R \in GL_m(\mathbb{F}), S \in GL_n(\mathbb{F}), K \in \mathbb{F}^{m \times n}$. Similarly, the observability indices are output injection invariants, i.e., for all $i = 1,\ldots,p$ and for all matrices $R \in GL_p(\mathbb{F}), S \in GL_n(\mathbb{F}), L \in \mathbb{F}^{n \times p}$, the equality

$$\lambda_i(RCS^{-1}, S(A+LC)S^{-1}) = \lambda_i(C,A)$$

is valid.

Proof. It is obvious, with S,R invertible maps in the state space and input space, respectively, that

$$\mathcal{V}_i(S(A+BK)S^{-1}, SBR^{-1}) = S\mathcal{V}_i(A,B)$$

for all $i \geq 0$. While the spaces $\mathcal{V}_i$ defined in (6.8) change under the action of an element of the feedback group, their dimensions do not, i.e., they are invariant. This shows that the $v_i(A,B)$ are state feedback invariant, as are, thus, the reachability indices $\kappa_i(A,B)$. The proof is similar for the observability indices. ∎

The preceding definition of reachability indices was introduced in state-space terms. We now show the connection to invariants defined in terms of coprime factorizations of $(zI - A)^{-1}B$.

Theorem 6.8. *Let $(A,B) \in \mathbb{F}^{n \times n} \times \mathbb{F}^{n \times m}$ be a reachable pair, with $\operatorname{rank} B = m$. Let $N(z)D(z)^{-1}$ be a right coprime factorization of $(zI - A)^{-1}B$, and let $\kappa_1 \geq \cdots \geq \kappa_m > 0$ be the reachability indices of the pair (A,B) as defined in (6.10). Then:*

1. *The reachability indices of the pair (A,B) are equal to the minimal column indices of the submodule $D(z)\mathbb{F}[z]^m \subset \mathbb{F}[z]^m$;*
2. *The reachability indices of the pair (A,B) are equal to the left Wiener–Hopf factorization indices of $D(z)$;*
3. *The reachability indices of the pair (A,B) are equal to the minimal column indices of the submodule $\operatorname{Ker}(zI - A, -B) \subset \mathbb{F}[z]^{n+m}$, defined by the linear multiplication operator*

$$(zI - A, -B) : \mathbb{F}[z]^{n+m} \longrightarrow \mathbb{F}[z]^n.$$

Proof. Assume $\eta_1 \geq \cdots \geq \eta_m$ are the minimal column indices of $D(z)$. Let $\Delta(z) = \mathrm{diag}(z^{\eta_1}, \ldots, z^{\eta_m})$. Then there exist a unimodular polynomial matrix $U(z)$ and a biproper matrix $\Gamma(z)$ such that

$$D(z)U(z) = \Gamma(z)\Delta(z).$$

By Theorem 6.2, the pairs (S_D, π_D) and (S_Δ, π_Δ) are feedback equivalent and, hence, have the same reachability indices. Thus, it suffices to prove the theorem for $A = S_\Delta, B = \pi_\Delta$. The reachability indices of (S_Δ, π_Δ) are easily computed as follows. With $e_1, \ldots, e_m$ the standard basis elements of $\mathbb{F}^m$, clearly, by our assumption that B has full column rank, we get $\mathrm{Im}\,\pi_\Delta = \mathrm{span}\{e_1, \ldots, e_m\}$. On the other hand, the equality between the coprime factorizations $(zI - A)^{-1}B = N(z)D(z)^{-1}$ implies, using the shift realization, that (A, B) is similar to (S_D, π_D), so it is feedback equivalent to (S_Δ, π_Δ). Consider now the subspaces $\mathscr{V}_i$, defined in (6.8), that correspond to the pair (S_Δ, π_Δ). Clearly, $\dim \mathscr{V}_1 = \#\{\eta_i > 0\} = m$ and $\dim \mathscr{V}_k = \dim \mathscr{V}_{k-1} + \#\{\eta_i \geq k\}$. So $v_k = \dim \mathscr{V}_k - \dim \mathscr{V}_{k-1} = \#\{\eta_i \geq k\}$. Thus $\eta_1 \geq \cdots \geq \eta_m$ are the dual indices to the v_i, but so are the reachability indices $\kappa_1, \ldots, \kappa_m$. Hence, necessarily, $\eta_i = \kappa_i$.

By part 1, the column indices of $D(z)$ are equal to the reachability indices of (S_D, π_D), i.e., to $\kappa_1, \ldots, \kappa_m$. Therefore, there exists a unimodular polynomial matrix $V(z)$ for which $D(z)V(z)$ is column proper with column indices $\kappa_1, \ldots, \kappa_m$. Writing $D(z)V(z) = \Gamma(z)\Delta(z)$, where $\Delta(z) = \mathrm{diag}(z^{\kappa_1}, \ldots, z^{\kappa_m})$, $\Gamma(z)$ is necessarily biproper because the leading term of $\Gamma(z)$ is $[DV]_{hc}$, which is nonsingular. This implies that, with $U(z) = V(z)^{-1}$, the left Wiener–Hopf factorization $D(z) = \Gamma(z)\Delta(z)U(z)$.

The equality $(zI - A)N(z) = BD(z)$ can be rewritten as

$$(zI - A, -B) \begin{pmatrix} N(z) \\ D(z) \end{pmatrix} = 0.$$

Using the coprimeness assumption (Theorem 2.27), it follows that

$$\mathrm{Ker}\,(zI - A, -B) = \begin{pmatrix} N(z) \\ D(z) \end{pmatrix} \mathbb{F}[z]^m.$$

Now $N(z)D(z)^{-1}$ is strictly proper, and thus the minimal column indices of $\begin{pmatrix} N(z) \\ D(z) \end{pmatrix}$ are equal to those of $D(z)$. By part 2, they coincide with the reachability indices of (A, B). ∎

The next result, which is a straightforward consequence of Theorem 6.8, characterizes the reachability and observability indices of an observable pair in terms of Wiener–Hopf factorization indices.

Corollary 6.9. *1. Let (A,B) be a reachable pair, and let $N(z)D(z)^{-1} = (zI - A)^{-1}B$ be a right coprime factorization. Then the reachability indices of (A,B) are equal to the left Wiener–Hopf indices of $D(z)$.*
2. Let (C,A) be an observable pair, and let $D_\ell(z)^{-1}N_\ell(z) = C(zI - A)^{-1}$ be a left coprime factorization. Then the observability indices of (C,A) are equal to the right Wiener–Hopf indices of $D_\ell(z)$.

If $G(z)$ is a proper, rational transfer function, then its Wiener–Hopf factorization indices must have a system-theoretic interpretation. This is indeed the case, and a system-theoretic interpretation of the factorization indices of the denominators in coprime matrix fraction representations of $G(z)$ can be derived.

Theorem 6.10. *Let $G(z) \in \mathbb{F}[z]^{p \times m}$ be a proper rational function admitting the coprime matrix fraction representations*

$$G(z) = N_r(z)D_r(z)^{-1} = D_\ell(z)^{-1}N_\ell(z), \tag{6.11}$$

and let (A,B,C,D) be a reachable and observable realization of $G(z)$. Then the reachability indices of the realization are equal to the left Wiener–Hopf indices of $D_r(z)$ and the observability indices are equal to the right Wiener–Hopf indices of $D_\ell(z)$.

Proof. By the state-space isomorphism theorem, the pair (A,B) is isomorphic to the pair (S_{D_r}, π_{D_r}). By Proposition 2.19, there exists a unimodular matrix $U(z)$ such that $D_r(z)U(z)$ is column proper with column indices $\kappa_1 \geq \kappa_2 \geq \cdots \geq \kappa_m$. Clearly,

$$D_r(z)U(z) = \Gamma(z)\Delta(z), \tag{6.12}$$

with $\Delta(z) = \text{diag}(z^{\kappa_1}, \ldots, z^{\kappa_m})$ and Γ biproper. By Theorem 6.2, this implies that (S_{D_r}, π_{D_r}) and (S_Δ, π_Δ) are feedback equivalent pairs. However, the reachability indices of (S_Δ, π_Δ) are easily seen to be equal to $\kappa_1, \ldots, \kappa_m$ (see the proof of the subsequently stated Theorem 6.14). Finally, (6.12) can be rewritten as

$$D_r(z) = G_-(z)\Delta(z)G_+(z),$$

with $G_-(z) = \Gamma(z)$ and $G_+(z) = U(z)^{-1}$. This is a left Wiener–Hopf factorization of $D_r(z)$. The statement concerning observability indices follows by duality. ∎

In Corollary 6.9 and Theorem 6.10, it was shown that the reachability indices of a pair (A,B) coincide with the left Wiener–Hopf indices of the nonsingular polynomial matrix $D(z)$ appearing in a coprime factorization

$$(zI - A)^{-1}B = N(z)D(z)^{-1}.$$

One attempts to extend this analysis to Wiener–Hopf factorizations of strictly proper transfer functions $G(z) = C(zI - A)^{-1}B$. For simplicity, our focus will be on strictly proper transfer functions, although an extension to proper transfer functions is possible.

Definition 6.11. A reachable and observable system (A, B, C) is called **state feedback irreducible** if and only if $(S(A + BK)S^{-1}, SBR^{-1}, CS^{-1})$ is reachable and observable for all state feedback matrices $(S, K, R) \in \mathscr{F}_{n,m}$.

Of course, while the reachability of a system is always preserved by state feedback, this is no longer true of observability. It is a simple observation that transfer functions of the form

$$(zI - A)^{-1}B \quad \text{or} \quad C(zI - A)^{-1},$$

with (A, B) reachable or (C, A) observable, are feedback irreducible. Thus feedback irreducibility is an extension of the situation discussed previously.

To begin with the analysis of feedback irreducibility, one considers the single-input single-output case. Let

$$g(z) = \frac{p(z)}{q(z)} \in \mathbb{F}[z]$$

denote a scalar strictly proper transfer function of degree n, given by a coprime factorization, with $q(z)$ monic and $\deg p(z) < \deg q(z) = n$. Let (A, b, c) denote a minimal realization of $g(z)$. Without loss of generality, one can assume that (A, b) is in Brunovsky canonical form, i.e.,

$$\begin{pmatrix} 0 & 1 & & \\ & \ddots & \ddots & \\ & & 0 & 1 \\ 0 & \dots & \dots & 0 \end{pmatrix}, \quad b = \begin{pmatrix} 0 \\ \vdots \\ 0 \\ 1 \end{pmatrix}, \quad c = \begin{pmatrix} c_0 & c_1 & \cdots & c_{n-1} \end{pmatrix},$$

with transfer function

$$g(z) = \frac{p(z)}{q(z)} = \frac{c_0 + \dots + c_{n-1} z^{n-1}}{z^n}.$$

Thus the system (A, b, c) is feedback irreducible if and only if the pair

$$A + bk = \begin{pmatrix} 0 & 1 & & \\ & \ddots & \ddots & \\ & & 0 & 1 \\ k_0 - q_0 & \dots & \dots & k_{n-1} - q_{n-1} \end{pmatrix}, \quad c = \begin{pmatrix} c_0 & c_1 & \cdots & c_{n-1} \end{pmatrix}$$

is observable for all state feedback matrices $k = (k_0, \dots, k_{n-1})$. This in turn is equivalent to $p(z) = \sum_{j=0}^{n-1} c_j z^j$ being coprime to all monic polynomials of the form $q_k(z) = z^n + k_{n-1} z^{n-1} + \cdots + k_0$, i.e., that $p(z) = c_0 \neq 0$ is a nonzero constant polynomial. Thus a scalar strictly proper transfer function $g(z)$ is feedback irreducible if and only if it has no finite zeros, i.e., if and only if the relative degree $\deg q - \deg p$ of $g(z)$ is equal to n. This analysis is now extended to the matrix case, beginning with the following lemma.

Lemma 6.12. *Let* $G(z) \in \mathbb{F}(z)^{m \times m}$ *be proper with the right Wiener–Hopf factorization* $G(z) = U(z)\Delta(z)\Gamma(z)$, *with* $U(z) \in \mathbb{F}[z]^{p \times p}$ *unimodular, and* $\Gamma(z)$ *biproper. Let* $N(z) \in \mathbb{F}[z]^{p \times m}$ *be a right prime polynomial matrix, with* $p \geq m$. *Then*

$$N(z)G(z) = U_1(z) \begin{pmatrix} \Delta(z) \\ 0 \end{pmatrix} \Gamma(z).$$

In particular, $G(z)$ *and* $N(z)G(z)$ *have the same right factorization indices.*

Proof. Since $N(z)$ is right prime, there exist unimodular matrices $V(z), W(z)$, with

$$N(z) = V(z) \begin{pmatrix} I_m \\ 0 \end{pmatrix} W(z).$$

Thus,

$$N(z)G(z) = V(z) \begin{pmatrix} I_m \\ 0 \end{pmatrix} W(z)U(z)\Delta(z)\Gamma(z) = U_1(z) \begin{pmatrix} \Delta(z) \\ 0 \end{pmatrix} \Gamma(z),$$

where $U_1(z) = V(z)\mathrm{diag}\,(W(z)U(z), I)$.   ∎

Theorem 6.13. *Let* (A,B,C) *be a reachable and observable realization of a strictly proper transfer function* $G(z)$ *with right coprime factorization* $G(z) = N(z)D(z)^{-1}$. *Assume that* $G(z)$ *has full column rank. Then:*

1. (A,B,C) *is feedback irreducible if and only if* $N(z)$ *is right prime (i.e., left invertible);*
2. *Assume that* (A,B,C) *is feedback irreducible. Then the negatives of the reachability indices of* (A,B) *coincide with the right Wiener–Hopf indices and the negatives of the observability indices of* (C,A)) *coincide with the left Wiener–Hopf indices of the transfer function* $G(z) = C(zI - A)^{-1}B$.

Proof. Recall that every state feedback transformation $(A,B,C) \mapsto (A + BK, B, C)$ acts on transfer functions by right multiplication with a biproper rational function, that is,

$$C(zI - A + BK)^{-1}B = C(zI - A)^{-1}B(I_m + K(zI - A)^{-1}B)^{-1}.$$

Moreover, each right coprime factorization of $(zI - A)^{-1}B = H(z)D(z)^{-1}$ implies the intertwining relation $BD(z) = (zI - A)H(z)$. This induces the factorization

$$G(z) = C(zI - A)^{-1}B = N(z)D(z)^{-1},$$

with $N(z) = CH(z)$. This is summarized in the system equivalence relation

$$\begin{pmatrix} B & 0 \\ 0 & I \end{pmatrix} \begin{pmatrix} D(z) & -I \\ N(z) & 0 \end{pmatrix} = \begin{pmatrix} zI-A & -B \\ C & 0 \end{pmatrix} \begin{pmatrix} H(z) & 0 \\ 0 & I \end{pmatrix},$$

with $B, zI - A$ left coprime and $D(z), H(z)$ right coprime, which implies the following equivalence:

$$\begin{pmatrix} D(z) & -I \\ N(z) & 0 \end{pmatrix} \simeq_{FSE} \begin{pmatrix} zI-A & -B \\ C & 0 \end{pmatrix}.$$

Using the Shift Realization Theorem 4.26, it follows that the minimality of (A,B,C) implies the right coprimeness of $N(z), D(z)$. Similarly, from the factorization $(zI - A - BK)^{-1}B = H(z)(D(z) - KH(z))^{-1}$ follows the intertwining relation

$$\begin{pmatrix} B & 0 \\ 0 & I \end{pmatrix} \begin{pmatrix} D(z)+KH(z) & -I \\ N(z) & 0 \end{pmatrix} = \begin{pmatrix} zI-A+BK & -B \\ C & 0 \end{pmatrix} \begin{pmatrix} H(z) & 0 \\ 0 & I \end{pmatrix}.$$

Here $B, zI - A + BK$ and $D(z) + KH(z), H(z)$ are left coprime and right coprime, respectively. In particular,

$$\begin{pmatrix} D(z)+KH(z) & -I \\ N(z) & 0 \end{pmatrix} \simeq_{FSE} \begin{pmatrix} zI-A+BK & -B \\ C & 0 \end{pmatrix}$$

for each K. This shows that (A,B,C) is feedback irreducible if and only if the polynomial matrices $N(z)$ and $D(z) + KH(z)$ are right coprime for each K.

Next, it will be shown that this condition is equivalent to the right primeness of $N(z)$. Clearly, the right primeness of $N(z)$ implies for each state feedback matrix K the right coprimeness of $N(z)$ and $D(z) - KH(z)$. Thus $N(z)$ right prime implies feedback irreducibility. To prove the converse implication, let us assume that $N(z)$ is not right prime, i.e., there exists a polynomial factorization $N(z) = N'(z)F(z)$ with $N'(z)$ right prime and $F(z) \in \mathbb{F}[z]^{m \times m}$ nonsingular and nonunimodular. Applying Lemma 6.12, it follows that $G(z) = N(z)D(z)^{-1}$ and $F(z)D(z)^{-1}$ have the same right Wiener–Hopf indices. Let

$$F(z)D(z)^{-1} = U(z)\Delta(z)^{-1}\Gamma(z)$$

be the right Wiener–Hopf factorization, with $\Delta(z) = \text{diag}(z^{\kappa_1}, \cdots, z^{\kappa_m})$. Then $E(z) := \Delta(z)U(z)^{-1}$ is a nonsingular polynomial matrix and

$$D_1(z) := E(z)F(z) = \Gamma(z)D(z)$$

is a nonsingular polynomial matrix with $\deg\det D_1(z) = \deg\det D(z)$. Computing

$$G(z)\Gamma(z)^{-1} = N(z)D(z)^{-1}\Gamma(z)^{-1} = N(z)D_1(z)^{-1}$$
$$= N'(z)E(z)^{-1}$$

yields a nontrivial factorization. Thus, the McMillan degrees of $G(z)\Gamma(z)^{-1}$ and $G(z)$ are related as

$$\delta(G\Gamma^{-1}) \le \deg\det E(z) < \deg\det D_1(z) = \deg\det D(z) = \delta(G).$$

This shows that $G(z)$ is feedback reducible and completes the proof of the first claim of the theorem.

By the first part, a full column rank coprime factorization $G(z) = N(z)D(z)^{-1}$ is feedback irreducible if and only if $N(z)$ is right prime. But then Lemma 6.12 implies that $G(z)$ and $D(z)^{-1}$ have the same right Wiener–Hopf indices. Thus the right Wiener–Hopf indices of $G(z)$ are equal to the negative of the left Wiener–Hopf indices of $D(z)$, which by Theorem 6.10 coincide with the reachability indices of (A,B). This completes the proof of the second claim 2.   ∎

Our attention turns now to the question of constructing a canonical form for reachable pairs under the action of the state feedback group.

Theorem 6.14. *1. Let $(A,B) \in \mathbb{F}^{n \times n} \times \mathbb{F}^{n \times m}$ be a reachable pair with reachability indices $\kappa_1 \ge \ldots \ge \kappa_m$. Then (A,B) is feedback equivalent to the block matrix representation*

$$\left(\begin{pmatrix} A_1 & & \\ & \ddots & \\ & & A_m \end{pmatrix}, \begin{pmatrix} B_1 & & \\ & \ddots & \\ & & B_m \end{pmatrix} \right), \qquad (6.13)$$

with the matrices $A_j \in \mathbb{F}^{\kappa_j \times \kappa_j}$ and $B_j \in \mathbb{F}^{\kappa_j \times 1}$ defined by

$$A_j = \begin{pmatrix} 0 & & & \\ 1 & \ddots & & \\ & \ddots & \ddots & \\ & & 1 & 0 \end{pmatrix}, \qquad B_j = \begin{pmatrix} 1 \\ 0 \\ \vdots \\ 0 \end{pmatrix}. \qquad (6.14)$$

*We will refer to $D(z) = \text{diag}(z^{\kappa_1}, \ldots, z^{\kappa_m})$ as the **polynomial Brunovsky form** and to (6.13) and (6.14) as the **Brunovsky canonical form**.*

2. Two reachable pairs $(A,B),(\overline{A},\overline{B}) \in \mathbb{F}^{n \times (n+m)}$ are state feedback equivalent if and only if they have the same reachability indices

$$\kappa_1(A,B) = \kappa_1(\overline{A},\overline{B}), \quad \ldots, \quad \kappa_m(A,B) = \kappa_m(\overline{A},\overline{B}).$$

Proof. The direct sum $X_\Delta = X_{z^{\kappa_1}} \oplus \cdots \oplus X_{z^{\kappa_m}}$ is a consequence of $\Delta(z)$ being diagonal. Let $\{e_1, \ldots, e_m\}$ be the standard basis in $\mathbb{F}^m$; then the vectors

$$\{z^i e_j \mid 1 \leq j \leq m, 0 \leq i < \kappa_j - 1\}$$

form a basis for X_Δ. Relative to these bases in $\mathbb{F}^m$ and X_Δ, the pair (S_Δ, π_Δ) has the matrix representation (6.13)–(6.14).

It is a trivial consequence of the Brunovsky canonical form that the reachability indices define a complete set of invariants for state feedback of reachable pairs. ■

For a reachable pair (A, B), the group

$$\mathrm{Stab}(A, B) = \left\{ \begin{pmatrix} S & 0 \\ K & R \end{pmatrix} \mid (S(A + BK)S^{-1}, SBR^{-1}) = (A, B) \right\} \subset GL_{n+m}(\mathbb{F})$$

of all elements of the feedback group that leave (A, B) invariant is called the **state feedback stabilizer group of** (A, B). Clearly, the stabilizers of feedback equivalent pairs are isomorphic. As a consequence, it suffices to study the stabilizer group for systems in Brunovsky canonical form. It follows that the structure of the stabilizer depends only on the reachability indices of the reachable pair (A, B). The relation between the state feedback stabilizer subgroup and the left factorization group introduced in Theorem 2.37 can be stated as follows.

Theorem 6.15. *Let (A, B) be a reachable pair, and let $N(z)D(z)^{-1}$ be a right coprime factorization of $(zI - A)^{-1}B$. Then the state feedback stabilizer group of (A, B) is isomorphic to the left factorization group of $D(z)$.*

Proof. The pair (A, B) is isomorphic to (S_D, π_D) and, in turn, state feedback equivalent to the polynomial Brunovsky form (S_Δ, π_Δ), with $\Delta(z) = \mathrm{diag}(z^{\kappa_1}, \ldots, z^{\kappa_m})$. It suffices, therefore, to study the state feedback stabilizer at (S_Δ, π_Δ). However, by Theorem 6.2, this is equivalent to finding all solutions of the equation

$$\Gamma(z)\Delta(z) = \Delta(z)U(z), \tag{6.15}$$

with $U(z) \in \mathbb{F}[z]^{m \times m}$ unimodular and $\Gamma(z) \in \mathbb{F}[[z^{-1}]]^{m \times m}$ biproper. Equation (6.15) is equivalent to $\gamma_{ij} z^{\kappa_j} = z^{\kappa_i} u_{ij}$, which in turn implies

$$\deg u_{ij} = \begin{cases} 0 & \kappa_i > \kappa_j \\ \leq \kappa_j - \kappa_i & \kappa_j \geq \kappa_i . \end{cases} \tag{6.16}$$

Conversely, if $U(z)$ is unimodular and satisfies (6.16), then it is easily seen that equation (6.15) is solvable with a biproper $\Gamma(z)$. Thus the unimodular matrices $U(z)$ of (6.15) have a block triangular structure. By Theorem 2.37, the set of such unimodular matrices $U(z)$ coincides with the left factorization group of $\Delta(z)$. This structure is reflected in $\Gamma(z)$, which is uniquely determined by $U(z)$ and $\Delta(z)$. This completes the proof. ■

From the preceding discussion it is clear that the existence of a Wiener–Hopf factorization of a nonsingular polynomial matrix is equivalent to the existence of Brunovsky's canonical form for a reachable pair. Next, in a purely state-space-oriented manner, a refinement of the Brunovsky canonical form is derived. Recall that the Kronecker indices of a state space pair $(A, B = (b_1, \ldots, b_m))$ are defined by the following deletion process on the columns of the reachability matrix. Let $\leq$ denote the lexicographical ordering on $\{0, \ldots, n-1\} \times \{1, \ldots, m\}$ defined as

$$(i, j) \leq (k, \ell) \iff \begin{cases} i < k \\ \text{or} \\ i = k, j \leq \ell. \end{cases}$$

While going from left to right in the list

$$\left(b_1, \ldots, b_m, Ab_1, \ldots, Ab_m, \ldots, A^{n-1}b_1, \ldots, A^{n-1}b_m\right)$$

of mn vectors in $\mathbb{F}^n$, delete all vectors $A_k b_\ell$ that are linearly dependent on the set of preceding vectors $\{A_i b_j | (i, j) \leq (k, \ell)\}$.

It is easily seen that the remaining vectors constitute a list of the form

$$(b_1, Ab_1, \ldots, A^{k_1-1}b_1, \ldots, b_m, \ldots, A^{k_m-1}b_m), \qquad (6.17)$$

for unique nonnegative integers $k_1, \ldots, k_m$, called the **Kronecker indices**. Note that the Kronecker indices define an m-tuple of integers and not a set of numbers. By construction, the vectors in (6.17) form a basis of the reachable set $\mathscr{R}$ of (A, B). Thus (A, B) is reachable if and only if the Kronecker indices satisfy $k_1 + \cdots + k_m = n$. An important difference that distinguishes the Kronecker indices $\mathbf{k} = (k_1, \ldots, k_m)$ from the reachability indices $\kappa = (\kappa_1, \ldots, \kappa_m)$ is that the Kronecker indices are not ordered by magnitude. Thus $(2, 0, 3)$ and $(3, 2, 0)$ are Kronecker indices of different systems (A, B). It is easily seen that $\mathbf{k} = (k_1, \ldots, k_m)$ are Kronecker indices of a system (A, B), with $k_i \geq 1$ for all i; then the reachability indices of (A, B) arise by reordering the Kronecker indices in decreasing form. However, this is not true if one of the Kronecker indices is zero.

Let $\mathscr{U}_m$ denote the subgroup of $GL_m(\mathbb{F})$ consisting of all $m \times m$ upper triangular matrices U with identical entries $u_{11} = \cdots = u_{mm} = 1$ on the diagonal. The **restricted state feedback group** is then defined by all state feedback transformations

$$\begin{pmatrix} S & 0 \\ K & U \end{pmatrix},$$

with $S \in GL_n(\mathbb{F})$, $K \in \mathbb{F}^{m \times n}$, and $U \in \mathscr{U}_m$. Two linear systems $(A, B), (\overline{A}, \overline{B})$ are called **restricted state feedback equivalent** if $(\overline{A}, \overline{B}) = (S(A - BK)S^{-1}, SBU^{-1})$ is satisfied for a restricted state feedback transformation (S, K, U). We proceed to show that the Kronecker indices are feedback invariants.

Lemma 6.16. *Let* $(A, B) \in \mathbb{F}^{n \times (n+m)}$ *be reachable with Kronecker indices* $\mathbf{k} = (k_1, \ldots, k_m)$. *For each* $S \in GL_n(\mathbb{F})$, $U \in \mathscr{U}_m$, *and* $K \in \mathbb{F}^{m \times n}$, *the pairs* (A, B) *and* $(\overline{A}, \overline{B}) = (S(A - BK)S^{-1}, SBU^{-1})$ *have the same Kronecker indices* $\mathbf{k} = (k_1, \ldots, k_m)$.

Proof. It is easily seen that the reachability matrix

$$R(A, B) = (B, \ldots, A^{n-1}B)$$

satisfies

$$R(\overline{A}, \overline{B}) = SR(A - BK, BU) = SR(A, B)V$$

for a suitable invertible upper triangular matrix $V \in GL_{nm}(\mathbb{F})$ with diagonal blocks $V_{11} = \cdots = V_{nn} = I_m$. This implies that the Kronecker indices of (A, B) and $(\overline{A}, \overline{B})$ coincide. This completes the proof. ∎

The following result will be needed.

Lemma 6.17. *Let* $(A, B) \in \mathbb{F}^{n \times (n+m)}$ *be reachable with Kronecker indices* $\mathbf{k} = (k_1, \ldots, k_m)$. *Then there exists a unipotent matrix* $U \in \mathscr{U}_m$ *such that* $\overline{B} = (\overline{b}_1, \ldots, \overline{b}_m) = BU$ *satisfies for each* $j = 1, \ldots, m$

$$A^{k_j}\overline{b}_j \in \operatorname{Im}B + \cdots + A^{k_j-1}\operatorname{Im}B,$$

$$A^{k_j}\overline{b}_j \notin \operatorname{Im}B + \cdots + A^{k_j-2}\operatorname{Im}B.$$

Proof. By construction of the Kronecker indices there exist $c_{ij} \in \mathbb{F}, i < j$, and $z_j \in \operatorname{Im}B + \cdots + A^{k_j-1}\operatorname{Im}B$ such that

$$A^{k_j}b_j = z_j + \sum_{i=1}^{j-1} c_{ij}A^{k_j}b_i, \tag{6.18}$$

$$A^{k_j-1}b_j \notin \operatorname{Im}B + \cdots + A^{k_j-2}\operatorname{Im}B + A^{k_j-1}\operatorname{span}\{b_1, \ldots, b_{j-1}\}$$

holds. Define $\overline{b}_j = b_j - \sum_{i=1}^{j-1} c_{ij}b_i$ and

$$U = \begin{pmatrix} 1 & -c_{12} & \cdots & -c_{1m} \\ & \ddots & \ddots & \vdots \\ & & \ddots & -c_{m-1,m} \\ & & & 1 \end{pmatrix} \in \mathscr{U}_m.$$

Then $A^{k_j}\bar{b}_j = z_j \in \mathrm{Im}\,B + \cdots + A^{k_j-1}\mathrm{Im}\,B$. Suppose

$$A^{k_j-1}\bar{b}_j \in \mathrm{Im}\,B + \cdots + A^{k_j-2}\mathrm{Im}\,B.$$

Then $A^{k_j-1}b_j = A^{k_j-1}\bar{b}_j + \sum_{i=1}^{j-1} c_{ij}A^{k_j-1}b_i$, in contradiction to (6.18). This completes the proof. ∎

Using the preceding lemmas, it will be shown next that the Kronecker indices define a complete set of invariants for the restricted state feedback equivalence of reachable pairs.

Theorem 6.18. *1. Let $(A,B) \in \mathbb{F}^{n\times(n+m)}$ be a reachable pair with Kronecker indices $\mathbf{k} = (k_1,\ldots,k_m)$. Then (A,B) is restricted state feedback equivalent to the block matrix*

$$\left(\begin{pmatrix} A_1 & \\ & \ddots & \\ & & A_m \end{pmatrix}, \begin{pmatrix} B_1 & \\ & \ddots & \\ & & B_m \end{pmatrix} \right) \tag{6.19}$$

$$A_j = \begin{pmatrix} 0 & & & \\ 1 & \ddots & & \\ & \ddots & \ddots & \\ & & 1 & 0 \end{pmatrix}, \quad B_j = \begin{pmatrix} 1 \\ 0 \\ \vdots \\ 0 \end{pmatrix} \tag{6.20}$$

if $k_j \geq 1$. If $k_j = 0$, then the block A_j is absent and $B_j = 0$.
2. Two reachable pairs $(A,B),(\bar{A},\bar{B}) \in \mathbb{F}^{n\times(n+m)}$ are restricted state feedback equivalent if and only if their Kronecker indices coincide.

Proof. Choose U as in Lemma 6.17. Then there exist elements $\beta_{1,j},\ldots,\beta_{k_j,j} \in \mathrm{Im}\,B$ such that

$$A^{k_j}\bar{b}_j - A^{k_j-1}\beta_{1,j} - \cdots - \beta_{k_j,j} = 0. \tag{6.21}$$

For $j = 1,\ldots,m$ define the state vectors

$$x_{1,j} = \bar{b}_j, \quad x_{2,j} = A\bar{b}_j - \beta_{1,j}, \quad \ldots, \quad x_{k_j,j} = A^{k_j-1}\bar{b}_j - A^{k_j-2}\beta_{1j} - \cdots - \beta_{k_j-1,j}.$$

Let $\mathscr{X} \subset \mathbb{F}^n$ denote the span of the vectors $\{x_{i,j} \mid 1 \leq i \leq k_j, j = 1,\ldots,m\}$. Clearly, $\mathrm{Im}\,B \subset \mathscr{X}$. Using (6.21), it follows that $Ax_{k_j,j} = \beta_{k_j,j} \in \mathrm{Im}\,B \subset \mathscr{X}$. Thus $\mathscr{X}$ is A-invariant. Thus the reachability of (A,B) implies that $\{x_{i,j} \mid 1 \leq i \leq k_j, j = 1,\ldots,m\}$ defines a basis of $\mathbb{F}^n$. Choose $u_{i,j} \in \mathbb{F}^m$ with $\bar{B}u_{i,j} = \beta_{i,j}$, $i = 1,\ldots,k_j$, $j = 1,\ldots,m$. Then the feedback transformation $K : \mathbb{F}^n \longrightarrow \mathbb{F}^m$ defined by

$$Kx_{i,j} = -u_{i,j}, \quad i = 1,\ldots,k_j, j = 1,\ldots,m,$$

satisfies, for each $j = 1, \ldots, m$,

$$(A - \overline{B}K)x_{i,j} = Ax_{i,j} - \beta_{i,j} = x_{i+1,j}, \quad 1 \leq i \leq k_j - 1,$$
$$(A - \overline{B}K)x_{K_j,j} = Ax_{K_j,j} - \beta_{k_j,j} = 0.$$

By choosing $S = (x_{1,1}, \ldots, x_{k_1,1}, \ldots, x_{1,m}, \ldots, x_{k_m,m})$, K, and U as above, one sees that S is invertible and $(S(A - BK)S^{-1}, SBU)$ has the form (6.19), (6.20). This completes the proof of the first part. The second follows easily from the first part, together with Lemma 6.16.   ∎

6.4 Pole Assignment

The study of the effects of state feedback on closed-loop dynamics begins with an analysis of the simple case of a single-input reachable system. It will be shown how, by the use of state feedback, the dynamics of the system, determined by its characteristic polynomial, can be arbitrarily assigned. This indicates the tremendous importance of feedback. In fact, as long as reachability is fulfilled, the original system can be flexibly modified by the use of feedback. In particular, every reachable system can be stabilized through feedback. The subsequent results are presented in an unashamedly matrix-oriented manner, beginning with the single-input case, where the analysis becomes particularly simple.

Theorem 6.19. *Let $(A,b) \in \mathbb{F}^{n \times n} \times \mathbb{F}^n$ be a reachable system with the n-dimensional state space $\mathbb{F}^n$. Let $f(z) = f_0 + \cdots + f_{n-1}z^{n-1} + z^n$ be a monic polynomial of degree n. Then there exists a unique feedback transformation $K \in \mathbb{F}^{1 \times n}$ such that $A - bK$ has $f(z)$ as its characteristic polynomial.*

Proof. Let $q(z) = q_0 + \cdots + q_{n-1}z^{n-1} + z^n$ denote the characteristic polynomial of A. Since (A,b) is reachable, the pair (A,b) is state space equivalent to the reachable shift realization (S_q, π_q) on X_q. Thus, without loss of generality, one can identify (A,b) with the pair (S_q, π_q) and, by the choice of basis in X_q, one can assume that the pair (A,b) has the **control canonical form**

$$A = \begin{pmatrix} 0 & 1 & & \\ & \ddots & \ddots & \\ & & 0 & 1 \\ -q_0 & -q_1 & \cdots & -q_{n-1} \end{pmatrix}, \quad b = \begin{pmatrix} 0 \\ \vdots \\ 0 \\ 1 \end{pmatrix}. \tag{6.22}$$

This can be done by taking a right coprime factorization $N(z)q(z)^{-1}$ of $(zI - A)^{-1}b$, with $q(z)$ the characteristic polynomial of A, and choosing in X_q the **control basis** $\mathscr{B}_{co} := \{\gamma_1(z), \ldots, \gamma_n(z)\}$, where, for $i = 0, \ldots, n$,

$$\gamma_i(z) = z^{n-i} + q_{n-1}z^{n-i-1} + \cdots + q_i.$$

A straightforward computation shows that the shift operator S_q acts on these basis vectors via

$$S_q(\gamma_i) = \gamma_{i-1}(z) - q_{i-1}.$$

Therefore, (A,b) in (6.22) is just the basis representation of (S_q, π_q) with respect to the control basis. Let $K = (k_0, \ldots, k_{n-1})$ be the feedback map, i.e.,

$$Kx = k_0 x_1 + \cdots + k_{n-1} x_n;$$

then

$$A - bK = \begin{pmatrix} 0 & 1 & & \\ & \ddots & \ddots & \\ & & 0 & 1 \\ -k_0 - q_0 & \cdots \cdots & -k_{n-1} - q_{n-1} \end{pmatrix}.$$

The unique choice $k_i = -q_i + f_i$ then yields $\det(zI - A + bK) = f(z)$. ∎

The following two results present explicit formulas for the feedback gain K.

Theorem 6.20 (Ackermann Formula). *Let* $(A,b) \in \mathbb{F}^{n \times n} \times \mathbb{F}^n$ *be reachable, and let* $f(z) = \sum_{j=0}^{n} f_j z^j, f_n = 1$, *be a monic polynomial of degree n. Let* $R(A,b) = (b, \ldots, A^{n-1}b) \in GL_n(\mathbb{F})$ *denote the reachability matrix. Then*

$$K = (0, \ldots, 0, 1) R(A,b)^{-1} f(A)$$

is the unique element $K \in \mathbb{F}^{1 \times n}$, *with* $\det(zI - A + bK) = f(z)$.

Proof. By Theorem 6.19, there exists a unique $K \in \mathbb{F}^{1 \times n}$ that satisfies $\det(zI - A + bK) = f(z)$. Applying the Cayley–Hamilton theorem, one obtains

$$\sum_{j=0}^{n} f_j (A - bK)^j = f(A - bK) = 0,$$

and therefore

$$f(A) = -\sum_{j=0}^{n} f_j \left((A - bK)^j - A^j \right).$$

There exist row vectors $k_{j,\ell} \in \mathbb{F}^{1 \times n}$, $k_{j,j-1} = -K$, with

$$(A - bK)^j - A^j = \sum_{\ell=0}^{j-1} A^\ell b k_{j,\ell}.$$

Thus

$$f(A) = -\sum_{j=0}^{n}\sum_{\ell=0}^{j-1} A^{\ell} b f_j k_{j,\ell} = -\sum_{i=0}^{n-1} A^{ib}\xi_i,$$

with $\xi_i = \sum_{j>i} f_j k_{j,i}$ and $\xi_{n-1} = f_n k_{n,n-1} = -K$. Defining the matrix

$$\xi = \begin{pmatrix} \xi_0 \\ \vdots \\ \xi_{n-1} \end{pmatrix} \in \mathbb{F}^{n\times n}$$

we obtain

$$f(A) = -R(A,b)\xi,$$

and hence $K = -e_n^{\top}\xi = e_n^{\top} R(A,b)^{-1} f(A)$.   ∎

We now turn to the analysis of state feedback in the general case $m \geq 1$.

Lemma 6.21. *Assume (A,B) is reachable and $b = Bv \neq 0$. Then there exist $u_0 = v, u_1, \ldots, u_{n-1} \in \mathbb{F}^m$ such that $(x_1, \ldots, x_n)$, recursively defined as*

$$x_1 = b, \quad x_k = Ax_{k-1} + Bu_k, \quad k = 2, \ldots, n,$$

is a basis of $\mathbb{F}^n$.

Proof. One constructs the input vectors u_k recursively, starting from $u_0 = v$. Suppose that $u_1, \ldots, u_{k-1}$ are such that $x_1, \ldots, x_k$ are linearly independent, satisfying ($x_0 := 0$) $x_j = Ax_{j-1} + Bu_{j-1}$ for $j = 1, \ldots, k$ and $k < n$. Let $L \subset \mathbb{F}^n$ denote the k-dimensional linear subspace spanned by $x_1, \ldots, x_k$. Then one chooses $u_k \in \mathbb{F}^m$ such that $x_{k+1} := Ax_k + Bu_k \notin L$. Such a vector u_k always exists, thereby proving the induction step that $\{x_1, \ldots, x_{k+1}\}$ is linearly independent. In fact, otherwise

$$Ax_k + Bu \in L$$

is true for all $u \in \mathbb{F}^m$. This implies $Ax_k \in L$, and therefore also $\operatorname{Im}B \subset L$, and, in turn, $Ax_j = x_{j+1} - Bu_j \in L$ for $j = 1, \ldots, k-1$. This shows that L is an A-invariant linear subspace that contains $\operatorname{Im}B$. The reachability of (A,B) thus implies $L = \mathbb{F}^n$, in contradiction to $\dim L = k < n$.   ∎

The preceding result has an interesting consequence for state feedback control.

Lemma 6.22 (Heymann). *Let $(A,B) \in \mathbb{F}^{n\times n} \times \mathbb{F}^{n\times m}$ and $b = Bv \neq 0$. Then there exists $K \in \mathbb{F}^{m\times n}$ such that $(A + BK, b)$ is reachable. In particular, for each reachable pair (A,B) there exists a feedback gain $K \in \mathbb{F}^{m\times n}$ such that $A + BK$ is cyclic.*

Proof. Choose $u_0, \ldots, u_{n-1} \in \mathbb{F}^m$ and basis vectors $x_1, \ldots, x_n$ of $\mathbb{F}^n$, as in Lemma 6.21. For each element $u_n \in \mathbb{F}^m$ there exists a unique $K \in \mathbb{F}^{m \times n}$, with

$$Kx_j = u_j, \quad j = 1, \ldots, n.$$

This shows that

$$(A + BK)x_j = Ax_j + Bu_j = x_{j+1}$$

for $j = 1, \ldots, n-1$. Since $b = x_1$, we obtain

$$(A + BK)^{jb} = x_{j+1}$$

for $j = 1, \ldots, n-1$. Thus $(b, (A + BK)b, \ldots, (A + BK)^{n-1}b)$ is a basis of $\mathbb{F}^n$, completing the proof. ∎

It is easy to see that, for nonzero $b \in \operatorname{Im} B$, the set of all such feedback gains K forms a Zariski-open subset of $\mathbb{F}^{m \times n}$. The celebrated pole-shifting theorem of M. Wonham is proved next. The reason for the name is due to the fact that poles of the rational function $(zI - A + BK)^{-1}B$ correspond to the eigenvalues of $A - BK$.

Theorem 6.23 (Pole-Shifting Theorem). *A linear system* $(A, B) \in \mathbb{F}^{n \times n} \times \mathbb{F}^{n \times m}$ *is reachable if and only if for every monic polynomial* $f(z) \in \mathbb{F}[z]$ *of degree n there exists* $K \in \mathbb{F}^{m \times n}$, *with*

$$\det(zI - A + BK) = f(z). \tag{6.23}$$

Proof. Suppose (A, B) is reachable. Choose a nonzero vector $b = Bv$ in the image space of B. By Heymann's Lemma 6.22, there exists $F \in \mathbb{F}^{m \times n}$ such that $(A + BF, b)$ is reachable. Thus, using Theorem 6.19, there exists a row vector $L \in \mathbb{F}^{1 \times n}$ such that

$$\det(zI - A - BF + bL) = f(z).$$

This proves (6.23) for $K = -F + vL$.

To prove the converse, the Kalman decomposition is used. Thus, assume that (A, B) is a system with k-dimensional reachable subspace

$$\mathscr{R} = \operatorname{Im} B + A \operatorname{Im} B + \cdots + A^{n-1} \operatorname{Im} B.$$

Choose the basis vectors $v_1, \ldots, v_k$ of $\mathscr{R}$ and extend them to a basis $v_1, \ldots, v_n$ of the state space $\mathbb{F}^n$. Then the matrix $S = (v_1, \ldots, v_n) \in \mathbb{F}^{n \times n}$ is invertible. Since $\mathscr{R}$ is an A-invariant linear subspace, this implies that the state-space equivalent system $(S^{-1}AS, S^{-1}B)$ has the structure

$$\begin{pmatrix} A_1 & A_2 \\ 0 & A_3 \end{pmatrix}, \quad \begin{pmatrix} B_1 \\ 0 \end{pmatrix}, \tag{6.24}$$

which is referred to as the **Kalman decomposition**; it exists for every $(A,B) \in$ $\mathbb{F}^{n \times (n+m)}$. Note that (A_1, B_1) is uniquely determined up to a similarity transformation and is reachable. Moreover, the eigenvalues of A_3 are uniquely determined by the similarity orbit of (A,B). Thus, for the converse, one can assume, without loss of generality, that (A,B) is given by (6.24), with $k < n$. Hence, for each feedback matrix $K = (K_1, K_2)$,

$$A - BK = \begin{pmatrix} A_1 - B_1 K_1 & A_2 - B_1 K_2 \\ 0 & A_3 \end{pmatrix},$$

with the characteristic polynomial $\det(zI - A + BK) = \det(zI - A_1 + B_1 K_1)$ $\det(zI - A_3)$. This implies that the characteristic polynomials $\det(zI - A + BK)$ of nonreachable pairs (A,B) all contain the same factor $\det(zI - A_3)$ and thus cannot be arbitrarily assigned. This completes the proof. ∎

There is a simple, inductive proof of Wonham's theorem that works over an algebraically closed field $\mathbb{F} = \overline{\mathbb{F}}$. We learned the following argument from Carsten Scherer. Without loss of generality, assume that $\operatorname{rk} B = m$ and (A,B) is of the form

$$A = \begin{pmatrix} A_{11} & A_{12} \\ A_{21} & A_{22} \end{pmatrix}, \quad B = \begin{pmatrix} I_m \\ 0 \end{pmatrix}.$$

Then, using the Hautus test, one sees that the reachability of (A,B) implies that of (A_{22}, A_{21}). Consider a monic polynomial $f(z) = f_1(z)f_2(z)$, with monic factors $f_1(z), f_2(z)$ of degrees $m, n - m$, respectively. Applying the induction hypothesis, there exists $K_2 \in \mathbb{F}^{m \times (n-m)}$ such that

$$\det(zI_{n-m} - A_{22} + A_{21}K_2) = f_2(z).$$

Let C be a matrix with the characteristic polynomial $f_1(z)$. With

$$S = \begin{pmatrix} I & K_2 \\ 0 & I \end{pmatrix}$$

and a suitable matrix X, one obtains

$$SAS^{-1} = \begin{pmatrix} A_{11} + K_2 A_{21} & X \\ A_{21} & A_{22} - A_{21}K_2 \end{pmatrix}, \quad SB = B = \begin{pmatrix} I \\ 0 \end{pmatrix}.$$

Therefore, it follows that, with $F = (C - A_{11} - K_2 A_{21}, -X)$, one obtains

$$SAS^{-1} + SBF = \begin{pmatrix} C & 0 \\ A_{21} & A_{22} - A_{21}K_2 \end{pmatrix},$$

which has the characteristic polynomial $f_1(z)f_2(z)$. This completes the proof.

6.5 Rosenbrock's Theorem

We turn now to the question of finding the extent to which the dynamics of a system can be modified by state feedback. Of course, as far as eigenvalue assignment of $A - BK$ is concerned, it would suffice to know whether the eigenvalues of $A - BK$ could be freely assigned. A deeper question concerns the ability to alter the entire eigenstructure of $A - BK$, i.e., the Jordan canonical form. Rosenbrock showed, in a subtle analysis, that the invariant factors of $A - BK$ can be freely assigned subject only to a finite set of constraints arising from the reachability indices. Our aim in this section is to prove this fundamental result. In view of Theorem 6.14, the only invariants of a reachable pair (A, B) under the action of a feedback group are the reachability indices. On the other hand, the invariant factors of a nonsingular polynomial matrix $D(z)$ are invariant under left and right multiplication by unimodular polynomial matrices. Now if $N(z)D(z)^{-1}$ is a right coprime factorization of the input to state transfer function $(zI - A)^{-1}B$, then the column indices of $D(z)$, which are the minimal indices of the submodule $M = D(z)\mathbb{F}[z]^m$, are equal to the reachability indices of the pair (A, B). Thus it suffices to see how, starting with a polynomial matrix $\mathrm{diag}\,(\psi_1, \ldots, \psi_m)$, where the ψ_i satisfy $\psi_i | \psi_{i-1}$ for $i = 2, \ldots, m$, that the minimal indices of M can be changed by left and right multiplication by unimodular matrices. Our starting point is the following.

Lemma 6.24. *Let $D(z) = (d_1(z), \ldots, d_m(z)) \in \mathbb{F}[z]^{m \times m}$ be a nonsingular, column proper polynomial matrix with its columns $d_1(z), \ldots, d_m(z)$ of degrees $\lambda_1 \geq \cdots \geq \lambda_m$. Assume, without loss of generality, that the highest column coefficient matrix is $[D]_{hc} = I_m$. If $1 \leq j, k \leq m$, with $\deg d_j < \deg d_k$, then there exist elementary row and column operations transforming $D(z)$ into $D'(z) = (d_1'(z), \ldots, d_m'(z))$, with*

$$
\deg d_i' = \begin{cases} \deg d_i & i \neq j, k, \\ \deg d_j + 1 & i = j, \\ \deg d_k - 1 & i = k, \end{cases}
$$

and $[D']_{hc} = I_m$.

Proof. Adding the jth row, multiplied by z, to the kth row of $D(z)$, one gets a matrix $D^{(1)}(z)$ with columns $d_i^{(1)}(z)$, with

$$
\deg d_i^{(1)} \begin{cases} = \deg d_i, & i \neq j, k, \\ = \deg d_j + 1, & i = j, \\ \leq \deg d_k, & i = k. \end{cases}
$$

Next, add a suitable multiple of the jth column to the kth column to obtain a $D^{(2)}(z)$, with

$$\deg d_i^{(2)} \begin{cases} = \deg d_i, & i \neq j,k, \\ = \deg d_j + 1, & i = j, \\ \leq \deg d_k, & i = k. \end{cases}$$

Since $\det D^{(2)}(z) = \det D(z)$, one necessarily has $\deg d_k^{(2)} = \deg d_k - 1$, and the highest column coefficient matrix satisfies $\det[D^{(2)}]_{hc} \neq 0$. Thus the matrix $D'(z) = [D^{(2)}]_{hc}^{-1} D^{(2)}(z)$ has the required properties.   ∎

As an example of the process, taken from Rosenbrock (1970), consider the nonsingular, column proper polynomial matrix

$$D(z) = \begin{pmatrix} z^2+2 & z^3 & z^5+z+1 \\ 2z+1 & z^4+3z+1 & 2z^2 \\ z+2 & 2z^2+1 & z^6-2z^4 \end{pmatrix}.$$

The column indices are $2,4,6$, and we will reduce the degree of the last column and increase the degree of the first. The successive stages are

$$D^{(1)}(z) = \begin{pmatrix} z^2+2 & z^3 & z^5+z+1 \\ 2z+1 & z^4+3z+1 & 2z^2 \\ z^3+3z+2 & z^4+2z^2+1 & 2z^6-2z^4+z^2+z \end{pmatrix},$$

$$D^{(2)}(z) = \begin{pmatrix} z^2+2 & z^3 & -z^5-4z^3+z+1 \\ 2z+1 & z^4+3z+1 & -4z^4-2z^3+2z^2 \\ z^3+3z+2 & z^4+2z^2+1 & -8z^4-4z^3+z^2+z \end{pmatrix},$$

$$[D^{(2)}]_{hc} = \begin{pmatrix} 0 & 0 & -1 \\ 0 & 1 & 0 \\ 1 & 1 & 0 \end{pmatrix}, \quad [D^{(2)}]_{hc}^{-1} = \begin{pmatrix} 0 & -1 & 1 \\ 0 & 1 & 0 \\ -1 & 0 & 0 \end{pmatrix},$$

and, finally,

$$D'(z) = \begin{pmatrix} z^3+z+1 & 2z^2-3z & -4z^4-2z^3-z^2+z \\ 2z+1 & z^4+3z+1 & -4z^4-2z^3+2z^2 \\ -z^2-2 & -z^3 & z^5+4z^3-z-1 \end{pmatrix}.$$

Proposition 6.25. *Let $M \subset \mathbb{F}[z]^m$ be a full submodule with minimal indices $\lambda_1 \geq \cdots \geq \lambda_m$, and let $\kappa_1 \geq \cdots \geq \kappa_m$ be a sequence of nonnegative integers. If the conditions*

$$\begin{aligned} \Sigma_{i=1}^j \lambda_i &\geq \Sigma_{i=1}^j \kappa_i, \qquad j = 1, \ldots, m-1, \\ \Sigma_{i=1}^m \lambda_i &= \Sigma_{i=1}^m \kappa_i, \end{aligned}$$

are satisfied, then there exists a submodule $N \subset \mathbb{F}[z]^m$, unimodularly equivalent to M, with minimal indices $\kappa_1 \geq \cdots \geq \kappa_m$.

Proof. The proof is by a purely combinatorial argument on partitions, applying Lemma 6.24. Recall that a partition of n is a decreasing sequence of integers $\kappa_1 \leq \cdots \leq \kappa_m$, with $\kappa_1 + \cdots + \kappa_m = n$. Define a partial order, the so-called **dominance order**, on partitions $\kappa = (\kappa_1, \ldots, \kappa_m)$ and $\lambda = (\lambda_1, \ldots, \lambda_m)$ of $\sum_{i=1}^m \kappa_i = n = \sum_{i=1}^m \lambda_i$ as

$$\kappa \preceq \lambda \iff \sum_{i=1}^j \kappa_i \leq \sum_{i=1}^j \lambda_i, \quad j = 1, \ldots, m-1.$$

A partition $\lambda \neq \kappa$ is called a cover of κ whenever λ is the smallest element in the dominance order that satisfies $\kappa \preceq \lambda$. The covers for the dominance order are characterized as follows. There exists $j < k$, with

$$\lambda_i = \begin{cases} \kappa_i & i, \neq j, k, \\ \kappa_j + 1, & i = j, \\ \kappa_k - 1, & i = k. \end{cases}$$

It is a simple and well-known combinatorial exercise to verify that two arbitrary partitions $\kappa \preceq \lambda$ are connected through a chain of covers, i.e.,

$$\kappa = \lambda^{(1)} \preceq \cdots \preceq \lambda^{(k)} = \lambda,$$

where $\lambda^{(i)}$ is a cover of $\lambda^{(i-1)}$, $i = 2, \ldots, k$. From this and Lemma 6.24 the result follows, as every product of elementary row and column operations is achieved by multiplying with appropriate unimodular matrices. ∎

To prove Rosenbrock's theorem, one can start from the coprime factorization

$$(zI - A)^{-1} B = N(z)D(z)^{-1}$$

and try to modify the invariant factors of $D(z)$, keeping the reachability indices invariant. This is a difficult process, though conceptually more natural. Much easier, at the cost of being somewhat indirect, is to start from a polynomial matrix with the required invariant factors and modify the reachability indices, without changing the invariant factors.

Theorem 6.26 (Rosenbrock). *Let $(A, B) \in \mathbb{F}^{n \times n} \times \mathbb{F}^{n \times m}$ be a reachable pair with reachability indices $\kappa_1 \geq \cdots \geq \kappa_m$. Let $\psi_i \in \mathbb{F}[z]$ be such that $\psi_{i+1} \mid \psi_i$ for $i = 1, \ldots, m-1$. Then a necessary and sufficient condition for the existence of a matrix K such that the invariant factors of $A - BK$ are $\psi_1, \ldots, \psi_m$ is that*

$$\sum_{i=1}^{d} \deg \psi_i \geq \sum_{i=1}^{d} \kappa_i, \qquad d = 1, \ldots, m-1,$$

$$(6.25)$$

$$\sum_{i=1}^{m} \deg \psi_i = \sum_{i=1}^{m} \kappa_i.$$

Proof. Let $(zI - A)^{-1}B = N(z)D(z)^{-1}$ be a right coprime factorization such that $D(z)$ is column proper. Then $D(z)$ is a minimal-basis matrix for the full submodule $D(z)\mathbb{F}[z]^m \subset \mathbb{F}[z]^m$ with minimal indices $\kappa_1 \geq \cdots \geq \kappa_m$ and Smith form $\mathrm{diag}\,(\psi_1, \ldots, \psi_m)$. Let g_k denote the degree of the g.c.d. $\delta_k(D)$ of all $k \times k$ minors of $D(z)$. Since $D(z)$ is a minimal basis, a $k \times k$ principal minor of $D(z)$ has degree $\kappa_m + \cdots + \kappa_{m-k+1}$, and therefore $g_k \leq \kappa_m + \cdots + \kappa_{m-k+1}$ for $k = 1, \ldots, m$. Similarly, since $\delta_k(D) = \psi_m \cdots \psi_{m-k+1}$, we obtain

$$g_k = \sum_{i=m-k+1}^{m} \deg \psi_i \leq \sum_{i=m-k+1}^{m} \kappa_i,$$

$$\sum_{i=1}^{m} \deg \psi_i = \sum_{i=1}^{m} \kappa_i,$$

which is equivalent to (6.25). This shows necessity.

Conversely, assume that conditions (6.25) are in force. By Proposition 6.25, the submodule M with the minimal-basis matrix $D_\psi := \mathrm{diag}\,(\psi_1, \ldots, \psi_m)$ is unimodularly equivalent to a full submodule $D(z)\mathbb{F}^m[z]$ with indices $\kappa_1 \geq \cdots \geq \kappa_m$. Thus there exist unimodular polynomial matrices $U(z), V(z)$ with $D(z) = U(z)D_\psi V(z)$, and $D(z)$ has invariant factors $\psi_1, \ldots, \psi_m$ and minimal indices $\kappa_1 \geq \cdots \geq \kappa_m$. Consider the shift realization (S_D, π_D). By Theorem 6.8, the reachability indices of (S_D, π_D) are $\kappa_1 \geq \cdots \geq \kappa_m$, and the invariant factors of S_D are equal to the Smith invariants of $D(z)$, i.e., they are $\psi_1, \ldots, \psi_m$. Now consider reachable pairs (A, B) with reachability indices $\kappa_1 \geq \cdots \geq \kappa_m$. By Theorem 6.1, the pair (A, B) is feedback equivalent to the pair (S_D, π_D), where S_D has invariant factors $\psi_1, \ldots, \psi_m$. This completes the proof. ∎

6.6   Stabilizability

Recall that a discrete-time linear dynamical system

$$x_{t+1} = Ax_t \tag{6.26}$$

on $\mathbb{F}^n$ is called **stable** whenever the sequence $x_t = A_t x_0$ converges to zero for all initial conditions $x_0 \in \mathbb{F}^n$. Similarly, a linear control system

$$x_{t+1} = Ax_t + Bu_t \tag{6.27}$$

in the state space $\mathbb{F}^n$ is called **open-loop stabilizable** if, for all initial conditions $x_0 \in \mathbb{F}^n$, there exists an input sequence $u_t \in \mathbb{F}^m$, with

$$\lim_{t \to \infty} x_t = 0.$$

Of course, these notions require specifying a topology on $\mathbb{F}^n$, and there are various ways to do that. This issue came up already in Chapter 5 in the discussion of the stability of linear systems. As in Chapter 5, we consider only two possibilities, depending on whether or not $\mathbb{F}$ is a subfield of the complex number field $\mathbb{C}$:

1. The Euclidean distance topology on subfields $\mathbb{F} \subset \mathbb{C}$;
2. The discrete topology on any other field $\mathbb{F}$.

Recall that **the discrete topology** is a unique topology on $\mathbb{F}$ whose open (and closed) subsets are subsets of $\mathbb{F}$. Thus every **finite field** is compact with respect to this topology. Moreover, if $\mathbb{F}$ is endowed with the discrete topology, then the stability of (6.26) means that the trajectories of (6.26) eventually become constant, i.e., $x_{t+T} = x_T, t \geq 0$, for a sufficiently large $T \in \mathbb{N}$. Equivalently, A is nilpotent. In contrast, if $\mathbb{F} \subset \mathbb{C}$ is endowed with the Euclidean topology, then the asymptotic stability of (6.26) is satisfied if and only if all eigenvalues λ of A have absolute value $|\lambda| < 1$. In this case, one says that A is **Schur stable**. The stability properties of linear systems (6.26) are thus summarized as follows.

Proposition 6.27. *Let $\mathbb{F}$ denote a field. A discrete-time dynamical system (6.26) is asymptotically stable if and only if*

1. *A is Schur stable whenever $\mathbb{F} \subset \mathbb{C}$;*
2. *A is nilpotent for the discrete topology on $\mathbb{F}$.*

For the remaining parts of this section, let us assume that $\mathbb{F} \subset \mathbb{C}$ is satisfied, so that one is dealing with the standard notion of stability. The standard **stability domain** for the discrete-time system (6.26) is the open unit disc in the complex plane

$$\mathbb{D} := \{z \mid |z| < 1\}.$$

In contrast, for continuous-time systems $\dot{x} = Ax$ it will be the open left half-plane $\mathbb{C}_- = \{z \mid \mathrm{Re}(z) < 0\}$. In more generality, one may consider a subset Λ of the complex field $\mathbb{C}$ and refer to it as a region of stability.

Definition 6.28. Let $\mathbb{F} \subset \mathbb{C}$ be a subfield.

1. A nonsingular polynomial matrix $T(z) \in \mathbb{F}[z]^{r \times r}$ will be called Λ-**stable**, with respect to a region of stability Λ, if $\det T(z)$ has all its zeros in Λ. If $\Lambda = \mathbb{D}$, then the polynomial matrix $T(z)$ is called **stable**.
2. The pair $(A, B) \in \mathbb{F}^{n \times n} \times \mathbb{F}^{n \times m}$ is called Λ-**feedback stabilizable** if there exists a state feedback gain $K \in \mathbb{F}^{m \times n}$ such that $A - BK$ has all eigenvalues in Λ.

For discrete-time systems (6.27) and $\Lambda = \mathbb{D}$, one refers simply to **feedback stabilizable** rather than to $\mathbb{D}$-feedback stabilizable.

Next, it is shown that the notions of feedback stabilizability and open-loop stabilizability are equivalent.

Proposition 6.29. *Let $\mathbb{F} \subset \mathbb{C}$ be a subfield.*

1. *A linear system (6.27) is open-loop stabilizable if and only if it is feedback stabilizable.*
2. *Let $\mathbb{F} = \mathbb{C}$. A linear system (6.27) is reachable if and only if it is Λ-feedback stabilizable for all nonempty subsets $\Lambda \subset \mathbb{C}$.*

Proof. Clearly, feedback stabilizability implies stabilizability. Suppose (A, B) is stabilizable. As a result of the pole-shifting theorem, it follows that the reachability of (A, B) is sufficient for feedback stabilizability. If (A, B) is not reachable, then, after applying a suitable similarity transformation $(A, B) \mapsto (SAS^{-1}, SB)$ by an invertible matrix $S \in GL_n(\mathbb{F})$, one can assume without loss of generality that (A, B) is in the Kalman decomposition form

$$A = \begin{pmatrix} A_{11} & A_{12} \\ 0 & A_{22} \end{pmatrix}, \quad B = \begin{pmatrix} B_1 \\ 0 \end{pmatrix},$$

where (A_{11}, B_1) is reachable. By the pole-shifting theorem, there exists a feedback matrix $K = (K_1, K_2) \in \mathbb{F}^{m \times n}$ such that $A_{11} - B_1 K_1$ is stable. Since (A, B) is stabilizable, the matrix A_{22} must be stable. This implies that $A - BK$ is stable, i.e., (A, B) is feedback stabilizable.

Again, using the pole-shifting theorem, the eigenvalues of $A - BK$ for reachable pairs (A, B) can be placed arbitrarily in the complex plane. If (A, B) is not reachable, then the unreachable modes, i.e., the eigenvalues of A_{22} in the Kalman decomposition, are in the spectrum of $A - BK$ for every feedback matrix K. This proves the converse. ∎

The same argument as in the preceding proof shows that systems (6.27) over the field of real numbers $\mathbb{R}$ are reachable if and only if they are Λ-feedback stabilizable for all nonempty self-conjugate subsets $\Lambda \subset \mathbb{C}$. However, for other subfields of $\mathbb{C}$ such as, for example, the algebraic number fields, such simple characterizations cannot be expected.

A polynomial characterization of stabilizability is our next objective.

Theorem 6.30. *Let $\mathbb{F} \subset \mathbb{C}$ be a subfield, and let $\Lambda \subset \mathbb{C}$ be a nonempty subset. Let $G(z) = V(z)T(z)^{-1}U(z) + W(z) \in \mathbb{F}(z)^{p \times m}$ be proper rational, and let (A, B, C, D) be the associated shift realization (4.20) defined over $\mathbb{F}$. Then the following conditions are equivalent:*

1. *The shift realization is Λ-feedback stabilizable.*
2. *The g.c.l.d. $E(z)$ of $T(z)$ and $U(z)$ is stable.*
3. *$\left(T(z) \ \ U(z) \right)$ has full row rank for every $z \notin \Lambda$.*
4. *$\left(zI - A \ \ B \right)$ has full row rank for every $z \notin \Lambda$.*

Proof. By Theorem 4.26, the reachability subspace of the shift realization is $\mathscr{R} = EX_{T'}$. Let $\mathscr{F}$ be a complementary subspace to $EX_{T'}$ in X_T, i.e., $X_T = EX_{T'} + \mathscr{F}$ and $\mathscr{F} \cap EX_{T'} = \{0\}$. Then $\mathscr{F} \simeq X_T/EX_{T'}$, and this in turn is isomorphic to X_E. To see this, consider the projection map $\pi_E : X_T \longrightarrow X_E$. From the intertwining relation $T(z) = E(z)T'(z)$ it follows, using Theorem 3.21, that this map is surjective and its kernel is $EX_{T'}$. Thus the isomorphism is proved. In terms of this direct sum, the pair (A, B) has the Kalman decomposition

$$\begin{pmatrix} A_{11} & A_{12} \\ 0 & A_{22} \end{pmatrix}, \quad \begin{pmatrix} B_1 \\ 0 \end{pmatrix},$$

with (A_{11}, B_1) reachable and $A_{22} \simeq S_E$. Thus the pair (A, B) is feedback stabilizable if and only if A_{22} is stable or, equivalently, if and only if $E(z)$ is a stable polynomial matrix. This proves the equivalence of the first two conditions. The last two conditions are equivalent to the assertion that the g.c.l.d. $E(z)$ is invertible for all $z \notin \Lambda$, i.e., to $\det E(z)$ being Λ-stable. In turn, this is equivalent to the matrix $(zI - A, B)$ being full row rank for all $z \notin \Lambda$. This completes the proof. ∎

A different way of stating this for the field $\mathbb{F} = \mathbb{C}$ is as follows. For $\Lambda \subset \mathbb{C}$ and $A \in \mathbb{C}^{n \times n}$ we let

$$X_\Lambda(A) = \bigoplus_{\lambda \in \Lambda} \operatorname{Ker}(\lambda I - A)^n$$

denote the direct sum of the generalized eigenspaces of A with respect to the eigenvalues $\lambda \in \Lambda$. If Λ_- is a stability region and Λ_+ is its complement in $\mathbb{C}$, then every polynomial $p(z)$ has a factorization $p(z) = p_+(z)p_-(z)$, with $p_-(z)$ stable and $p_+(z)$ antistable. Clearly,

$$X_{\Lambda_+}(A) = \operatorname{Ker} d_+(A).$$

For each stability region Λ_- and its complement Λ_+ we will also write

$$X_\pm = X_{\Lambda_\pm}(A).$$

The preceding result can also be stated in a state-space representation.

Theorem 6.31. *A pair $(A, B) \in \mathbb{C}^{n \times n} \times \mathbb{C}^{n \times m}$ is Λ_--stabilizable if and only if one of the following equivalent conditions is satisfied:*

1. *$X_+(A)$ is contained in the reachable set of (A, B).*
2. *$\operatorname{rk}(zI - A, B) = n$ is valid for all $z \in \Lambda_+$.*

Proof. The first condition is often expressed by saying that the unstable modes are reachable. The equivalence of stabilizability with the first condition thus follows from the Kalman decomposition of (A, B). The equivalence of stabilizability with the second condition follows at once from Theorem 6.30. ∎

This construction can be extended to a field $\mathbb{F}$ as follows. Let $\mathbb{F}_-[z] \subset \mathbb{F}[z]$ denote a multiplicatively closed subset of nonzero polynomials such that, with an element $f(z) \in \mathbb{F}_-[z]$, all its prime factors are also contained in $\mathbb{F}_-[z]$, and further assume that $1 \in \mathbb{F}_-[z]$. Such subsets are called **saturated**. Denote by $\mathbb{F}_+[z]$ the set of all polynomials that are coprime with all elements of $\mathbb{F}_-[z]$. Elements of $\mathbb{F}_-[z]$ will be called **stable** polynomials and those of $\mathbb{F}_+[z]$ **antistable**. It is a consequence of the primary decomposition theorem that every polynomial $p(z)$ has a factorization $p(z) = p_-(z)p_+(z)$, with $p_-(z)$ stable and $p_+(z) \in \mathbb{F}_+[z]$. As an example, take $\mathbb{F} = \mathbb{R}$ and $\mathbb{R}_-[z]$ the set of all monic Schur polynomials of arbitrary degree (including 1). Alternatively, one considers the set of all monic real Hurwitz polynomials, including 1. As another example, one may consider $\mathbb{F}_-[z] := \{z^n \mid n \in \mathbb{N}_0\}$.

Let $A : \mathbb{F}^n \longrightarrow \mathbb{F}^n$ be a linear transformation and $d(z) = \det(zI - A)$ its characteristic polynomial, and let $d(z) = d_-(z)d_+(z)$ be its factorization into stable and antistable factors. In Chapter 3, it was shown that such a factorization induces essentially unique factorizations

$$zI - A = \overline{S}_+(z)S_-(z) = \overline{S}_-(z)S_+(z),$$

with $\overline{S}_-(z), S_-(z)$ stable and $\overline{S}_+(z), S_+(z)$ antistable. This leads to the spectral decomposition

$$\mathbb{F}^n = X_{zI-A} = \overline{S}_+(z)X_{S_-} \oplus \overline{S}_-(z)X_{S_+} = X_-(A) \oplus X_+(A),$$

where the subspaces $X_-(A), X_+(A)$ are the generalized eigenspaces associated with the sets of stable and antistable eigenvalues, respectively. With these constructions in our hands, Theorem 6.31 generalizes as follows; the proof is by a straightforward modification of the arguments for Theorem 6.31 and is omitted.

Theorem 6.32. *Let $\mathbb{F}_-[z]$ be a saturated subset of nonzero polynomials. For $(A, B) \in \mathbb{F}^{n \times n} \times \mathbb{F}^{n \times m}$, the following conditions are equivalent:*

1. *There exists $K \in \mathbb{F}^{m \times n}$ with $\det(zI - A + BK) \in \mathbb{F}_-[z]$.*
2. *$X_+(A)$ is contained in the reachable set (A, B).*
3. *$\operatorname{rank}(zI - A \ \ B) = n$ is satisfied for all roots $z \in \overline{\mathbb{F}}$ of all irreducible polynomials in $\mathbb{F}_+[z]$.*

The dual notion to open-loop stabilizability is that of detectability. While we will introduce this concept in Chapter 7 in a larger context, here we confine ourselves to a more specialized situation.

Definition 6.33. The system

$$x_{t+1} = Ax_t + Bu_t$$
$$y_t = Cx_t \tag{6.28}$$

is called **detectable**, provided all pairs of trajectories $(x_t), (\bar{x}_t)$ with the same input sequence (u_t) and identical output sequences $(Cx_t) = (C\bar{x}_t)$ satisfy

$$\lim_{t \to \infty} (x_t - \bar{x}_t) = 0.$$

Using linearity, it is easily seen that the detectability of (6.28) is satisfied if and only if all state trajectories (x_t) of the input-free or, equivalently, **autonomous** system

$$x_{t+1} = Ax_t,$$
$$y_t = Cx_t,$$

(6.29)

with $Cx_t = 0$ for $t \geq 0$, satisfy

$$\lim_{t \to \infty} x_t = 0.$$

Thus (6.28) is detectable if and only if (6.29) is detectable. The system-theoretic interpretation of detectability is clarified by the following result.

Proposition 6.34. *Let $\mathbb{F} \subset \mathbb{C}$. The following conditions are equivalent:*

1. *System (6.28) is detectable.*
2. *The unobservable states*

$$\mathscr{O}_* = \bigcap_{i=0}^{n-1} Ker\, CA^{i-1}$$

   satisfy

$$\mathscr{O}_* \subset \{x_0 \in \mathbb{F}^n \mid \lim_{t \to \infty} x_t = 0\}.$$

3. *The unobservable modes $\lambda \in \mathbb{C}$ of (C,A) are all stable, i.e., satisfy $|\lambda| < 1$.*
4. *The dual system $(A^\top, C^\top)$ is stabilizable.*
5. *There exists an output injection transformation $L \in \mathbb{F}^{n \times p}$ such that $A - LC$ is Schur stable.*

Proof. The equivalence of statements (1) and (2) follows directly from the definition. Equivalently, the dual Kalman decomposition of (C,A) is of the form

$$SAS^{-1} = \begin{pmatrix} A_{11} & 0 \\ A_{21} & A_{22} \end{pmatrix}, \quad CS^{-1} = (C_1 \; 0),$$

with (C_1, A_{11}) observable and A_{22} stable. Equivalently, the Kalman decomposition of $(A^\top, C^\top)$ is

$$\begin{pmatrix} A_{11}^\top & A_{21}^\top \\ 0 & A_{22}^\top \end{pmatrix}, \quad \begin{pmatrix} C_1^\top \\ 0 \end{pmatrix},$$

with $(A_{11}^\top, C_1^\top)$ reachable and $A_{22}^\top$ stable. Thus statement (2) is both equivalent to statement (3) and to $(A^\top, C^\top)$ being stabilizable. By Proposition 6.29, this is equivalent to the existence of $K \in \mathbb{F}^{n \times p}$ such that $A^\top + C^\top K$ is stable. With $L = K^\top$ this shows the equivalence of statements (4) and (5). ∎

Using this simple proposition, all mentioned results on state feedback stabilizability dualize to corresponding results on detectability. For a subset $\Lambda \subset \mathbb{C}$, a realization (A, B, C, D) defined over a subfield $\mathbb{F} \subset \mathbb{C}$ is called Λ-**output injection stabilizable**, if there exists $L \in \mathbb{F}^{n \times p}$ such that $\det(zI - A + LC)$ has all its roots in Λ. With this notation, the dual result to Theorem 6.30 is stated as follows.

Theorem 6.35. *Let $\mathbb{F} \subset \mathbb{C}$ be a subfield, and let $\Lambda \subset \mathbb{C}$ be a nonempty subset. Let $G(z) = V(z)T(z)^{-1}U(z) + W(z) \in \mathbb{F}(z)^{p \times m}$ be a proper rational matrix, and let (A, B, C, D) be the associated shift realization (4.20), defined over $\mathbb{F}$. Then the following conditions are equivalent:*

1. *The shift realization is Λ-output injection stabilizable.*
2. *The g.c.l.d. $E(z)$ of $T(z)$ and $V(z)$ is stable.*
3. $\begin{pmatrix} V(z) \\ T(z) \end{pmatrix}$ *has full column rank for every $z \notin \Lambda$.*
4. $\begin{pmatrix} C \\ zI - A \end{pmatrix}$ *has full column rank for every $z \notin \Lambda$.*

Proof. By transposing the transfer function $G(z)$, one obtains the factorization $G(z)^\top = U(z)^\top T(z)^{-\top} V(z)^\top + W(z)^\top$, having the shift realization $(A^\top, C^\top, B^\top, D^\top)$. Moreover, (C, A) is Λ-output injection stabilizable if and only if $(A^\top, C^\top)$ is Λ-feedback stabilizable. Thus the result follows by applying Theorem 6.30 to $G(z)^\top$. ∎

The next result is valid over an arbitrary field. It is obtained by dualizing Theorem 6.32; the straightforward arguments are omitted.

Theorem 6.36. *Let $\mathbb{F}_-[z]$ be a saturated subset of nonzero polynomials, and let $(A, B) \in \mathbb{F}^{n \times n} \times \mathbb{F}^{n \times m}$; then the following conditions are equivalent:*

1. *There exists $L \in \mathbb{F}^{n \times p}$ with $\det(zI - A + LC) \in \mathbb{F}_-[z]$.*
2. *The unobservable states $\mathcal{O}_*$ of (C, A) satisfy*

$$\mathcal{O}_* \subset X_-(A).$$

3. *The full rank condition*

$$\mathrm{rk} \begin{pmatrix} C \\ zI - A \end{pmatrix} = n$$

is fulfilled for all roots $z \in \overline{\mathbb{F}}$ of all irreducible polynomials in $\mathbb{F}_+[z]$.

6.7 Dynamic Output Feedback Stabilization

Next, let us consider the task of characterizing the notions of internal stability and stabilizability for the standard **output feedback connection** of a system with transfer function $G(z)$ and a controller with transfer function $K(z)$. This is a generalization of the situation discussed so far, insofar as it refers to **dynamic output feedback** rather than to static state feedback or output injection. For simplicity, we will focus on continuous-time systems described over the field of real numbers; the discrete-time case runs similarly. Schematically, the following diagram describes the feedback connection:

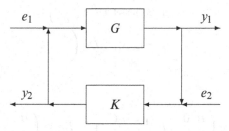

Our assumption is that both transfer functions $G(z) \in \mathbb{R}(z)^{p \times m}$ and $K(z) \in \mathbb{R}(z)^{m \times p}$ are proper rational. The full system equations are then given by

$$G(e_1 + y_2) = y_1,$$
$$K(e_2 + y_1) = y_2. \tag{6.30}$$

Equivalently, the closed-loop feedback interconnection Σ_{cl} is described using state-space realizations

$$\dot{x} = Ax + Bu_1,$$
$$y_1 = Cx + Du_1, \tag{6.31}$$

$$\dot{\xi} = A_c \xi + B_c u_2,$$
$$y_2 = C_c \xi + D_c u_2, \tag{6.32}$$

for $G(z)$ and $K(z)$, respectively, together with the coupling equations

$$u_1 = e_1 + y_2, \quad u_2 = e_2 + y_1.$$

The feedback interconnection Σ_{cl} is called **well-posed**, provided $I - DD_c$ is invertible or, equivalently, if the transfer function $I - G(z)K(z)$ is properly invertible. This condition is easily seen to be equivalent to the $(m+p) \times (m+p)$-matrix

$$F = \begin{pmatrix} I_m & -D_c \\ -D & I_p \end{pmatrix}$$

being invertible. Thus the feedback interconnection of a strictly proper transfer function $G(z)$ with a proper controller $K(z)$ is well posed. The assumption of well-posedness allows one to eliminate the internal input variables u_1, u_2 via

$$F \begin{pmatrix} u_1 \\ u_2 \end{pmatrix} = \begin{pmatrix} 0 & C_c \\ C & 0 \end{pmatrix} \begin{pmatrix} x \\ \xi \end{pmatrix} + \begin{pmatrix} e_1 \\ e_2 \end{pmatrix}$$

from the state equations (6.31) and (6.32). The closed-loop state space representation Σ_{cl} of a well-posed feedback interconnection then takes the form

$$\dot{x}_{cl} = A_{cl} x_{cl} + B_{cl} \begin{pmatrix} e_1 \\ e_2 \end{pmatrix},$$

$$\begin{pmatrix} y_1 \\ y_2 \end{pmatrix} = C_{cl} x_{cl} + D_{cl} \begin{pmatrix} e_1 \\ e_2 \end{pmatrix}, \tag{6.33}$$

with system matrices

$$A_{cl} = \begin{pmatrix} A & 0 \\ 0 & A_c \end{pmatrix} + \begin{pmatrix} B & 0 \\ 0 & B_c \end{pmatrix} F^{-1} \begin{pmatrix} 0 & C_c \\ C & 0 \end{pmatrix}, \quad B_{cl} = \begin{pmatrix} B & 0 \\ 0 & B_c \end{pmatrix} F^{-1}$$

$$C_{cl} = \begin{pmatrix} 0 & I \\ I & 0 \end{pmatrix} F^{-1} \begin{pmatrix} 0 & C_c \\ C & 0 \end{pmatrix}, \quad D_{cl} = \begin{pmatrix} D & 0 \\ 0 & D_c \end{pmatrix} F^{-1}. \tag{6.34}$$

Thus (A_{cl}, B_{cl}, C_{cl}) is static output feedback equivalent to the direct sum system

$$\left(\begin{pmatrix} A & 0 \\ 0 & A_c \end{pmatrix}, \begin{pmatrix} B & 0 \\ 0 & B_c \end{pmatrix}, \begin{pmatrix} C & 0 \\ 0 & C_c \end{pmatrix} \right),$$

which shows that (6.34) is reachable and observable if and only if both (A,B,C,D) and (A_c,B_c,C_c,D_c) are reachable and observable. Thus the minimality of the realizations (A,B,C,D), (A_c,B_c,C_c,D_c) of $G(z)$ and $K(z)$ implies the minimality of the closed-loop system $(A_{cl},B_{cl},C_{cl},D_{cl})$.

Definition 6.37. The feedback interconnection Σ_{cl} is **internally stable** if and only if Σ_{cl} is well posed and the system matrix A_{cl} is stable, i.e., all eigenvalues of A_{cl} have negative real part.

Note that

$$\Phi(z) = D_{cl} + C_{cl}(zI - A_{cl})^{-1} B_{cl} \tag{6.35}$$

is the transfer function of the feedback interconnection (6.34).

Proposition 6.38. $\Phi(z)$, the closed-loop transfer function in (6.35) from $\begin{pmatrix} e_1 \\ e_2 \end{pmatrix}$ to $\begin{pmatrix} y_1 \\ y_2 \end{pmatrix}$, is given by

$$\Phi = \begin{pmatrix} I & -G \\ -K & I \end{pmatrix}^{-1} \begin{pmatrix} G & 0 \\ 0 & K \end{pmatrix} = \begin{pmatrix} G & 0 \\ 0 & K \end{pmatrix} \begin{pmatrix} I & -K \\ -G & I \end{pmatrix}^{-1}. \tag{6.36}$$

In particular, the closed-loop transfer function G_f from e_1 to y_1 is given by

$$G_f(z) = \begin{pmatrix} I & 0 \end{pmatrix} \Phi(z) \begin{pmatrix} I \\ 0 \end{pmatrix} \tag{6.37}$$

$$= (I - G(z)K(z))^{-1} G(z) = G(z)(I - K(z)G(z))^{-1}.$$

Proof. The system equations (6.30) can be written in matrix form as

$$\begin{pmatrix} I & -G \\ -K & I \end{pmatrix} \begin{pmatrix} y_1 \\ y_2 \end{pmatrix} = \begin{pmatrix} G & 0 \\ 0 & K \end{pmatrix} \begin{pmatrix} e_1 \\ e_2 \end{pmatrix}.$$

It is easily calculated that

$$\begin{pmatrix} I & -G \\ -K & I \end{pmatrix}^{-1} = \begin{pmatrix} (I - GK)^{-1} & G(I - KG)^{-1} \\ (I - KG)^{-1}K & (I - KG)^{-1} \end{pmatrix},$$

and hence

$$\begin{pmatrix} y_1 \\ y_2 \end{pmatrix} = \begin{pmatrix} (I - GK)^{-1}G & G(I - KG)^{-1}K \\ (I - KG)^{-1}KG & (I - KG)^{-1}K \end{pmatrix} \begin{pmatrix} e_1 \\ e_2 \end{pmatrix}. \tag{6.38}$$

The expression for the transfer function $G_f(z)$ from e_1 to y_1 follows easily from (6.38). ∎

The definition of internal stability is stated solely in state-space terms. If the realizations (A, B, C, D) and (A_c, B_c, C_c, D_c) of the plant $G(z)$ and the controller $K(z)$ are stabilizable and detectable, then we can reformulate this condition in terms of the two transfer functions.

Proposition 6.39. Assume that (A, B, C, D) and (A_c, B_c, C_c, D_c) are stabilizable and detectable. Then the feedback interconnection Σ_{cl} is internally stable if and only if $I - G(z)K(z)$ is properly invertible and the transfer function $\Phi(z)$ of Σ_{cl} is stable.

Proof. By assumption, the realization $(A_{cl}, B_{cl}, C_{cl}, D_{cl})$ is stabilizable and detectable. Thus the transfer function $\Phi(z)$ is stable if and only if A_{cl} is stable. Moreover, $I - G(\infty)K(\infty) = I - DD_c$. This completes the proof. ∎

To characterize internally stabilizing controllers $K(z)$ for a transfer function $G(z)$, it is convenient to replace the ring of polynomials by the ring of stable proper rational functions. We proceed to formulate the relevant results on coprime factorizations in this context.

1. Coprime Factorizations over RH_∞.

Let RH_∞ denote the set of all proper rational transfer functions $g(z) \in \mathbb{R}(z)$ with poles only in the left half-plane $\mathbb{C}_-$. It is easily seen that RH_∞ is a ring and, indeed, a principal ideal domain; see Chapter 2 for details. Since any real rational function in $\mathbb{R}(z)$ can be expressed as the quotient of two elements from RH_∞, we conclude that $\mathbb{R}(z)$ is the field of fractions of RH_∞. Many important properties of rational matrix functions remain valid over RH_∞. One says that a rational matrix valued function $G(z) \in \mathbb{R}(z)^{p \times m}$ is in $RH_\infty^{p \times m}$ if and only if $G(z)$ is proper and stable. Two results for rational matrix functions – of importance for us, namely the existence of coprime and doubly coprime factorizations over the ring RH_∞ – are stated next. The proofs of these theorems run verbatim with those of Theorem 2.29, Theorem 2.33, and Corollary 2.34 in Chapter 2; they are therefore omitted.

Theorem 6.40. *For $G(z) \in \mathbb{F}(z)^{p \times m}$ the following assertions are valid:*

1. *There exist right coprime matrices $P(z) \in RH_\infty^{p \times m}, Q(z) \in RH_\infty^{m \times m}$ with $\det Q \neq 0$ such that*

$$G(z) = P(z)Q(z)^{-1}. \tag{6.39}$$

If $P_1(z) \in RH_\infty^{p \times m}, Q_1(z) \in RH_\infty^{m \times m}$ are right coprime with $\det Q_1 \neq 0$ and

$$P_1(z)Q_1(z)^{-1} = P(z)Q(z)^{-1} = G(z),$$

then there exists a unique unimodular matrix $U \in GL_m(RH_\infty)$ with $(P_1(z), Q_1(z)) = (P(z)U(z), Q(z)U(z))$.
2. *There exist left coprime matrices $P_\ell(z) \in RH_\infty^{p \times m}, Q_\ell(z) \in RH_\infty^{p \times p}$ with $\det Q_\ell \neq 0$ such that*

$$G(z) = Q_\ell(z)^{-1}P_\ell(z). \tag{6.40}$$

If $P_{\ell,1}(z) \in RH_\infty^{p \times m}, Q_{\ell,1}(z) \in RH_\infty^{p \times p}$ are left coprime with $\det Q_{\ell,1} \neq 0$ and

$$Q_{\ell,1}(z)^{-1}P_{\ell,1}(z) = Q_{\ell,2}(z)^{-1}P_{\ell,2}(z) = G(z),$$

then $(P_{\ell,1}(z), Q_{\ell,1}(z)) = (U(z)P_{\ell,2}(z), U(z)Q_{\ell,2}(z))$ for a uniquely determined unimodular matrix $U \in GL_m(RH_\infty)$.

*Factorizations as in (6.39) and (6.40) are called **right and left coprime factorizations of G over RH_∞**.*

We now relate the coprimeness of factorizations over RH_∞ to the solvability of Bezout equations, i.e., to unimodular embeddings. Let

$$G(z) = Q_\ell(z)^{-1} P_\ell(z) = P_r(z) Q_r(z)^{-1}$$

be right and left coprime factorizations of $G(z) \in \mathbb{F}(z)^{p \times m}$ over the ring RH_∞, respectively, which implies the intertwining relation

$$P_\ell(z) Q_r(z) = Q_\ell(z) P_r(z).$$

Theorem 6.41 (Doubly Coprime Factorization). *Let $P_\ell(z) \in RH_\infty^{p \times m}$ and $Q_\ell(z) \in RH_\infty^{p \times p}$ be right coprime and $P_r(z) \in RH_\infty^{p \times m}$ and $Q_r(z) \in RH_\infty^{m \times m}$ be left coprime, with*

$$Q_\ell(z) P_r(z) = P_\ell(z) Q_r(z).$$

Then there exist stable proper rational matrices $X(z) \in RH_\infty^{m \times p}, \overline{X}(z) \in RH_\infty^{m \times p}, Y(z) \in RH_\infty^{m \times m}, \overline{Y}(z) \in RH_\infty^{p \times p}$, with

$$\begin{pmatrix} Y(z) & X(z) \\ -P_\ell(z) & Q_\ell(z) \end{pmatrix} \begin{pmatrix} Q_r(z) & -\overline{X}(z) \\ P_r(z) & \overline{Y}(z) \end{pmatrix} = \begin{pmatrix} I_m & 0 \\ 0 & I_p \end{pmatrix}.$$

In particular, every right coprime factorization $G(z) = P(z) Q(z)^{-1}$ of a proper rational function $G(z) \in \mathbb{R}(z)^{p \times m}$ admits an extension to a unimodular matrix

$$\begin{pmatrix} Q(z) & -\overline{X}(z) \\ P(z) & \overline{Y}(z) \end{pmatrix} \in GL_{m+p}(RH_\infty).$$

2. Characterization of stabilizing controllers.
The characterization of controllers that internally stabilize a transfer function $G(z)$ is our next task. For this, the following abstract extension of Lemma 2.28 is needed.

Lemma 6.42. *Let R denote a principal ideal domain, and let $A \in R^{r \times r}, B \in R^{r \times m}, C \in R^{p \times r}$ be such that $\det A \neq 0$, A and B are left coprime and C and A are right coprime. Then $CA^{-1}B \in R^{p \times m}$ if and only if $A \in GL_r(R)$ is unimodular.*

Proof. The condition is obviously sufficient. For the necessity part consider a solution $X \in R^{r \times r}, Y \in R^{r \times p}$ of the Bezout equation

$$XA + YC = I_r.$$

The solution exists since A and C are right coprime. Thus $A^{-1}B = XB + YCA^{-1}B \in R^{r \times m}$. Using the dual Bezout equation

$$A\tilde{X} + B\tilde{Y} = I$$

we conclude that $A^{-1} = \tilde{X} + A^{-1}B\tilde{Y} \in R^{r \times r}$. Thus A is unimodular, i.e., $A \in GL_r(R)$. ∎

Theorem 6.43. *Assume that $G(z) \in \mathbb{F}(z)^{p \times m}$ is proper and that $K(z) \in \mathbb{F}(z)^{m \times p}$ is a proper controller, with the assumption that $I - G(\infty)K(\infty)$ is invertible. Assume that $G(z)$ and $K(z)$ have the following coprime factorizations over RH_∞:*

$$G(z) = Q_\ell(z)^{-1} P_\ell(z) = P_r(z) Q_r(z)^{-1},$$

$$K(z) = S_\ell(z)^{-1} R_\ell(z) = R_r(z) S_r(z)^{-1}. \tag{6.41}$$

Then:

1. The transfer function $\Phi(z)$ of Σ_{cl} has the following coprime factorizations:

$$\Phi(z) = \begin{pmatrix} Q_\ell & -P_\ell \\ -R_\ell & S_\ell \end{pmatrix}^{-1} \begin{pmatrix} P_\ell & 0 \\ 0 & R_\ell \end{pmatrix} = \begin{pmatrix} P_r & 0 \\ 0 & R_r \end{pmatrix} \begin{pmatrix} Q_r & -R_r \\ -P_r & S_r \end{pmatrix}^{-1}. \tag{6.42}$$

2. For suitable units u_1, u_2, u_3 in RH_∞, the following equations are fulfilled:

$$\det \begin{pmatrix} Q_\ell & -P_\ell \\ -R_\ell & S_\ell \end{pmatrix} = u_1 \det(Q_\ell S_r - P_\ell R_r) = u_2 \det(S_\ell Q_r - R_\ell P_r)$$

$$= u_3 \det \begin{pmatrix} Q_r & -R_r \\ -P_r & S_r \end{pmatrix}.$$

3. $\Phi(z)$ is proper and stable if and only if

$$S_\ell Q_r - R_\ell P_r \in GL_m(RH_\infty) \quad \text{or} \quad Q_\ell S_r - P_\ell R_r \in GL_p(RH_\infty).$$

4. The closed-loop transfer function G_f from e_1 to y_1 has the following equivalent representations:

$$G_f(z) = S_r(z)(Q_\ell(z) S_r(z) - P_\ell(z) R_r(z))^{-1} P_\ell(z)$$

$$= P_r(z)(S_\ell(z) Q_r(z) - R_\ell(z) P_r(z))^{-1} S_\ell(z)$$

$$= (P_r \ 0) \begin{pmatrix} Q_r & -R_r \\ -P_r & S_r \end{pmatrix}^{-1} \begin{pmatrix} I \\ 0 \end{pmatrix} \tag{6.43}$$

$$= (I \ 0) \begin{pmatrix} Q_\ell & -P_\ell \\ -R_\ell & S_\ell \end{pmatrix}^{-1} \begin{pmatrix} P_\ell \\ 0 \end{pmatrix}.$$

More generally, the closed-loop transfer function is

$$\Phi(z) = \begin{pmatrix} 0 & 0 \\ -I & 0 \end{pmatrix} + \begin{pmatrix} P_r \\ Q_r \end{pmatrix} (S_\ell Q_r - R_\ell P_r)^{-1} (S_\ell \ R_\ell)$$

$$= \begin{pmatrix} 0 & -I \\ 0 & 0 \end{pmatrix} + \begin{pmatrix} S_r \\ R_r \end{pmatrix} (Q_\ell S_r - P_\ell R_r)^{-1} (P_\ell \ Q_\ell).$$

Proof. The representations (6.42) follow by substituting the coprime factorizations (6.41) into equation (6.36). The right coprimeness of the factorization

$$\begin{pmatrix} P_r & 0 \\ 0 & R_r \end{pmatrix} \begin{pmatrix} Q_r & -R_r \\ -P_r & S_r \end{pmatrix}^{-1}$$

is equivalent to the right primeness of the matrix

$$\begin{pmatrix} P_r & 0 \\ 0 & R_r \\ Q_r & -R_r \\ -P_r & S_r \end{pmatrix},$$

i.e., after suitable elementary row operations, to the right primeness of

$$\begin{pmatrix} P_r & 0 \\ Q_r & 0 \\ 0 & R_r \\ 0 & S_r \end{pmatrix}.$$

In turn, this is equivalent to our assumption that P_r, Q_r and R_r, S_r are right coprime, respectively. A similar argument shows left coprimeness.

2. First, note that from the coprime factorizations (6.45) it follows that $\det Q_\ell = c \det Q_r$ and $\det S_\ell = d \det S_r$ for suitable units $c, d \in RH_\infty$. Next, computing

$$\begin{pmatrix} I & 0 \\ R_\ell Q_\ell^{-1} & I \end{pmatrix} \begin{pmatrix} Q_\ell & -P_\ell \\ -R_\ell & S_\ell \end{pmatrix} = \begin{pmatrix} Q_\ell & -P_\ell \\ 0 & S_\ell - R_\ell Q_\ell^{-1} P_\ell \end{pmatrix} = \begin{pmatrix} Q_\ell & -P_\ell \\ 0 & (S_\ell Q_r - R_\ell P_r) Q_r^{-1} \end{pmatrix},$$

and applying the multiplication rule of determinants, it follows that

$$\det \begin{pmatrix} Q_\ell & -P_\ell \\ -R_\ell & S_\ell \end{pmatrix} = \det Q_\ell \cdot \det(S_\ell Q_r - R_\ell P_r) \cdot \det Q_r^{-1}$$

$$= c \cdot \det(S_\ell Q_r - R_\ell P_r).$$

The other equalities are derived analogously.

4. Substituting representations (6.41) into (6.37), the closed-loop transfer function $G_f(z)$ has the following representations:

$$G_f(z) = S_r(z)(Q_\ell(z) S_r(z) - P_\ell(z) R_r(z))^{-1} P_\ell(z)$$

$$= P_r(z)(S_\ell(z) Q_r(z) - R_\ell(z) P_r(z))^{-1} S_\ell(z).$$

To obtain the third representation in (6.43), the coprime factorizations (6.42) are used to compute

$$G_f(z) = (I \ 0) \begin{pmatrix} P_r & 0 \\ 0 & R_r \end{pmatrix} \begin{pmatrix} Q_r & -R_r \\ -P_r & S_r \end{pmatrix}^{-1} \begin{pmatrix} I \\ 0 \end{pmatrix}$$

$$= (P_r \ 0) \begin{pmatrix} Q_r & -R_r \\ -P_r & S_r \end{pmatrix}^{-1} \begin{pmatrix} I \\ 0 \end{pmatrix}.$$

The last representation in (6.43) is similarly derived. The formulas for the closed-loop transfer function $\Phi(z)$ are similarly derived using formula (6.38). ∎

The preceding result leads to the first main characterization of all stabilizing controllers.

Theorem 6.44. *The following assertions are equivalent for proper transfer functions $G(z) \in \mathbb{R}(z)^{p \times m}$ and $K(z) \in \mathbb{R}(z)^{m \times p}$ that define a well-posed feedback interconnection:*

1. *K internally stabilizes G.*
2. *The matrix*

$$\begin{pmatrix} Q_\ell & -P_\ell \\ -R_\ell & S_\ell \end{pmatrix} \in RH_\infty^{(p+m) \times (p+m)}$$

   is invertible over RH_∞.
3. *The matrix*

$$\begin{pmatrix} Q_r & -P_r \\ -R_r & S_r \end{pmatrix} \in RH_\infty^{(p+m) \times (p+m)}$$

   is invertible over RH_∞.
4. *$\det(Q_\ell S_r - P_\ell R_r)$ is a biproper stable rational function with stable inverse.*
5. *$\det(S_\ell Q_r - R_\ell P_r)$ is a biproper stable rational function with stable inverse.*

Proof. By Proposition 6.39, the controller K internally stabilizes K if and only if the transfer function $\Phi \in RH_\infty^{(p+m) \times (p+m)}$. By Lemma 6.42 this is true if and only if

$$\begin{pmatrix} Q_\ell & -P_\ell \\ -R_\ell & S_\ell \end{pmatrix}^{-1} \in RH_\infty^{(p+m) \times (p+m)}.$$

This in turn is equivalent to

$$\begin{pmatrix} Q_r & -P_r \\ -R_r & S_r \end{pmatrix}^{-1} \in RH_\infty^{(p+m) \times (p+m)}.$$

By Theorem 6.43, this is true if and only if $\det(Q_\ell S_r - P_\ell R_r)$ [or $\det(S_\ell Q_r - R_\ell P_r)$] is a biproper stable rational function with stable inverse. This proves the equivalence of parts (1)–(5). ∎

3. The Youla–Kucera Parameterization.

The characterization of stabilizing controllers via unimodular embeddings is precise but has the disadvantage of not leading to an easily manageable parameterization of all such controllers. The Youla–Kucera parameterization resolves this issue by giving a complete parameterization of all stabilizing controllers via linear fractional transformations.

Theorem 6.45 (Youla-Kucera). *Let*

$$G(z) = Q_\ell(z)^{-1} P_\ell(z) = P_r(z) Q_r(z)^{-1} \in \mathbb{R}(z)^{p \times m}$$

be a proper rational, stable, and coprime factorization over RH_∞. Let

$$\begin{pmatrix} Y(z) & X(z) \\ P_\ell(z) & Q_\ell(z) \end{pmatrix} \begin{pmatrix} Q_r(z) & -\overline{X}(z) \\ -P_r(z) & \overline{Y}(z) \end{pmatrix} = \begin{pmatrix} I_m & 0 \\ 0 & I_p \end{pmatrix} \qquad (6.44)$$

be a unimodular embedding for this doubly coprime factorization over RH_∞. Then all proper rational and internally stabilizing controllers for $G(z)$ are parameterized by

$$K = (\overline{X} - Q_r \Gamma)(\overline{Y} - P_r \Gamma)^{-1} = (Y - \Gamma P_\ell)^{-1}(X - \Gamma Q_\ell). \qquad (6.45)$$

Here Γ denotes an arbitrary element of $(RH_\infty)^{m \times p}$ such that the matrices

$$\overline{Y}(\infty) - P_r(\infty) \Gamma(\infty) \quad \text{and} \quad Y(\infty) - \Gamma(\infty) P_\ell(\infty)$$

are invertible.

Proof. By Theorem 6.44, all internally stabilizing controllers $K = R_r S_r^{-1}$ are such that the matrix

$$\begin{pmatrix} Q_r(z) & -R_r(z) \\ -P_r(z) & S_r(z) \end{pmatrix}$$

is unimodular over RH_∞. Thus there exists a unimodular matrix

$$U = \begin{pmatrix} U_{11} & U_{12} \\ U_{21} & U_{22} \end{pmatrix} \in GL_{m+p}(RH_\infty),$$

with

$$\begin{pmatrix} Q_r(z) & -R_r(z) \\ -P_r(z) & S_r(z) \end{pmatrix} = \begin{pmatrix} Q_r(z) & -\overline{X}(z) \\ -P_r(z) & \overline{Y}(z) \end{pmatrix} \begin{pmatrix} U_{11} & U_{12} \\ U_{21} & U_{22} \end{pmatrix} \in GL_{m+p}(RH_\infty).$$

This implies $U_{11} = I$, $U_{21} = 0$, and U_{22} is unimodular over RH_∞. Thus $\Gamma := U_{12} U_{22}^{-1}$ exists in RH_∞ and

$$\begin{pmatrix} -R_r(z) \\ S_r(z) \end{pmatrix} = \begin{pmatrix} Q_r(z) & -\overline{X}(z) \\ -P_r(z) & \overline{Y}(z) \end{pmatrix} \begin{pmatrix} \Gamma U_{22} \\ U_{22} \end{pmatrix}$$

follows, implying

$$K = R_r S_r^{-1} = (\overline{X} - Q_r \Gamma)(\overline{Y} - P_r \Gamma)^{-1}.$$

By the unimodular embedding (6.44), we see that

$$\begin{pmatrix} Y - \Gamma P_\ell & X - \Gamma Q_\ell \\ P_\ell & Q_\ell \end{pmatrix} \begin{pmatrix} Q_r & -R_r \\ -P_r & S_r \end{pmatrix} = \begin{pmatrix} I & -\Gamma \\ 0 & I \end{pmatrix} \begin{pmatrix} Y & X \\ P_\ell & Q_\ell \end{pmatrix} \begin{pmatrix} Q_r & -\overline{X} \\ -P_r & \overline{Y} \end{pmatrix} \begin{pmatrix} I & \Gamma \\ 0 & I \end{pmatrix}$$

$$= \begin{pmatrix} I & 0 \\ 0 & I \end{pmatrix}.$$

Defining $R_\ell = X - \Gamma Q_\ell, S_\ell := Y - \Gamma P_\ell$, one obtains $S_\ell R_r = R_\ell S_r$, and therefore $S_\ell^{-1} R_\ell = R_r S_r^{-1}$, which completes the proof. ∎

As a consequence of the Youla–Kucera parameterization, it is now shown that the closed-loop transfer function depends **affinely** on the stabilizing parameter Γ. This fact is very important in robust controller design and opens the way to applying convex optimization techniques to robust controller design.

Theorem 6.46. *Let*

$$G(z) = Q_\ell(z)^{-1} P_\ell(z) = P_r(z) Q_r(z)^{-1} \in \mathbb{R}(z)^{p \times m}$$

be a proper rational, stable, and coprime factorization over RH_∞. *Let*

$$K_0(z) = Y(z)^{-1} X(z) = \overline{X}(z) \overline{Y}(z)^{-1} \in \mathbb{R}(z)^{m \times p}$$

denote a proper rational, stable, and coprime factorization over RH_∞ *of a stabilizing controller* $K_0(z)$ *of* $G(z)$. *Then the closed-loop transfer function* $\Phi_K(z) =$

$$\begin{pmatrix} P_r(z) & 0 \\ 0 & \overline{X}(z) \end{pmatrix} \begin{pmatrix} Q_r(z) & -\overline{X}(z) \\ -P_r(z) & \overline{Y}(z) \end{pmatrix}^{-1} - \begin{pmatrix} P_r(z) \\ Q_r(z) \end{pmatrix} \begin{pmatrix} 0 & \Gamma(z) \end{pmatrix} \begin{pmatrix} Q_r(z) & -\overline{X}(z) \\ -P_r(z) & \overline{Y}(z) \end{pmatrix}^{-1}$$

of all proper stabilizing controllers $K(z)$ *of* $G(z)$ *depends affinely on the stabilizing parameter* Γ, *where* $\Gamma(z) \in RH_\infty$ *is such that the matrices*

$$\overline{Y}(\infty) - P_r(\infty) \Gamma(\infty) \quad and \quad Y(\infty) - \Gamma(\infty) P_\ell(\infty)$$

are invertible. In particular, the closed-loop transfer function from e_1 *to* y_1 *has the affine parametric form*

$$G_f(z) = (I - P_\ell \Gamma \overline{Y}^{-1}) G (I - K_0 G)^{-1}.$$

Proof. By Theorem 6.43, the transfer function for the feedback interconnection of G with $K = \overline{ML}^{-1}$ is

$$G_f(z) = (P_r\ 0) \begin{pmatrix} Q_r & -\overline{M} \\ -P_r & \overline{L} \end{pmatrix}^{-1} \begin{pmatrix} I \\ 0 \end{pmatrix}.$$

From the Youla–Kucera parameterization it follows that

$$\begin{pmatrix} Q_r & -\overline{M} \\ -P_r & \overline{L} \end{pmatrix} = \begin{pmatrix} Q_r & -\overline{X} \\ -P_r & \overline{Y} \end{pmatrix} \begin{pmatrix} I & \Gamma \\ 0 & I \end{pmatrix}.$$

Moreover,

$$\begin{pmatrix} Q_r & -\overline{X} \\ -P_r & \overline{Y} \end{pmatrix} = \begin{pmatrix} I & -K_0 \\ -G & I \end{pmatrix} \begin{pmatrix} Q_r & 0 \\ 0 & \overline{Y} \end{pmatrix}.$$

Using (6.38), we compute

$$G_f(z) = (P_r\ 0) \begin{pmatrix} I & -\Gamma \\ 0 & I \end{pmatrix} \begin{pmatrix} Q_r & -\overline{X} \\ -P_r & \overline{Y} \end{pmatrix}^{-1} \begin{pmatrix} I \\ 0 \end{pmatrix}$$

$$= (P_r\ -P_r\Gamma) \begin{pmatrix} Q_r^{-1} & 0 \\ 0 & \overline{Y}^{-1} \end{pmatrix} \begin{pmatrix} I & -K_0 \\ -G & I \end{pmatrix}^{-1} \begin{pmatrix} I \\ 0 \end{pmatrix}$$

$$= G(I - K_0 G)^{-1} - P_r \Gamma \overline{Y}^{-1} G (I - K_0 G)^{-1},$$

which completes the proof. ∎

To present a state-space representation of the Youla–Kucera parameterization, state-space formulas for the involved transfer functions are derived. Let (A, B, C, D) and (A_c, B_c, C_c, D_c) be realizations of the transfer functions $G(z)$ and $K(z)$, respectively. Choosing a state feedback F such that $A + BF$ is stable, the transfer functions

$$Q_r(z) = \left[\begin{array}{c|c} A+BF & B \\ \hline F & I \end{array} \right], \quad P_r(z) = \left[\begin{array}{c|c} A+BF & B \\ \hline C+DF & D \end{array} \right] \qquad (6.46)$$

are in RH_∞ and define a right coprime factorization of $G(z) = P_r(z) Q_r(z)^{-1}$. Similarly, for J such that $A - JC$ is stable, the transfer functions

$$Q_\ell(z) = \left[\begin{array}{c|c} A-JC & J \\ \hline -C & I \end{array} \right], \quad P_\ell(z) = \left[\begin{array}{c|c} A-JC & B-JD \\ \hline C & D \end{array} \right]$$

are in RH_∞ and define a left coprime factorization of

$$G(z) = Q_\ell(z)^{-1} P_\ell(z).$$

A solution of the unimodular embedding (6.44) is constructed as follows. Define

$$\overline{Y}(z) = \left[\begin{array}{c|c} A+BF & J \\ \hline C+DF & I \end{array}\right], \quad \overline{X}(z) = \left[\begin{array}{c|c} A+BF & J \\ \hline F & 0 \end{array}\right].$$

Similarly,

$$Y(z) = \left[\begin{array}{c|c} A-JC & B-JD \\ \hline -F & I \end{array}\right], \quad X(z) = \left[\begin{array}{c|c} A-JC & J \\ \hline F & 0 \end{array}\right]. \tag{6.47}$$

The state-space representation of the stabilizing controller then has the form

$$\dot{\xi} = A\xi + Bu + J(y - C\xi - Du)$$

$$u = F\xi - \Gamma(\frac{d}{dt})(y - C\xi - Du), \tag{6.48}$$

where $\Gamma(z)$ denotes a proper rational and stable matrix function such that the feedback system is well posed. One observes that the transfer function of (6.48) from y to u is exactly the controller transfer function $K(z)$ in Youla–Kucera form (6.45). By choosing a state-space realization of $\Gamma(z)$ as

$$\Gamma(z) = D_c + C_c(zI - A_c)B_c,$$

we obtain the (implicit) first-order representation of the stabilizing controller as

$$\dot{\xi} = A\xi + Bu + J(y - C\xi - Du)$$

$$\dot{\xi}_c = A_c\xi_c + B_c(y - C\xi - Du)$$

$$u = F\xi - C_c\xi_c - D_c(y - C\xi - Du).$$

For $D = 0$, this leads to the input/output representation of all stabilizing controllers:

$$\dot{\xi} = (A + BF - (J - D_c)C)\xi - B_c C_c \xi_c + Jy$$

$$\dot{\xi}_c = -B_c C\xi + A_c \xi_c + B_c y$$

$$u = (F + D_c C)\xi - C_c \xi_c - D_c y.$$

6.8   Controlled Invariant Subspaces

The next two sections are devoted to a presentation of basic concepts of geometric control, i.e., the notions of controlled and conditioned invariant subspaces. Geometric control was developed in the early 1970s, by Francis, Wonham, and Morse on the one hand and by Basile and Marro on the other, as a tool for better understanding the structure of linear control systems within the context of state-space theory. Controlled and conditioned invariant subspaces generalize the class of invariant subspaces of a linear operator into a system-theoretic context; we refer the reader to Basile and Marro (1992) and Wonham (1979) for comprehensive accounts of the theory. The emphasis of geometric control is on clear, conceptual statements rather than a reliance on complex matrix manipulations. The term *geometry* refers to the basis-free study of classes of subspaces of the state space of a system realization. This development led to (sometimes iterative) design procedures based on elegant vector space geometric considerations.

However, it turned out that even the most fundamental problem of linear control – characterizing the limitations of pole placement by state feedback – was first solved by Rosenbrock using polynomial algebra. This brought about the need to find a bridge between the elegance of state-space problem formulations on the one hand and the computational effectiveness of polynomial algebra on the other.

The availability of the shift realizations allows us, given a system in a polynomial matrix description, to interpret the relation between the elements of the associated polynomial system matrix in state-space terms. Conversely, constructs coming from geometric control theory can, in the case of the shift realization, be completely characterized in polynomial terms. This leads to the most direct connection between abstract module theory, polynomial matrix descriptions, and state-space theory. In particular, the most basic objects of geometric control theory, namely, controlled and conditioned invariant subspaces, have very nice characterizations in terms of the zero structure of linear systems. Space limitations prevent us from delving deeper into the subject.

For a linear transformation A in $\mathscr{X}$ and an A-invariant subspace $\mathscr{V}$, denote by $A|_{\mathscr{V}}$ the restriction of A to $\mathscr{V}$. By a slight abuse of notation, an **induced map**, i.e., a map induced by A in the quotient space $\mathscr{X}/\mathscr{V}$, will be denoted by $A|_{\mathscr{X}/\mathscr{V}}$. Controlled invariant subspaces are introduced in state-space terms, and functional characterizations are derived.

Definition 6.47. 1. For an input pair (A,B), a subspace $\mathscr{V} \subset \mathscr{X}$ is called a **controlled invariant subspace**, or an (A,B)**-invariant subspace**, if

$$A\mathscr{V} \subset \mathscr{V} + \operatorname{Im} B. \tag{6.49}$$

2. Let $\mathscr{V}$ be a controlled invariant subspace for the pair (A,B). A feedback map $K : \mathscr{X} \longrightarrow \mathscr{U}$ that satisfies $(A - BK)\mathscr{V} \subset \mathscr{V}$ is called a **friend** of $\mathscr{V}$. For a controlled invariant subspace $\mathscr{V}$, denote by $\mathscr{F}(\mathscr{V})$ the set of all friends of $\mathscr{V}$.

3. One says that a family $\{\mathcal{V}_\alpha\}$ of controlled invariant subspaces is **compatible** if $\bigcap_\alpha \mathcal{F}(\mathcal{V}_\alpha) \neq \emptyset$, i.e., if there exists a single feedback map K such that

$$(A - BK)\mathcal{V}_\alpha \subset \mathcal{V}_\alpha$$

if fulfilled for all α.

4. A controlled invariant subspace $\mathcal{V}$ is called a **reachability subspace** if for each monic polynomial $q(z)$ of degree equal to $\dim \mathcal{V}$ there exists a friend $K \in \mathcal{F}(\mathcal{V})$ such that $q(z)$ is the characteristic polynomial of $(A - BK)|_{\mathcal{V}}$.

If $\mathcal{V} \subset \mathcal{X}$ is a controlled invariant subspace for the pair (A, B), and if $K \in \mathcal{F}(\mathcal{V})$, the notation $(A - BK)|_{\mathcal{V}}$ and $(A - BK)|_{\mathcal{X}/\mathcal{V}}$ will be used for restricted and induced maps, respectively. The next proposition is basic to all subsequent characterizations of controlled invariant subspaces.

Proposition 6.48. *For an input pair (A, B), the following statements are equivalent:*

1. *$\mathcal{V}$ is a controlled invariant subspace.*
2. *There exists a state feedback map $K : \mathcal{X} \longrightarrow \mathcal{U}$ such that the subspace $\mathcal{V}$ is $(A - BK)$-invariant.*
3. *For each $x_0 \in \mathcal{V}$, there exists an infinite sequence of inputs (u_t) such that the state trajectory (x_t) stays in $\mathcal{V}$.*

Proof. Assume $\mathcal{V}$ is controlled invariant. We choose a basis $\{v_1, \ldots, v_m\}$ for $\mathcal{V}$. By our assumption, $Av_i = w_i + Bu_i$, with $w_i \in \mathcal{V}$ and $u_i \in U$. Define a linear map K on $\mathcal{V}$ by $Kv_i = u_i$, and arbitrarily extend it to all of $\mathcal{X}$. Thus there exists a linear map $K : \mathcal{X} \longrightarrow U$ such that $(A - BK)v_i = w_i$, i.e., $(A - BK)\mathcal{V} \subset \mathcal{V}$, and (1) implies (2).

Let $x_0 \in \mathcal{V}$. It suffices to show that there exists a control u such that $x_1 = Ax_0 + Bu \in \mathcal{V}$. Since $(A - BK)x_0 = x_1 \in \mathcal{V}$, we simply choose $u = -Kx_0$, and (2) implies (3). To show that (3) implies (1), consider $x_0 \in \mathcal{X}$. It suffices to show that there exists a u such that $Ax_0 = x_1 - Bu$. By our assumption, there exists a $u \in U$ such that $x_1 = Ax_0 + Bu \in \mathcal{V}$.   ∎

The following are simple, yet very useful, characterizations.

Proposition 6.49. *Let (A, B) be an input pair in the state space $\mathcal{X}$. Then:*

1. *A q-dimensional subspace $\mathcal{V} \subset \mathcal{X}$ is controlled invariant if and only if there exists a linear map $F : \mathcal{V}_0 \longrightarrow \mathcal{V}_0$ on a q-dimensional subspace $\mathcal{V}_0$ and an injective map $Z : \mathcal{V}_0 \longrightarrow \mathcal{X}$, with $\mathrm{Im}\, Z = \mathcal{V}$, so that for some K*

$$ZF = (A - BK)Z. \tag{6.50}$$

2. *A q-dimensional subspace $\mathcal{R} \subset \mathcal{X}$ is a reachability subspace if and only if there exists a reachable pair (F, G) in a q-dimensional state space $\mathcal{V}_0$ and an injective map $Z : \mathcal{V}_0 \longrightarrow \mathcal{X}$, with $\mathrm{Im}\, Z = \mathcal{R}$, so that for some K, L*

$$ZF = (A - BK)Z,$$
$$ZG = BL. \tag{6.51}$$

Proof. If $\mathcal{V} \subset \mathcal{X}$ is controlled invariant, choose $\mathcal{V}_0 = \mathcal{V}$, $F = (A - BK)|\mathcal{V}$, and Z to be the embedding map of $\mathcal{V}$ into $\mathcal{X}$. Conversely, if F and Z exist and satisfy (6.50), then clearly $\mathcal{V} = \operatorname{Im} Z$ satisfies $(A + BK)\mathcal{V} \subset \mathcal{V}$, i.e., $\mathcal{V}$ is controlled invariant.

To prove the second claim, observe that $BL \subset \operatorname{Im} Z$ implies the existence of G. If $\mathcal{R}$ is a reachability subspace, it is in particular controlled invariant. So, with F and Z defined as previously, the first equation of (6.51) was derived. By the injectivity of Z, G is uniquely determined. Thus, the second equation in (6.51) follows. Conversely, if (6.51) holds, then, as previously, $\mathcal{R} = \operatorname{Im} Z$ is controlled invariant. Now, with $k = \dim \mathcal{X}_0$ and using the reachability of (F, G), we compute

$$\sum_{i=0}^{k-1} (A - BK)^i \operatorname{Im} BL = Z \sum_{i=0}^{k-1} F^i \operatorname{Im} G = Z \mathcal{X}_0 = \mathcal{R},$$

which shows that $\mathcal{R}$ is a reachability subspace. ∎

Stated next are some elementary properties of controlled invariant subspaces.

Proposition 6.50. *Let (A, B) be an input pair in the state space $\mathcal{X}$. Then:*

1. *The set of controlled invariant subspaces is closed under sums;*
2. *For each subspace $\mathcal{K} \subset \mathcal{X}$ there exists a maximal controlled invariant subspace contained in $\mathcal{K}$ that is denoted by $\mathcal{V}^*(\mathcal{K})$.*

Proof. The first claim follows directly from (6.49). For the second claim note that the set of all controlled invariant subspaces contained in $\mathcal{K}$ is closed under sums and is nonempty because the zero subspace is included. $\mathcal{V}^*(\mathcal{K})$ is the sum of all these subspaces. ∎

There exists a simple subspace algorithm to compute $\mathcal{V}^*$; see Wonham (1979).

Theorem 6.51 ($\mathcal{V}^*$-Algorithm). *Let $(A, B) \in \mathbb{F}^{n \times n} \times \mathbb{F}^{n \times m}$ and a linear subspace $W \subset \mathbb{F}^n$. Define a sequence of linear subspaces V_i, recursively constructed as follows:*

$$V_0 = W$$

$$V_{i+1} = W \cap A^{-1}(V_i + \operatorname{Im} B).$$

Then $V_0 \supset V_1 \supset V_2 \supset \cdots$ and, for a suitable $k \leq \dim W$, the equality $V_k = \mathcal{V}^*(W)$ is valid. In particular,

$$\mathcal{V}^*(W) = W \cap A^{-1}(W) \cap \cdots \cap A^{-k+1}(W) \cap A^{-k}(W + \operatorname{Im} B).$$

Proof. The construction of V_i implies $V_i \subset W$ for all i. Suppose that $V_i \subset V_{i-1}$ is satisfied for some i. Then

$$V_{i+1} = W \cap A^{-1}(V_i + \operatorname{Im} B) \subset W \cap A^{-1}(V_{i-1} + \operatorname{Im} B) = V_i.$$

Thus $V_j \subset V_{j-1}$ is satisfied for all $j \geq 1$. Moreover, by the same argument, $V_i = V_{i-1}$ implies $V_r = V_{i-1}$ for all $r \geq i$. Thus there exists $k \leq \dim W$ with $V_k = V_{k+r}$ for all $r \geq 1$. Since $V_{k+1} = W \cap A^{-1}(V_k + \operatorname{Im} B) = V_k$, the inclusions $V_k \subset W$ and $AV_k \subset V_k + \operatorname{Im} B$ follow. Thus V_k is a controlled invariant subspace contained in W. If $V \subset W$ is a controlled invariant subspace, then $V \subset V_0$. Let r be the largest nonnegative number with $V \subset V_r$. Then V being controlled invariant implies the inclusion $V \subset A^{-1}(V + \operatorname{Im} B)$, and therefore $V_{r+1} = W \cap A^{-1}(V_r + \operatorname{Im} B) \supset W \cap A^{-1}(V + \operatorname{Im} B) \supset V$. This proves that $V \subset V_i$ for all i, and therefore $V \subset V_k$. Thus V_k is the largest controlled invariant subspace contained in W. ∎

In applications, for example, to disturbance decoupling, it is often useful to extend the notion of controlled invariant subspaces by taking the output of the system under consideration. To this end, new geometric objects are introduced.

Definition 6.52. Let (A, B, C, D) be a state-space system in the state space $\mathscr{X}$.

1. A subspace $\mathscr{V} \subset \mathscr{X}$ is called **output nulling** if there exists a state feedback map K that satisfies

$$(A - BK)\mathscr{V} \subset \mathscr{V} \subset \operatorname{Ker}(C - DK). \tag{6.52}$$

2. Let $\mathscr{V}$ be a controlled invariant subspace for the pair (A, B). A feedback map K that satisfies (6.52) is called an **output nulling friend** of $\mathscr{V}$. Denote by $\mathscr{F}_{ON}(\mathscr{V})$ the set of all output nulling friends of $\mathscr{V}$.
3. A subspace $\mathscr{R}$ of the state space is called an **output nulling reachability subspace** if for each monic polynomial $q(z)$ of degree equal to $\dim \mathscr{R}$ there exists a friend $K \in \mathscr{F}_{ON}(\mathscr{V})$ such that $q(z)$ is the characteristic polynomial of $(A - BK)|_{\mathscr{V}}$.

Thus a subspace is output nulling if and only if, for each initial state in $\mathscr{V}$, one can find a state feedback controller that keeps the state in $\mathscr{V}$ while keeping the output zero. Usually, for an output nulling space $\mathscr{V}$, there exist also some external inputs that may be output nulled. Thus there exists a linear map L for which

$$(A - BK)\mathscr{V} \subset \mathscr{V} \subset \operatorname{Ker}(C - DK)$$

$$\operatorname{Im} BL \subset \mathscr{V}$$

$$DL = 0.$$

This is equivalent to nulling the output using a feedback law of the form

$$u = Kx + Lv.$$

Such an L, in fact a maximal one, can be constructed by considering the subspace

$$\mathscr{L} = \{\xi \mid B\xi \in \mathscr{V}, \quad D\xi = 0\},$$

and choosing L to be a basis matrix for this subspace.

In addition to the preceding basic definitions, a number of further useful classes of controlled invariant subspaces are linked with certain stability properties of the restriction and induced operators. Refer to Section 6.6 for the class of stable polynomials defined by a multiplicatively closed subset $\mathbb{F}_-[z] \subset \mathbb{F}[z]$.

Definition 6.53. Let (A,B,C,D) be a state-space system acting in the state space $\mathscr{X}$. A controlled invariant subspace $\mathscr{V}$ for the pair (A,B) is called stabilizable, or **inner stabilizable**, if there exists a friend $K \in \mathscr{F}(\mathscr{V})$ such that $(A - BK)|_{\mathscr{V}}$ is $\mathbb{F}_-[z]$-stable.

Analogously, a controlled invariant subspace $\mathscr{V}$ is called **outer stabilizable** if there exists a friend $K \in \mathscr{F}(\mathscr{V})$ such that $(A - BK)|_{\mathscr{X}/\mathscr{V}}$ is $\mathbb{F}_-[z]$-stable. One defines **inner antistabilizable** subspaces similarly.

If $\mathscr{V}$ is controlled invariant with respect to the reachable pair (A,B) and $K \in \mathscr{F}(\mathscr{V})$, then the pair induced by $(A - BK, B)$ in $\mathscr{X}/\mathscr{V}$ is also reachable; hence, $\mathscr{V}$ is both outer stabilizable and antistabilizable. From this point of view, it is more interesting to study inner stabilizability and antistabilizability.

1. Polynomial Characterization of Controlled Invariant Subspaces

By Theorem 3.14, the study of invariant subspaces, in the polynomial model context, is directly related to factorization theory. So it is natural to try and extend this correspondence to the study of controlled and conditioned invariant subspaces as well as other classes of subspaces.

A consequence of the shift realization procedure is the compression of information. All information, up to state-space isomorphism, of a reachable pair (A,B) is encoded in one, nonsingular, polynomial matrix. To see this, recall that reachability is equivalent to the left coprimeness of $zI - A, B$. Taking coprime factorizations

$$(zI - A)^{-1}B = N(z)D(z)^{-1}, \tag{6.53}$$

the isomorphism of the input pairs (A,B) and (S_D, π_D) follows and is described by the following diagram:

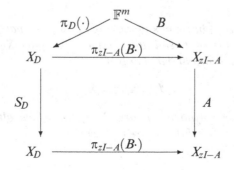

Let (A_i, B_i), $i = 1, 2$, be state feedback equivalent pairs and $D_i(z)$ the respective denominators in the coprime factorizations (6.53). Then, by Theorem 6.2, the feedback equivalence is expressed by the fact that $D_2(z)D_1(z)^{-1}$ is biproper. Since a controlled invariant subspace is an invariant subspace for a feedback equivalent pair, we obtain the following theorem.

Theorem 6.54. *Let $D(z) \in \mathbb{F}[z]^{m \times m}$ be nonsingular. Then a subspace $\mathcal{V} \subset X_D$ is a controlled invariant subspace, i.e., an (S_D, π_D)-invariant subspace, if and only if there exist necessarily nonsingular, polynomial matrices $E_1(z)$, $F_1(z) \in \mathbb{F}[z]^{m \times m}$ such that*

1. $D(z)D_1(z)^{-1}$ is biproper for

$$D_1(z) := E_1(z)F_1(z). \tag{6.54}$$

2. The subspace $\mathcal{V}$ has the representation

$$\mathcal{V} = \pi_D T_{DD_1^{-1}}(E_1 X_{F_1}). \tag{6.55}$$

Proof. Assume there exist $D_1(z), E_1(z)$ and $F_1(z) \in \mathbb{F}[z]^{m \times m}$ such that (1) and (2) are satisfied. Then (6.54) implies that $E_1 X_{F_1}$ is an S_{D_1}-invariant subspace of X_{D_1}. From (1) it follows that the pairs (S_D, π_D) and (S_{D_1}, π_{D_1}) are feedback equivalent pairs, with the Toeplitz induced map $\pi_D T_{DD_1^{-1}} : X_{D_1} \longrightarrow X_D$ satisfying

$$S_D \pi_D T_{DD_1^{-1}} - \pi_D T_{DD_1^{-1}} S_{D_1} = \pi_D K \tag{6.56}$$

for some $K : X_{D_1} \longrightarrow \mathbb{F}^m$. This implies that $\mathcal{V}$ is a controlled invariant subspace.

Conversely, assume that $\mathcal{V} \subset X_D$ is a controlled invariant subspace of X_D. By the definition of controlled invariant subspaces, $\mathcal{V}$ is an invariant subspace of a feedback equivalent pair. This pair can be taken, without loss of generality, to be (S_{D_1}, π_{D_1}). This implies that $D(z)D_1(z)^{-1}$ is biproper. The map from X_{D_1} to X_D that exhibits the feedback, i.e., that satisfies (6.56), is simply an induced Toeplitz map. Since S_{D_1}-invariant subspaces of X_{D_1} are of the form $E_1 X_{F_1}$ for some factorization (6.54), it follows that $\mathcal{V}$ has the required representation (6.55). ∎

We next aim at a characterization of controlled invariant subspaces of X_D in terms of rational models.

Theorem 6.55. *Let $D(z)$ be an $m \times m$ nonsingular polynomial matrix. Let the pair (A, B) be defined by the shift realization in the state space X_D. Then a subspace $\mathcal{V} \subset X_D$ is controlled invariant if and only if*

$$\mathcal{V} = \pi_D \pi_+ D X^{F_1}$$

for some nonsingular polynomial matrix $F_1(z)$ for which all left Wiener–Hopf factorization indices of $D(z)F_1(z)^{-1}$ are nonnegative.

Proof. Assume such an $F_1(z)$ exists. This implies the existence of a nonsingular polynomial matrix E_1 for which $DF_1^{-1}E_1^{-1}$ is biproper. We define

$$D_1(z) = E_1(z)F_1(z). \tag{6.57}$$

By Proposition 3.50, the map $Y : X_{D_1} \longrightarrow X_D$, defined, for $f(z) \in X_{D_1}$, by $Yf = \pi_D T_{DD_1^{-1}} f$, is invertible. Now, the factorization (6.57) implies that $E_1 X_{F_1}$ is an S_{D_1}-invariant subspace, so its image under Y is a controlled invariant subspace of X_D. Computing

$$\begin{aligned}
\mathscr{V} &= Y(E_1 X_{F_1}) = \pi_D T_{DD_1^{-1}} E_1 X_{F_1} = D\pi_- D^{-1}\pi_+ DF_1^{-1} E_1^{-1} E_1 X_{F_1} \\
&= D\pi_- D^{-1}\pi_+ DX^{F_1} = \pi_D \pi_+ DX^{F_1} = D\pi^D X^{F_1},
\end{aligned}$$

it is a consequence of Theorem 3.49 that the map $Z : X^F \longrightarrow X_D$, defined, for $h(z) \in X^F$, by

$$Zh = \pi_D \pi_+ Dh,$$

is injective. This shows that $\mathrm{Im}\, Z = \pi_D \pi_+ DX^F$ is indeed controlled invariant for the pair (A, B) defined by the shift realization.

Conversely, assume that $\mathscr{V} \subset X_D$ is a controlled invariant subspace. Every pair that is feedback equivalent to the pair $(S_D, \pi_D(\cdot))$ can be assumed, up to similarity, to be of the form $(S_{D_1}, \pi_{D_1}(\cdot))$, with $D(z)D_1(z)^{-1}$ biproper. An invariant subspace of X_{D_1} is of the form $\mathscr{V}_1 = E_1 X_{F_1}$ for a factorization $D_1(z) = E_1(z)F_1(z)$ into nonsingular factors. The biproperness assumption on $D(z)D_1(z)^{-1}$ implies that all left Wiener–Hopf factorization indices of $D(z)F_1(z)^{-1}$ are nonnegative. The Toeplitz induced map $\overline{T}_{DD_1^{-1}} : X_{D_1} \longrightarrow X_D$, defined for $f \in X_{D_1}$ by $\overline{T}_{DD_1^{-1}} f = \pi_D \pi_+ DD_1^{-1} f$, is therefore, by Theorem 3.49, invertible, and hence its restriction to $E_1 X_{F_1}$ is injective. Moreover, it satisfies $S_D \overline{T}_{DD_1^{-1}} - \overline{T}_{DD_1^{-1}} S_{D_1} = \pi_D K$ for some feedback map $K : X_{D_1} \longrightarrow \mathbb{F}^m$. Indeed, for $f \in X_{D_1}$, setting $g_+ = \pi_+ DD_1^{-1} f$ and $g_+ = \pi_+ DD_1^{-1} f$, we compute

$$\begin{aligned}
(S_D \overline{T}_{DD_1^{-1}} - \overline{T}_{DD_1^{-1}} S_{D_1})f &= \pi_D z\pi_D \pi_+ DD_1^{-1} f - \pi_D \pi_+ DD_1^{-1} \pi_{D_1} zf \\
&= \pi_D z\pi_+ DD_1^{-1} f - \pi_D \pi_+ DD_1^{-1} D_1 \pi_- D_1^{-1} zf \\
&= \pi_D \left\{ z\pi_+ DD_1^{-1} f - \pi_+ D\pi_- D_1^{-1} zf \right\} \\
&= \pi_D \left\{ z\pi_+ DD_1^{-1} f - \pi_+ D(I - \pi_+) D_1^{-1} zf \right\} \\
&= \pi_D \left\{ z\pi_+ DD_1^{-1} f - \pi_+ zDD_1^{-1} f \right\} \\
&= \pi_D \left\{ zg_+ - \pi_+ z(g_+ + g_-) \right\} = -\pi_D \pi_+ zg_- = -\pi_D \xi,
\end{aligned}$$

with $\xi = \pi_+ z g_- \in \mathbb{F}^m$. Thus

$$\mathcal{V} = T\mathcal{V}_1 = \pi_D \pi_+ DD_1^{-1}(E_1 X_{F_1}) = \pi_D \pi_+ DF_1^{-1} E_1^{-1}(E_1 X_{F_1}) = \pi_D \pi_+ DX^{F_1}.$$

∎

Using Theorem 3.32, i.e., the isomorphism between polynomial and rational models, one can restate Theorem 6.54 in terms of the rational functional model. The trivial proof of the next result is omitted.

Theorem 6.56. *Let $D(z)$ be a nonsingular polynomial matrix. With the (A, B) defined by the shift realization (4.26), a subspace $\mathcal{V} \subset X^D$ is a controlled invariant subspace if and only if there exist nonsingular polynomial matrices $E_1(z)$, $F_1(z) \in \mathbb{F}[z]^{m \times m}$ such that*

1. $D(z)D_1(z)^{-1}$ *is biproper for $D_1(z) = E_1(z)F_1(z)$;*
2. $\mathcal{V} = \pi^D X^{F_1}$.

Proposition 6.57. *Let $D(z) \in \mathbb{F}[z]^{m \times m}$ be nonsingular. Then a subspace $\mathcal{V} \subset X^D$ is controlled invariant with respect to the shift realization $\Sigma^{\bullet D^{-1}}$ if and only if it has a (not necessarily unique) representation of the form*

$$\mathcal{V} = \pi^D X^F$$

for some nonsingular polynomial matrix $F(z) \in \mathbb{F}[z]^{m \times m}$.

Proof. Using the isomorphism of the shift realizations $\Sigma_{\bullet D^{-1}}$ and $\Sigma^{\bullet D^{-1}}$, it follows from Theorem 6.54 that the controlled invariant subspace $\mathcal{V}$ has the representation

$$\mathcal{V} = D^{-1}\pi_D \pi_+ DX^F = D^{-1}D\pi_- D^{-1}\pi_+ DX^F$$
$$= \pi_- D^{-1}\pi_+ DX^F = \pi^D X^F.$$

Conversely, assume $\mathcal{V} = \pi^{D_\ell} X^F$ for some nonsingular $F(z) \in \mathbb{F}[z]^{m \times m}$. To prove that $\mathcal{V}$ is controlled invariant, one must show that for each $h \in \mathcal{V}$ there exist $h' \in X^F$ and $\xi \in \mathbb{F}^m$ such that $S^D \pi^D h = \pi^D h' + \pi_- D^{-1}\xi$. This is done by choosing $h' = S_- h$ and $\xi = (Dh)_{-1}$. To this end, one computes

$$S^D \pi^D h - \pi^D S_- h = \pi_- z \pi_- D^{-1}\pi_+ Dh - \pi_- D^{-1}\pi_+ D\pi_- zh$$
$$= \pi_- z D^{-1}\pi_+ Dh - \pi_- D^{-1}\pi_+ D\pi_- zh$$
$$= \pi_- D^{-1}\{z\pi_+ Dh - \pi_+ zDh\} = \pi_- D^{-1}\xi.$$

∎

Theorem 6.54 can be strengthened to yield a particularly clean representation of controlled invariant subspaces, a representation that has no direct reference to specific factorizations or to particular representations of submodules.

Theorem 6.58. *Let $D(z) \in \mathbb{F}[z]^{m \times m}$ be nonsingular. With respect to the realization (4.26) in the state space X^D, a subspace $\mathcal{V} \subset X^D$ is controlled invariant if and only if*

$$\mathcal{V} = \pi^D \mathcal{L} \tag{6.58}$$

for some submodule $\mathcal{L} \subset z^{-1} \mathbb{F}[[z^{-1}]]^m$.

Proof. In view of Theorem 6.54, all that needs to be proven is that the image under the projection π^D of a submodule of $\mathcal{L} \subset z^{-1} \mathbb{F}[[z^{-1}]]^m$ is a controlled invariant subspace of X^D. Equivalently, one must show that if $h \in \mathcal{L}$, then there exist $h_1 \in \mathcal{L}$ and $\xi \in \mathbb{F}^m$ such that

$$S^D \pi^{Dh} = \pi_D h_1 + \pi_- D^{-1} \xi. \tag{6.59}$$

We will prove (6.59), with $h_1 = S_- h$ and $\xi = (Dh)_{-1}$. In this case,

$$S^D \pi^{Dh} - \pi_D S_- h = \pi_- z \pi_- D^{-1} \pi_+ Dh - \pi_- D^{-1} \pi_+ D \pi_- zh$$
$$= \pi_- z D^{-1} \pi_+ Dzh - \pi_- D^{-1} \pi_+ Dzh$$
$$= \pi_- D^{-1} z \pi_+ Dh - \pi_+ zDh = \pi_- D^{-1} \xi.$$

∎

For the rational model characterization of controlled invariant subspaces as in Theorem 6.58, the shift realization $\Sigma^{\bullet D^{-1}}$ was used, with $D(z)$ as a right denominator. However, when analyzing output nulling subspaces, it turns out to be more convenient to work with the polynomial model shift realization $\Sigma_{T^{-1}V}$. To state the relevant characterization, one first extends the definition of a polynomial model from square nonsingular polynomial matrices to rectangular polynomial matrices $U(z) \in \mathbb{F}[z]^{p \times m}$ as

$$X_U := U(z)(z^{-1} \mathbb{F}[[z^{-1}]]^m) \cap \mathbb{F}[z]^p. \tag{6.60}$$

We refer to this space as the **rectangular polynomial model**. It is emphasized that X_U is certainly an $\mathbb{F}$-vector space of polynomials; however, unless U is nonsingular square, it is not a module over $\mathbb{F}[z]$.

Proposition 6.59. *Let $G(z)$ be a $p \times m$ proper rational matrix function with matrix fraction representation $G(z) = T(z)^{-1} U(z)$.*

1. Assume $U(z) = E_1(z) U_1(z)$, with $E_1(z) \in \mathbb{F}[z]^{p \times p}$ a nonsingular polynomial matrix. Then

$$\mathcal{V} = E_1 X_{U_1}$$

is an output nulling subspace of the shift realization $\Sigma_{T^{-1}U}$ in the state space X_T, and the following inclusions are valid:

$$E_1 X_{U_1} \subset X_U \subset X_T.$$

2. *A subspace $\mathcal{V} \subset X_T$ is output nulling if and only if $\mathcal{V} = E_1 X_{U_1}$, with $U(z) = E_1(z)U_1(z)$ and $E_1(z) \in \mathbb{F}[z]^{p \times p}$ nonsingular.*

Proof. Let $f \in E_1 X_{U_1}$, that is, $f = E_1 g$, with $g = U_1 h \in \mathbb{F}[z]^p$ for a strictly proper power series h. Since $T^{-1}U$ is proper and h is strictly proper, it follows that $T^{-1}f = (T^{-1}U)h$ is strictly proper. Thus $f \in X_T$, which shows $E_1 X_{U_1} \subset X_T$. Next we show that if h is strictly proper with $g = U_1 h \in \mathbb{F}[z]^p$, then also $U_1(S_-h) \in \mathbb{F}[z]^p$. Denoting by $\eta = h_{-1} \in \mathbb{F}^m$ the residue term of h, the equality $zh(z) = \eta + S_-h$ follows. This implies

$$U_1(S_-h) = U_1(zh) - U_1\eta = z(U_1h) - U_1\eta = zg - U_1\eta \in \mathbb{F}[z]^p.$$

To show that $\mathcal{V} = E_1 X_{U_1}$ is controlled invariant, let (A,B,C,D) denote the shift realization $\Sigma_{T^{-1}U}$. For $f \in E_1 X_{U_1}$, i.e., $f = E_1 g$ and $g = U_1 h$, one computes

$$S_{Tf} = \pi_T(zf) = \pi_T(zE_1U_1h) = \pi_T(E_1U_1zh) = \pi_T(E_1U_1(\eta + S_-h))$$
$$= \pi_T(U\eta) + \pi_T(E_1U_1S_-h) = \pi_T(U\eta) + E_1U_1S_-h.$$

Since we assume $T^{-1}U$ to be proper, there exists a representation $U(z) = T(z)D + V(z)$, with $T^{-1}V$ strictly proper. Hence,

$$\pi_T(U\eta) = \pi_T(TD+V)\eta = V\eta,$$

and therefore

$$Af = S_{Tf} = E_1U_1(S_-h) + V\eta.$$

As $E_1U_1S_-h \in E_1 X_{U_1}$ and $V\eta \in \operatorname{Im}B$, this proves that $\mathcal{V}$ is controlled invariant. Next, compute

$$Cf = (T^{-1}f)_{-1} = (T^{-1}Uh)_{-1} = (T^{-1}(TD+V)h)_{-1}$$
$$= (Dh)_{-1} = Dh_{-1} = D\eta.$$

Since η depends linearly on f, there exists a linear transformation K such that $\eta = Kf$. Thus $(A - BK)f \in \mathcal{V}$ and $(C - DK)f = 0$. This completes the proof of (1). The proof of the second claim is omitted.   ∎

The preceding result leads us to prove the following polynomial characterization of the maximal output nulling subspace.

Theorem 6.60. *With respect to the shift realization $\Sigma_{T^{-1}U}$ in the state space X_T, the maximal output nulling subspace is given as*

$$\mathscr{V}^* = X_U.$$

Proof. By Proposition 6.59 (1), it follows that X_U is output nulling. Moreover, by part (2) of the same proposition, each output nulling subspace $\mathscr{V}$ of X_T is of the form $\mathscr{V} = E_1 X_{U_1}$, with $U = E_1 U_1$ and E_1 nonsingular. Using Proposition 6.59 (1) once more, we obtain $\mathscr{V} \subset X_U$, which implies that X_U is the maximal output nulling subspace. ∎

A purely module-theoretic characterization of $\mathscr{R}^*$ is presented without proof.

Theorem 6.61. *Let $G(z)$ be a $p \times m$ proper rational matrix function with the left coprime matrix fraction representation $G(z) = T(z)^{-1} U(z)$. Then, with respect to the shift realization on X_T, the maximal output nulling reachability subspace is given by*

$$\mathscr{R}^* = X_U \cap U\mathbb{F}[z]^m.$$

6.9 Conditioned Invariant Subspaces

We begin by introducing basic concepts from geometric control theory that are relevant to observer design, i.e., conditioned invariant subspaces and related subspaces. The theory of such subspaces is dual to that of controlled invariant subspaces and thus can be developed in parallel.

Definition 6.62. 1. For an output pair (C, A), a subspace $\mathscr{V} \subset \mathscr{X}$ is called **conditioned invariant** if

$$A(\mathscr{V} \cap \operatorname{Ker} C) \subset \mathscr{V}.$$

2. For a conditioned invariant subspace $\mathscr{V}$ of the pair (C, A), an output injection map J for which $(A - JC)\mathscr{V} \subset \mathscr{V}$ is called a **friend** of $\mathscr{V}$. Denote by $\mathscr{G}(\mathscr{V})$ the set of all friends of $\mathscr{V}$.

3. For a pair (C, A), a conditioned invariant subspace $\mathscr{V} \subset \mathscr{X}$ is called **tight** if it satisfies

$$\mathscr{V} + \operatorname{Ker} C = \mathscr{X}.$$

4. A set of conditioned invariant subspaces $\mathscr{V}_\alpha$ is **compatible** if $\bigcap_\alpha \mathscr{F}(\mathscr{V}_\alpha) \neq \emptyset$.

5. A conditioned invariant subspace $\mathcal{V}$ will be called an **observability subspace** if, for each monic polynomial $q(z)$ of degree equal to codim $\mathcal{V}$, there exists a friend $J \in \mathscr{G}(\mathcal{V})$ such that $q(z)$ is the characteristic polynomial of $(A - JC)|_{\mathcal{X}/\mathcal{V}}$, the map induced on the quotient space $\mathcal{X}/\mathcal{V}$ by $A - JC$.

Some elementary properties of conditioned invariant subspaces are stated next.

Proposition 6.63. *Let (C,A) be an output pair acting in the state space $\mathcal{X}$. Then:*

1. *The set of conditioned invariant subspaces is closed under intersections;*
2. *For every subspace $\mathscr{L} \subset \mathcal{X}$, there exists a minimal conditioned invariant subspace containing $\mathscr{L}$; this subspace is denoted by $\mathcal{V}_*(\mathscr{L})$;*
3. *$\mathcal{V}$ is a conditioned invariant subspace if and only if there exists an output injection map $J : \mathbb{F}^p \longrightarrow \mathcal{X}$ such that $\mathcal{V}$ is $(A - JC)-$ invariant.*

Proof. The set of all conditioned invariant subspaces containing $\mathscr{L}$ is closed under intersections and is nonempty because $\mathcal{X}$ is included. $\mathcal{V}_*(\mathscr{L})$ is the intersection of all these subspaces. The proof of the last claim runs parallel to that of Proposition 6.48. Explicitly, choose a basis $\{v_1, \ldots, v_r\}$ of $\mathcal{V} \cap \operatorname{Ker} C$, and extend it to a basis $\{v_1, \ldots, v_q\}$ of $\mathcal{V}$. Then $Cv_{r+1}, \ldots, Cv_q \subset \mathbb{F}^p$ are linearly independent, with $q - r \leq p$. Thus there exists a linear transformation $J : \mathbb{F}^p \longrightarrow \mathcal{X}$ that maps $Cv_{r+1}, \ldots, Cv_q$ to $Av_{r+1}, \ldots, Av_q \in \mathcal{V}$. Thus $(A - JC)v_i = Av_i$ for $1 \leq i \leq r$ and $(A - JC)v_i = 0$ for $r + 1 \leq i \leq q$. Thus $(A - JC)\mathcal{V} \subset \mathcal{V}$. The converse is obvious. ∎

It may be instructive to see how the last claim can also be deduced from Proposition 6.48 by reasons of duality. In fact, assume for simplicity that $\mathcal{X} = \mathbb{F}^n$ and, thus, $A \in \mathbb{F}^{n \times n}, B \in \mathbb{F}^{n \times m}$ and $C \in \mathbb{F}^{p \times n}$. Consider a linear subspace $\mathcal{V} \subset \mathbb{F}^n$. Let $\mathcal{V}^\perp \subset \mathbb{F}^n$ denote the orthogonal complement with respect to the standard nondegenerate bilinear form $< x, y > = x^\top y$ on $\mathbb{F}^n$. Then

$$A\mathcal{V} \subset \mathcal{V} + \operatorname{Im} B$$

is satisfied if and only if

$$\mathcal{V}^\perp \cap \operatorname{Ker} B^\top = \mathcal{V}^\perp \cap (\operatorname{Im} B)^\perp = (\mathcal{V} + \operatorname{Im} B)^\perp \subset (A\mathcal{V})^\perp = (A^\top)^{-1}(\mathcal{V}^\perp),$$

i.e., if and only if

$$A^\top(\mathcal{V}^\perp \cap \operatorname{Ker} B^\top) \subset \mathcal{V}^\perp.$$

Thus $\mathcal{V}$ is (A,B)-invariant if and only if $\mathcal{V}^\perp$ is $(B^\top, A^\top)$-invariant. In this way, most results on controlled invariant subspaces for a linear system (A, B, C) can be equivalently reformulated as results on conditioned invariant subspaces for the dual system $(A^\top, C^\top, B^\top)$.

The dual notions to inner and outer stabilizable controlled invariant subspaces are introduced next.

Definition 6.64. Let (A, B, C, D) be a state-space system acting in the state space $\mathscr{X}$. A subspace $\mathscr{V} \subset \mathscr{X}$ is called **inner detectable** if there exists a friend $J \in \mathscr{G}(\mathscr{V})$ such that $(A - JC)|_{\mathscr{V}}$ is stable. A subspace $\mathscr{V}$ is **outer detectable** if there exists a $J \in \mathscr{G}(\mathscr{V})$ such that $(A - JC)|_{\mathscr{X}/\mathscr{V}}$ is stable. Again, the concepts of **inner and outer antidetectability** are naturally defined.

Let $\mathscr{V}$ be a conditioned invariant subspace for the observable pair (C, A), and let $J \in \mathscr{G}(\mathscr{V})$; then $(C, A - JC)$ has a restriction to $\mathscr{V}$, which is observable. Thus, if $\mathscr{V}$ is conditioned invariant, then it is both inner detectable and antidetectable. By standard duality considerations, one expects the notions of inner stabilizability to be related to outer detectability, and this indeed is the case.

1. Polynomial Characterizations of Conditioned Invariant Subspaces

We next derive polynomial characterizations of conditioned invariant subspaces.

Theorem 6.65. *Let $T(z) \in \mathbb{F}[z]^{p \times p}$ be nonsingular, and let (C, A) be the associated observable pair obtained via the shift realization $\Sigma_{T^{-1}}$. Then a subspace $\mathscr{V} \subset X_T$ is a conditioned invariant subspace, i.e., a $((T \cdot)_{-1}, S_T)$-invariant subspace, if and only if there exist nonsingular polynomial matrices $E_1(z)$, $F_1(z) \in \mathbb{F}[z]^{p \times p}$ such that:*

1. $T_1(z)^{-1} T(z)$ is normalized biproper for

$$T_1(z) = E_1(z) F_1(z); \tag{6.61}$$

2. In terms of the factorization (6.61), the representation

$$\mathscr{V} = E_1 X_{F_1}$$

is obtained.

Proof. Two proofs of this theorem are given.

Proof I:
$\mathscr{V}$ is a conditioned invariant subspace if and only if it is invariant for $A_1 = A - JC$. By Theorem 6.5, if the pair (C, A) is associated with the matrix fraction $T(z)^{-1} U(z)$, then the pair (C, A_1) is associated with the matrix fraction $T_1(z)^{-1} U(z)$, where $T_1(z)^{-1} T(z)$ is biproper. Since X_T and X_{T_1} are equal as sets, $\mathscr{V}$ is an S_{T_1}-invariant subspace of X_{T_1}. These subspaces are, by Theorem 3.11, of the form $\mathscr{V} = E_1 X_{F_1}$ with $T_1(z) = E_1(z) F_1(z)$.

Proof II:
In this proof, duality and the characterization of controlled invariant subspaces given in Theorem 6.54 will be used. The subspace $\mathscr{V} \subset X_T$ is conditioned invariant if and only if $\mathscr{V}^{\perp} \subset X_{T^{\top}}$ is controlled invariant, i.e., an $(S_{T^{\top}}, \pi_{T^{\top}})$-invariant subspace. By Theorem 6.54, there exists a polynomial matrix $T_1(z) \in \mathbb{F}[z]^{p \times p}$ such that $T^{\top}(z) T_1^{\top}(z)^{-1}$ is biproper and

$$\mathscr{V}^\perp = \pi_{T^\top} T_{T^\top T_1^{-\top}} (F_1^\top X_{E_1^\top}),$$

where $T_1(z) = E_1(z)F_1(z)$, and hence also $T_1^\top(z) = F_1^\top(z)E_1^\top(z)$. By the elementary properties of dual maps, $(\pi_{T^\top} T_{T^\top T_1^{-\top}})^* \mathscr{V} = \mathscr{V}_1 \subset X_{T_1}$ and $\mathscr{V}_1^\perp = F_1^\top X_{E_1^\top}$. Applying Theorem 3.11, one obtains $\mathscr{V}_1 = E_1 X_{F_1}$, and since $(\pi_{T^\top} T_{T^\top T_1^{-\top}})^* : X_T \longrightarrow X_{T_1}$ acts as the identity map, it follows that $\mathscr{V} = E_1 X_{F_1}$. ∎

In view of Theorems 6.56 and 6.65, it is of considerable interest to characterize the factorizations appearing in these theorems. The key to this are Wiener–Hopf factorizations at infinity.

Proposition 6.66. *1. Let $D(z), F_1(z) \in \mathbb{F}[z]^{m \times m}$ be nonsingular. Then there exist $E_1(z) \in \mathbb{F}[z]^{m \times m}$, and $D_1(z) := E_1(z)F_1(z)$ such that $D(z)D_1(z)^{-1}$ is biproper if and only if all the left Wiener–Hopf factorization indices at infinity of $D(z)F_1(z)^{-1}$ are nonnegative.*
2. Let $T(z), E_1(z) \in F[z]^{p \times p}$ be nonsingular. Then there exist polynomial matrices $F_1(z) \in \mathbb{F}[z]^{p \times p}$, and $T_1(z) := E_1(z)F_1(z)$ such that $T_1(z)^{-1}T(z)$ is biproper if and only if the right Wiener–Hopf factorization indices at infinity of $E_1(z)^{-1}T(z)$ are nonnegative.

Proof. Define $D_1(z) = E_1(z)F_1(z)$. If $\Gamma(z) = D(z)D_1(z)^{-1}$ is biproper, then $D(z)F_1(z)^{-1} = \Gamma(z)E_1(z)$. Now let $E_1(z) = \Omega(z)\Delta(z)U(z)$ be a left Wiener–Hopf factorization of $E_1(z)$. Then necessarily the factorization indices of $E_1(z)$ are nonnegative, being the reachability indices of the input pair (S_{E_1}, π_{E_1}). It follows that

$$D(z)F_1(z)^{-1} = (\Gamma(z)\Omega(z))\Delta(z)U(z), \tag{6.62}$$

i.e., $D(z)F_1(z)^{-1}$ has nonnegative left factorization indices. Conversely, if (6.62) holds with $\Delta(z) = \mathrm{diag}(z^{\kappa_1}, \ldots, z^{\kappa_m})$, and $\kappa_1 \geq \cdots \geq \kappa_m \geq 0$, then, defining $E_1(z) = \Delta(z)U(z)$, it follows that $D(z) = \Omega(z)D_1(z)$, with $D_1(z) = E_1(z)F_1(z)$ and $\Omega(z)$ biproper.

The proof of the second claim follows the lines of the proof of part 1 or can be derived from that theorem by duality. ∎

The characterizations appearing in Theorems 6.54 and 6.65 are factorization-dependent. The following proposition makes this unnecessary.

Proposition 6.67. *Let $D_\ell(z) \in \mathbb{F}[z]^{p \times p}$ be nonsingular. A subspace $\mathscr{V} \subset X_{D_\ell}$ is conditioned invariant with respect to the shift realization $\Sigma_{D_\ell^{-1}}$ if and only if it has a (not necessarily unique) representation of the form*

$$\mathscr{V} = \mathrm{Ker}\, \pi_T | X_{D_\ell} = X_{D_\ell} \cap T(z)\mathbb{F}[z]^p,$$

where $T(z) \in \mathbb{F}[z]^{p \times p}$ is a nonsingular polynomial matrix.

Proof. By Theorem 6.65, there exists a representation $\mathcal{V} = E_1 X_{F_1}$, with $D_\ell^{-1} E_1 F_1$ biproper. Assume $f \in \mathcal{V} = E_1 X_{F_1}$; then $f \in X_{D_\ell}$ and $f = E_1 g$, so $\mathcal{V} \subset X_{D_\ell} \cap E_1 \mathbb{F}[z]^p$.

Conversely, if $f \in X_{D_\ell} \cap E_1 \mathbb{F}[z]^p$, then $f = X_{D_\ell}$ and $f = E_1 g$. Let $T_1 := E_1 F_1$; then $T_1^{-1} D_\ell$ is biproper. Applying these facts, it follows that $D_\ell^{-1} f$ is strictly proper, and the biproperness of $T_1^{-1} D_\ell$ implies that also $T_1^{-1} f = T_1^{-1} E_1 g = F_1^{-1} E_1^{-1} E_1 g = F_1^{-1} g$ is strictly proper. But this shows that $g \in X_{F_1}$, and the inclusion $X_{D_\ell} \cap E_1 \mathbb{F}[z]^p \subset \mathcal{V}$ is proved. ∎

Theorems 6.54 and 6.65 can be strengthened to yield a particularly clean representation of controlled and conditioned invariant subspaces that has no direct reference to specific factorizations or to particular representations of submodules.

Theorem 6.68. *Let $D_\ell(z) \in \mathbb{F}[z]^{p \times p}$ be nonsingular. With respect to the realization (4.24) in the state space X_{D_ℓ}, a subspace $\mathcal{V} \subset X_{D_\ell}$ is conditioned invariant if and only if*

$$\mathcal{V} = X_{D_\ell} \cap \mathcal{M} \tag{6.63}$$

for some submodule $\mathcal{M} \subset \mathbb{F}[z]^p$.

Proof. The *only if* part was proved in Theorem 6.65. To prove the *if* part, assume $\mathcal{V} = X_{D_\ell} \cap \mathcal{M}$, where $\mathcal{M}$ is a submodule of $\mathbb{F}[z]^p$. If $f \in \mathcal{V} \cap \operatorname{Ker} C$, then $(D_\ell^{-1} f)_{-1} = 0$, which implies that $S_{D_\ell} f = \pi_{D_\ell} z f = z f$. But $S_{D_\ell} f \in X_{D_\ell}$ and $S_{D_\ell} f = z f \in \mathcal{M}$. Therefore, $S_{D_\ell} f \in X_{D_\ell} \cap \mathcal{M}$ follows, which shows that $\mathcal{V}$ is a conditioned invariant subspace, thus proving the theorem. ∎

The availability of the representation (6.63) of conditioned invariant subspaces allows us to give different proofs of the closure of the sets of controlled/conditioned invariant subspaces under sums/intersections, respectively.

Proposition 6.69. *1. Let $D(z) \in \mathbb{F}[z]^{m \times m}$ be nonsingular. Let $\mathcal{V}_i$ be controlled invariant subspaces of X^D, with respect to the shift realization $\Sigma^{\bullet D^{-1}}$, and having the representations $\mathcal{V}_i = \pi^D \mathcal{L}_i$ for submodules $\mathcal{L}_i \subset z^{-1} \mathbb{F}[[z^{-1}]]^m$. Then $\mathcal{V} = \sum_i \mathcal{V}_i$ is controlled invariant with the representation $\mathcal{V} = \pi^D \sum_i \mathcal{L}_i$.*

2. Let $D_\ell(z) \in \mathbb{F}[z]^{p \times p}$ be nonsingular. Let $\mathcal{V}_i$ be conditioned invariant subspaces of X_{D_ℓ}, with respect to the shift realization $\Sigma_{D_\ell^{-1} \bullet}$, and having the representations $\mathcal{V}_i = X_{D_\ell} \cap \mathcal{M}_i$ for submodules $\mathcal{M}_i \subset \mathbb{F}[z]^p$. Then $\mathcal{V} = \cap_i \mathcal{V}_i$ is conditioned invariant and has the representation $\mathcal{V} = X_{D_\ell} \cap (\cap_i \mathcal{M}_i)$.

Proof. The proof is obvious. ∎

The representation formula (6.63) is at the heart of the analysis of conditioned invariant subspaces and opens up several interesting questions, some of which will be described later. Note first that, for unimodular polynomial matrices $U(z)$, one has $T \mathbb{F}[z]^p = TU \mathbb{F}[z]^p$ and $X^{UT} = X^T$. Thus it is not important to distinguish between representing polynomial matrices up to an appropriate, one-sided unimodular factor. The representations of controlled and conditioned invariant subspaces that appear in Theorem 6.65 have the advantage of using nonsingular polynomial matrices in the representations. The disadvantage is the nonuniqueness of the representing polynomial matrices. However, uniqueness modulo unimodular factors can be recovered by switching to the use of rectangular polynomial matrices. This is done next.

As noted already, the submodule $\mathscr{M}$ in a representation of the form $\mathscr{V} = X_{D_\ell} \cap \mathscr{M}$ is, in general, not unique. To get a unique representation, one needs to associate with a conditioned invariant subspace of X_{D_ℓ} a unique submodule, and none is more natural than the submodule of $\mathbb{F}[z]^p$ generated by $\mathscr{V}$.

Proposition 6.70. *Let $D_\ell(z) \in \mathbb{F}[z]^{p \times p}$ be nonsingular, and let $\mathscr{V} \subset X_{D_\ell}$ be a conditioned invariant subspace. Let $< \mathscr{V} >$ be the submodule of $\mathbb{F}[z]^p$ generated by $\mathscr{V}$, which is the smallest submodule of $\mathbb{F}[z]^p$ that contains $\mathscr{V}$. Then*

$$\mathscr{V} = X_{D_\ell} \cap < \mathscr{V} >.$$

Proof. Assume $\mathscr{V} = X_{D_\ell} \cap \mathscr{M}$ for some submodule of $\mathbb{F}[z]^p$. Clearly, $\mathscr{V} \subset \mathscr{M}$, and hence $< \mathscr{V} > \subset \mathscr{M}$, and so $\mathscr{V} \subset < \mathscr{V} > \subset \mathscr{M}$, which in turn implies

$$\mathscr{V} \quad \subset \quad X_{D_\ell} \cap < \mathscr{V} > \quad \subset \quad X_{D_\ell} \cap \mathscr{M} = \mathscr{V}.$$

∎

Corollary 6.71. *For each subset $E \subset X_{D_\ell}$, the intersection $X_{D_\ell} \cap < E >$ is the smallest conditioned invariant subspace of X_{D_ℓ} that contains E.*

Proof. $X_{D_\ell} \cap < E >$ is a conditioned invariant subspace and contains E. Let $\mathscr{W}$ be another conditioned invariant subspace containing E. Then $< E > \subset < \mathscr{W} >$, and hence

$$X_{D_\ell} \cap < E > \quad \subset \quad X_{D_\ell} \cap < \mathscr{W} > = \mathscr{W}.$$

∎

Finally, we arrive at a very useful characterization of conditioned invariant subspaces.

Theorem 6.72. *A subspace $\mathscr{V} \subset X_{D_\ell}$ is a conditioned invariant subspace if and only if it has a representation of the form*

$$\mathscr{V} = X_{D_\ell} \cap H(z)\mathbb{F}[z]^k,$$

where $H(z)$ is a full column rank $p \times k$ polynomial matrix whose columns are in $\mathscr{V}$. $H(z)$ is uniquely determined up to a right $k \times k$ unimodular factor.

Proof. Follows from Theorem 6.68 and the basis representation of submodules of $\mathbb{F}[z]^p$ by full column rank polynomial matrices.   ∎

6.10 Zeros and Geometric Control

In this section we clarify the connection between the analysis of zeros based on module theory and that based on geometric control concepts. Recalling the definition of the zero module given in (4.43), we proceed with the following computational result. For any rectangular polynomial matrices we define the **rectangular polynomial model** X_U as

$$X_U := U(z)(z^{-1}\mathbb{F}[[z^{-1}]]^m) \cap \mathbb{F}[z]^m \tag{6.64}$$

and the **rectangular rational model** X^U as

$$X^U := \{h \in z^{-1}\mathbb{F}[[z^{-1}]]^m \mid \pi_-(Uh) = 0\}. \tag{6.65}$$

In particular, both identities $X^U = \operatorname{Ker} U(\sigma)$ and $U X^U = X_U$ are satisfied.

Proposition 6.73. *Let $G(z)$ be a strictly proper, $p \times m$ transfer function, with the left coprime factorization*

$$G(z) = T(z)^{-1} U(z). \tag{6.66}$$

Then:

1. Viewed as linear multiplication maps from $\mathbb{F}(z)^m$ to $\mathbb{F}(z)^p$,

$$\operatorname{Ker} G = \operatorname{Ker} U;$$

2.

$$\pi_- G^{-1}(\mathbb{F}[z]^p) = X^U = \operatorname{Ker} U(\sigma) \tag{6.67}$$

and

$$U\pi_- G^{-1}(\mathbb{F}[z]^p) = X_U, \tag{6.68}$$

where X_U and X^U are defined by (6.64) and (6.65), respectively;

3.

$$U\pi_- \operatorname{Ker} G = X_U \cap U\mathbb{F}[z]^m. \tag{6.69}$$

Proof. 1. Obvious.

2. Assume $h \in G^{-1}(\mathbb{F}[z]^p)$, i.e., $g = T^{-1}Uh \in \mathbb{F}[z]^p$. Defining $h_\pm = \pi_\pm h$, this implies $Tg = Uh = Uh_- + Uh_+$, or $Uh_- = Tg - Uh_+ \in \mathbb{F}[z]^p$ and, in turn, $h_- \in \operatorname{Ker} U(\sigma) = X^U$.

Conversely, if $h_- \in \operatorname{Ker} U(\sigma)$, then $Uh_- \in \mathbb{F}[z]^p$. By the left coprimeness of $T(z), U(z)$, there exist polynomial vectors $g(z), h_+(z)$ for which $Uh_- = Tg - Uh_+$. Therefore, with $h = h_- + h_+$ we have $g = T^{-1}Uh$, i.e., $h \in G^{-1}(\mathbb{F}[z]^p)$. From the equality $Uh_- = Tg - Uh_+$ it follows that $\pi_- Uh_- = 0$, i.e., (6.67) is proven.

3. Clearly, $G^{-1}(\{0\}) = \operatorname{Ker} G = \operatorname{Ker} U$. For $h = h_- + h_+ \in \operatorname{Ker} U$, we have $Uh_- = -Uh_+$, which implies $U(\sigma)h_- = \pi_- Uh_- = 0$, that is, $h_- \in X^U$ as well as $Uh_- = -Uh_+ \in U\mathbb{F}[z]^m$, and the inclusion $U\pi_- \operatorname{Ker} G \subset X_U \cap U\mathbb{F}[z]^m$ follows.

Conversely, if $Uh \in X_U \cap U\mathbb{F}[z]^m$, then there exist $h_+ \in \mathbb{F}[z]^m$ and $h_- \in z^{-1}\mathbb{F}[[z^{-1}]]^m$ for which $Uh = Uh_- = -Uh_+$. From this it follows that $\pi_-(h_- + h_+) = h_-$ and $U(h_- + h_+) = 0$, or $(h_- + h_+) \in \operatorname{Ker} G$. This implies the inclusion $X_U \cap U\mathbb{F}[z]^m \subset U\pi_- \operatorname{Ker} G$. The two inclusions imply (6.69). ∎

Following Wyman, Sain, Conte and Perdon (1989), we define

$$Z_\Gamma(G) = \frac{G(z)^{-1}(\mathbb{F}[z]^p)}{G(z)^{-1}(\mathbb{F}[z]^p) \cap \mathbb{F}[z]^m}, \tag{6.70}$$

$$Z_0(G) = \frac{\operatorname{Ker} G}{\operatorname{Ker} G \cap \mathbb{F}[z]^m}, \tag{6.71}$$

and recall the definition of the zero module, given in (4.43), namely,

$$Z(G) = \frac{G(z)^{-1}\mathbb{F}[z]^p + \mathbb{F}[z]^m}{\operatorname{Ker} G(z) + \mathbb{F}[z]^m}. $$

Theorem 6.74. *Let $G(z)$ be a (strictly) proper, $p \times m$ transfer function, with the left coprime factorization (6.66). Then:*

$$Z_\Gamma(G) \simeq \pi_- G^{-1}(\mathbb{F}[z]^p) = X^U, \tag{6.72}$$

$$Z_0(G) \simeq \pi_- \operatorname{Ker} G. \tag{6.73}$$

For the zero module, defined by (4.43), the following isomorphism is true:

$$Z(G) \simeq \frac{X_U}{X_U \cap U\mathbb{F}[z]^m} = \frac{\mathscr{V}^*}{\mathscr{R}^*}. \tag{6.74}$$

Here $\mathscr{V}^*$ and $\mathscr{R}^*$ are defined by Theorems 6.60 and 6.61, respectively.

Proof. For the proof we will use the following standard module isomorphisms Lang (1965). Assuming M, N, M_i are submodules of a module X over a commutative ring R,

$$\frac{M+N}{N} \simeq \frac{M}{M \cap N}$$

and, assuming additionally the inclusions $M_0 \subset M_1 \subset M_2$,

$$\frac{M_2}{M_1} \simeq \frac{M_2}{M_0} \bigg/ \frac{M_2}{M_1}.$$

Using these, we have the isomorphisms

$$Z_\Gamma(G) = \frac{G(z)^{-1}(\mathbb{F}[z]^p)}{G(z)^{-1}(\mathbb{F}[z]^p) \cap \mathbb{F}[z]^m} \simeq \frac{G(z)^{-1}(\mathbb{F}[z]^p) + \mathbb{F}[z]^m}{\mathbb{F}[z]^m}$$

and

$$Z_0(G) = \frac{\mathrm{Ker}\,G}{\mathrm{Ker}\,G \cap \mathbb{F}[z]^m} \simeq \frac{\mathrm{Ker}\,G + \mathbb{F}[z]^m}{\mathbb{F}[z]^m}.$$

Clearly, the inclusion $\mathrm{Ker}\,G \subset G(z)^{-1}(\mathbb{F}[z]^p)$ implies the inclusions $\mathbb{F}[z]^m \subset \mathrm{Ker}\,G + \mathbb{F}[z]^m \subset G(z)^{-1}(\mathbb{F}[z]^p) + \mathbb{F}[z]^m$. Again, we obtain the isomorphism

$$Z(G) = \frac{G(z)^{-1}(\mathbb{F}[z]^p) + \mathbb{F}[z]^m}{\mathrm{Ker}\,G(z) + \mathbb{F}[z]^m} \simeq \frac{G(z)^{-1}(\mathbb{F}[z]^p) + \mathbb{F}[z]^m}{\mathbb{F}[z]^m} \bigg/ \frac{\mathrm{Ker}\,G + \mathbb{F}[z]^m}{\mathbb{F}[z]^m}.$$
(6.75)

Note that (6.68) implies $U\pi_-(G(z)^{-1}(\mathbb{F}[z]^p) + \mathbb{F}[z]^m) = X_U$, and, similarly, (6.69) implies $U\pi_-(\mathrm{Ker}\,G + \mathbb{F}[z]^m) = X_U \cap U\mathbb{F}[z]^m$. Furthermore,

$$\mathrm{Ker}\,U\pi_-|(G(z)^{-1}(\mathbb{F}[z]^p) + \mathbb{F}[z]^m) = \mathrm{Ker}\,G + \mathbb{F}[z]^m,$$

and hence the isomorphism (6.74) follows. ∎

The isomorphism (6.74) shows that the zero module is directly related to the transmission zeros Morse (1973). The modules $Z_\Gamma(G)$ and $Z_0(G)$ also have system-theoretic interpretations, but this is beyond the scope of the present monograph.

6.11 Exercises

1. Let $R(A,B) = (B, AB, \ldots, A^{n-1}B)$ denote the reachability matrix of $(A,B) \in \mathbb{F}^{n \times (n+m)}$. Prove that for a feedback $K \in \mathbb{F}^{m \times n}$ there exists an upper triangular block matrix $U \in \mathbb{F}^{nm \times nm}$ with diagonal blocks $U_{11} = \cdots = U_{nn} = I_m$ and

$$R(A+BK, B) = R(A, B)U.$$

2. Prove that the reachability indices of a pair (A,B) coincide with the Kronecker indices, arranged in decreasing order.

3. Assume that the reachable pair $(A,b) \in \mathbb{F}^{n \times n} \times \mathbb{F}^n$ is in Jordan canonical form,

$$A = \begin{pmatrix} \lambda_1 & \cdots & 0 \\ \vdots & \ddots & \vdots \\ 0 & \cdots & \lambda_n \end{pmatrix}, \quad A = \begin{pmatrix} 1 \\ \vdots \\ 1 \end{pmatrix},$$

with eigenvalues $\lambda_1, \ldots, \lambda_n \in \mathbb{F}$. Let $\mu_1, \ldots, \mu_n \in \mathbb{F}$. Prove the formula by Mayne-Murdoch, i.e., that the feedback gain $K = (k_1, \ldots, k_n) \in \mathbb{F}^{1 \times n}$,

$$k_i = \frac{\prod_j (\lambda_i - \mu_j)}{\prod_j (\lambda_i - \lambda_j)}, \quad i = 1, \ldots, n,$$

satisfies

$$\det(zI - A + bK) = \prod_{j=1}^n (z - \mu_j).$$

4. (a) Let the pair $(A,B) \in \mathbb{F}^{n \times (n+m)}$ be reachable. For a monic polynomial $f(z) \in \mathbb{F}[z]$ of degree $m+n$, show the existence of matrices $X \in \mathbb{F}^{m \times n}$ and $Y \in \mathbb{F}^{m \times n}$ such that $f(z)$ is the characteristic polynomial of

$$M = \begin{pmatrix} A & B \\ X & Y \end{pmatrix}.$$

   (b) Let

$$A = \begin{pmatrix} 1 & 0 & 0 & 0 \\ 0 & 0 & 1 & 0 \\ 0 & 0 & 0 & 0 \\ 1 & 0 & 0 & 0 \end{pmatrix}, \quad B = \begin{pmatrix} 1 & 0 \\ 1 & 0 \\ 0 & 1 \\ 0 & 0 \end{pmatrix}.$$

   Determine the matrices X and Y such that the matrix M is nilpotent.

5. Let $\kappa = (\kappa_1, \ldots, \kappa_m)$ denote a partition of n. Prove that the set of pairs $(A,B) \in \mathbb{F}^{n \times n} \times \mathbb{F}^{n \times m}$ whose reachability indices $\lambda = (\lambda_1, \ldots, \lambda_m)$ satisfy $\sum_{j=1}^r \kappa_j \le \sum_{j=1}^r \lambda_j$, $r = 1, \ldots, m$, forms a Zariski-closed subset of $\mathbb{F}^{n \times n} \times \mathbb{F}^{n \times m}$. Let $n = km + \ell$ with $0 \le \ell < m$. Deduce that the set of pairs $(A,B) \in \mathbb{F}^{n \times n} \times \mathbb{F}^{n \times m}$ with reachability indices $\kappa = (k+1, \ldots, k+1, k, \ldots, k)$ is a nonempty Zariski-open subset in $\mathbb{F}^{n \times n} \times \mathbb{F}^{n \times m}$.

6. Prove that a linear subspace $\mathcal{V} \subset \mathbb{F}^n$ is simultaneously (A,B)- and (C,A)-invariant if and only if there exists an output feedback gain $K \in \mathbb{F}^{m \times p}$ that satisfies $(A + BKC)\mathcal{V} \subset \mathcal{V}$.

7. Let (A,B,C) be reachable and observable. Then the following conditions are equivalent:

   (a) (A,B,C) is state feedback irreducible.
   (b) $(A + BK, B, C)$ is observable for all K.
   (c) $\mathcal{V}^*(\operatorname{Ker} C) = \{0\}$.

8. Consider the scalar real rational transfer function $G(s) = \frac{1}{(s-1)(s-2)}$ of a continuous-time linear system.

   (a) Show that $G(s) = P(s)Q(s)^{-1}$, with $P(s) = \frac{1}{(s+1)^2}$ and $Q(s) = \frac{(s-1)(s-2)}{(s+1)^2}$, is a coprime factorization over RH_∞.
   (b) Determine all stabilizing controllers of $G(s)$.

9. Let $G(s) \in RH_\infty$ be a $p \times m$ proper rational stable transfer function. Show that all stabilizing controllers of G are of the form

$$K(s) = \Gamma(s)(I - G(s)\Gamma(s))^{-1},$$

with $\Gamma \in RH_\infty$ and $(I - G(s)\Gamma(s))^{-1}$ proper. Is K always in RH_∞?

6.12 Notes and References

A module-theoretic approach to the study of state feedback was initiated by Hautus and Heymann (1978); see also Fuhrmann (1979). The characterization in Theorem 6.13 of reachability indices via Wiener–Hopf indices for feedback irreducible systems can be extended to transfer functions that are not full column rank; see Fuhrmann and Willems (1979). A closely related characterization of feedback irreducibility is due to Heymann (1975).

The Brunovsky canonical form is due to Brunovsky (1970). The Kronecker indices appear first in the work by Popov (1972) and Wang and Davison (1976) on state-space canonical forms. Their characterization in Theorem 6.18 as complete invariants for restricted state feedback equivalence was shown by Helmke (1985). The dimension of the state feedback orbit of a reachable pair was expressed by Brockett (1977) via the reachability indices $\kappa_1 \geq \ldots \geq \kappa_m$ as

$$n^2 + nm + m^2 - \sum_{i,j=1}^{m} \max(\kappa_i - \kappa_j + 1, 0).$$

This formula is a straightforward consequence of Theorem 6.15. For a study of the feedback group, see also Tannenbaum (1981). For an analysis of the topological closure of each state feedback orbit in terms of the dominance order on partitions, see Hazewinkel and Martin (1983) and Helmke (1985).

The connection between reachability indices of a reachable pair and the Wiener–Hopf factorization indices is due to Gohberg, Lerer and Rodman (1978); see also Fuhrmann and Willems (1979). Rosenbrock's theorem is one important instance where polynomial arguments become much simpler than a pure state-space proof. The polynomial proof of Rosenbrock's theorem by Münzner and Prätzel-Wolters (1979) shows the elegancy of polynomial algebra arguments.

The Youla–Kucera parameterization of stabilizing controllers, together with a general theory of coprime factorizations, is nicely revealed in the book by Vidyasagar (1987). The state-space formulas (6.46)–(6.47) and (6.48) for the Youla–Kucera parameterization are taken from Kucera (2011). A more difficult problem is that of strong stabilizability that deals with the issue of finding a stable controller that stabilizes a plant. In Chapter 5.3 of Vidyasagar (1987), strong stabilizability is characterized in terms of parity interlacing conditions on the poles of plants with respect to unstable blocking zeros. A natural generalization of strong stabilizability is the simultaneous stabilization problem of N plants by a single controller, which has been studied by many researchers, including, for example, Blondel (1994); Ghosh and Byrnes (1983); Vidyasagar (1987).

For an early connection between state feedback pole placement and matrix extension problems, see Wimmer (1974b). The problem of pole placement and stabilization by static output feedback is considerably more difficult than that of state feedback control. Pole placement by constant output feedback is equivalent to an intersection problem in a Grassmann manifold, first solved, over the complex numbers, by Schubert in 1886. The connection to intersection theory in the Grassmann manifold was first revealed in the paper by Brockett and Byrnes (1981); see also Byrnes (1989). A striking result that generalized all of the preceding ones is due to Wang (1992), who showed that the pole-placement problem is generically solvable over the reals if $mp > n$. An interesting extension of the pole-placement problem to finite fields $\mathbb{F}$ is due to Gorla and Rosenthal (2010).

Geometric control was initiated by Francis and Wonham (1976) and Basile and Marro (1992); see also Wonham (1979). The mutual relations between the various subspaces are summarized by the so-called Morse relations and the Morse diamond; see Morse (1973). Readers who want to delve somewhat deeper into geometric control theory must master these important contributions. For a nice exposition and further results, we refer the reader to Aling and Schumacher (1984). Geometric control concepts can be applied, for example, to disturbance decoupling with measurement feedback Willems and Commault (1981) and to noninteracting control Falb and Wolovich (1967); Morse and Wonham (1971). A very important problem, which can be successfully dealt with in the framework of geometric control, is the so-called servomechanism or output regulation problem; see Francis (1977).

The polynomial model approach to the characterization of controlled and conditioned invariant subspaces yields clean module-theoretic representations. In particular, (6.58) is due to Fuhrmann and Willems (1980), whereas (6.63) was proved in Fuhrmann (1981). Closely related, and more general, characterizations of controlled and conditioned invariant subspaces in terms of transfer function representations are due to Hautus (1980) and Özgüler (1994). For the case of strictly proper rational transfer functions, the polynomial characterization of $\mathscr{V}^*$ in Theorem 6.60 is due to Emre and Hautus (1980) and Fuhrmann and Willems (1980). The characterization of the maximal output nulling reachability subspace $\mathscr{R}^*$ in Theorem 6.61 is due to Fuhrmann (1981) and Khargonekar and Emre (1982).

Chapter 7
Observer Theory

Observer theory is one of the most basic, and important, aspects of linear systems theory. The problem addressed in this chapter is that of indirect observation, or partial state estimation. It arises from the fact that in a control system Σ_{sys}, the observed variables are not necessarily the variables one needs to estimate for control, or other, purposes. A standard situation often encountered is that of partial state estimation, where a few, or all, state variables are to be estimated from the output variables. Of course, if one can estimate the state, then one automatically has the ability to estimate a function of the state. However, especially in a large and complex system, estimating the full state may be a daunting task and more than what is needed. The task of state estimation is also instrumental for practical implementations of state feedback control using estimates of the unknown state functions. More generally, our aim is to find a mechanism, called an **observer**, that allows us to use information on observed variables y and the inputs u in order to estimate linear functions z of the state variables. Loosely speaking, an observer for the system is itself a linear system Σ_{est}, which is driven by the variables u and y and whose output is the desired estimate ζ of z, with the estimation error being $e = z - \zeta$. The following diagram describes the observation process:

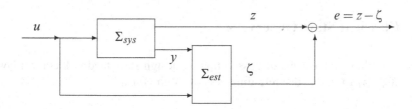

The error trajectory depends on the system transfer function, the observer transfer function, and on the initial conditions of both Σ_{sys} and Σ_{est}. There is great freedom in the choice of the observer, the only constraint being the compatibility with

© Springer International Publishing Switzerland 2015   355
P.A. Fuhrmann, U. Helmke, *The Mathematics of Networks*
of Linear Systems, Universitext, DOI 10.1007/978-3-319-16646-9_7

the signals u, y, z. Of course, if the observer is not chosen appropriately, the error trajectory may be large, which makes the estimate useless. Our aim is to characterize the properties of the observer in terms of these transfer functions. Even of greater importance is, whenever possible, the construction of observers having desired properties.

The issues of observation and estimation play a crucial role in analyzing observation processes for networks and are therefore of paramount importance for all questions concerning fault detection, monitoring, and measurement processes. Observer theory has a long history laden with vagueness, imprecision, and incomplete or even false proofs; see Trumpf (2013) for a short list of these. The principal reason for the difficulty in clarifying the structure theory of observers, functional observers in particular, seems to be that a full understanding of the problems requires the ability to integrate many topics and viewpoints that cover most of algebraic systems theory. These include state-space theory (including the dual Brunovsky form, realizations and partial realizations, Sylvester equations, and some old results of Roth and Halmos), polynomial and rational model theory, geometric control (conditioned invariant subspaces, as well as detectability and outer observability subspaces), and Hankel matrices. Another interesting point of view on observers is the behavioral approach, as developed and presented in the papers by Valcher and Willems (1999) and Fuhrmann (2008). However, to keep the exposition within reasonable limits, this direction will not be pursued. We will draw heavily on Fuhrmann and Helmke (2001a) and Trumpf (2002, 2013). Of course, the theory of observers depends strongly on the interpretation of what a linear system is and how it is represented. The state-space representation of a finite-dimensional, time-invariant linear system is chosen as our starting point. In addition, we focus on discrete-time systems because this simplifies matters when comparing trajectory-based formulations with statements for rational functions. Moreover, this enables us to state several results for systems over fields more general than the field $\mathbb{R}$ of real numbers. Because most of the statements remain true for continuous-time systems (and the field of real numbers), this restriction to discrete-time systems presents no real loss of generality.

7.1 Classical State Observers

In the preceding chapter, we discussed how to design state feedback control laws $u_t = -Kx_t + v_t$ for linear discrete-time systems of the form

$$
\begin{aligned}
x_{t+1} &= Ax_t + Bu_t, \\
y_t &= Cx_t,
\end{aligned}
\tag{7.1}
$$

where $A \in \mathbb{F}^{n \times n}, B \in \mathbb{F}^{n \times m}, C \in \mathbb{F}^{p \times n}$, and $\mathbb{F}$ is a field. Obviously, to implement control laws such as $u_t = -Kx_t + v_t$, one needs to know the state x_t, or at least an approximation to the state. State observers are designed precisely to fulfill such a purpose and are thus indispensable for practical applications of state feedback control. In this section, the classical construction of a full state observer, due to Luenberger (1964), is described; see also Luenberger (1971). To investigate standard stability properties of observers, it will be assumed that $\mathbb{F}$ is a subfield of the field $\mathbb{C}$ of complex numbers.

A **state observer** for system (7.1) is an n-dimensional linear control system

$$z_{t+1} = Fz_t + Gy_t + Hu_t, \tag{7.2}$$

with matrices $F \in \mathbb{F}^{n \times n}, G \in \mathbb{F}^{n \times p}, H \in \mathbb{F}^{n \times m}$, such that, for each initial condition $x_0, z_0 \in \mathbb{F}^n$ and every input sequence $u = (u_t)$, $\lim_{t \to \infty}(x_t - z_t) = 0$. A state observer is therefore a dynamical system that is driven by the input and output of (7.1) and whose state vectors z_t will asymptotically converge to the state vectors of (7.1).

How can one construct such an observer? Luenberger's ingenious idea was to consider systems with $F = A - LC, G = L, H = B$, i.e.,

$$\begin{aligned} z_{t+1} &= Az_t + Bu_t + L(y_t - \hat{y}_t), \\ \hat{y}_t &= Cz_t, \end{aligned} \tag{7.3}$$

with an **observer gain matrix** $L \in \mathbb{F}^{n \times p}$. Thus (7.3) consists of an identical copy of a system that is driven by the **innovations** $y_t - \hat{y}_t$. System (7.3) is often called the **Luenberger observer**. It has only one free design parameter, i.e., the observer gain matrix. To see how to choose the observer gain in order to achieve a state observer, one must consider the evolution of the **estimation error**

$$e_t = x_t - z_t, \quad t \in \mathbb{N}.$$

The dynamics of the estimation error is

$$e_{t+1} = (A - LC)e_t.$$

Thus the estimation error converges to zero if and only if L is chosen such that $A - LC$ has all its eigenvalues in the open unit disc. This leads to the following classical result on state observers.

Theorem 7.1. *The Luenberger observer (7.3) is a state observer for system (7.1) if and only if the observer gain $L \in \mathbb{F}^{n \times p}$ is such that $A - LC$ is asymptotically stable. Such an observer gain exists if and only if (C, A) is detectable.*

Proof. The observer condition for (7.3) is equivalent to $\lim_{t \to \infty} e_t = 0$ for all initial conditions $e_0 \in \mathbb{F}^n$. Thus (7.3) defines a state observer if and only if $A - LC$ is asymptotically stable, i.e., has all eigenvalues in the open unit disc. There exists such a stabilizing observer gain L if and only if (C, A) is detectable. $\blacksquare$

Having a state observer at hand, how can one use it for the purpose of state feedback control? Here we consider the closed-loop control system of the form

$$x_{t+1} = (A - B\mathscr{F})x_t + Bu_t,$$
$$y_t = Cx_t,$$

(7.4)

where $\mathscr{F} \in \mathbb{F}^{m \times n}$ is a desired state feedback gain. For example, $\mathscr{F}$ may be chosen such that the closed-loop characteristic polynomial $\det(zI - A + B\mathscr{F})$ is a prescribed monic polynomial of degree n. How must one choose the observer gain L? A beautiful simple result, the **separation principle**, provides a solution. It states that the designs of the state feedback and observer gain matrices can be done separately. But even then there is a problem because implementing (7.4) requires knowledge of the feedback term $-\mathscr{F}x_t$. This can be resolved by replacing $-\mathscr{F}x_t$ with the observer estimate $-\mathscr{F}z_t$. This then leads to the composed controller/observer dynamics with joint state variables $\xi = \mathrm{col}(\hat{x}, z)$:

$$\hat{x}_{t+1} = A\hat{x}_t - B\mathscr{F}z_t + Bu_t,$$
$$z_{t+1} = Az_t - B\mathscr{F}z_t + Bu_t + L(C\hat{x}_t - Cz_t),$$
$$y_t = C\hat{x}_t.$$

(7.5)

Written in matrix form we obtain

$$\xi_{t+1} = \mathscr{A}_c\xi_t + \mathscr{B}_c u_t,$$

with

$$\mathscr{A}_c = \begin{pmatrix} A & -B\mathscr{F} \\ LC & A - LC - B\mathscr{F} \end{pmatrix}, \qquad \mathscr{B}_c = \begin{pmatrix} B \\ B \end{pmatrix}.$$

(7.6)

The fundamental result for a combined controller and observer design is stated next.

Theorem 7.2 (Separation Principle). *Let $\mathscr{A}_c$ be defined by (7.6).*

1. The identity

$$\det(zI - \mathscr{A}_c) = \det(zI - A + B\mathscr{F})\det(zI - A + LC)$$

is true. In particular, for each state feedback gain $\mathscr{F}$ and every output injection gain $L \in \mathbb{F}^{n \times p}$ such that $A - LC$ is asymptotically stable, the composed controller/observer dynamics (7.5) satisfies

$$\lim_{t \to \infty}(z_t - \hat{x}_t) = 0$$

for arbitrary initial states $\hat{x}_0, z_0$ and input sequences (u_t).

2. *The transfer function from u to y of (7.5) is*

$$G(z) = C(zI - A + B\mathscr{F})^{-1}B.$$

More generally, the Z-transforms of (u_t) and (y_t) are related as

$$y(z) = C(zI - A + B\mathscr{F})^{-1}Bu(z)$$
$$+ C(zI - A + B\mathscr{F})^{-1}\hat{x}_0 + C(zI - A + B\mathscr{F})^{-1}B\mathscr{F}(zI - A + LC)^{-1}(z_0 - \hat{x}_0).$$

3. *Assuming that $A - B\mathscr{F}$ is stable, the system*

$$x_{t+1} = (A - B\mathscr{F})x_t + Bu_t,$$
$$y_t = Cx_t$$

satisfies

$$\lim_{t \to \infty}(x_t - \hat{x}_t) = 0$$

for all initial states $\hat{x}_0, z_0$ and input sequences (u_t).

Proof. For the invertible matrix

$$S = \begin{pmatrix} I & 0 \\ -I & I \end{pmatrix}$$

we compute

$$S\mathscr{A}_c S^{-1} = \begin{pmatrix} A - B\mathscr{F} & -B\mathscr{F} \\ 0 & A - LC \end{pmatrix}, \quad S\mathscr{B}_c = \begin{pmatrix} B \\ 0 \end{pmatrix}, \quad (C\ 0)S^{-1} = (C\ 0).$$

The transfer function of (7.5) is thus equal to $C(zI - A + B\mathscr{F})^{-1}B$. For the remaining parts, we proceed to consider the dynamics of the error term $\varepsilon_t := z_t - \hat{x}_t$, which is given as

$$\varepsilon_{t+1} = (A - LC)\varepsilon_t.$$

Our assumption on L implies $\lim_{t \to \infty}(z_t - \hat{x}_t) = 0$. For the last claim, consider the error sequence $e_t := x_t - \hat{x}_t$ with associated error dynamics. It satisfies

$$e_{t+1} = (A - B\mathscr{F})e_t + B\mathscr{F}(z_t - \hat{x}_t).$$

Since $A - B\mathscr{F}$ is stable and $\lim_{t \to \infty}(z_t - \hat{x}_t) = 0$, we conclude that $\lim_{t \to \infty}(x_t - \hat{x}_t) = 0$. This completes the proof. ∎

Of course, the preceding results on full state observers are only the starting point for a deeper theory of observers that enables one to estimate a finite number of linear state functionals. This more general observer theory is developed in subsequent sections.

7.2 Observation Properties

One of Kalman's major achievements has been the introduction of the concepts of reachability and observability, as distinct from compensator or observer design. This separation is reflected in Valcher and Willems (1999), where the observability or detectability of one set of system variables from another is studied before observer design is attempted. Such an approach is adopted in this section. Clearly, observers depend on the observability properties of a system, and a few gradations of observability will be introduced. Naturally, one expects that the stronger the observation properties of a system are, the better behaved should be the corresponding observers. How the observation properties of the system are reflected in the corresponding observers will be examined in Subsection 7.3.

To state the subsequent definitions and results over a field $\mathbb{F}$, the meaning of convergence in a finite-dimensional vector space $\mathbb{F}^n$ must be clarified. Here we proceed as in Chapter 6 with respect to the dual situation, i.e., that of state feedback stabilization. A general field $\mathbb{F}$ is endowed with the discrete topology, i.e., the unique topology on $\mathbb{F}$ whose open (and closed) subsets are defined by subsets of $\mathbb{F}$. In contrast, for a subfield $\mathbb{F} \subset \mathbb{C}$ we introduce the Euclidean topology on $\mathbb{F}$, which is defined by the standard Euclidean distance $|x - y|$ of complex numbers $x, y \in \mathbb{C}$. In either case, a discrete-time dynamical system $x_{t+1} = Ax_t$ on $\mathbb{F}^n$ is called stable whenever the sequence $x_t = A^t x_0$ converges to zero for all initial conditions $x_0 \in \mathbb{F}^n$. Thus a discrete-time dynamical system $x_{t+1} = Ax_t$ is asymptotically stable if and only if either A is Schur stable, for the Euclidean topology on $\mathbb{F} \subset \mathbb{C}$, or if A is nilpotent, for the discrete topology on $\mathbb{F}$. Let (e_t) denote a sequence of points in $\mathbb{F}^m$ that defines a proper rational function

$$e(z) = \sum_{t=0}^{\infty} e_t z^{-t} \in z^{-1}\mathbb{F}[[z^{-1}]]^m,$$

and let $(A, B, C, D) \in \mathbb{F}^{n \times n} \times \mathbb{F}^{n \times 1} \times \mathbb{F}^{m \times n} \times \mathbb{F}^m$ denote a minimal realization of

$$e(z) = D + C(zI - A)^{-1}B.$$

Thus $e_0 = D$, $e_t = CA^{t-1}B$, $t \geq 1$, is the sequence of Markov parameters of $e(z)$.

Proposition 7.3. *Let $e(z) \in \mathbb{F}(z)^m$ be proper rational. The sequence of Markov parameters (e_t) satisfies $\lim_{t \to \infty} e_t = 0$ if and only if:*

1. *All poles of $e(z)$ have absolute value < 1 or, equivalently, A is Schur stable. This assumes that $\mathbb{F} \subset \mathbb{C}$ carries the Euclidean topology;*
2. *All poles of $e(z)$ are in $z = 0$ or, equivalently, A is nilpotent. This assumes that $\mathbb{F}$ carries the discrete topology.*

Proof. Since $e_t = CA^{t-1}B$ for $t \geq 1$, and, by the minimality of (A, B, C), the observability and reachability matrices

$$\begin{pmatrix} C \\ \vdots \\ CA^{n-1} \end{pmatrix}, \quad (B \cdots A^{n-1}B)$$

have full column rank and full row rank, respectively. Thus the sequence of Hankel matrices

$$\begin{pmatrix} e_{nt+1} & \cdots & e_{n(t+1)} \\ \vdots & & \vdots \\ e_{n(t+1)} & \cdots & e_{n(t+2)-1} \end{pmatrix} = \begin{pmatrix} C \\ \vdots \\ CA^{n-1} \end{pmatrix} A^{tn} (B \cdots A^{n-1}B), \quad t \in \mathbb{N},$$

converges to zero if and only if $\lim_{t \to \infty} A^{tn} = 0$. If the field $\mathbb{F}$ carries the discrete topology, this is equivalent to A being nilpotent, while for the Euclidean topology on $\mathbb{F} \subset \mathbb{C}$ this is equivalent to the eigenvalues of A being in the open complex unit disc. $\blacksquare$

The next characterization will be useful later on.

Proposition 7.4. *Let $Q(z) \in \mathbb{F}[z]^{m \times m}$ be nonsingular, and let X^Q denote the associated rational model.*

1. *Let $\mathbb{F} \subset \mathbb{C}$ be endowed with the Euclidean topology. The following statements are equivalent:*

   (a) *All elements $h(z) \in X^Q$ are stable, i.e., the coefficients h_t of $h(z)$ satisfy $\lim_{t \to \infty} h_t = 0$.*
   (b) *$\det Q(z)$ is a Schur polynomial, i.e., all its roots are in the open unit disc.*

2. *Let $\mathbb{F}$ be endowed with the discrete topology. The following statements are equivalent:*

   (a) *All elements $h(z) \in X^Q$ are stable, i.e., the coefficients h_t of $h(z)$ satisfy $h_t = 0$ for t sufficiently large.*
   (b) *$\det Q(z)$ is a monomial, i.e., all its roots are equal to 0.*

Proof. Choosing a polynomial basis matrix $P(z) \in \mathbb{F}[z]^{m \times n}$ for the finite-dimensional polynomial model X_Q implies $Q(z)^{-1}P(z)$ is a basis matrix for the rational model X^Q, and therefore the elements of X^Q are of the form $h(z) = Q(z)^{-1}P(z)\xi$ for unique vectors $\xi \in \mathbb{F}^n$. By Proposition 4.36, there exists an observable pair $(C,A) \in \mathbb{F}^{m \times n} \times \mathbb{F}^{n \times n}$, with

$$C(zI - A)^{-1} = Q(z)^{-1}P(z).$$

Thus the coefficients of $Q(z)^{-1}P(z)\xi$ converge to zero for all choices of ξ if and only if $\lim_{t \to \infty} CA^t = 0$. This is equivalent to (C,A) being detectable. By Proposition 6.34, and applying the observability of (C,A), this is equivalent to $\det(zI - A)$ being a Schur polynomial, i.e., to all eigenvalues of A being in the open unit disc. Part (1) follows by observing the identity $\det Q(z) = \det(zI - A)$. Part (2) is proven similarly.

Next, some of the important observation properties of a system are introduced, e.g., the extent to which the observed variables y determine the relevant, or to-be-estimated, variables z.

Definition 7.5. Let Σ_{sys} be a linear system with the state-space representation

$$\Sigma_{sys} := \begin{cases} x_{t+1} = Ax_t + Bu_t, \\ \quad\;\; y_t = Cx_t, \\ \quad\;\; z_t = Kx_t, \end{cases} \tag{7.7}$$

with x_t, y_t, u_t, z_t taking values in $\mathbb{F}^n, \mathbb{F}^p, \mathbb{F}^m, \mathbb{F}^k$, respectively.

1. The variable z is **T-trackable** from (y,u) if there exists a nonnegative integer T such that for every two solutions $(x,y,u,z), (\bar{x},y,u,\bar{z})$ of (7.7) the condition $\bar{z}_t = z_t$ for $0 \le t \le T$ implies $\bar{z} = z$. The smallest such T is called the **tracking index** and is denoted by τ.
2. The variable z is **detectable** from (y,u) if each pair of solutions (x,y,u,z) and $(\bar{x},y,u,\bar{z})$ of (7.7) satisfies $\lim_{t \to \infty}(z_t - \bar{z}_t) = 0$.
3. The variable z is **reconstructible** from (y,u) if for each pair of solutions (x,y,u,z) and $(\bar{x},y,u,\bar{z})$ of (7.7) there exists a nonnegative integer T such that $z_t - \bar{z}_t = 0$ for $t > T$. The smallest such T is called the **reconstructibility index**.
4. The variable z is **observable** from (y,u) if each pair of solutions (x,y,u,z) and $(\bar{x},y,u,\bar{z})$ of (7.7) satisfies $\bar{z} = z$.

One says that a system Σ_{sys}, given by (7.7), is **T-trackable** if z is T-trackable from y. We similarly define **detectability, reconstructibility**, and **observability**.

In view of the Cayley–Hamilton theorem, it is obvious that every linear state function $z = Kx$ is T-trackable from the output y with $u = 0$ of an n-dimensional linear system

$$\begin{aligned} x_{t+1} &= Ax_t \\ y_t &= Cx_t, \end{aligned}$$

provided $T \ge n$. Thus a finite-dimensional linear system is always trackable, unless one requires an a priori bound on the tracking index. Therefore, defining trackability without imposing a constraint on the tracking index is meaningless. For reconstructibility, the situation becomes slightly different because not every linear system is reconstructible. Of course, the definition of detectability strongly depends on the topology of the field $\mathbb{F}$. If the field $\mathbb{F}$ is finite, or more generally if $\mathbb{F}$ carries the discrete topology, detectability is equivalent to reconstructibility. However, for a subfield $\mathbb{F} \subset \mathbb{C}$ with the standard Euclidean topology, this is no longer true. In fact, for the standard Euclidean topology on a subfield $\mathbb{F} \subset \mathbb{C}$, the detectability of z from y is equivalent to the condition that for an unobservable state x_0 the rational function $K(zI - A)^{-1}x_0$ has only poles in the open unit disc. Intuitively, it is clear that, since perfect knowledge of the system is assumed, the effect of the input variable on the estimate can be removed without affecting the observation properties. The following simple proposition is stated, with its trivial proof omitted.

Proposition 7.6. *Let Σ be a linear system with the state-space representation (7.7), together with the associated system Σ' given by*

$$(\Sigma') \quad \begin{cases} x_{t+1} = Ax_t, \\ \quad y_t = Cx_t, \\ \quad z_t = Kx_t. \end{cases} \tag{7.8}$$

Then:

1. *The following conditions are equivalent:*

   (a) *z is T-trackable from (y, u) with respect to Σ,*
   (b) *z is T-trackable from y with respect to Σ',*
   (c) *For all initial conditions x_0 such that $y_t = 0$ and $u_t = 0$ are satisfied for all t, the condition $z_0 = \cdots = z_T = 0$ implies $z_t = 0$ for all t;*

2. *The following conditions are equivalent:*

   (a) *z is detectable from (y, u) with respect to Σ,*
   (b) *z is detectable from y with respect to Σ',*
   (c) *For all initial conditions x_0 such that $y_t = 0$ and $u_t = 0$ are satisfied for all t, $\lim_{t \to \infty} z_t = 0$;*

3. *The following conditions are equivalent:*

   (a) *z is reconstructible from (y, u) with respect to Σ,*
   (b) *z is reconstructible from y with respect to Σ',*
   (c) *For all initial conditions x_0 that satisfy $y_t = 0$ and $u_t = 0$ for all t, then $z_t = 0$ for all $t > T$;*

4. *The following conditions are equivalent:*

   (a) *z is observable from (y, u) with respect to Σ,*
   (b) *z is observable from y with respect to Σ',*
   (c) *For all initial conditions x_0 such that $y_t = 0$ and $u_t = 0$ are satisfied for all t, $z_t = 0$ for all t.*

Next, the invariance of the preceding notions with respect to **output injection equivalence** is explored.

Proposition 7.7. *Let $S \in GL_n(\mathbb{F}), L \in \mathbb{F}^{n \times p}, R \in GL_p(\mathbb{F}), U \in GL_m(\mathbb{F})$ be output injection transformations. A system (A,B,C,K) is T-trackable, reconstructible, detectable, or observable, respectively, if and only if the output injection equivalent system $(S(A+LC)S^{-1}, SBU^{-1}, RCS^{-1}, KS^{-1})$ is.*

Proof. It is obvious that state-space similarity transformations and invertible coordinate changes in the input and output spaces, respectively, do not change the aforementioned properties. Thus it suffices to prove the result for output injection transformations. But the invariance of the notions under output injection $(A,B,C,K) \mapsto (A+LC,B,C,K)$ is obvious from Proposition 7.6. ∎

It is convenient, for our further analysis, to first transform the system into a simple normal form by state-space similarity. This is done next and depends on an observability condition for the pair $\left(\binom{C}{K}, A \right)$. A priori, there is no reason to assume that the pair $\left(\binom{C}{K}, A \right)$ is observable, but the following proposition shows that this entails no great loss of generality.

Proposition 7.8. *1. Every linear system (7.8) can be reduced to the case that the pair $\left(\binom{C}{K}, A \right)$ is observable.*

2. If $\left(\binom{C}{K}, A \right)$ is observable but (C,A) is not, then the system Σ' has a state-space equivalent representation of the form

$$A = \begin{pmatrix} A_{11} & 0 \\ A_{21} & A_{22} \end{pmatrix},$$

$$C = (C_1 \ 0),$$

$$K = (K_1 \ K_2),$$

(7.9)

with both pairs (C_1, A_{11}) and (K_2, A_{22}) observable.

Proof. 1. If the pair $\left(\binom{C}{K}, A \right)$ is not observable, the system can be reduced to an observable one. Letting $\mathcal{V} = \bigcap_{j \geq 0} \mathrm{Ker} \left(\binom{C}{K} \right) A^j$ be the unobservable subspace for the pair $\left(\binom{C}{K}, A \right)$ and $\mathcal{W}$ be a complementary subspace leads to the direct sum decomposition of the state space into $\mathcal{X} = \mathcal{W} \oplus \mathcal{V}$. Writing $x = \binom{x_1}{x_2}$, with $x_1 \in \mathcal{W}$ and $x_2 \in \mathcal{V}$, implies the following block matrix representations:

$$A = \begin{pmatrix} A_{11} & 0 \\ A_{21} & A_{22} \end{pmatrix}, \quad \begin{pmatrix} C \\ K \end{pmatrix} = \begin{pmatrix} C_1 & 0 \\ K_1 & 0 \end{pmatrix}, \quad B = \begin{pmatrix} B_1 \\ B_2 \end{pmatrix}.$$

Necessarily, $\left(\left(\begin{smallmatrix} C_1 \\ K_1 \end{smallmatrix}\right), A_{11}\right)$ is an observable pair and (7.7) can be replaced by

$$x_{t+1} = A_{11}x_t + B_1 u_t,$$
$$y_t = C_1 x_t,$$
$$z_t = K_1 x_t,$$

since C_2 and K_2 are both zero, and hence x_2 plays no role.

2. If (C,A) is not an observable pair, then let $\mathscr{O}_* = \mathscr{O}_*(C,A) = \bigcap_{j \geq 0} \mathrm{Ker}\, CA^j \subset \mathbb{F}^n$ be the unobservable subspace of (C,A). Let $\mathscr{W}$ be a complementary subspace to $\mathscr{O}_*$. With respect to the direct sum decomposition

$$\mathbb{F}^n = \mathscr{W} \oplus \mathscr{O}_*, \tag{7.10}$$

we obtain the block matrix representation (7.9). By construction, the pair (C_1, A_{11}) is observable. Also, A_{22} is similar to $A|\mathscr{O}_*$. Our assumption that the pair $\left(\left(\begin{smallmatrix} C \\ K \end{smallmatrix}\right), A\right)$ is observable implies that the pair (K_2, A_{22}) is also observable. ∎

Coprime factorizations are the most effective tool in bridging the gap between frequency-domain and state-space methods. This is done using the shift realization. The following results, split into two separate theorems, examine the corresponding functional characterizations. Consider a state-space system

$$\Sigma_{sys} := \begin{cases} x_{t+1} = Ax_t + Bu_t, \\ \quad\ y_t = Cx_t, \\ \quad\ z_t = Kx_t, \end{cases}$$

with $A \in \mathbb{F}^{n \times n}, B \in \mathbb{F}^{n \times m}, C \in \mathbb{F}^{p \times n}, K \in \mathbb{F}^{k \times n}$.

Theorem 7.9. *Assume that* $\left(\left(\begin{smallmatrix} C \\ K \end{smallmatrix}\right), A\right)$ *is observable and has the representation (7.9).*

1. *There exists a left coprime factorization of the state-to-output transfer function of* Σ_{sys} *of the form*

$$\begin{pmatrix} C_1 & 0 \\ K_1 & K_2 \end{pmatrix} \begin{pmatrix} zI - A_{11} & 0 \\ -A_{21} & zI - A_{22} \end{pmatrix}^{-1} = \begin{pmatrix} D_{11}(z) & 0 \\ D_{21}(z) & D_{22}(z) \end{pmatrix}^{-1} \begin{pmatrix} \Theta_{11}(z) & 0 \\ \Theta_{21}(z) & \Theta_{22}(z) \end{pmatrix}, \tag{7.11}$$

with $D_{11}(z) \in \mathbb{F}[z]^{p \times p}$ *and* $D_{22}(z) \in \mathbb{F}[z]^{k \times k}$ *nonsingular,* $D_{21}(z) \in \mathbb{F}[z]^{k \times p}, \Theta_{11}(z) \in \mathbb{F}[z]^{p \times (n-r)}, \Theta_{21}(z) \in \mathbb{F}[z]^{k \times (n-r)}, \Theta_{22}(z) \in \mathbb{F}[z]^{k \times r}$, *for which:*

(a) $D_{11}(z)^{-1}\Theta_{11}(z)$ and $D_{22}(z)^{-1}\Theta_{22}(z)$ are left coprime factorizations of the transfer functions $C_1(zI - A_{11})^{-1}$ and $K_2(zI - A_{22})^{-1}$, respectively. $D_{21}(z)$ and $\Theta_{21}(z)$ satisfy the polynomial equation

$$D_{22}(z)K_1 + \Theta_{22}(z)A_{21} = -D_{21}(z)C_1 + \Theta_{21}(z)(zI - A_{11}); \qquad (7.12)$$

(b) $D_{11}(z)$ and $D_{22}(z)$ are row proper;
(c) $D_{21}(z)D_{11}(z)^{-1}$ and $D_{22}(z)^{-1}D_{21}(z)D_{11}(z)^{-1}$ are strictly proper.

One refers to the coprime factorization (7.11), satisfying parts (a)–(c), as an **adapted coprime factorization**.
2. Assume that (7.11) is an adapted coprime factorization. The following properties are in force:

(a) $n = \deg\det(zI - A) = \deg\det D_{11}(z) + \deg\det D_{22}(z)$.
(b) $D_{22}(z)$ is a nonsingular polynomial matrix.
(c) The linear map

$$\psi : \mathscr{O}_*(C,A) \longrightarrow X_{D_{22}}, \quad \psi(x) = \Theta_{22}x,$$

is bijective, satisfying

$$\psi A_{22} = S_{D_{22}}\psi. \qquad (7.13)$$

This implies the isomorphism

$$S_{D_{22}} \simeq A_{22} = A|_{\mathscr{O}_*(C,A)}. \qquad (7.14)$$

Proof. The first claim of statement 1 is proved first. Applying Proposition 7.8, with respect to the direct sum decomposition (7.10), yields the block matrix representation (7.9), with the pairs (C_1,A_{11}) and (K_2,A_{22}) observable. Let $D_{11}(z)^{-1}\Theta_{11}(z)$ and $D_{22}(z)^{-1}\Theta_{22}(z)$ be left coprime factorizations of $C_1(zI - A_{11})^{-1}$ and $K_2(zI - A_{22})^{-1}$, respectively. Since a left coprime factorization is unique only up to a common left unimodular factor, we will assume, without loss of generality, that $D_{11}(z)$ and $D_{22}(z)$ are both row proper. So (*b*) holds by construction. Thus the (11)- and (22)-terms on both sides of (7.11) are equal. Comparing the (21)-terms of both sides of equation (7.11), multiplying by $D_{22}(z)$ on the left and by $(zI - A_{11})$ on the right, we obtain (7.12). Thus $D_{21}(z),\Theta_{21}(z)$ fulfill (7.11) if and only if (7.12) is satisfied. By the observability of the pair (C_1,A_{11}), the existence of a polynomial solution $X(z),Y(z)$ of the Bezout equation $X(z)C_1 + Y(z)(zI - A_{11}) = I$ follows. Consequently, taking into consideration the general, polynomial, solution of the homogeneous equation, we obtain the parameterization

$$D_{21}(z) = -(D_{22}(z)K_1 + \Theta_{22}(z)A_{21})X(z) - Q(z)D_{11}(z),$$
$$\Theta_{21}(z) = (D_{22}(z)K_1 + \Theta_{22}(z)A_{21})Y(z) + Q(z)\Theta_{11}(z),$$

where $Q(z) \in \mathbb{F}[z]^{k \times p}$. Choosing $Q(z) = -\pi_+ \left((D_{22}K_1 + \Theta_{22}A_{21})XD_{11}^{-1} \right)$ guarantees that $D_{21}(z)D_{11}(z)^{-1}$ is strictly proper. This does not change the row properness of $D_{11}(z)$ and $D_{22}(z)$. That $D_{22}(z)^{-1}D_{21}(z)D_{11}(z)^{-1}$ is strictly proper follows from the strict properness of $D_{21}(z)D_{11}(z)^{-1}$ and the fact that the nonsingular, row proper polynomial matrix $D_{22}(z)$ has a proper inverse.

Proof of 2. From the observability assumption on the pair $\left(\binom{C}{K}, A \right)$ and the left coprime factorization (7.11), we conclude that

$$n = \deg \det(zI - A) = \deg \det \begin{pmatrix} D_{11}(z) & 0 \\ D_{21}(z) & D_{22}(z) \end{pmatrix}$$
$$= \deg \det D_{11}(z) + \deg \det D_{22}(z),$$

which proves the first claim. The nonsingularity of $D_{22}(z)$ follows from the nonsingularity of $\begin{pmatrix} D_{11}(z) & 0 \\ D_{21}(z) & D_{22}(z) \end{pmatrix}$.

To prove the last claim, note that the coprime factorization

$$D_{22}(z)^{-1}\Theta_{22}(z) = K_2(zI - A_{22})^{-1} \tag{7.15}$$

is equivalent to the intertwining relation

$$\Theta_{22}(z)(zI - A_{22}) = D_{22}(z)K_2.$$

Applying Theorems 3.20 and 3.21 proves the intertwining relation (7.13) as well as the invertibility of the map ψ defined by (7.23). The isomorphism (7.14) follows from (7.13) and the invertibility of ψ. ∎

Theorem 7.10. *Assume that (7.11) is an adapted coprime factorization.*

1. Define strictly proper rational matrices $Z_K(z), Z_C(z)$ by

$$\begin{aligned} Z_C(z) &= C(zI - A)^{-1}, \\ Z_K(z) &= K(zI - A)^{-1}. \end{aligned} \tag{7.16}$$

The general rational solutions of the equation

$$Z_K(z) = Z_1(z)Z_C(z) + Z_2(z) \tag{7.17}$$

are

$$\begin{aligned} Z_1(z) &= -D_{22}(z)^{-1}D_{21}(z) + W(z)D_{11}(z), \\ Z_2(z) &= \left(D_{22}(z)^{-1}\Theta_{21}(z) - W(z)\Theta_{11}(z), \ D_{22}(z)^{-1}\Theta_{22}(z) \right), \end{aligned} \tag{7.18}$$

where $W(z) \in \mathbb{F}(z)^{k \times p}$ is a rational function. Let $Z_1(z)$ be proper rational. Using an adapted left coprime factorization (7.11), there exists a proper rational $W(z)$ such that $Z_1(z)$ is given by (7.18) and $Z_2(z)$ is strictly proper.

2. (a) $D_{22}(z)$ is a stable matrix if and only if the pair (C,A) is detectable.
   (b) $D_{22}(z)$ is a unimodular matrix if and only if the pair (C,A) is observable. In this case, that is, where $\mathcal{O}_*(C,A) = \{0\}$, the coprime factorization (7.11) reduces to

$$\begin{pmatrix} C \\ K \end{pmatrix} (zI - A)^{-1} = \begin{pmatrix} D_{11}(z) & 0 \\ D_{21}(z) & I \end{pmatrix}^{-1} \begin{pmatrix} \Theta_{11}(z) \\ \Theta_{21}(z) \end{pmatrix}, \tag{7.19}$$

with $D_{11}^{-1}\Theta_{11}$ a left coprime factorization of $C(zI - A)^{-1}$ and $D_{21}(z), \Theta_{21}(z)$ determined from the equation

$$K = -D_{21}(z)C + \Theta_{21}(z)(zI - A). \tag{7.20}$$

In this case, the general solution of equation (7.17) is given by

$$\begin{aligned} Z_1(z) &= -D_{21}(z) + W(z)D_{11}(z), \\ Z_2(z) &= \Theta_{21}(z) - W(z)\Theta_{11}(z). \end{aligned} \tag{7.21}$$

3. (a) Let $\phi : \mathbb{F}^n \longrightarrow X^{zI-A}$ be defined by $\phi(x) = (zI - A)^{-1}x$. Then ϕ is injective, with

$$\phi(\mathcal{O}_*) = \{0\} \oplus X^{zI-A_{22}}, \tag{7.22}$$

and thus induces an isomorphism $\phi : \mathcal{O}_* \longrightarrow X^{zI-A_{22}}$.
   (b) The map $\psi : \mathcal{O}_* \longrightarrow X_{D_{22}}$, defined by

$$\psi(x) = \Theta_{22}(z)x, \tag{7.23}$$

is an isomorphism satisfying $\psi(Ax) = S_{D_{22}}\psi(x)$.
   (c) The map $\Psi : \mathcal{O}_* \longrightarrow X^{D_{22}}$ defined by $x \mapsto D_{22}(z)^{-1}\Theta_{22}(z)x$ is an isomorphism satisfying $\Psi(Ax) = S^{D_{22}}\Psi(x)$, i.e., $A|\mathcal{O}_* \simeq S^{D_{22}}$.

Proof. Proof of 1. Using (7.11) and computing

$$Z_K(z) = (K_1 \ K_2) \begin{pmatrix} (zI - A_{11})^{-1} & 0 \\ (zI - A_{22})^{-1}A_{21}(zI - A_{11})^{-1} & (zI - A_{22})^{-1} \end{pmatrix}$$

$$= \left(K_1(zI - A_{11})^{-1} + K_2(zI - A_{22})^{-1}A_{21}(zI - A_{11})^{-1}, \ K_2(zI - A_{22})^{-1} \right)$$

$$= \left(-D_{22}(z)^{-1}D_{21}(z)D_{11}(z)^{-1}\Theta_{11}(z) + D_{22}(z)^{-1}\Theta_{21}(z), \ D_{22}(z)^{-1}\Theta_{22}(z) \right)$$

$$= Z_1(z)Z_C(z) + Z_2(z)$$

$$(7.24)$$

leads to a particular solution of (7.17), i.e.,

$$Z_1(z) = -D_{22}(z)^{-1}D_{21}(z),$$

$$Z_2(z) = \left(D_{22}(z)^{-1}\Theta_{21}(z), \ D_{22}(z)^{-1}\Theta_{22}(z) \right).$$

To obtain the general rational solution, one needs to add to (Z_1, Z_2) the general rational solution (Y_1, Y_2) of the homogeneous equation $Y_1(z)Z_C(z) + Y_2(z) = 0$ or, equivalently, $Y_1(z)C + Y_2(z)(zI - A) = 0$. Noting that $Z_C(z) = \left(D_{11}(z)^{-1}\Theta_{11}(z) \ 0 \right)$ and writing $Y_2(z) = \left(Y_2'(z), \ Y_2''(z) \right)$ implies $Y_1(z) = W(z)D_{11}(z), \ Y_2'(z) = -W(z)\Theta_{11}(z))$, and $Y_2''(z) = 0$, with $W(z)$ a free, rational parameter. This proves (7.18).

For the second claim, choose

$$W(z) = (Z_1(z) + D_{22}(z)^{-1}D_{21}(z))D_{11}(z)^{-1}$$
$$= Z_1(z)D_{11}(z)^{-1} + D_{22}(z)^{-1}D_{21}(z)D_{11}(z)^{-1},$$

and note that the properness of $W(z)$ follows from the assumed properness of $Z_1(z)$ and the assumption that (7.11) is an adapted coprime factorization.

Proof of 2. The pair (C, A) is detectable if and only if $A_{22} \simeq A|\bigcap_{i=0}^{\infty} \mathrm{Ker}\, CA^i$ is stable. By the isomorphism (7.14), this is equivalent to the stability of $D_{22}(z)$, which proves the first claim. The pair (C, A) is observable if and only if the equality of degrees $n = \deg\det(zI - A) = \deg\det D_{11}(z)$ is satisfied. This is equivalent to $\deg\det D_{22}(z) = 0$, i.e., to $D_{22}(z)$ being unimodular. Equation (7.20) is a special case of (7.12). Similarly, the parameterization (7.21) is a special case of (7.18).

Proof of 3. Since (C, A) is assumed to be in Kalman decomposition form, $x \in \mathcal{O}_*$ if and only if $x = \mathrm{col}\,(0, x_2)$. Thus $\phi(\mathcal{O}_*) = \{(zI - A)^{-1}x|x \in \mathcal{O}_*\} = \{0\} \times X^{zI - A_{22}}$, which proves (7.22). The coprime factorizations $D_{22}(z)^{-1}\Theta_{22}(z) = K_2(zI - A_{22})^{-1}$ yields the intertwining relation

$$\Theta_{22}(z)(zI - A_{22}) = D_{22}(z)K_2.$$

Thus the map $\psi : X_{zI - A_{22}} \longrightarrow X_{D_{22}}$ defined by $\psi(x) = \pi_{D_{22}}\Theta_{22}x = \Theta_{22}x$ is an isomorphism. Note that $\{0\} \times X_{zI - A_{22}} = \mathcal{O}_*$.

Since autonomous behaviors are equal to rational models (Theorem 3.36), we conclude that $X^{D_{22}} = \mathrm{Ker}\, D_{22}(\sigma)$ is true for the backward shift operator σ. Now the multiplication map $D_{22}^{-1} : X_{D_{22}} \longrightarrow X^{D_{22}}$ is an $\mathbb{F}[z]$-module isomorphism, and therefore the composed map $\Psi = D_{22}^{-1} \psi$ is also an $\mathbb{F}[z]$-module isomorphism from $\mathcal{O}_*$ onto $X^{D_{22}}$. This proves the last claim.   ∎

For unobservable states $x \in \mathcal{O}_*$, using (7.15), one computes

$$K_2 \phi(x) = K_2(zI - A_{22})^{-1}x = D_{22}(z)^{-1}\Theta_{22}(z)x = \Psi x = D_{22}(z)^{-1}\psi(x),$$

which implies that the following diagram is a commutative diagram of $\mathbb{F}$-vector space isomorphisms:

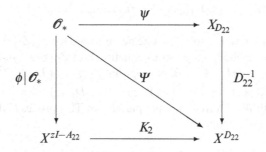

It is of principal interest to find characterizations of the observation properties introduced in Definition 7.5. This depends very much on the functional relation between the observed variables y and the to-be-estimated variables z. Of course, in the state-space representation (7.8) of the system Σ', this relation is indirect. To get a direct relation, one needs to eliminate the state variable x from (7.8). This is best done in a behavioral setting but is avoided here.

Thus, avoiding the explicit use of behaviors, we proceed by characterizing the tracking index of linear systems. First let us note, as an immediate consequence of the definition, that the trackability of (7.7) with tracking index τ is satisfied if and only if τ is the smallest number such that, for every initial state $x_0 \in \mathbb{F}^n$ with $C(zI - A)^{-1}x_0 = 0$, the implication

$$Kx_0 = \cdots = Kx_\tau = 0 \implies Kx_t = 0, \quad \forall t \geq 0,$$

follows. Note that if (C,A) is observable, then the tracking index of an output functional $z = Kx$ is $\tau = 0$.

Proposition 7.11. *A linear system (7.7) with $\left(\binom{C}{K}, A \right)$ observable and having the representation (7.9), has a tracking index τ if and only if the largest observability index of (K_2, A_{22}) is equal to τ. In particular, T-trackability is fulfilled for every T that is greater than or equal to the degree of the minimal polynomial of A.*

Proof. Using the representation of states with respect to the direct sum (7.10), let an initial state have the representation $x_0 = \begin{pmatrix} u \\ v \end{pmatrix}$. The initial condition x_0 is unobservable, that is, $C(zI - A)^{-1}x_0 = 0$ if and only if $u = 0$. For each such x_0 one has $KA^i x_0 = K_2 A_{22}^i v$ for all $i \geq 0$, implying the system is T-trackable if and only if the implication $K_2 A_{22}^i v = 0, \quad i = 0, \ldots, T \implies K_2 A_{22}^i v = 0 \quad \forall i \geq 0$ is valid for all vectors $v \in \mathbb{F}^{n_2}$. Equivalently, this says that

$$\mathrm{Ker} \begin{pmatrix} K_2 \\ \vdots \\ K_2 A_{22}^T \end{pmatrix} \subset \mathrm{Ker} \begin{pmatrix} K_2 \\ \vdots \\ K_2 A_{22}^{n-1} \end{pmatrix}.$$

In turn, this is equivalent to saying that all observability indices of (K_2, A_{22}) are less than or equal to T, which implies the result. $\blacksquare$

The following lemma will be needed.

Lemma 7.12. *Let $Q(z) \in \mathbb{F}[z]^{r \times r}$ be nonsingular with degree ℓ, i.e., $Q(z) = Q_0 + Q_1 z + \cdots + Q_\ell z^\ell$. Assuming $h(z) = \sum_{j=1}^{\infty} \dfrac{h_j}{z^j} \in X^Q$ and $h_1 = \cdots = h_\ell = 0$ implies $h(z) = 0$.*

Proof. Since $X^Q = \mathrm{Im}\,\pi^Q$, it follows that $h \in X^Q$ if and only if $h = \pi^Q h$. However, under our assumptions, $\pi^Q h = \pi_- Q^{-1} \pi_+ (Qh) = \pi_- Q^{-1} \pi_+ (Qz^{-\ell} z^\ell h)$. Clearly, $Q(z) z^{-\ell}$ is proper, whereas $z^\ell h$ is strictly proper, so the product is strictly proper with $\pi_+(Qh) = 0$. This implies $h(z) = 0$. $\blacksquare$

Lemma 7.12 leads us to a simple polynomial characterization of the tracking index.

Proposition 7.13. *Assume that (7.7) has the representation (7.9) such that the pair $\left(\begin{pmatrix} C \\ K \end{pmatrix}, A \right)$ is observable. Letting $K_2(zI - A_{22})^{-1} = D_{22}(z)^{-1} \Theta_{22}(z)$ be a left coprime factorization, with $D_{22}(z)$ row proper, implies the degree of $D_{22}(z)$ coincides with the tracking index τ of (7.7). If $D_{22}(z)$ is not row proper, the degree of $D_{22}(z)$ provides an upper bound for the tracking index.*

Proof. By Proposition 7.11, the minimal tracking index is equal to the maximal observability index of (K_2, A_{22}). In view of Corollary 6.9, the observability indices of (K_2, A_{22}) coincide with the right Wiener–Hopf indices of $D_{22}(z)$. Since $D_{22}(z)$ is assumed to be in row proper form, the row indices of $D_{22}(z)$ coincide with the right Wiener–Hopf indices. In particular, the degree of $D_{22}(z)$ is the largest right Wiener–Hopf index, i.e., it is equal to the largest observability index of (K_2, A_{22}). If $D_{22}(z)$ is not row proper, Lemma 7.12 can be applied to see that the degree of $D_{22}(z)$ gives an upper bound for the tracking index. $\blacksquare$

The preceding analysis leads to the following explicit characterizations of the different observation properties.

Theorem 7.14. *Assume that (7.7) has the representation (7.9) such that* $\left(\begin{pmatrix} C \\ K \end{pmatrix}, A \right)$ *is observable. Let*

$$K_2(zI - A_{22})^{-1} = D_{22}(z)^{-1}\Theta_{22}(z),$$

with $D_{22}(z), \Theta_{22}(z)$ *left coprime.*

1. *Suppose* $D_{22}(z)$ *is row proper. The following conditions are equivalent:*

   (a) *(7.7) has tracking index* τ.
   (b) *The largest observability index of* (K_2, A_{22}) *is equal to* τ.
   (c) *The degree of* $D_{22}(z)$ *is equal to* τ.

2. *The following conditions are equivalent:*

   (a) z *is detectable from* (y, u) *in (7.7).*
   (b) A_{22} *is stable.*
   (c) $\det D_{22}(z)$ *is a Schur polynomial.*
   (d) *All elements of* $X^{D_{22}}$ *have their poles in the open unit disc.*

3. *The following conditions are equivalent:*

   (a) z *is reconstructible from* (y, u) *in (7.7).*
   (b) A_{22} *is nilpotent.*
   (c) $D_{22}(z)$ *is a monomic polynomial matrix.*
   (d) *All elements of* $X^{D_{22}}$ *have their poles in zero.*

4. *The following conditions are equivalent:*

   (a) z *is observable from* (y, u) *in (7.7).*
   (b) $D_{22}(z)$ *is unimodular.*
   (c) (C, A) *is observable.*

Proof. Part 1 has already been shown.

Part 2. By Proposition 7.6, detectability is satisfied if and only if $\lim_{t \to \infty} z_t = 0$, whenever $(y_t) = 0$ and $(u_t) = 0$. The Z-transform of (z_t) is equal to $K(zI - A)^{-1}x_0$, where x_0 is in the unobservable subspace $\mathcal{O}_*$. By Theorem 7.10,

$$\{K(zI - A)^{-1}x_0 \mid x_0 \in \mathcal{O}_*\} = K_2 X^{zI - A_{22}} = X^{D_{22}}$$

is an autonomous behavior. Applying Proposition 7.4, we conclude that detectability is satisfied if and only if $\det D_{22}(z)$ is a Schur polynomial. Since $\det D_{22}(z)$ is equal to $\det(zI - A_{22})$, this is equivalent to A_{22} being stable.

For part 3 one can argue similarly. In fact, reconstructibility is equivalent to all elements of $X^{D_{22}}$ being stable for the discrete topology of $\mathbb{F}$, which just says that all elements of $X^{D_{22}}$ are of the form $z^{-N}p(z)$ for polynomials of degree $< N$ or, equivalently, that $D_{22}(z)$ is monomic. Applying Proposition 7.4, one sees that this in turn is equivalent to $\det D_{22}(z) = cz^{n_2}$ being a monomial or, equivalently, that all roots of $\det D_{22}(z) = 0$ are equal to zero, or that A_{22} is nilpotent.

Finally, the observability of z from (y, u) is fulfilled if and only if $(y_t) = 0$ and $(u_t) = 0$ implies $(z_t) = 0$. This is equivalent to

$$\{K(zI - A)^{-1}x_0 \mid x_0 \in \mathcal{O}_*\} = K_2 X^{zI - A_{22}} = \{0\}.$$

By the observability of (K_2, A_{22}), this is possible only if (C, A) is observable. This completes the proof.   ∎

It may be surprising to note that the characterizations of detectability, reconstructibility, and observability for a linear functional K in Theorem 7.14 are identical with the corresponding ones for $K = I_n$. This is stated as a corollary.

Corollary 7.15. *The same assumptions are used as in Theorem 7.14. For a linear system (7.7), the output $z = Kx$ is detectable, reconstructible, or observable from y if and only if the pair (C, A) is detectable, reconstructible, or observable, respectively.*

7.3 Functional State Observers

When dealing with complex systems, one may want to track only a sample of the state variables. However, these variables of interest may be impossible to observe directly, so one must have recourse to estimation procedures that utilize only the available observations. In certain cases this can be achieved even if the system is not completely observable. As indicated in the introduction to this chapter and emphasized by the observer diagram appearing there, an observer is itself a dynamical system driven by inputs and observations and whose output is an estimate ζ of the relevant variable z. This leads us to the following definition of functional state observers, which broadly extends the class of full state Luenberger observers. In the sequel, the principal results will be stated and proved only for discrete-time systems; these results hold, mutatis mutandis, also in the continuous time case.

Let

$$\Sigma_{sys} := \begin{cases} x_{t+1} = Ax_t + Bu_t \\ y_t = Cx_t \\ z_t = Kx_t \end{cases} \tag{7.25}$$

be a linear system, with A, B, C, K in $\mathbb{F}^{n \times n}, \mathbb{F}^{n \times m}, \mathbb{F}^{p \times n}, \mathbb{F}^{k \times n}$, respectively. Let another system,

$$\Sigma_{est} := \begin{cases} \xi_{t+1} = F\xi_t + Gy_t + Hu_t, \\ \zeta_t = J\xi_t + Ey_t, \end{cases} \tag{7.26}$$

be given with system matrices F, G, H, J, E in $\mathbb{F}^{q \times q}, \mathbb{F}^{q \times p}, \mathbb{F}^{q \times m}, \mathbb{F}^{k \times q}, \mathbb{F}^{k \times p}$ respectively, and driven by the input u and output y of (7.25). It will always be assumed that J is of full row rank, which presents no restriction of generality, as well as that both C and K have full row rank and that the pair $\left(\begin{pmatrix} C \\ K \end{pmatrix}, A \right)$ is observable.

Define the **estimation error** or **error trajectory** e by

$$e_t = z_t - \zeta_t = Kx_t - J\xi_t - Ey_t = Kx_t - \begin{pmatrix} J & E \end{pmatrix} \begin{pmatrix} x_t \\ y_t \end{pmatrix}.$$

The error trajectory defines the strictly proper power series

$$e = e(z) = \sum_{t=0}^{\infty} e_t z^{-t-1}. \tag{7.27}$$

As usual, the error trajectory will often be identified with the equivalent formal power series expansion (7.27) that it defines. We will refer to (7.26) as a **functional observer** because it is designed to estimate a function of the state rather than the state itself.

Definition 7.16. Consider the linear system (7.25). The system Σ_{est} defined by (7.26) will be called

1. a **finitely determined observer** for K if there exists a $T \in \mathbb{N}$ such that $e_t = 0$ for $t < T$ implies $e = 0$;
2. a **tracking observer** for K if for every $x_0 \in \mathbb{F}^n$ there exists a $\xi_0 \in \mathbb{F}^q$ such that, for all input functions u, the solutions x_t and ξ_t of (7.25) and (7.26), respectively, satisfy $e_t = z_t - \zeta_t = 0$ for all $t \geq 0$;
3. a **strongly tracking observer** for K if $e_0 = z_0 - \zeta_0 = 0$ implies $e_t = 0$ for all input functions u and $t \geq 0$;
4. an **asymptotic observer** for K if, for all initial conditions of the states x and ξ and all inputs u, $\lim_{t \to \infty} e_t = \lim_{t \to \infty} (z_t - \zeta_t) = 0$; an observer is called an **asymptotic tracking observer** for K if it is both a tracking observer and an asymptotic observer;
5. **spectrally assignable** if, given a polynomial $p(z)$ of degree q, there exists an observer in the family for which the characteristic polynomial of F is $p(z)$.

In all cases, q will be called the **order** of the observer.

Naturally, there are two fundamental problems that present themselves, namely, given system (7.25), how does one obtain a characterization of observers and how does one show the existence of observers of the various types, together with a computational procedure for observer construction? Note further that, in general, the initial value of the state of a system is not known, which is at the core of the estimation/observation problem. Even if a tracking observer exists, there will be a nonzero tracking error whenever the initialization of the observer is incorrect. This points out the importance of asymptotic observers as well as, even more, spectrally assignable observers where one also controls the rate of convergence to zero of the error. Several further remarks are in order.

1. Incorporating a feedthrough term in the observer, as in (7.26), improves our ability to construct reduced-order functional observers. An example of this is the case of observing Kx, where $\operatorname{Ker} C \subset \operatorname{Ker} K$. This implies the existence of a map E for which $K = EC$, which leads to a zero-order, i.e., nondynamic, observer given by $\zeta_t = Ey_t$. Note that incorporating a feedthrough term in the observer is not new; it already appeared in Luenberger (1971) in the construction of an observer for a single functional of the state.
2. The definition of a tracking observer clearly implies that the set of the trajectories to be estimated is included in the set of outputs of the tracking observer.
3. A strongly tracking observer is at the same time a tracking observer. This follows from our assumption that J has full row rank. Thus $e_0 = Kx_0 - J\xi_0$ can always be made zero by an appropriate choice of ξ_0. Note also that a strongly tracking observer is finitely determined, with $T = 1$.

It was already observed that the trackability of an output of a finite-dimensional linear system is always satisfied. In the same vein, it is always possible to construct tracking observers by inspection. For instance, taking the copy of a system as

$$\xi_{t+1} = A\xi_t + Bu_t,$$
$$\zeta_t = K\xi_t$$

obviously leads to a tracking observer of system (7.25). Note that this observer has the same dimension n as (7.25). More generally, for a matrix $L \in \mathbb{F}^{n \times p}$, the system

$$\xi_{t+1} = (A - LC)\xi_t + LCx_t + Bu_t,$$
$$\zeta_t = K\xi_t$$

is a tracking observer for (7.25). Therefore, the main issue is not the existence of tracking observers (they always exist), but whether or not tracking observers with prescribed dimension $q \leq n$ exist or, even better, whether a minimal-order observer can be constructed. This problem will be addressed in Theorem 7.27, but first, a characterization of functional observers is derived.

Our starting point is the derivation of a state-space characterization, in terms of matrix Sylvester equations, for the classes of observers introduced in Definition 7.16. Thus we consider a linear system Σ_{sys}:

$$\Sigma_{sys} := \begin{cases} x_{t+1} = Ax_t + Bu_t, \\ \quad y_t = Cx_t, \\ \quad z_t = Kx_t, \end{cases} \tag{7.28}$$

with state space $\mathbb{F}^n$ and the estimator system

$$\Sigma_{est} := \begin{cases} \xi_{t+1} = F\xi_t + Gy_t + Hu_t, \\ \zeta_t = J\xi_t + Ey_t, \end{cases} \tag{7.29}$$

in the state space $\mathbb{F}^q$.

Theorem 7.17. *Assume that both* $(\begin{pmatrix} C \\ K \end{pmatrix}, A)$ *and* (J, F) *are observable.*

1. *System (7.29) is a tracking observer for K if and only if there exists a solution $Z \in \mathbb{F}^{q \times n}$ of the **observer Sylvester equations***

$$ZA = \begin{pmatrix} F & G \end{pmatrix} \begin{pmatrix} Z \\ C \end{pmatrix},$$

$$H = ZB, \tag{7.30}$$

$$K = \begin{pmatrix} J & E \end{pmatrix} \begin{pmatrix} Z \\ C \end{pmatrix}.$$

The solution Z of (7.30) is uniquely determined.
2. *Let Z be the unique solution to the observer Sylvester equations (7.30). Defining $Z_K(z) = K(zI - A)^{-1}$ and $Z_C(z) = C(zI - A)^{-1}$, the equation*

$$Z_K(z) = Z_1(z)Z_C(z) + Z_2(z) \tag{7.31}$$

is solvable with

$$Z_1(z) = E + J(zI - F)^{-1}G, \quad Z_2(z) = J(zI - F)^{-1}Z. \tag{7.32}$$

3. *Defining an auxiliary variable ε by*

$$\varepsilon = Zx - \xi, \tag{7.33}$$

the observer error dynamics with the initial condition $\varepsilon_0 = Zx_0 - \xi_0$ are

$$\begin{aligned} \varepsilon_{t+1} &= F\varepsilon_t, \\ e_t &= J\varepsilon_t, \end{aligned} \tag{7.34}$$

i.e., the error trajectory is the output of an autonomous linear system. The set $\mathscr{B}_{err}$ of all error trajectories is an autonomous behavior of the form

$$\mathscr{B}_{err} = X^Q, \tag{7.35}$$

where

$$Q(z)^{-1}P(z) = J(zI - F)^{-1} \qquad (7.36)$$

are coprime factorizations.
4. *The following conditions are equivalent:*

  (a) *System (7.29) is an asymptotic tracking observer for K.*
  (b) *There exists a linear transformations Z, with F stable, such that (7.30) holds.*

5. *The following conditions are equivalent:*

  (a) *System (7.29) represents a family of spectrally assignable tracking observers for K.*
  (b) *With the characteristic polynomial of F preassigned, there exists a linear transformation Z that satisfies the observer Sylvester equations (7.30).*

Proof. 1.
For initial conditions x_0 for Σ_{sys} and ξ_0 for Σ_{est}, the Z-transforms of the solutions to equations (7.28), (7.29), and (7.34) are given by

$$x = (zI - A)^{-1}x_0 + (zI - A)^{-1}Bu,$$

$$y = C(zI - A)^{-1}x_0 + C(zI - A)^{-1}Bu,$$

$$z = K(zI - A)^{-1}x_0 + K(zI - A)^{-1}Bu,$$

$$\xi = (zI - F)^{-1}\xi_0 + (zI - F)^{-1}GC(zI - A)^{-1}x_0 + (zI - F)^{-1}(H + GC(zI - A)^{-1}B)u,$$

$$\zeta = J(zI - F)^{-1}\xi_0 + J(zI - F)^{-1}GC(zI - A)^{-1}x_0 + EC(zI - A)^{-1}x_0$$

$$+ J(zI - F)^{-1}(H + GC(zI - A)^{-1}B)u + EC(zI - A)^{-1}x_0 + EC(zI - A)^{-1}Bu,$$

$$e = z - \zeta = \left[K - J(zI - F)^{-1}GC - EC\right](zI - A)^{-1}x_0 - J(zI - F)^{-1}\xi_0,$$

$$+ \left[(K - EC)(zI - A)^{-1}B - J(zI - F)^{-1}(H + GC(zI - A)^{-1}B)\right]u. \qquad (7.37)$$

To begin, one takes $u = 0$. The trackability assumption translates into the following statement. For each vector $x_0 \in \mathbb{F}^n$, there exists a vector $\xi_0 \in \mathbb{F}^q$ such that

$$J(zI - F)^{-1}\xi_0 + J(zI - F)^{-1}GC(zI - A)^{-1}x_0 + EC(zI - A)^{-1}x_0 - K(zI - A)^{-1}x_0 = 0.$$

This implies that ξ_0 is a linear function of x_0. Because x_0 is unrestricted, this means that there exists a $Z \in \mathbb{F}^{q \times n}$ for which $\xi_0 = Zx_0$. This leads to the identity

$$J(zI - F)^{-1}Z + J(zI - F)^{-1}GC(zI - A)^{-1} + EC(zI - A)^{-1} - K(zI - A)^{-1} = 0.$$

Equating residues, one obtains

$$K = JZ + EC = \begin{pmatrix} J & E \end{pmatrix} \begin{pmatrix} Z \\ C \end{pmatrix}. \tag{7.38}$$

Using this identity one computes

$$
\begin{aligned}
0 &= J(zI - F)^{-1}Z + J(zI - F)^{-1}GC(zI - A)^{-1} + EC(zI - A)^{-1} - K(zI - A)^{-1} \\
&= J(zI - F)^{-1}Z + J(zI - F)^{-1}GC(zI - A)^{-1} + (K - JZ)(zI - A)^{-1} - K(zI - A)^{-1} \\
&= J(zI - F)^{-1}[Z(zI - A) + GC - (zI - F)Z](zI - A)^{-1} \\
&= J(zI - F)^{-1}[-ZA + GC + FZ](zI - A)^{-1}.
\end{aligned}
$$

The nonsingularity of $(zI - A)$ and the observability of the pair (J, F) imply the identity

$$ZA - FZ = GC, \tag{7.39}$$

which can be rewritten as

$$ZA = \begin{pmatrix} F & G \end{pmatrix} \begin{pmatrix} Z \\ C \end{pmatrix}.$$

By inserting identities (7.38) and (7.39), together with $\xi_0 = Zx_0$, back into the representation of e in (7.37), one gets

$$
\begin{aligned}
0 &= J\left[Z - (zI - F)^{-1}GC\right](zI - A)^{-1}Bu - J(zI - F)^{-1}Hu \\
&= J(zI - F)^{-1}\left[(zI - F)Z - GC\right](zI - A)^{-1}Bu - J(zI - F)^{-1}Hu \\
&= J(zI - F)^{-1}\left[Z(zI - A)\right](zI - A)^{-1}Bu - J(zI - F)^{-1}Hu \\
&= J(zI - F)^{-1}(ZB - H)u.
\end{aligned}
$$

Choosing constant inputs and using the observability of (J, F), this implies $H = ZB$. Thus the observer Sylvester equations (7.30) hold.

To show the uniqueness of the solution to the observer Sylvester equations (7.30), assume there exist two maps Z', Z'' satisfying them. Setting $Z = Z'' - Z'$ yields $ZA = FZ$ and $JZ = 0$. The intertwining relation implies that $ZA^k = F^k Z$, for all $k \geq 0$, and hence $JF^k Z = JZA^k = 0$, i.e., $\operatorname{Im} Z \subset \bigcap_{k \geq 0} \operatorname{Ker} JF^k$. The observability of the pair (J, F) implies now $Z = 0$, i.e., $Z'' = Z'$.

Conversely, assume the observer Sylvester equations (7.30) are satisfied. For a control u and an initial condition x_0 for Σ', we choose $\xi_0 = Zx_0$. By (7.37), using the Sylvester equations, the error trajectory is given by

$$e = \left[K - J(zI - F)^{-1}GC \right](zI - A)^{-1}x_0 - J(zI - F)^{-1}Zx_0$$
$$+ \left[JZ(zI - A)^{-1}B - J(zI - F)^{-1}(H + GC(zI - A)^{-1}B) \right]u$$
$$= J(zI - F)^{-1}\left[(zI - F)Z - Z(zI - A) - GC \right](zI - A)^{-1}x_0$$
$$+ J(zI - F)^{-1}\left[(zI - F)Z - Z(zI - A) - GC \right](zI - A)^{-1}Bu(z) = 0.$$

This shows that Σ_{est} is a tracking observer for Σ_{sys}.

2. From equation (7.39) it follows that

$$Z(zI - A) - (zI - F)Z = -GC,$$

and hence

$$(zI - F)^{-1}Z - Z(zI - A)^{-1} = -(zI - F)^{-1}GC(zI - A)^{-1}.$$

Using (7.38), this leads to

$$J(zI - F)^{-1}Z + J(zI - F)^{-1}GC(zI - A)^{-1} = JZ(zI - A)^{-1} = (K - EC)(zI - A)^{-1},$$

which proves the statement.

3. To determine the error dynamics, one computes, using the observer Sylvester equations (7.30),

$$\varepsilon_{t+1} = Zx_{t+1} - \xi_{t+1}$$
$$= ZAx_t + ZBu_t - F\xi_t - GCx_t - Hu_t$$
$$= ZAx_t + ZBu_t - \left[F(Zx_t - \varepsilon_t) + GCx_t - ZBu_t \right]$$
$$= F\varepsilon_t,$$
$$e_t = Kx_t - J\xi_t = J\varepsilon_t.$$

This proves (7.34). The error behavior, i.e., the space of error trajectories, is given by $\mathscr{B}_{err} = \{J(zI - F)^{-1}\xi \,|\, \xi \in \mathbb{F}^q\}$. Applying Proposition 4.36 to the coprime factorizations (7.36) leads to the representation $\{J(zI - F)^{-1}\xi \,|\, \xi \in \mathbb{F}^q\} = X^Q$.

4. $(a) \Leftrightarrow (b)$.

Assume that (7.29) is an asymptotic tracking observer for (7.28). By part 1, there exists a uniquely determined linear transformation Z that satisfy the Sylvester equations (7.30). Since the error dynamics is given by (7.34), and (J, F) is an observable pair by assumption, the convergence $e_t \to 0$ always implies $\varepsilon_t \to 0$. Thus the error dynamics (7.34) are stable, which shows that F is stable.

Conversely, the existence of a map Z solving the Sylvester equations (7.30) implies Σ_{est} is a tracking observer. The assumed stability of F implies, using the error dynamics (7.34), that Σ_{est} is actually an asymptotic tracking observer.

5. $(a) \Leftrightarrow (b)$.
   Follows directly from the definition of a spectrally assignable family of tracking observers, together with part 1.

∎

It is of interest to relate the error trajectories to the proper, rational solutions $Z_1(z), Z_2(z)$ of the equation $Z_K(z) = Z_1(z)Z_C(z) + Z_2(z)$. Here $Z_K(z)$ and $Z_C(z)$ are defined by (7.16) and $Z_1(z)$ and $Z_2(z)$ by (7.32). Note that $Z_1(z)$ is the transfer function of the observer from y to ζ, while $Z_2(z)$ is related to the error estimate. Choosing the initial condition of the observer as $\xi_0 = Zx_0$ would make the error trajectory zero. However, the initial state x_0 is unknown to the observer and, in the absence of that initial state information, the challenge is to obtain an error estimate. This is provided by the following proposition.

Proposition 7.18. *Let the initial conditions for system (7.25) and the observable tracking observer (7.26) be x_0 and ξ_0, respectively. The error trajectory is given by*

$$e = J(zI - F)^{-1}(Zx_0 - \xi_0) = Z_2(z)x_0 - J(zI - F)^{-1}\xi_0. \qquad (7.40)$$

In particular, e does not depend on the input u.

Proof. Computing, using equation (7.37), together with the observer Sylvester equations (7.30) and (7.32),

$$
\begin{aligned}
e &= \left[K - J(zI - F)^{-1}GC - EC \right](zI - A)^{-1}x_0 - J(zI - F)^{-1}\xi_0 \\
&\quad + \left[(K - EC)(zI - A)^{-1}B - J(zI - F)^{-1}(H + GC(zI - A)^{-1}B) \right]u \\
&= \left[Z_K(z) - Z_1(z)Z_C(z) \right]x_0 - J(zI - F)^{-1}\xi_0 \\
&\quad + \left[Z_K - Z_1(z)Z_C(z) - Z_2(z) \right]Bu \\
&= Z_2(z)x_0 - J(zI - F)^{-1}\xi_0
\end{aligned}
$$

completes the proof.   ∎

How the existence of tracking observers is preserved under the action of the output injection group $\mathcal{G}$ is shown in the next proposition. This is done by showing how the observer Sylvester equations (7.30) transform under the same group action.

Proposition 7.19. *Let the linear system*

$$\Sigma_{sys} := \begin{cases} x_{t+1} = Ax_t + Bu_t, \\ \quad y_t = Cx_t, \\ \quad z_t = Kx_t \end{cases} \qquad (7.41)$$

act in the state space $\mathbb{F}^n$. It is assumed that C and K have full row rank and that $\left(\binom{C}{K}, A\right)$ is observable. Assume that

$$\Sigma_{est} := \begin{cases} \xi_{t+1} = F\xi_t + Gy_t + Hu_t, \\ \zeta_t = J\xi_t + Ey_t \end{cases}$$

is an observable tracking observer with state space $\mathbb{F}^q$ that satisfies the observer Sylvester equations

$$ZA = (F\ G)\binom{Z}{C},$$
$$H = ZB, \tag{7.42}$$
$$K = (J\ E)\binom{Z}{C}.$$

Equations (6.3) are extended to the action of the output injection group $\mathscr{G}$ on quadruples A, B, C, K by

$$\begin{pmatrix} \overline{A} & \overline{B} \\ \overline{C} & 0 \\ \overline{K} & 0 \end{pmatrix} = \begin{pmatrix} R & L & 0 \\ 0 & S & 0 \\ 0 & 0 & I \end{pmatrix} \begin{pmatrix} A & B \\ C & 0 \\ K & 0 \end{pmatrix} \begin{pmatrix} R & 0 \\ 0 & I \end{pmatrix}^{-1}.$$

Under this action, the observer Sylvester equations transform as follows:

$$\overline{Z}\overline{A} = (\overline{F}\ \overline{G})\binom{\overline{Z}}{\overline{C}},$$
$$\overline{H} = \overline{Z}\overline{B}, \tag{7.43}$$
$$\overline{K} = (\overline{J}\ \overline{E})\binom{\overline{Z}}{\overline{C}},$$

where

$$\overline{A} = (RA - LC)R^{-1}, \quad \overline{B} = RB, \quad \overline{C} = SCR^{-1}, \quad \overline{K} = KR^{-1},$$
$$\overline{Z} = ZR^{-1}, \quad \overline{L} = LS^{-1},$$
$$\overline{F} = F, \quad \overline{G} = G - \overline{Z}L, \quad \overline{H} = H, \quad \overline{J} = J, \quad \overline{E} = ES^{-1}.$$

Proof. Equation (7.43) can be rewritten as

$$(ZR^{-1})(R(A - LS^{-1}C)R^{-1}) = (F\ (G - ZLS^{-1}))\binom{ZR^{-1}}{CR^{-1}},$$
$$H = (ZR^{-1})(RB),$$
$$KR^{-1} = (J\ E)\binom{ZR^{-1}}{CR^{-1}}.$$

This in turn is equivalent to the Sylvester equations (7.30).

Corollary 7.20. *Consider system (7.28). The observer (7.29) is finitely determined if and only if it is tracking.*

Proof. Follows from the error dynamics given by (7.35) and Lemma 7.12.   ■

The matrix Z that solves the observer Sylvester equations is not necessarily of full rank. Next, it is shown how the maps F,G,H,J,E can always be modified in such a way that a full rank solution $\overline{Z}$ to the Sylvester equations of the modified observer exists.

Proposition 7.21. *Suppose there exists a q-dimensional observable tracking observer for K. Then there exists a tracking observer $\overline{F},\overline{G},\overline{H},\overline{J},\overline{E}$ of dimension $\overline{q} \leq q$ together with a full row rank solution $\overline{Z}$ to the observer Sylvester equations*

$$\overline{Z}A - \overline{F}\overline{Z} = \overline{G}C,$$
$$\overline{H} = \overline{Z}B, \tag{7.44}$$
$$K = \overline{J}\overline{Z} + \overline{E}C.$$

Furthermore, the pair $(\overline{J},\overline{F})$ can be chosen to be observable.

Proof. Let F,G,H,J,E denote an observable tracking observer for system (7.25), with Z the solution to the observer Sylvester equation (7.30). If Z is not surjective, this implies that, in an appropriate basis, one has $Z = \begin{pmatrix} \overline{Z} \\ 0 \end{pmatrix}$, with $\overline{Z}$ surjective, i.e., of full row rank. The corresponding representations are as follows:

$$F = \begin{pmatrix} F_{11} & F_{12} \\ F_{21} & F_{22} \end{pmatrix}, \quad G = \begin{pmatrix} G_1 \\ G_2 \end{pmatrix}, \quad H = \begin{pmatrix} H_1 \\ H_2 \end{pmatrix}, \quad J = \begin{pmatrix} J_1 & J_2 \end{pmatrix}.$$

Equations (7.44) can now be rewritten as

$$\begin{pmatrix} \overline{Z} \\ 0 \end{pmatrix} A = \begin{pmatrix} F_{11} & F_{12} \\ F_{21} & F_{22} \end{pmatrix} \begin{pmatrix} \overline{Z} \\ 0 \end{pmatrix} + \begin{pmatrix} G_1 \\ G_2 \end{pmatrix} C,$$

$$\begin{pmatrix} H_1 \\ H_2 \end{pmatrix} = \begin{pmatrix} \overline{Z} \\ 0 \end{pmatrix} B,$$

$$K = \begin{pmatrix} J_1 & J_2 \end{pmatrix} \begin{pmatrix} \overline{Z} \\ 0 \end{pmatrix} + EC.$$

This in turn implies

$$\overline{Z}A - F_{11}\overline{Z} = G_1C,$$

$$H_1 = \overline{Z}B,$$

$$K = J_1\overline{Z} + EC.$$

The observability of the pair (J,F) implies the observability of the pair (J_1,F_{11}). Therefore,

$$\xi_{t+1} = F_{11}\xi_t + G_1 y_t + H_1 u_t,$$
$$\zeta_t = J_1 \xi_t + EC$$

is a tracking observer for (7.28). ∎

Before proceeding to establish the basic connections between observation properties and observer constructions, the following simple lemma is stated.

Lemma 7.22. *Let* (F,G,H,J,E) *be a tracking observer for* (A,B,C,K).

1. *If Z is a solution to the observer Sylvester equations, then, for each output injection map L,* $(F,G-ZL,H,J,E)$ *is a tracking observer for* $(A-LC,B,C,K)$.
2. *If Z is a solution to the observer Sylvester equations and P is nonsingular, then PZ solves the Sylvester equations for the tracking observer defined by* $(PFP^{-1},PG,PH,JP^{-1},E)$.

Proof. 1. Theorem 7.17 is applied to conclude that there exists a solution Z to the Sylvester equations (7.30). The first equation, $ZA = FZ + GC$, implies $Z(A - LC) = FZ + (G - ZL)C$, while the other two equations remain untouched. This implies that the observer defined by $(F,G-ZL,H,J,E)$ is a tracking observer for the system defined by $(A-LC,B,H,C,K)$.
2. Using the observer Sylvester equations (7.30), one computes

$$(PZ)A = (PFP^{-1})(PZ) + (PG)C,$$
$$K = (JP^{-1})(PZ) + EC.$$

The identity of transfer functions

$$\left[\begin{array}{c|cc} PFP^{-1} & PG & PZ \\ \hline JP^{-1} & E & 0 \end{array}\right] = \left[\begin{array}{c|cc} F & G & Z \\ \hline J & E & 0 \end{array}\right]$$

is easily checked. ∎

An interesting question is to analyze the extent of our control over the error dynamics. In particular, one might want to clarify the following question: under what conditions can we preassign the error dynamics? In one direction, this is easily resolved using observability subspaces. This is studied in Fuhrmann and Trumpf (2006). For further studies, one would first have to extend Definition 7.16 to the notion of spectral assignability and derive results characterizing the existence of such observers.

7.4 Existence of Observers

Having studied the observation properties of linear systems in Section 7.2 and introduced several classes of observers in Section 7.3, it will come as no great surprise that there is a natural correspondence between observation properties and observers of linear systems. This correspondence is addressed next.

Theorem 7.23. *Let the linear system*

$$\Sigma_{sys} := \begin{cases} x_{t+1} = Ax_t + Bu_t, \\ \quad y_t = Cx_t, \\ \quad z_t = Kx_t \end{cases} \tag{7.45}$$

act in the state space $\mathbb{F}^n$. *It is assumed that* $\left(\begin{pmatrix} C \\ K \end{pmatrix}, A \right)$ *is observable and is of the form (7.9).*

1. *Let* Σ_{sys} *be trackable with minimal tracking index* τ. *Then the maximal observability index of every tracking observer for K is greater than or equal to* τ.
2. *The following conditions are equivalent:*

   (a) *There exists an asymptotic tracking observer for K.*
   (b) *The pair* (C,A) *is detectable.*

3. *The following conditions are equivalent:*

   (a) *There exists a spectrally assignable family of tracking observers for K.*
   (b) *The pair* (C,A) *is observable.*

Proof. Part 1.

Let Σ_{est} be a tracking observer (7.26) with maximal observability index equal to τ_*. Choose an initial condition $x_0 \in \mathbb{F}^n$ with $CA^t x_0 = 0$ for all t. Assume $Kx_0 = \cdots = Kx_{\tau_*} = 0$. Since (7.26) is a tracking observer, there exists ξ_0 such that $Kx_t = J\xi_t$ is true for all t. Since $y_t = CA^t x_0 = 0$ for all t, one concludes $J\xi_t = JF^t \xi_0$ for all t. In particular, $J\xi_0 = \cdots = JF^{\tau_*-1}\xi_0 = 0$. By the observability of J and F, we have that $\xi_0 = 0$, and thus $Kx_t = 0$ for all t. This implies the bound $\tau \le \tau_*$ for the minimal tracking index.

Part 2. $(a) \Leftrightarrow (b)$

Assume that an asymptotic tracking observer

$$\xi_{t+1} = F\xi_t + Gy_t + Hu_t,$$
$$\zeta_t = J\xi_t + Ey_t \tag{7.46}$$

for z exists; then F is necessarily stable. One must show that all unobservable modes of (C,A) are stable. Choose $u = 0$, and pick an unobservable initial state $x_0 \in \mathcal{O}_*$. Then $\zeta_t = JF^t \xi_0$ is true for all t and all ξ_0. The stability of F implies

$\lim_{t\to\infty} JF^t \xi_0 = 0$. Since (7.46) is a tracking observer, there exists ξ_0 such that $\zeta_t = z_t$ for all $t \geq 0$. Thus $\lim_{t\to\infty} KA^t x_0 = \lim_{t\to\infty} z_t = 0$ for all $x_0 \in \mathcal{O}_*$. This is equivalent to $\lim_{t\to\infty} K_2 A_{22}^t = 0$ and, therefore, by the observability of (K_2, A_{22}), to the stability of A_{22}. Thus (C, A) is detectable.

Conversely, assume (C, A) is detectable. By Theorem 7.14, the unobservable modes of A are stable and there exists an output injection matrix L such that $A - LC$ is stable. Therefore, the full state Luenberger observer

$$\begin{aligned} \xi_{t+1} &= (A - LC)\xi_t + Ly_t + Bu_t, \\ \zeta_t &= K\xi_t \end{aligned} \tag{7.47}$$

is an asymptotic tracking observer for K.

Part 3. $(a) \Leftrightarrow (b)$

If (C, A) is observable, then one can find L such that $A - LC$ has a preassigned characteristic polynomial. Thus the Luenberger observer (7.47) yields a spectrally assignable tracking observer for K. This shows that $(b) \Rightarrow (a)$. For the converse, assume (7.46) is a spectrally assignable tracking observer. Thus F can be chosen with a preassigned characteristic polynomial. To show that (C, A) is observable, suppose, to obtain a contradiction, that there exists a nonzero unobservable state $x_0 \in \mathcal{O}_*$. Choose $u = 0$ and ξ_0 such that $z_t = \zeta_t$ for all t. Then, for each v, there exists ξ_0 with $z_t = K_2 A_{22}^t v = JF^t \xi_0$, or, equivalently, there exists a matrix Z that satisfies

$$K_2 A_{22}^t = JF^t Z$$

for all $t \geq 0$. Equivalently,

$$K_2(zI - A_{22})^{-1} = J(zI - F)^{-1} Z. \tag{7.48}$$

Since (7.46) is spectrally assignable, one can choose F so that the minimal polynomial $b(z)$ of F is coprime to the minimal polynomial $a(z)$ of A_{22}. Then the poles of the rational functions on both sides of (7.48) are disjoint. Therefore, both sides must be zero, i.e.,

$$K_2(zI - A_{22})^{-1} = 0 = J(zI - F)^{-1} Z,$$

in contradiction to the observability of (K_2, A_{22}). This completes the proof. ∎

Our next objective is the characterization of several classes of observers

$$\Sigma_{est} = \begin{cases} \xi_{t+1} = F\xi_t + Gy_t + Hu_t, \\ \zeta_t = J\xi_t + Ey_t, \end{cases} \tag{7.49}$$

defined in the state space $\mathbb{F}^q$. This includes the characterization of minimal-order observers as a special case, relating, for example, minimal-order q tracking observers (7.49) with maximal conditioned invariant subspaces of codimension q. It is tacitly assumed throughout the subsequent arguments that (7.49) is observable. Our analysis is done in essentially two ways, geometrically and functionally. For the case of detectability, it will always be assumed that the underlying field is a subfield of $\mathbb{C}$. Our results remain in force for all fields $\mathbb{F}$ endowed with the discrete topology. In that case, detectability is equivalent to reconstructibility.

Theorem 7.24. *Let the linear system*

$$\Sigma_{sys} := \begin{cases} x_{t+1} = Ax_t + Bu_t, \\ \quad y_t = Cx_t, \\ \quad z_t = Kx_t \end{cases} \qquad (7.50)$$

act in the state space $\mathbb{F}^n$. It is assumed that the pair $\left(\begin{pmatrix} C \\ K \end{pmatrix}, A \right)$ is observable and C and K are both of full row rank. Let $Z_C(z), Z_K(z)$ be defined by (7.16).

1. *The following conditions are equivalent:*

    (a) There exists a tracking observer for K of order q.

    (b) There exists a conditioned invariant subspace $\mathcal{V} \subset \mathbb{F}^n$, of codimension q, satisfying

    $$\mathcal{V} \cap \operatorname{Ker} C \subset \operatorname{Ker} K. \qquad (7.51)$$

    (c) There exist proper rational functions $Z_1(z), Z_2(z)$, with McMillan degree of $\left(Z_1(z) \ Z_2(z) \right)$ less than or equal to q, that solve

    $$\left(Z_1(z) \ Z_2(z) \right) \begin{pmatrix} C \\ zI - A \end{pmatrix} = K \qquad (7.52)$$

    or the equivalent equation

    $$Z_K(z) = Z_1(z)Z_C(z) + Z_2(z). \qquad (7.53)$$

2. *The following statements are equivalent:*

    (a) There exists an order q asymptotic tracking observer for K.

    (b) There exists an outer detectable subspace, with $\operatorname{codim} \mathcal{D} = q$, satisfying

    $$\mathcal{D} \cap \operatorname{Ker} C \subset \operatorname{Ker} K. \qquad (7.54)$$

    (c) There exist strictly proper, stable rational functions $Z_1(z), Z_2(z)$, with McMillan degree of $\left(Z_1(z) \ Z_2(z) \right)$ equal to q, that solve (7.53).

3. *The following conditions are equivalent:*

   (a) *There exists an order-q, spectrally assignable family of tracking observers for K.*
   (b) *There exists an outer observability subspace $\mathcal{O}$ of codimension q satisfying*

$$\mathcal{O} \cap \operatorname{Ker} C \subset \operatorname{Ker} K. \tag{7.55}$$

   (c) *There exist polynomial matrices $P_1(z)$ and $P_2(z)$ that solve (7.53).*

Proof. Part 1. $(a) \Leftrightarrow (b)$.

Assume that a q-dimensional tracking observer exists and is given by (7.49). By Theorem 7.17, there exists a solution to the observer Sylvester equations (7.30). By Proposition 7.21, it can be assumed without loss of generality that Z is of full row rank. For minimal tracking observers this is automatically satisfied. Otherwise, the observer order could be further reduced, contradicting the assumption that the observer has minimal order. Define now $\mathcal{V} = \operatorname{Ker} Z$. The equation

$$ZA = (F \ G) \begin{pmatrix} Z \\ C \end{pmatrix}$$

implies

$$A\operatorname{Ker} \begin{pmatrix} Z \\ C \end{pmatrix} \subset \operatorname{Ker} Z$$

or, equivalently,

$$A(\mathcal{V} \cap \operatorname{Ker} C) \subset \mathcal{V},$$

which shows that $\mathcal{V}$ is a conditioned invariant subspace. By Theorem 7.17, there exist F and Z that satisfy the Sylvester equations (7.30). In particular, $ZA = FZ + GC$ shows that $\mathcal{V} = \operatorname{Ker} Z$ is conditioned invariant. Moreover, the equation $K = JZ + EC$ implies the inclusion

$$\mathcal{V} \cap \operatorname{Ker} C = \operatorname{Ker} \begin{pmatrix} Z \\ C \end{pmatrix} \subset \operatorname{Ker} K.$$

Z having row rank implies $\operatorname{rank} Z = q$. The equality $\dim \operatorname{Ker} Z = n - \operatorname{rank} Z$ implies $\operatorname{codim} \mathcal{V} = q$.

To prove the converse, assume there exists a conditioned invariant subspace $\mathcal{V}$ of codimension q satisfying the inclusion (7.51). Let $Z \in \mathbb{F}^{q \times n}$ be of full row rank q such that $\operatorname{Ker} Z = \mathcal{V}$. The inclusion (7.51) implies the factorization

$$K = (J\ E) \begin{pmatrix} Z \\ C \end{pmatrix}.$$

Since $\mathscr{V}$ is a conditioned invariant subspace, there exists an output injection L such that $(A - LC)\mathrm{Ker}\,Z \subset \mathrm{Ker}\,Z$. This inclusion implies the existence of $F \in \mathbb{F}^{q \times q}$ for which $Z(A - LC) = FZ$. Defining $G = ZL$ and $H = ZB$, the Sylvester equations (7.30) and an order q observer are obtained.

$(a) \Leftrightarrow (c)$.

Assume that Σ_{est}, given by (7.49), is an order-q tracking observer. Define the transfer function

$$(Z_1(z)\ Z_2(z)) = \left[\begin{array}{c|cc} F & G & Z \\ \hline J & E & 0 \end{array}\right],\tag{7.56}$$

that is,

$$Z_1(z) = J(zI - F)^{-1}G + E,$$
$$Z_2(z) = J(zI - F)^{-1}Z.$$

Using the Sylvester equations (7.30), we compute $Z_1(z)Z_C(z) + Z_2(z) =$

$$= (J(zI - F)^{-1}G + E)C(zI - A)^{-1} + J(zI - F)^{-1}Z$$
$$= J(zI - F)^{-1}GC(zI - A)^{-1} + (K - JZ)(zI - A)^{-1} + J(zI - F)^{-1}Z$$
$$= K(zI - A)^{-1} + J(zI - F)^{-1}[GC - (zI - F)Z + Z(zI - A)](zI - A)^{-1}$$
$$= K(zI - A)^{-1} + J(zI - F)^{-1}[GC + FZ - ZA](zI - A)^{-1}$$
$$= K(zI - A)^{-1} = Z_K(z),$$

i.e., we obtain a proper solution of (7.53), of McMillan degree q. Note that the equivalence of the solvability of equations (7.52) and (7.53) is trivial.

Conversely, assume that, with $Z_C(z), Z_K(z)$ defined in (7.16), $Z_1(z), Z_2(z)$ is a proper solution of equation (7.53), of McMillan degree q. Note that, since $Z_K(z)$ and $Z_1(z)Z_C(z)$ are both strictly proper, necessarily $Z_2(z)$ is strictly proper, too. Therefore, a minimal realization of $(Z_1(z)\ Z_2(z))$ has the form $\left(\begin{array}{c|cc} F & G & Z \\ \hline J & E & 0 \end{array}\right)$, which has dimension q. Then

$$0 = K(zI - A)^{-1} - (J(zI - F)^{-1}G + E)C(zI - A)^{-1} - J(zI - F)^{-1}Z.$$

By inspection of the residue term, this implies $K = JZ + EC$. Substituting this back into the previous equation, we compute

$$0 = (JZ + EC)(zI - A)^{-1} - (J(zI - F)^{-1}G + E)C(zI - A)^{-1} - J(zI - F)^{-1}Z$$
$$= JZ(zI - A)^{-1} - J(zI - F)^{-1}GC(zI - A)^{-1} - J(zI - F)^{-1}Z$$
$$= J(zI - F)^{-1}[(zI - F)Z - GC - Z(zI - A)](zI - A)^{-1}$$
$$= J(zI - F)^{-1}[-FZ - GC + ZA](zI - A)^{-1}.$$

By the nonsingularity of $zI - A$ and the observability of the pair (J, F), we conclude that $ZA - FZ - GC = 0$. Defining $H = ZB$, it follows that (7.29) is a tracking observer of dimension at most q.

Proof of the equivalence $(a) \Leftrightarrow (b)$ in part 4.

Assuming there exists an outer detectable subspace $\mathscr{D}$ that satisfies (7.54), there exists an output injection map L for which $(A - LC)\mathscr{D} \subset \mathscr{D}$ and the induced map $(A - LC)|_{\mathscr{X}/\mathscr{D}}$ is stable. As in part 1, we set $\mathscr{D} = \operatorname{Ker}Z$ for some surjective linear transformation Z; then, by Lemma 3.8, there exists a map F that satisfies $Z(A - LC) = FZ$, i.e., $ZA - FZ = GC$ with $G = ZL$, as well as maps J and E, with $K = JZ + EC$. The stability of the induced map $(A - LC)|_{\mathscr{X}/\mathscr{D}}$ and the isomorphism $F \simeq (A - LC)|_{\mathscr{X}/\mathscr{D}}$ imply the stability of F. Finally, we define $H = ZB$. Thus, equations (7.30) have been derived.

Conversely, assume Σ_{est} is an asymptotic tracking observer for Σ_{sys}. By Proposition 7.21, there exists a reduced-order observer for which the Sylvester equations (7.30) are satisfied, with Z of full row rank $\bar{q} \le q$. By the surjectivity of Z, there exists an L for which $G = ZL$, and hence $Z(A - LC) = FZ$ holds. Moreover, since $K = JZ + EC$, $\mathscr{D} = \operatorname{Ker}Z$ is a conditioned invariant subspace of codimension $\bar{q} \le q$ that satisfies (7.54). By the surjectivity of Z, the map $(A - LC)|_{\mathbb{F}^n/\mathscr{D}}$ is isomorphic to F. Since F is stable, $\mathscr{D}$ is an outer detectability subspace of codimension $\bar{q} \le q$.

$(a) \Leftrightarrow (c)$

Assume the Sylvester equations (7.30) hold, with F stable. Then $Z_1(z)$ and $Z_2(z)$, as defined in (7.56), are necessarily proper and stable. Conversely, assume equation (7.52) is solvable with strictly proper and stable $Z_1(z)$ and $Z_2(z)$. Choose a minimal realization

$$\left(\begin{array}{c|cc} F & G & Z \\ \hline J & E & 0 \end{array} \right)$$

of $\left(Z_1(z)\ Z_2(z) \right)$. Necessarily, F is stable. By part 1, equations (7.30) are satisfied with F stable.

Proof of the equivalence $(a) \Leftrightarrow (b)$ in part 5.

Assume there exists a spectrally assignable family of observers (7.29). The subspace $\mathscr{O} = \operatorname{Ker}Z$ is a conditioned invariant subspace such that, for each polynomial $f(z)$ of degree equal to $\operatorname{codim}\mathscr{O}$, there exists a friend L of $\mathscr{O}$ for which the characteristic polynomial of F is $f(z)$. Necessarily, $\mathscr{O}$ is an outer observability subspace.

Conversely, assume that $\mathcal{O}$ is an observability subspace of codimension q that satisfies (7.55). Let $\mathcal{O}$ have the kernel representation $\mathcal{O} = \mathrm{Ker}\, Z$, with Z surjective. By the definition of observability subspaces, for each polynomial $f(z)$ of degree q there exists a friend L of $\mathcal{O}$ for which $Z(A - LC) = FZ$ and F has $f(z)$ as its characteristic polynomial. From the inclusion (7.55) it follows that there exist unique J and E for which $K = JZ + EC$. Finally, by defining $G = ZL$, the required family of observers is obtained.

$(c) \Leftrightarrow (b)$

Let $\mathcal{O}_*$ denote the unobservability subspace, i.e., the smallest outer observability subspace of (C, A). Assume that there exists an outer observability subspace $\mathcal{O}$ that is contained in the kernel of K. Thus $\mathcal{O}_* \subset \mathrm{Ker}\, K$, i.e., with respect to the representation (7.9), one has $K_2 = 0$. Since (C_1, A_{11}) is observable, there exist polynomial matrices $P_1(z)$ and $P_2(z)$, with $P_1(z)C_1 + P_2(z)(zI - A_{11}) = K_1$. Thus

$$
\begin{pmatrix} P_1(z) & P_2(z) & 0 \end{pmatrix} \begin{pmatrix} C_1 & 0 \\ zI - A_{11} & 0 \\ -A_{21} & zI - A_{22} \end{pmatrix} = \begin{pmatrix} K_1 & 0 \end{pmatrix},
$$

and a polynomial solution of (7.53) was constructed.

Conversely, let $P_1(z), P_2(z)$ denote a polynomial solution to (7.53). Then

$$
P_1 C(zI - A)^{-1} + P_2(z) = K(zI - A)^{-1}.
$$

Choose an unobservable state $x \in \mathcal{O}_*$. Thus $C(zI - A)^{-1}x = 0$, and therefore

$$
P_2(z)x = K(zI - A)^{-1}x.
$$

Since the left-hand side is polynomial and the right-hand side is strictly proper, we conclude $K(zI - A)^{-1}x = 0$ and $P_2(z)x = 0$ for all $x \in \mathcal{O}_*$. This implies $Kx = 0$, i.e., $\mathcal{O}_* \subset \mathrm{Ker}\, K$. This proves the converse. Moreover, $x = \mathrm{col}\,(u, v)$, with $u = 0$ and $K_2(zI - A_{22})^{-1}v = K(zI - A)^{-1}x = 0$. Thus the observability of (K_2, A_{22}) implies $\mathcal{O}_* = \{0\}$. This shows the observability of (C, A).   ∎

Remarks: 1. It has already been noted that trackability is a weak concept. Therefore, one expects that a tracking observer for Σ_{sys}, given by (7.45), should always exist. This is indeed the case. One can define the observer as

$$
\begin{aligned}
\xi_{t+1} &= A\xi_t + Bu_t, \\
\zeta_t &= K\xi_t,
\end{aligned}
\tag{7.57}
$$

and check that it is a tracking observer. Also, note that one strictly proper solution of (7.52) is given by $\begin{pmatrix} Z_1(z) & Z_2(z) \end{pmatrix} = \begin{pmatrix} 0 & K(zI - A)^{-1} \end{pmatrix}$. This also leads to the observer (7.57). Finally, the zero subspace is a conditioned invariant subspace for (C, A) and is contained in $\mathrm{Ker}\, K$. This allows us to take $Z = I$ and, hence, from the Sylvester equations, show that $F = A$ and $J = K$. So, once again, we are back to the observer (7.57). Such an observer is of course totally useless because it disregards all the observed data y.

2. Note that the existence of fixed-order tracking observers with preassignable spectra does not necessarily imply the existence of a suitable observability subspace, not even in the minimal-order case.

3. Clearly, the existence of a spectrally assignable family of observers implies the existence of an asymptotic observer. In particular, part 3 (c) of Theorem 7.24 should imply part 2 (c), and in the same way, part 3 (d) should imply part 2 (d). This can be verified directly using partial realization theory, a topic that will not be discussed in this book.

It may be of interest to understand the conditions under which the observer equations (7.49) can be simplified to the form $\zeta_{t+1} = F\zeta_t + Gy_t + Hu_t$. The following proposition addresses this question and gives a geometric characterization for the existence of strongly tracking observers.

Proposition 7.25. *Consider the system*

$$\Sigma_{sys} := \begin{cases} x_{t+1} = Ax_t + Bu_t, \\ y_t = Cx_t, \\ z_t = Kx_t. \end{cases}$$

It is assumed that both C and K have full row rank and that $\left(\begin{pmatrix} C \\ K \end{pmatrix}, A \right)$ is observable.

1. A tracking observer

$$\Sigma_{est} := \begin{cases} \xi_{t+1} = F\xi_t + Gy_t + Hu_t, \\ \zeta_t = J\xi_t, \end{cases} \tag{7.58}$$

with (J,F) observable, is a strongly tracking observer if and only if J is nonsingular. In that case, we may assume without loss of generality that the observer is given by

$$\zeta_{t+1} = F\zeta_t + Gy_t + Hu_t. \tag{7.59}$$

2. A strongly tracking observer of the form (7.59) exists if and only if $\text{Ker} K$ is a conditioned invariant subspace. In this case, the error dynamics are given by

$$e_{t+1} = Fe_t. \tag{7.60}$$

Proof. 1. Assume J in (7.58) is nonsingular. The error dynamics are given by (7.34), and hence

$$e_{t+1} = J\varepsilon_{t+1} = JF\varepsilon_t = JFJ^{-1}J\varepsilon_t = JFJ^{-1}e_t.$$

This shows that $e_t = (JFJ^{-1})^{t-1}e_0$, and hence $e_0 = 0$ implies $e_t = 0$, i.e., Σ_{est} is a strongly tracking observer.

Conversely, assume that Σ_{est} is a strongly tracking observer. The error dynamics are given by (7.34), and hence $e_t = JF^{t-1}\varepsilon_0$. By the property of strong tracking, $e_0 = 0$ implies $e_t = 0$ for all $t \geq 0$, i.e., $\varepsilon_0 \in \bigcap \operatorname{Ker} JF^{t-1}$. By the observability of the pair (J, F), we conclude that $e_0 = 0$ implies $\varepsilon_0 = 0$. This shows that J is injective and, hence, since it was assumed that J has full row rank, actually invertible. Substituting $\xi = J^{-1}\zeta$ into the observer equation and multiplying from the left by J, $\zeta_{t+1} = (JFJ^{-1})\zeta_t + (JG)y_t + (JH)u_t$ follows. Modifying appropriately the definitions of F, G, H, equation (7.59) is proved.

2. Assume (7.59) is a strongly tracking observer. By Theorem 7.17, there exists a map Z satisfying the following Sylvester equations:

$$ZA = FZ + GC,$$
$$H = ZB,$$
$$K = Z.$$

Letting $x \in \operatorname{Ker} K \cap \operatorname{Ker} C$ implies $K(Ax) = 0$, i.e., $Ax \in \operatorname{Ker} K$, so $\operatorname{Ker} K$ is a conditioned invariant subspace.

Conversely, assume $\operatorname{Ker} K$ is a conditioned invariant subspace. Letting $Z = K$, there exists a map L such that $(A - LC)\operatorname{Ker} K \subset \operatorname{Ker} K$, and using, once again Lemma 3.8, we infer that $K(A - LC) = FK$ for some L. Thus $KA - FK = GC$, with $G = KL$. Setting $J = I$ and defining $H = KB$, we are done. That the error dynamics are given by (7.60) follows from (7.34) and the fact that $J = I$. ∎

This section ends with some ideas on state-space constructions of functional observers. Theorem 7.24 contains equivalent characterizations for minimal-order observers but does not give the minimal order of a tracking observer or a way of computing such an observer. From a practical point of view, it is important to have a systematic way of constructing observers, and, for computational efficiency, it is important to have the order of the observer minimal. These questions are addressed next. To this end, assume Σ_{est}, defined by (7.49), is a minimal-order tracking observer for system (7.50). By Theorem 7.17, there exists a solution Z of the observer Sylvester equations (7.30), and, by Proposition 7.21, it may be assumed without loss of generality that Z has full row rank. Define now $\mathscr{V} = \operatorname{Ker} Z$. The first equation in (7.30) means that $A(\mathscr{V} \cap \operatorname{Ker} C) \subset \mathscr{V}$, i.e., that $\mathscr{V}$ is a conditioned invariant subspace, whereas the last equation in (7.30) means that $\mathscr{V} \cap \operatorname{Ker} C \subset \operatorname{Ker} K$. In view of the geometric characterizations given in Theorem 7.24, to find minimal-order observers for system (7.28), one must find all maximal dimensional conditioned invariant subspaces $\mathscr{V}$ that satisfy $\mathscr{V} \cap \operatorname{Ker} C \subset \operatorname{Ker} K$. Since the set of all conditioned invariant subspaces is closed under intersections but not under sums, one must approach the minimality question differently. In spirit, we follow Michelangelo's dictum: "Carving is easy, you just go down to the skin and stop."

Therefore, to get a minimal-order tracking observer, one must choose a minimal rank extension $\begin{pmatrix} z \\ c \end{pmatrix}$ for which $\mathscr{V} = \operatorname{Ker} Z$ is conditioned invariant and

$$\mathscr{V} \cap \operatorname{Ker} C = \operatorname{Ker} \begin{pmatrix} Z \\ C \end{pmatrix} \subset \operatorname{Ker} K. \tag{7.61}$$

By the minimality of such an extension, Z necessarily has full row rank. It is easy to fulfill the second requirement, simply by choosing $Z = K$, which implies

$$\operatorname{Ker} \begin{pmatrix} Z \\ C \end{pmatrix} = \operatorname{Ker} \begin{pmatrix} K \\ C \end{pmatrix} \subset \operatorname{Ker} K.$$

There are two problems with this choice, of which the first one is minor. Since, in the case where $\operatorname{Ker} K$ and $\operatorname{Ker} C$ have a nontrivial intersection, K can be reduced modulo C to get a lower rank extension K'. The second problem is due to the fact that, in general, there is no reason why $\mathscr{V} = \operatorname{Ker} K$ should be conditioned invariant. (But, as we shall see in Example 7.32, it may.) One way to overcome this is to add, if necessary, additional terms to Z. An easy way to do this is to set

$$Z = \begin{pmatrix} KA^{n-1} \\ \vdots \\ KA \\ K \end{pmatrix}.$$

Clearly, $\operatorname{Ker} Z$ is not only a conditioned invariant subspace but actually an invariant one. However, in general, the constructed Z does not have full row rank and thus would lead to a nonminimal observer. The remedy to these two problems is to maintain a fine balance between increasing the rank of Z sufficiently so that $\operatorname{Ker} Z$ is conditioned invariant and (7.61) being satisfied, but small enough to preserve the maximality of the dimension of $\operatorname{Ker} Z$ and, hence, leading to a minimal-order functional observer. This will be treated in Theorem 7.27.

To formalize the reduction process, the following proposition, which is of interest on its own, is stated and proved.

Proposition 7.26. *1. Let $S \in \mathbb{F}^{n \times q}$ and $T \in \mathbb{F}^{n \times p}$ be of full column rank. Then there exists $S' \in \mathbb{F}^{n \times r}$, of full column rank, such that*

$$\operatorname{Im} S' \subset \operatorname{Im} S,$$

$$\operatorname{Im} S' + \operatorname{Im} T = \operatorname{Im} S + \operatorname{Im} T,$$

$$\operatorname{Im} S' \cap \operatorname{Im} T = \{0\},$$

implying the direct sum representation $\operatorname{Im} S + \operatorname{Im} T = \operatorname{Im} S' \oplus \operatorname{Im} T$.

2. *Let $K \in \mathbb{F}^{q \times n}$ and $C \in \mathbb{F}^{p \times n}$ be of full row rank. Then there exists $K' \in \mathbb{F}^{r \times n}$, of full row rank, such that*

$$\operatorname{Ker} K' \supset \operatorname{Ker} K,$$

$$\operatorname{Ker} K' \cap \operatorname{Ker} C = \operatorname{Ker} K \cap \operatorname{Ker} C,$$

$$\operatorname{Ker} K' + \operatorname{Ker} C = \mathbb{F}^n,$$

i.e., $\operatorname{Ker} K \cap \operatorname{Ker} C$ is the transversal intersection of $\operatorname{Ker} K'$ and $\operatorname{Ker} C$, which implies the direct sum representation

$$\mathbb{F}^n / (\operatorname{Ker} K' \cap \operatorname{Ker} C) = \operatorname{Ker} K' / (\operatorname{Ker} K' \cap \operatorname{Ker} C) \oplus \operatorname{Ker} C / (\operatorname{Ker} K' \cap \operatorname{Ker} C).$$

Moreover,

$$r = \operatorname{rank} K' = \operatorname{codim} \operatorname{Ker} K'.$$

Proof. 1. Noting that S is assumed to be of full column rank, it follows that, with S_i being the columns of S, the set $\mathcal{B} = \{S_1, \ldots, S_q\}$ is a basis for $\operatorname{Im} S$. Let $\{R_1, \ldots, R_r\}$ be a basis for $\operatorname{Im} S \cap \operatorname{Im} T$. By the basis exchange theorem, there exist r elements of $\mathcal{B}$, which without loss of generality one can take to be the first r, for which $\{R_1, \ldots, R_r, S_{r+1}, \ldots, S_q\}$ is a basis for $\operatorname{Im} S$. Defining $S' = (S_{r+1} \ldots S_q)$, we are done.

2. The first assertion follows from the first part by duality considerations, while the second one follows from the identity $\dim \operatorname{Ker} K' + \dim \operatorname{Im} K' = n$. ∎

Let

$$\Sigma_{sys} := \begin{cases} x_{t+1} = Ax_t + Bu_t, \\ \phantom{x_{t+1}} y_t = Cx_t, \\ \phantom{x_{t+1}} z_t = Kx_t \end{cases}$$

be a linear system acting in the state space $\mathbb{F}^n$. Assume that $(\begin{pmatrix} C \\ K \end{pmatrix}, A)$ is observable and that both C and K are of full row rank. Define the *i*th **partial observability matrix** $\mathcal{O}_i(K,A)$ by

$$\mathcal{O}_i(K,A) := \begin{pmatrix} KA^{i-1} \\ \vdots \\ KA \\ K \end{pmatrix}.$$

Define, inductively, a sequence of full row rank matrices $\{Z_i\}$ as follows:
Set $Z_0 = 0$, and proceed inductively. Assume $Z_0, \ldots, Z_i$ are constructed. If $\operatorname{Ker} Z_i \cap \operatorname{Ker} C$ is conditioned invariant, then one sets $Z = Z_i$ and stops. Otherwise, Proposition 7.26 is applied to construct Z_{i+1}, which satisfies

$$
\begin{cases}
\operatorname{Ker} Z_{i+1} \supset \operatorname{Ker} \mathcal{O}_{i+1}(K,A), \\
\operatorname{Ker} \begin{pmatrix} Z_{i+1} \\ C \end{pmatrix} = \operatorname{Ker} \begin{pmatrix} KA^i \\ Z_i \\ C \end{pmatrix} = \operatorname{Ker} \begin{pmatrix} \mathcal{O}_{i+1}(K,A), \\ C \end{pmatrix} \\
\operatorname{Ker} Z_{i+1} + \operatorname{Ker} C = \mathbb{F}^n.
\end{cases} \tag{7.62}
$$

Theorem 7.27. *Let Z_i be constructed as above. The following assertions are true:*

1. *For all i, $\operatorname{Ker} \begin{pmatrix} Z_{i+1} \\ C \end{pmatrix} \subset \operatorname{Ker} \begin{pmatrix} Z_i \\ C \end{pmatrix}$.*
2. *Let ν be the smallest index with*

$$
\operatorname{Ker} \begin{pmatrix} Z_{i+1} \\ C \end{pmatrix} = \operatorname{Ker} \begin{pmatrix} Z_i \\ C \end{pmatrix}. \tag{7.63}
$$

Setting $Z = Z_\nu$, the subspace $\mathcal{V} = \operatorname{Ker} Z$ is a maximal conditioned invariant subspace that satisfies (7.61).
3. *There exist matrices $F \in \mathbb{F}^{q \times q}$, $G \in \mathbb{F}^{q \times p}$, $J \in \mathbb{F}^{k \times q}$, and $E \in \mathbb{F}^{k \times p}$ for which the following Sylvester equations are satisfied:*

$$
ZA = FZ + GC,
$$

$$
K = JZ + EC.
$$

Then the system

$$
\Sigma_{est} := \begin{cases} \xi_{t+1} = F\xi_t + Gy_t + Hu_t, \\ \zeta_t = J\xi_t + Ey_t \end{cases}
$$

is an order-q tracking observer for K, where

$$
q = \operatorname{rank} Z = \operatorname{codim} \operatorname{Ker} Z. \tag{7.64}
$$

Proof. 1. Follows from the equality $\operatorname{Ker} \begin{pmatrix} Z_{i+1} \\ C \end{pmatrix} = \operatorname{Ker} \begin{pmatrix} KA^i \\ Z_i \\ C \end{pmatrix}$.

2. The equalities (7.62) and (7.63) imply the equality

$$\text{Ker}\left(\begin{array}{c}\mathscr{O}_i(K,A)\\C\end{array}\right)=\text{Ker}\left(\begin{array}{c}\mathscr{O}_{i+1}(K,A)\\C\end{array}\right),$$

which in turn implies the inclusion

$$A\text{Ker}\left(\begin{array}{c}\mathscr{O}_i(K,A)\\C\end{array}\right)\subset\text{Ker}\left(\begin{array}{c}\mathscr{O}_{i+1}(K,A)\\C\end{array}\right).$$

Using this, one computes

$$A(\mathscr{V}\cap\text{Ker}\,C)=A\text{Ker}\left(\begin{array}{c}Z_v\\C\end{array}\right)=A\text{Ker}\left(\begin{array}{c}\mathscr{O}_v(K,A)\\C\end{array}\right)\subset\text{Ker}\left(\begin{array}{c}\mathscr{O}_{v+1}(K,A)\\C\end{array}\right)$$
$$=\text{Ker}\left(\begin{array}{c}Z_{v+1}\\C\end{array}\right)=\text{Ker}\left(\begin{array}{c}Z_v\\C\end{array}\right)=\mathscr{V},$$

showing that $\mathscr{V}$ is conditioned invariant. Since, for $i\geq 1$, one has $\text{Ker}\,\mathscr{O}_i(K,A)\subset\text{Ker}\,K$, it follows that

$$\mathscr{V}\cap\text{Ker}\,C=\text{Ker}\left(\begin{array}{c}Z_v\\C\end{array}\right)=\text{Ker}\left(\begin{array}{c}\mathscr{O}_v(K,A)\\C\end{array}\right)\subset\text{Ker}\,K,$$

i.e., (7.61) holds.
3. Follows from Theorem 7.24.   ∎

The decision whether or not q, defined in (7.64), is the minimal order of tracking observers is left as an open problem. Next, we consider a few special cases of Theorem 7.27.

Case I: $\text{Ker}\,C\subset\text{Ker}\,K$. This means that unobserved states are not to be estimated, or, equivalently stated, states to be estimated are directly observed. From the inclusion $\text{Ker}\,C\subset\text{Ker}\,K$ one deduces that there exists an E for which $K=EC$. This implies $Z=0$, $F=0$, $G=0$, and $J=0$. Thus, as intuitively expected, the existence of a zero-order or, equivalently, a nondynamic, observer for K is obtained.

Case II: $K=I$, i.e., tracking the state. To this end, one needs a map Z for which $\text{Ker}\left(\begin{array}{c}Z\\C\end{array}\right)\subset\text{Ker}\,K=\{0\}$. The easiest choice, though not necessarily the minimal one, is to take $Z=I$ and $E=0$ and $G=0$. This implies $J=I$ and $F=A$. Thus, a tracking observer is given by

$$\Sigma_{est}:=\left\{\begin{array}{l}\sigma\xi=A\xi,\\\zeta=\xi,\end{array}\right.$$

and the Luenberger state observer has been rederived. To obtain a reduced-order state observer, Z is chosen so that $\left(\begin{array}{c}z\\c\end{array}\right)$ is nonsingular.

Case III: $C = 0$, i.e., no observations are available. By our assumption of the observability of the pair $\left(\binom{C}{K}, A\right)$, (K, A) is necessarily an observable pair. Let $K_1, \ldots, K_q$ be the rows of K. Since K is assumed to be of full row rank, the rows are linearly independent. The **dual Kronecker indices** are defined by the following deletion process:

Starting from the top of the observability matrix

$$\text{col}\,(K_1, \ldots, K_q, K_1 A, \ldots, K_q A, \ldots, K_1 A^{n-1}, \ldots, K_q A^{n-1}),$$

delete all row vectors that are linearly dependent on the set of preceding ones. In this way one obtains, up to a permutation of the rows, a matrix of the form

$$Z = \text{col}\,(K_1, K_1 A, \ldots, K_1 A^{\nu_1 - 1}, \ldots, K_q, \ldots, K_q A^{\nu_q - 1}).$$

The observability of (K, A) implies $\sum_{i=1}^{q} \nu_i = n$ and, hence, the invertibility of Z. Defining $F = ZAZ^{-1}$, $J = KZ^{-1}$, then

$$\Sigma_{est} := \begin{cases} \sigma\xi = F\xi, \\ \zeta = J\xi \end{cases}$$

is a tracking observer. For tracking, given an initial condition x_0 of the state system, one chooses $\xi_0 = Zx_0$.

Example 7.28. This example is taken from Fernando, Trinh, Hieu and Jennings (2010). One takes A, C, K as follows:

$$A = \begin{pmatrix} -1 & 0 & 0 \\ 0 & -1 & 0 \\ 0 & 0 & 2 \end{pmatrix}, \quad C = (1\ 1\ 0), \quad K = (1\ 2\ 0).$$

Note that the rows of C and K are linearly independent, so $Z_1 = K$. Computing $KA = (-1\ -2\ 0)$, which is linearly dependent on K, we conclude that $Z = Z_1 = K$. Since $\text{Ker} Z + \text{Ker} C = \mathbb{F}^3$, $\mathscr{V} = \text{Ker} Z$ is a tight conditioned invariant subspace. It is easily checked that the observer Sylvester equations have a unique solution given by $F = -1, G = 0, J = 1, E = 0$, which gives an asymptotic observer.

Example 7.29. Let A, C, K be as follows:

$$A = \left(\begin{array}{ccc|cc} -1 & 0 & 0 & 0 & 0 \\ 1 & -1 & 0 & 0 & 0 \\ 0 & 1 & -1 & 0 & 0 \\ \hline 0 & 0 & 0 & 1 & 0 \\ 0 & 0 & 0 & 1 & 1 \end{array} \right), \quad C = (0\ 0\ 1|0\ 0), \quad K = (0\ 1\ 0|0\ 1).$$

In this example, the pair (C,A) is not observable, but $\left(\begin{pmatrix} C \\ K \end{pmatrix}, A\right)$ is. Since C and K are linearly independent, one gets $Z_1 = K$. Computing

$$\mathscr{O}_5(K,A) = \begin{pmatrix} KA^4 \\ KA^3 \\ KA^2 \\ KA \\ K \end{pmatrix} = \begin{pmatrix} -4 & 1 & 0 & 4 & 1 \\ 3 & -1 & 0 & 3 & 1 \\ -2 & 1 & 0 & 2 & 1 \\ 1 & -1 & 0 & 1 & 1 \\ 0 & 1 & 0 & 0 & 1 \end{pmatrix},$$

it is seen that the bottom four rows are linearly independent, but the top row depends linearly on them. Therefore, $Z = \mathscr{O}_4(K,A)$ and $\mathrm{Ker}\, Z = \{\mathrm{col}\,(0,0,\gamma,0,0) | \gamma \in \mathbb{R}\}$. Since

$$A \begin{pmatrix} 0 \\ 0 \\ 1 \\ 0 \\ 0 \end{pmatrix} = \begin{pmatrix} 0 \\ 0 \\ -1 \\ 0 \\ 0 \end{pmatrix},$$

$\mathrm{Ker}\, Z$ is actually an A-invariant subspace. It is easily checked that $\mathrm{Ker}\, Z + \mathrm{Ker}\, C = \mathbb{R}^5$, hence $\mathscr{V} = \mathrm{Ker}\, Z$ is a tight conditioned invariant subspace, which means that the corresponding tracking observer is uniquely determined up to similarity. Using the observer Sylvester equations, one obtains

$$F = \begin{pmatrix} 0 & 2 & 0 & -1 \\ 1 & 0 & 0 & 0 \\ 0 & 1 & 0 & 0 \\ 0 & 0 & 1 & 0 \end{pmatrix}, \quad G = \begin{pmatrix} 0 \\ 0 \\ 0 \\ 0 \end{pmatrix}, \quad J = (0\ 0\ 0\ 1), \quad E = (0).$$

7.5 Construction of Functional Observers

In Theorem 7.24, characterizations of various classes of observers in terms of conditioned invariant, outer detectability, and outer observability subspaces appear. Recalling that these subspaces have nice functional representations, given in Proposition 6.67 and Theorem 6.72, it is only natural to attempt observer construction using these representations. Key ingredients in the analysis are the coprime factorization (7.19), the parameterizations (7.18) and (7.21), and the shift realization (4.20).

Consider the system

$$x_{t+1} = Ax_t + Bu_t,$$
$$y_t = Cx_t, \qquad\qquad (7.65)$$
$$z_t = Kx_t,$$

with $A \in \mathbb{F}^{n \times n}, B \in \mathbb{F}^{n \times m}, C \in \mathbb{F}^{p \times n}, K \in \mathbb{F}^{k \times n}$, assuming that $\left(\begin{pmatrix} C \\ K \end{pmatrix}, A \right)$ is observable. Let C, K, A have the representation (7.9), with respect to the direct sum representation (7.10), and let

$$\begin{pmatrix} C_1 & 0 \\ K_1 & K_2 \end{pmatrix} \begin{pmatrix} zI - A_{11} & 0 \\ -A_{21} & zI - A_{22} \end{pmatrix}^{-1} = \begin{pmatrix} D_{11}(z) & 0 \\ D_{21}(z) & D_{22}(z) \end{pmatrix}^{-1} \begin{pmatrix} \Theta_{11}(z) & 0 \\ \Theta_{21}(z) & \Theta_{22}(z) \end{pmatrix}$$

be an adapted coprime factorization, i.e., satisfying conditions (1a)–(1c) of Theorem 7.9. Under these assumptions the following assertions are true.

Theorem 7.30. *1. The map*

$$\Theta : X_{\begin{pmatrix} zI - A_{11} & 0 \\ -A_{21} & zI - A_{22} \end{pmatrix}} \longrightarrow X_{\begin{pmatrix} D_{11}(z) & 0 \\ D_{21}(z) & D_{22}(z) \end{pmatrix}} \qquad (7.66)$$

$$\begin{pmatrix} f_1 \\ f_2 \end{pmatrix} = \Theta \begin{pmatrix} \xi_1 \\ \xi_2 \end{pmatrix} = \begin{pmatrix} \Theta_{11}(z) & 0 \\ \Theta_{21}(z) & \Theta_{22}(z) \end{pmatrix} \begin{pmatrix} \xi_1 \\ \xi_2 \end{pmatrix}$$

is an $\mathbb{F}[z]$-*isomorphism. Defining, via the shift realization,*

$$\bar{A} := S_{\begin{pmatrix} D_{11} & 0 \\ D_{21} & D_{22} \end{pmatrix}}, \qquad \bar{C} \begin{pmatrix} f_1 \\ f_2 \end{pmatrix} := \left((D_{11}^{-1} \ 0) \begin{pmatrix} f_1 \\ f_2 \end{pmatrix} \right)_{-1} = (D_{11}^{-1} f_1)_{-1},$$

$$\bar{K} \begin{pmatrix} f_1 \\ f_2 \end{pmatrix} := \left((-D_{22}^{-1} D_{21} D_{11}^{-1} \ D_{22}^{-1}) \begin{pmatrix} f_1 \\ f_2 \end{pmatrix} \right)_{-1} = (-D_{22}^{-1} D_{21} D_{11}^{-1} f_1 + D_{22}^{-1} f_2)_{-1},$$

$$(7.67)$$

the intertwining relations

$$\Theta A = \bar{A} \Theta, \quad C = \bar{C} \Theta, \quad K = \bar{K} \Theta \qquad (7.68)$$

are satisfied.
2. For the parameterization (7.18) of the set of rational solutions of (7.17), we can choose a proper rational matrix $W(z)$, with left coprime factorization

$$W(z) = T(z)^{-1} L(z), \qquad (7.69)$$

such that $Z_1(z)$ is proper. Let $Q(z)$ be the l.c.l.m.$(T(z),D_{22}(z))$, and let $R_D(z)$ and $R_L(z)$ be left coprime polynomial matrices with

$$Q(z) = R_D(z)D_{22}(z) = R_T(z)T(z). \tag{7.70}$$

Then the parameterization (7.18) can be rewritten as

$$Z_1(z) = Q(z)^{-1}(-R_D(z)D_{21}(z) + R_T(z)L(z)D_{11}(z)),$$
$$Z_2(z) = Q(z)^{-1}\left(-R_T(z)L(z)\Theta_{11}(z) + R_D(z)\Theta_{21}(z) \ R_D(z)\Theta_{22}(z).\right) \tag{7.71}$$

3. *Define maps*

$$Z:X_{\begin{pmatrix} D_{11} & 0 \\ D_{21} & D_{22} \end{pmatrix}} \longrightarrow X_Q, \quad F:X_Q \longrightarrow X_Q, \quad G:\mathbb{F}^p \longrightarrow X_Q,$$

$$H:\mathbb{F}^m \longrightarrow X_Q, \quad J:X_Q \longrightarrow \mathbb{F}^k, \quad E:\mathbb{F}^p \longrightarrow \mathbb{F}^k$$

by

$$Z = \pi_Q\left(-R_TL \ R_D,\right)|X_{\begin{pmatrix} D_{11} & 0 \\ D_{21} & D_{22} \end{pmatrix}}$$
$$F = S_Q,$$
$$G = -\pi_Q\left((-R_TL \ R_D,)\begin{pmatrix} D_{11} \\ D_{21} \end{pmatrix}(\cdot)\right) \tag{7.72}$$
$$H = ZB,$$
$$J = \left(Q^{-1}(\cdot)\right)_{-1},$$
$$E = \pi_+Q^{-1}(-R_DD_{21} + R_TLD_{11}).$$

Then the system

$$\Sigma_{est} = \begin{cases} \xi_{t+1} = F\xi_t + Gy_t + Hu_t, \\ \zeta_t = J\xi_t + Ey_t, \end{cases} \tag{7.73}$$

defined in the state space X_Q by these maps, is an observable tracking observer for K with transfer functions

$$Z_1(z) = \left[\begin{array}{c|c} F & G \\ \hline J & E \end{array}\right], \quad Z_2(z) = \left[\begin{array}{c|c} F & Z \\ \hline J & 0 \end{array}\right]$$

of (F,G,J,E) and $(F,Z,J,0)$, respectively. The realization (F,G,J,E) of $Z_1(z)$ is always observable. It is reachable if and only if the polynomial matrices $Q(z)$ and $-R_T(z)L(z)D_{11}(z) + R_D(z)D_{21}(z)$ are left coprime.

Proof. The adapted coprime factorization leads to the intertwining relation

$$\begin{pmatrix} \Theta_{11}(z) & 0 \\ \Theta_{21}(z) & \Theta_{22}(z) \end{pmatrix} \begin{pmatrix} zI - A_{11} & 0 \\ -A_{21} & zI - A_{22} \end{pmatrix} = \begin{pmatrix} D_{11}(z) & 0 \\ D_{21}(z) & D_{22}(z) \end{pmatrix} \begin{pmatrix} C_1 & 0 \\ K_1 & K_2 \end{pmatrix}.$$

Applying Theorem 3.21 shows that the map Θ, defined by (7.68), is an $\mathbb{F}[z]$-module isomorphism. The intertwining relations (7.66) are easily verified. This proves part 1.

Part 2 follows from a straightforward substitution of (7.69) in the parameterization (7.18), using the identities $D_{22}(z)^{-1} = Q(z)^{-1}R_D(z)$ and $T(z)^{-1} = Q(z)^{-1}R_T(z)$.

Proof of part 3. Recall from Theorem 3.30 that

$$S_{\begin{pmatrix} D_{11} & 0 \\ D_{21} & D_{22} \end{pmatrix}} \begin{pmatrix} f_1 \\ f_2 \end{pmatrix} = z \begin{pmatrix} f_1 \\ f_2 \end{pmatrix} - \begin{pmatrix} D_{11} & 0 \\ D_{21} & D_{22} \end{pmatrix} \begin{pmatrix} \xi_1 \\ \xi_2 \end{pmatrix},$$

where $\begin{pmatrix} \xi_1 \\ \xi_2 \end{pmatrix} = \left(\begin{pmatrix} D_{11} & 0 \\ D_{21} & D_{22} \end{pmatrix}^{-1} \begin{pmatrix} f_1 \\ f_2 \end{pmatrix} \right)_{-1}$. For $f = \begin{pmatrix} f_1 \\ f_2 \end{pmatrix} \in X_{\begin{pmatrix} D_{11} & 0 \\ D_{21} & D_{22} \end{pmatrix}}$,

using (7.67), one computes $(Z\overline{A} - F Z)f =$

$$= \pi_Q \left(-R_T L \; R_D \right) S_{\begin{pmatrix} D_{11} & 0 \\ D_{21} & D_{22} \end{pmatrix}} \begin{pmatrix} f_1 \\ f_2 \end{pmatrix} - S_Q \pi_Q \left(-R_T L \; R_D \right) \begin{pmatrix} f_1 \\ f_2 \end{pmatrix}$$

$$= \pi_Q \left(-R_T L \; R_D \right) \left(z \begin{pmatrix} f_1 \\ f_2 \end{pmatrix} - \begin{pmatrix} D_{11} & 0 \\ D_{21} & D_{22} \end{pmatrix} \begin{pmatrix} \xi_1 \\ \xi_2 \end{pmatrix} \right) - \pi_Q \left(-R_T L \; R_D \right) z \begin{pmatrix} f_1 \\ f_2 \end{pmatrix}$$

$$= -\pi_Q \left(-R_T L \; R_D \right) \begin{pmatrix} D_{11} & 0 \\ D_{21} & D_{22} \end{pmatrix} \begin{pmatrix} \xi_1 \\ \xi_2 \end{pmatrix} = -\pi_Q \left(-R_T L \; R_D \right) \begin{pmatrix} D_{11} \\ D_{21} \end{pmatrix} \xi_1.$$

It was shown that, with $\overline{C}, \overline{A}$ defined by (7.67) and Z, F, G by (7.72), the observer Sylvester equation $Z\overline{A} = FZ + G\overline{C}$ is satisfied. The equation $\overline{K} = JZ + E\overline{C}$ can be verified similarly. This completes the proof. ∎

A few remarks are in order.

1. The construction of the tracking observer (7.73) works for every left coprime pair of polynomials $T(z), L(z)$, with $T(z)$ nonsingular. In particular, one does not need to impose a properness assumption on

$$Z_1 = -(TD_{22})^{-1}(TD_{21} - LD_{11}).$$

However, if one wants to obtain the observer from a shift realization of $Z_1(z)$, then $W(z) = T(z)^{-1}L(z)$ must be chosen such that Z_1 is proper.

2. The unobservable subspace $\mathcal{O}_*(C,A)$ of system (7.65) has the representation

$$\mathcal{O}_*(C,A) = \begin{pmatrix} zI - A_{11} & 0 \\ -A_{21} & I \end{pmatrix} X \begin{pmatrix} I & 0 \\ 0 & zI - A_{22} \end{pmatrix}$$

and dimension $\deg\det(zI - A_{22})$. Its image under Θ is given by

$$\begin{pmatrix} D_{11}(z) & 0 \\ D_{21}(z) & I \end{pmatrix} X \begin{pmatrix} I & 0 \\ 0 & D_{22}(z) \end{pmatrix},$$

which has dimension $\deg\det D_{22} = \deg\det(zI - A_{22})$.

3. The constructed tracking observer

$$Z_1(z) = \left[\begin{array}{c|c} F & G \\ \hline J & E \end{array}\right]$$

can be written as in (7.71), with the shift realization defined in the state space $X_{TD_{22}}$. Clearly, $\dim X_{TD_{22}} = \deg\det(T) + \deg\det(D_{22})$, and the term $\deg\det(D_{22})$ is the price of tracking the unobservable subspace $\mathcal{O}_*(C,A)$.

4. The choice of the rational matrix $W(z)$ in Theorem 7.30 is closely related to partial realizations; however, we will not follow this path in this book and instead refer the reader to Fuhrmann (2008) for some of the details.

From Theorem 7.30 we deduce several special cases as corollaries.

Corollary 7.31. *Consider system (7.65), with* $\left(\begin{pmatrix} C \\ K \end{pmatrix}, A\right)$ *observable.*

1. Assume that (C,A) is observable.

(a) The coprime factorization (7.5) reduces to

$$\begin{pmatrix} C \\ K \end{pmatrix}(zI - A)^{-1} = \begin{pmatrix} D_{11}(z) & 0 \\ D_{21}(z) & I \end{pmatrix}^{-1}\begin{pmatrix} \Theta_{11}(z) \\ \Theta_{21}(z) \end{pmatrix}.$$

(b) Choose a proper rational matrix $W(z)$ with left coprime factorization $W(z) = T(z)^{-1}L(z)$ so that $Z_1(z)$ in (7.18) is proper. Define maps

$$Z : X_{D_{11}} \longrightarrow X_T, \quad F : X_T \longrightarrow X_T, \quad G : \mathbb{F}^p \longrightarrow X_T,$$
$$H : \mathbb{F}^m \longrightarrow X_T, \quad J : X_T \longrightarrow \mathbb{F}^k, \quad E : \mathbb{F}^p \longrightarrow \mathbb{F}^k$$

by

$$Z = -\pi_T L | X_{D_{11}},$$
$$F = S_T,$$
$$G = -\pi_T L D_{11}(\cdot),$$
$$H = ZB,$$
$$J = (T^{-1}(\cdot))_{-1},$$
$$E = \pi_+(-D_{21} + T^{-1} L D_{11}).$$

(7.74)

Then system (7.73), defined by F, G, H, J, E, is a tracking observer for K.

2. Assume $C = 0$, i.e., there are no observations at all.

(a)   The pair (K, A) is observable, and the coprime factorization (7.5) reduces to

$$K(zI - A)^{-1} = D_{22}(z)^{-1} \Theta_{22}(z).$$

(b)   The map $Z : X_{zI-A} \longrightarrow X_{D_{22}}$, defined as $Z\xi = \Theta_{22}(z)\xi$, is an isomorphism, and the Sylvester equation (7.42) reduces to $ZA = FZ$, with Z invertible and $K = JZ$.

(c)   The pair (J, F), defined in the state space $X_{D_{22}}$ by

$$F = S_{D_{22}},$$
$$Jg = (D_{22}^{-1} g)_{-1},$$

is a tracking observer.

3. Assume the pair (C, A) is detectable.

(a)   In the coprime factorization (7.5), the polynomial matrix $D_{22}(z)$ is stable.
(b)   In the parameterization (7.18) of the set of rational solutions of (7.17), we can choose the rational matrix $W(z)$ with coprime factorization $W(z) = T(z)^{-1} L(z)$ so that $Z_1(z)$ is proper and stable.
(c)   System (7.73), defined by the F, G, H, J, E in (7.72), is an asymptotic tracking observer for K.

Proof.   Basically, this follows from Theorem 7.30. However, we add a few remarks. A simple computation yields

$$X_{\begin{pmatrix} D_{11} & 0 \\ D_{21} & I \end{pmatrix}} = \left\{ \begin{pmatrix} f(z) \\ 0 \end{pmatrix} \Big| f(z) \in X_{D_{11}} \right\}.$$

Note that the intertwining relation

$$(I\,0)\begin{pmatrix} D_{11}(z)\ 0 \\ D_{21}(z)\ I \end{pmatrix} = D_{11}(z)\,(I\,0),$$

taken together with the trivial associated coprimeness relations, implies that the map $\Phi : X_{\begin{pmatrix} D_{11}\ 0 \\ D_{21}\ I \end{pmatrix}} \longrightarrow X_{D_{11}}$, defined by $\begin{pmatrix} f(z) \\ 0 \end{pmatrix} \mapsto f(z)$, is an $\mathbb{F}[z]$-isomorphism. Using this, the maps in (7.72) have the simplified form (7.74). ∎

It should be pointed out that, to compensate for the total absence of observations, i.e., for the case $C = 0$, the order of a minimal tracking observer is necessarily equal to the dimension of the state space of the system.

To clarify the issues of observer characterization and construction, we present an example from two points of view: the state-space formulation on the one hand and the functional model formulation on the other. Each formulation has its own advantages and insights.

Example 7.32. Let A, C, K be as follows:

$$A = \begin{pmatrix} 0\ 0\ 0 \\ 1\ 0\ 0 \\ 0\ 1\ 0 \end{pmatrix}, \quad C = (0\ 1\ 0), \quad K = (0\ 0\ 1).$$

Computing $Z_1 = K = (0\ 0\ 1)$, $KA = (0\ 1\ 0)$, clearly,

$$\mathrm{Ker}\begin{pmatrix} K \\ C \end{pmatrix} = \mathrm{Ker}\begin{pmatrix} 0\ 0\ 1 \\ 0\ 1\ 0 \end{pmatrix} \subset \mathrm{Ker}\,(0\ 1\ 0) = \mathrm{Ker}\,KA,$$

which shows that $\mathcal{V} = \mathrm{Ker}\,K$ is a conditioned invariant subspace satisfying $\mathcal{V}\,\cap\,\mathrm{Ker}\,C \subset \mathrm{Ker}\,K$. Obviously, it is a maximal such subspace. The Sylvester equations lead to an observer defined by

$$\begin{aligned} F &= (0), \quad G = (1) \\ J &= (1), \quad E = (0). \end{aligned}$$

Since the matrix $\begin{pmatrix} K \\ C \end{pmatrix}$ has full row rank, this representation is uniquely determined.

Computing further $KA^2 = (1\ 0\ 0)$, and setting $Z = \begin{pmatrix} 1\ 0\ 0 \\ 0\ 1\ 0 \\ 0\ 0\ 1 \end{pmatrix}$, one has $ZA = \begin{pmatrix} 0\ 0\ 0 \\ 1\ 0\ 0 \\ 0\ 1\ 0 \end{pmatrix}$, and the equation $ZA = FZ + GC$ implies

$$F = \begin{pmatrix} 0 & 0 & -t_0 \\ 1 & 0 & -t_1 \\ 0 & 1 & -t_2 \end{pmatrix}, G = \begin{pmatrix} t_0 \\ t_1 \\ t_2 \end{pmatrix}. \tag{7.75}$$

The nonuniqueness of the constructed minimal-order observer is a consequence of the inequality $\dim(\operatorname{Ker} Z + \operatorname{Ker} C) = 2 < 3 = \dim \mathbb{F}^3$.

From a polynomial point of view, one has $\mathcal{X} = X_{z^3}$ and $A = [S_{z^3}]$. Since $\operatorname{Ker} K = \{\xi_0 + \xi_1 z\}$, the maximal conditioned invariant subspace $\mathcal{V}$ contained in $\operatorname{Ker} K$ is the zero subspace; hence, $\mathcal{V} = X_{z^3} \cap t(z)\mathbb{F}[z]$, with $t(z)$ a polynomial of degree greater than or equal to 3. To obtain minimal-order observers, we take $\deg t(z) = 3$. Setting $t(z) = t_0 + t_1 z + t_2 z^2 + z^3$, we compute $F = [\pi_t | X_{z^3}]$ and $G = -[\pi_t z^3]$. Simple computations lead to (7.75).

7.6 Exercises

1. Consider the undamped harmonic oscillator

$$\dot{x}_1(t) = x_2(t),$$
$$\dot{x}_2(t) = -\omega^2 x_1(t) + u(t),$$
$$y(t) = x_2(t).$$

(a) Determine an observer

$$\dot{z}(t) = (A - LC)z(t) + Ly(t) + Bu(t),$$
$$u(t) = Fz(t) + v(t),$$
$$y(t) = x_2(t),$$

such that the eigenvalues of the closed-loop system

$$\begin{pmatrix} A & B\mathscr{F} \\ LC & A - LC + B\mathscr{F} \end{pmatrix}$$

are $-\omega \pm \sqrt{-1}\omega, -\omega, -\omega$.

(b) Determine a one-dimensional reduced observer.

2. Consider the third-order system $\dot{x} = Ax + bu, y = cx$, with $\alpha < 0, \beta \neq 0, k \neq 0$, and

$$A = \begin{pmatrix} 0 & 1 & 0 \\ 0 & -\beta & 1 \\ 0 & 0 & \alpha \end{pmatrix}, \quad b = \begin{pmatrix} 0 \\ k \\ 0 \end{pmatrix}, \quad c = \begin{pmatrix} 1 & 0 & 0 \end{pmatrix}.$$

(a) Prove that the system is stabilizable and observable.
(b) Find an observer of order 3 such that the eigenvalues of the closed-loop system are $-1, -2, -3, -4, -5, \alpha$.

3. Let $(A, B) \in \mathbb{R}^{n \times n} \times \mathbb{R}^{n \times m}$. Assume that the unreachable modes $\lambda \in \mathbb{C}$ of

$$x_{t+1} = Ax_t + Bu_t$$

are unstable, i.e., satisfy $|\lambda| \geq 1$. Then, for every $C \in \mathbb{R}^{p \times n}$ with $C \neq 0$, there exists an initial state x_0 and an input sequence (u_t) such that $Cx_t \not\to 0$ for $t \to \infty$.

4. Assume that $G(z) = D + C(zI - A)^{-1}B$ is a $p \times m$ proper rational function such that $G(z)u(z)$ is stable for all proper rational functions $u(z) \in \mathbb{F}(z)^m$. Show that $D = 0$ and $C(zI - A)^{-1}B = 0$.

5. Assume that the output sequence $\{y_t\}$ of the linear system

$$x_{t+1} = Ax_t + Bu_t,$$
$$y_t = Cx_t + Du_t$$

satisfies $\lim_{t \to \infty} y_t = 0$ for all initial conditions $x_0 \in \mathbb{F}^n$ and all input sequences $u = (u_t)$. Assume further that all unreachable modes of A are unstable. Show that $C = 0$ and $D = 0$.

6. Let (J, F) be an observable pair, and let (A, B) be reachable. Prove the following result from Fuhrmann and Helmke (2001a): for matrices M and N one has

$$J(zI - F)^{-1}(zM + N)(zI - A)^{-1}B = 0$$

if and only if there exist constant matrices X and Y with $JY = 0$ and $XB = 0$ satisfying

$$zM + N = X(zI - A) - (zI - F)Y.$$

7.7 Notes and References

Probably the first application of modern control theory was that of system stabilization (Chapter 6), and this was done by state feedback. Since the state of a system is hardly ever available, the immediate question arises of how to estimate the state from measurements. This question immediately leads to the fields of optimal filtering and observer theory. Although attempts at state estimation were made earlier, it is generally accepted that the origin of observer theory can be traced to Luenberger (1964). What is surprising is that, over the years, the analysis and synthesis of functional observers, i.e., observer theory, did not attract the appropriate attention from the control community that it so rightly deserves. Moreover, in the system literature, there are several gaps, faulty proofs, and lack of insights that only now are beginning to be filled in. Refer to Trumpf (2013) for details on these gaps in the development of observer theory. In this connection we mention also, for example, Fuhrmann and Helmke (2001) for a fairly complete account of asymptotic observers and to the Ph.D. thesis by Trumpf (2002), which focuses on certain geometric properties that relate to observer theory.

In recent years, the behavioral approach, an approach that avoids the input/output point of view, has been initiated and developed by Willems (1986, 1991) and coworkers. For a study of observers in the behavioral context we refer to the work by Valcher and Willems (1999) and Trumpf, Trentelmann and Willems (2014); see also Fuhrmann (2008), who pointed out how conventional state observer theory fits into the behavioral framework. A full study of the connections between conventional and behavioral observer theories has not yet been undertaken.

The concept of reconstructibility is important for cases dealing with dead-beat observers, a case that will not be addressed in this book. For a treatment of dead-beat observers, see Bisiacco, Valcher and Willems (2006) and Fuhrmann and Trumpf (2006). The parameterization results, given by (7.18) when the pair (C,A) is not observable and by (7.21) when it is, relate to partial realizations; they are also reminiscent of the Youla–Kucera parameterization as outlined in Chapter 6. In fact, from the first equation of (7.21) it follows that $W = -D_{21}D_{11}^{-1} - Z_1 D_{11}^{-1}$, which shows that $W(z)$ is a solution to a nice partial realization problem induced by $-D_{21}D_{11}^{-1}$. Minimal McMillan degree solutions to (7.17) can be obtained from minimal McMillan degree solutions of the partial realization problem. We refer the reader to Fuhrmann (2008) for a full analysis of the observable case.

The linear equation $Z_K(z) = Z_1(z)Z_C(z) + Z_2(z)$ in proper rational functions $Z_1(z), Z_2(z)$ plays a central role in our approach to functional observers. Since the space of proper rational functions is a valuation ring, and hence a local ring, this task amounts to studying linear matrix equations over a local ring. This fact may be useful in developing solution algorithms for $Z_K(z) = Z_1(z)Z_C(z) + Z_2(z)$.

In Definition 7.16, we made a distinction between asymptotic observers and asymptotic tracking observers. Theorem 7.17 left open the question of how to characterize asymptotic observers. A natural question arises as to how to specify extra conditions such that an asymptotic observer is also tracking. This issue has been addressed by Trumpf (2013) for continuous-time systems. Some preparatory results from Trumpf (2013) appear as Exercises 4–6. In the absence of a full proof, this is stated as a conjecture.

Conjecture 7.33. Let the linear system (7.28) act in the state space $\mathbb{R}^n$. Assuming that all unreachable modes of the system are unstable, the system, defined in the state space $\mathbb{R}^q$ by

$$\xi_{t+1} = F\xi_t + Gy_t + Hu_t,$$
$$\zeta_t = J\xi_t + Ey_t,$$

is an observable, asymptotic observer for K if and only if F has all its eigenvalues in the open unit disc, (J,F) is observable, and there exists a matrix $Z \in \mathbb{R}^{q \times n}$ such that

$$ZA - FZ = GC, \quad H = ZB, \quad K = JZ + EC.$$

In particular, under such an assumption, asymptotic observers are automatically asymptotic tracking observers.

Under the assumption that the pair (A,B) is reachable, this was proved in Fuhrmann and Helmke (2001a). See also Fernando, Jennings and Trinh (2011) for a claim toward Conjecture 7.33. For a characterization of asymptotic observers in the behavioral framework, see Trumpf, Trentelmann and Willems (2014).

Part III
Networks of Linear Systems

Chapter 8
Nonnegative Matrices and Graph Theory

The interconnections and coupling patterns of dynamical systems are best described in terms of graph theory. This chapter serves the purpose of summarizing the main results and tools from matrix analysis and graph theory that will be important for the analysis of interconnected systems in subsequent chapters. This includes a proof of the Perron–Frobenius theorem for irreducible nonnegative matrices using a contraction mapping principle on convex cones due to Birkhoff (1957). We introduce adjacency matrices and Laplacians associated to a weighted directed graph and study their spectral properties. The analysis of eigenvalues and eigenvectors for graph adjacency matrices and Laplacians is the subject of spectral graph theory, which is briefly summarized in this chapter; see the book by Godsil and Royle (2001) for a comprehensive presentation. Explicit formulas for the eigenvalues and eigenvectors of Laplacians are derived for special types of graphs such as cycles and paths. These formulas will be used later on, in Chapter 9, in an examination of homogeneous networks. The technique of graph compression is briefly discussed owing to its relevance for the model reduction of networks. Properties of graphs are increasingly important for applications to, for example, formation control and molecular geometry. Therefore, a brief section is included on graph rigidity and the characterization of Euclidean distance matrices.

We begin by establishing some notation to be used subsequently and presenting some basic facts on Kronecker products of matrices over a field $\mathbb{F}$. For rectangular matrices $A \in \mathbb{F}^{m \times n}, B \in \mathbb{F}^{k \times l}$, the **Kronecker product** is defined as the $mk \times nl$ matrix

$$A \otimes B = \begin{pmatrix} a_{11}B & \cdots & a_{1n}B \\ \vdots & \ddots & \vdots \\ a_{m1}B & \cdots & a_{mn}B \end{pmatrix}.$$

By this definition, the Kronecker product of an upper triangular matrix A with a rectangular matrix B is block-upper triangular. In particular, the Kronecker product

© Springer International Publishing Switzerland 2015   411
P.A. Fuhrmann, U. Helmke, *The Mathematics of Networks*
of Linear Systems, Universitext, DOI 10.1007/978-3-319-16646-9_8

$B \otimes I_N$ is of the form

$$B \otimes I_N = \begin{pmatrix} b_{11}I_N & \cdots & b_{1l}I_N \\ \vdots & \ddots & \vdots \\ b_{k1}I_N & \cdots & b_{kl}I_N \end{pmatrix},$$

while

$$I_N \otimes A = \mathrm{diag}\,(A,\ldots,A) = \begin{pmatrix} A & \cdots & 0 \\ \vdots & \ddots & \vdots \\ 0 & \cdots & A \end{pmatrix}.$$

If A and B are invertible $n \times n$ and $m \times m$ matrices, respectively, then the Kronecker product $A \otimes B$ is invertible and

$$(A \otimes B)^{-1} = A^{-1} \otimes B^{-1}.$$

The eigenvalues of $A \otimes B$ are the products $\lambda_i(A)\lambda_j(B)$ of the eigenvalues $\lambda_i(A)$ and $\lambda_j(B)$ of A and B, respectively. Therefore, the trace and determinant of $A \otimes B$ of matrices A and B are $\mathrm{tr}\,(A \otimes B) = \mathrm{tr}\,(A)\mathrm{tr}\,(B)$ and $\det(A \otimes B) = \det(A)^m \det(B)^n$. Similarly, the eigenvalues of $A \otimes I_m + I_n \otimes B$ are the sums $\lambda_i(A) + \lambda_j(B)$. The following rules for the Kronecker product are easily verified:

$$(A \otimes B)(C \otimes D) = AC \otimes BD,$$

$$(A \otimes B)^\top = A^\top \otimes B^\top.$$

Let $\mathrm{vec}(A) \in \mathbb{F}^{mn}$ denote a column vector obtained by stacking the second column of A under the first, then the third under the second, and so on. This leads to the following important identity:

$$\mathrm{vec}(ABC) = \left(C^\top \otimes A\right)\mathrm{vec}(B).$$

8.1   Nonnegative Matrices and Contractions

Let $\mathbb{R}_+$ denote the subset of all nonnegative real numbers. A matrix $A \in \mathbb{R}^{n \times n}$ is called **nonnegative (positive)** if all entries a_{ij} of A are nonnegative (positive) real numbers. The notation for nonnegative and positive matrices

$$A \geq 0 \iff a_{ij} \geq 0, \quad i,j = 1,\ldots,n,$$

$$A > 0 \iff a_{ij} > 0, \quad i,j = 1,\ldots,n$$

should not be confused with the notion of positive definiteness for symmetric matrices. To distinguish these notions from each other, the **positive (semi-) definiteness** of real symmetric matrices $A = A^\top \in \mathbb{R}^{n \times n}$ is denoted by

$$A \succeq 0 \iff x^\top A x \geq 0 \quad \text{for all } x \in \mathbb{R}^n,$$

$$A \succ 0 \iff x^\top A x > 0 \quad \text{for all } x \in \mathbb{R}^n \setminus \{0\}.$$

The sum and product of finitely many nonnegative matrices is nonnegative. Moreover, the scalar product λA of a nonnegative matrix A with a nonnegative real number λ is a nonnegative matrix. Thus the set of nonnegative matrices forms a closed convex cone $\mathbb{R}_+^{n \times n}$ in the matrix space $\mathbb{R}^{n \times n}$, which is multiplicatively closed.

The set of nonnegative matrices forms the largest class of matrices that leave the cone $\mathbb{R}_+^n$ invariant. The Hilbert metric allows for a natural generalization of this situation. We refer to the papers by Birkhoff (1957), Bushell (1986), and Kohlberg and Pratt (1982) for additional background. Let $C \subset \mathbb{R}^n$ denote a closed convex cone that is **pointed**, i.e., C has nonempty interior and satisfies $C \cap (-C) = \{0\}$. We use the notation $x \geq 0$ whenever $x \in C$ and $x > 0$ whenever x is an interior point of C. Recall that the dual cone of C is defined as

$$C^* = \{\lambda \in \mathbb{R}^{1 \times n} \mid \lambda(x) \geq 0\}.$$

Clearly, the dual cone $(\mathbb{R}_+^n)^*$ of $\mathbb{R}_+^n$ is equal to $\mathbb{R}_+^{1 \times n}$. We mention, without proof, the well-known fact that for pointed closed convex sets the interior of the dual cone C^* is nonempty. Note further that for C a closed pointed convex cone, every linear functional λ in the interior of C^* satisfies $\lambda(x) > 0$ for all nonzero $x \in C$. This implies the following lemma.

Lemma 8.1. *Let C be a closed convex and pointed cone in $\mathbb{R}^n$. Then the subset $C_1 = \{x \in C \mid \lambda(x) = 1\}$ is compact for all interior points λ of the dual cone C^*.*

Proof. Clearly, $C_1 \mathbb{R}^n$. Suppose C_1 is unbounded. Then there exists a sequence $x_k \in C_1$, with $\|x_k\| \to \infty$ and $\lambda(x_k) = 1$ for all k. Thus $\lambda(\frac{x_k}{\|x_k\|})$ converges to 0. By the compactness of the unit sphere in $\mathbb{R}^n$, there exists an infinite subsequence $y_m, m \in \mathbb{N}$ of $\frac{x_k}{\|x_k\|}$ that converges to a unit vector $y \in C$. Thus $\lambda(y) = \lim_{m \to \infty} \lambda(y_m) = 0$. But λ is in the interior of C^*, and therefore $\lambda(x) > 0$ for all $x \in C \setminus \{0\}$. Thus $y = 0$, which is a contradiction. $\blacksquare$

A **projective metric** on C is a map $d : C \times C \longrightarrow \mathbb{R} \cup \{\infty\}$ such that for all $x, y, z \in C$ (and $r \leq \infty$, $r + \infty = \infty = \infty + \infty$ for all real r):

1. $d(x, y) = d(y, x)$;
2. $d(x, y) \geq 0$, $d(x, y) = 0$ if and only if $x = \lambda y$ for some real $\lambda > 0$;
3. $d(x, z) \leq d(x, y) + d(y, z)$.

Conditions 1–3 imply the identity

4. $d(\lambda x, \mu y) = d(x, y)$ for all $\lambda > 0, \mu > 0$.

Let

$$M(x,y) = \inf\{\lambda \geq 0 \mid x \leq \lambda y\},$$

$$m(x,y) = \sup\{\lambda \geq 0 \mid x \geq \lambda y\} = \frac{1}{M(y,x)}.$$

The following properties are easily seen:

$$0 \leq m(x,y) \leq M(x,y) \leq \infty,$$

$$m(x,y)y \leq x \leq M(x,y)y.$$

Definition 8.2. The **Hilbert metric** on C is a projective metric defined by

$$d(x,y) = \log \frac{M(x,y)}{m(x,y)} = \log M(x,y) + \log M(y,x).$$

Here, $d(0,0) = 0$; $d(x,0) = d(0,y) = \infty$ for $x,y \in C$.

The preceding definitions are illustrated by the following examples.

Example 8.3. (a) Let $C = \mathbb{R}^n_+$. Then for all $x > 0, y > 0$

$$m(x,y) = \min_{i=1,\dots,n} \frac{x_i}{y_i}, \quad M(x,y) = \max_{i=1,\dots,n} \frac{x_i}{y_i}.$$

Thus the Hilbert metric on $\mathbb{R}^n_+$ is

$$d(x,y) = \max_{1 \leq i,j \leq n} \log \frac{x_i y_j}{x_j y_i}, \quad \text{for } x > 0, y > 0.$$

(b) Let $C = \{X \in \mathbb{R}^{n \times n} \mid X = X^\top \succeq 0\}$ denote the closed convex cone of positive semidefinite real symmetric matrices. Let $\lambda_{\min}(X)$ and $\lambda_{\max}(X)$ denote the smallest and largest eigenvalues of a symmetric matrix X, respectively. For positive definite matrices $X \succ 0, Y \succ 0$ then

$$m(X,Y) = \lambda_{\min}(XY^{-1}), \quad M(X,Y) = \lambda_{\max}(XY^{-1}).$$

Thus the Hilbert metric of two positive definite matrices $X \succ 0, Y \succ 0$ is

$$d(X,Y) = \log \frac{\lambda_{\max}(XY^{-1})}{\lambda_{\min}(XY^{-1})}.$$

For the proof of the following result see Kohlberg and Pratt (1982).

Theorem 8.4. *Let C be a closed convex cone in $\mathbb{R}^n$ that is pointed. Let $\lambda \in C^*$ be an interior point of C^*. Then the following properties are satisfied:*

1. *The Hilbert metric is a projective metric on C. The distance satisfies $d(x,y) < \infty$ if and only if x and y are interior points of $L(x,y) \cap C$. In particular, $d(x,y) < \infty$ for all interior points x,y of C.*
2. *Let $C_1 := \{x \in C \mid \lambda(x) = 1\}$. Then (C_1,d) is a compact metric space, and there exists a constant $\gamma > 0$ with*

$$\|x - y\| \leq \gamma d(x,y) \quad \forall x,y \in C_1.$$

3. *Let $x,y \in C$ be such that $d(x,y) < \infty$. Then there exist $a,b \in C$ such that*

$$d(x,y) = \log \frac{\|a - y\| \|x - b\|}{\|a - x\| \|y - b\|}.$$

A linear map $A : \mathbb{R}^n \longrightarrow \mathbb{R}^n$ is called C-monotonic whenever $AC \subset C$. Let d denote the Hilbert metric on C. Then

$$k(A) = \inf\{k \geq 0 \mid d(Ax,Ay) \leq kd(x,y) \; \forall x,y \in C\}$$

denotes the contraction constant of A. The operator A is called a **contraction** if $k(A) < 1$. There is a beautiful formula for the contraction constant of a C-monotonic linear map A.

Theorem 8.5 (Birkhoff (1957)). *A linear C-monotonic map $A : \mathbb{R}^n \longrightarrow \mathbb{R}^n$ is a contraction if and only if*

$$\delta = \sup\{d(Ax,Ay) \mid x,y \in C\} < \infty.$$

Whenever this is satisfied, the contraction constant is equal to

$$k(A) = \frac{e^{\delta/2} - 1}{e^{\delta/2} + 1}.$$

For a linear contraction on a closed convex and pointed cone C the Banach fixed-point theorem applies. The following result extends the classical Perron–Frobenius theorem to monotonic maps on a closed convex pointed cone.

Theorem 8.6 (Contraction Mapping Theorem). *Let $C \subset \mathbb{R}^n$ denote a closed convex and pointed cone, $\lambda \in C^*$ an interior point of the dual cone C^*, and $C_1 := \{x \in C \mid \lambda(x) = 1\}$. Let $\mu \geq 0$ be a nonnegative real number. Let $N \in \mathbb{N}$ and $A : \mathbb{R}^n \longrightarrow \mathbb{R}^n$ be a linear map such that $(\mu I + A)^N$ maps C_1 into the interior of C.*

1. *Then $(\mu I + A)^N$ is a contraction on C and A has a unique eigenvector $x_* \in C_1$. The vector x_* is contained in the interior of C_1 with a positive eigenvalue $r_* > 0$.*
2. *The discrete dynamical system*

$$x_{t+1} = \frac{(\mu I + A)_t^{Nx}}{\lambda((\mu I + A)_t^{Nx})}, \quad x_t \in C_1 \tag{8.1}$$

converges to x_* from each initial condition $x_0 \in C_1$.

Proof. Let $B = (\mu I + A)^N$. By Lemma 8.1, C_1 is compact, and therefore the image $K := B(C_1)$ is a compact subset of the interior of C. Thus

$$\sup\{d(Bx, By) \mid x, y \in C\} = \sup\{d(Bx, By) \mid x, y \in C_1\}$$
$$\leq \delta(K) < \infty,$$

where $\delta(K)$ denotes the diameter of the compact set K. The Birkhoff Theorem 8.5 therefore implies that B is a contraction on the complete metric space C_1. Consider the discrete-dynamical system $x_{t+1} = f(x_t)$ on C_1, defined by iterating the map

$$f : C_1 \longrightarrow C_1, \quad f(x) = \frac{Bx}{\lambda(Bx)}.$$

By our assumption on A, the map f is well defined and satisfies $d(f(x), f(y)) = d(Bx, By)$. Since B is a contraction on C_1, so is f. Thus there exists $0 \leq k < 1$ with $d(f(x), f(y)) \leq kd(x, y)$ for all $x, y \in C_1$. Therefore, one can apply the Banach fixed-point theorem to f and conclude that there exists a unique fixed point $x_* \in C_1$ of f. Moreover, the dynamical system (8.1) converges to x_* from every initial point $x_0 \in C_1$. This shows that $x_* \in C_1$ is an eigenvector of B, with $Bx_* = \sigma x_*$. Since B maps C_1 into the interior of C, $\sigma x_* = Bx_*$ must be an interior point of C. But this implies that $\sigma > 0$ as well as that x_* is an interior point of C_1. By projective invariance of the Hilbert metric,

$$d(Ax_*, x_*) = d(A(Bx_*), Bx_*)$$
$$= d(B(Ax_*), Bx_*) \leq kd(Ax_*, x_*),$$

and therefore $d(Ax_*, x_*) = 0$. But this implies $Ax_* = r_* x_*$ for some $r > 0$. The result follows. ∎

One can give a steepest-descent interpretation of the preceding arguments that will be useful for the proof of the Perron–Frobenius theorem. Consider the continuous function

$$R_A : K \longrightarrow \mathbb{R}_+, \quad R_A(x) = m(Ax, x). \tag{8.2}$$

It is instructive to compute this function in the special case $C = \mathbb{R}_+^n$, where for $x > 0$ one has

$$R_A(x) = \min_{1 \leq i \leq n} \frac{(Ax)_i}{x_i}.$$

This form is reminiscent of the Rayleigh quotient function.

Proposition 8.7 (Generalized Rayleigh Quotient). *The same notation is used here as in Theorem 8.6. The function (8.2) has x_* as its unique maximum with $R_A(x_*) = r_*$. Moreover, $R_A(x_t)$, $t \geq 0$, is monotonically increasing in t and converges to $R_A(x_*) = r_*$ for each sequence of points $(x_t)_{t \geq 0}$ that are generated by the power iterations (8.1).*

Proof. We use the notation from the proof of Theorem 8.6. Note that R_A is well defined and continuous, because $K = B(C_1)$ is a compact subset of the interior of C. Since $m(cx, cy) = m(x,y)$ for $c > 0$, one obtains for all $x \in K$

$$R_A(f(x)) = m(ABx, Bx) = \sup\{\lambda \mid ABx \geq \lambda Bx\}$$
$$= \sup\{\lambda \mid B(Ax - \lambda x) \geq 0\}$$
$$\geq \sup\{\lambda \mid Ax - \lambda x \geq 0\} = R_A(x).$$

Since $Bv > 0$ for all nonzero vectors $v \in C$, this shows $B(Ax - \lambda x) > 0$ for all $x \in K$ that are not eigenvectors of A. This implies that $\sup\{\lambda \mid B(Ax - \lambda x) \geq 0\} > \sup\{\lambda \mid Ax - \lambda x \geq 0\}$ is a strict inequality, unless $x \in K$ is an eigenvector of A. Thus $R_A(x_{t+1}) = R_A(f(x_t)) < R_A(x_t)$, unless $x_t = f(x_t)$ is a fixed point of f. By Theorem 8.6, the eigenvector x_* is the only fixed point of f and satisfies $R_A(x_*) = m(Ax_*, x_*) = m(r_* x_*, x_*) = r_*$. This completes the proof. ∎

The next result yields an explicit form for the contraction constant for positive matrices $A \in \mathbb{R}^{n \times n}$, $A > 0$. Here $C = \mathbb{R}^n_+$.

Corollary 8.8. *Every positive matrix $A \in \mathbb{R}^{n \times n}$ is a contraction with respect to the Hilbert metric on $\mathbb{R}^n_+$. The contraction constant is*

$$k(A) = \frac{\sqrt{\gamma} - 1}{\sqrt{\gamma} + 1},$$

where

$$\gamma = \max_{i,j,k,l} \frac{a_{ki} a_{lj}}{a_{kj} a_{li}}.$$

8.2 Perron–Frobenius Theorem

The Perron–Frobenius theorem establishes a surprising and deep connection between the spectral properties of nonnegative matrices and the properties of the associated graph. Let us begin by deriving the theorem using the contraction mapping theorem on closed convex cones. For other approaches and further details see, for example, the beautiful books by Fiedler (2008) and Sternberg (2010).

The notion of the irreducibility of nonnegative matrices plays a central role in the subsequent analysis. In Section 8.6, a simple graph-theoretic characterization of irreducibility is derived.

Definition 8.9. A matrix $A \in \mathbb{R}^{n \times n}$ is called **reducible** if either $(n = 1, A = 0)$ or $n \geq 2$ and there exists a permutation matrix $P \in \mathbb{R}^{n \times n}$, $1 \leq r \leq n - 1$, and the matrices $B \in \mathbb{R}^{r \times r}, C \in \mathbb{R}^{r \times (n-r)}, D \in \mathbb{R}^{(n-r) \times (n-r)}$, with

$$P^\top A P = \begin{pmatrix} B & C \\ 0 & D \end{pmatrix}.$$

Otherwise, A is called **irreducible**.

Irreducible nonnegative matrices have useful properties.

Lemma 8.10. *Let $A \in \mathbb{R}^{n \times n}$ be a nonnegative irreducible matrix and $x \in \mathbb{R}^n$ a vector with nonnegative components. Then $Ax = 0$ implies $x = 0$.*

Proof. After a suitable permutation of the entries of x (and an induced similarity transformation on A), one may assume that $x = (\xi^\top, 0)^\top$, with $\xi = (x_1, \ldots, x_r)^\top$ and $x_1 > 0, \ldots, x_r > 0$. Suppose $r \geq 1$. Partition the matrix A accordingly as

$$A = \begin{pmatrix} A_{11} & A_{12} \\ A_{21} & A_{22} \end{pmatrix}, \tag{8.3}$$

with $A_{11} \in \mathbb{R}^{r \times r}$, and so forth. Thus $Ax = 0$ is equivalent to $A_{11}\xi = 0, A_{12}\xi = 0$, which implies $A_{11} = 0, A_{21} = 0$. This is a contradiction to A being irreducible. Therefore, $Ax = 0$ implies $x = 0$.   ∎

Using this lemma, we next prove a basic existence and uniqueness result for positive eigenvectors of nonnegative irreducible matrices. Let

$$\rho(A) = \max\{|\lambda| \, | \, \det(\lambda I - A) = 0\} \tag{8.4}$$

denote the **spectral radius** of matrix A. Suppose that A has exactly h eigenvalues with absolute value $\rho(A)$. Then h is called the **index** of A.

Theorem 8.11. *Let $A \in \mathbb{R}^{n \times n}$ be a nonnegative irreducible matrix. Let*

$$e = (1, \ldots, 1)^\top \in \mathbb{R}^n, \quad C_1 = \{x \in \mathbb{R}_+^n \mid e^\top x = 1\}.$$

1. *Then A has a unique nonnegative eigenvector $x_* \in \mathbb{R}_+^n$, with $e^\top x_* = 1$, called the **Perron vector**. Both the Perron vector and the associated eigenvalue are positive.*
2. *If A is positive, then the sequence of power iterates*

$$x_{t+1} = \frac{Ax_t}{e^\top A x_t}, \quad x_t \in C_1 \tag{8.5}$$

converges to x_* from each initial condition $x_0 \in C_1, x_0 > 0$.

Proof. C_1 is a compact convex subset of the closed convex and pointed cone $C = \mathbb{R}_+^n$. By Lemma 8.10, matrix A maps $C \setminus \{0\}$ into itself. Moreover, $\mathbf{e}^\top A x > 0$ for all $x \in C \setminus \{0\}$. Since A is irreducible, the matrix $(I + A)^{n-1}$ is positive (see the subsequent Theorem 8.26) and therefore maps C_1 into the interior of C. Thus one can apply Theorem 8.6, with $N = n - 1$ and $\mu = 1$, to conclude that A possesses a unique eigenvector x_* in C_1. Moreover, x_* is contained in the interior of C_1 with positive eigenvalue $r > 0$. In particular, $x_* > 0$. The second part follows again from Theorem 8.6. ∎

After this first step we can state and prove the Perron–Frobenius theorem.

Theorem 8.12 (Perron–Frobenius). *Let $A \in \mathbb{R}^{n \times n}$ denote an irreducible nonnegative matrix. Then the spectral radius $\rho(A)$ is a simple positive eigenvalue of A and there exists a positive eigenvector $x > 0$ for $\rho(A)$. No eigenvector corresponding to other eigenvalues of A is positive.*

Proof. Suppose, $x \in \mathbb{R}_+^n$ is a nonnegative eigenvector of A. By Theorem 8.11, then $x = x_*$, where $x_* > 0$ denotes the unique Perron vector of A; thus $A x_* = r_* x_*$ and $r_* > 0$. Since $\mathbb{R} x_*$ is the only eigenspace that intersects $\mathbb{R}_+^n$, the eigenvalue r_* has an algebraic multiplicity of one. It suffices, therefore, to show that $\rho(A)$ coincides with the eigenvalue r_* for x_*. In fact, let $\lambda \in \mathbb{C}$ be an eigenvalue of A so that $A z = \lambda z$ for some complex vector $z = (z_1, \ldots, z_n)^\top$. Let $|z| = (|z_1|, \ldots, |z_n|) \in \mathbb{R}_+^n$ be the associated nonnegative vector of absolute values. From the triangle inequality one obtains $|\lambda| \, |z| \le A |z|$. Let

$$m(x,y) = \sup\{\lambda \ge 0 \mid x \ge \lambda y\}$$

be the order function for $\mathbb{R}_+^n$. Applying Proposition 8.7, then

$$|\lambda| = m(\lambda |z|, |z|) \le m(A|z|, |z|) \le r_*.$$

Thus r_* is an eigenvalue with eigenvector x_* and is equal to the spectral radius. This completes the proof. ∎

The following perturbation result is of independent interest.

Proposition 8.13 (Wielandt). *Let $A \in \mathbb{R}^{n \times n}$ be an irreducible nonnegative matrix. Let $B \in \mathbb{C}^{n \times n}$ be a complex matrix with*

$$|b_{ij}| \le a_{ij} \quad \text{for all } i, j = 1, \ldots n.$$

Then $\rho(B) \le \rho(A)$. If $\rho(B) = \rho(A)$ and $\rho(B) e^{\phi \sqrt{-1}}$ is an eigenvalue of B, then there exists a diagonal matrix $D = \operatorname{diag}(z_1, \ldots, z_n) \in \mathbb{C}^{n \times n}$ with $|z_1| = \ldots = |z_n| = 1$ such that

$$B = e^{\phi \sqrt{-1}} D A D^{-1}.$$

In particular, $|b_{ij}| = a_{ij}$ for all $i, j = 1, \ldots n$.

Proof. In the preceding proof of the Perron–Frobenius theorem it was shown that the maximum $r_* = R_A(x_*)$ of the function $R_A : \mathbb{R}^n_+ \longrightarrow \mathbb{R} \cup \{\infty\}, R_A(x) = m(Ax, x)$ exists and $r_* = \rho(A)$. Let $z \in \mathbb{C}^n, z \neq 0$, and $\lambda \in \mathbb{C}$, with $Bz = \lambda z$. Then, using the triangle inequality, one obtains $|\lambda| |z| = |Bz| \leq A|z|$, and therefore

$$|\lambda| = m(|Bz|, |z|) \leq m(A|z|, |z|) \leq r_* = \rho(A). \tag{8.6}$$

This shows that $\rho(B) \leq \rho(A)$. Assume that $\rho(B) = \rho(A) = r_*$ and λ is an eigenvalue of B, with $|\lambda| = r_*$. Then (8.6) implies that $r_* = |\lambda| = m(A|z|, |z|)$. The Perron–Frobenius theorem therefore implies $A|z| = r_*|z|$ and $|z| > 0$. Similarly, for $|B| = (|b_{ij}|)$ one obtains

$$r_*|z| = |\lambda z| = |Bz| \leq |B| \cdot |z| \leq A|z| = r_*|z|,$$

and therefore $|B| \cdot |z| = A \cdot |z|$. Since $A - |B|$ is a nonnegative matrix and $|z| > 0$, this implies $A = |B|$.

Define

$$D = \text{diag}\left(\frac{z_1}{|z_1|}, \ldots, \frac{z_n}{|z_n|}\right).$$

Then $D|z| = z$ and $BD|z| = Bz = \lambda D|z|$. Thus $C := e^{\phi\sqrt{-1}} D^{-1} AD$ satisfies $C|z| = A|z|$ and $|C| = |B| = A$. Split the complex matrix $C = \text{Re}\, C + \sqrt{-1}\,\text{Im}\, C$ into real and imaginary parts $\text{Re}\, C$ and $\text{Im}\, C$, respectively. Since A is real, $C|z| = A|z|$ implies $\text{Re}\, C|z| = A|z|$. From $\text{Re}\, C \leq |C| = A$ it follows that $A - \text{Re}\, C$ is nonnegative, with $(A - \text{Re}\, C)|z| = 0$. Since $|z| > 0$, this implies $|C| = A = \text{Re}\, C$, and therefore $C = A$. This completes the proof. ∎

Let $A \in \mathbb{R}^{n \times n}$ be nonnegative and irreducible. For $i = 1, \ldots, n$, the **period** $p(i)$ is defined as the greatest common divisor of all $m \in \mathbb{N}$ satisfying $(A^m)_{ii} > 0$. By a theorem of Romanovsky, $p(1) = \cdots = p(n)$ for all irreducible nonnegative matrices A. The common value $p(A) := p(1) = \cdots = p(n)$ is called the **period** of A. A nonnegative matrix A of period 1 is called **aperiodic**. We now state, without providing full proof details, a full characterization of the structure of irreducible nonnegative matrices. A stronger form of the subsequent result and its proof appeared as Theorem 4.3.1 in the book by Fiedler (2008).

Theorem 8.14. *Let $A \in \mathbb{R}^{n \times n}$ be irreducible and nonnegative, and let $\lambda_0, \ldots, \lambda_{k-1}$ denote the eigenvalues of A with absolute value equal to the spectral radius $\rho(A)$. The following statements are true:*

1. $\lambda_0, \ldots, \lambda_{k-1}$ are simple eigenvalues of A and satisfy

$$\lambda_j = e^{\frac{2\pi\sqrt{-1}j}{k}} \rho(A), \quad j = 0, \ldots, k-1.$$

2. The spectrum of A is invariant under rotations with angle $\frac{2\pi}{k}$.

3. *If $k > 1$, then there exists a permutation matrix P such that*

$$PAP^\top = \begin{pmatrix} 0 & B_{12} & 0 & \cdots & 0 \\ 0 & 0 & B_{23} & & \vdots \\ \vdots & & \ddots & \ddots & 0 \\ 0 & & & 0 & B_{k-1,k} \\ B_{k1} & 0 & \cdots & 0 & 0 \end{pmatrix},$$

   with block matrices B_{ij} of suitable sizes.
4. *The index k of A in (1) coincides with the period of A.*
5. *A is primitive if and only if the spectral radius $\rho(A)$ is the only eigenvalue λ of A with $|\lambda| = \rho(A)$.*

Proof. Only the first two statements are shown; we refer the reader to Fiedler (2008) for a proof of the remaining claims. Let $\lambda_j = \rho e^{2\pi\phi_j\sqrt{-1}}, \phi_0 = 0$, denote the eigenvalues of absolute value $\rho = \rho(A)$. Applying Proposition 8.13 with $B = A$ one obtains

$$A = e^{2\pi\phi_j\sqrt{-1}}D_j A D_j^{-1}, \quad j = 0, \cdots, k-1$$

for suitable unitary diagonal matrices $D_0, \ldots, D_{k-1}$. Thus the spectrum of A is invariant under multiplications by $e^{2\pi\phi_j\sqrt{-1}}, j = 0, \cdots, k-1$. Since the spectral radius $\lambda_0 = \rho$ is a simple eigenvalue of A, $\lambda_0, \ldots, \lambda_{k-1}$ are simple eigenvalues of $e^{2\pi\phi_j\sqrt{-1}}D_j A D_j^{-1} = A$. For all $0 \le r, s \le k-1$ and $D_{rs} = D_r D_s$,

$$A = e^{2\pi\phi_r\sqrt{-1}}D_r A D_r^{-1} = e^{2\pi\phi_r\sqrt{-1}}e^{2\pi\phi_s\sqrt{-1}}D_r D_s A D_r^{-1} D_s^{-1}$$

$$= e^{2\pi(\phi_r+\phi_s)\sqrt{-1}}D_{rs} A D_{rs}^{-1}.$$

Thus $e^{2\pi(\phi_r+\phi_s)\sqrt{-1}}\rho$ are eigenvalues of A for all $0 \le r, s \le k-1$. This implies that $\{1, e^{2\pi\phi_1\sqrt{-1}}, \ldots, e^{2\pi\phi_{k-1}\sqrt{-1}}\}$ is a multiplicative subgroup of $S^1 = \{z \in \mathbb{C} \mid |z| = 1\}$ of order k. Thus

$$\lambda_j = e^{\frac{2\pi j\sqrt{-1}}{k}}, \quad j = 0, \ldots, k-1.$$

In particular, the spectrum of A is invariant under rotations by $\frac{2\pi}{k}$. This completes the proof for the first two items. ∎

The preceding result allows for an interesting dynamical interpretation of irreducible nonnegative matrices A in terms of discrete-time periodic linear systems. In fact, if A is not primitive, then the dynamical system $x_{t+1} = Ax_t$ is permutation equivalent to a **periodic** time-varying system $x_{t+1} = A_{[t]}x_t$, with local states $x_t \in \mathbb{R}^{n_t}$ and a periodic sequence of matrices $A_{[0]} = B_{12}, A_{[1]} = B_{23}, \ldots, A_{[k-1]} = B_{k1}, A_{[k]} = A_{[0]}$, and so forth.

8.3 Stochastic Matrices and Markov Chains

We present a simple version of the ergodic theorem in the context of finite-dimensional matrix algebras. Recall that a norm $\|\cdot\|$ on the matrix space $\mathbb{C}^{n\times n}$ is called submultiplicative if $\|AB\| \le \|A\|\|B\|$ for all matrices $A, B \in \mathbb{C}^{n\times n}$. Standard examples of submultiplicative matrix norms include the 1-norm $\|A\|_1 = \sum_{i,j=1}^{n} |a_{ij}|$, the Frobenius norm $\|A\|_F = \sqrt{\sum_{i,j=1}^{n} |a_{ij}|^2}$, and the operator norm $\|A\| = \sup_{\|x\|=1} \|Ax\|$. Let $\rho(A)$ denote the spectral radius of A.

Proposition 8.15. *Let $\rho(A)$ denote the spectral radius of a matrix $A \in \mathbb{C}^{n\times n}$. Then the following assertions are true:*

1. If $\rho(A) < 1$, then

$$\lim_{k\to\infty} A^k = \lim_{k\to\infty} \frac{1}{k} \sum_{i=0}^{k-1} A^i = 0.$$

2. If $\rho(A) > 1$, then both sequences (A^k) and $\frac{1}{k}\sum_{i=0}^{k-1} A^i$ diverge.
3. Let $\rho(A) = 1$. The limit $\lim_{k\to\infty} A^k$ exists if and only if 1 is the only eigenvalue of A with absolute value 1 and all Jordan blocks for 1 are 1×1.
4. Let $\rho(A) = 1$. The limit $\lim_{k\to\infty} \frac{1}{k}\sum_{i=0}^{k-1} A^i$ exists if and only if all eigenvalues λ of A with absolute value $|\lambda| = 1$ have a geometric multiplicity of one.

Proof. The simple proofs of assertions 1–3 are omitted. To prove assertion 4, assume that the limit $\lim_{k\to\infty} \frac{1}{k}\sum_{i=0}^{k-1} A^i$ exists. From assertion it follows that each eigenvalue λ of A must satisfy $|\lambda| \le 1$. Suppose $|\lambda| = 1$. Without loss of generality, one can assume that $A = \lambda I + N$ is a Jordan block. Then

$$\frac{1}{k}\sum_{i=0}^{k-1} A^i = \frac{1}{k}\sum_{i=0}^{k-1} \lambda^i I + \frac{1}{k}\sum_{i=0}^{k-1} i\lambda^{i-1} N + \ldots + \frac{1}{k}\sum_{i=0}^{k-1} N^i$$

diverges whenever

$$\frac{1}{k}\sum_{i=0}^{k-1} i\lambda^{i-1} = \begin{cases} \frac{k\lambda^{k-1}(\lambda-1)-(\lambda^k-1)}{k(\lambda-1)^2} & \text{if} \quad \lambda \ne 1 \\ \frac{k-1}{2} & \text{if} \quad \lambda = 1 \end{cases}$$

diverges. This completes the proof. ∎

Theorem 8.16 (Ergodic Theorem). *Let $\|\cdot\|$ be a submultiplicative matrix norm on $\mathbb{C}^{n\times n}$ and $A \in \mathbb{C}^{n\times n}$ with $\|A\| \le 1$. Then:*

1. The limit

$$P = \lim_{k\to\infty} \frac{1}{k}\sum_{i=0}^{k-1} A^i$$

exists and satisfies $P^2 = P = PA = AP$. Moreover,

$$\mathrm{Im}\, P = \mathrm{Ker}\,(A - I)$$

$$\mathrm{Ker}\, P = \bigoplus_{\lambda \neq 1} \mathrm{Ker}\,(A - \lambda I)^n;$$

2. *If 1 is the only eigenvalue of A with an absolute value of one, then*

$$P = \lim_{k \to \infty} A^k$$

exists.

Proof. Assume that the limit

$$P = \lim_{k \to \infty} \frac{1}{k} \sum_{i=0}^{k-1} A^i \qquad (8.7)$$

exists. Proposition 8.15 then implies $\rho(A) \leq 1$. For complex numbers λ with $|\lambda| \leq 1$,

$$\lim_{k \to \infty} \frac{1}{k} \sum_{i=0}^{k-1} \lambda^i = \lim_{k \to \infty} \frac{1}{k} \frac{\lambda^k - 1}{\lambda - 1} = \begin{cases} 0 & \text{if } \lambda \neq 1, \\ 1 & \text{if } \lambda = 1. \end{cases}$$

Thus, by decomposing the Jordan canonical form $J = \mathrm{diag}\,(J_1, J_2)$ of A, where J_1 and J_2 have eigenvalues $\lambda = 1$ and $|\lambda| < 1$, respectively, one obtains $\lim_{k \to \infty} \frac{1}{k} \sum_{i=0}^{k-1} J_2^i = I$ and $\lim_{k \to \infty} \frac{1}{k} \sum_{i=0}^{k-1} J_1^i = 0$. Thus $P = \lim_{k \to \infty} \frac{1}{k} \sum_{i=0}^{k-1} A^i$ is a projection operator that commutes with A. Therefore, $P^2 = P = AP = PA$. The preceding argument also implies the formulas for the image space $\mathrm{Im}\, P$ and kernel of P. The second claim follows from Proposition 8.15. Thus it remains to show that the limit (8.7) exists.

For all complex matrices and submultiplicative norms the inequality $\rho(A) \leq \|A\|$ is valid. By Proposition 8.15, it is enough to show that the Jordan blocks for the eigenvalues λ with $|\lambda| = 1$ are 1×1. Assume that $A = SJS^{-1}$, where $J = \mathrm{diag}\,(J_1, \ldots, J_r)$ are in Jordan canonical form. Let $J_1, \ldots, J_\nu$ be the Jordan blocks for the eigenvalues λ with an absolute value of one. For all $m \in \mathbb{N}$, therefore, $\|J^m\| \leq \|S\|\|S^{-1}\|\|A^m\| \leq \|S\|\|S^{-1}\|$. Since all norms on $\mathbb{C}^{n \times n}$ are equivalent, one obtains, for $i = 1, \ldots, \nu$,

$$\|J_i^m\|_1 \leq \|J^m\|_1 \leq \gamma \|J^m\| \leq \gamma \|S\|\|S^{-1}\|.$$

For every Jordan block $J_i = \lambda I + N, i = 1, \ldots, \nu$, of size s and $m \geq 1$, one has the estimate

$$\|J_i^m\|_1 = \|(\lambda I + N)^m\|_1 \geq \begin{cases} 1 & s = 1, \\ m+1 & s > 1 \end{cases}.$$

Thus the sequence $(\|J^m\|_1)_m$ grows unbounded if there exists an eigenvalue λ of geometric multiplicity greater than one and $|\lambda| = 1$. This completes the proof.   ∎

We establish a concrete form of the ergodic theorem for doubly stochastic matrices.

Definition 8.17. A nonnegative matrix $A \in \mathbb{R}^{n \times n}$ is called **stochastic** if

$$\sum_{j=1}^{n} a_{ij} = 1, \quad i = 1, \ldots, n.$$

A is called **doubly stochastic** if it is nonnegative and satisfies

$$\sum_{j=1}^{n} a_{ij} = 1, \quad \sum_{l=1}^{n} a_{lj} = 1, \quad i, l = 1, \ldots, n. \tag{8.8}$$

Let

$$e_n = \begin{pmatrix} 1 \\ \vdots \\ 1 \end{pmatrix}$$

denote a vector in $\mathbb{R}^n$ with all components equal to one. Then a nonnegative matrix A is stochastic (or doubly stochastic) if and only if $Ae_n = e_n$ (or $Ae_n = e_n, e_n^\top A = e_n^\top$).

Theorem 8.18. *Let $A \in \mathbb{R}^{n \times n}$ be a stochastic matrix. Then*

1. *1 is an eigenvalue of A and A has a spectral radius equal to 1;*
2. *The limit*

$$P = \lim_{k \to \infty} \frac{1}{k} \sum_{i=0}^{k-1} A^i$$

exists and is a stochastic matrix that satisfies $P^2 = P = PA = AP$;
3. *If 1 is the only eigenvalue of A with an absolute value of one, then*

$$P = \lim_{k \to \infty} A^k$$

exists. In particular, this is case if A is primitive;
4. *If A is irreducible, then*

$$\lim_{k \to \infty} \frac{1}{k} \sum_{i=0}^{k-1} A^i = e_n y^\top.$$

Here y is a uniquely determined positive vector with

$$y^\top A = y^\top, \quad y_1 + \cdots + y_n = 1.$$

Proof. The matrix norm $\|A\| = \max_{1 \le i \le n} \sum_{j=1}^{n} |a_{ij}|$ is submultiplicative, and every stochastic matrix A satisfies $\|A\| = 1$. The vector $\mathbf{e}_n$ is an eigenvector of A with eigenvalue 1. Therefore, the spectral radius satisfies $1 \le \rho(A) \le \|A\| = 1$, i.e.,

$$\rho(A) = \|A\| = 1.$$

This proves claim 1. Moreover, claim 2 follows from the Ergodic Theorem 8.16, together with the observation that A being stochastic implies that for each k the matrix $\frac{1}{k} \sum_{i=0}^{k-1} A^i$ is nonnegative, with $\frac{1}{k} \sum_{i=0}^{k-1} A^i \mathbf{e}_n = \frac{1}{k} \sum_{i=0}^{k-1} \mathbf{e}_n = \mathbf{e}_n$. The first claim in statement 3 follows from Theorem 8.16, while the second claim follows from the Perron–Frobenius theorem. To prove the last claim, one applies the Perron–Frobenius theorem to the irreducible matrix $A^\top$. Thus, there exists a unique vector $y \in \mathbb{R}^n, y > 0$, with $y^\top A = y^\top$ and $\mathbf{e}^\top y = 1$. Moreover, 1 is a simple eigenvalue of A. Thus Theorem 8.16 implies that $\mathrm{rk} P = \dim \mathrm{Ker}(A - I) = 1$, and therefore P is of the form $P = bc^\top$ for unique nonzero vectors $b, c \in \mathbb{R}^n$, with $c^\top \mathbf{1} = 1$. Since $Pe = \mathbf{e}$, $b = \mathbf{e}$. Moreover, $y^\top A = y^\top$ implies $y^\top P = y^\top$, and therefore $y^\top = y^\top P = y^\top \mathbf{e} c^\top = c^\top$, since $y^\top \mathbf{e} = 1$. This completes the proof. $\blacksquare$

Corollary 8.19. *For every irreducible, doubly stochastic matrix* $A \in \mathbb{R}^{n \times n}$,

$$\lim_{k \to \infty} \frac{1}{k} \sum_{i=0}^{k-1} A^i = \frac{1}{n} \mathbf{e} \mathbf{e}^\top.$$

It is straightforward to apply the preceding results to Markov chains. Consider random variables with values in the finite alphabet $\{1, \ldots, n\}$ and associated probabilities $\pi_i = P(X = i), i = 1, \ldots, n$. Thus $\pi = (\pi(1), \ldots, \pi(n))^\top \in \mathbb{R}^n$ satisfies $\pi \ge 0$ and $\pi(1) + \cdots + \pi(n) = 1$. One can easily generalize this simple static model to a dynamic one by considering a stochastic process (X_t) defined by a sequence of random variables $X_t, t \in \mathbb{N}_0$, with values in $\{1, \ldots, n\}$. Let $(\pi_t) \in \mathbb{R}^n$ denote the associated vector of probabilities. Markov chains are special stochastic processes where for each time t the vector of probabilities π_t depends only on π_{t-1}. More precisely, it is assumed that for each $i, j \in \{1, \ldots, n\}$ the conditioned probabilities

$$p_{ij} := P(X_{t+1} = j | X_t = i), \quad t \in \mathbb{N}_0,$$

are independent of time t. Therefore, one can describe the transition probabilities between states i and j by a matrix

$$A = (p_{ij}) \in \mathbb{R}^{n \times n}$$

of real numbers $p_{ij} \ge 0$, with $\sum_{j=1}^{n} p_{ij} = 1$ for $i = 1, \ldots, n$. Thus $A \in \mathbb{R}^{n \times n}$ is a stochastic matrix of transition probabilities, with $A\mathbf{e} = \mathbf{e}$.

Definition 8.20. A **Markov chain** on the finite state space $\{1, \ldots, n\}$ is a discrete dynamical system

$$\pi_{t+1}^\top = \pi_t^\top A, \quad \pi_0 = \pi, \quad t \in \mathbb{N}_0 \tag{8.9}$$

defined by a stochastic matrix A. Here the initial probability distribution π is allowed to be an arbitrary vector of nonnegative numbers $p_1, \ldots, p_n$, $p_1 + \cdots + p_n = 1$.

The preceding results on stochastic matrices can be reformulated as follows.

Theorem 8.21. *Let (A, π) be a Markov chain on $\{1, \ldots, n\}$ with initial probability distribution π. Let $\pi_t^\top = (\pi_t(1), \ldots, \pi_t(n))$ denote the probability distributions that evolve according to the Markov chain (8.9).*

1. If A is irreducible, then there exists a unique stationary probability distribution $\pi_\infty^\top = (\pi_\infty(1), \ldots, \pi_\infty(n)) \in \mathbb{R}^{1 \times n}$ satisfying

$$\pi_\infty > 0, \quad \pi_\infty^\top A = \pi_\infty^\top, \quad e^\top \pi_\infty = 1.$$

Moreover,

$$\lim_{k \to \infty} \frac{1}{k} \sum_{i=0}^{k-1} A^i = e \pi_\infty^\top,$$

which implies that

$$\lim_{k \to \infty} \frac{1}{k} \sum_{t=0}^{k-1} \pi_t^\top = \pi_\infty^\top.$$

2. Assume A is primitive, i.e., $A^m > 0$ for some $m \in \mathbb{N}$. Then the following limits exist:

$$\lim_{k \to \infty} A^k = e \pi_\infty^\top,$$

$$\lim_{t \to \infty} E(X_t) = \sum_{i=1}^{n} i \pi_\infty(i).$$

Here the expectation value of X_t is defined as $E(X_t) := \sum_{i=1}^{n} i \pi_t(i)$.

Example 8.22. We discuss the Ehrenfest diffusion model from statistical mechanics. Assume a domain Ω is partitioned into two regions, Ω_1 and Ω_2. Assume further that Ω contains exactly n particles that may move around in Ω, passing from one region to the other. Let $X_t \in \{0, \ldots, n\}$ denote the number of particles that are in region Ω_1 at time t. Assume that the probability for a change of a particle from region Ω_1 to region Ω_2, or vice versa, is exactly $\frac{1}{n}$. The transition probability matrix, then, is the $(n+1) \times (n+1)$ tridiagonal matrix

$$A = \begin{pmatrix} 0 & 1 & 0 & \cdots & \cdots & 0 \\ \frac{1}{n} & 0 & \frac{n-1}{n} & & & \vdots \\ 0 & \frac{2}{n} & 0 & \frac{n-2}{n} & & \vdots \\ \vdots & & \ddots & \ddots & \ddots & 0 \\ \vdots & & & \ddots & \ddots & \frac{1}{n} \\ 0 & \cdots & \cdots & 0 & 1 & 0 \end{pmatrix}.$$

Note that A is an irreducible stochastic matrix. Therefore, a unique stationary probability distribution π_∞ of X_t exists and satisfies $\pi_\infty^\top A = \pi_\infty^\top$. Define $y = (y_0, \ldots, y_n)^\top$, with $y_j = 2^{-j}\binom{n}{j}$. A straightforward computation shows that $y^\top A = y^\top, y^\top e = 1$. Therefore, the stationary probabilities are

$$\pi_\infty(j) = \frac{1}{2^j}\binom{n}{j}, \quad j = 0, \ldots, n.$$

In particular, the expectation value of the number of particles in region Ω_1 is equal to

$$E(X_t) = \sum_{j=0}^{n} j\pi_\infty(j) = \sum_{j=0}^{n} j\frac{1}{2^j}\binom{n}{j} = \frac{n}{2},$$

as expected. One can show that the eigenvalues of A are the real numbers $1 - \frac{2k}{n}$, $k = 0, \ldots, n$. The convergence rate of the Markov chain is dependent on the second largest eigenvalue of A, i.e., it is equal to $1 - \frac{2}{n}$. Thus, for large numbers n of particles, the Markov chain will converge quite slowly to the equilibrium distribution.

8.4 Graphs and Matrices

We now introduce some of the basic notions from graph theory and relate the graph concepts to the structure of nonnegative matrices. A **directed graph (digraph)** $\Gamma = (V, E)$ consists of a finite set $V = \{v_1, \ldots, v_N\}$ of **vertices**, together with a finite subset $E \subset V \times V$ of pairs of vertices called **edges**. Thus each edge of a graph is a pair (v, w) of vertices v and w, which are called the initial and terminal vertices of e, respectively. This leads to well-defined maps $\iota, \tau : E \longrightarrow V$ that assign to each edge the initial vertex $\iota(v, w) = v$ and terminal vertex $\tau(v, w) = w$, respectively. We refer to the pair ι, τ as the **canonical orientation** on a digraph (Figures 8.1 and 8.2).

Each vertex element v in a digraph has two kinds of neighborhoods, the **in-neighborhood**,

$$\mathcal{N}^i(v) = \{u \in V \mid (u, v) \in E\},$$

Fig. 8.1 Directed graph

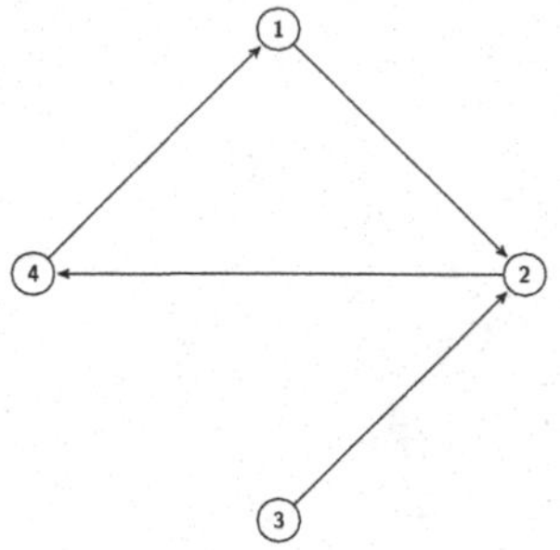

Fig. 8.2 Spanning tree

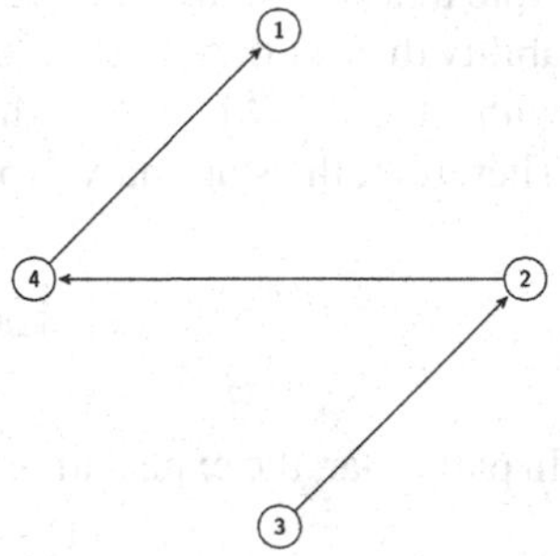

and the **out-neighborhood**,

$$\mathcal{N}^o(v) = \{w \in V \mid (v,w) \in E\}.$$

The cardinalities $d_i(v) = |\mathcal{N}^i(v)|$ and $d_o(v) = |\mathcal{N}^o(v)|$ are called the **in-degree** and **out-degree** of v, respectively. A **subgraph** of a digraph $\Gamma = (V,E)$ is a digraph $\Gamma' = (V',E')$, with $V' \subset V$ and $E' \subset E$. It is called a **spanning subgraph** if $V' = V$ and $E' \subset E$. An **induced subgraph** of $\Gamma = (V,E)$ is a subgraph $\Gamma' = (V',E')$ that contains all edges in E between pairs of vertices in V'. A **walk** in a directed graph Γ of length $r-1$ is a finite sequence of vertices $(v_1,\ldots,v_r)$ such that (v_i,v_{i+1}) are edges for $i = 1,\ldots,r-1$. A walk is **cyclic** if $v_1 = v_r$. A path is a walk where all vertices $v_1,\ldots,v_r$ are distinct. Thus a path cannot be cyclic. A directed graph is called **acyclic** if it does not contain a cycle.

An important topological concept in graph theory is that of connectivity. A digraph Γ is called **strongly connected** if there exists a directed path between all pairs $(u,v) \in V \times V$ of distinct vertices. Γ is called **connected** if, for each $(u,v) \in V \times V$, there exists a directed path from u to v or from v to u. A **strong component** of a digraph is a maximal, strongly connected induced subgraph (Figure 8.3).

For a proof of the following characterization of strong components we refer the reader to Fiedler (2008).

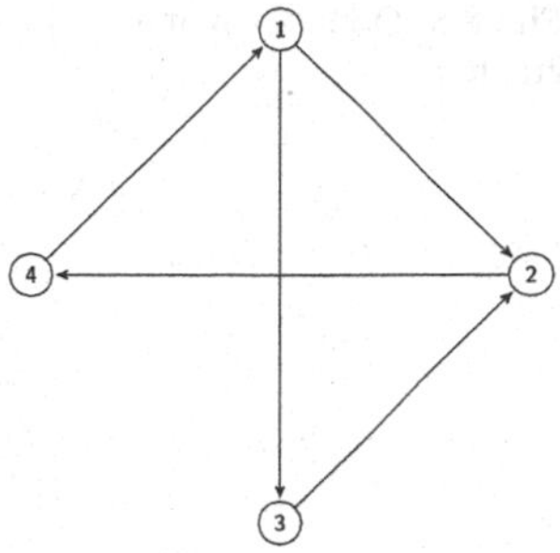

Fig. 8.3 Strongly connected
graph

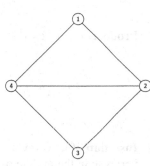

Fig. 8.4 Undirected graph

Proposition 8.23. *Let $\Gamma = (V,E)$ be a digraph.*

1. *Every vertex v is contained in a unique strong component.*
2. *Two distinct strong components have disjoint sets of vertices.*
3. *Two distinct vertices $u,v \in V$ belong to the same strong component if and only if there are directed paths from u to v and from v to u.*

An **undirected graph** $\Gamma = (V,E)$, often simply called a **graph**, consists of a finite set $V = \{v_1, \ldots, v_N\}$ of **vertices**, together with a finite set $E = \{\{v_i, v_j\} \mid (i,j) \in I\}$ of **edges**. Here I denotes a finite subset of $\{1, \ldots, N\} \times \{1, \ldots, N\}$. Thus the edges of an undirected graph are unordered pairs of vertices. Frequently, self-loops are excluded from the definition. A graph Γ is **oriented** if there exist maps $\iota, \tau : E \longrightarrow V$ such that for each edge, $e = \{\iota(e), \tau(e)\}$. Thus, $\hat{\Gamma} = (V,E)$ is a directed graph with a set of edges $E = \{(\iota(e), \tau(e)) \mid e \in E\}$. In many situations concerning graphs one often assumes that an underlying orientation of the graph is specified. A directed graph $\Gamma = (V,E)$ carries a natural orientation by defining $\iota(v,w) = v, \tau(v,w) = w$ for all vertices (Figures 8.4 and 8.5).

We briefly mention a number of elementary operations one can perform with graphs. Let $\Gamma = (V,E)$ and $\Gamma' = (V',E')$ denote graphs with a disjoint set of vertices, i.e., $V \cap V' = \emptyset$. Then the following operations yield new graphs:

1. **Union:** $\Gamma \cup \Gamma' := (V \cup V', E \cup E')$.
2. **Join:** $\Gamma + \Gamma' := (V \cup V', < E \cup E' >)$, where

$$< E \cup E' > := E \cup E' \cup \{\{v,v'\} \mid v \in V, \, v' \in V'\}.$$

Fig. 8.5 Orientation of a
graph

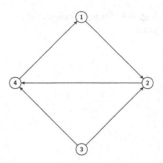

3. **Product:** $\Gamma \times \Gamma' := (V \times V', \hat{E})$, where the set of edges $\hat{E}$ is defined as

$$\{(v,v'),(w,w')\} \in \hat{E} \iff \begin{cases} \{v,w\} \in E \text{ and } v' = w' \\ \text{or} \\ \{v',w'\} \in E' \text{ and } v = w. \end{cases}$$

4. **Insertion of a vertex into an edge:** Geometrically this means one places a new
   vertex w in the middle of an edge uv and replaces the old edge uv with the two
   new ones uw, wv. This operation does not change the topology of the graph. Thus,
   for every edge $\{u,v\} \in E$, a new graph $\Gamma' = (V',E')$ is defined as

$$V' = V \cup \{\{u,v\}\}$$
$$E' = E \setminus \{\{u,v\}\} \cup \{\{u,w\},\{w,v\}\}.$$

5. **Contraction of an edge:** Here one replaces an edge $\{u,v\}$ with a new vertex w
   and adds new edges to all neighboring vertices of u, v.

   The preceding definitions and constructions for digraphs carry over to undirected
graphs in an obvious way. For a vertex $v \in V$, let

$$N(v) = \{w \in V \mid \{v,w\} \in E\}$$

denote the neighborhood of v in the graph. The degree of v is the number of all
neighboring vertices, i.e., it is equal to $|N(v)|$. A graph is called k-regular if all
vertices have the same degree k. A subgraph of Γ is defined by a pair $\Gamma' = (V',E')$
such that $V' \subset V$ and $E' \subset E$. A **spanning subgraph** of Γ is a subgraph $\Gamma' = (V',E')$
with the same set of vertices $V' = V$. A **path** in Γ of length $r-1$ is a finite sequence
of vertices $(v_0,\dots,v_r)$ such that $e_i = \{v_{i-1},v_i\}$ are edges of Γ for $i = 1,\dots,r$.
One says that the path connects the vertices v_0 and v_r. If $v_r = v_0$, then the path
is called closed or a **cycle**. A graph is called **connected** if two distinct vertices
$v \neq w$ are always connected through a suitable path in Γ. A maximal, connected,
induced subgraph of a graph is called a **connected component**. The counterpart
to Proposition 8.23 is true, too, i.e., each vertex is contained in a unique connected

component, and the connected components of a graph form a disjoint decomposition of the set of vertices. A **tree** is a connected graph Γ without cycles. This is easily seen to be equivalent to $|V| = |E| + 1$. A **forest** is a graph whose connected components are trees. A **spanning tree** in a graph $\Gamma = (V, E)$ is a spanning subgraph $\Gamma' = (V', E')$, which is a tree. The number of spanning trees in a graph can be counted by the so-called Matrix-Tree Theorem 8.43.

Weighted Digraphs and Matrices. Nonnegative matrices are associated with digraphs in various ways. A digraph $\Gamma = (V, E)$ is called **weighted** if for each edge $(v_i, v_j) \in E$ one specifies a nonzero real number $a_{ij} \in \mathbb{R}$. For $(v_i, v_j) \notin E$ set $a_{ij} = 0$. Thus, using a labeling $\{1, \ldots, N\} \longrightarrow V$, one associates with the graph a real $N \times N$ matrix $A(\Gamma) = (a_{ij}) \in \mathbb{R}^{N \times N}$. We refer to $A(\Gamma)$ as the **weighted adjacency matrix**. The labelings of the set of vertices differ from each other by a permutation π on $\{1, \ldots N\}$. Thus the associated adjacency matrix changes by a similarity transformation $\pi A(\Gamma) \pi^{-1}$. Conversely, if A denotes a real $N \times N$ matrix, then let $\Gamma_A = (V_A, E_A)$ denote the associated finite directed graph with vertex set $V_A = \{1, \ldots, N\}$. A pair $(i, j) \in V_A \times V_A$ is an edge of Γ_A if and only if $a_{ij} \neq 0$. Then A is the weighted adjacency matrix of Γ_A. Similarly, weighted undirected graphs are defined by specifying for each edge $\{v_i, v_j\}$ a real number a_{ij} and $a_{ij} = 0$ for $\{v_i, v_j\} \notin E$. Thus the weight matrix $A = (a_{ij})$ of an undirected graph is always a real symmetric matrix and therefore has only real eigenvalues.

Every digraph can be considered in a canonical way as a weighted digraph by defining the weight matrix with $0, 1$ entries as

$$\mathfrak{A} = (a_{ij}) \in \{0, 1\}^{N \times N}, \quad \text{with} \quad a_{ij} = \begin{cases} 1 & (v_i, v_j) \in E, \\ 0 & \text{otherwise.} \end{cases}$$

We refer to the digraph $\Gamma_{\mathfrak{A}}$ as a **canonically weighted** digraph.

Example 8.24. A simple example of digraphs with nonnegative weights arises in Euclidean distance geometry and shape analysis. Thus, consider an arbitrary directed graph $\Gamma = (V, E)$ with vertex set $V = \{v_1, \ldots, v_N\} \subset \mathbb{R}^m$. Using the Euclidean distance $\|v - w\|$ between two vertices, define the weights as $a_{ij} = \|v_i - v_j\|$ if and only if $(v_i, v_j) \in E$, and $a_{ij} = 0$ otherwise. Then the weighted adjacency matrix contains all the mutual distances between ordered pairs of points (v_i, v_j) that are specified by the edges of the graph. Thus this matrix contains very interesting information on the geometric configuration of the vertex points.

One can express the classical adjacency matrices of a graph in terms of basic graph operations. Let Γ and Γ' be graphs on m and n vertices, respectively. The classical adjacency matrices for unions, sums, and products are

$$\mathfrak{A}_{\Gamma \cup \Gamma'} = \mathrm{diag}\,(\mathfrak{A}_\Gamma, \mathfrak{A}_{\Gamma'}), \quad \mathfrak{A}_{\Gamma \times \Gamma'} = \mathfrak{A}_\Gamma \otimes I_n + I_m \otimes \mathfrak{A}_{\Gamma'},$$

$$\mathfrak{A}_{\Gamma + \Gamma'} = \begin{pmatrix} \mathfrak{A}_\Gamma & J \\ J^\top & \mathfrak{A}_{\Gamma'} \end{pmatrix}.$$

Here J denotes the $m \times n$ matrix with all entries equal to 1.

The analysis of the structural properties of matrices is closely related to graph theory. The basic connectivity properties of digraphs are reflected in the associated properties of the weighted adjacency matrix. There is a simple graph-theoretic characterization of irreducible matrices.

Proposition 8.25. *The following conditions are equivalent for a matrix* $A \in \mathbb{R}^{N \times N}$:

1. A is irreducible.
2. The digraph Γ_A is strongly connected.

Proof. Assume that Γ_A is not strongly connected. Then there exist vertices $w \neq j \in V_A$ such that there exists no directed path from w to j. Let $V' \subset V_A$ denote the set of vertices v such that there exists a directed path from v to j. Define $V_1 = V' \cup \{j\}$ and $V_2 = V_A \setminus V_1$. By construction of these sets, there does not exist a path from a vertex w in V_2 to some vertex v in V_1; otherwise, one could concatenate the path from w to v with the path from v to j to obtain a path from w to j. This is a contradiction, since $V_2 \cap V_1 = \emptyset$. By assumption on w, j, one has $j \in V_1 \neq \emptyset, w \in V_2$. After a suitable renumbering of vertices one can assume, without loss of generality, that $V_1 = \{1, \ldots, r\}, V_2 = \{r+1, \ldots, N\}, 1 \leq r \leq N-1$. Thus there exists a permutation matrix P such that

$$P^\top A P = \begin{pmatrix} B & C \\ 0 & D \end{pmatrix}.$$

Therefore, A is reducible. Conversely, if A is reducible, then there exists no path from the set $V_2 = \{r+1, \ldots, N\}$ to $V_1 = \{1, \ldots, r\}$. Thus the graph is not strongly connected. This completes the proof. ∎

For nonnegative matrices a stronger form of Proposition 8.25 is valid.

Theorem 8.26. *The following conditions are equivalent for a nonnegative $A \in \mathbb{R}^{N \times N}$.*

1. A is irreducible.
2. For every pair of indices $i, j \in \{1, \ldots, N\}$ there exists $m \in \mathbb{N}$, with $(A^m)_{ij} \neq 0$.
3. $(I_N + A)^{N-1} > 0$.
4. The digraph Γ_A is strongly connected.

Proof. The equivalence of conditions 1 and 4 has already been shown. For $m \in \mathbb{N}$, the sum

$$(A^m)_{ij} = \sum_{k_1, \ldots, k_{m-1}} a_{ik_1} a_{k_1 k_2} \cdots a_{k_{m-1} j} = 0$$

is zero if and only if each summand is zero. But this is equivalent to the property that there exists no walk from i to j of length m. Thus condition 2 is equivalent to condition 4.

Fig. 8.6 Cyclic graph

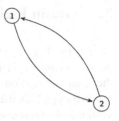

Assume condition 3. Since A is nonnegative, the entries of the matrix

$$(I+A)^{N-1} = \sum_{j=0}^{N-1} \binom{N-1}{j} A^j$$

are all positive. Thus, for each i, j, the ij entry of A^m is positive, for $m = 1, \ldots, N-1$. Thus condition 2 is satisfied and condition 3 implies condition 2 and, therefore, condition 4. Conversely, assume that Γ_A is strongly connected. Between every two distinct vertices there exists, then, a path of length $\leq N-1$ that connects them. Thus the off-diagonal entries of $(I+A)^{N-1}$ are positive. Since the diagonal entries of $(I+A)^{N-1}$ are positive, too, this implies condition 3. This completes the proof. ∎

The adjacency matrix of the cyclic graph in Figure 8.6 is

$$A = \begin{pmatrix} 0 & 1 \\ 1 & 0 \end{pmatrix}$$

and satisfies, for each m,

$$A^{2m+1} = \begin{pmatrix} 0 & 1 \\ 1 & 0 \end{pmatrix}, \quad A^{2m} = \begin{pmatrix} 1 & 0 \\ 0 & 1 \end{pmatrix}.$$

Thus A is an example of a nonnegative matrix that satisfies assertion 2 in Theorem 8.26 but does not satisfy $A^m > 0$ for some m. Such phenomena are explained by the Perron–Frobenius theorem.

Definition 8.27. A nonnegative matrix $A \in \mathbb{R}^{N \times N}$ is called **primitive** if $A^m > 0$ for some $m \in \mathbb{N}$. The smallest such integer m is called the **primitivity index** $\gamma(A)$.

Thus a primitive nonnegative matrix A is irreducible, but the converse does not hold. In general, the primitivity index does not satisfy $\gamma(A) \leq N$. One can show that the primitivity index always satisfies the sharp bound

$$\gamma(A) \leq N^2 - 2N + 2.$$

8.5   Graph Rigidity and Euclidean Distance Matrices

Graph rigidity is an important notion from Euclidean distance geometry that plays a central role in diverse areas such as civil engineering Henneberg (1911), the characterization of tensegrity structures, molecular geometry and 2D-NMR spectroscopy Havel and Wüthrich (1985), and formation shape control of multiagent systems Anderson et. al. (2007). Formations of specified shape are useful in control for sensing and localizing objects, and formations of fixed shape can be contemplated for moving massive objects placed upon them. To steer formations of points from one location to another, steepest-descent methods are used to optimize a suitable cost function. Typically, the smooth cost function

$$V(X) = \frac{1}{4} \sum_{ij \in E} (\|x_i - x_j\|^2 - d_{ij}^2)^2$$

on the space of all formations $X = (x_1, \ldots, x_N)$ of N points x_i in $\mathbb{R}^m$ is used. The gradient flow of V is

$$\dot{x}_i = \sum_{j:\, ij \in E} (\|x_i - x_j\|^2 - d_{ij}^2)(x_i - x_j), \quad i = 1, \ldots, N,$$

and can be shown to converge from every initial condition pointwise to a single equilibrium point. It thus provides a simple computational approach to find a formation that realizes a specified set of distances $d_{ij} > 0, ij \in E$, indexed by the edges of a graph. The characterization of such critical formations and the analysis of their local stability properties in terms of the properties of the graph are among the open research problems in this field. Such research depends crucially on a deeper understanding of Euclidean distance geometry and associated graph-theoretic concepts, such as rigidity. We next turn to a brief description of such methods.

In Euclidean distance geometry one considers a finite tuple of points $x_1, \ldots, x_N$ in Euclidean space $\mathbb{R}^m$, called a formation, together with an undirected, connected graph $\Gamma = (V, E)$ on the set of vertices $V = \{1, \ldots, N\}$, with prescribed distances $d_{ij} = \|x_i - x_j\|$ for each edge $ij \in E$ of Γ. Conversely, by assigning positive real numbers d_{ij}^* to the edges ij of a graph Γ, one is asked to find a formation $x_1, \ldots, x_N \in \mathbb{R}^m$ that realizes the d_{ij}^* as distances $\|x_i - x_j\|$. In heuristic terms (see below for a more formal definition), a formation $(x_1, \ldots, x_N)$ is then called rigid whenever there does not exist a nontrivial perturbed formation $(x_1', \ldots, x_N')$ near $(x_1, \ldots, x_N)$ that realizes the same prescribed distances.

Let us associate with a vertex element $i \in V$ a point x_i in Euclidean space $\mathbb{R}^m$. The $m \times N$ matrix $X = (x_1, \ldots, x_N)$ then describes a **formation** X of N points labeled by the set of vertices of Γ. With this notation at hand, consider the smooth distance map

$$\mathscr{D} : \mathbb{R}^{m \times N} \longrightarrow \mathbb{R}^M, \quad \mathscr{D}(X) = (\|x_i - x_j\|^2)_{(i,j) \in E}.$$

The image set

$$CM_m(\Gamma) = \{\mathscr{D}(X) \mid X \in \mathbb{R}^{m \times N}\}$$

is called the **Cayley–Menger variety**. Being the image of a real polynomial map, the Cayley–Menger variety defines a semialgebraic subset of $\mathbb{R}^M$, which is in fact closed. It is a fundamental geometric object that is attached to the set of all realizations of a graph in $\mathbb{R}^m$. For simplicity, let us focus on the **complete graph** K_N with a set of vertices $V = \{1, \ldots, N\}$ and a set of edges $E = V \times V$. Then the elements of the Cayley–Menger variety $C_m(K_N)$ are in bijective correspondence with the set of $N \times N$ **Euclidean distance matrices**

$$D(x_1, \ldots, x_N) = \begin{pmatrix} 0 & \|x_1 - x_2\|^2 & \cdots & \|x_1 - x_N\|^2 \\ \|x_1 - x_2\|^2 & 0 & \cdots & \|x_2 - x_N\|^2 \\ \vdots & \ddots & \ddots & \vdots \\ \|x_1 - x_{N-1}\|^2 & & 0 & \|x_{N-1} - x_N\|^2 \\ \|x_1 - x_N\|^2 & \cdots & \|x_{N-1} - x_N\|^2 & 0 \end{pmatrix}$$

defined by $x_1 \ldots, x_N \in \mathbb{R}^m$. Thus,

$$D(x_1, \ldots, x_N) = -2X^\top X + \mathbf{x}\mathbf{e}^\top + \mathbf{e}\mathbf{x}^\top.$$

Here $\mathbf{x} = \mathrm{col}(\|x_1\|^2, \ldots, \|x_N\|^2) \in \mathbb{R}^N$ and $X^\top X = (x_i^\top x_j) \in \mathbb{R}^{N \times N}$ denote the Gramian matrix associated with $x_1, \ldots, x_N$. In particular, $D(x_1, \ldots, x_N)$ is a rank two perturbation of the rank $\leq m$ Gramian matrix $X^\top X$. This observation implies that Euclidean distance matrices of N points in $\mathbb{R}^m$ have rank $\leq m + 2$, while for generic choices of $x_1, \ldots, x_N$ the rank is equal to $m + 2$. To characterize the set of Euclidean distance matrices, one needs a simple lemma from linear algebra.

Lemma 8.28. *Let $A = A^\top \in \mathbb{R}^{n \times n}$, and assume that $B \in \mathbb{R}^{r \times n}$ has full row rank r. Let Q_A denote the quadratic form $x^\top A x$ defined on the kernel $\mathrm{Ker}\, B$ of B. Then the symmetric matrix*

$$C = \begin{pmatrix} 0 & B^\top \\ B & A \end{pmatrix}$$

satisfies the equations for rank and signature

$$\mathrm{rk}\, C = \mathrm{rk}\, Q_A + 2r,$$

$$\mathrm{sign}\, C = \mathrm{sign}\, Q_A.$$

Proof. Let R and L be invertible $r \times r$ and $n \times n$ matrices, with $LBR = (0, I_r)$, respectively. Then

$$\begin{pmatrix} R^\top & 0 \\ 0 & L \end{pmatrix} \begin{pmatrix} 0 & B^\top \\ B & A \end{pmatrix} \begin{pmatrix} R & 0 \\ 0 & L^\top \end{pmatrix} = \begin{pmatrix} 0 & (LBR)^\top \\ LBR & LAL^\top \end{pmatrix},$$

and after such a suitable transformation one can assume without loss of generality that $B = (I_r, 0)$. Partition the matrix A

$$A = \begin{pmatrix} A_{11} & A_{12} \\ A_{12}^\top & A_{22} \end{pmatrix},$$

where $A_{11}, A_{12},$ and A_{22} have sizes $r \times r$, $r \times (n-r)$, and $(n-r) \times (n-r)$. Applying elementary row and column operations we obtain

$$C = \begin{pmatrix} 0 & I_r & 0 \\ I_r & A_{11} & A_{12} \\ 0 & A_{12}^\top & A_{22} \end{pmatrix} = \begin{pmatrix} I & 0 & 0 \\ \frac{1}{2}A_{11} & I & 0 \\ A_{12}^\top & 0 & I \end{pmatrix} \begin{pmatrix} 0 & I_r & 0 \\ I_r & 0 & 0 \\ 0 & 0 & A_{22} \end{pmatrix} \begin{pmatrix} I & \frac{1}{2}A_{11} & A_{12} \\ 0 & I & 0 \\ 0 & 0 & I \end{pmatrix}.$$

Thus the inertia theorem of Sylvester implies that

$$\mathrm{rk}\ C = \mathrm{rk}\ A_{22} + 2r = \mathrm{rk}\ Q_A + 2r,$$

$$\mathrm{sign}\ C = \mathrm{sign}\ A_{11} = \mathrm{sign}\ Q_A.$$

∎

A classical result by Menger (1928), see also Blumenthal (1953), asserts that the Euclidean distance matrices $D(x_1 \ldots, x_N)$ of N points in $\mathbb{R}^m$ have nonpositive **Cayley–Menger determinants**

$$\det \begin{pmatrix} 0 & 1 & 1 & \cdots & 1 \\ 1 & 0 & -\frac{1}{2}\|x_1 - x_2\|^2 & \cdots & -\frac{1}{2}\|x_1 - x_k\|^2 \\ 1 & -\frac{1}{2}\|x_1 - x_2\|^2 & 0 & \cdots & -\frac{1}{2}\|x_2 - x_k\|^2 \\ \vdots & \vdots & \vdots & \ddots & \vdots \\ 1 & -\frac{1}{2}\|x_1 - x_k\|^2 & -\frac{1}{2}\|x_2 - x_k\|^2 & \cdots & 0 \end{pmatrix} \le 0 \qquad (8.10)$$

for each $k \le N$ (and are equal to zero for $k > m+1$). One can easily deduce this condition from the following more general characterization of Euclidean distance matrices.

Theorem 8.29. *Let $A = (a_{ij}) \in \mathbb{R}_+^{N \times N}$ be a nonnegative symmetric matrix, with $a_{11} = \cdots = a_{NN} = 0$. The following assertions are equivalent:*

(a) A is a Euclidean distance matrix of N points in $\mathbb{R}^m$.
(b) There exists a nonnegative vector $a \in \mathbb{R}_+^N$ that satisfies the linear matrix inequality with rank constraint

$$-A + ae^\top + ea^\top \succeq 0, \quad \mathrm{rk}(-A + ae^\top + ea^\top) \leq m.$$

(c) There exists a positive semidefinite matrix S of rank $\leq m$, with

$$-\frac{1}{2}A = S - \frac{1}{2}\left(\mathrm{diag}\,(S)ee^\top + ee^\top \mathrm{diag}\,(S)\right). \tag{8.11}$$

(d) The matrix

$$S_A := -\frac{1}{2}\left(I_N - \frac{1}{N}ee^\top\right)A\left(I_N - \frac{1}{N}ee^\top\right)$$

is positive semidefinite of rank $\leq m$.

(e) The restriction of the quadratic form $x^\top Ax$ on $(\mathbb{R}e)^\perp$ is negative semidefinite and has rank $\leq m$.

*(f) The **Cayley–Menger matrix***

$$CM(A) := \begin{pmatrix} 0 & 1 & 1 & \cdots & 1 \\ 1 & 0 & -\frac{1}{2}a_{12} & \cdots & -\frac{1}{2}a_{1N} \\ 1 & -\frac{1}{2}a_{12} & 0 & \cdots & -\frac{1}{2}a_{2N} \\ \vdots & \vdots & \vdots & \ddots & \vdots \\ 1 & -\frac{1}{2}a_{1N} & -\frac{1}{2}a_{2N} & \cdots & 0 \end{pmatrix}$$

has exactly one negative eigenvalue and at most $m + 1$ positive eigenvalues.

Proof. Using the identity $\|x_i - x_j\|^2 = \|x_i\|^2 + \|x_j\|^2 - 2x_i^\top x_j$, we see that $A = D(x_1, \ldots, x_N)$ for some points $x_1, \ldots, x_N \in \mathbb{R}^m$ if and only if

$$-\frac{1}{2}A = X^\top X - \frac{1}{2}ae^\top - \frac{1}{2}ea^\top, \tag{8.12}$$

where $a = \mathrm{col}\,(\|x_1\|^2, \ldots, \|x_N\|^2)$. Equivalently, A is a Euclidean distance matrix in $\mathbb{R}^m$ if and only if there exists a positive semidefinite matrix S of rank $\leq m$ with

$$-\frac{1}{2}A = S - \frac{1}{2}\left(\mathrm{diag}\,(S)ee^\top + ee^\top \mathrm{diag}\,(S)\right).$$

Here $\mathrm{diag}\,S$ is a diagonal matrix with the same diagonal entries as S. This completes the proof of the equivalence of the first three conditions. It is easily seen that

$$S := -\frac{1}{2}\left(I_N - \frac{1}{N}ee^\top\right)A\left(I_N - \frac{1}{N}ee^\top\right)$$

satisfies (8.11). Thus (d) implies (c) and hence also (a). Conversely, (a) implies (8.12), and therefore

$$S_A = -\frac{1}{2}\left(I_N - \frac{1}{N}\mathbf{e}\mathbf{e}^\top\right)X^\top X\left(I_N - \frac{1}{N}\mathbf{e}\mathbf{e}^\top\right)$$

is positive semidefinite of rank $\leq m$. This shows the equivalence of (a) and (d). The equivalence of (a) with (e) and (f) follows from Lemma 8.28 by noting that S_A in (d) satisfies $x^\top S_A x = -\frac{1}{2}x^\top A x$ for all $x \in (\mathbb{R}\mathbf{e})^\perp$. This completes the proof. ∎

There are two simple facts about Cayley–Menger determinants that are worth mentioning. First, for $N = 3$ points in $\mathbb{R}^2$, one has the expression for the determinant of the distance matrix (with $a_{ij} := \|x_i - x_j\|$)

$$\det D(x_1, x_2, x_3)$$

$$= -\frac{1}{4}(a_{12} + a_{13} + a_{23})(a_{12} + a_{13} - a_{23})(a_{12} - a_{13} + a_{23})(-a_{12} + a_{13} + a_{23}).$$

This relates to the familiar triangle inequalities that characterize a triple of nonnegative real numbers d_1, d_2, d_3 as the side lengths of a triangle. Second, a well-known formula for the volume Vol (Σ) of the simplex Σ defined by $N + 1$ points $x_0, \ldots, x_N$ in $\mathbb{R}^N$ asserts that

$$\text{Vol}(\Sigma) = \frac{1}{N!}|\det(x_1 - x_0, \ldots, x_N - x_0)|.$$

From the translational invariance of the norm, the distance matrix $A := D(x_0, \ldots, x_N) = D(0, p_1, \ldots, p_N)$, with $p_i := x_i - x_0$ for $i = 1, \ldots, N$. Applying (8.12) to $d := \text{col}(0, \|p_1\|^2, \ldots, \|p_N\|^2)$, $P := (p_1, \ldots, p_N) \in \mathbb{R}^{d \times N}$, we obtain

$$\text{Vol}^2(\Sigma) = \left(\frac{1}{N!}\right)^2 |\det P|^2 = \det\begin{pmatrix} I_2 & 0 \\ 0 & P^\top P \end{pmatrix}$$

$$= -\det\begin{pmatrix} 0 & 1 & 1 & \cdots & 1 \\ 1 & 0 & 0 & \cdots & 0 \\ 1 & 0 & p_1^\top p_1 & \cdots & p_1^\top p_N \\ \vdots & \vdots & \vdots & \ddots & \vdots \\ 1 & 0 & p_N^\top p_1 & \cdots & p_N^\top p_N \end{pmatrix}$$

$$= -\det\begin{pmatrix} 0 & \mathbf{e}^\top \\ \mathbf{e} & -\frac{1}{2}D \end{pmatrix}.$$

Thus we obtain the formula for the squared volume of the simplex in terms of the mutual distances as

$$\text{Vol}^2(\Sigma) = -\left(\frac{1}{N!}\right)^2 \det CM,$$

where CM is the Cayley–Menger matrix of the distance matrix $D(x_0,\ldots,x_N)$.

Returning to the situation of formations defined over a graph, the **rigidity matrix** of a formation is defined as the $M \times mN$ Jacobi matrix $R(X) = Jac_{\mathscr{D}}(X)$ whose ijth row $(ij \in E)$ is

$$R(X)_{ij} = (e_i - e_j)^\top \otimes (x_i - x_j)^\top.$$

A formation X is called **regular** whenever

$$\mathrm{rk}\, R(X) = \max_{Z \in \mathbb{R}^{m \times N}} \mathrm{rk}\, R(Z).$$

The regular formations form an open and dense subset in the space of all formations. The geometry of formations is closely connected with the standard action of the Euclidean group on $\mathbb{R}^m$. Let $O(m)$ denote the compact matrix Lie group of all real orthogonal $m \times m$ matrices. The Euclidean group $E(m)$ then parameterizes all Euclidean group transformations of the form $p \mapsto gp + v$, where $g \in O(m)$ and $v \in \mathbb{R}^m$ denotes a translation vector. Thus $E(m)$ is a Lie group of dimension $\frac{m(m+1)}{2}$, which is in fact a semidirect product of $O(m)$ and $\mathbb{R}^m$. Since $\mathscr{D}$ is invariant under orthogonal rotations, i.e., $\mathscr{D}(SX) = \mathscr{D}(X)$ for $S \in O(m)$, the tangent space to such a group orbit is contained in the kernel of the rigidity matrix $R(X)$.

Lemma 8.30. *The kernel of the rigidity map $R(X)$ contains the tangent space $T_X(O(m) \cdot X)$. Suppose the linear span of the columns $x_1,\ldots,x_N$ of X has dimension r. Then the kernel of $R(X)$ has at least dimension $\frac{1}{2} r(2m - r - 1)$.*

Proof. The first statement is a simple consequence of the invariance of $\mathscr{D}$ under the group of orthogonal transformations $X \mapsto SX$. Note that the stabilizer group $O(m)_X$ of X coincides with the subgroup of $O(m)$ that leaves the elements of the linear span $< x_1,\ldots,x_N >$ pointwise invariant. Thus a straightforward computation reveals that the dimension of $O(m)_X$ is equal to $\frac{1}{2}(m - r)(m - r - 1)$. Therefore, the dimension of the group orbit $O(m) \cdot X$ is equal to $\frac{1}{2} m(m - 1) - \frac{1}{2}(m - r)(m - r - 1) = \frac{1}{2} r(2m - r - 1)$. This completes the proof.   ∎

A formation is called **infinitesimally rigid** if the kernel of the rigidity matrix coincides with the tangent space $T_X(O(m) \cdot X)$. Equivalently, infinitesimal rigidity holds if and only if the following rank condition is satisfied:

$$\mathrm{rk}\, R(X) = m(N - 1) - \frac{1}{2} \mathrm{rk}\, X (2m - \mathrm{rk}\, X - 1).$$

Note that from the structure of $R(X)$ one can easily check that $\mathrm{rk}\, R(X) \le N - 1$ for $r = 1$. Likewise one can check that $\mathrm{rk}\, R(X) \le 2N - 3$ for $r = 2$ and all d with $M \ge 2N - 3$. With the aid of these bounds, one can then verify using the rank condition that a formation of N points in the plane $\mathbb{R}^2$ is infinitesimally rigid if and only if $r = 2$ and the rank of $R(X)$ is equal to $2N - 3$. Similarly, a formation of $N \ge 4$ points in $\mathbb{R}^3$ is infinitesimally rigid if and only if $r = 3$ and the rank of $R(X)$ is equal

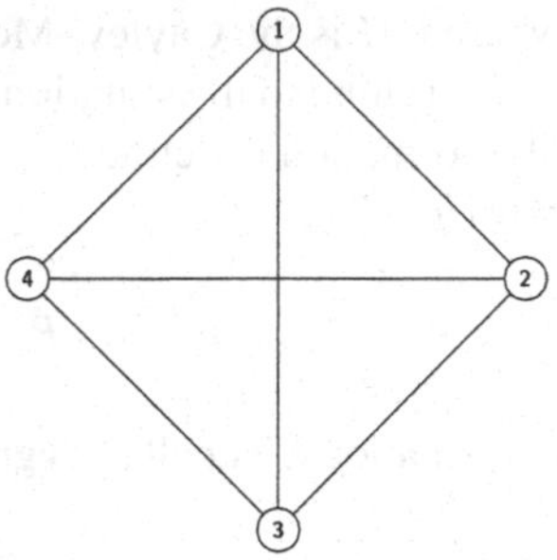

Fig. 8.7 Complete graphs
are rigid

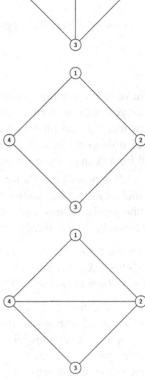

Fig. 8.8 Nonrigid graph

Fig. 8.9 Minimally rigid
graph

to $3N - 6$. A formation X is called **rigid** whenever the orbit $SO(m) \cdot X$ is isolated in the fiber $\mathscr{D}^{-1}(\mathscr{D}(X))$. Every infinitesimally rigid formation is rigid, but the converse does not hold. In fact, regular formations are infinitesimally rigid if and only if they are rigid (Figures 8.7 and 8.8).

A **rigid graph** in $\mathbb{R}^m$ is one for which almost every $X \in \mathbb{R}^m$ is infinitesimally rigid. Thus Γ is rigid in $\mathbb{R}^m$ if and only if the rigidity matrix $R(X)$ has generic rank equal to $mN - \frac{m(m+1)}{2}$. A rigid graph is called **minimally rigid** if it has exactly $mN - \frac{m(m+1)}{2}$ edges (Figure 8.9). An example of a rigid graph is the complete graph K_N that has an edge between each pair of the N vertices. K_N is minimally rigid if and only if $N = 2, 3$, but not if $N \geq 4$. In contrast, a graph with 4 vertices and 5 edges realized in $\mathbb{R}^2$ is minimally rigid.

Rigid graphs in $\mathbb{R}^2$ are characterized in a combinatorial manner by the so-called Laman condition stated in the next theorem.

Theorem 8.31 (Laman (1970)). *A graph Γ with M edges and N vertices is minimally rigid in $\mathbb{R}^2$ if and only if the following two conditions are satisfied:*

(a) $M = 2N - 3$.
(b) *Every subgraph Γ' of Γ with N' vertices and M' edges satisfies $M' \leq 2N' - 3$.*

An explicit combinatorial characterization of rigid graphs in $\mathbb{R}^3$ is unknown.

8.6 Spectral Graph Theory

Definition 8.32. The **spectrum of a weighted graph** is defined as the set of eigenvalues of the adjacency matrix $A(\Gamma)$, counted with their multiplicities. The characteristic polynomial of Γ is defined as the characteristic polynomial

$$\det(zI - A(\Gamma)) = z^N + c_1 z^{N-1} + \cdots + c_N.$$

The field of spectral graph theory is concerned with attempting to characterize the properties of graphs using information on the spectrum. Typically, the graphs are not weighted, and thus the adjacency matrix considered is the classical adjacency matrix of a graph. The first three coefficients of the characteristic polynomial of an unweighted graph (without self-loops) are easily characterized as follows:

1. $c_1 = 0$,
2. $c_2 = -|E|$,
3. $c_3 = -2\delta$, where δ denotes the number of triangles in Γ.

We refer the reader to Cvetkovic, Rowlinson and Simic (2010) for a detailed study of graph spectra.

Laplacian matrices are constructed from the adjacency matrix of a weighted graph through the notion of oriented incidence matrices. Their spectral properties dominate the present theory of consensus and synchronization, as will be further explained in Chapter 11. Let $\Gamma = (V, E)$ be an oriented (directed or undirected) weighted graph with associated maps $\iota : E \ \rangle \ V$ and $\tau : E \ \rangle \ V$ on the initial and terminal points, respectively. Here and in the sequel we will always use the canonical orientation on a digraph. Assume that Γ has n vertices $\{v_1, \ldots, v_N\}$ and M edges $\{e_1, \ldots, e_M\}$. Thus, for an edge $e \in E$ with initial and terminal points $v_i = \iota(e), v_j = \tau(e)$, respectively, there are associated weights $a_e = a_{ij} \geq 0$. Let $A \in \mathbb{R}_+^{N \times N}$ denote the weight adjacency matrix of $\Gamma = (V, E)$. Equivalently, one can present the weights as the diagonal matrix

$$W = \operatorname{diag}(a_e)_{e \in E} \in \mathbb{R}^{M \times M}.$$

The **oriented incidence matrix** is defined as

Fig. 8.10 Orientation
labeling of a graph

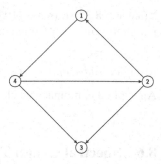

$$B = (b_{ij}) \in \mathbb{R}^{N \times M}, \quad b_{ij} = \begin{cases} 1 & \text{if } v_i = \tau(e_j) \neq \iota(e_j), \\ -1 & \text{if } v_i = \iota(e_j) \neq \tau(e_j), \\ 0 & \text{otherwise.} \end{cases} \tag{8.13}$$

Thus every incidence matrix has in each of its columns a single entry 1 and a single entry -1. All other entries in the column are zero. If B, B' are incidence matrices of two identical graphs but with different orientations, then $B' = BS$ for a unique diagonal matrix $S = \text{diag}(s_1, \ldots, s_M)$ and $s_i = \pm 1$. Thus the product $B'W(B')^\top = BWB^\top$ is independent of the orientation. If the graph Γ is strongly connected, then the incidence matrix B is known to have full row rank $N - 1$; see, for example, Fiedler (2008). This shows that B has full row rank $N - 1$ and the kernel of B has dimension $M - N + 1$. Each vector in Ker B describes a cycle in the graph Γ. Thus there are exactly $M - N + 1$ linearly independent cycles for a (directed or undirected) graph defined by a basis of the kernel of B with integer coefficients (Figure 8.10).

Let $\overleftarrow{\Gamma} = (V, \overleftarrow{E})$ denote the so-called **reverse graph**, i.e., Γ and $\overleftarrow{\Gamma}$ have the same set V of vertices and for each pair of vertices u, v:

$$(u, v) \in E \iff (v, u) \in \overleftarrow{E}.$$

Then the adjacency matrix of $\overleftarrow{\Gamma}$ is the transpose of that of Γ, i.e.,

$$A(\overleftarrow{\Gamma}) = A(\Gamma)^\top.$$

Definition 8.33 (Laplacian). Let $\Gamma = (V, E)$ be a weighted (directed or undirected) graph with nonnegative weight matrix $A(\Gamma) \in \mathbb{R}_+^{N \times N}$. Let $D_\Gamma = \text{diag}(d_1, \ldots, d_N)$ denote the real diagonal matrix with entries $d_i := \sum_{j=1}^N a_{ij}$. The **Laplacian** of the weighted digraph is defined as

$$L(\Gamma) = D(\Gamma) - A(\Gamma) \in \mathbb{R}^{N \times N}.$$

The **canonical Laplacian** $\mathscr{L}(\Gamma) \in \mathbb{Z}^{N \times N}$ of a (directed or undirected) graph is defined as a Laplacian with the canonical $0,1$ adjacency weight matrix $\mathfrak{A}(\Gamma)$.

Proposition 8.34. *Let $\Gamma = (V, E)$ be an oriented directed graph with nonnegative weight matrix A_Γ. Then*

$$BWB^\top = L(\Gamma) + L(\overleftarrow{\Gamma}).$$

If Γ is undirected, then

$$BWB^\top = L(\Gamma).$$

Proof. Let $e_1, \ldots, e_M$ denote the edges of Γ. The ij entry of $BWB^\top$ is equal to $\sum_{k=1}^M a_{e_k} b_{ik} b_{jk}$. Note that $b_{ik} b_{jk} = 0$ if and only if either i or j is not a vertex of the edge e_k. This shows that

$$(BWB^\top)_{ij} = \begin{cases} -(a_{ij} + a_{ji}) & \text{for } i \neq j \\ \sum_{r=1}^N (a_{ir} + a_{ri}) & \text{for } i = j. \end{cases}$$

This shows the formula for directed graphs. For undirected graphs, the edges ij and ji appear only once. This accounts for the factor of $\frac{1}{2}$. This completes the proof. ∎

The Laplacian of an undirected, weighted graph is always a real symmetric matrix and therefore has only real eigenvalues. Of course, for a directed graph this does not need to be true. However, there are important constraints on the spectrum of a Laplacian that are specified by the next theorem.

For $\xi = \text{col}(\xi_1, \ldots, \xi_N) \in \mathbb{R}^N$, introduce the diagonal matrix $\text{diag}(\xi_i - \xi_j)_{(i,j) \in E} \in \mathbb{R}^{M \times M}$. The oriented incidence matrix then satisfies the useful identity

$$B^\top \xi = \text{diag}(\xi_i - \xi_j)\mathbf{e}.$$

The preceding lemma then implies the explicit representation of the associated quadratic form as

$$Q_\Gamma(\xi) = \xi^\top (L(\Gamma) + L(\overleftarrow{\Gamma}))\xi = \xi^\top BWB^\top \xi$$

$$= \sum_{i,j=1}^n a_{ij}(\xi_i - \xi_j)^2. \tag{8.14}$$

Most spectral properties of Laplacian matrices can be derived via the **canonical quadratic form** $Q_\Gamma : \mathbb{R}^{N \times N} \longrightarrow \mathbb{R}$ associated with the graph.

Lemma 8.35. *The quadratic form Q_Γ vanishes on $\mathbb{R}\mathbf{e}$. It is exactly zero on $\mathbb{R}\mathbf{e}$ if the graph Γ is strongly connected.*

Proof. The first claim is obvious. By (8.14), it follows that $Q_\Gamma(\xi) = 0$ if and only if $a_{ij}(\xi_i - \xi_j)^2 = 0$ for all i, j. This implies $\xi_i = \xi_j$ for all edges $(i, j) \in E$. Thus, if Γ is strongly connected, this implies $\xi_1 = \cdots = \xi_n$. ∎

Theorem 8.36. *Let $L(\Gamma)$ denote the Laplacian of a weighted graph Γ on N vertices with nonnegative weights. Let $e = (1, \ldots, 1)^\top \in \mathbb{R}^N$. Then $L(\Gamma)e = L(\overleftarrow{\Gamma})e = 0$. Moreover:*

1. *The eigenvalues of $L(\Gamma)$ have nonnegative real part;*
2. *If Γ is strongly connected, then $L(\Gamma)$ has rank $N - 1$, i.e.,*

$$\mathrm{Ker}\, L(\Gamma) = \mathbb{R}e.$$

   Thus, 0 is a simple eigenvalue of $L(\Gamma)$, and all other eigenvalues of L_Γ have positive real part.
3. *The quadratic form Q_Γ is positive semidefinite if and only if $e^\top L(\Gamma) = 0$.*

Proof. The claim $L(\Gamma)e = L(\overleftarrow{\Gamma})e = 0$ is obvious from the definition of Laplacians. Note that $e^\top L(\Gamma) = 0$ if and only if $\sum_{j=1}^N a_{ij} = \sum_{j=1}^N a_{ji}$ for all $i = 1, \ldots, N$. This in turn is equivalent to $L(\overleftarrow{\Gamma}) = L(\Gamma)^\top$. For simplicity, we only prove claims 1 and 2 for symmetric weights, i.e., for $L(\overleftarrow{\Gamma}) = L(\Gamma)^\top$. See Bullo, Cortés and Martínez (2009), Theorem 1.32, for the proof in the general case.

Claims 1 and 2. Assume $Lx = \lambda x$ for a nonzero complex vector x and $\lambda \in \mathbb{C}$. By Proposition 8.34, then $L + L^\top = BWB^\top$, and therefore, using $x^* = \bar{x}^\top$,

$$2\mathrm{Re}\lambda \|x\|^2 = x^* L x + x^* L^\top x = x^* BWB^\top x \geq 0.$$

Thus $\mathrm{Re}\lambda \geq 0$. Now suppose that Γ is strongly connected. By Lemma 8.35, the symmetric matrix $L(\Gamma) + L(\Gamma)^\top$ is positive semidefinite and degenerates exactly on $\mathbb{R}e$. This proves the claim.

Claim 3. If $e^\top L(\Gamma) = 0$, then $L(\Gamma) + L(\Gamma)^\top = L(\Gamma) + L(\overleftarrow{\Gamma}) = BWB^\top$ is positive semidefinite. Conversely, assume that $L(\Gamma) + L(\Gamma)^\top$ is positive semidefinite. Then

$$0 = e^\top (L(\Gamma) + L(\Gamma)^\top)e.$$

By the positive semidefiniteness of $L(\Gamma) + L(\Gamma)^\top$, thus $(L(\Gamma)^\top e = L(\Gamma) + L(\Gamma)^\top)e = 0$. This proves (3). ∎

A classical result on the eigenvalues of Hermitian matrices is the Courant–Fischer minimax principle, which characterizes the eigenvalues of Hermitian matrices.

Theorem 8.37 (Courant–Fischer). *Let $A = A^* \in \mathbb{C}^{n \times n}$ be Hermitian with eigenvalues $\lambda_1 \geq \cdots \geq \lambda_n$. Let E_i denote the direct sum of the eigenspaces corresponding to the eigenvalues $\lambda_1, \ldots, \lambda_i$. Then*

$$\lambda_i = \min_{\dim S = n-i+1} \max_{0 \neq x \in S} \frac{x^* A x}{x^* x} = \min_{0 \neq x \in E_i} \frac{x^* A x}{x^* x}.$$

The minimax principle can be extended to establish bounds for the eigenvalues of sums of Hermitian matrices, such as the Weyl inequalities. The **Weyl inequality** asserts that the eigenvalues of the sum $A + B$ of Hermitian matrices $A, B \in \mathbb{C}^{n \times n}$, ordered decreasingly, satisfy, for $1 \leq i + j - 1 \leq n$, the inequality

$$\lambda_{i+j-1}(A+B) \leq \lambda_i(A) + \lambda_j(B).$$

More refined eigenvalue estimates are obtained from generalizations of the Weyl inequalities such as the Freede–Thompson inequality. We next state a straightforward consequence of the minimax principle to Laplacian matrices.

Corollary 8.38. *Let $L(\Gamma)$ denote the Laplacian of a weighted, directed graph on N vertices with nonnegative weights. Assume that Γ is strongly connected and satisfies $e^\top L = 0$. Then the eigenvalue λ_{N-1} of $L(\Gamma)$ with smallest positive real part satisfies*

$$\mathrm{Re}(\lambda_{N-1}) \geq \min_{0 \neq x \in (\mathbb{R}e)^\perp} \frac{x^* L x}{x^* x} > 0. \tag{8.15}$$

Moreover, if $L(\Gamma)$ is symmetric, then λ_{N-1} is real, and equality in (8.15) holds.

Proof. By Theorem 8.36 the quadratic form $Q_\Gamma(x) = 2x^* L x = x^*(L + L^*)x$ is positive semidefinite and degenerates exactly on $\mathbb{R}e$. Moreover, all nonzero eigenvalues σ of L have positive real part. Thus the eigenvectors v of L with $Lv = \sigma v$ satisfy $v^* e = 0$ and $v^*(L+L^*)v = 2\mathrm{Re}(\sigma)\|v\|^2$. Thus the result follows from the minimax principle applied to $A := L + L^*$. ∎

Let us briefly mention a coordinate-free approach to Laplacian matrices. Let $S(N)$ denote the vector space of real symmetric $N \times N$ matrices. For an $N \times N$ matrix S, let $\delta(S) = (s_{11}, s_{22}, \ldots, s_{NN})^\top$ denote the column vector defined by the diagonal entries of S. Moreover, let $\mathrm{diag}(S)$ denote the diagonal matrix obtained from S by setting all off-diagonal entries equal to zero. Define a linear map $\mathscr{D} : S(N) \longrightarrow S(N)$ as

$$\mathscr{D}(S) = S - \frac{1}{2}(\delta(S)\mathbf{e}^\top + \mathbf{e}\delta(S)^\top),$$

where $\mathbf{e} = (1, \ldots, 1)^\top \in \mathbb{R}^N$.

Lemma 8.39.

$$\mathrm{Ker}\,\mathscr{D} = \{a\mathbf{e}^\top + \mathbf{e}a^\top\}.$$

Proof. The inclusion $\subset$ follows from the definition of $\mathscr{D}(S)$ with $a = \frac{1}{2}\delta(S)$. For the other direction let $S = a\mathbf{e}^\top + \mathbf{e}a^\top$. Then $\delta(S) = 2a$, and thus $\mathscr{D}(S) = 0$. ∎

It is easily seen that the adjoint operator of $\mathscr{D}$ with respect to the Frobenius inner product $< S_1, S_2 >:= \mathrm{tr}(S_1 S_2)$ is $\mathscr{D}^* : S(N) \longrightarrow S(N)$, with

$$\mathscr{D}^*(S) = S - \frac{1}{2}\mathrm{diag}\left(\delta(S)\mathbf{e}\mathbf{e}^\top + \mathbf{e}\mathbf{e}^\top \delta(S)^\top\right).$$

The **Laplacian operator** is the linear map $L = \mathscr{D}^* \circ \mathscr{D}$. Obviously, a symmetric matrix S satisfies

$$\mathrm{tr}(L(S)S) = \mathrm{tr}(\mathscr{D}(S)\mathscr{D}(S)) = \|\mathscr{D}(S)\|^2 \geq 0,$$

and therefore $\mathrm{tr}(L(S)S) = 0$ if and only if $\mathscr{D}(S) = 0$. Note further that $\mathscr{D} \circ \mathscr{D}^* = \mathscr{D}^*$ and the operators $\mathscr{D}$ and $\mathscr{D}^*$ commute on the space of symmetric matrices with zero diagonal entries. A brute force calculation shows

$$L(S) = S - \frac{1}{2}(\delta(S)\mathbf{e}^\top + \mathbf{e}\delta(S)^\top) - \frac{1}{2}\mathrm{diag}\left(S\mathbf{e}\mathbf{e}^\top + \mathbf{e}\mathbf{e}^\top S\right) + \frac{N}{2}\mathrm{diag}\left(S\right) + \frac{1}{2}\mathrm{tr}(S)I_N.$$

Using the preceding formula for the Laplacian operator one concludes for each symmetric matrix S that

$$L_{ij}(S) = \begin{cases} s_{ij} - \frac{s_{ii}+s_{jj}}{2} & \text{if } i \neq j \\ -\sum_{j=1}^N (s_{ik} - \frac{s_{ii}+s_{kk}}{2}) & \text{if } i = j. \end{cases}$$

In particular, $L(S)\mathbf{e} = 0$ for all S. This explicit formula implies the following corollary.

Corollary 8.40. *For $(x_1,\ldots,x_N) \in \mathbb{R}^{m\times N}$, define the distance matrix $D(x_1,\ldots,x_N)$ $= (\|x_i - x_j\|^2)$. Then the Laplacian of $S = X^\top X$ is*

$$L(X^\top X) = -\frac{1}{2}(D(x_1,\ldots,x_N) - \mathrm{diag}\left(D(x_1,\ldots,x_N)\mathbf{e}\right).$$

In particular, if the matrix $D(x_1,\ldots,x_N)$ is irreducible, then

$$\mathrm{Ker}\, L(X^\top X) = \mathbb{R}\mathbf{e}.$$

Laplacian operators share an important monotonicity property.

Proposition 8.41. *L is a monotonic operator, i.e., if $S_1 - S_2$ is positive semidefinite, then so is $L(S_1) - L(S_2)$. The kernel of L is equal to the kernel of $\mathscr{D}$.*

Proof. The second claim is obvious from the definition of $L = \mathscr{D}^* \circ \mathscr{D}$. For the first claim it suffices to show that L maps positive semidefinite matrices into positive semidefinite matrices. If S is positive semidefinite, then there exists a full column row matrix $X = (x_1, \ldots, x_N)$, with $S = X^\top X$. Therefore, $L_{ij}(X^\top X) = -\|x_i - x_j\|^2$ for $i \neq j$ and $L_{ii}(X^\top X) = \sum_{j=1}^N \|x_i - x_j\|^2$. Thus, for a vector ξ,

$$\xi^\top L(X^\top X)\xi = \frac{1}{2}\sum_{i<j} \|x_i - x_j\|^2 (\xi_i - \xi_j)^2 \geq 0.$$

This completes the proof. ∎

A different version of the Laplacian matrix of a graph that is frequently of interest in applications is the **normalized Laplacian**, or the **flocking matrix**

$$\mathscr{L} = D^{-1}A.$$

Here A denotes the weighted adjacency matrix and $D = \mathrm{diag}(Ae)$. We list a few spectral properties of the normalized Laplacian for undirected graphs.

Theorem 8.42. *The normalized Laplacian $\mathscr{L}$ of an undirected, weighted, connected graph Γ has the following properties:*

1. *$\mathscr{L}$ is a stochastic matrix with only real eigenvalues $-1 \leq \lambda \leq 1$;*
2. *1 is a simple eigenvalue of $\mathscr{L}$ with eigenspace $\mathbb{R}e$. Moreover, -1 is not an eigenvalue of Γ if and only if Γ is not bipartite;*
3. *If A has at least one positive entry on the diagonal, then -1 is not an eigenvalue of $\mathscr{L}$.*

Proof. $\mathscr{L}$ is similar to the real symmetric matrix

$$D^{\frac{1}{2}}\mathscr{L}D^{-\frac{1}{2}} = D^{-\frac{1}{2}}AD^{-\frac{1}{2}} = I - D^{-\frac{1}{2}}LD^{-\frac{1}{2}}$$

and therefore has only real eigenvalues. Moreover, $D^{-1}Ax = x$ if and only if $Lx = (D - A)x = 0$. Theorem 8.36 implies that 1 is a simple eigenvalue of $\mathscr{L}$ with eigenspace equal to the kernel of L, i.e., it coincides with $\mathbb{R}e$. $\mathscr{L}$ is nonnegative, with $\mathscr{L}e = e$, and therefore a stochastic matrix. Thus Theorem 8.18 implies that $\mathscr{L}$ has spectral radius 1. Moreover, the irreducibility of the adjacency matrix A implies that $\mathscr{L}$ is irreducible. Suppose that -1 is an eigenvalue of $\mathscr{L}$. Applying Theorem 8.23 we conclude that $\mathscr{L}$ and, hence, A are permutation equivalent to a matrix of the form

$$\begin{pmatrix} 0 & B_1 \\ B_1^\top & 0 \end{pmatrix}. \tag{8.16}$$

But this is equivalent to the graph being bipartite. Conversely, assume that A is the adjacency matrix of a bipartite graph. Then $\mathscr{L}$ is permutation equivalent to (8.16). Thus the characteristic polynomial of $\mathscr{L}$ is even, and therefore -1 is an eigenvalue. This proves the first two claims. Now suppose that A and, hence, $\mathscr{L}$ have at least one positive diagonal entry. Then $\mathscr{L}$ cannot be permutation equivalent to a matrix of the form (8.16) because diagonal entries remain on the diagonal under permutations. Thus Theorem 8.23 implies that -1 cannot be an eigenvalue of $\mathscr{L}$. This completes the proof.   ∎

Finally, we prove the classical matrix-tree theorem for weighted graphs, which plays an important role in algebraic combinatorics. We use the following notation. For each edge $e \in E$ let $a_e > 0$ denote the associated weight. For a subset $E' \subset E$ define

$$a_{E'} = \prod_{e \in E'} a_e.$$

Note that the classical adjoint of a matrix A is the transposed matrix adj $(A$ of cofactors, i.e., adj $A_{ij} = (-1)^{i+j} \det A_{ji}$.

Theorem 8.43 (Matrix-Tree Theorem). *Let Γ be an undirected weighted graph and $L(\Gamma)$ the associated Laplacian with real eigenvalues $0 = \lambda_1 \leq \lambda_2 \leq \cdots \leq \lambda_N$. Then:*

1. The $(N-1) \times (N-1)$ leading principal minor of $L(\Gamma)$ is equal to

$$\kappa(\Gamma) = \sum_{|E'|=N-1} a_{E'}, \tag{8.17}$$

where the sum is over all spanning subtrees (V, E') of Γ;
2. The adjoint matrix of $L(\Gamma)$ is

$$\mathrm{adj}\, L(\Gamma) = \kappa(\Gamma) ee^{\top}; \tag{8.18}$$

3.

$$\kappa(\Gamma) = \frac{\lambda_2 \cdots \lambda_N}{N}. \tag{8.19}$$

Proof. By Lemma 8.34, the 11-minor of the Laplacian is equal to $\det(B_1 W B_1^{\top})$, where $B_1 \in \mathbb{R}^{(N-1) \times M}$ denotes the $(N-1) \times M$ submatrix of the oriented incidence matrix B formed by the first $N-1$ rows. By the Cauchy–Binet formula,

$$\det(B_1 W B_1^{\top}) = \sum_{|E'|=N-1} \det^2(B_{E'}) \det W_{E'}, \tag{8.20}$$

where the summation is over all subsets of edges $E' \subset E \cap (\{1,\ldots,N-1\} \times \{1,\ldots,N-1\})$ of cardinality $N-1$. One can always assume without loss of generality that the subgraphs $(\{1,\ldots,N-1\}, E')$ are connected, because otherwise $B_{E'}$ would contain a zero row (or zero column), and thus $\det^2(B_{E'}) = 0$. Assume that $(\{1,\ldots,N-1\}, E')$ contains a cycle of length $r \leq N-1$. Then, after a suitable permutation of rows and columns, $B_{E'}$ would be of the form

$$\begin{pmatrix} B_{11} & B_{12} \\ 0 & B_{22} \end{pmatrix},$$

with

$$B_{11} = \begin{pmatrix} 1 & 0 & \cdots & 0 & -1 \\ -1 & 1 & 0 & & 0 \\ 0 & \ddots & \ddots & \ddots & \vdots \\ \vdots & & -1 & 1 & 0 \\ 0 & \cdots & 0 & -1 & 1 \end{pmatrix}$$

a circulant matrix whose last column is the negative of the sum of the previous ones. Thus $\det B_{11} = 0$, and therefore the corresponding summand in (8.20) vanishes. Thus only the cycle-free subgraphs $\Gamma' = (\{1,\dots,N-1\}, E')$ contribute. These are exactly the spanning subtrees of $(\{1,\dots,N-1\}, E')$ with determinant $\det B_{E'} = \pm 1$. This proves (8.20).

For the second claim note that Γ is connected if and only if the rank of the Laplacian is $n - 1$. Thus, if Γ is not connected, then both sides of (8.19) are zero. Hence, one can assume that Γ is connected. From $L(\Gamma)\mathrm{adj}\, L(\Gamma) = \det L(\Gamma)I = 0$ we conclude that every column of $\mathrm{adj}\, L(\Gamma)$ is in the kernel of $L(\Gamma)$, i.e., is a scalar multiple of $\mathbf{e}$. Since $L(\Gamma)$ is symmetric, so is $\mathrm{adj}\, L(\Gamma)$. Thus $\mathrm{adj}\, L(\Gamma)$ is a multiple of $\mathbf{e}\mathbf{e}^\top$.

The last claim follows by taking traces in (8.19). Thus $N\kappa(\Gamma) = \mathrm{tr}\,\mathrm{adj}\, L(\Gamma)$ coincides with the sum of eigenvalues of $\mathrm{adj}\, L(\Gamma)$. If $\lambda_1,\dots,\lambda_N$ denote the eigenvalues of a matrix M, then the eigenvalues of the adjoint $\mathrm{adj}\, M$ are $\prod_{j\neq i}\lambda_j$, $j = 1,\dots,N$. Thus we obtain $\mathrm{tr}\,\mathrm{adj}\, L(\Gamma) = \lambda_2\cdots\lambda_N$. This completes the proof. ∎

As a consequence of the matrix-tree theorem one can derive an explicit formula for the number of spanning trees in a graph.

Corollary 8.44 (Temperley (1964)). *Let $L := L(\Gamma)$ be the Laplacian of an undirected weighted graph on N vertices and $J = \mathbf{e}\mathbf{e}^\top$. Then*

$$\kappa(\Gamma) = \frac{\det\,(J+L)}{N^2}.$$

Proof. The identities $NJ = J^2, JL = 0$ imply $(NI - J)(J + L) = NL$. By taking adjoints, therefore, $\mathrm{adj}\,(J+L)\mathrm{adj}\,(NI - J) = \mathrm{adj}\,(NL) = N^{N-1}\mathrm{adj}\, L$. Thus, using $\mathrm{adj}\,(NI - J) = N^{N-2}J$, the matrix-tree theorem implies

$$N^{N-1}\kappa(\Gamma)J = N^{N-1}\mathrm{adj}\, L = N^{N-2}\mathrm{adj}\,(J+L)J.$$

Thus $N\kappa(\Gamma)J = \mathrm{adj}\,(J+L)J$, and therefore

$$\det\,(J+L)J = (J+L)\mathrm{adj}\,\,(J+L)J = \mathrm{adj}\,\,(J+L)(J+L)J = N\mathrm{adj}\,\,(J+L)J$$
$$= N^2 \kappa(\Gamma).$$

■

For the complete graph K_N on N vertices, the classical graph Laplacian is $\mathscr{L} = NI - J$. This implies the well-known formula

$$\kappa(K_N) = \frac{\det\,(J+\mathscr{L})}{N^2} = \frac{N^N}{N^2} = N^{N-2}$$

for the number of spanning trees in K_N.

8.7 Laplacians of Simple Graphs

We determine spectral information for important classes of classical Laplacians and adjacency matrices for directed and undirected graphs Γ. For simplicity we focus on unweighted graphs, i.e., on classical Laplacians and adjacency matrices. Note that both Laplacians and adjacency matrices of undirected, weighted graphs satisfy the following properties.

$$L(\Gamma \cup \Gamma')) = \mathrm{diag}\,(L(\Gamma), L(\Gamma')),$$
$$L(\Gamma \times \Gamma') = L(\Gamma) \otimes I_n + I_m \otimes L(\Gamma')\,.$$

In particular, the eigenvalues of the Laplacian of the direct product graph $\Gamma \times \Gamma'$ are the sums $\lambda_i + \mu_j$ of the eigenvalues λ_i and μ_j of Γ and Γ', respectively.

1. Simple Path Graph. Our first example is the simple directed path graph Γ_n on the vertex set $V = \{1,\dots,N\}$ and edges $E = \{(1,2),(2,3),\dots,(N-1,N)\}$ (Figure 8.11).

The adjacency matrix and Laplacian of Γ_N are respectively

$$\mathfrak{A} = \begin{pmatrix} 0 & \cdots & \cdots & 0 \\ 1 & \ddots & & \vdots \\ \vdots & \ddots & \ddots & \vdots \\ 0 & \cdots & 1 & 0 \end{pmatrix}, \quad L = \begin{pmatrix} 0 & \cdots & \cdots & 0 \\ -1 & 1 & & \vdots \\ \vdots & \ddots & \ddots & \vdots \\ 0 & \cdots & -1 & 1 \end{pmatrix}.$$

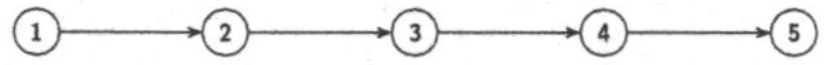

Fig. 8.11 Directed simple path

Fig. 8.12 Undirected simple path

Thus $\mathfrak{A}$ is the standard cyclic nilpotent matrix, while L has 0 as a simple eigenvalue and the eigenvalue 1 has a geometric multiplicity of one (and an algebraic multiplicity of $N-1$) (Figure 8.12).

More interesting is the associated undirected graph with a set of edges $E = \{\{1,2\},\{2,3\},\ldots,\{N-1,N\}\}$ and graph adjacency and Laplacian matrices, respectively,

$$\mathfrak{A}_N = \begin{pmatrix} 0 & 1 & & & \\ 1 & 0 & 1 & & \\ & 1 & \ddots & \ddots & \\ & & \ddots & \ddots & 1 \\ & & & 1 & 0 \end{pmatrix}, \quad L_N = \begin{pmatrix} 1 & -1 & & & \\ -1 & 2 & -1 & & \\ & -1 & \ddots & \ddots & \\ & & \ddots & \ddots & -1 \\ & & -1 & 2 & -1 \\ & & & -1 & 1 \end{pmatrix}.$$

We begin with a spectral analysis of $\mathfrak{A}_N$. This is a classical exercise from analysis.

Theorem 8.45. *1. The eigenvalues of $\mathfrak{A}_N$ are the distinct real numbers $\lambda_k(\mathfrak{A}_N) = 2\cos\frac{k\pi}{N+1}$, $k = 1,\ldots,N$.*

2. The unique eigenvector $x^{(k)} = (\xi_0^{(k)},\ldots,\xi_{N-1}^{(k)})^\top$, normalized as $\xi_0^{(k)} := 1$, for the eigenvalue $2\cos\frac{k\pi}{N+1}$ is

$$\xi_v^{(k)} = \frac{\sin k(\frac{v+1}{N+1})\pi}{\sin(\frac{k\pi}{N+1})}, \quad v = 0,\ldots,N-1.$$

In particular, the coordinates of the eigenvector $x^{(k)}$ are reflection symmetric, that is, they satisfy $\xi_v^{(k)} = \xi_{n-1-v}^{(k)}$.

Proof. Let

$$e_N(z) = \det(zI - \mathfrak{A}_N)$$

be the characteristic polynomial of $\mathfrak{A}_N$. Expanding by the first row leads to the three-term recursion

$$e_N(z) = ze_{N-1}(z) - e_{N-2}(z). \tag{8.21}$$

For the 1-norm $\|\mathfrak{A}\| \leq 2$, and hence it follows that the eigenvalues satisfy $|\lambda| \leq 2$ and, as eigenvalues of a real symmetric matrix, they are all real. So one can set

$\lambda = 2\cos x$. From (8.21) it follows that λ is an eigenvalue if and only if $e_N(\lambda) = 0$. The difference equation (8.21) can be written now as

$$e_N(2\cos x) = 2\cos x \cdot e_{N-1}(2\cos x) - e_{N-2}(2\cos x). \qquad (8.22)$$

We try an exponential solution to this difference equation, i.e., we put $e_N = A\zeta_1^N + B\zeta_2^N$, where ζ_1 and ζ_2 are the two roots of the characteristic polynomial $\zeta^2 - 2\zeta\cos x + 1 = 0$. This leads to $\zeta = e^{\pm\sqrt{-1}x}$. The initial conditions for the difference equation (8.21) are $e_0(z) = 1$ and $e_1(z) = z$. Setting $e_N(2\cos x) = Ae^{\sqrt{-1}Nx} + Be^{-\sqrt{-1}Nx}$ leads to the pair of equations

$$A + B = 1, \quad Ae^{\sqrt{-1}x} + Be^{-\sqrt{-1}x} = 2\cos x.$$

Solving and substituting back in (8.22) one obtains

$$e_N(2\cos x) = \frac{\sin(N+1)x}{\sin x}.$$

The right-hand side vanishes for $x = \frac{k\pi}{N+1}$, $k = 1,\dots,N$, and therefore the eigenvalues of $\mathfrak{A}_N$ are $2\cos\frac{k\pi}{N+1}$.

We proceed now to the computation of the eigenvectors of $\mathfrak{A}_N$. Let $x^{(k)} = (\xi_0^{(k)},\dots,\xi_{N-1}^{(k)})$ be the eigenvector corresponding to the eigenvalue $\lambda_k = 2\cos(\frac{k\pi}{N+1})$. The characteristic equation $Tx^{(k)} = \lambda_k x^{(k)}$ is equivalent to the system

$$\xi_1^{(k)} = \lambda_k \xi_0^{(k)}$$

$$\xi_0^{(k)} + \xi_2^{(k)} = \lambda_k \xi_1^{(k)}$$

$$\vdots \qquad\qquad\qquad (8.23)$$

$$\xi_{N-3}^{(k)} + \xi_{N-1}^{(k)} = \lambda_k \xi_{N-2}^{(k)}$$

$$\xi_{N-2}^{(k)} = \lambda_k \xi_{N-1}^{(k)}.$$

The coordinates of the eigenvector $x^{(k)}$ satisfy the recursion

$$\xi_\nu^{(k)} = 2\cos(\frac{k\pi}{N+1})\xi_{\nu-1}^{(k)} - \xi_{\nu-2}^{(k)}.$$

Normalize the eigenvector by requiring $\xi_0^{(k)} = 1$. The second coordinate is determined by the first equation in (8.23) and $\xi_1^{(k)} = 2\cos(\frac{k\pi}{N+1})$. As is the case with eigenvalues, one solves the difference equation using a linear combination of two exponential solutions. Thus

$$\xi_v^{(k)} = A e^{\frac{\sqrt{-1}v\pi}{N+1}} + B e^{\frac{-\sqrt{-1}v\pi}{N+1}}.$$

The initial conditions determine the ξ_v, and we obtain the explicit formula

$$\xi_v^{(k)} = \frac{\sin k(\frac{v+1}{N+1})\pi}{\sin(\frac{k\pi}{N+1})}.$$

∎

For later use we formulate a similar result for the matrices

$$M_N = \begin{pmatrix} 1 & -1 & & & & \\ -1 & 2 & -1 & & & \\ & -1 & \ddots & \ddots & & \\ & & \ddots & \ddots & -1 & \\ & & & -1 & 2 & -1 \\ & & & & -1 & 2 \end{pmatrix} \qquad L_N = \begin{pmatrix} 1 & -1 & & & & \\ -1 & 2 & -1 & & & \\ & -1 & \ddots & \ddots & & \\ & & \ddots & \ddots & -1 & \\ & & & -1 & 2 & -1 \\ & & & & -1 & 1 \end{pmatrix}.$$

Expanding the characteristic polynomial of M_N

$$g_N(z) = \det(zI - M_N)$$

by the last row one obtains the recursion

$$g_N(z) = (z-2)g_{N-1}(z) - g_{N-2}(z),$$

with initial conditions $g_0(z) = 1, g_1(z) := z - 1, g_2(z) = (z-1)(z-2) - 1$. Note that $\gamma_N(z) = e_N(z-2)$ satisfies the same recursion, but with different initial conditions. For a proof of the next result we refer to Yueh (2005); see Willms (2008) for further eigenvalue formulas for tridiagonal matrices.

Theorem 8.46. *1. The eigenvalues of M_N are distinct and are*

$$\lambda_k(M_N) = 2 - 2\cos\frac{(2k-1)\pi}{2N+1}, \quad k = 1, \dots, N.$$

An eigenvector $x^{(k)} = (\xi_0^{(k)}, \dots, \xi_{N-1}^{(k)})^\top$ for the eigenvalue $\lambda_k(M_N)$ is

$$\xi_v^{(k)} = \sin\left(\frac{(2k-1)(v+N+1)\pi}{2N+1}\right), \quad v = 0, \dots, N-1.$$

2. The eigenvalues of L_N are distinct

$$\lambda_k(L_N) = 2 - 2\cos\frac{(k-1)\pi}{N}, \quad k = 1, \dots, N.$$

Fig. 8.13 Directed cycle
graph

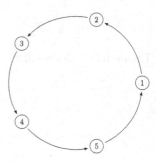

The unique eigenvector $x^{(k)} = (\xi_0^{(k)}, \ldots, \xi_{N-1}^{(k)})^\top$, normalized as $\xi_v^{(1)} := 1$, for the
eigenvalue $\lambda_k(L_N)$ is

$$\xi_v^{(k)} = \cos\left(\frac{(k-1)(2v+1)\pi}{2N}\right), \quad v = 0, \ldots, N-1.$$

2. Simple Cycle Graph. In this case the set of edges of the digraph is $E = \{(1,2), \ldots, (N-1,N), (N,1)\}$ and in the undirected case

$$E = \{\{1,2\}, \ldots, \{N-1,N\}, \{1,N\}\}.$$

Consider first the directed graph case (Figure 8.13). Then the adjacency matrix
is the circulant matrix

$$C_N = \begin{pmatrix} 0 & & & 1 \\ 1 & \ddots & & \\ & \ddots & \ddots & \\ & & 1 & 0 \end{pmatrix}. \tag{8.24}$$

Being a circulant matrix, C is diagonalized by the Fourier matrix. Explicitly, let

$$\Phi = \frac{1}{\sqrt{N}} \begin{pmatrix} 1 & 1 & 1 & \cdots & 1 \\ 1 & \omega & \omega^2 & \cdots & \omega^{N-1} \\ 1 & \omega^2 & \omega^4 & \cdots & \omega^{2N-2} \\ \vdots & & & & \vdots \\ 1 & \omega^{N-1} & \omega^{2N-2} & \cdots & \omega^{(N-1)^2} \end{pmatrix} \tag{8.25}$$

denote the Fourier matrix, where $\omega = e^{2\pi\sqrt{-1}/N}$ denotes a primitive Nth root of
unity. Notice that Φ is both a unitary and a symmetric matrix:

$$C_N = \Phi \operatorname{diag}(1, \omega, \ldots, \omega^{N-1}) \Phi^*.$$

Fig. 8.14 Undirected cycle graph

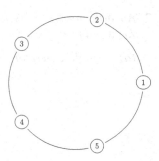

This proves the following theorem.

Theorem 8.47. *The eigenvalues of C_N are distinct and are the Nth roots of unity:*

$$\lambda_k(C_N) = \omega^k = e^{\frac{2k\sqrt{-1}\pi}{N}}, \quad k = 1, \ldots, N.$$

An eigenvector $x^{(k)} = (\xi_0^{(k)}, \ldots, \xi_{N-1}^{(k)})^\top$ for the eigenvalue $\lambda_k(C_N)$ is

$$x^{(k)} = \Phi e_k = \sum_{j=0}^{N-1} \omega^{(k-1)j} e_{j+1}, \quad k = 1, \ldots, N.$$

The associated Laplacian matrix is equal to $L_N = I_N - C_N$. Thus the eigenvalues and eigenvector are trivially related to those of C_N (Figure 8.14).

The undirected case is more interesting. The adjacency matrix and Laplacian matrices are

$$\mathfrak{A}_N = \begin{pmatrix} 0 & 1 & & & 1 \\ 1 & 0 & 1 & & \\ & 1 & \ddots & \ddots & \\ & & \ddots & \ddots & 1 \\ 1 & & & 1 & 0 \end{pmatrix}, \quad L_N = 2I_N - \mathfrak{A}_N = \begin{pmatrix} 2 & 1 & & & -1 \\ -1 & 2 & -1 & & \\ & -1 & \ddots & \ddots & \\ & & \ddots & \ddots & -1 \\ -1 & & & -1 & 2 \end{pmatrix}. \quad (8.26)$$

Theorem 8.48. *1. The eigenvalues of $\mathfrak{A}_N$ and L_N defined in (8.26) are*

$$\lambda_k(\mathfrak{A}_N) = 2\cos(\frac{2k\pi}{N}), \quad k = 1, \ldots, N,$$

$$\lambda_k(L_N) = 2 - 2\cos(\frac{2k\pi}{N}), \quad k = 1, \ldots, N.$$

In either case, $\lambda_k = \lambda_l$ for $1 \leq k, l \leq N$ if and only if $l = N - k$. For $N = 2m$ even, λ_m and λ_N are simple and λ_k has a multiplicity of two for all other k. For $N = 2m + 1$ odd, λ_N is simple and all other eigenvalues have a multiplicity of two.

2. *An orthonormal basis for the eigenspaces of $\mathfrak{A}_N$ and L_N for the eigenvalue $\lambda_k(\mathfrak{A}_N)$ and $\lambda_k(L_N)$, respectively, is as follows:*

   (a) *A single generator*

$$\frac{1}{\sqrt{N}}\begin{pmatrix} 1 \\ 1 \\ 1 \\ \vdots \\ 1 \end{pmatrix}, \qquad \frac{1}{\sqrt{N}}\begin{pmatrix} 1 \\ -1 \\ 1 \\ \vdots \\ 1 \end{pmatrix}$$

   for $k = N$ or $k = m, N = 2m$, respectively.
   (b) *Otherwise, a basis of two orthonormal vectors*

$$x^{(k)} = \frac{1}{\sqrt{N}}\begin{pmatrix} 1 \\ \cos(\frac{2k\pi}{N}) \\ \cos(\frac{4k\pi}{N}) \\ \vdots \\ \cos(\frac{2(N-1)k\pi}{N}) \end{pmatrix}, \qquad y^{(k)} = \frac{1}{\sqrt{N}}\begin{pmatrix} 1 \\ \sin(\frac{2k\pi}{N}) \\ \sin(\frac{4k\pi}{N}) \\ \vdots \\ \sin(\frac{2(N-1)k\pi}{N}) \end{pmatrix}.$$

Proof. Since

$$\mathfrak{A}_N = C_N + C_N^\top = \Phi\,\mathrm{diag}\,(1,\omega,\dots,\omega^{N-1})\Phi^* + \Phi\,\mathrm{diag}\,(1,\overline{\omega},\dots,\overline{\omega}^{N-1})\Phi^*,$$

the eigenvalues of $C + C^\top$ are equal to $\mathrm{Re}(\omega^k + \overline{\omega}^k) = 2\cos(\frac{2k\pi}{N})$. Moreover, the complex eigenvectors of $\mathfrak{A}_N$ are simply the columns

$$\phi_k = \frac{1}{\sqrt{N}}\begin{pmatrix} 1 \\ \omega^k \\ \omega^{2k} \\ \vdots \\ \omega^{(N-1)k} \end{pmatrix}$$

of the Fourier matrix Φ_N. Thus the real and imaginary parts

$$x^{(k)} = \frac{1}{2}(\phi_k + \overline{\phi_k}), \quad y^{(k)} = \frac{1}{2i}(\phi_k - \overline{\phi_k})$$

form a real basis of the corresponding eigenspaces. Writing $x^{(k)} = \frac{1}{\sqrt{N}}(\xi_0^{(k)},\dots,$ $\xi_{N-1}^{(k)})^\top$ and $y^{(k)} = \frac{1}{\sqrt{N}}(\eta_0^{(k)},\dots,\eta_{N-1}^{(k)})^\top$ one obtains for each $k = 1,\dots,N$

$$x_v^{(k)} = \cos \frac{2kv\pi}{N}, \quad y_v^{(k)} = \sin \frac{2kv\pi}{N}, \quad k = 1, \ldots, N, v = 0, \ldots, N-1.$$

This completes the proof for $\mathfrak{A}_N$. The result on the Laplacian follows trivially as $L_N = 2I_N - \mathfrak{A}_N$. ∎

8.8 Compressions and Extensions of Laplacians

We begin by recalling the definition of the Schur complement. Let M be an $N \times N$ matrix and $I, J \subset \{1, \ldots, N\}$. Then M_{IJ} denotes the submatrix of M with row indices in I and column indices in J, respectively.

Definition 8.49. Let M be an $N \times N$ matrix and $I \subset \{1, \ldots, N\}$ such that M_{II} is invertible. Let $J = \{1, \ldots, N\} \setminus I \neq \emptyset$. Then

$$M/M_{II} := M_{JJ} - M_{JI}M_{II}^{-1}M_{IJ}$$

is called the **Schur complement**.

The Schur complement has some basic properties that are easily established, as follows.

Proposition 8.50. *Let M be an $N \times N$ matrix and $I \subset \{1, \ldots, N\}$ such that M_{II} is invertible. The Schur complement M/M_{II} has the following properties:*

1. $\operatorname{rk} M = \operatorname{rk} M_{II} + \operatorname{rk} M/M_{II}$;
2. *Let M be Hermitian; then M/M_{II} is Hermitian with signature*

$$\operatorname{sign}(M) = \operatorname{sign}(M_{II}) + \operatorname{sign}(M/M_{II}).$$

Proof. Without loss of generality, assume that $I = \{1, \ldots, r\}, 1 \leq r < n$ and M is partitioned as

$$M = \begin{pmatrix} M_{11} & M_{12} \\ M_{21} & M_{22} \end{pmatrix}.$$

The result follows easily from the identity

$$\begin{pmatrix} I & 0 \\ -M_{21}M_{11}^{-1} & I \end{pmatrix} \begin{pmatrix} M_{11} & M_{12} \\ M_{21} & M_{22} \end{pmatrix} \begin{pmatrix} I & -M_{11}^{-1}M_{12} \\ 0 & I \end{pmatrix} = \begin{pmatrix} M_{11} & 0 \\ 0 & M_{22} - M_{21}M_{11}^{-1}M_{12} \end{pmatrix}.$$

∎

The 2×2 matrix

$$\begin{pmatrix} 1 & -2 \\ 2 & -3 \end{pmatrix}$$

shows that the Schur complement of a Hurwitz matrix need not be a Hurwitz matrix. Information about the spectral properties of the Schur complement is provided by the next result. For the proof we refer the reader to Fiedler (2008).

Theorem 8.51. *Let M be a real $N \times N$ matrix with nonpositive off-diagonal entries $m_{ij} \leq 0, i \neq j$. Let $I \subset \{1, \ldots, N\}$, with M_{II} invertible. Suppose there exists a vector $x \geq 0$ with $Mx > 0$. Then:*

1. *All eigenvalues of M have positive real part;*
2. *The eigenvalues of M_{II} and M/M_{II} have positive real parts, respectively;*
3. *The off-diagonal entries of M_{II} and the Schur complement M/M_{II} are both nonpositive. The inverses M_{II}^{-1} and $(M/M_{II})^{-1}$ exist and are nonnegative matrices.*

Let L denote the Laplacian matrix of an undirected, weighted graph Γ. Assume that $\Delta = \text{diag}(\delta_1, \ldots, \delta_N)$ denotes a diagonal matrix with nonnegative entries $\delta_1 \geq 0, \ldots, \delta_N \geq 0$. Then the matrix

$$\mathscr{L} = L + \Delta$$

is called a **generalized Laplacian** for Γ. Thus the generalized Laplacians $\mathscr{L}$ are characterized as those matrices with nonpositive off-diagonal entries that satisfy $\mathscr{L}\mathbf{e} \geq 0$. Let A denote the weighted adjacency matrix of Γ. The **loopy Laplacian** is then the generalized Laplacian $Q = L + \Delta$ defined by setting $\delta_i := a_{ii}$ for $i = 1, \ldots, N$.

We now prove a central result on the submatrices of generalized Laplacians.

Theorem 8.52. *Let Γ be an undirected, weighted graph with generalized Laplacian matrix $\mathscr{L}$.*

1. *Γ is connected if and only if $\mathscr{L}$ is irreducible.*
2. *Every eigenvalue and every principal minor of $\mathscr{L}$ are nonnegative.*
3. *For each $I \neq \{1, \ldots, N\}$, $\mathscr{L}_{II}$ is a positive definite matrix and its inverse $(\mathscr{L}_{II})^{-1}$ is a nonnegative matrix.*
4. *Let Γ be connected. Then, for each $I \neq \{1, \ldots, N\}$, both $\mathscr{L}_{II}$ and the Schur complement $\mathscr{L}/\mathscr{L}_{II}$ are generalized Laplacians.*

Proof. Since the off-diagonal entries of $-\mathscr{L}$ coincide with those of the graph adjacency matrix A of Γ, it follows that $\mathscr{L}$ is irreducible if and only if A is irreducible. But this is equivalent to Γ being connected.

By Proposition 8.34, every principal submatrix of $\mathscr{L}$ is of the form

$$\mathscr{L}_{II} = \Delta_{II} + B_I W B_I^\top,$$

where B_I denotes the submatrix of the incidence matrix B formed by the rows that are indexed by I. The matrix Δ_{II} is a diagonal matrix with nonnegative entries. Thus

$\mathscr{L}_{II}$ has nonnegative off-diagonal terms and is positive semidefinite. This proves claim 2. Assume that Γ is connected. Then for each proper subset $I \subset \{1,\dots,N\}$, the matrix $B_I W B_I^\top$ is positive definite. In fact, $x^\top B_I W B_I^\top x = 0$ implies $B_I^\top x = 0$. Extend x to $z \in \mathbb{R}^N$ by adding zeros, so that $B^\top z = B_I^\top x$. Since $\operatorname{Ker} B^\top = \mathbb{R}e$, therefore $z = \lambda e$. Since at least one entry of z is zero, we obtain $\lambda = 0$. Thus $x = 0$, which proves positive definiteness of $B_I W B_I^\top$. In particular, $\mathscr{L}_{II} = \Delta_{II} + B_I W B_I^\top$ is positive definite for all proper index sets I. Moreover, $\mathscr{L}_{II}$ is a generalized Laplacian matrix because the off-diagonal entries are all nonpositive and $\mathscr{L}_{II}e \geq 0$. Let A denote a real matrix with nonpositive off-diagonal entries such that all eigenvalues of A have positive real part. By Theorem 5.2.1 in Fiedler (2008), one obtains that A^{-1} is a nonnegative matrix. Applying this result to $A = \mathscr{L}_{II}$ we conclude that $\mathscr{L}_{II}^{-1}$ is nonnegative. This completes the proof of claim 3. Since $\mathscr{L}_{II}$ is invertible, the Schur complement $\mathscr{L}/\mathscr{L}_{II} = \mathscr{L}_{JJ} - \mathscr{L}_{JI}\mathscr{L}_{II}^{-1}\mathscr{L}_{IJ}$ exists. Moreover, $\mathscr{L}_{II}^{-1}$ is nonnegative and the entries of $\mathscr{L}_{JI}, \mathscr{L}_{IJ}$ are nonpositive. Thus all entries of $-\mathscr{L}_{JI}\mathscr{L}_{II}^{-1}\mathscr{L}_{IJ}$ are ≤ 0. Since the off-diagonal entries of $\mathscr{L}_{JJ}$ are all ≤ 0, this shows that the off-diagonal entries of $\mathscr{L}/\mathscr{L}_{II}$ are nonpositive. Thus it remains to show that the diagonal entries of the Schur complement are nonnegative. To this end, we simplify the notation by assuming that $I = \{1,\dots,r\}$. Then diagonal entries of $\mathscr{L}/\mathscr{L}_{II}$ are of the form

$$v^\top \left(-\mathscr{L}_{21}\mathscr{L}_{11}^{-1}\ I\right) \begin{pmatrix} \mathscr{L}_{11} & \mathscr{L}_{12} \\ \mathscr{L}_{21} & \mathscr{L}_{22} \end{pmatrix} \begin{pmatrix} -\mathscr{L}_{11}^{-1}\mathscr{L}_{12} \\ I \end{pmatrix} v = w^\top \mathscr{L} w,$$

for suitable choices of v, w. By claim 2, then $w^\top \mathscr{L} w \geq 0$, and the result follows. ∎

We now explore in more detail the underlying graphs that are associated with forming submatrices and Schur complements. Let $\Gamma = (V, E)$ be an undirected weighted graph and $V' \subset V$ a nonempty subset of r vertices in V. Let $\Gamma_{V'} = (V', E \cap (V' \times V'))$ denote an induced graph with the induced weight adjacency matrix A'. The relation between the Laplacians of $\Gamma_{V'}$ and Γ is established by the following result, whose easy proof is omitted.

Proposition 8.53. *Let* $I = \{i_1 < \dots < i_r\} \subset \{1,\dots,N\}$ *and* $V' = \{v_{i_1},\dots,v_{i_r}\}$ *denote the corresponding set of vertices in V. Let*

$$L(\Gamma)_{II} = (L_{ab})_{a,b \in I}$$

denote the $r \times r$ principal submatrix of the Laplacian $L(\Gamma) = (L_{ij})$, with row and column indices in I. Then

$$L(\Gamma)_{II} = L(\Gamma_{V'}) + D_{V'},$$

where $D_{V'} = \operatorname{diag}(\delta_1,\dots,\delta_r)$ is a diagonal matrix with nonnegative entries $\delta_i = \sum_{j \notin I}(a_{ij} + a_{ji})$. In particular, the submatrix $L(\Gamma)_{II}$ of the Laplacian L is a generalized Laplacian of the induced graph $\Gamma_{V'} = (V', E \cap (V' \times V'))$.

For the Schur complement we introduce the following notion; see Fiedler (2008) and Horn and Johnson (1990).

Definition 8.54. Let $\Gamma = (V,E)$ be an undirected weighted graph on a vertex set $V = \{1,\dots,N\}$ and $I \subset V$. The **Schur complement**, or the **Kron reduction**, on J is the graph $\Gamma_J = (J,E_J)$ with the set of vertices $J := V \setminus I$. Between vertices $i,j \in J$ an edge $(i,j) \in E'$ is defined if and only if there exists a path from i to j such that all its interior vertices (i.e., those of the path that differ from i and j) belong to W.

The Kron reduction of an undirected graph is an undirected graph on a subset of vertices; however, it may contain self-loops even if $\Gamma = (V,E)$ does not have self-loops. The Kron reduction graph has some appealing properties that are partially stated in the next result.

Theorem 8.55. *Let $\Gamma = (V,E)$ be an undirected weighted graph that is connected. The Kron reduction of Γ is connected. The Schur complement $\mathscr{L}/\mathscr{L}_{II}$ of a generalized Laplacian of Γ is a generalized Laplacian of the Kron reduction graph $\Gamma_J = (V_J, E_J)$.*

Proof. For a proof that the Kron reduction is connected, we refer to Doerfler and Bullo (2013). By the preceding reasoning, $L(J) := \mathscr{L}/\mathscr{L}_{II}$ is a generalized Laplacian on the vertex set J. Thus it remains to show that the off-diagonal entries $L_{ij}(J)$ are nonzero if and only if ij is an edge of the Kron reduction $\Gamma_J = (J,E_J)$. This is shown in Theorem 14.1.2 by Fiedler (2008) in the case where $\mathscr{L}$ possesses a vector $x \geq 0$ with $\mathscr{L}x > 0$. In Theorem 3.4 by Doerfler and Bullo (2013), this is shown for the so-called loopy Laplacian matrix of a graph. ∎

The Courant–Fischer minimax principle has important implications for the characterization of the eigenvalues of submatrices of Hermitian matrices via interlacing conditions. We state one of the simplest known results here, which is often attributed to Cauchy and Weyl.

Theorem 8.56 (Eigenvalue Interlacing Theorem). *Let M be a Hermitian $n \times n$ matrix and $I \subset \{1,\dots,n\}$ a subset of cardinality r. Assume that the eigenvalues of Hermitian matrices A are ordered increasingly as $\lambda_1(A) \leq \cdots \leq \lambda_n(A)$. Then, for $1 \leq k \leq r$,*

$$\lambda_k(M) \leq \lambda_k(M_{II}) \leq \lambda_{k+n-r}(M). \qquad (8.27)$$

If in addition M_{II} is positive definite, then for $1 \leq k \leq n-r$,

$$\lambda_k(M) \leq \lambda_k(M/M_{II}) \leq \lambda_k(M_{JJ}) \leq \lambda_{k+r}(M).$$

Proof. For the proof of the first inequality we refer to Theorem 4.3.15 in Horn and Johnson (1990). For the second claim note that the positive definiteness of M_{II} implies that of $M_{JI}M_{II}^{-1}M_{IJ}$. Therefore, $M_{JJ} \succeq M_{JJ} - M_{JI}M_{II}^{-1}M_{IJ}$, and thus $\lambda_k(M/M_{II}) \leq \lambda_k(M_{JJ})$ for all $1 \leq k \leq n$. Applying (8.27) to the submatrix M_{JJ} gives the result. ∎

We next describe inequalities between the eigenvalues of the Laplacians of a graph and induced subgraph.

Theorem 8.57. *Let $\Gamma = (V,E)$ be an undirected weighted graph and $\Gamma_{V'} = (V', E \cap (V' \times V'))$ an induced subgraph on $V' \subset V$. Let $0 = \lambda_1 \leq \lambda_2 \leq \cdots \leq \lambda_n$, $\lambda_1' \leq \lambda_2' \leq \cdots \leq \lambda_r'$, and $0 = \mu_1 \leq \mu_2 \leq \cdots \leq \mu_r$ denote the eigenvalues of $L(\Gamma)$, $L(\Gamma)_{V'}$, and $L(\Gamma_{V'})$, respectively. Then for all $1 \leq k \leq r$,*

$$\lambda_k \leq \lambda_k' \leq \lambda_{N-r+k},$$

$$\sum_{j=1}^{k} \lambda_j' \geq \sum_{j=1}^{k} \mu_j + \sum_{j=1}^{k} \delta_j.$$

In particular,

$$\lambda_N + \cdots + \lambda_{N-r+1} \geq \sum_{i=1}^{r} \sum_{j \neq i} (a_{ij} + a_{ji}).$$

Proof. The first inequality follows from the interlacing theorem for eigenvalues of nested Hermitian matrices; see, for example, Horn, Rhee and So (1998). The second estimate follows from a standard eigenvalue inequality for sums of Hermitian matrices. The last inequality follows from the other two. In fact, by the first inequality, $\lambda_n + \cdots + \lambda_{n-r+1} \geq \mathrm{tr}(L(\Gamma)_{V'}) = \mathrm{tr} L(\Gamma_{V'}) + D_{V'}$. This completes the proof. ∎

Similar eigenvalue inequalities exist for the Schur complement of generalized Laplacians. The straightforward proof of the next theorem is omitted.

Theorem 8.58. *Let $\mathscr{L}$ denote a generalized Laplacian of an undirected, connected graph, and let $\mathscr{L}_{\mathrm{red}} = \mathscr{L}/\mathscr{L}_{II}$ denote the Schur complement, $|I| = r$. Then the following interlacing conditions for eigenvalues are satisfied:*

$$\lambda_k(\mathscr{L}) \leq \lambda_k(\mathscr{L}_{\mathrm{red}}) \leq \lambda_k(\mathscr{L}_{JJ}) \leq \lambda_{k+r}(\mathscr{L}) \quad \text{for } 1 \leq k \leq N-r.$$

8.9 Exercises

1. Let $\lambda_1, \ldots, \lambda_n$ and $\mu_1, \ldots, \mu_m$ be the eigenvalues of the matrices $A \in \mathbb{R}^{n \times n}$ and $B \in \mathbb{R}^{m \times m}$, respectively. Prove that the eigenvalues of the Kronecker product $A \otimes B$ and of $A \otimes I_m + I_n \otimes B$ are $\lambda_i \mu_j$ and $\lambda_i + \mu_j$, respectively, for $i = 1 \ldots, n; j = 1, \ldots, m$. Deduce that the Sylvester operator $A \otimes I_m - I_n \otimes B$ is invertible if and only if A and B have disjoint spectra.

2. Let $\lambda_1, \ldots, \lambda_n$ and $\mu_1, \ldots, \mu_m$ be the eigenvalues of the matrices $A \in \mathbb{R}^{n \times n}$ and $B \in \mathbb{R}^{m \times m}$, respectively. Let $p(x,y) = \sum_{i,j} c_{ij} x^i y^j$ denote a real polynomial in

two commuting variables. Generalize the preceding exercise by showing that the eigenvalues of

$$\sum_{ij} c_{ij} A^i \otimes B^j$$

are equal to $p(\lambda_k, \mu_l)$.

3. The *Hadamard product* of two matrices $A, B \in \mathbb{R}^{n \times n}$ is defined as the $n \times n$ matrix $A * B = (a_{ij} b_{ij})$. Prove that $A * B$ is a principal submatrix of $A \otimes B$. Deduce that the Hadamard product $A * B$ of two positive definite symmetric matrices A and B is again positive definite.

4. Prove that the set of matrices $A \in \mathbb{R}^{n \times n}$, with $e^\top A = e^\top$ and $Ae = e$, forms an affine space of dimension $(n-1)^2$.

5. Prove Birkhoff's theorem, stating that the set of $n \times n$ doubly stochastic matrices form a convex polyhedron whose $n!$ vertices are permutation matrices.

6. Let $A \in \mathbb{C}^{n \times n}$ be unitary. Prove that the $n \times n$ matrix $(|a_{ij}|^2)$ is doubly stochastic.

7. Let $A \in \mathbb{R}^{n \times n}$ be irreducible and $D \in \mathbb{R}^{n \times n}$ be diagonal with $AD = DA$. Prove that $D = \lambda I_n$ is suitable for $\lambda \in \mathbb{R}$.

8. Let $A \in \mathbb{R}^{n \times n}$ be irreducible and $D_1, \ldots, D_{N-1} \in \mathbb{R}^{n \times n}$ diagonal, with

$$A = e^{2\pi\sqrt{-1}k/N} D_k A D_k^{-1}, \quad k = 1, \ldots, N-1.$$

Then there exists $\lambda_k \in \mathbb{C}$, with $D_k = \lambda_k D_1^k$ for $k = 1, \ldots, N-1$.

9. A connected graph with N vertices, without loops and multiple edges, has at least $N-1$ edges. If the graph has more than $N-1$ edges, then it contains a polygon as a subgraph.

10. Prove that a graph is connected if and only if it has a spanning tree.

11. Consider the directed graph Γ on the vertex set $\mathcal{V} = \{1,2,3,4,5,6\}$ with adjacency matrix

$$A = \begin{pmatrix} 0 & 1 & 0 & 0 & 0 & 0 \\ 1 & 0 & 0 & 0 & 1 & 0 \\ 0 & 1 & 0 & 0 & 0 & 0 \\ 1 & 0 & 1 & 0 & 0 & 0 \\ 0 & 0 & 0 & 1 & 0 & 1 \\ 0 & 0 & 1 & 0 & 1 & 0 \end{pmatrix}.$$

  (a) Prove that Γ is strongly connected.
  (b) Prove that there exists a cycle of period two through vertex 1 and 1 has no cycle of odd period.
  (c) Prove that the period of A is 2.
  (d) Compute the eigenvalues of A.

12. Let $A \in \mathbb{R}^{n \times n}$ be nonnegative and irreducible. Show that $(A + \varepsilon I)^{n-1} > 0$ for all $\varepsilon > 0$.

13. Consider the matrices

$$A_1 = \begin{pmatrix} 0 & 1 & 0 & 0 \\ 0 & 0 & 1 & 0 \\ 0 & 0 & 0 & 1 \\ 1 & 0 & 0 & 0 \end{pmatrix}, \quad A_2 = \begin{pmatrix} 0 & 1 & 0 & 0 \\ 0 & 0 & 1 & 0 \\ 0 & 0 & 0 & 1 \\ 1 & 1 & 0 & 0 \end{pmatrix}.$$

Check for the primitivity of the matrices and, if possible, determine the smallest $m \in \mathbb{N}$ such that $A_i^m > 0$.

14. Show that the contraction constant for the Hilbert metric of

$$A = \begin{pmatrix} 1 & \frac{1}{2} \\ \frac{1}{2} & \frac{1}{3} \end{pmatrix}$$

is equal to

$$k(A) = \frac{2 - \sqrt{3}}{2 + \sqrt{3}},$$

while the eigenvalues of A are $\lambda_{\pm} = \frac{4 \pm \sqrt{13}}{6}$. Deduce that $k(A)$ is strictly smaller than the convergence rate for the power iteration defined by A.

15. The primitivity index $\gamma(A)$ of a nonnegative matrix A is defined as the smallest $m \subset \mathbb{N}$, with $A^m > 0$. Prove that the $n \times n$ Wielandt matrix

$$A = \begin{pmatrix} 0 & 1 & \cdots & 0 \\ \vdots & \ddots & \ddots & \vdots \\ 0 & & \ddots & 1 \\ 1 & 1 & \cdots & 0 \end{pmatrix}$$

is primitive with primitivity index $\gamma(A) = n^2 - 2n + 2$.

16. Prove that every nonnegative irreducible matrix $A \in \mathbb{R}^{n \times n}$ with at least one positive diagonal element is primitive.

17. Consider a real $n \times n$ tridiagonal matrix

$$A = \begin{pmatrix} a_1 & b_1 & \cdots & & 0 \\ c_1 & \ddots & \ddots & & \vdots \\ \vdots & \ddots & \ddots & & b_{n-1} \\ 0 & \cdots & & c_{n-1} & a_n \end{pmatrix}$$

with spectral radius $r(A)$. Prove:

(a) If $b_j c_j \geq 0$ for all j, then A has only real eigenvalues.

(b) If $b_j c_j > 0$ for all j, then A has only real simple eigenvalues.
(c) Assume $b_j > 0$, $c_j > 0$, and $a_j \geq 0$ for all j. Then A is irreducible. Matrix A is primitive if at least one $a_j > 0$. If $a_1 = \ldots = a_n = 0$, then $-r(A)$ is an eigenvalue of A.

18. Let $\Gamma = (V = \{1,\ldots,N\}, E)$ be a finite directed graph and $d_o(j) = |\mathcal{N}^o(j)| = |\{i \in V \mid (j,i) \in \Gamma\}|$ the out-degree of vertex j. For a real number $0 \leq \alpha < 1$ define the $N \times N$ **Google matrix** $\mathcal{G} = (g_{ij})$ of the digraph Γ as

$$g_{ij} := \begin{cases} \frac{\alpha}{d_o(j)} + \frac{1-\alpha}{N} & i \in \mathcal{N}^o(j) \neq \emptyset \\ \frac{1-\alpha}{N} & i \notin \mathcal{N}^o(j) \neq \emptyset \\ \frac{1}{N} & \mathcal{N}^o(j) = \emptyset. \end{cases}$$

(a) Prove that $\mathcal{G}$ is column stochastic and primitive,
(b) Prove that the largest eigenvalue of $\mathcal{G}$ is $\lambda_1 = 1$ and the second largest eigenvalue of $\mathcal{G}$ is $\lambda_2 = \alpha$.

19. The **Leslie matrix** is a nonnegative matrix of the form

$$A = \begin{pmatrix} a_1 & a_2 & \cdots & a_n \\ b_1 & 0 & \cdots & 0 \\ \vdots & \ddots & \ddots & \vdots \\ 0 & \cdots & b_{n-1} & 0 \end{pmatrix}.$$

We assume $a_n > 0$ and $b_1 > 0, \ldots, b_n > 0$.

(a) Show that A is irreducible.
(b) Show that A is primitive whenever there exists i with $a_i > 0$ and $a_{i+1} > 0$.
(c) Show that A is not primitive if $n = 3$ and $a_1 = a_2 = 0$.

20. Prove that the Cayley–Menger determinants of a formation $x_1, \ldots, x_N \in \mathbb{R}^m$ are nonpositive for $k \leq N$ and are zero for $k > m+1$.

8.10   Notes and References

Classical references for nonnegative matrices, Markov chains, and the Perron–Frobenius theorem include Gantmacher (1959), Horn and Johnson (1990), and Seneta (1981). We also mention the excellent book by Fiedler (2008), which provides a useful collection of results on special matrices and connections to graph theory. Part of the material on stochastic matrices and the ergodic theorem in Section 8.3 was inspired by the book of Huppert (1990). Infinite-dimensional generalizations of the Perron–Frobenius theory can be found in the work of Jentzsch (1912), Krein and Rutman (1950), and Krasnoselskii (1964). A special case of the

Contraction Mapping Theorem 8.6 was applied by Pollicot and Yuri (1998) to prove
the existence of a unique Perron vector for aperiodic $\{0,1\}$ matrices. For positive
matrices A, the existence of the Perron vector in Theorem 8.11 is well known and
easily deduced from purely topological arguments. In fact, the standard simplex C_1
is homeomorphic to the closed unit ball, on which

$$x \mapsto \frac{Ax}{\mathbf{e}^\top Ax}$$

defines a continuous map. Thus the Brouwer fixed-point theorem implies the
existence of an eigenvector $x \in C_1$ with positive eigenvalue. The papers by Bushell
(1986) and Kohlberg and Pratt (1982) provide further background on the Hilbert
metric and the analysis of positive operators. The sequence of power iterates (8.5)
to compute the Perron vector is reminiscent of the well-known power method from
numerical linear algebra,

$$x_{t+1} = \frac{Ax_t}{\|Ax_t\|},$$

for computing dominant eigenvectors of a matrix A. The convergence speed of
the general power method depends on the ratio $\frac{|\lambda_1|}{|\lambda_2|}$ of the largest and second
largest eigenvalues. See Parlett and Poole (1973) and Helmke and Moore (1994)
for convergence proofs of the power method on projective spaces and Grassmann
manifolds.

Graph-theoretic methods have long been used for studying the convergence
properties of random walks on a graph, for analyzing synchronization and clustering
phenomena in physical systems, and for algorithms in distributed computing,
formation control, and networked control systems; we refer the reader to the
monographies by Bullo, Cortés and Martínez (2009) and Meshbahi and Egerstedt
(2010) for extensive background material and further references. Boyd, Diaconis
and Xiao (2004) developed linear matrix inequalities characterizing the fastest
Markov chain on a graph, while Xiao and Boyd (2004) studied linear iterations for
distributed averaging and consensus in networks. Ottaviani and Sturmfels (2013)
studied the problem of finding weights in a complete graph such that the associated
Markov chain has a stationary probability distribution that is contained in a specified
linear subspace. This problem is equivalent to characterizing the Laplacian matrices
A of a graph such that the pair (C,A) is not observable. This problem seems widely
open, but in a special situation (complete graph; weights are complex numbers),
Ottaviani and Sturmfels (2013) successfully computed the degree of the variety of
unobservable pairs.

The literature on formation shape control via distances and graph concepts
includes early work by Olfati-Saber, Fax and Murray (2007) and Doerfler and
Francis (2010). For characterizations of rigid graphs see Asimov and Roth (1978)
and Connelly (1993). References on Euclidean distance geometry and applications
include Crippen and Havel (1988), Dress and Havel (1993), and Blumenthal (1953).

The nonpositivity condition (8.10) for the Cayley–Menger determinants yields a simple determinantal condition that is necessary for a nonnegative symmetric matrix A with zero diagonal entries to be a Euclidean distance matrix. Blumenthal (1953), Chapter IV, p. 105, has shown that every such matrix A is a Euclidean distance matrix if and only if the Cayley–Menger determinants for all $k \times k$ principal submatrices of A are nonpositive, $k = 1, \ldots, N$. The conditions of parts (d) and (e) in Theorem 8.29 are due to Gower (1985) and Schoenberg (1935), respectively.

A reference for the proof of Theorem 8.45 and related material is Grenander and Szegö (1958). The spectral properties of circulant matrices are well studied. An important fact is that the set of all circulant matrices is simultaneously diagonalized by the Fourier matrix (8.25). Further information on circulant matrices can be found in Davis (1979) and the recent survey by Kra and Simanca (2012). For a statement and proof of the Courant–Fischer minimax principle, see Horn and Johnson (1990). A generalization is Wielandt's minimax theorem on partial sums of eigenvalues. The eigenvalue inequalities appearing in Theorem 8.57 are the simplest of a whole series of eigenvalue inequalities, which can be derived from eigenvalue inequalities on sums of Hermitian matrices. For a derivation of such eigenvalue inequalities via Schubert calculus on Grassmann manifolds, see Helmke and Rosenthal (1995). The full set of eigenvalue inequalities for sums of Hermitian matrices has been characterized by Klyachko; his work is nicely summarized by Fulton (2000). Such results should be useful in deriving sharp eigenvalue bounds for the Schur complement of Laplacian matrices.

Chapter 9
Interconnected Systems

The system-theoretic study of interconnected systems is not new. It started with the work by Gilbert (1963) on controllability and observability for generic classes of systems in parallel, series, and feedback interconnections. Complete characterizations for multivariable linear systems were obtained by Callier and Nahum (1975) for series and feedback interconnections and in a short note by Fuhrmann (1975) for parallel interconnections. We refer the reader to Chapter 10 for a proof of these classical characterizations using the techniques developed here. However, the interconnection structures of most complex systems are generally not of the series, parallel, or feedback type. Thus, one needs to pass from the standard interconnections to more complex ones, where the interconnection pattern between the node systems is described by a weighted directed graph. This will be done in the first part of this chapter. The main tool used is the classical concept of strict system equivalence. This concept was first introduced by Rosenbrock in the 1970s for the analysis of higher-order linear systems and was subsequently developed into a systematic tool for realization theory through the work of Fuhrmann. Rosenbrock and Pugh (1974) provided an extension of this notion toward a permanence principle for networks of linear systems. Section 9.2 contains a proof of a generalization of this permanence principle for dynamic interconnections. From this principle we then derive our main results on the reachability and observability of interconnected systems. This leads to very concise and explicit characterizations of reachability and observability for homogeneous networks consisting of identical SISO systems. Further characterizations of reachability are obtained for special interconnection structures, such as paths, cycles, and circulant structures.

Before we delve into the technical details of this chapter, let us discuss some simple examples on the reachability of interconnected systems (Fig. 9.1). The first example is defined by a continuous-time consensus algorithm on a path graph whose nodes are first-order scalar systems. Thus, for the path graph $\Gamma_N = (\mathcal{V}, \mathcal{E})$ with

© Springer International Publishing Switzerland 2015

P.A. Fuhrmann, U. Helmke, *The Mathematics of Networks of Linear Systems*, Universitext, DOI 10.1007/978-3-319-16646-9_9

Fig. 9.1 Interconnection graph of (9.1) for $N = 5$

vertex set $\mathscr{V} = \{1,\dots,N\}$ and set of edges $\mathscr{E} = \{\{1,2\},\{2,3\},\dots,\{N-1,N\}\}$, consider the autonomous dynamical system

$$
\begin{aligned}
\dot{z}_1(t) &= z_2(t) - z_1(t), \\
\dot{z}_i(t) &= z_{i+1}(t) - 2z_i(t) + z_{i-1}(t), \quad i = 2,\dots,N-1, \\
\dot{z}_N(t) &= z_{N-1}(t) - z_N(t).
\end{aligned}
\tag{9.1}
$$

Equivalently, the system can be written in matrix form as $\dot{z} = -L_N z$, where

$$
L_N = \begin{pmatrix}
1 & -1 & & & & \\
-1 & 2 & -1 & & & \\
& -1 & \ddots & \ddots & & \\
& & \ddots & \ddots & -1 & \\
& & & -1 & 2 & -1 \\
& & & & -1 & 1
\end{pmatrix}
$$

denotes the Laplacian matrix of the graph Γ_N. We emphasize that system (9.1) achieves consensus in the sense that all solutions satisfy $\lim_{t\to\infty}(z_i(t) - z_j(t)) = 0$. More generally, consider an autonomous system of linear differential equations

$$
\begin{aligned}
\dot{z}_1 &= a_{11}z_1 + \cdots + a_{1N}z_N \\
&\;\;\vdots \\
\dot{z}_N &= a_{N1}z_1 + \cdots + a_{NN}z_N
\end{aligned}
\tag{9.2}
$$

with a matrix of coefficients $\mathfrak{A} = (a_{ij}) \in \mathbb{R}^{N \times N}$. To study the influence of a node on the evolution of the remaining system variables, select, say, the last variable z_N and consider it a control variable. This leads to the linear control system

$$
\dot{x}(t) = Ax(t) + bu(t)
\tag{9.3}
$$

in the variable $x = \mathrm{col}\,(z_1,\dots,z_{N-1})$, where

$$
A = \begin{pmatrix}
a_{11} & \cdots & a_{1N-1} \\
\vdots & \ddots & \vdots \\
a_{N-11} & \cdots & a_{N-1N-1}
\end{pmatrix}, \quad
b = \begin{pmatrix}
a_{1N} \\
\vdots \\
a_{N-1N}
\end{pmatrix}.
$$

In particular, from (9.1) one obtains the reachable system

$$A = \begin{pmatrix} 1 & -1 & & & & \\ -1 & 2 & -1 & & & \\ & -1 & \ddots & \ddots & & \\ & & \ddots & \ddots & -1 & \\ & & & -1 & 2 & -1 \\ & & & & -1 & 2 \end{pmatrix}, \quad b = \begin{pmatrix} 0 \\ 0 \\ \vdots \\ 0 \\ 0 \\ -1 \end{pmatrix}.$$

More generally, one can select a finite number of state variables in an interconnected autonomous system and replace them by free input variables that act on the remaining system. Thus, starting from a system of ordinary differential equations, one obtains a control system, and one can study its reachability properties. This leads to the topic of **pinning control**. Now, suppose that certain of the entries a_{ij} are set to zero, i.e., assume that the coefficients a_{ij} are defined by the adjacency matrix of an undirected graph on N nodes. Then the reduced control system (9.3) has an induced graph structure and clearly defines an interconnected linear system. One can then ask about the extent to which the graph structure of the autonomous system (9.2) impacts the reachability properties of the network (9.3). Such pinning reachability questions have been considered in the past few years by a number of researchers, including Tanner (2004), Liu, Slotine and Barabasi (2011), and Parlangeli and Notarstefano (2012). The techniques that we will develop in this chapter can be applied to such problems.

To further illustrate the issue of pinning control, let us consider some examples of networks studied by Tanner (2004) and Parlangeli and Notarstefano (2012). The first example is perhaps a bit surprising because it shows that a complete graph can lead to unreachability.

Example 9.1. Suppose that $\mathfrak{A}$ is an (unweighted) adjacency matrix of the complete graph K_N on N vertices. Then (9.3) is equal to

$$A = \begin{pmatrix} 1 & \cdots & 1 \\ \vdots & \ddots & \vdots \\ 1 & \cdots & 1 \end{pmatrix}, \quad b = \begin{pmatrix} 1 \\ \vdots \\ 1 \end{pmatrix}.$$

Obviously, the system (A, b) is unreachable for $N \geq 2$. The same conclusion is true if $\mathfrak{A}$ is the Laplacian of K_N.

Example 9.2. Here $\mathfrak{A}$ is the Laplacian matrix of the path graph Γ_N on N vertices. The pinned system is then

Fig. 9.2 Cycle graph

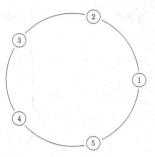

$$A = \begin{pmatrix} 1 & -1 & & & & \\ -1 & 2 & -1 & & & \\ & -1 & \ddots & \ddots & & \\ & & \ddots & \ddots & -1 & \\ & & & -1 & 2 & -1 \\ & & & & -1 & 2 \end{pmatrix}, \quad b = \begin{pmatrix} 0 \\ 0 \\ \vdots \\ 0 \\ 0 \\ -1 \end{pmatrix},$$

which is reachable. One could extend this example by replacing the rth state variable z_r by an input. This situation has been analyzed by Parlangeli and Notarstefano (2012).

Example 9.3. Now assume that $\mathfrak{A}$ is the symmetric adjacency matrix of the cycle graph on N vertices and edges $\{1,2\},\{2,3\},\ldots,\{N-1,N\},\{N,1\}$ (Fig. 9.2).

Thus $\mathfrak{A}$ is the tridiagonal matrix

$$\mathfrak{A}_N = \begin{pmatrix} 0 & 1 & & & 1 \\ 1 & 0 & 1 & & \\ & 1 & \ddots & \ddots & \\ & & \ddots & \ddots & 1 \\ 1 & & & 1 & 0 \end{pmatrix},$$

and therefore one obtains the pinned system for z_N as

$$A = \begin{pmatrix} 0 & 1 & & & \\ 1 & 0 & 1 & & \\ & 1 & \ddots & \ddots & \\ & & \ddots & \ddots & 1 \\ & & & 1 & 0 \end{pmatrix}, \quad b = \begin{pmatrix} 1 \\ 0 \\ \vdots \\ 0 \\ 0 \\ 1 \end{pmatrix},$$

which is not reachable. Likewise, by pinning the variable z_{N-1}, one obtains the unreachable system

$$A = \begin{pmatrix} 0 & 1 & & & & 1 \\ 1 & 0 & 1 & & & \\ & 0 & 1 & \ddots & \ddots & \\ & & \ddots & \ddots & 1 & 0 \\ & & & 1 & 0 & 0 \\ 1 & & & 0 & 0 & 0 \end{pmatrix}, \quad b = \begin{pmatrix} 0 \\ \vdots \\ \vdots \\ 0 \\ 1 \\ 1 \end{pmatrix}.$$

One can also consider pinning control problems where the adjacency matrix $\mathfrak{A}$ is replaced by the associated Laplacian matrix. In that situation, one encounters more interesting reachability phenomena, which have been studied by Parlangeli and Notarstefano (2012).

The preceding examples show that the reachability of the pinned system depends in a nontrivial way on the underlying graph structure of the system, as well as on the selection of the pinned control variables. In Section 9.5, we will study the reachability of such networks in greater generality. In fact, system (9.2) can be interpreted as a system of n integrators,

$$\dot{z}_1(t) = u_1(t)$$

$$\vdots$$

$$\dot{z}_N(t) = u_N(t),$$

with feedback terms $u_i = \sum_{j=1}^{N} a_{ij}z_j$. If one replaces the integrator dynamics with a general first-order systems $\dot{z}_i = \alpha_i z_i + \beta_i u_i$ with local state variables $z_i \in \mathbb{R}^{n_i}$, system matrices $\alpha_i, \beta_i \in \mathbb{F}^{n_i \times n_i}$, and using the same coupling terms $u_i = \sum_{j=1}^{N} a_{ij}z_j$, then the closed-loop system is

$$\dot{z}_1(t) = \alpha_1 z_1(t) + \beta_1 \sum_{j=1}^{N} a_{1j}z_j(t)$$

$$\vdots$$

$$\dot{z}_N(t) = \alpha_N z_N(t) + \beta_N \sum_{j=1}^{N} a_{Nj}z_j(t).$$

By pinning the last variable z_N, one obtains the control system

$$\dot{z}_1(t) = \alpha_1 z_1(t) + \beta_1 \sum_{j=1}^{N-1} a_{1j} z_j(t) + \beta_1 a_{1N} u(t)$$

$$\vdots$$

$$\dot{z}_{N-1}(t) = \alpha_{N-1} z_{N-1}(t) + \beta_{N-1} \sum_{j=1}^{N-1} a_{N-1,j} z_j(t) + \beta_{N-1} a_{N-1,N} u(t) .$$

Thus one can ask when such a system is reachable and how one can relate reachability to the graph properties that define the structure of the matrix of coupling parameters $\mathfrak{A}$. We will now develop systematic tools for the reachability and observability analysis of such systems.

9.1   Interconnection Models

State-Space Representations. We present a state-space formulation of the situation we are interested in and introduce notation to be used subsequently. Consider N discrete-time linear systems, which we refer to as **node systems** $\Sigma_i, i = 1,\dots,N$,

$$\begin{aligned} x_i(t+1) &= \alpha_i x_i(t) + \beta_i v_i(t), \\ w_i(t) &= \gamma_i x_i(t). \end{aligned} \tag{9.4}$$

Here $\alpha_i \in \mathbb{F}^{n_i \times n_i}$, $\beta_i \in \mathbb{F}^{n_i \times m_i}$, and $\gamma_i \in \mathbb{F}^{p_i \times n_i}$ are the associated system matrices, and $\mathbb{F}$ denotes a field. Assume that each system is reachable and observable. To interconnect the node systems, apply static coupling laws

$$v_i(t) = \sum_{j=1}^{N} A_{ij} w_j(t) + B_i u(t) \in \mathbb{F}^{m_i}$$

with constant matrices $A_{ij} \in \mathbb{F}^{m_i \times p_j}$ and $B_i \in \mathbb{F}^{m_i \times m}$, although more complex dynamic interconnections laws are possible, too, and will be considered later on. The interconnected output is

$$y(t) = \sum_{i=1}^{N} C_i w_i(t) + D u(t), \quad \text{with } C_i \in \mathbb{F}^{p \times p_i}, \ i = 1,\dots,N.$$

To express the closed-loop system in compact matrix form, define $\bar{n} := n_1 + \cdots + n_N$, $\bar{m} := m_1 + \cdots + m_N$, $\bar{p} := p_1 + \cdots + p_N$. Moreover,

$$A := (A_{ij})_{ij} \in \mathbb{F}^{\bar{m} \times \bar{p}}, \quad C := (C_1, \ldots, C_N) \in \mathbb{F}^{p \times \bar{p}}, \quad B := \begin{pmatrix} B_1 \\ \vdots \\ B_N \end{pmatrix} \in \mathbb{F}^{\bar{m} \times m}, \quad D \in \mathbb{F}^{p \times m}$$

and

$$\alpha := \begin{pmatrix} \alpha_1 & & \\ & \ddots & \\ & & \alpha_N \end{pmatrix} \in \mathbb{F}^{\bar{n} \times \bar{n}}, \quad \beta := \begin{pmatrix} \beta_1 & & \\ & \ddots & \\ & & \beta_N \end{pmatrix} \in \mathbb{F}^{\bar{n} \times \bar{m}},$$

$$\gamma := \begin{pmatrix} \gamma_1 & & \\ & \ddots & \\ & & \gamma_N \end{pmatrix} \in \mathbb{F}^{\bar{p} \times \bar{n}}, \quad x(t) := \begin{pmatrix} x_1(t) \\ \vdots \\ x_N(t) \end{pmatrix} \in \mathbb{F}^{\bar{n}}.$$

Thus, the global state-space representation of the node systems Σ_i is

$$x(t+1) = \alpha x(t) + \beta v(t),$$
$$w(t) = \gamma x(t),$$

and the interconnection is

$$v(t) = Aw(t) + Bu(t),$$
$$y(t) = Cw(t) + Du(t) .$$

Here $u(t)$ is the external input and $y(t)$ the external output of the network. The restriction to strictly proper rather than proper node systems is not crucial and is done here only to simplify some of the subsequent expressions. Thus the network dynamics has the state-space form

$$x(t+1) = \mathscr{A} x(t) + \mathscr{B} u(t), \tag{9.5}$$
$$y(t) = \mathscr{C} x(t) + Du(t),$$

with

$$\mathscr{A} := \alpha + \beta A \gamma \in \mathbb{F}^{\bar{n} \times \bar{n}}, \quad \mathscr{B} := \beta B \in \mathbb{F}^{\bar{n} \times m}, \quad \mathscr{C} := C\gamma \in \mathbb{F}^{p \times \bar{n}}. \tag{9.6}$$

It is convenient to describe an interconnected system in terms of the transfer functions of the node systems. The ith node transfer function is defined as a strictly proper transfer function of McMillan degree n_i and is given in state-space form as

$$G_i(z) = \gamma_i (zI - \alpha_i)^{-1} \beta_i. \tag{9.7}$$

Define the **node transfer function** as

$$G(z) := \text{diag}(G_1(z), \ldots, G_N(z)) = \gamma(zI - \alpha)^{-1}\beta.$$

In the case where $\bar{m} = \bar{p}$, the **interconnection transfer function** is defined as

$$\mathcal{N}(z) = C(zI - A)^{-1}B + D. \tag{9.8}$$

The global **network transfer function** is then defined as

$$\mathcal{N}_G(z) = \mathscr{C}(zI - \mathscr{A})^{-1}\mathscr{B} + D;$$

thus, explicitly,

$$\mathcal{N}_G(z) = C\gamma(zI - \alpha - \beta A\gamma)^{-1}\beta B + D. \tag{9.9}$$

A network of systems (9.5) is called **homogeneous** if the transfer functions of the node systems (9.7) are identical scalar rational functions that are strictly proper. The reachability and observability analysis of homogeneous networks is particularly easy, as is subsequently shown.

Polynomial Matrix Descriptions. A general class of higher-order system representations, the so-called **polynomial matrix descriptions** (PMD), was introduced by Rosenbrock (1970). In this case, Σ_i is defined in terms of systems of higher-order difference equations:

$$\begin{aligned} T_i(\sigma)\xi_i &= U_i(\sigma)v_i, \\ w_i &= V_i(\sigma)\xi_i + W_i(\sigma)v_i, \end{aligned} \tag{9.10}$$

with transfer functions

$$G_i(z) = V_i(z)T_i(z)^{-1}U_i(z) + W_i(z).$$

Here, as well as in other parts of this book, σ denotes the **backward shift operator** (4.8), defined for discrete-time systems. For continuous-time systems, σ denotes the differentiation operator. Let

$$T(z) := \text{diag}(T_1(z), \ldots, T_N(z)) \in \mathbb{F}[z]^{\bar{r} \times \bar{r}},$$

and similarly define $V(z), U(z), W(z)$. Here $\bar{r} = \sum_{i=1}^{N} r_i$. Using this notation, (9.10) can be rewritten as the polynomial matrix representation

$$\begin{pmatrix} 0 \\ I \end{pmatrix} w = \begin{pmatrix} T(\sigma) & -U(\sigma) \\ V(\sigma) & W(\sigma) \end{pmatrix} \begin{pmatrix} \xi \\ v \end{pmatrix}.$$

The transfer function of the decoupled system is

$$V(z)T(z)^{-1}U(z) + W(z).$$

The interconnections are

$$v = Aw + Bu,$$
$$y = Cw + Du.$$

The network transfer function (for $W = 0$) is then

$$\mathcal{N}_G(z) = CV(z)(T(z) - U(z)AV(z))^{-1}U(z)B + D.$$

Matrix Fraction Systems. A special, but interesting, class of polynomial matrix representations is described as **matrix fraction descriptions** (MFD). Here, the node systems Σ_i, $i = 1, \ldots, N$, are given by polynomial matrix descriptions:

$$D_{\ell,i}(\sigma)\xi_i = N_{\ell,i}(\sigma)v_i,$$
$$w_i = \xi_i,$$

with transfer function representations

$$G_i(z) = D_{\ell,i}(z)^{-1}N_{\ell,i}(z) = N_{r,i}(z)D_{r,i}(z)^{-1}.$$

Define the polynomial matrices $D_\ell(z) \in \mathbb{F}[z]^{\bar{p} \times \bar{p}}, N_\ell(z) \in \mathbb{F}[z]^{\bar{p} \times \bar{m}}$ by

$$D_\ell(z) = \mathrm{diag}\,(D_{\ell,1}(z), \ldots, D_{\ell,N}(z)), \quad N_\ell(z) = \mathrm{diag}\,(N_{\ell,1}(z), \ldots, N_{\ell,N}(z)),$$

and similarly for $D_r(z), N_r(z)$. If the interconnections are given by

$$v(t) = Aw(t) + Bu(t),$$
$$y(t) = Cw(t) + Du(t),$$

then the network transfer function is

$$\begin{aligned}\mathcal{N}_G(z) &= C(D_\ell(z) - N_\ell(z)A)^{-1}N_\ell(z)B + D \\ &= CN_r(z)(D_r(z) - AN_r(z))^{-1}B + D.\end{aligned} \tag{9.11}$$

9.2 Equivalence of Interconnected Systems

We next treat the mixed case, where several different models for decoupled node systems are possible, namely, state space, left and right matrix fractions, and polynomial system matrices. For each of the polynomial-based representations one can associate a state-space realization via the shift realization described in Theorem 4.26. The following theorem shows that the similarity of shift realizations associated with different representations of a decoupled system is preserved for interconnected systems. This is true despite the fact that, as a result of inter-connection, the reachability and observability properties of the uncoupled node systems may have been lost. In view of Definition 4.30, proving a similarity of the realizations associated with polynomial system matrices is equivalent to showing that the polynomial system matrices are strictly system equivalent. This is the case even when the associated realizations are not minimal, and hence the state-space isomorphism theorem is not applicable. This constitutes a great simplification because strict system equivalence can be verified without computing the realizations.

Special emphasis will be placed on polynomial matrix descriptions because they cover all system representations of interest to us. Thus, we assume the node systems have the polynomial matrix descriptions

$$T_i(z)\xi_i(z) = U_i(z)v_i,$$
$$w_i = V_i(z)\xi_i + W_i(z)v_i ,$$

(9.12)

with transfer function $G_i(z) = V_i(z)T_i(z)^{-1}U_i(z) + W_i(z)$. The system interconnections are given by

$$v_i = \sum_{j=1}^{N} A_{ij}w_j + B_i u,$$

$$y = \sum_{j=1}^{N} C_j w_j + Du.$$

(9.13)

Let

$$T(z) = \mathrm{diag}\,(T_1(z),\ldots,T_r(z)) \in \mathbb{F}[z]^{\bar{r}\times\bar{r}},$$

and similarly for $V(z), U(z), W(z)$. Using this notation, (9.12) can be rewritten in matrix form as

$$\begin{pmatrix} 0 \\ I \end{pmatrix} w = \begin{pmatrix} T(z) & -U(z) \\ V(z) & W(z) \end{pmatrix} \begin{pmatrix} \xi \\ v \end{pmatrix}.$$

(9.14)

Similarly, equation (9.13) can be rewritten as

$$v = Aw + Bu,$$
$$y = Cw + Du.$$

More generally, we allow for dynamic interconnections described by first-order difference equations of the form

$$E(\sigma)v = A(\sigma)w + B(\sigma)u,$$
$$y = C(\sigma)w + Du. \tag{9.15}$$

Here $E(z)$ is a square nonsingular polynomial matrix, $A(z), B(z), C(z)$ are appropriately sized polynomial matrices, and D is a constant feedthrough matrix. Assuming that the rational function $E(z)^{-1}(A(z), B(z))$ is proper, and by ignoring the output part, consider a first-order shift realization for

$$E(\sigma)v = A(\sigma)w + B(\sigma)u$$

as

$$\zeta(t+1) = F\zeta(t) + G_1 w(t) + G_2 u(t),$$
$$v(t) = H\zeta(t) + J_1 w(t) + J_2 u(t). \tag{9.16}$$

Therefore, one obtains the strict system equivalence

$$\left(\begin{array}{c|cc} E(z) & -A(z) & -B(z) \\ \hline I & 0 & 0 \end{array} \right) \simeq_{FSE} \left(\begin{array}{c|cc} zI - F & -G_1 & -G_2 \\ \hline H & J_1 & J_2 \end{array} \right).$$

Clearly, ζ, v, and w are latent variables, whereas u and y are manifest variables. Thus equations (9.14) and (9.15) can be combined to yield a polynomial matrix description of the following closed-loop interconnected system:

$$\begin{pmatrix} 0 \\ 0 \\ 0 \\ I \end{pmatrix} y = \left(\begin{array}{ccc|c} T(z) & -U(z) & 0 & 0 \\ V(z) & W(z) & -I & 0 \\ 0 & E(z) & -A(z) & -B(z) \\ 0 & 0 & C(z) & D \end{array} \right) \begin{pmatrix} \xi \\ v \\ w \\ u \end{pmatrix}.$$

Similarly, for $C(z) = C$ constant, the closed-loop interconnected system has the first-order representation

$$z(t+1) = \mathscr{A}_c(t) + \mathscr{B}_c u(t),$$
$$y(t) = \mathscr{C}_c z(t) + Du(t), \tag{9.17}$$

with system matrices

$$\mathscr{A}_c = \left(\begin{array}{c|c} \alpha + \beta J_1 \gamma & \beta H \\ \hline G_1 \gamma & F \end{array}\right), \mathscr{B}_c = \left(\begin{array}{c} \beta J_2 \\ G_2 \end{array}\right), \mathscr{C}_c = (C\gamma \ 0).$$

Theorem 9.4. *Consider two pairs of N node systems with polynomial system matrices*

$$\begin{pmatrix} T_i^{(v)}(z) & -U_i^{(v)}(z) \\ V_i^{(v)}(z) & W_i^{(v)}(z) \end{pmatrix}, v = 1, 2, \ i = 1, \ldots, N.$$

Assume that, for all i,

$$\begin{pmatrix} T_i^{(1)}(z) & -U_i^{(1)}(z) \\ V_i^{(1)}(z) & W_i^{(1)}(z) \end{pmatrix} \simeq_{FSE} \begin{pmatrix} T_i^{(2)}(z) & -U_i^{(2)}(z) \\ V_i^{(2)}(z) & W_i^{(2)}(z) \end{pmatrix}.$$

Defining $T^{(v)}(z) = \text{diag}(T_1^{(v)}(z), \ldots, T_N^{(v)}(z))$, *and similarly for the other matrices, then*

$$\begin{pmatrix} T^{(1)}(z) & -U^{(1)}(z) \\ V^{(1)}(z) & W^{(1)}(z) \end{pmatrix} \simeq_{FSE} \begin{pmatrix} T^{(2)}(z) & -U^{(2)}(z) \\ V^{(2)}(z) & W^{(2)}(z) \end{pmatrix}.$$

Assume that each of the two systems is connected by the same interconnection rule (9.15). Then

$$\left(\begin{array}{ccc|c} T^{(1)}(z) & -U^{(1)}(z) & 0 & 0 \\ V^{(1)}(z) & W^{(1)}(z) & -I & 0 \\ 0 & E(z) & -A(z) & -B(z) \\ \hline 0 & 0 & C(z) & D \end{array}\right) \simeq_{FSE} \left(\begin{array}{ccc|c} T^{(2)}(z) & -U^{(2)}(z) & 0 & 0 \\ V^{(2)}(z) & W^{(2)}(z) & -I & 0 \\ 0 & E(z) & -A(z) & -B(z) \\ \hline 0 & 0 & C(z) & D \end{array}\right).$$

$$(9.18)$$

Assume that C is constant, with

$$\begin{pmatrix} T(z) & -U(z) \\ V(z) & W(z) \end{pmatrix} \simeq_{FSE} \begin{pmatrix} zI - \alpha & -\beta \\ \gamma & 0 \end{pmatrix}$$

and

$$\left(\begin{array}{c|cc} E(z) & -A(z) & -B(z) \\ \hline I & 0 & 0 \end{array}\right) \simeq_{FSE} \left(\begin{array}{c|cc} zI - F & -G_1 & -G_2 \\ \hline H & J_1 & J_2 \end{array}\right).$$

Then

$$
\left(\begin{array}{ccc|c}
T(z) & -U(z) & 0 & 0 \\
V(z) & W(z) & -I & 0 \\
0 & E(z) & -A(z) & -B(z) \\
0 & 0 & C & D
\end{array}\right)
\simeq_{FSE}
\left(\begin{array}{cc|c}
zI-\alpha-\beta J_1\gamma & -\beta H & -\beta J_2 \\
-G_1\gamma & zI-F & -G_2 \\
C\gamma & 0 & D
\end{array}\right).
$$

Proof. By our assumption, there exist polynomial matrices $M(z), N(z), X(z), Y(z)$, with $M(z), T^{(2)}(z)$ left coprime and $T^{(1)}(z), N(z)$ right coprime, for which

$$
\begin{pmatrix} M(z) & 0 \\ -X(z) & I \end{pmatrix}
\begin{pmatrix} T^{(1)}(z) & -U^{(1)}(z) \\ V^{(1)}(z) & W^{(1)}(z) \end{pmatrix}
=
\begin{pmatrix} T^{(2)}(z) & -U^{(2)}(z) \\ V^{(2)}(z) & W^{(2)}(z) \end{pmatrix}
\begin{pmatrix} N(z) & Y(z) \\ 0 & I \end{pmatrix}.
$$

In turn, this implies

$$
\begin{pmatrix} M(z) & 0 & 0 & 0 \\ -X(z) & I & 0 & 0 \\ 0 & 0 & I & 0 \\ 0 & 0 & 0 & I \end{pmatrix}
\left(\begin{array}{ccc|c}
T^{(1)}(z) & -U^{(1)}(z) & 0 & 0 \\
V^{(1)}(z) & W^{(1)}(z) & -I & 0 \\
0 & E(z) & -A(z) & -B(z) \\
0 & 0 & C(z) & D
\end{array}\right)
$$

$$
=
\left(\begin{array}{ccc|c}
T^{(2)}(z) & -U^{(2)}(z) & 0 & 0 \\
V^{(2)}(z) & W^{(2)}(z) & -I & 0 \\
0 & E(z) & -A(z) & -B(z) \\
0 & 0 & C(z) & D
\end{array}\right)
\begin{pmatrix} N(z) & Y(z) & 0 & 0 \\ 0 & I & 0 & 0 \\ 0 & 0 & I & 0 \\ 0 & 0 & 0 & I \end{pmatrix}.
$$

The left coprimeness of

$$
\begin{pmatrix} M(z) & 0 & 0 & 0 \\ -X(z) & I & 0 & 0 \\ 0 & 0 & I & 0 \\ 0 & 0 & 0 & I \end{pmatrix}
\quad \text{and} \quad
\left(\begin{array}{ccc|c}
T^{(2)}(z) & -U^{(2)}(z) & 0 & 0 \\
V^{(2)}(z) & W^{(2)}(z) & -I & 0 \\
0 & E(z) & -A(z) & -B(z) \\
0 & 0 & C(z) & D
\end{array}\right)
$$

follows from the left coprimeness of $M(z)$ and $T^{(2)}(z)$, and similarly for right coprimeness. Thus (9.18) follows. For the remaining part, observe that the following strict equivalences are valid:

$$
\left(\begin{array}{ccc|c}
T(z) & -U(z) & 0 & 0 \\
V(z) & W(z) & -I & 0 \\
0 & E(z) & -A(z) & -B(z) \\
0 & 0 & C & D
\end{array}\right)
\simeq_{FSE}
\left(\begin{array}{cccc|c}
T(z) & -U(z) & 0 & 0 & 0 \\
V(z) & W(z) & 0 & -I & 0 \\
0 & E(z) & 0 & -A(z) & -B(z) \\
0 & 0 & I & 0 & 0 \\
0 & 0 & C & 0 & D
\end{array}\right)
\simeq_{FSE}
$$

$$
\begin{pmatrix}
T(z) & -U(z) & 0 & 0 & 0 \\
V(z) & W(z) & 0 & -I & 0 \\
0 & E(z) & E(z) & -A(z) & -B(z) \\
0 & 0 & I & 0 & 0 \\
\hline
0 & 0 & C & 0 & D
\end{pmatrix}
\simeq_{FSE}
\begin{pmatrix}
T(z) & -U(z) & 0 & 0 & 0 \\
V(z) & W(z) & 0 & -I & 0 \\
0 & 0 & E(z) & -A(z) & -B(z) \\
0 & -I & I & 0 & 0 \\
\hline
0 & 0 & C & 0 & D
\end{pmatrix}
$$

$$
\simeq_{FSE}
\begin{pmatrix}
zI - \alpha & -\beta & 0 & 0 & 0 \\
\gamma & 0 & 0 & -I & 0 \\
0 & 0 & zI - F & -G_1 & -G_2 \\
0 & -I & H & J_1 & J_2 \\
\hline
0 & 0 & 0 & C & D
\end{pmatrix}
\simeq_{FSE}
\begin{pmatrix}
zI - \alpha - \beta J_1 \gamma & -\beta H & -\beta J_2 \\
-G_1 \gamma & zI - F & -G_2 \\
\hline
C\gamma & 0 & D
\end{pmatrix}.
$$

∎

We emphasize that the formulation of this theorem for polynomial matrix descriptions covers several cases of interest, including nonminimal state-space descriptions and (not necessarily coprime) matrix fraction descriptions. The next result for constant interconnections is a straightforward consequence of Theorem 9.4; however, the proof is more specific in exhibiting the required isomorphisms. For ease of exposition we assume that the node transfer function $G(z) = V(z)T(z)^{-1}U(z) + W(z)$ is strictly proper, with $W(z) = 0$.

Theorem 9.5. *Assume the strictly proper transfer function $G(z)$ of decoupled node systems has the following, minimal, representations:*

$$
\begin{aligned}
G(z) &= \gamma(zI - \alpha)^{-1}\beta \\
&= D_\ell(z)^{-1}N_\ell(z) = N_r(z)D_r(z)^{-1} = V(z)T(z)^{-1}U(z).
\end{aligned}
\tag{9.19}
$$

Let A, B, C, D be interconnection matrices.

1. *The shift realizations associated with the interconnected polynomial system matrices*

$$
\begin{pmatrix}
T(z) - U(z)AV(z) & -U(z)B \\
CV(z) & D
\end{pmatrix}, \quad
\begin{pmatrix}
D_r(z) - AN_r(z) & -B \\
CN_r(z) & D
\end{pmatrix}
$$

are similar.

2. *The shift realizations associated with the interconnected polynomial system matrices*

$$
\begin{pmatrix}
T(z) - U(z)AV(z) & -U(z)B \\
CV(z) & D
\end{pmatrix}, \quad
\begin{pmatrix}
D_\ell(z) - N_\ell(z)A & -N_\ell(z)B \\
C & D
\end{pmatrix}
$$

are similar.

3. *The shift realizations associated with the interconnected polynomial system matrices*

$$\begin{pmatrix} D_\ell(z) - N_\ell(z)A & -N_\ell(z)B \\ C & D \end{pmatrix}, \quad \begin{pmatrix} D_r(z) - AN_r(z) & -B \\ CN_r(z) & D \end{pmatrix}$$

are similar.

4. *The realizations associated with the interconnected polynomial system matrices*

$$\begin{pmatrix} T(z) - U(z)AV(z) & -U(z)B \\ CV(z) & D \end{pmatrix}, \quad \begin{pmatrix} zI - \alpha - \beta A\gamma & -\beta B \\ C\gamma & D \end{pmatrix}$$

are similar.

Proof. Without loss of generality, one can assume that $D = 0$.

1. By our assumption of minimality, using the state-space isomorphism theorem and the definition of FSE, it follows that all polynomial system matrices

$$\begin{pmatrix} \alpha & -\beta \\ \gamma & 0 \end{pmatrix}, \begin{pmatrix} D_\ell(z) & -N_\ell(z) \\ I & 0 \end{pmatrix}, \begin{pmatrix} D_r(z) & -I \\ N_r(z) & 0 \end{pmatrix}, \text{and} \begin{pmatrix} T(z) & -U(z) \\ V(z) & 0 \end{pmatrix}$$

are system equivalent. Our plan is to show that all polynomial system matrices of the connected system, namely,

$$\begin{pmatrix} zI - \alpha - \beta A\gamma & -\beta \\ \gamma & 0 \end{pmatrix}, \begin{pmatrix} D_\ell(z) - N_\ell(z)A & -N_\ell(z) \\ I & 0 \end{pmatrix}, \begin{pmatrix} D_r(z) - AN_r(z) & -I \\ N_r(z) & 0 \end{pmatrix}, \text{and}$$

$$\begin{pmatrix} T(z) - U(z)AV(z) & -U(z) \\ V(z) & 0 \end{pmatrix},$$

are also system equivalent. Noting that the transfer function of the unconnected system has the representation (9.19), it follows that there exists a polynomial matrix $S(z)$ for which $T(z)^{-1}U(z) = S(z)D_r(z)^{-1}$, and hence both the intertwining relation

$$U(z)D_r(z) = T(z)S(z)$$

and

$$N_r(z) = V(z)S(z)$$

hold. Since $UAN_r = UAVS$, the identity

$$\begin{pmatrix} U(z) & 0 \\ 0 & I \end{pmatrix} \begin{pmatrix} D_r(z) & -I \\ N_r(z) & 0 \end{pmatrix} = \begin{pmatrix} T(z) & -U(z) \\ V(z) & 0 \end{pmatrix} \begin{pmatrix} S(z) & 0 \\ 0 & I \end{pmatrix}$$

implies

$$\begin{pmatrix} U(z) & 0 \\ 0 & I \end{pmatrix}\begin{pmatrix} D_r(z)-AN_r(z) & -I \\ N_r(z) & 0 \end{pmatrix} = \begin{pmatrix} T(z)-U(z)AV(z) & -U(z) \\ V(z) & 0 \end{pmatrix}\begin{pmatrix} S(z) & 0 \\ 0 & I \end{pmatrix}.$$

In turn, this implies

$$\begin{pmatrix} D_r(z)-AN_r(z) & -I \\ N_r(z) & 0 \end{pmatrix} \simeq_{FSE} \begin{pmatrix} T(z)-U(z)AV(z) & -U(z) \\ V(z) & 0 \end{pmatrix}.$$

2. The proof is similar.
3. The equality $D_\ell(z)^{-1}N_\ell(z) = N_r(z)D_r(z)^{-1}$ leads to the intertwining relation

$$\begin{pmatrix} N_\ell(z) & 0 \\ 0 & I \end{pmatrix}\begin{pmatrix} D_r(z)-AN_r(z) & -I \\ N_r(z) & 0 \end{pmatrix} = \begin{pmatrix} D_\ell(z)-N_\ell(z)A & -N_\ell(z) \\ I & 0 \end{pmatrix}\begin{pmatrix} N_r(z) & 0 \\ 0 & I \end{pmatrix}.$$

In turn, this implies

$$\begin{pmatrix} D_r(z)-AN_r(z) & -I \\ N_r(z) & 0 \end{pmatrix} \simeq_{FSE} \begin{pmatrix} D_\ell(z)-N_\ell(z)A & -N_\ell(z) \\ I & 0 \end{pmatrix}, \tag{9.20}$$

which proves the similarity of the associated shift realizations.
4. By the state-space isomorphism theorem, the shift realizations associated with the polynomial system matrices

$$\begin{pmatrix} zI-\alpha & -\beta \\ \gamma & 0 \end{pmatrix} \quad \text{and} \quad \begin{pmatrix} T(z) & -U(z) \\ V(z) & 0 \end{pmatrix}$$

are similar. Thus, applying Definition 4.30,

$$\begin{pmatrix} zI-\alpha & -\beta \\ \gamma & 0 \end{pmatrix} \simeq_{FSE} \begin{pmatrix} T(z) & -U(z) \\ V(z) & 0 \end{pmatrix}.$$

Our next step is to go from a state-space representation to a minimal right matrix fraction representation. To this end, let $H(z)D_r(z)^{-1}$ be a right coprime factorization of $(zI-\alpha)^{-1}\beta$. This implies the intertwining relation $\beta D_r(z) = (zI-\alpha)H(z)$. Moreover, define $N_r(z) = \gamma H(z)$. Similarity implies system equivalence, but one can actually write down the equivalence explicitly, namely,

$$\begin{pmatrix} \beta & 0 \\ 0 & I \end{pmatrix}\begin{pmatrix} D_r(z) & -I \\ N_r(z) & 0 \end{pmatrix} = \begin{pmatrix} zI-\alpha & -\beta \\ \gamma & 0 \end{pmatrix}\begin{pmatrix} H(z) & 0 \\ 0 & I \end{pmatrix}.$$

Thus

$$\begin{pmatrix} D_r(z) & -I \\ N_r(z) & 0 \end{pmatrix} \simeq_{FSE} \begin{pmatrix} zI - \alpha & -\beta \\ \gamma & 0 \end{pmatrix}.$$

Passing on to the polynomial system matrices associated with a coupled system, an easy computation shows that

$$\begin{pmatrix} \beta & 0 \\ 0 & I \end{pmatrix} \begin{pmatrix} D_r(z) - AN_r(z) & -I \\ N_r(z) & 0 \end{pmatrix} = \begin{pmatrix} zI - \alpha - \beta A\gamma & -\beta \\ \gamma & 0 \end{pmatrix} \begin{pmatrix} H(z) & 0 \\ 0 & I \end{pmatrix}.$$

This implies

$$\begin{pmatrix} zI - \alpha - \beta A\gamma & -\beta \\ \gamma & 0 \end{pmatrix} \simeq_{FSE} \begin{pmatrix} D_r(z) - AN_r(z) & -I \\ N_r(z) & 0 \end{pmatrix}. \tag{9.21}$$

Using equations (9.20) and (9.21) and the transitivity of FSE, one obtains

$$\begin{pmatrix} zI - \alpha - \beta A\gamma & -\beta \\ \gamma & 0 \end{pmatrix} \simeq_{FSE} \begin{pmatrix} T(z) - U(z)AV(z) & -U(z) \\ V(z) & 0 \end{pmatrix}.$$

So far, the required system equivalences were shown for $C = I$ and $B = I$. However, it is easily seen that if two polynomial system matrices

$$\begin{pmatrix} T_1(z) & -U_1(z) \\ V_1(z) & 0 \end{pmatrix}, \quad \begin{pmatrix} T_2(z) & -U_2(z) \\ V_2(z) & 0 \end{pmatrix}$$

are FSE, then for matrices B and C the polynomial matrices

$$\begin{pmatrix} T_1(z) & -U_1(z)B \\ CV_1(z) & 0 \end{pmatrix}, \quad \begin{pmatrix} T_2(z) & -U_2(z)B \\ CV_2(z) & 0 \end{pmatrix}.$$

are also system equivalent. This completes the proof. ∎

One can reformulate this result as follows.

Corollary 9.6 (Permanence Principle). *Suppose strictly proper input-output systems $\Sigma_1, \ldots, \Sigma_N$ are coupled by interconnection matrices A, B, C to define a network Σ. Assume that the systems Σ_i are system equivalent to systems $\hat{\Sigma}_i$, $i = 1, \ldots, N$. Let $\hat{\Sigma}$ be a network obtained by interconnecting $\hat{\Sigma}_1, \ldots, \hat{\Sigma}_N$ using the identical interconnections A, B, C. Then Σ is system equivalent to $\hat{\Sigma}$.*

We list an obvious consequence of this result for poles and zeros of systems that was introduced in Chapter 4.7. The proof is left to the reader. Recall that $\overline{\mathbb{F}}$ denotes the algebraic closure of $\mathbb{F}$.

Corollary 9.7. *Assume that the strictly proper transfer function $G(z)$ of the decoupled node systems has the following, minimal, representations:*

$$G(z) = \gamma(zI - \alpha)^{-1}\beta,$$

$$= D_\ell(z)^{-1}N_\ell(z) = N_r(z)D_r(z)^{-1},$$

$$= V(z)T(z)^{-1}U(z).$$

Consider the interconnection matrices A, B, C, D. Then:

1. *The interconnected system $(\alpha + \beta A\gamma, \beta B, C\gamma, D)$ has a finite zero at $z \in \overline{\mathbb{F}}$ if and only if*

$$\text{rk}\begin{pmatrix} T(z) - U(z)AV(z) & -U(z)B \\ CV(z) & D \end{pmatrix} < \bar{r} + grk;$$

   here grk denotes the generic rank of $\mathcal{N}_G(z)$;
2. *$(\alpha + \beta A\gamma, \beta B, C\gamma, D)$ has a pole at $z \in \overline{\mathbb{F}}$ if and only if*

$$\det(T(z) - U(z)AV(z)) = 0;$$

3. *Assume $\mathbb{F} = \mathbb{R}$. Then $(\alpha + \beta A\gamma, \beta B, C\gamma, D)$ is discrete-time (continuous-time) asymptotically stable if and only if $\det(T(z) - U(z)AV(z))$ is a Schur (Hurwitz) polynomial.*

9.3   Reachability and Observability of Networks of Systems

The permanence principle is our main tool for analyzing the reachability and observability properties of interconnected systems. The results will depend on the type of interconnections, i.e., whether they are static or dynamic.

1. Static Interconnections. The question we are interested in is to decide when an interconnected system in state-space form (9.5) is reachable or observable. Of course, if the input and output interconnection matrices B, C are identity matrices, then the effect of the interconnection matrix A is simply by the action of static output feedback on the decoupled, block-diagonal system α, β, γ. In particular, reachability and observability would be preserved. However, except for this trivial case, it is more difficult to characterize reachability and observability. A naive approach might be to compute the $\bar{n} \times \bar{n}m$-Kalman reachability matrix (or, equivalently, the Hautus test) for system (9.5) and check its rank. But this requires checking the rank of a potentially huge matrix. Moreover, owing to the additive perturbation structure of $\mathcal{A}$, the impact of interconnection parameters on the reachability properties is hard to assess. Therefore, one searches for an alternative reachability characterization that exhibits interconnection parameters and node dynamics in a more direct form.

The characterization of the reachability and observability of the shift realization associated with a polynomial system matrix, as in Theorem 4.26, allows us to easily derive such characterizations for interconnected systems. Since reachability and observability are preserved by FSE, Theorem 9.5 implies the following result.

Theorem 9.8. *With the same assumptions as in Corollary 9.7, let*

$$\mathscr{A} = \alpha + \beta A \gamma, \quad \mathscr{B} = \beta B, \quad \mathscr{C} = C \gamma.$$

1. The transfer function $\mathscr{N}_G(z)$ of an interconnected system has the following representations:

$$\mathscr{N}_G(z) = \mathscr{C}(zI - \mathscr{A})^{-1}\mathscr{B} + D$$

$$= C(D_\ell(z) - N_\ell(z)A)^{-1}N_\ell(z)B + D = CN_r(z)(D_r(z) - AN_r(z))^{-1}B + D$$

$$= CV(z)(T(z) - U(z)AV(z))^{-1}U(z)B + D.$$

2. The following statements are equivalent:

  (a) The system $(\mathscr{A}, \mathscr{B}, \mathscr{C})$ is reachable.
  (b) $D_\ell(z) - N_\ell(z)A$ and $N_\ell(z)B$ are left coprime.
  (c) $D_r(z) - AN_r(z)$ and B are left coprime.
  (d) $T(z) - U(z)AV(z)$ and $U(z)B$ are left coprime.

3. The following statements are equivalent:

  (a) The system $(\mathscr{A}, \mathscr{B}, \mathscr{C})$ is observable.
  (b) $D_\ell(z) - N_\ell(z)A$ and C are right coprime.
  (c) $D_r(z) - AN_r(z)$ and $CN_r(z)$ are right coprime.
  (d) $CV(z)$ and $T(z) - U(z)AV(z)$ are right coprime.

4. The system $(\mathscr{A}, \mathscr{B}, \mathscr{C})$ is minimal if and only if $D_r(z) - AN_r(z)$ and B are left coprime and $D_r(z) - AN_r(z)$ and $CN_r(z)$ are right coprime.

Proof. By Theorem 9.5, the triple $(\mathscr{A}, \mathscr{B}, \mathscr{C})$ is similar to the shift realizations of each of three polynomial matrix representations $(D_\ell(z) - N_\ell(z)A, N_\ell(z)B, C)$, $(D_r(z) - AN_r(z), B, CN_r(z))$, and $(T(z) - U(z)AV(z), U(z)B, CV(z))$. The Shift Realization Theorem 4.26 then implies that the reachability and observability of these realizations is equivalent to left coprimeness and right coprimeness, respectively. The result follows. ∎

Corollary 9.9. *A necessary condition for the reachability/observability of an interconnected system is the reachability/observability of all node systems.*

Proof. By Theorem 9.8, an interconnected system is reachable if and only if the polynomial matrices

$$
\begin{pmatrix} T(z) & -U(z) & 0 \\ V(z) & 0 & -I \\ 0 & I & A \end{pmatrix} \text{ and } \begin{pmatrix} 0 \\ 0 \\ B \end{pmatrix}
$$

are left coprime. Clearly, for this, the left coprimeness of $T(z)$ and $U(z)$ is necessary. Because of the diagonal nature of both polynomial matrices, this is equivalent to the left coprimeness of $T_i(z)$ and $U_i(z)$ for all i. One argues similarly for observability.   ∎

Let us point out that a similar result, phrased in terms of decoupling zeros, appeared in Rosenbrock and Pugh (1974). At that time, the connection between decoupling zeros and the properties of reachability and observability had not yet been clarified.

While Corollary 9.9 provides a simple necessary condition for reachability, the condition is in general not sufficient. As a simple consequence of Theorem 9.8 one obtains the following Hautus-type characterization of the reachability and observability of networks. Let $\overline{\mathbb{F}}$ denote the algebraic closure of the field $\mathbb{F}$.

Theorem 9.10. *(a)* $(\mathscr{A}, \mathscr{B})$ *is reachable if and only if*

$$
\mathrm{rk}\left(T(z) - U(z)AV(z)\; U(z)B \right) = \bar{r}, \quad \forall z \in \overline{\mathbb{F}}.
$$

(b) $(\mathscr{C}, \mathscr{A})$ *is observable if and only if*

$$
\mathrm{rk}\left(\begin{matrix} T(z) - U(z)AV(z) \\ CV(z) \end{matrix} \right) = \bar{r}, \quad \forall z \in \overline{\mathbb{F}}.
$$

The preceding result exhibits, in a clear way, how the different components of the network contribute to reachability and observability. In comparison with the Kalman reachability matrix, the size is reduced to $\bar{r} \times (\bar{r} + m)$. Note that for homogeneous networks with scalar node functions, the matrices $T(z), U(z), V(z)$ become scalar multiples of the identity matrix. Therefore, Theorem 9.10 implies that the reachability and observability properties of a homogeneous network are actually independent of the choice of the strictly proper node function. Thus, for homogeneous networks and scalar nodes, the network realization $(\mathscr{A}, \mathscr{B})$ is reachable if and only if (A, B) is reachable. This greatly simplifies the analysis of scalar homogeneous networks; see Section 9.6 for further details and applications. For homogeneous networks with multivariable node transfer functions, the result is not true without further assumptions. We will now extend Theorem 9.8 to dynamical interconnection laws and analyze in detail some special interconnection schemes.

2. Dynamic Interconnections. We next consider more general dynamical coupling laws between the various node systems. This is important for network control applications, where one wants to allow for possible delays in interconnections, thus modeling potential communication delays between subsystems. Let

$$
\begin{aligned}
x(t+1) &= \alpha x(t) + \beta v(t) \\
w(t) &= \gamma x(t)
\end{aligned}
\tag{9.22}
$$

denote the uncoupled array of node systems. As before, assume that (α, β, γ) is reachable and observable with right and left coprime factorizations of the block-diagonal transfer function

$$
\gamma(zI - \alpha)^{-1}\beta = D_\ell(z)^{-1}N_\ell(z) = N_r(z)D_r(z)^{-1}.
$$

Let $\gamma(zI - \alpha)^{-1}\beta = V(z)T(z)^{-1}U(z) + W(z)$ be a polynomial matrix fraction decomposition, with $V(z), T(z)$ right coprime and $T(z), U(z)$ left coprime. Consider the dynamic interconnection law (9.15) via

$$
E(\sigma)v = A(\sigma)w + B(\sigma)u,
$$

$$
y = Cw + Du.
$$

Here $E(z)$ is a square nonsingular polynomial matrix, $A(z)$ and $B(z)$ are appropriately sized polynomial matrices, and C and D are constant matrices. Assuming that the rational function $E(z)^{-1}(A(z), B(z))$ is proper, there exists a proper, first-order shift realization as

$$
\begin{aligned}
\zeta(t+1) &= F\zeta(t) + G_1 w(t) + G_2 u(t), \\
v(t) &= H\zeta(t) + J_1 w(t) + J_2 u(t).
\end{aligned}
\tag{9.23}
$$

Equations (9.22) and (9.23) can be combined to yield the first-order representation

$$
\begin{aligned}
z(t+1) &= \mathscr{A}_c z(t) + \mathscr{B}_c u(t), \\
y(t) &= \mathscr{C}_c z(t) + Du(t),
\end{aligned}
\tag{9.24}
$$

with system matrices

$$
\mathscr{A}_c = \left(\begin{array}{c|c} \alpha + \beta J_1 \gamma & \beta H \\ \hline G_1 \gamma & F \end{array} \right), \quad \mathscr{B}_c = \left(\begin{array}{c} \beta J_2 \\ G_2 \end{array} \right), \quad \mathscr{C}_c = (C\gamma \; 0).
$$

The interconnected system then has the following PMD representation:

$$
\begin{pmatrix} 0 \\ 0 \\ 0 \\ I \end{pmatrix} y = \left(\begin{array}{ccc|c} T(z) & -U(z) & 0 & 0 \\ V(z) & W(z) & -I & 0 \\ 0 & E(z) & -A(z) & -B(z) \\ 0 & 0 & C & D \end{array} \right) \begin{pmatrix} \xi \\ v \\ w \\ u \end{pmatrix}.
$$

The special choice $T(z) = D_r(z), U(z) = I, V(z) = N_r(z), W(z) = 0$ leads to the following result.

Theorem 9.11. *Given N node strictly proper systems, with right and left coprime matrix fraction representations $D_\ell(z)^{-1}N_\ell(z) = N_r(z)D_r(z)^{-1}$, respectively, the following strict system equivalences are fulfilled:*

$$
\left(\begin{array}{cc|c}
zI - \alpha - \beta J_1 \gamma & -\beta H & -\beta J_2 \\
-G_1 \gamma & zI - F & -G_2 \\
\hline
C\gamma & 0 & D
\end{array}\right) \simeq_{FSE}
$$

$$
\left(\begin{array}{ccc|c}
D_r(z) & -I & 0 & 0 \\
N_r(z) & 0 & -I & 0 \\
0 & E(z) & -A(z) & -B(z) \\
\hline
0 & 0 & C & D
\end{array}\right) \simeq_{FSE}
\left(\begin{array}{cc|c}
I & 0 & 0 \\
0 & -A(z)N_r(z) + E(z)D_r(z) & -B(z) \\
\hline
0 & CN_r(z) & D
\end{array}\right).
$$

Proof. The first equivalence follows from Theorem 9.4. The FSE representations

$$
\left(\begin{array}{cc|c}
N_\ell(z) & -D_\ell(z) & 0 & 0 \\
0 & 0 & I & 0 \\
0 & 0 & 0 & I
\end{array}\right)
\left(\begin{array}{ccc|c}
D_r(z) & -I & 0 & 0 \\
N_r(z) & 0 & -I & 0 \\
0 & E(z) & -A(z) & -B(z) \\
0 & 0 & C & D
\end{array}\right) =
\left(\begin{array}{ccc|c}
0 & -N_\ell(z) & D_\ell(z) & 0 \\
0 & E(z) & -A(z) & -B(z) \\
0 & 0 & C & D
\end{array}\right)
$$

as well as

$$
\left(\begin{array}{cc|c}
0 & 0 & 0 \\
0 & I & 0 \\
0 & 0 & I
\end{array}\right)
\left(\begin{array}{cc|c}
I & 0 & 0 \\
0 & -A(z)N_r(z) + E(z)D_r(z) & -B(z) \\
0 & CN_r(z) & D
\end{array}\right)
$$
$$
= \left(\begin{array}{ccc|c}
0 & -N_\ell(z) & D_\ell(z) & 0 \\
0 & E(z) & -A(z) & -B(z) \\
0 & 0 & C & D
\end{array}\right)
\left(\begin{array}{ccc|c}
I & 0 & 0 \\
0 & D_r(z) & 0 \\
0 & N_r(z) & 0 \\
0 & 0 & I
\end{array}\right)
$$

are satisfied. It is easily seen that these representations define FSE transformations. The result follows. ∎

We conclude that the reachability of the node systems and left coprimeness of $(E(z), A(z), B(z))$ are necessary conditions for the reachability of a dynamically interconnected network. The next theorem characterizes the reachability and observability properties of interconnected systems. The proof is an obvious consequence of the preceding equivalence result and therefore omitted.

Theorem 9.12. *Assume that (α, β, γ) are reachable and observable and N_r and D_r are right coprime, with $N_r(z)D_r(z)^{-1} = \gamma(zI - \alpha)^{-1}\beta$. Assume further that*

$(F,(G_1,G_2),H)$ *is reachable and observable with left coprime factorization* $E(z)^{-1}(A(z),B(z)) = (J_1,J_2) + H(zI - F)^{-1}(G_1,G_2)$. *Let*

$$\mathscr{A}_c = \left(\begin{array}{c|c} \alpha + \beta J_1 \gamma & \beta H \\ \hline G_1\gamma & F \end{array} \right), \mathscr{B}_c = \left(\begin{array}{c} \beta J_2 \\ G_2 \end{array} \right), \mathscr{C}_c = (C\gamma \ 0)$$

denote the realization of the dynamically interconnected system (9.24).

1. $(\mathscr{A}_c, \mathscr{B}_c, \mathscr{C}_c)$ *is reachable if and only if*

$$\mathrm{rk}\left(A(z)N_r(z) - E(z)D_r(z), B(z) \right) = \bar{n}, \quad \forall z \in \overline{\mathbb{F}}.$$

2. $(\mathscr{A}_c, \mathscr{B}_c, \mathscr{C}_c)$ *is observable if and only if*

$$\mathrm{rk}\left(\begin{array}{c} A(z)N_r(z) - E(z)D_r(z) \\ CN_r(z) \end{array} \right) = \bar{n}, \quad \forall z \in \overline{\mathbb{F}}.$$

As a special case of the general dynamical coupling law one can characterize reachability and observability for delayed interconnection schemes of the form

$$v_i(t) = \sum_{j=1}^{N} A_{ij} w_j(t - L_{ij}) + B_i u(t)$$

for nonnegative integers L_{ij}. This network is described as follows using the interconnection law (9.15). We use the notations $L_i := \max_{j=1,\dots,N} L_{ij}$, $i = 1,\dots,N$, and $\bar{L}_{ij} = L_i - L_{ij} \geq 0$. Define the polynomial matrices $E_L(z) = \mathrm{diag}(z^{L_1},\cdots,z^{L_N})$, $A_L(z) = (A_{ij}z^{\bar{L}_{ij}})_{i,j=1,\dots,N} \in \mathbb{F}^{\bar{p} \times \bar{m}}[z]$, $B(z) := E_L(z)B$, and $C(z) = C$. One obtains the following characterization of reachability and observability for delayed networks.

Theorem 9.13. *The shift realization of a delayed network is reachable if and only if*

$$\mathrm{rk}\left(A_L(z)N_r(z) - E_L(z)D_r(z), E_L(z)B \right) = \bar{n}, \quad \forall z \in \overline{\mathbb{F}}.$$

The shift realization of a delayed network is observable if and only if

$$\mathrm{rk}\left(\begin{array}{c} A_L(z)N_r(z) - E_L(z)D_r(z) \\ CN_r(z) \end{array} \right) = \bar{n}, \quad \forall z \in \overline{\mathbb{F}}.$$

As a special case, consider homogeneous networks of identical SISO systems with node transfer functions $G_i(z) = \frac{p(z)}{q(z)}$ satisfying $p(0) \neq 0$. In this situation the preceding result implies the following corollary.

Corollary 9.14. *Consider a network of identical node transfer functions* $G_i(z) = \frac{p(z)}{q(z)}$ *satisfying* $p(0) \neq 0$. *Assume that A is invertible and* (A,B,C) *is reachable (observable). Assume further that all delays* L_{ij} *are identical and* equal to $L \geq 1$. Then the delayed network $(\mathscr{A}_c, \mathscr{B}_c, \mathscr{C}_c)$ is reachable (observable), independently of the value of L.

Proof. Let $A = (A_{ij})$. Since $L_{ij} = L$, one obtains $\bar{L}_{ij} = 0$, and thus $A(z) = A$. From Theorem 9.12 we conclude that reachability is equivalent to $(p(z)A - z^{Lq}(z)I, z^L B)$ having full row rank. For $z = 0$ this is true since $p(0) \neq 0$ and A is invertible. For $z \neq 0$ this is equivalent to $(A - \frac{z^{Lq}(z)}{p(z)}I, B)$ having full row rank, which again follows from the reachability of (A, B). For observability one argues similarly. ∎

9.4 Homogeneous Networks

Clearly, the simplest classes of networks are the homogeneous ones, defined by interconnections of identical linear systems with SISO node transfer function $g(z)$. Thus, assume that the dynamics of the identical node systems in a linear network are described by a single scalar strictly proper transfer function

$$g(z) = \gamma(zI_n - \alpha)^{-1}\beta,$$

with $\alpha \in \mathbb{F}^{n \times n}, \beta \in \mathbb{F}^n, \gamma \in \mathbb{F}^{1 \times n}$ reachable and observable. This is a special case of (9.7). Define $h(z) = 1/g(z)$. Let $\mathscr{N}(z) = C(zI - A)^{-1}B$ denote the $p \times m$ interconnection transfer function. Then the network transfer function $\mathscr{N}_G(z)$ of the homogeneous network is $\mathscr{N}_G(z) = C(h(z)I - A)^{-1}B$, i.e., it is the composition of rational functions $\mathscr{N} \circ h$. Let $(\mathscr{A}, \mathscr{B}, \mathscr{C})$ be the shift realization of the network transfer function $\mathscr{N}_G(z)$ associated with a minimal factorization of $g(z) = p/q$. Then the matrices in (9.6) are represented in Kronecker product form as

$$\mathscr{A} = I_N \otimes \alpha + A \otimes \beta\gamma \in \mathbb{F}^{nN \otimes nN},$$

$$\mathscr{B} = B \otimes \beta \in \mathbb{F}^{nN \times m}, \tag{9.25}$$

$$\mathscr{C} = C \otimes \gamma \in \mathbb{F}^{p \times nN}.$$

The following test for the reachability of homogeneous networks is a simple consequence of Theorem 9.10. It was first stated and proved by Hara, Hayakawa and Sugata (2009). Our proof is quite different and avoids complicated state-space canonical form arguments.

Theorem 9.15. *The shift realization $(\mathscr{A}, \mathscr{B}, \mathscr{C})$ of $\mathscr{N}_G(z)$, defined by (9.25), is reachable (observable) if and only if the realization (A, B, C) of the interconnection transfer function $\mathscr{N}(z)$ is reachable (observable). In particular, the reachability of $(\mathscr{A}, \mathscr{B}, \mathscr{C})$ is independent of the choice of the node transfer function $g(z)$, as long as $g(z)$ is scalar rational and strictly proper.*

Proof. By Theorem 9.8, $(\mathscr{A},\mathscr{B},\mathscr{C})$ is reachable if and only if $Q(z) - P(z)A, P(z)B$ are left coprime. The coprime factorization of the decoupled system $Q(z)^{-1}P(z) = g(z)I_N$ is $Q(z) = q(z)I_N$, $P(z) = p(z)I_N$, with $g(z) = p(z)/q(z)$ coprime. Thus $(\mathscr{A},\mathscr{B})$ is reachable if and only if

$$\mathrm{rk}(q(z)I_N - p(z)A, p(z)B) = N \qquad (9.26)$$

for all $z \in \mathbb{F}$. If $p(z) = 0$, then, by coprimeness, $q(z) \neq 0$, and (9.26) is satisfied. The fundamental theorem of algebra implies that for all $w \in \mathbb{F}$ there exists $z \in \mathbb{F}$, with $p(z) \neq 0$ and $w = \frac{q(z)}{p(z)}$. Dividing by $p(z)$, it follows that the left coprimeness condition is equivalent to $\mathrm{rk}(wI_N - A, B) = N$ for all $w \in \mathbb{F}$. Thus the reachability of $(\mathscr{A},\mathscr{B})$ is equivalent to the reachability of (A,B), and we are done. One argues similarly for observability and minimality. ∎

The amazing consequence of the preceding theorem is that reachability can be analyzed completely independently of the choice of node function. We will now make this even more explicit by relating coprime factorizations of $\mathscr{N}_G$ and $\mathscr{N}$. Assume that $(A,B) \in \mathbb{F}^{N \times (N+m)}$ is reachable with reachability indices $\kappa_1 \geq \cdots \geq \kappa_m$. Choose a right coprime factorization

$$(zI - A)^{-1}B = N(z)D(z)^{-1}$$

by $N \times m$ and $m \times m$ polynomial matrices $N(z)$ and $D(z)$, respectively. Therefore, $\det D(z) = \det(zI - A)$. Without loss of generality, one can assume that $D(z)$ is in column proper form, i.e., the leading coefficient matrix of $D(z)$ is $D_0\Delta(z)$, with D_0 invertible and

$$\Delta(z) = \mathrm{diag}(z^{\kappa_1}, \cdots, z^{\kappa_m}).$$

Let $g(z)$ denote a strictly proper, scalar rational transfer function with coprime factorization $g(z) = p(z)/q(z)$ and McMillan degree n. Define the homogenizations

$$N_g(z) = N(\frac{q(z)}{p(z)})\Delta(p(z)), \quad D_g(z) = D(\frac{q(z)}{p(z)})\Delta(p(z)).$$

Proposition 9.16. *Under the preceding assumptions, the following assertions are true:*

1. *$N_g(z)$ and $D_g(z)$ are right coprime polynomial matrices and $D_g(z)$ is in column proper form.*
2. *$\det D_g(z) = \det(q(z)I - p(z)A)$.*
3. *The reachability indices of the shift realization of $N_g(z)D_g(z)^{-1}$ are equal to $(n\kappa_1, \ldots, n\kappa_m)$.*

Proof. That $N_g(z)$ and $D_g(z)$ are polynomials follows easily from $N(z)$ and $D(z)$ being in column proper form. For right coprimeness, one must show that

$$\mathrm{rk}\begin{pmatrix} N_g(z) \\ D_g(z) \end{pmatrix} = m \tag{9.27}$$

for all z in the algebraic closure of the field $\mathbb{F}$. If z is not a zero of p, then $\Delta(p(z))$ is invertible. Therefore, the rank condition (9.27) follows from the corresponding rank condition for $N(z)$ and $D(z)$ at the point $h(z) = \frac{q(z)}{p(z)}$. If $p(z) = 0$, then $D_g(z) = D_0\Delta(q(z))$ is invertible. Moreover, by the strict properness of $N(z)D(z)^{-1}$, at such a point z one has $N_g(z) = 0$. This proves right coprimeness. The column properness of $D_g(z)$ follows from the fact that the leading term of $D_g(z)$ is $D_0\Delta(q(z))$. This also implies part 3, i.e., that the reachability indices of the shift realization of $N_g(z)D_g(z)^{-1}$ are $n\kappa_i$ for $i = 1,\ldots,m$.

2. It suffices to verify the formula for the transfer function for all z that are not zeros of $p(z)$. Note that

$$N_g(z)D_g(z)^{-1} = N(h(z))D(h(z))^{-1} = C(h(z)I - A)^{-1}B = C(q(z)I - p(z)A)^{-1}p(z)B.$$

Finally, for each z that is not a zero of p,

$$\det D_g(z) = \det D(h(z))\det\Delta(p(z)) = \det D(h(z))p(z)^n = \det(q(z)I - p(z)A)).$$

This completes the proof.   ∎

9.5   Special Coupling Structures

Many coupling patterns in interconnected systems arise by specifying linear dependency relations among the coefficients of the coupling matrices A,B,C. Thus, for example, one may consider $0 - *$ patterns in which the entries of A,B,C are either 0 or free independent variables. Other examples include block upper triangular matrices, symmetric matrices, and Toeplitz matrices A. In this section, we will explore the reachability task for some of these interconnection structures. A more systematic approach would require tools from graph theory. Here we pursue modest goals and focus on the analysis of special cases such as path graphs and circular structures.

1. Paths. Path graphs or, more generally, trees are among the simplest hierarchical interconnection patterns. Certainly, the easiest example of a coupling pattern that comes from a path is the nearest-neighbor interconnection scheme with controls at the first node. Thus, consider N node systems Σ_i with reachable and observable state-space representations $\alpha_i \in \mathbb{F}^{n\times n}$, $\beta_i \in \mathbb{F}^{n\times m}$, and $\gamma_i \in \mathbb{F}^{p\times n}$. For $i = 1,\ldots,N$, let

$$\gamma_i(zI - \alpha_i)^{-1}\beta_i = D_{\ell,i}(z)^{-1}N_{\ell,i}(z) = N_{r,i}(z)D_{r,i}(z)^{-1} \tag{9.28}$$

denote left and right coprime factorizations of the associated transfer functions. For simplicity assume $m = p$. Consider the state interconnection matrices

$$
A = \begin{pmatrix} 0 & \cdots\cdots & 0 \\ I_m & \ddots & \vdots \\ \vdots & \ddots & \ddots & \vdots \\ 0 & \cdots & I_m & 0 \end{pmatrix}, \quad B = \begin{pmatrix} 0 \\ \vdots \\ I_m \\ \vdots \\ 0 \end{pmatrix}, \tag{9.29}
$$

the I_m-component of B being at position $1 \le r \le N$. Clearly, (9.29) represents a nearest-neighbor interaction of N systems, with the external controls entering at node r. The closed-loop system matrix then has the lower bidiagonal form

$$
\mathscr{A} = \begin{pmatrix} \alpha_1 \\ \beta_2 \gamma_1 & \alpha_2 \\ & \beta_3 \gamma_2 & \ddots \\ & & \ddots & \ddots \\ & & & \beta_N \gamma_{N-1} & \alpha_N \end{pmatrix}, \quad \mathscr{B} = \begin{pmatrix} 0 \\ \vdots \\ \beta_r \\ 0 \\ \vdots \\ 0 \end{pmatrix}. \tag{9.30}
$$

Note that, for $r = 1$, the network (9.30) is simply the series connection $\Sigma_1 \wedge \ldots \wedge \Sigma_N$ of N systems and thus is reachable if and only if the $(N-1)m \times Nm$ polynomial Sylvester-type matrix

$$
\begin{pmatrix} N_{r,1}(z) & D_{r,2}(z) \\ & \ddots & \ddots \\ & & \ddots & \ddots \\ & & & N_{r,N-1}(z) & D_{r,N}(z) \end{pmatrix}
$$

is left prime. Applying Theorem 9.8, one observes that the system is not reachable for $r > 1$.

The situation becomes more interesting for symmetric couplings defined by the interconnection matrices $A = J \otimes I_m$ and $B = e_k \otimes I_m$, where

$$
J = \begin{pmatrix} 0 & 1 \\ 1 & 0 & 1 \\ & 1 & \ddots & \ddots \\ & & \ddots & \ddots & 1 \\ & & & 1 & 0 \end{pmatrix}. \tag{9.31}
$$

The interconnected system is then

$$
\mathcal{A} = \begin{pmatrix} \alpha_1 & \beta_1\gamma_2 & & & \\ \beta_2\gamma_1 & \alpha_2 & \ddots & & \\ \vdots & \ddots & \ddots & \beta_{N-1}\gamma_N & \\ 0 & . & \beta_N\gamma_{N-1} & \alpha_N \end{pmatrix}, \quad \mathcal{B} = \begin{pmatrix} 0 \\ . \\ \beta_k \\ . \\ 0 \end{pmatrix}. \tag{9.32}
$$

Again, applying Theorem 9.8 to the coprime factorization (9.28), we conclude that (9.32) is reachable if and only if the polynomial matrices

$$
\begin{pmatrix} D_{r,1}(z) & N_{r,1}(z) & & & \\ N_{r,2}(z) & D_{r,2}(z) & \ddots & & \\ \vdots & \ddots & \ddots & N_{r,N-1}(z) & \\ 0 & . & N_{r,N}(z) & D_{r,N}(z) \end{pmatrix}, \quad \begin{pmatrix} 0 \\ . \\ I_m \\ . \\ 0 \end{pmatrix}
$$

are left prime. Equivalently, the polynomial matrix

$$
\begin{pmatrix} D_{r,1}(z) & N_{r,1}(z) & & & & \\ N_{r,2}(z) & D_{r,2}(z) & N_{r,2}(z) & & & \\ & \ddots & \ddots & \ddots & & \\ & & N_{r,k-1}(z) & D_{r,k-1}(z) & N_{r,k-1}(z) & \\ & & & N_{r,k+1}(z) & D_{r,k+1}(z) & N_{r,k+1}(z) \\ & & & & \ddots & \ddots & \ddots \\ & & & & & N_{r,N-1}(z) & D_{r,N-1}(z) & N_{r,N-1}(z) \\ & & & & & & N_{r,N}(z) & D_{r,N}(z) \end{pmatrix}
$$

is left prime.

For identical node systems with $D(z) := D_{r,1}(z) = \ldots = D_{r,N}(z)$ and $N(z) := N_{r,1}(z) = \ldots = N_{r,N}(z)$, more explicit results can be obtained using the spectral information on (9.31). By Theorem 8.45, matrix J has N distinct real eigenvalues $2\cos\frac{k\pi}{N+1}, k = 1, \ldots, N$, with eigenvectors given by the columns of

$$
T = \sqrt{\frac{2}{N+1}} \begin{pmatrix} \sin\frac{\pi}{N+1} & \cdots & \sin\frac{N\pi}{N+1} \\ \vdots & \ddots & \vdots \\ \sin\frac{N\pi}{N+1} & \cdots & \sin\frac{N^2\pi}{N+1} \end{pmatrix}.
$$

Note that the column vectors $x^{(k)}$ of T are pairwise orthogonal with Euclidean norm

$$\frac{N+1}{2}\|x^{(k)}\|^2 = \sum_{j=1}^{N} \sin^2\left(\frac{kj\pi}{N+1}\right) = \frac{N}{2} - \frac{1}{2}\sum_{j=1}^{N}\cos\frac{2kj\pi}{N+1}$$

$$= \frac{N}{2} - \frac{1}{2}\sum_{j=1}^{N}\mathrm{Re}(\omega^{kj}) = \frac{N+1}{2},$$

where $\omega = e^{\frac{2\pi\sqrt{-1}}{N+1}}$. Thus $T = \sqrt{\frac{2}{N+1}}(\sin\frac{kl\pi}{N+1})_{k,l}$ is a real orthogonal matrix such that $T^{-1}JT$ is diagonal. Then $T \otimes I_N$ diagonalizes A with eigenvalues $2\cos\frac{k\pi}{N+1}, k = 1,\ldots,N$, each one occurring with multiplicity m. Moreover,

$$T^{-1}e_r = \sqrt{\frac{N+1}{2}}\begin{pmatrix} \sin\frac{r\pi}{N+1} \\ \vdots \\ \sin\frac{Nr\pi}{N+1} \end{pmatrix}$$

has a zero entry if and only if $N+1$ divides rk for some $1 \le k \le N$, i.e., if and only if $N+1$ and r are not coprime. This leads to an explicit characterization of reachability that is independent of the node system.

Theorem 9.17. *The interconnected system (9.32) with identical nodes is reachable if and only if $N+1$ and r are coprime.*

Proof. The matrix $T \otimes I_m (I_N \otimes D(z) - J \otimes N(z)) T^{-1} \otimes I_m$ is block-diagonal with block-diagonal entries $D(z) - 2\cos\frac{k\pi}{N+1}N(z), 1 \le k \le N$. Therefore, the pair $(I_N \otimes D(z) - J \otimes N(z), T^{-1}e_r \otimes I_m)$ is left coprime if and only if $T^{-1}e_r$ has no zero entry, i.e., if and only if r and $N+1$ are coprime. ∎

2. Simple Circulant Structures. Here the reachability problem for linear systems with special circulant interconnection structures are explored. We refer the reader to Brockett and Willems (1974) and Lunze (1986) for earlier work on circulant systems. Further motivation derives from the observation that such systems present the simplest kind of systems with symmetries; see, for example, Hazewinkel and Martin (1983). Symmetric systems also arise in recent studies on spatially invariant systems; see, for example, Bamieh, Paganini and Dahleh (2002).

Consider now N nodes with transfer functions $G_i(z)$ coupled circularly. Specifically, in terms of minimal state-space realizations (9.4), one has the following state-space equations describing the individual nodes ($i = 1,\ldots,N$):

$$\sigma x_i = \alpha_i x_i(t) + \beta_i v_i(t),$$
$$w_i(t) = \gamma_i x_i(t).$$

This by itself is not sufficient to describe an interconnected system. We need to describe the cross influence between the nodes and the way in which the external input influences the nodes. There are many options. The cross influence between the nodes can be one- or two-sided nearest-neighbor interactions up to interactions between all nodes in the same way. Then one must consider how the external input influences the individual nodes. The two extreme cases are, first, that the external input is applied only to one node and, second, that it is applied directly to all nodes. Similar cases of interest exist for the global output of an interconnected system. We briefly review some of the options.

2a. Unidirectional Nearest-Neighbor Coupling and One-Node External Control. The coupling information is described by the matrices

$$
A = \begin{pmatrix} 0 & \cdots & 0 & I \\ I & \ddots & & 0 \\ & \ddots & \ddots & \vdots \\ & & I & 0 \end{pmatrix}, \quad B = \begin{pmatrix} I \\ 0 \\ \vdots \\ 0 \end{pmatrix}, \quad C = \begin{pmatrix} I & 0 & \cdots & 0 \end{pmatrix}. \tag{9.33}
$$

The coupled system has the following representation:

$$
\mathscr{A} = \begin{pmatrix} \alpha_1 & 0 & \cdots & 0 & \beta_1\gamma_N \\ \beta_2\gamma_1 & \alpha_2 & \ddots & & 0 \\ & \beta_3\gamma_2 & \ddots & \ddots & \vdots \\ & & \ddots & \ddots & 0 \\ & & & \beta_N\gamma_{N-1} & \alpha_N \end{pmatrix}, \quad \mathscr{B} = \begin{pmatrix} \beta_1 \\ 0 \\ \vdots \\ \vdots \\ 0 \end{pmatrix}, \quad \mathscr{C} = \begin{pmatrix} \gamma_1 & 0 & \cdots & 0 \end{pmatrix}. \tag{9.34}
$$

2b. Bidirectional Nearest-Neighbor Coupling and One-Node External Control. The coupling in (9.33) is unidirectional. Alternatively, one can use the more symmetric, nearest-neighbor, coupling described by

$$
A = \begin{pmatrix} 0 & I & & & I \\ I & \ddots & \ddots & & \\ & \ddots & \ddots & \ddots & \\ & & \ddots & \ddots & I \\ I & & & I & 0 \end{pmatrix}, \quad B = \begin{pmatrix} I \\ 0 \\ \vdots \\ 0 \end{pmatrix}, \quad C = \begin{pmatrix} I & 0 & \cdots & 0 \end{pmatrix}.
$$

In this case, the coupled system has the following representation:

$$
\mathscr{A} = \begin{pmatrix} \alpha_1 & \beta_1\gamma_2 & 0 & \cdots & \beta_1\gamma_N \\ \beta_2\gamma_1 & \alpha_2 & \ddots & & 0 \\ 0 & \beta_3\gamma_2 & \ddots & \ddots & \vdots \\ \vdots & & \ddots & \ddots & \beta_{N-1}\gamma_N \\ \beta_N\gamma_1 & 0 & \cdots & \beta_N\gamma_{N-1} & \alpha_N \end{pmatrix}, \quad \mathscr{B} = \begin{pmatrix} \beta_1 \\ 0 \\ \vdots \\ \vdots \\ 0 \end{pmatrix}, \quad \mathscr{C} = \begin{pmatrix} \gamma_1 & 0 & \cdots & 0 \end{pmatrix}. \tag{9.35}
$$

2c. Full Coupling and One-Node External Control. The coupling is described by

$$
A = \begin{pmatrix} 0 & I & & & I \\ I & \ddots & \ddots & & \\ & \ddots & \ddots & \ddots & \\ & & \ddots & \ddots & I \\ I & & & I & 0 \end{pmatrix}, \quad B = \begin{pmatrix} I \\ 0 \\ \vdots \\ 0 \end{pmatrix}, \quad C = (I \; 0 \; \cdots \; 0).
$$

In this case, the coupled system has the following representation:

$$
\mathscr{A} = \begin{pmatrix} \alpha_1 & \beta_1\gamma_2 & 0 & \cdots & & \beta_1\gamma_N \\ \beta_2\gamma_1 & \alpha_2 & \ddots & & & 0 \\ 0 & \beta_3\gamma_2 & \ddots & \ddots & & \vdots \\ \vdots & & \ddots & \ddots & & \beta_{N-1}\gamma_N \\ \beta_N\gamma_1 & 0 & \cdots & & \beta_N\gamma_{N-1} & \alpha_N \end{pmatrix}, \quad \mathscr{B} = \begin{pmatrix} \beta_1 \\ 0 \\ \vdots \\ \vdots \\ 0 \end{pmatrix}, \quad \mathscr{C} = (\gamma_1 \; 0 \; \cdots \; 0).
$$

$$(9.36)$$

If all nodes have the same state-space representation, that is, in the homogeneous case, then the matrices $\mathscr{A}$ in the representations (9.34), (9.35), and (9.36) will all have a block-circulant structure; see the next subsection for a discussion of general block-circulant structures.

For our purposes, it is advantageous to obtain matrix fraction representations of the various interconnections. However, because the computations are similar, we restrict ourselves to a single case, for which a characterization of reachability is obtained, which is summarized by the following theorem. This should be compared with the criteria for the reachability of series connections derived subsequently in (10.5).

Theorem 9.18. *Consider the node systems Σ_i, $i = 1,\ldots,N$, with coprime matrix fraction representations as in (9.7). The circular interconnection system (9.34) is reachable if and only if the polynomial matrix*

$$
\begin{pmatrix} N_{r,1}(z) & D_{r,1}(z) & & & \\ & N_{r,2}(z) & D_{r,2}(z) & & \\ & & \ddots & \ddots & \\ & & & N_{r,N-1}(z) & D_{r,N}(z) \end{pmatrix}
\tag{9.37}
$$

is left prime.

Proof. Applying Theorem 9.8, one sees that (9.34) is reachable if and only if the pair of polynomial matrices

$$
(D_r(z) - AN_r(z), B) = \left(\begin{pmatrix} D_{r,1} & & & -N_{r,N} \\ -N_{r,1} & D_{r,2} & & \\ & \ddots & \ddots & \\ & & -N_{r,N-1} & D_{r,N} \end{pmatrix}, \begin{pmatrix} I_m \\ 0 \\ \vdots \\ 0 \end{pmatrix} \right)
$$

is left coprime. After a simple column operation, this is equivalent to the left primeness of (9.37).   ∎

3. Block Circulant Structures. Following Brockett and Willems (1974), we begin by presenting a state-space formulation of the situation we are interested in and introduce our subsequent notation. Since Fourier transform techniques will be applied, we restrict ourselves to the field $\mathbb{C}$ of complex numbers. An $N \times N$ **block-circulant matrix** has the form

$$
\mathbf{A} = \begin{pmatrix}
A_0 & A_1 & \cdots & A_{N-2} & A_{N-1} \\
A_{N-1} & A_0 & \cdots & & A_{N-2} \\
A_{N-2} & \ddots & \ddots & & \vdots \\
\vdots & \ddots & \ddots & \ddots & \vdots \\
A_1 & \cdots & A_{N-2} & A_{N-1} & A_0
\end{pmatrix},
$$

where A_i denotes an $n \times n$ matrix with complex coefficients. Similarly, let **B** and **C** denote block-circulant matrices, where the block matrices are of the form $B_i \in \mathbb{C}^{n \times m}$ and $C_i \in \mathbb{C}^{p \times n}$. Consider the input and output matrices, respectively,

$$
\beta = \begin{pmatrix} \beta_1 \\ \beta_2 \\ \vdots \\ \beta_N \end{pmatrix}, \quad \gamma = \begin{pmatrix} \gamma_1 & \gamma_2 & \cdots & \gamma_N \end{pmatrix}.
$$

Here, the submatrices satisfy $\beta_i \in \mathbb{C}^{m \times r}$ and $\gamma_i \in \mathbb{C}^{s \times p}$, respectively. Consider N interconnected discrete-time block-circulant linear systems

$$
\begin{aligned}
x(t+1) &= \mathbf{A}x(t) + \mathbf{B}\beta u(t), \\
y(t) &= \gamma \mathbf{C}x(t).
\end{aligned}
\tag{9.38}
$$

We are interested in characterizing when such systems are reachable. Let $\omega := \exp(2\pi\sqrt{-1}/N)$ denote the primitive Nth root of unity, and let

$$
\Phi = \frac{1}{\sqrt{N}} \left(\omega^{(k-1)(\ell-1)} \right)_{k,\ell}
\tag{9.39}
$$

denote the $N \times N$ **Fourier matrix**. Note that the Fourier matrix Φ is the reachability matrix of the reachable pair

$$\Delta(\omega) = \begin{pmatrix} 1 & & & \\ & \omega & & \\ & & \ddots & \\ & & & \omega^{N-1} \end{pmatrix}, \quad \mathbf{1} = \frac{1}{\sqrt{N}} \begin{pmatrix} 1 \\ 1 \\ \vdots \\ 1 \end{pmatrix}.$$

It is easily seen that the block-circulant matrix is exactly of the form $\mathbf{A} = \sum_{i=0}^{N-1} S^i \otimes A_i$, where

$$S = \Phi\Delta(\omega)\Phi^* = \begin{pmatrix} 0 & 1 & & & \\ & 0 & 1 & & \\ & & \ddots & \ddots & \\ & & & \ddots & 1 \\ 1 & & & & 0 \end{pmatrix} \tag{9.40}$$

denotes the standard $N \times N$ circulant matrix. In particular, finite sums and products of block-circulant matrices with square blocks are block-circulant. Moreover, the Fourier matrix is unitary and $\Phi \otimes I_n$ block-diagonalizes all block-circulant matrices $\mathbf{A} = \sum_{i=0}^{N-1} S^i \otimes A_i$. Block-circulant matrices $\mathbf{A}$ are best analyzed in terms of the associated matrix polynomial

$$A(z) := \sum_{k=0}^{N-1} A_k z^k \in \mathbb{C}[z]^{n \times n}.$$

Thus

$$\mathbf{A} = (\Phi \otimes I_n) \begin{pmatrix} A(1) & & & \\ & A(\omega) & & \\ & & \ddots & \\ & & & A(\omega^{N-1}) \end{pmatrix} (\Phi^* \otimes I_n),$$

and similarly for $\mathbf{B}$ and $\mathbf{C}$. This shows that one has full knowledge on the eigenstructure of block-circulant matrices. Explicitly, the eigenvalues of block-circulant matrices are the eigenvalues of $A(1), \ldots, A(\omega^{N-1})$, respectively, while the eigenvectors of $\mathbf{A}$ are equal to $(\Phi \otimes I_N)v$ for the eigenvectors v of $\mathrm{diag}\,(A(1), \ldots, A(\omega^{N-1}))$.

Define $b \in \mathbb{C}^{Nm \times r}, c \in \mathbb{C}^{s \times Np}$ as

$$b = (\Phi \otimes I_m)\beta = \begin{pmatrix} b_N \\ b_1 \\ \vdots \\ b_{N-1} \end{pmatrix}, \quad c = \gamma(\Phi^* \otimes I_p) = \begin{pmatrix} c_N & c_1 \cdots & c_{N-1} \end{pmatrix}.$$

Thus, in the discrete-time case (and similarly for continuous-time systems), the block-circulant system (9.38) is state-space equivalent to the parallel connected system

$$x_k(t+1) = A(\omega^k)x_k(t) + B(\omega^k)b_k u(t),$$
$$y_k(t) = c_k C(\omega^k)x_k(t), \quad k = 1,\ldots,N. \tag{9.41}$$

We emphasize that this is simply the parallel sum of N systems $(c_k C(\omega^k), A(\omega^k), B(\omega^k)b_k)$. The $s \times r$ transfer function of system (9.41) is

$$\mathcal{N}_G(z) = c\,\mathrm{diag}\,(G_1(z), \cdots, G_N(z))b,$$

where $G_k(z) = C(\omega^k)(zI - A(\omega^k))^{-1}B(\omega^k)$. Let

$$(zI - A(\omega^k))^{-1}B(\omega^k)b_k := N_k(z)D_k(z)^{-1}$$

denote a right coprime factorization into polynomial matrices $N_k(z), D_k(z)$, $k \in 1,\ldots,N$. Thus $N_k(z)$ and $D_k(z)$ are $n \times r$ and $r \times r$ polynomial matrices, respectively. From Theorem 10.4 one arrives at the following theorem.

Theorem 9.19. *Assume that $(A(\omega^k), B(\omega^k)b_k)$ are reachable for $k = 1, \cdots, N$. The block-circulant system (9.38) is reachable if and only if the N polynomial matrices $D_k(z) \in \mathbb{C}[z]^{r \times r}, k = 1,\ldots,N$ are mutually left coprime. In particular, for $r = 1$ and reachable pairs $(A(\omega^k), B(\omega^k)b_k), k = 1,\ldots,N$, system (9.38) is reachable if and only if the polynomials $\det(zI - A(\omega^k))$ are pairwise coprime for $k = 1, \cdots, N$.*

The preceding result generalizes previous results by Lunze (1986) and Brockett and Willems (1974). These authors considered block-circulant control systems in which each subsystem was controlled independently. Thus, they effectively assumed that b was the identity matrix. This excludes several interesting cases, such as leader–follower networks. The more general case treated here is motivated by the more recent work of Brockett (2010), in which the inputs are broadcasted to all nodes of a network. The next example has been studied by Lunze (1986).

Example 9.20. Consider the circulant system

$$x(t+1) = \mathbf{A}x(t) + \mathbf{B}u(t)$$

with independent controls. Let $\mathbf{A}$ and $\mathbf{B}$ be circulant matrices with $A(z) = A_0 - A_1 + A_1(1 + \cdots + z^{N-1})$ and $B(z) = B_0$. For $z = 1$ one obtains $A(1) = A_0 + (N-1)A_1$, and $A(\omega^i) = A_0 - A_1$ for $i = 1,\ldots,N-1$. Therefore, $(A(z), B(z))$ is reachable for all Nth roots of unity z if and only if the two systems $(A_0 - A_1, B_0)$ and $(A_0 + (N-1)A_1, B_0)$ are reachable. This coincides with the result by Lunze (1986). In contrast, if one replaces $\mathbf{B}$ with $\mathbf{B}\beta$, with β a column vector, then Theorem 9.19 implies that the system is not reachable for $N > 2$.

4. Periodic Interconnections. We proceed to discuss briefly an extension to periodic interconnection laws for discrete-time systems. We refer the reader to Bittanti and Colaneri (2010) for background material on periodic linear systems. Thus, consider N reachable and observable discrete-time decoupled node systems

$$x_k(t+1) = \alpha_k x_k(t) + \beta_k v_k(t),$$

$$w_k(t) = \gamma_k x_k(t), \quad k = 1,\dots,N,$$

with system matrices $\alpha_k \in \mathbb{F}^{n_k \times n_k}, \beta_k \in \mathbb{F}^{n_k \times m_k}, \gamma_k \in \mathbb{F}^{p_k \times n_k}$. Equivalently, introducing the global state vectors $x = \mathrm{col}\,(x_1,\dots,x_N) \in \mathbb{F}^n$, and similarly for the input and output vectors v, w, one obtains the global decoupled system

$$x(t+1) = \alpha x(t) + \beta v_k(t),$$

$$w(t) = \gamma x(t),$$

where $\alpha = \mathrm{diag}\,(\alpha_1,\dots,\alpha_N)$, and one argues similarly for β, γ. We emphasize at this point that (α, β, γ) is assumed to be a time-invariant reachable and observable system. One could also investigate periodic node systems, but we will not do so here. Let $G(z) = N_r(z)D_r(z)^{-1} = \gamma(zI - \alpha)^{-1}\beta$ be a right coprime factorization of the global node transfer function. Consider the periodic interconnection law

$$v(t) = A_t w(t) + B_t u(t),$$

$$y(t) = C_t w(t),$$

with $A_t = A_{t+\tau}, B_t = B_{t+\tau}, C_t = C_{t+\tau}$ time-varying matrices of period $\tau \in \mathbb{N}$. The closed-loop, first-order system is then the τ-periodic system

$$\begin{aligned} x(t+1) &= \mathscr{A}_t x(t) + \mathscr{B}_t u(t), \\ y(t) &= \mathscr{C}_t x(t), \end{aligned} \tag{9.42}$$

with $\mathscr{A}_t = \alpha + \beta A_t \gamma, \mathscr{B}_t = \beta B_t, \mathscr{C}_t = C_t \gamma$. For simplicity, let us focus on reachability. Define

$$\bar{A} = \begin{pmatrix} 0 & & & A_\tau \\ A_1 & \ddots & & \\ & \ddots & \ddots & \\ & & A_{\tau-1} & 0 \end{pmatrix}, \quad \bar{B} = \begin{pmatrix} B_\tau & & & \\ & B_1 & & \\ & & \ddots & \\ & & & B_{\tau-1} \end{pmatrix}, \quad \bar{C} = \begin{pmatrix} 0 & & & C_\tau \\ C_1 & \ddots & & \\ & \ddots & \ddots & \\ & & C_{\tau-1} & 0 \end{pmatrix}.$$

The reachability properties of a periodic system are characterized by the reachability properties of the so-called lifted system. We refer the reader to Bittanti and Colaneri (2010) for a discussion on the reachability properties of periodic systems and a proof of the equivalence of the reachability of periodic systems and of the reachability of lifted systems. Let S denote the standard $\tau \times \tau$ circulant matrix defined in (9.40).

Proposition 9.21. *The closed-loop periodic system (9.42) is reachable (observable) if and only if the time-invariant **lifted system***

$$\mathscr{A}_e = S^\top \otimes \alpha + (I \otimes \beta)\overline{A}(I \otimes \gamma), \quad \mathscr{B}_e = (I \otimes \beta)\overline{B}, \quad \mathscr{C}_e = \overline{C}(I \otimes \gamma)$$

is reachable (observable).

Since we are using the Fourier transform, let us assume from now on that the systems are defined over the field $\mathbb{F} = \mathbb{R}$ of real numbers or over the field $\mathbb{F} = \mathbb{C}$ of complex numbers. Define $\omega = e^{2\pi\sqrt{-1}/\tau}, \overline{\omega} = e^{-2\pi\sqrt{-1}/\tau}$, and let Φ denote the $\tau \times \tau$ Fourier matrix. Then

$$(\Phi^* \otimes I)\mathscr{A}_e(\Phi \otimes I) = \Delta(\overline{\omega}) \otimes \alpha + (I \otimes \beta)\hat{A}(I \otimes \gamma), \quad (\Phi^* \otimes I)\mathscr{B}_e = (I \otimes \beta)\hat{B},$$

where $\Delta(\overline{\omega}) = \mathrm{diag}(1, \overline{\omega}, \dots, \overline{\omega}^{\tau-1})$, $\hat{A} = (\Phi^* \otimes I)\overline{A}(\Phi \otimes I)$, and $\hat{B} = (\Phi^* \otimes I)\overline{B}$. Thus $(\mathscr{A}_e, \mathscr{B}_e)$ is reachable if and only if $(\Delta(\overline{\omega}) \otimes \alpha + (I \otimes \beta)\hat{A}(I \otimes \gamma), (I \otimes \beta)\hat{B})$ is reachable. The latter system is obtained by interconnecting the decoupled node system

$$\hat{\alpha} = \mathrm{diag}(\alpha, \overline{\omega}\alpha, \dots, \overline{\omega}^{\tau-1}\alpha), \quad \hat{\beta} = I \otimes \beta, \quad \hat{\gamma} = I \otimes \gamma$$

with the interconnection matrices $\hat{A}$ and $\hat{B}$. The transfer function of the system $(\hat{\alpha}, \hat{\beta}, \hat{\gamma})$ is easily computed from the right coprime factorization $\overline{P}(z)\overline{Q}(z)^{-1}$ of $\gamma(zI - \alpha)^{-1}\beta$ as

$$\hat{G}(z) := (I \otimes \gamma)(zI - \hat{\alpha})^{-1}(I \otimes \beta) = \hat{P}(z)\hat{Q}(z)^{-1},$$

with right coprime factors

$$\hat{Q}(z) = \mathrm{diag}\left(D_r(z), D_r(\omega z), \dots, D_r(\omega^{\tau-1}z)\right),$$

$$\hat{P}(z) = \mathrm{diag}\left(\omega N_r(\omega z), \omega N_r(\omega z), \dots, \omega^{\tau-1}N_r(\omega^{\tau-1}z)\right).$$

Applying Theorem 9.10 we arrive at the following corollary.

Corollary 9.22. *A periodically interconnected network of linear systems (9.42) is reachable if and only if the matrix $\left(\hat{Q}(z) - \hat{A}\hat{P}(z), \hat{B}\right)$ has full row rank for all z.*

9.6 Exercises

1. For matrix fraction representations (9.11), prove that the map

$$Z : X_D \longrightarrow X_{D-NA}, \quad Zf = \pi_{D-NA}(f)$$

defines an isomorphism of $\mathbb{F}[z]$-modules.

2. Let $(\kappa_1, \ldots, \kappa_m)$ denote the reachability indices of (A, B), and let (α, β, γ) denote a minimal realization of a McMillan degree n, SISO transfer function. Show that the reachability indices of the homogeneous network $\mathscr{A} = I_N \otimes \alpha + A \otimes \beta\gamma$, $\mathscr{B} = B \otimes \beta$ are $(n\kappa_1, \ldots, n\kappa_m)$. Deduce that every state x of the network can be reached from 0 in at most $n\kappa_1$ steps.

3. Deduce from Proposition 9.16 the formula

$$\delta(\mathscr{N} \circ h) = \delta(\mathscr{N})\delta(g)$$

for the McMillan degree of the network transfer function of a homogeneous network. Apply this to obtain a new proof of Theorem 9.15.

4. Let $\lambda \in \mathbb{C}$ be nonzero and $p_C(z) := \sum_{j=0}^{N-1} c_{jz}^j \in \mathbb{C}[z]$. A complex λ-**circulant matrix** is a Toeplitz matrix of the form

$$C_\lambda = \begin{pmatrix} c_0 & c_1 & \cdots & c_{N-1} \\ \lambda c_{N-1} & \ddots & \ddots & \vdots \\ \vdots & \ddots & \ddots & c_1 \\ \lambda c_1 & \cdots & \lambda c_{N-1} & c_0 \end{pmatrix}, \quad S_\lambda := \begin{pmatrix} 0 & 1 & \cdots & 0 \\ 0 & \ddots & \ddots & \vdots \\ \vdots & \ddots & \ddots & 1 \\ \lambda & \cdots & 0 & 0 \end{pmatrix}.$$

   a. Prove that a λ-circulant is equal to $p_C(S_\lambda)$. Conversely, each such matrix is a λ-circulant. Deduce that the set of λ-circulants is an Abelian algebra.
   b. Let γ denote an Nth root of λ, i.e., $\gamma^N = \lambda$. Prove that the eigenvectors of a λ-circulant matrix C_λ are the columns of the matrix $\mathrm{diag}(1, \gamma, \ldots, \gamma^{N-1})\Phi$. What are the eigenvalues?

5. Extend Theorem 9.19 to λ-circulant interconnection matrices $\mathbf{A}, \mathbf{B}, \mathbf{C}$.
6. Compute the eigenvalues and eigenvectors of the $N \times N$ circulant matrix

$$\mathbf{A} = \begin{pmatrix} a & & & b \\ b & \ddots & & \\ & \ddots & \ddots & \\ & & b & a \end{pmatrix}.$$

For which $\beta \in \mathbb{R}^N$ is $\dot{x}(t) = \mathbf{A}x(t) + \beta u(t)$ reachable?

9.7 Notes and References

A natural question in network analysis is that of structural controllability, i.e., the classification of all networks that are reachable for a generic choice of coupling parameters. We refer the reader to Dion, Commault and van der Woude (2013) for a survey on this topic. Liu, Slotine and Barabasi (2011) characterized all graphs such that (9.3) is structurally reachable, i.e., if the network is reachable for a generic choice of the nonzero coefficients in the coupling parameters A, b. Their work is based on the characterization by Lin (1974) on structural controllability and makes it possible to estimate the percentage of so-called driver nodes in a network, i.e., those state variables that, after pinning, lead to structurally reachable systems. We emphasize that the work by Liu, Slotine and Barabasi (2011) deals only with the very special situation in which the node systems are first-order integrators $\dot{x}_i = u_i$. For networks of a more general type of node system their conclusions on structural controllability need not hold. Theorem 9.15 enables one to take the first steps in that direction, i.e., to extend the graph-theoretic characterization of the structural controllability of linear systems to homogeneous networks. Characterizing structural controllability for heterogeneous networks is an open problem. Linear systems theory as developed over abstract fields is useful in several areas, including coding theory. This provides further motivation for the algebraic approach taken in this book. Observability and state estimation tasks for linear systems over finite fields are studied in Sundaram and Hadjicostis (2013).

Most of the present chapter is based on Fuhrmann and Helmke (2013). Our central tool for the structural analysis of networks is the equivalence theorem of Rosenbrock and Pugh (1974), extended in Theorem 9.4 to include dynamic couplings. All the subsequent results proven in this chapter follow directly from Theorem 9.4. Theorem 9.15 has been proven by Hara, Hayakawa and Sugata (2009) using complicated canonical form arguments. The early paper by Sontag (1979), which proves the same result in larger generality, has apparently been overlooked.

The key to a deeper understanding of homogeneous networks is the fact that the network transfer functions $\mathcal{N}_G = \mathcal{N} \circ h$, see (9.9), is the composition of the interconnection transfer function $\mathcal{N}(z)$, as defined by (9.8), with the reciprocal $h(z) = 1/g(z)$ of the scalar node transfer function $g(z)$. This simple observation in fact characterizes the transfer functions of homogeneous networks. The problem of characterizing the transfer functions of homogeneous networks is thus equivalent to the question, first raised by J.F. Ritt, of which rational functions can be written as a composition of two rational functions. Ritt (1922) proved that a complex scalar rational function f is the composition of two scalar rational functions if and only if the Galois group (or monodromy group) of f is imprimitive. This shows that a rational function f is the transfer function of a homogeneous network if and only if the Galois group of f is imprimitive. Ritt also solved the decomposition problem for complex polynomials; we refer the reader to Müller (1995) for a classification of the Galois groups of indecomposable polynomials. Of course, a full classification of Galois groups defined by rational functions is difficult and refers to the so-called inverse problem of Galois theory. Even more so, the characterization of imprimitive Galois groups of rational functions remains an open problem. We refer the reader to Brockett (1983) for related work on Galois groups attached to linear feedback systems. Algebraic-geometric characterizations of decomposable rational functions in terms of root loci or associated Bezoutian curves have been obtained by Pakovich (2011).

The necessary conditions for transfer functions of homogeneous networks are easily described in terms of fundamental system invariants. For example, the McMillan degree of rational functions is multiplicative, i.e., $\delta(f_1 \circ f_2) = \delta(f_1)\delta(f_2)$ is satisfied for rational functions f_1, f_2. A similar property holds for the Cauchy index of a rational function. If $F(z) = F(z)^\top$ is a real symmetric $m \times m$ proper rational function, then the **matrix Cauchy index** $CI(F)$, see Bitmead and Anderson (1977), of $F(z)$ is defined and the matrix Cauchy index of the composition $F \circ h$ with $h = 1/g$ satisfies

$$CI(F \circ h) = CI(F) \cdot CI(g) . \tag{9.43}$$

Consequently, this imposes a constraint on the Cauchy index of homogeneous networks. Formula (9.43) follows easily from the well-known interpretation of the Cauchy–Maslov index as the topological degree of the associated rational curve in the Lagrange–Grassmann manifold; see, for example, Byrnes and Duncan (1981). See Helmke (1989) for a generalization of formula (9.43) in terms of Bezoutian matrices.

The question of characterizing homogeneous networks is reminiscent of, but not equivalent to, the classical synthesis problem for electrical circuits, tackled by Brune (1931) and Bott and Duffin (1949). An amazing M.S. thesis by Ladenheim (1948) presents a catalog of 108 circuits that are claimed to realize all biquadratic positive real transfer functions. Despite these efforts, and those of many others, this fundamental circuit synthesis problem remains unsolved to date, but it has attracted some attention lately; see, for example, Kalman (2010), Smith (2002), Jiang and Smith (2011), and Hughes and Smith (2013).

Another interesting topic is the model reduction of networks of systems. There exist several options for doing this, either by reducing the number of nodes and coupling parameters or by order reduction of the node systems. For homogeneous networks, the situation becomes particularly nice. Mullis and Roberts (1976) have shown that if the discrete-time node transfer function $g(z)$ is allpass with respect to the unit circle, then the discrete-time reachability Gramian satisfies $W_c(\mathscr{A}, \mathscr{B}) = W_c(A, B) \otimes W_c(\alpha, \beta)$, and similarly for the observability Gramian. This has been generalized by Koshita, Abe and Kawamata (2007) for bounded real transfer functions $g(z)$ and leads to useful techniques for model reduction by balanced truncation.

Chapter 10
Control of Standard Interconnections

This chapter is devoted to several computational problems from Chapters 4 and 9 that are related to the concepts of reachability and observability. Having characterized the reachability of a linear system, or a network, we turn to the more difficult problem of the computation of controls that steer a system from rest to a prescribed state or, alternatively, steer a system to the zero state. Steering the system to a specified state is not the only option. In many cases, one is satisfied in steering a system to a function of the state, that is, to a partial state. This happens, in particular and in a very natural way, when dealing with systems defined by higher-order difference, or differential, equations.

In principle, Theorem 4.47 solves the problem of computing a steering controller. Its drawback emerges in situations where the pair (A, B) is a result of coupling a large number of nodes of much smaller McMillan degree. In such situations, one would like to make intelligent use of the reachability information of the individual nodes in order to reduce the complexity of computing the inverse of the reduced reachability map $\mathscr{R}$. This topic will be taken up in Section 10.2 for the case of parallel connections.

Some of these computational problems will be studied by focusing on special networks of linear systems. In principle, every algorithm for finding open-loop inputs can be applied to networks of arbitrary size. However, such generic solution approaches quickly face considerable numerical problems. A way out would be to find decentralized computational methods that make use of local information. In such generality, this is a widely open, exciting, and very relevant research area. We restrict ourselves to some of the more common types of interconnection, namely, parallel and series interconnections, and show how these results tie in to polynomial interpolation theory. The solution to these problems involves the embedding of left (right) prime polynomial matrices in unimodular ones. Such unimodular embeddings have been encountered already in computing the inverses of reachability maps. Hence, it comes as no surprise that such embeddings play a central role in the study of open-loop controls.

© Springer International Publishing Switzerland 2015   507
P.A. Fuhrmann, U. Helmke, *The Mathematics of Networks*
of Linear Systems, Universitext, DOI 10.1007/978-3-319-16646-9_10

10.1 Standard Interconnections

It is well known that all Boolean operations can be reduced to a combination of simple ones, such as AND, OR, and NEGATION. Similarly, the operations of parallel, series, and feedback interconnection can be regarded as the building blocks for designing large networks of control systems. These operations are studied next in more detail, in particular, concerning their effects on reachability and observability.

1. Parallel Interconnection. Parallel coupling is certainly one of the easiest ways to interconnect linear systems. The reason for this is that, in this case, there is no direct interaction between the nodes, and the only coupling is via the external input and output. However, controlling such a network is not that easy. Thus, for a finite number N of linear control systems

$$\dot{x}_k(t) = A_k x_k(t) + B_k u_k(t), \quad k = 1,\ldots,N, \tag{10.1}$$

with system matrices $(A_k, B_k) \in \mathbb{R}^{n_k \times (n_k+m)}$, consider the **parallel interconnected system**

$$\dot{x}_1(t) = A_1 x_1(t) + B_1 u(t)$$

$$\vdots \tag{10.2}$$

$$\dot{x}_N(t) = A_N x_N(t) + B_N u(t).$$

The subsystems (A_k, B_k) are referred to as the **node systems** of system (10.2). Of course, the difference between the two systems (10.1) and (10.2) is that the first one is completely decoupled with independent input functions while in the second one all subsystems share the same input. One can also say that the input $u(t) \in \mathbb{R}^m$ is **broadcasted** to all systems. When is such a system reachable and how does one control the system? The first question is of course a classical one whose answer goes back to the beginnings of linear state-space theory in the 1970s. Let us begin with the formal definition of parallel connection of two input-output systems.

Definition 10.1. Consider systems Σ_i with proper rational transfer functions $G_i(z) = \left(\begin{array}{c|c} A_i & B_i \\ \hline C_i & D_i \end{array} \right) \in \mathbb{F}(z)^{p \times m}$ and state spaces $\mathscr{X}_i$, $i = 1, 2$. The **parallel coupling**, $\Sigma_1 \vee \Sigma_2$, of the two systems is defined in the state space $\mathscr{X}_1 \oplus \mathscr{X}_2$ by the system

$$\begin{pmatrix} x_{t+1}^{(1)} \\ x_{t+1}^{(2)} \end{pmatrix} = \begin{pmatrix} A_1 & 0 \\ 0 & A_2 \end{pmatrix} \begin{pmatrix} x_t^{(1)} \\ x_t^{(2)} \end{pmatrix} + \begin{pmatrix} B_1 \\ B_2 \end{pmatrix} u_t,$$

$$y_t = \begin{pmatrix} C_1 & C_2 \end{pmatrix} \begin{pmatrix} x_t^{(1)} \\ x_t^{(2)} \end{pmatrix} + (D_1 + D_2) u_t.$$

Schematically, this is expressed by the following diagram:

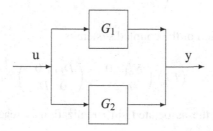

The characterization of the reachability and observability properties of the parallel connection of linear systems, stated below, is due to Fuhrmann (1975).

Theorem 10.2. *Let $G_1(z), G_2(z) \in \mathbb{F}(z)^{p \times m}$ be proper transfer functions with minimal realizations $G_i(z) = \left[\begin{array}{c|c} A_i & B_i \\ \hline C_i & D_i \end{array}\right]$ and associated state spaces $\mathcal{X}_i$, $i = 1, 2$.*

1. For parallel coupling the transfer function is

$$G(z) = G_1(z) + G_2(z) = \left[\begin{array}{cc|c} A_1 & 0 & B_1 \\ 0 & A_2 & B_2 \\ \hline C_1 & C_2 & D_1 + D_2 \end{array}\right]. \tag{10.3}$$

2. Let $G_1(z)$ and $G_2(z)$ be proper rational matrices admitting the following coprime factorizations:

$$G_i(z) = N_{r,i}(z) D_{r,i}(z)^{-1} = D_{\ell,i}(z)^{-1} N_{\ell,i}(z), \qquad i = 1, 2. \tag{10.4}$$

(a) The parallel coupling of the shift realizations associated with $N_{r,2}(z) D_{r,2}(z)^{-1}$ and $N_{r,1}(z) D_{r,1}(z)^{-1}$ is reachable if and only if $D_{r,1}(z)$ and $D_{r,2}(z)$ are left coprime.

(b) The parallel coupling of the shift realizations associated with $D_{\ell,2}(z)^{-1} N_{\ell,2}(z)$ and $D_{\ell,1}(z)^{-1} N_{\ell,1}(z)$ is observable if and only if $D_{\ell,1}(z)$ and $D_{\ell,2}(z)$ are right coprime.

(c) The parallel coupling of the shift realizations associated with $D_{\ell,2}(z)^{-1} N_{\ell,2}(z)$ and $D_{\ell,1}(z)^{-1} N_{\ell,1}(z)$ is minimal if and only if $D_{\ell,1}(z)$ and $D_{\ell,2}(z)$ are right coprime and $D_{r,1}(z)$ and $D_{r,2}(z)$ are left coprime. Equivalently,

$$\delta(G_1 + G_2) = \delta(G_1) + \delta(G_2).$$

Proof. 1. The realization (10.3) is obvious.
     Proof of 2.

(a) The coupling matrices are

$$A = \begin{pmatrix} 0 & 0 \\ 0 & 0 \end{pmatrix}, B = \begin{pmatrix} I \\ I \end{pmatrix}, C = \begin{pmatrix} I & I \end{pmatrix}.$$

The transfer function of the coupled system is

$$G(z) = \begin{pmatrix} I & I \end{pmatrix} \begin{pmatrix} N_{r,1} & 0 \\ 0 & N_{r,2} \end{pmatrix} \begin{pmatrix} D_{r,1} & 0 \\ 0 & D_{r,2} \end{pmatrix}^{-1} \begin{pmatrix} I \\ I \end{pmatrix}.$$

By Theorem 9.8, the associated shift realization is reachable if and only if $\begin{pmatrix} D_{r,1} & 0 \\ 0 & D_{r,2} \end{pmatrix}$ and $\begin{pmatrix} I \\ I \end{pmatrix}$ are left coprime. Observing that

$$\begin{pmatrix} D_{r,1} & 0 & I \\ 0 & D_{r,2} & I \end{pmatrix} = \begin{pmatrix} I & 0 \\ I & I \end{pmatrix} \begin{pmatrix} D_{r,1} & 0 & I \\ -D_{r,1} & D_{r,2} & 0 \end{pmatrix},$$

this is the case if and only if $D_{r,1}(z)$ and $D_{r,2}(z)$ are left coprime. This proves the claim.

(b) The proof is analogous to that of the previous part. It also follows from (a) by duality considerations.

(c) The parallel coupling is minimal if and only if it is both reachable and observable, which is equivalent to both coprimeness conditions. This proves the first claim. The parallel coupling of the shift realizations of $G_1(z)$ and $G_2(z)$ has a state space of dimension $\delta(G_1) + \delta(G_2)$. Since this realization is both reachable and observable, it has dimension $\delta(G_1 + G_2)$, and the result follows. ∎

Next, the analysis of the parallel connection of two nodes is extended to the case of the parallel coupling of N node systems Σ_i, where the node transfer functions have the coprime factorizations

$$G_i(z) = D_{\ell,i}(z)^{-1} N_{\ell,i}(z) = N_{r,i}(z) D_{r,i}(z)^{-1}.$$

For this, one needs the concept of mutual coprimeness introduced in Definition 3.13. Note that, in the matrix case, mutual left coprimeness is a stronger condition than pairwise left coprimeness. We prove the following resultant-type characterization of mutual left coprimeness; a similar characterization is valid for mutual right coprimeness.

Proposition 10.3. *Nonsingular polynomial matrices $D_i(z) \in \mathbb{F}[z]^{m \times m}$, $i = 1,\ldots,N$, are mutually left coprime if and only if the polynomial matrix*

$$\mathscr{D}_N(z) := \begin{pmatrix} -D_1(z) & D_2(z) \\ & \ddots & \ddots \\ & & -D_{N-1}(z) & D_N(z) \end{pmatrix} \tag{10.5}$$

is left prime.

Proof. For $N = 2$ the result is obvious, so let us assume that $N \geq 3$. We first prove sufficiency. If $\mathscr{D}_N(z)$ is left prime, then so is $\mathscr{D}_{N-1}(z)$. Thus, by induction, we can assume that $D_1(z), \ldots, D_{N-1}(z)$ are mutually left coprime. We next show that $D_N(z)$ is left coprime with the least common right multiple of $D_1(z), \ldots, D_{N-1}(z)$. To verify this, partition $\mathscr{D}_N$ into

$$\mathscr{D}_N = \begin{pmatrix} \mathscr{D}_{N-1} & 0 \\ D'(z) & D_N(z) \end{pmatrix},$$

with $D'(z) = \begin{pmatrix} 0 & \cdots & 0 & -D_{N-1}(z) \end{pmatrix}$. By the left coprimeness of $\mathscr{D}_{N-1}(z)$, there exists a unimodular polynomial matrix $U(z) \in GL_{m(N-1)}(\mathbb{F}[z])$, with $\mathscr{D}_{N-1}U(z) = (I_{m(N-2)}, 0)$. Let $U'(z) := \mathrm{col}\,(U_1(z), \cdots, U_{N-1}(z))$ denote the last block column vector of $U(z)$, with $U_i(z) \in \mathbb{F}[z]^{m \times m}$. The entries of $U'(z)$ satisfy

$$\Delta(z) := D_1(z)U_1(z) = \cdots = D_{N-1}(z)U_{N-1}(z).$$

In particular, $\Delta(z)$ is a common left multiple of $D_1(z), \ldots, D_{N-1}(z)$. Since $U'(z)$ is right prime, $\Delta(z) = l.c.r.m.\{D_1(z), \ldots, D_{N-1}(z)\}$ is in fact a least common right multiple of $D_1(z), \ldots, D_{N-1}(z)$. One easily checks $D'(z)U(z) = (B(z), -\Delta(z))$ for a suitable polynomial matrix $B(z) \in \mathbb{F}[z]^{m \times m(N-2)}$. This implies the identity

$$\begin{pmatrix} \mathscr{D}_{N-1}(z) & 0 \\ D'(z) & D_N(z) \end{pmatrix} \begin{pmatrix} U(z) & 0 \\ 0 & I_m \end{pmatrix} = \begin{pmatrix} I_{m(N-2)} & 0 & 0 \\ B(z) & -\Delta(z) & D_N(z) \end{pmatrix}.$$

After an elementary row operation on both sides of this equation one can reduce $B(z)$ to the zero matrix. This shows that $\mathscr{D}_N$ is unimodularly equivalent to

$$\begin{pmatrix} I_{m(N-2)} & 0 & 0 \\ 0 & -\Delta(z) & D_N(z) \end{pmatrix}.$$

Thus, by the left primeness of $\mathscr{D}_N(z)$, the polynomial matrices $D_N(z)$ and $\Delta(z)$ are left prime. In the same way one verifies that $D_i(z)$ and $l.c.r.m.\{D_j(z); j \neq i\}$ are left prime. This completes the proof that $D_1(z), \ldots, D_N(z)$ are mutually left coprime.

For the converse direction, assume that $D_1(z), \ldots, D_N(z)$ are mutually left coprime. Suppose that $\mathscr{D}_N$ is not left prime, i.e., there exists $\lambda \in \overline{\mathbb{F}}$ and a nonzero row vector $\xi = (\xi_1, \ldots, \xi_{N-1}) \in \overline{\mathbb{F}}^{1 \times m(N-1)}$ such that $\xi \mathscr{D}_N(\lambda) = 0$. Equivalently, $\xi_1 D_1(\lambda) = 0$, $\xi_{N-1} D_N(\lambda) = 0$, and $(\xi_{i-1} - \xi_i)D_i(\lambda) = 0$ for $i = 2, \ldots, N-1$. Let $\Delta(z) := l.c.r.m.\{D_1(z), \ldots, D_{N-1}(z)\}$. Then there are polynomial matrices $\overline{D}_i(z)$ that satisfy $D_i(z)\overline{D}_i(z) = \Delta(z)$ for $i = 1, \ldots, N-1$. Thus we conclude

$$\xi_{N-1} D_N(\lambda) = 0 \quad \text{and} \quad \xi_i \Delta(\lambda) = 0, \quad \text{for} \quad i = 2, \ldots, N-1.$$

Since $D_N(z)$ and $\Delta(z)$ are left coprime, this implies $\xi_{N-1} = 0$. Proceeding induc-tively, using the mutual coprimeness of $D_1(z), \ldots, D_{N-1}(z)$, we conclude that $\xi_1 = \cdots = \xi_{N-2} = 0$, and therefore $\xi = 0$. But this is a contradiction.   ■

An independent, though indirect, proof of the previous resultantlike characteri-zation of mutual left coprimeness is given in Theorem 10.19 based on the analysis of parallel coupling of N systems, to which we turn next. Let $\Sigma_1 \vee \cdots \vee \Sigma_i$ denote the parallel coupling of the first i systems. Define the **parallel coupling** of the Σ_i inductively by

$$\Sigma_1 \vee \cdots \vee \Sigma_{i+1} = (\Sigma_1 \vee \cdots \vee \Sigma_i) \vee \Sigma_{i+1}.$$

Clearly, parallel coupling is an associative operation. In this case, the intercon-nection matrices are

$$A = \begin{pmatrix} 0 & \cdots & 0 \\ \vdots & \ddots & \vdots \\ 0 & \cdots & 0 \end{pmatrix}, \quad B = \begin{pmatrix} I \\ \vdots \\ I \end{pmatrix}, \quad C = (I \cdots I).$$

We apply Theorem 9.8 and assume the transfer functions of the node systems have the coprime factorizations (10.4). Let

$$D_r(z) = \begin{pmatrix} D_{r,1}(z) & & \\ & \ddots & \\ & & D_{r,N}(z) \end{pmatrix}, \quad N_r(z) = \begin{pmatrix} N_{r,1}(z) & & \\ & \ddots & \\ & & N_{r,N}(z) \end{pmatrix}.$$

The reachability of the parallel interconnected system is equivalent to the left primeness of

$$(D_r(z), B) = \begin{pmatrix} D_{r,1}(z) & & & I \\ & \ddots & & \vdots \\ & & D_{r,N}(z) & I \end{pmatrix}. \tag{10.6}$$

After suitable elementary row operations, it follows that this condition is in turn equivalent to the left primeness of the resultant-type matrix

$$\begin{pmatrix} -D_{r,1}(z) & D_{r,2}(z) & & \\ & \ddots & \ddots & \\ & & -D_{r,N-1}(z) & D_{r,N}(z) \end{pmatrix}. \tag{10.7}$$

Proceeding as in the case for $N = 2$ and using Proposition 10.3 we arrive at the following theorem.

Theorem 10.4. *Let Σ_i, $i = 1, \ldots, N$, be systems with proper rational transfer functions $G_i(z) \in \mathbb{F}(z)^{p \times m}$ and coprime factorizations (10.4). A necessary and sufficient condition for $\Sigma_1 \vee \cdots \vee \Sigma_N$ to be reachable is that the $D_{r,i}(z)$ are mutually left coprime.* ∎

2. Series Interconnection. Our next topic is the study of linear systems connected in series. Our approach is inductive; we first analyze the series connection of two systems and then move on to the general case. Schematically, this is expressed by the following diagram:

Definition 10.5. Let Σ_i, $i = 1, 2$, be systems with proper rational transfer functions $G_i(z) = \left[\begin{array}{c|c} A_i & B_i \\ \hline C_i & D_i \end{array}\right]$ and state spaces $\mathscr{X}_i$, $i = 1, 2$. Let $G_1(z) \in \mathbb{F}(z)^{k \times m}$ and $G_2(z) \in \mathbb{F}(z)^{p \times k}$. The **series coupling** $\Sigma_1 \wedge \Sigma_2$ of the two systems is defined in the state space $\mathscr{X}_1 \oplus \mathscr{X}_2$ as the system

$$\begin{pmatrix} x_{t+1}^{(1)} \\ x_{t+1}^{(2)} \end{pmatrix} = \begin{pmatrix} A_1 & 0 \\ B_2 C_1 & A_2 \end{pmatrix} \begin{pmatrix} x_t^{(1)} \\ x_t^{(2)} \end{pmatrix} + \begin{pmatrix} B_1 \\ B_2 D_1 \end{pmatrix} u_t,$$

$$y_t = \begin{pmatrix} D_2 C_1 & C_2 \end{pmatrix} \begin{pmatrix} x_t^{(1)} \\ x_t^{(2)} \end{pmatrix} + D_2 D_1 u_t.$$

The associated transfer function is

$$G(z) := G_2(z) G_1(z) = \left[\begin{array}{cc|c} A_1 & 0 & B_1 \\ B_2 C_1 & A_2 & B_2 D_1 \\ \hline D_2 C_1 & C_2 & D_2 D_1 \end{array}\right]. \tag{10.8}$$

It is not easy to read off from a state-space representation of the transfer function of a series connection meaningful conditions for reachability and observability. This changes dramatically if one uses coprime factorization representations of the transfer functions $G_i(z)$.

Theorem 10.6. *Let $G_1(z) \in \mathbb{F}(z)^{q \times m}$ and $G_2(z) \in \mathbb{F}(z)^{p \times q}$ be proper rational with minimal realizations $G_i(z) = \left[\begin{array}{c|c} A_i & B_i \\ \hline C_i & D_i \end{array}\right]$ in state spaces $\mathscr{X}_i$, $i = 1, 2$, respectively.*

1. Let $G_1(z) \in \mathbb{F}(z)^{k \times m}$ and $G_2(z) \in \mathbb{F}(z)^{p \times k}$ have the coprime matrix fraction representations

$$G_1(z) = D_{\ell,1}(z)^{-1} N_{\ell,1}(z) = N_{r,1}(z) D_{r,1}(z)^{-1},$$
$$G_2(z) = D_{\ell,2}(z)^{-1} N_{\ell,2}(z) = N_{r,2}(z) D_{r,2}(z)^{-1}.$$

Then

$$\delta(G_2 G_1) \leq \delta(G_1) + \delta(G_2). \tag{10.9}$$

2. *The series coupling of the shift realizations associated with $N_{r,2}(z)D_{r,2}(z)^{-1}$ and $N_{r,1}(z)D_{r,1}(z)^{-1}$ is reachable if and only if the polynomial matrices $N_{r,1}(z)$ and $D_{r,2}(z)$ are left coprime.*
3. *The series coupling of the shift realizations associated with $D_{\ell,2}(z)^{-1}N_{\ell,2}(z)$ and $D_{\ell,1}(z)^{-1}N_{\ell,1}(z)$ is observable if and only if the polynomial matrices $N_{\ell,2}(z)$ and $D_{\ell,1}(z)$ are right coprime.*
4. *The equality*

$$\delta(G_2 G_1) = \delta(G_1) + \delta(G_2)$$

is satisfied if and only if $N_{\ell,2}(z), D_{\ell,1}(z)$ are right coprime and $N_{r,1}(z)$ and $D_{r,2}(z)$ are left coprime.

Proof. 1. The realization (10.8) implies (10.9).
2. The proof is deduced from our results about the general interconnections of node systems. For the nodes $G_1(z)$ and $G_2(z)$, the coupling matrices are

$$A = \begin{pmatrix} 0 & 0 \\ I & 0 \end{pmatrix}, B = \begin{pmatrix} I \\ 0 \end{pmatrix}, C = \begin{pmatrix} 0 & I \end{pmatrix}.$$

By (9.11), the transfer function of the interconnected system is

$$G(z) = \begin{pmatrix} 0 & I \end{pmatrix} \begin{pmatrix} N_{r,1} & 0 \\ 0 & N_{r,2} \end{pmatrix} \left(\begin{pmatrix} D_{r,1} & 0 \\ 0 & D_{r,2} \end{pmatrix} - \begin{pmatrix} 0 & 0 \\ I & 0 \end{pmatrix} \begin{pmatrix} N_{r,1} & 0 \\ 0 & N_{r,2} \end{pmatrix} \right)^{-1} \begin{pmatrix} I \\ 0 \end{pmatrix}$$

$$= \begin{pmatrix} 0 & I \end{pmatrix} \begin{pmatrix} N_{r,1} & 0 \\ 0 & N_{r,2} \end{pmatrix} \begin{pmatrix} D_{r,1} & 0 \\ -N_{r,1} & D_{r,2} \end{pmatrix}^{-1} \begin{pmatrix} I \\ 0 \end{pmatrix}.$$

Applying Theorem 9.8, the associated shift realization is reachable if and only if the polynomial matrix

$$\begin{pmatrix} D_{r,1} & 0 & I \\ -N_{r,1} & D_{r,2} & 0 \end{pmatrix}$$

is left prime. This is clearly the case if and only if $N_{r,1}(z)$ and $D_{r,2}(z)$ are left coprime.
3. The proof is analogous to the proof of part 2.
4. Follows from parts 2 and 3.   ∎

The intuitive interpretation for, say, the left coprimenes of $N_{r,1}(z)$ and $D_{r,2}(z)$ is that there are no cancellations between the left zeros of $G_1(z)$ and the right poles of $G_2(z)$. We proceed next to examine the slightly more general problem of the series coupling of N node systems Σ_i in terms of the coprime factorizations of the nodes

$$G_i(z) = D_{\ell,i}(z)^{-1} N_{\ell,i}(z) = N_{r,i}(z) D_{r,i}(z)^{-1}. \tag{10.10}$$

First, let us introduce some notation. Let $\Sigma_1 \wedge \cdots \wedge \Sigma_i$ denote the series coupling of the first i systems. The series coupling of the systems Σ_i is inductively defined by

$$\Sigma_1 \wedge \cdots \wedge \Sigma_{i+1} = (\Sigma_1 \wedge \cdots \wedge \Sigma_i) \wedge \Sigma_{i+1}.$$

Clearly, series coupling is an associative operation. In this case, the interconnection matrices are

$$A = \begin{pmatrix} 0 & \cdots & \cdots & 0 \\ I & \ddots & & \vdots \\ & \ddots & \ddots & \vdots \\ & & I & 0 \end{pmatrix}, \quad B = \begin{pmatrix} I \\ 0 \\ \vdots \\ 0 \end{pmatrix}, \quad C = \begin{pmatrix} 0 & \cdots & 0 & I \end{pmatrix}.$$

Our intention is to apply Proposition 9.8. To this end, let

$$D_r(z) = \mathrm{diag}\,(D_{r,1}(z), \ldots, D_{r,N}(z)), \; N_r(z) = \mathrm{diag}\,(N_{r,1}(z), \ldots, N_{r,N}(z)).$$

The reachability of the series interconnected system is equivalent to the left primeness of $(D_r(z) - A N_r(z), B)$ or, equivalently, of

$$\begin{pmatrix} -N_{r,1}(z) & D_{r,2}(z) & & \\ & \ddots & \ddots & \\ & & -N_{r,N-1}(z) & D_{r,N}(z) \end{pmatrix}. \tag{10.11}$$

We conclude by stating the following theorem.

Theorem 10.7. *Let $G_i(z) \in \mathbb{F}[z]^{p \times m}$ be the transfer functions of the systems Σ_i, $i = 1, \ldots, N$, with coprime factorizations (10.10). A necessary and sufficient condition for $\Sigma_1 \wedge \cdots \wedge \Sigma_N$ to be reachable is that the matrix in (10.11) is left prime.* ∎

From this characterization one derives a necessary condition for the reachability of series connections, i.e., that all pairs $N_{r,i}(z), D_{r,i+1}(z)$ are left coprime for $i = 1, \ldots, N-1$. However, these individual coprimeness conditions are not sufficient because the zeros of a node might cancel a pole even if the nodes are not direct neighbors. To obtain workable necessary and sufficient conditions, we proceed recursively.

Theorem 10.8. *Consider the systems Σ_i, $i = 1,\ldots,N$, with right coprime factorizations $N_{r,i}(z)D_{r,i}(z)^{-1}$ of the node transfer functions. Let $\overline{\mathbf{N}}_i(z)\overline{\mathbf{D}}_i(z)^{-1}$ denote the right coprime factorization of the transfer function $\mathbf{G}_i(z) = G_i(z)\cdots G_1(z)$ of $\Sigma_1 \wedge \cdots \wedge \Sigma_i$, and let $\hat{\mathbf{N}}_i(z)\hat{D}_{i+1}(z)^{-1}$ be a right coprime factorization of $D_{r,i+1}(z)^{-1}\overline{\mathbf{N}}_i(z)$, $i < N$. Define, inductively,*

$$\overline{\mathbf{N}}_{i+1}(z) = N_{r,i+1}(z)\hat{\mathbf{N}}_i(z),$$

$$\overline{\mathbf{D}}_{i+1}(z) = \overline{\mathbf{D}}_i(z)\hat{D}_{i+1}(z).$$

Then a necessary and sufficient condition for $\Sigma_1 \wedge \cdots \wedge \Sigma_N$ to be reachable is that one of the following equivalent conditions is satisfied:

1. The polynomial matrix (10.11) is left prime.
2. The polynomial matrices $\overline{\mathbf{N}}_i(z)$ and $D_{r,i+1}(z)$ are left coprime, for $i = 1,\ldots,N-1$.

Proof. The proof is by induction. Clearly, $\Sigma_1 \wedge \cdots \wedge \Sigma_N$ is reachable if and only if $\Sigma_1 \wedge \cdots \wedge \Sigma_i$ is reachable for all $i = 1,\ldots N$. For $i = 1$ it is assumed. Assuming that $\Sigma_1 \wedge \cdots \wedge \Sigma_i$ is reachable and that $\overline{\mathbf{N}}_i(z)$ and $D_{r,i+1}(z)$ are left coprime, the reachability of $\Sigma_1 \wedge \cdots \wedge \Sigma_{i+1}$ follows by applying Theorem 10.6.

$\overline{\mathbf{N}}_i(z)$ and $\overline{D}_{i+1}(z)$ are embedded in a doubly coprime factorization as follows:

$$\begin{pmatrix} \mathbf{Y}_{i+1}(z) & \mathbf{X}_i(z) \\ -\overline{\mathbf{N}}_i(z) & D_{r,i+1}(z) \end{pmatrix} \begin{pmatrix} \hat{D}_{i+1}(z) & -\hat{\mathbf{X}}_i(z) \\ \hat{\mathbf{N}}_i(z) & \hat{Y}_{i+1}(z) \end{pmatrix} = \begin{pmatrix} I & 0 \\ 0 & I \end{pmatrix}.$$

Next, compute

$$\begin{pmatrix} -\overline{\mathbf{N}}_i(z) & D_{r,i+1}(z) & 0 \\ 0 & -N_{r,i+1}(z) & D_{r,i+2}(z) \end{pmatrix} \begin{pmatrix} \hat{D}_{i+1}(z) & -\hat{\mathbf{X}}_i(z) & 0 \\ \hat{\mathbf{N}}_i(z) & \hat{Y}_{i+1}(z) & 0 \\ 0 & 0 & I \end{pmatrix}$$

$$= \begin{pmatrix} 0 & I & 0 \\ -\overline{\mathbf{N}}_{i+1}(z) & * & D_{r,i+2}(z) \end{pmatrix}.$$

Clearly, the last polynomial matrix is left prime if and only if the polynomial matrices $\overline{\mathbf{N}}_{i+1}(z)$ and $D_{r,i+2}(z)$ are left coprime. The result follows by induction. ∎

A similar result can be obtained for the observability of the series connection using left coprime factorizations of the nodes' transfer functions. The details are omitted.

3. Feedback Interconnection. Finally, our aim is to characterize the reachability and observability of the **feedback connection** of a system with transfer function $G(z)$ and a controller with transfer function $K(z)$. This case was first treated by Callier and Nahum (1975). However, the subsequent analysis seems to be more complete. The full system equations are

$$G(e_1 + y_2) = y_1,$$
$$K(e_2 + y_1) = y_2. \tag{10.12}$$

The following diagram describes the feedback connection:

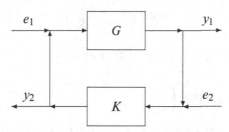

Proposition 10.9. *Let $G(z) \in \mathbb{F}(z)^{p \times m}$ be proper rational with a proper controller $K(z) \in \mathbb{F}(z)^{m \times p}$ such that $I - G(z)K(z)$ is properly invertible. Assume $G(z)$ and $K(z)$ have the following coprime factorizations:*

$$G(z) = D_\ell(z)^{-1} N_\ell(z) = N_r(z) D_r(z)^{-1},$$
$$K(z) = S_\ell(z)^{-1} R_\ell(z) = R_r(z) S_r(z)^{-1}. \tag{10.13}$$

Then:

1. The closed-loop transfer function $\Phi(z)$ from $\begin{pmatrix} e_1 \\ e_2 \end{pmatrix}$ to $\begin{pmatrix} y_1 \\ y_2 \end{pmatrix}$ is

$$\Phi = \begin{pmatrix} I & -G \\ -K & I \end{pmatrix}^{-1} \begin{pmatrix} G & 0 \\ 0 & K \end{pmatrix} = \begin{pmatrix} G & 0 \\ 0 & K \end{pmatrix} \begin{pmatrix} I & -K \\ -G & I \end{pmatrix}^{-1}. \tag{10.14}$$

The transfer function $\Phi(z)$ has the following coprime factorizations:

$$\Phi(z) = \begin{pmatrix} D_\ell & -N_\ell \\ -R_\ell & S_\ell \end{pmatrix}^{-1} \begin{pmatrix} N_\ell & 0 \\ 0 & R_\ell \end{pmatrix} = \begin{pmatrix} N_r & 0 \\ 0 & R_r \end{pmatrix} \begin{pmatrix} D_r & -R_r \\ -N_r & S_r \end{pmatrix}^{-1}. \tag{10.15}$$

2. The intertwining relation

$$\begin{pmatrix} N_\ell & 0 \\ 0 & R_\ell \end{pmatrix} \begin{pmatrix} D_r & -R_r \\ -N_r & S_r \end{pmatrix} = \begin{pmatrix} D_\ell & -N_\ell \\ -R_\ell & S_\ell \end{pmatrix} \begin{pmatrix} N_r & 0 \\ 0 & R_r \end{pmatrix}$$

is satisfied.
3. The polynomial system matrices associated with the coprime factorizations (10.15) are FSE.

4. *The following equations are true:*

$$\det\begin{pmatrix} D_\ell & -N_\ell \\ -R_\ell & S_\ell \end{pmatrix} = \det(D_\ell S_r - N_\ell R_r) = \det(S_\ell D_r - R_\ell N_r)$$

$$= \det\begin{pmatrix} D_r & -R_r \\ -N_r & S_r \end{pmatrix}.$$

5. *The closed-loop transfer function G_f from e_1 to y_1 is*

$$G_f(z) = \begin{pmatrix} I & 0 \end{pmatrix} \Phi(z) \begin{pmatrix} I \\ 0 \end{pmatrix}$$

$$= (I - G(z)K(z))^{-1}G(z) = G(z)(I - K(z)G(z))^{-1},$$

(10.16)

with the following equivalent representations:

$$G_f(z) = S_r(z)(D_\ell(z)S_r(z) - N_\ell(z)R_r(z))^{-1}N_\ell(z)$$

$$= N_r(z)(S_\ell(z)D_r(z) - R_\ell(z)N_r(z))^{-1}S_\ell(z)$$

$$= \begin{pmatrix} N_r & 0 \end{pmatrix} \begin{pmatrix} D_r & -R_r \\ -N_r & S_r \end{pmatrix}^{-1} \begin{pmatrix} I \\ 0 \end{pmatrix}$$

(10.17)

$$= \begin{pmatrix} I & 0 \end{pmatrix} \begin{pmatrix} D_\ell & -N_\ell \\ -R_\ell & S_\ell \end{pmatrix}^{-1} \begin{pmatrix} N_\ell \\ 0 \end{pmatrix}.$$

Proof. 1. The system equations (10.12) can be written in matrix form as

$$\begin{pmatrix} I & -G \\ -K & I \end{pmatrix} \begin{pmatrix} y_1 \\ y_2 \end{pmatrix} = \begin{pmatrix} G & 0 \\ 0 & K \end{pmatrix} \begin{pmatrix} e_1 \\ e_2 \end{pmatrix}.$$

It is easily calculated that

$$\begin{pmatrix} I & -G \\ -K & I \end{pmatrix}^{-1} = \begin{pmatrix} (I-GK)^{-1} & G(I-KG)^{-1} \\ (I-KG)^{-1}K & (I-KG)^{-1} \end{pmatrix},$$

and hence

$$\begin{pmatrix} y_1 \\ y_2 \end{pmatrix} = \begin{pmatrix} (I-GK)^{-1}G & G(I-KG)^{-1}K \\ (I-KG)^{-1}KG & (I-KG)^{-1}K \end{pmatrix} \begin{pmatrix} e_1 \\ e_2 \end{pmatrix}.$$

Representations (10.15) follow by substituting the coprime factorizations (10.13) into equation (10.14). It is easily checked that the factorizations in (10.15) are coprime factorizations.

2. Follows from (10.15).
3. Follows from

$$
\begin{pmatrix} N_\ell & 0 & 0 & 0 \\ 0 & R_\ell & 0 & 0 \\ 0 & 0 & I & 0 \\ 0 & 0 & 0 & I \end{pmatrix}
\begin{pmatrix} D_r & -R_r & -I & 0 \\ -N_r & S_r & 0 & -I \\ N_r & 0 & 0 & 0 \\ 0 & R_r & 0 & 0 \end{pmatrix}
=
\begin{pmatrix} D_\ell & -N_\ell & -N_\ell & 0 \\ -R_\ell & S_\ell & 0 & -R_\ell \\ I & 0 & 0 & 0 \\ 0 & I & 0 & 0 \end{pmatrix}
\begin{pmatrix} N_r & 0 & 0 & 0 \\ 0 & R_r & 0 & 0 \\ 0 & 0 & I & 0 \\ 0 & 0 & 0 & I \end{pmatrix} .
$$

The relevant coprimeness conditions follow from the assumed coprimeness of factorizations (10.13).

4. First, note that from the coprime factorizations (10.13) it follows that $\det D_\ell = \det D_r$ and $\det S_\ell = \det S_r$. Next, compute

$$
\begin{pmatrix} I & 0 \\ R_\ell D_\ell^{-1} & I \end{pmatrix}
\begin{pmatrix} D_\ell & -N_\ell \\ -R_\ell & S_\ell \end{pmatrix}
=
\begin{pmatrix} D_\ell & -N_\ell \\ 0 & S_\ell - R_\ell N_r D_r^{-1} \end{pmatrix}
=
\begin{pmatrix} D_\ell & -N_\ell \\ 0 & (S_\ell D_r - R_\ell N_r) D_r^{-1} \end{pmatrix} .
$$

By application of the multiplication rule of determinants, it follows that

$$
\det \begin{pmatrix} D_\ell & -N_\ell \\ -R_\ell & S_\ell \end{pmatrix} = \det D_\ell \cdot \det(S_\ell D_r - R_\ell N_r) \cdot \det D_r^{-1}
$$

$$
= \det(S_\ell D_r - R_\ell N_r).
$$

Similarly,

$$
\begin{pmatrix} D_\ell & -N_\ell \\ -R_\ell & S_\ell \end{pmatrix}
\begin{pmatrix} I & 0 \\ S_\ell^{-1} R_\ell & I \end{pmatrix}
=
\begin{pmatrix} D_\ell - N_\ell S_\ell^{-1} R_\ell & -N_\ell \\ 0 & S_\ell \end{pmatrix}
=
\begin{pmatrix} (D_\ell S_r - N_\ell R_r) S_r^{-1} & -N_\ell \\ 0 & S_\ell \end{pmatrix} ,
$$

which implies $\det \begin{pmatrix} D_\ell & -N_\ell \\ -R_\ell & S_\ell \end{pmatrix} = \det(D_\ell S_r - N_\ell R_r)$. The other equalities are derived analogously.

5. Substituting representations (10.13) into (10.16), one obtains for the closed-loop transfer function $G_f(z)$ the representations

$$
G_f(z) = S_r(z)(D_\ell(z)S_r(z) - N_\ell(z)R_r(z))^{-1} N_\ell(z)
$$
$$
= N_r(z)(S_\ell(z)D_r(z) - R_\ell(z)N_r(z))^{-1} S_\ell(z).
$$

To obtain the third representation in (10.17), we use (10.15) and compute

$$
G_f(z) = (I\ 0) \begin{pmatrix} N_r & 0 \\ 0 & R_r \end{pmatrix} \begin{pmatrix} D_r & -R_r \\ -N_r & S_r \end{pmatrix}^{-1} \begin{pmatrix} I \\ 0 \end{pmatrix}
$$

$$
= (N_r\ 0) \begin{pmatrix} D_r & -R_r \\ -N_r & S_r \end{pmatrix}^{-1} \begin{pmatrix} I \\ 0 \end{pmatrix} .
$$

The last representation in (10.17) is similarly derived. ∎

Next, we derive the basic characterizations for the reachability and observability of feedback interconnections.

Theorem 10.10. *Apply the same notation as in Proposition 10.9.*

1. *The following statements are equivalent:*

   (a) *The shift realization associated with each of the representations in (10.17) is reachable.*
   (b) $N_r(z)$ *and* $S_r(z)$ *are left coprime.*
   (c) $D_\ell(z)S_\ell(z)$ *and* $N_\ell(z)$ *are left coprime.*
   (d) $R_\ell(z)N_r(z)$ *and* $S_\ell(z)$ *are left coprime.*

2. *The following statements are equivalent:*

   (a) *The shift realization associated with each of the representations in (10.17) is observable.*
   (b) $N_\ell(z)$ *and* $S_\ell(z)$ *are right coprime.*
   (c) $S_\ell(z)D_r(z)$ *and* $N_r(z)$ *are right coprime.*
   (d) $N_\ell(z)R_r(z)$ *and* $S_r(z)$ *are right coprime.*

Proof. 1. Follows by applying Theorem 9.8 to representations (10.17). For example, the reachability of the shift realization is equivalent to the left coprimeness of

$$\begin{pmatrix} D_r(z) & -R_r(z) \\ -N_r(z) & S_r(z) \end{pmatrix} \text{ and } \begin{pmatrix} I \\ 0 \end{pmatrix}$$

or to the left primeness of

$$\begin{pmatrix} D_r(z) & -R_r(z) & I \\ -N_r(z) & S_r(z) & 0 \end{pmatrix}.$$

However, the last polynomial matrix is left prime if and only if the polynomial matrices $N_r(z)$ and $S_r(z)$ are left coprime. The equivalence of reachability to the other coprimeness conditions is similarly derived.
2. The proof proceeds on the same line as in part 1.   ∎

Even a cursory comparison of Theorem 9.8 and Theorems 10.9–10.10 shows a great similarity. This indicates that the dynamic output feedback problem is equivalent to the analysis of networks of systems under dynamic coupling.

4. Clouds. So far, reachability and observability have been characterized for the series and parallel connection of finitely many transfer functions. Our next aim is to extend this to cases where the transfer functions of subsystems are network transfer functions of homogeneous networks. Such systems of homogeneous networks, interconnected in series or parallel, are called **clouds**. The results are stated only for reachability; they also hold, mutatis mutandis, for observability.

4.1. Series Connection. Consider, for $i = 1, 2$, homogeneous networks Σ_i with scalar strictly proper rational node transfer functions $g_i(z) = p_i(z)/q_i(z)$ of McMillan degree n_i and strictly proper interconnection transfer functions

$$\mathcal{N}_1(z) = C_1(zI_{N_1} - A_1)^{-1}B_1 \in \mathbb{F}(z)^{p \times m},$$

$$\mathcal{N}_2(z) = C_2(zI_{N_2} - A_2)^{-1}B_2 \in \mathbb{F}(z)^{q \times p}.$$

The series connection of Σ_1 with Σ_2 then defines a heterogeneous network $\Sigma_1 \wedge \Sigma_2$ with interconnection transfer function $\mathcal{N}(z) = C(zI - A)^{-1}B$ and

$$A = \begin{pmatrix} A_1 & 0 \\ B_2C_1 & A_2 \end{pmatrix}, \quad B = \begin{pmatrix} B_1 \\ 0 \end{pmatrix}, \quad C = \begin{pmatrix} 0 & C_2 \end{pmatrix}.$$

The transfer function of the series connection network $\Sigma_1 \wedge \Sigma_2$ is then

$$\mathcal{N}_G(z) = C(H(z) - A)^{-1}B,$$

where

$$H(z) = \mathrm{diag}\left(\frac{q_1(z)}{p_1(z)}I_{N_1}, \frac{q_2(z)}{p_2(z)}I_{N_2}\right).$$

Applying Theorem 9.10, one concludes that the series network $\Sigma_1 \wedge \Sigma_2$ is reachable if and only if the matrix has rank

$$\mathrm{rank}\begin{pmatrix} q_1(z)I - p_1(z)A_1 & 0 & p_1(z)B_1 \\ -p_2(z)B_2C_1 & q_2(z) - p_2(z)A_2 & 0 \end{pmatrix} = N_1 + N_2$$

for all $z \in \overline{\mathbb{F}}$. One can deduce a more easily manageable sufficient condition for controllability. First, note that the reachability of $(A_1, B_1), (A_2, B_2)$ is clearly necessary for the reachability of $\Sigma_1 \wedge \Sigma_2$. Moreover, the coprimeness of the polynomials $p_1(z)$ and $\det(q_2(z) - p_2(z)A_2)$ is also a necessary condition. One can prove, under the restricted assumption that $p \leq m$, that $\Sigma_1 \wedge \Sigma_2$ is reachable provided the poles of Σ_2 are disjoint from the zeros of Σ_1.

Theorem 10.11. *Let $p \leq m$, and assume that (A_1, B_1) and (A_2, B_2) are reachable. The series connection $\Sigma_1 \wedge \Sigma_2$ is reachable, provided the matrix*

$$\begin{pmatrix} q_1(z)I - p_1(z)A_1 & -p_1(z)B_1 \\ C_1 & 0 \end{pmatrix}$$

has full row rank for all roots of $\det(q_2(z)I - p_2(z)A_2)$.

Proof. Let $N_i(z)D_i(z)^{-1}$ be right coprime factorizations of the transfer functions of Σ_i. By the coprimeness characterization of reachability for series connections, it follows that $\Sigma_1 \wedge \Sigma_2$ is reachable provided $N_1(z), D_2(z)$ are left coprime, i.e., the $p \times (p+m)$ matrix $(D_2(z), N_1(z))$ has full row rank for all $z \in \overline{\mathbb{F}}$. Since $p \le m$, a sufficient condition for this is that $N_1(z)$ has full row rank for all z that are zeros of $\det D_2(z)$. Proposition 9.16 implies that $\det D_2(z) = \det(q_2(z)I - p_2(z)A_2)$. Moreover, $N_1(z)$ has full row rank at z if and only if Σ_1 does not have a zero at z, i.e., the rank condition

$$\text{rank} \begin{pmatrix} q_1(z)I - p_1(z)A_1 & -p_1(z)B_1 \\ C_1 & 0 \end{pmatrix} = N_1 + p$$

is satisfied. The result follows. ∎

The preceding result is easily extended to the series connection of a finite number of homogeneous networks. The obvious, inductive, proof of the next theorem is omitted.

Theorem 10.12. *Let $p_i \le m_i$ for $i = 1, \ldots, N-1$, and assume that (A_i, B_i, C_i) are reachable for $i = 1, \ldots, N$. Let Σ_i denote a homogeneous network with strictly proper node transfer function $g_i(z) = p_i(z)/q_i(z)$ and interconnection transfer function $\mathcal{N}_i(z) = C_i(zI_{N_i} - A_i)^{-1}B_i$. The series connection $\Sigma_1 \wedge \cdots \wedge \Sigma_N$ is reachable provided*

$$\text{rank} \begin{pmatrix} q_i(z)I_{N_i} - p_i(z)A_i & -p_i(z)B_i \\ C_i & 0 \end{pmatrix} = N_i + p_i$$

for all z, with

$$\prod_{j=i+1}^{N} \det(q_j(z)I - p_j(z)A_j) = 0$$

and $i = 1, \ldots, N-1$.

4.2. Parallel and Feedback Connections. Similar results can be obtained for the parallel connection $\Sigma_1 \vee \cdots \vee \Sigma_N$ of homogeneous networks Σ_i. We use the same notation as was used earlier.

Theorem 10.13. *Assume that (A_i, B_i) are reachable for $i = 1, \ldots, N$. Let Σ_i denote a homogeneous network with strictly proper node transfer function $g_i(z) = p_i(z)/q_i(z)$ and interconnection transfer function $\mathcal{N}_i(z) = (zI_{N_i} - A_i)^{-1}B_i$. The parallel connection $\Sigma_1 \vee \cdots \vee \Sigma_N$ is reachable provided the polynomials $\det(q_j(z)I - p_j(z)A_j)$ are pairwise coprime, $j = 1, \ldots, N$.*

Proof. The parallel connection of N systems Σ_i is reachable if and only if $D_1, \cdots, D_N$ are mutually coprime. This condition is satisfied provided the polynomials $\det D_1, \cdots, \det D_N$ are pairwise coprime. The result follows from Proposition 9.16. ∎

Despite its rather restrictive coprimeness assumption, this result is quite useful in designing concrete heterogeneous networks of systems that are reachable. In fact, the theorem gives precise sufficient conditions for the construction of node transfer functions $g_1, \ldots, g_N$ such that the parallel sum $\Sigma_1 \vee \cdots \vee \Sigma_N$ is reachable for interconnections defined by identical reachable pairs $(A_1, B_1) = \cdots = (A_N, B_N)$.

The feedback interconnection of two homogeneous systems Σ_1, Σ_2 with node transfer functions g_1, g_2 is described by the interconnection transfer function $\mathcal{N}(z) = C(zI - A)^{-1}B$, with

$$
A = \begin{pmatrix} A_1 & B_1 C_2 \\ B_2 C_1 & A_2 \end{pmatrix}, \quad B = \begin{pmatrix} B_1 \\ 0 \end{pmatrix} \quad C = (C_1 \ 0).
$$

By Theorem 9.10, the feedback interconnection is reachable if and only if

$$
\mathrm{rank} \begin{pmatrix} q_1(z)I - p_1(z)A_1 & -p_1(z)B_1 C_2 & p_1(z)B_1 \\ -p_2(z)B_2 C_1 & q_2(z) - p_2(z)A_2 & 0 \end{pmatrix} = N_1 + N_2
$$

for all z. By adding a suitable multiple of the last block column to the second one, it follows that this condition is equivalent to

$$
\begin{pmatrix} q_1(z)I - p_1(z)A_1 & 0 & p_1(z)B_1 \\ -p_2(z)B_2 C_1 & q_2(z) - p_2(z)A_2 & 0 \end{pmatrix}
$$

being full row rank for all z. But this in turn is equivalent to the series connection $\Sigma_1 \wedge \Sigma_2$ being reachable. This proves the following theorem.

Theorem 10.14. *The feedback interconnection of two homogeneous networks Σ_1, Σ_2 is reachable if and only if the series connection $\Sigma_1 \wedge \Sigma_2$ is reachable.*

10.2   Open-Loop Controls for Parallel Connections

While Theorem 10.4 effectively solves the reachability problem for parallel interconnection of N systems

$$
\dot{x}_1(t) = A_1 x_1(t) + B_1 u(t)
$$

$$
\vdots \tag{10.18}
$$

$$
\dot{x}_N(t) = A_N x_N(t) + B_N u(t),
$$

the question remains as to how one can compute a control function $u : [0, T] \longrightarrow \mathbb{R}^m$ for the parallel connection of N systems that steers the zero state $x = 0$ in finite time $T > 0$ to a terminal state $x^* = (x_1^*, \ldots, x_N^*) \in \mathbb{R}^{nN}$. Even more, one would like to be able to compute such a control u for (10.18) using knowledge of local controls $u_k^* : [0, T] \longrightarrow \mathbb{R}^m$ that steer the node systems (10.1) from the zero state to the local terminal states x_k^*, $k = 1, \ldots, N$. Although such a problem looks daunting at first sight, we will develop in this chapter an approach that enables us to tackle such problems. To explain the basic idea of open-loop control of parallel interconnections, let us begin with a rather informal discussion of discrete-time, single-input, single-output (SISO) systems because these are much easier to analyze than the general multivariable case. We first establish some useful notation.

Single-Input Systems. Consider an nth-order, discrete-time, single-input system

$$x(t+1) = Ax(t) + bu(t), \quad t = 0, 1, 2, \ldots.$$

An input sequence $u_0, u_1, \ldots u_{M-1}$ steers the zero state of the system into a desired state $x(M) = x^* \in \mathbb{R}^n$ if and only if $u(A)b = x^*$ for the scalar polynomial

$$\mathbf{u}(z) = \sum_{k=0}^{M-1} u_{M-1-k} z^k.$$

We refer to $\mathbf{u}(z)$ as the **input polynomial** for x^*. The reachability of (A, b) is equivalent to the existence of such an input polynomial. Moreover, the unique input polynomial of minimal degree $n - 1$ is

$$\mathbf{u}^*(z) = (1, \ldots, z^{n-1})(b, Ab, \ldots, A^{n-1}b)^{-1} x^*.$$

Now consider the interconnected single-input systems

$$\begin{aligned} x_1(t+1) &= A_1 x_1(t) + b_1 u(t), \\ x_2(t+1) &= A_2 x_2(t) + b_2 u(t), \end{aligned} \tag{10.19}$$

where $A_i \in \mathbb{R}^{n_i \times n_i}$ and $b_i \in \mathbb{R}^{n_i}$ for $i = 1, 2$. Assume that system (10.19) is reachable, i.e., that (A_1, b_1) and (A_2, b_2) are both reachable and the characteristic polynomials

$$q_1(z) = \det(zI - A_1), \quad q_2(z) = \det(zI - A_2)$$

are coprime. For local state vectors $x_1^* \in \mathbb{R}^{n_1}, x_2^* \in \mathbb{R}^{n_2}$ there exist unique input polynomials $\mathbf{u}_1^*(z)$ and $\mathbf{u}_2^*(z)$ of degrees bounded by $n_1 - 1$ and $n_2 - 1$, respectively, with

$$\mathbf{u}_1^*(A_1)b_1 = x_1^*, \quad \mathbf{u}_2^*(A_2)b_2 = x_2^*.$$

By the coprimeness of the characteristic polynomials $q_1(z)$ and $q_2(z)$, there exist unique polynomials $c(z), d(z) \in \mathbb{R}[z]$ of degrees $\deg c(z) < n_2$ and $\deg d(z) < n_1$ that satisfy the Bezout identity

$$c(z)q_1(z) + d(z)q_2(z) = 1.$$

Note that, by the Cayley–Hamilton theorem, $q_i(A_i) = 0$ for $i = 1,2$. Thus $d(A_1)q_2(A_1) = I = c(A_2)q_1(A_2)$. Consider the polynomial

$$\mathbf{u}(z) = d(z)q_2(z)\mathbf{u}_1^*(z) + c(z)q_1(z)\mathbf{u}_2^*(z).$$

Substituting matrices A_i into $u(z)$ and using the Bezout equation one obtains

$$\mathbf{u}(A_1)b_1 = d(A_1)q_2(A_1)\mathbf{u}_1^*(A_1)b_1 = x_1^*,$$
$$\mathbf{u}(A_2)b_2 = c(A_2)q_1(A_2)\mathbf{u}_2^*(A_2)b_2 = x_2^*.$$

Thus the (reverse) coefficients of the polynomial $\mathbf{u}(z)$ yield an input sequence that steers (10.19) to the desired states.

The preceding solution approach can be extended in several directions. First, note that it is very closely related to the Chinese remainder theorem, stated in Chapter 2. Second, the degree of $\mathbf{u}(z)$ can be upper bounded by $n_1 + n_2$ by replacing the factors $d(z)\mathbf{u}_1^*(z)$ and $c(z)\mathbf{u}_2^*(z)$ with their remainders modulo $q_1(z)$ and $q_2(z)$, respectively. Third, the construction of the control can be easily carried out for finitely many parallel connected reachable SISO systems

$$x_1(t+1) = A_1 x_1(t) + b_1 u(t)$$

$$\vdots \tag{10.20}$$

$$x_N(t+1) = A_N x_N(t) + b_N u(t).$$

Assume that for each of the N local subsystems (A_j, b_j) local control sequences u_j are known that steer the zero state to a desired terminal state x_j^*. How can one compute from such local controls a single global input sequence u that steers all subsystems simultaneously to the desired terminal states? Ideally, one would like to obtain a formula such as $u = \sum_{j=1}^N f_j u_j$ that expresses the desired control as a weighted sum of local controls, where the weights f_j are suitable filter operators that act on the respective local input. To answer this question one needs just a little bit more notation.

Let $q_j(z) = \det(zI - A_j)$ denote the characteristic polynomial of A_j, and define $\hat{q}_j(z) = \prod_{i \neq j} q_i(z)$. Assume that the pairs (A_j, b_j) are reachable for each $j = 1, \ldots, N$. The reachability of the parallel connection (10.20) is then equivalent to the coprimeness of $q_j, \hat{q}_j$ for each $j = 1, \ldots, N$. Let the polynomials $c_j(z)$ and $d_j(z)$ denote the unique solutions of the Bezout equations

$$c_j(z)q_j(z) + d_j(z)\hat{q}_j(z) = 1, \quad j = 1, \ldots, N \tag{10.21}$$

with degrees $\deg d_j < n_j$. Proceeding as previously, one stores the input sequences $u_0, \ldots, u_{M-1}$ for controlling (10.20) as coefficients of the associated input polynomial $\mathbf{u}(z) = \sum_{j=0}^{M-1} u_{M-j-1} z^j$. Our basic control result for (10.20) is stated as follows.

Theorem 10.15. *Assume that (10.20) is reachable. For local state vectors $x_1^*, \ldots, x_N^*$, let $\mathbf{u}_1(z), \ldots, \mathbf{u}_N(z) \in \mathbb{R}[z]$ denote input polynomials satisfying*

$$\mathbf{u}_1(A_1)b_1 = x_1^*, \ \ldots, \ \mathbf{u}_N(A_N)b_N = x_N^*.$$

Then the input polynomial for (10.20)

$$\mathbf{u}(z) = \sum_{j=1}^{N} d_j(z)\hat{q}_j(z)\mathbf{u}_j(z) \tag{10.22}$$

satisfies $\mathbf{u}(A_j)b_j = x_j^*$ for all j.

Proof. From the Bezout equations (10.21) one obtains that $d_k(A_k)\hat{q}_k(A_k) = I$ and $d_j(A_k)\hat{q}_j(A_k) = 0$ for $j \neq k$. This implies

$$\mathbf{u}(A_k)b_k = \sum_{j=1}^{N} d_j(A_k)\hat{q}_j(A_k)\mathbf{u}_j(A_k)b_k = d_k(A_k)\hat{q}_k(A_k)\mathbf{u}_k(A_k)b_k = \mathbf{u}_k(A_k)b_k = x_k^*. \ \blacksquare$$

The preceding computations of open-loop controls for discrete-time systems are easily extended to continuous-time systems. Consider, for example, the parallel connection of higher-order systems

$$q_k(\frac{d}{dt})z_k(t) = u(t), \quad k = 1, \ldots, N, \tag{10.23}$$

with monic real polynomials $q_k(z)$ of degree n_k. Assume that the system is reachable, i.e., that the polynomials $q_k(z)$ and $\hat{q}_k(z) := \prod_{j \neq k} q_j(z)$ are coprime for each k. Then the product $q(z) = q_1(z) \cdots q_N(z)$ is the least common multiple and $n := \deg q(z) = n_1 + \cdots + n_N$. The next result illustrates the so-called **flatness approach** for solving open-loop control problems of (10.23).

Theorem 10.16. *Let $y(t) \in C^n([0,T])$ and $(z,u) := (z_1, \ldots, z_N, u) : [0,T] \longrightarrow \mathbb{R}^{n+1}$ be a pair of functions defined as*

$$z_k(t) = \hat{q}_k(\frac{d}{dt})y(t), \quad u(t) = q(\frac{d}{dt})y(t).$$

Then $z_k(t) \in C^{n_k}([0,T]), u(t) \in C([0,T])$ is a solution of (10.23). Conversely, all solutions of (10.23) are obtained in this way. The function y satisfies $y(0) = \cdots = y^{(n-1)}(0) = 0$ if and only if $z_k^{(j)}(0) = 0$ for $j = 0, \ldots, n_k - 1$; $k = 1, \ldots, N$.

Proof. From $q_k(\frac{d}{dt})z_k = q(\frac{d}{dt})y(t) = u(t)$ one obtains that $(z_k(t), u(t))$ solves (10.23). $\hat{q}_1, \ldots, \hat{q}_N$ are coprime. Thus $(\hat{q}_1(z), \ldots, z^{n_1-1}\hat{q}_1(z), \ldots, \hat{q}_N(z), \ldots,$ $z^{n_N-1}\hat{q}_N(z))$ is a basis of the space of polynomials of degree $< n$. This implies that the initial data conditions on the derivatives of z_k at $t = 0$ correspond uniquely, and linearly, to conditions on the first n derivatives of y at $t = 0$. For a solution z_k, u of (10.23) let $y \in C^n([0,T])$ denote the unique solution of $q(\frac{d}{dt})y = u$. Then both functions z_k and $\hat{q}_k(\frac{d}{dt})y$ satisfy (10.23) and have the same initial conditions. The result follows.   ∎

Example 10.17. Inspired by Rouchon (2005), we discuss the computation of open-loop controls for the parallel connection of harmonic oscillators

$$\ddot{z}_k(t) + \omega_k^2 z_k(t) = u(t), \quad k = 1, \ldots, N. \tag{10.24}$$

Suppose one wants to steer a system in finite time $T > 0$ from the zero initial state at $t = 0$ to $\mathrm{col}(Le_1, \ldots, Le_1)$, where $L > 0$ and e_1 denotes the first standard basis vector in $\mathbb{R}^2$. Assuming that the frequencies $0 < \omega_k^2 \neq \omega_l^2$ are pairwise distinct, it follows that (10.24) is reachable. To compute the control $u(t)$, consider the coprime polynomials

$$q_k(z) = z^2 + \omega_k^2, \quad \text{and} \quad \hat{q}_k(z) = \prod_{j \neq k}(z^2 + \omega_j^2),$$

with least common multiple $q(z) = \prod_{j=1}^N(z^2 + \omega_j^2)$. By coprimeness, there exist unique polynomials $c_k(z), d_k(z) \in \mathbb{R}[z]$ with $\deg d_k < 2$ and

$$c_k(z)q_k(z) + d_k(z)\hat{q}_k(z) = 1.$$

In fact, $d_k(z)$ is equal to the constant polynomial $d_k(z) = \prod_{j \neq k}(\omega_j^2 - \omega_k^2)^{-1}$. Next, for a $2N$-times continuously differentiable function $y : [0,T] \longrightarrow \mathbb{R}$, consider the functions

$$z_k(t) = \hat{q}_k(\frac{d}{dt})y(t), \quad u(t) = q(\frac{d}{dt})y(t). \tag{10.25}$$

It follows from $q_k(\frac{d}{dt})z_k(t) = q(\frac{d}{dt})y(t) = u(t)$ that $z_k(t)$ is a solution of (10.24). The control conditions $z_k(0) = \dot{z}_k(0) = 0$ and $z_k(T) = L, \dot{z}_k(T) = 0$ then easily translate as the interpolation conditions ($k = 1, \ldots, N$)

$$\hat{q}_k(\frac{d}{dt})y(0) = 0, \quad \hat{q}_k(\frac{d}{dt})\dot{y}(0) = 0, \quad \hat{q}_k(\frac{d}{dt})y(T) = L, \quad \hat{q}_k(\frac{d}{dt})\dot{y}(T) = 0. \tag{10.26}$$

The N even polynomials $\hat{q}_1(z), \ldots, \hat{q}_N(z)$ of degree $2N - 2$ are linearly independent and thus form a basis of the space of all even polynomials of degree $\leq 2N - 2$. Therefore, conditions (10.26) are equivalent to

$$y(0) = 0, \quad y(T) = L, \quad y^{(i)}(0) = y^{(i)}(T) = 0, \quad i = 1, \ldots, 2N - 1.$$

Thus every solution of this Hermite interpolation problem solves, via (10.25), the control problem.

Multivariable Systems. Theorem 10.15 presents a simple formula (10.22) for the global control of (10.20) in terms of a weighted sum of local controls for the decoupled systems (A_j, b_j). The polynomial weights $d_j(z)\hat{q}_j(z)$ are independent of the desired local states x_j^* and can be computed by solving N Bezout equations. Moreover, by replacing $\mathbf{u}(z)$ with the remainder polynomial $\mathbf{u}^*(z)$ obtained by division with remainders $\mathbf{u}(z) = m(z)q_1(z) \cdots q_N(z) + \mathbf{u}^*(z)$, one obtains a global control of minimal degree bounded by $n_1 + \cdots + n_N$. All this requires only elementary operations from polynomial algebra (Euclidean algorithm, division with remainders) for which efficient implementations are known. At this stage, the reader may wonder why we did not simply invert the reachability matrix of the parallel interconnection scheme (10.19) to compute the minimal-length input sequence. The reason for this is that we wanted to find controls that are computed from local controls u_i^*, and the inversion of the global reachability matrix does not allow one to do so. In fact, our solution, although possibly of higher degree than necessary, has an advantage in comparison with the global inversion method. The solutions $c_j(z), d_j(z)$ to the N Bezout equations must be computed only once. By choosing solutions of the Bezout equations in an appropriate way, for example by using Newton interpolation, may even pave the way toward recursive computations of $\mathbf{u}(z)$. This scheme is therefore favorable if the number of parallel connected systems changes or, more generally, if the effects of changing the coupling parameters in an interconnected system are of interest. The multivariable case poses more difficulties and will be treated next in full generality, including a discussion of related interpolation problems.

In accordance with the general philosophy that decomposing a complex problem into a finite number of easier ones reduces the complexity of the solution, we approach the terminal state problem for a parallel connection of node systems by using local minimal controllers as the building block for a global controller. Assume that $(A_i, B_i) \in \mathbb{F}^{n_i \times n_i} \times \mathbb{F}^{n_i \times m}, i = 1, \ldots, N$ are reachable pairs. The reachability map $\mathscr{R}_{(A_i, B_i)} : \mathbb{F}[z]^m \longrightarrow X_{zI-A_i}$ is defined as in (4.54). From

$$\operatorname{Ker} \mathscr{R}_{(A_i, B_i)} = D_i(z)\mathbb{F}[z]^m$$

one deduces the coprime factorization

$$(zI - A_i)^{-1}B_i = N_i(z)D_i(z)^{-1}. \tag{10.27}$$

The map $\mathscr{R}_i : X_{D_i} \longrightarrow X_{zI-A_i}$, i.e., the **reduced reachability map** for the ith node, is

$$\mathscr{R}_i u_i = \pi_{zI-A_i} B_i u_i.$$

For $\xi_i \in X_{zI-A_i}$, the polynomial $u_i = \mathscr{R}_i^{-1} \xi_i$ is the minimal steering controller. Our principal aim in what follows is to compute the minimal steering controller for the parallel connection of the nodes based on the availability of the minimal steering controllers for the individual nodes. The parallel connection of the node pairs $\{(A_i, B_i) \mid i = 1, \ldots, N\}$ is the pair (A, B), defined by

$$A = \begin{pmatrix} A_1 & & \\ & \ddots & \\ & & A_N \end{pmatrix}, \quad B = \begin{pmatrix} B_1 \\ \vdots \\ B_N \end{pmatrix}. \tag{10.28}$$

In general, the pair (A, B) need not be reachable. When it is, the computation of a control sequence steering the system from the origin to a terminal state follows Theorem 4.47. However, to significantly minimize computations, one wants to utilize all the available information on the individual nodes. The relevant assumptions on the node system are collected as follows.

Assumption A. Let $(A_i, B_i) \in \mathbb{F}^{n_i \times n_i} \times \mathbb{F}^{n_i \times m}, i = 1, \ldots, N$, be reachable pairs. Assume that $N_i(z)D_i(z)^{-1}$ is a right coprime factorization of $(zI - A_i)^{-1}B_i$. Assume further that the intertwining relation $B_i D_i(z) = (zI - A_i)N_i(z)$ is embedded in the doubly coprime factorization

$$\begin{pmatrix} Y_i(z) & X_i(z) \\ -B_i & zI - A_i \end{pmatrix} \begin{pmatrix} D_i(z) & -\overline{X}_i(z) \\ N_i(z) & \overline{Y}_i(z) \end{pmatrix} = \begin{pmatrix} I & 0 \\ 0 & I \end{pmatrix}. \tag{10.29}$$

Define the pair (A, B) by (10.28) and

$$\begin{aligned}
D(z) &= l.c.r.m.\{D_i(z) | i = 1, \ldots, N\}, \\
L_i(z) &= D_i^{-1}(z)D(z), \\
\overline{L}_i(z) &= l.c.r.m.\{D_j(z) | j \neq i\}, \\
\overline{D}_i(z) &= \overline{L}_i(z)^{-1}D(z).
\end{aligned} \tag{10.30}$$

To enhance readability, we split our main result into three theorems.

Theorem 10.18. *Under assumption A, the following assertions are true:*

1. The factorizations

$$D(z) = D_i(z)L_i(z) = \overline{L}_i(z)\overline{D}_i(z) \tag{10.31}$$

are satisfied by $L_1(z), \ldots, L_N(z)$ and $\overline{L}_1(z), \ldots, \overline{L}_N(z)$ right coprime and left coprime, respectively.

2. *Defining*

$$H(z) = \begin{pmatrix} H_1(z) \\ \vdots \\ H_N(z) \end{pmatrix} = \begin{pmatrix} N_1(z)L_1(z) \\ \vdots \\ N_N(z)L_N(z) \end{pmatrix},$$

the intertwining relation

$$BD(z) = (zI - A)H(z)$$

is satisfied by

$$(zI - A)^{-1}B = H(z)D(z)^{-1}$$

coprime factorizations.
3. *The direct sum decomposition*

$$X_{zI-A} = X_{zI-A_1} \oplus \cdots \oplus X_{zI-A_N}$$

holds.
4. *There exist uniquely determined polynomial matrices* $\Phi_i(z)$, $i = 0,\ldots,N$, *for which*

$$I = D(z)\Phi_0(z) + \sum_{i=1}^{N} \overline{L}_i(z)\Phi_i(z). \qquad (10.32)$$

5. *The intertwining relation (10.31) is embeddable in the following doubly coprime factorization:*

$$\begin{pmatrix} \overline{E}_i(z) & F_i(z) \\ -\overline{L}_i(z) & D_i(z) \end{pmatrix} \begin{pmatrix} \overline{D}_i(z) & -\overline{F}_i(z) \\ L_i(z) & E_i(z) \end{pmatrix} = \begin{pmatrix} I & 0 \\ 0 & I \end{pmatrix}. \qquad (10.33)$$

Proof. 1. Factorizations (10.31) follow from (10.30). The right coprimeness of the $L_i(z)$ follows from the fact that $D(z)$ is the least common right multiple of the $D_i(z)$.
2. Since

$$(zI - A_i)^{-1}B_i = N_i(z)D_i(z)^{-1} \qquad (10.34)$$

are coprime factorizations, this implies the intertwining relation

$$B_iD_i(z) = (zI - A_i)N_i(z), \qquad (10.35)$$

which by the coprimeness assumption implies the embeddability of the inter-twining relation in the doubly coprime factorization (10.29). Without loss of generality, one can assume that $X_i(z)(zI - A_i)^{-1}$ and, hence, $D_i(z)^{-1}\overline{X}_i(z)$, are strictly proper. In particular, X_i is constant. Multiplying (10.35) on the right by $L_i(z)$, one obtains

$$BD(z) = (zI - A)H(z). \tag{10.36}$$

The right coprimeness of $N_i(z), D_i(z)$ implies the right coprimeness of $H(z), D(z)$. To see this, one observes that, by construction, $N_i(z)$ and $D_i(z)$ are right coprime. Hence, there exist polynomial matrices $X_i(z)$ and $Y_i(z)$ for which

$$X_i(z)N_i(z) + Y_i(z)D_i(z) = I, \qquad i = 1, \ldots, N. \tag{10.37}$$

By the right coprimeness of the $L_i(z)$, there exist polynomial matrices $Z_i(z)$ for which

$$\sum_{i=1}^{N} Z_i(z)L_i(z) = I. \tag{10.38}$$

From the Bezout equations (10.37) and (10.38), and recalling that $H_i(z) = N_i(z)L_i(z)$, one obtains $L_i(z) = X_i(z)H_i(z) + Y_i(z)D(z)$. Hence,

$$I = \sum_{i=1}^{N} Z_i(z)L_i(z) = \sum_{i=1}^{N} Z_i(z)(X_i(z)H_i(z) + Y_i(z)D(z))$$

$$= \sum_{i=1}^{N} (Z_i(z)X_i(z))H_i(z) + (\sum_{i=1}^{N} Z_i(z)Y_i(z))D(z),$$

which shows the right coprimeness of $H(z)$ and $D(z)$.
3. Follows from the fact that A has the diagonal representation (10.28).
4. From the direct sum decomposition $\mathbb{F}[z]^m = X_D \oplus D\mathbb{F}[z]$ it follows, using (3.18), that

$$\mathbb{F}[z]^m = \bigoplus_{i=1}^{N} \overline{L}_i X_{\overline{D}_i} \oplus D\mathbb{F}[z]^m.$$

This implies the existence of polynomial matrices $\Phi_i(z)$, $i = 0, \ldots, N$ that satisfy (10.32).
5. Follows from the fact that $\overline{D}_i(z)$ and $L_i(z)$ are left coprime and $\overline{L}_i(z)$ and $D_i(z)$ are right coprime. ∎

Theorem 10.19. *Under assumption A, the following statements are equivalent:*

1. The pair (A, B), defined in (10.28), is reachable.
2. $zI - A, B$ are left coprime.

3. *The* **reduced reachability map** $\mathscr{R} : X_D \longrightarrow X_{zI-A}$, *defined by*

$$\mathscr{R}u = \pi_{zI-A}Bu, \qquad u(z) \in X_D, \tag{10.39}$$

is an $\mathbb{F}[z]$-*isomorphism.*
4. $\dim X_D = n = \sum_{i=1}^{N} n_i.$
5. *Factorizations (10.31) are skew-prime factorizations.*
6. *The polynomial matrices* $D_i(z)$, $i = 1, \ldots, N$, *are mutually left coprime.*
7. *The polynomial matrices* $\overline{D}_i(z)$, $i = 1, \ldots, N$, *are mutually right coprime.*
8. *There are direct sum representations*

$$X^D = X^{\overline{D}_1} \oplus \cdots \oplus X^{\overline{D}_N}$$

and

$$X_D = \overline{L}_1 X_{\overline{D}_1} \oplus \cdots \oplus \overline{L}_N X_{\overline{D}_N}. \tag{10.40}$$

9. *The polynomial resultant-type matrix (10.5) is left prime.*

Proof. $1 \Leftrightarrow 2$:   Follows from Corollary 4.12.
$1 \Leftrightarrow 3$:   Since (A,B) is assumed to be a reachable pair, the reachability map $\mathscr{R}$ is a surjective homomorphism. This implies that $\dim X_D \geq n = \sum_{i=1}^{N} n_i = \sum_{i=1}^{N} \dim X_{D_i}$. On the other hand, since $D(z) = l.c.r.m.\{D_i(z) | i = 1, \ldots, N\}$, one must have $\dim X_D \leq \sum_{i=1}^{N} \dim X_{D_i}$. The two equalities, taken together, imply the equality

$$\dim X_D = \sum_{i=1}^{N} \dim X_{D_i}.$$

$3 \Leftrightarrow 4$: $\mathscr{R}$ is, by the left coprimeness of $(zI - A), B$, a surjective homomorphism. Since $\dim X_D = n$, it is also injective and, hence, an isomorphism.
     The equivalence of statements 4, 5, 6, 7, and 8 follows from Theorem 3.19.
$1 \Leftrightarrow 9$:   Follows from (10.6).   ∎

Theorem 10.20. *Assume that* $F_i(z)D_i(z)^{-1}$ *and* $\overline{D}_i(z)^{-1}\overline{F}_i(z)$, *arising from the doubly coprime factorization (10.33), are strictly proper. Under assumption A, the following assertions are true:*

1. *Define the maps* $\mathscr{U}_i : X_{\overline{D}_i} \longrightarrow X_{D_i}$, $\mathscr{R}_i : X_{D_i} \longrightarrow X_{zI-A_i}$ *and* $\mathscr{V}_i : X_{\overline{D}_i} \longrightarrow X_{zI-A_i}$, $i = 1, \ldots, N$, *by*

$$\mathcal{U}_i f_i = \pi_{D_i} \overline{L}_i f_i,$$

$$\mathcal{R}_i g_i = \pi_{zI-A_i} B_i g_i,$$

$$\mathcal{V}_i f_i = \pi_{zI-A_i} B_i \overline{L}_i f_i.$$

All three maps are $\mathbb{F}[z]$-*isomorphisms, and the following diagram is commutative:*

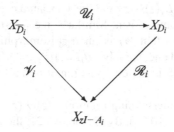

The inverses of these maps are

$$\mathcal{U}_i^{-1} g_i = \pi_{\overline{D}_i} \overline{F}_i g_i,$$

$$\mathcal{R}_i^{-1} \xi_i = \pi_{D_i} \overline{X}_i \xi_i, \qquad (10.41)$$

$$\mathcal{V}_i^{-1} \xi_i = \pi_{\overline{D}_i} \overline{F}_i \overline{X}_i \xi_i.$$

2. *Let the reduced reachability map* $\mathcal{R} : X_D \longrightarrow X_{zI-A}$ *be defined by (10.39). Let* $u(z) \in X_D$ *have the representation* $u(z) = \sum_{j=1}^{N} \overline{L}_j \pi_{\overline{D}_j} \overline{F}_j u_j(z)$ *with respect to the direct sum representation (10.40). Then*

$$\mathcal{R}u = \mathcal{R} \sum_{j=1}^{N} \overline{L}_j \pi_{\overline{D}_j} \overline{F}_j u_j(z) = \begin{pmatrix} \xi_1 \\ \vdots \\ \xi_N \end{pmatrix} = \begin{pmatrix} \pi_{zI-A_1} B_1 u_1 \\ \vdots \\ \pi_{zI-A_N} B_N u_N \end{pmatrix} = \begin{pmatrix} \mathcal{R}_1 u_1 \\ \vdots \\ \mathcal{R}_N u_N \end{pmatrix},$$

i.e.,

$$\mathcal{R} \simeq \mathcal{R}_1 \oplus \cdots \oplus \mathcal{R}_N. \qquad (10.42)$$

3. *The intertwining relation*

$$(\overline{F}_i \overline{X}_i)(zI - A_i) = \overline{D}_i(F_i X_i) \qquad (10.43)$$

exists and is embeddable in the following doubly coprime factorization:

$$\begin{pmatrix} \overline{D}_i(z) & \overline{F}_i(z)\overline{X}_i(z) \\ N_i(z)L_i(z) & \overline{Y}_i(z) + N_i(z)E_i(z)\overline{X}_i(z) \end{pmatrix} \begin{pmatrix} \overline{E}_i(z) + F_i(z)Y_i(z)\overline{L}_i(z) & -F_i(z)X_i(z) \\ -B_i\overline{L}_i(z) & zI - A_i \end{pmatrix}$$

$$= \begin{pmatrix} I & 0 \\ 0 & I \end{pmatrix}. \tag{10.44}$$

Proof. 1. The proof follows from Theorems 3.20 and 3.21. The intertwining relation $\overline{L}_i(z)\overline{D}_i(z) = D_i(z)L_i(z)$ and the associated coprimeness conditions imply that $\mathscr{U}_i$ is an $\mathbb{F}[z]$-isomorphism. Similarly, the intertwining relation $B_iD_i(z) = (zI - A_i)N_i(z)$ implies that $\mathscr{R}_i$ is an $\mathbb{F}[z]$-isomorphism. The two intertwining relations imply a third one, namely, $(B_i\overline{L}_i(z))\overline{D}_i(z) = (zI - A_i)(N_i(z)L_i(z))$, which implies that $\mathscr{V}_i$, which is the composition of two $\mathbb{F}[z]$-isomorphisms, is also an $\mathbb{F}[z]$-isomorphism.

Next, embed the intertwining relation $\overline{L}_i(z)\overline{D}_i(z) = D_i(z)L_i(z)$ in the doubly coprime factorization (10.33). By Theorem 3.21, the map $\mathscr{U}_i^{-1} : X_{D_i} \longrightarrow X_{\overline{D}_i}$ is

$$\mathscr{U}_i^{-1}g_i = \pi_{\overline{D}_i}\overline{F}_ig_i.$$

Similarly, embed the intertwining relation $B_iD_i(z) = (zI - A_i)N_i(z)$ in the doubly coprime factorization (10.29). Applying Theorem 3.21 once more, the map $\mathscr{R}_i^{-1} : X_{zI-A_i} \longrightarrow X_{D_i}$ is

$$\mathscr{R}_i^{-1}\xi_i = \pi_{D_i}\overline{X}_i\xi_i.$$

Finally, compute

$$\mathscr{V}_i^{-1}\xi_i = \mathscr{U}_i^{-1}\mathscr{R}_i^{-1}\xi_i = \pi_{D_i}\overline{F}_i\pi_{D_i}\overline{X}_i\xi_i = \pi_{\overline{D}_i}\overline{F}_i\overline{X}_i\xi_i.$$

2. Using (10.28) we compute

$$\mathscr{R}f = \pi_{(zI-A)}Bf = (zI - A)\pi_-\left(zI - A)^{-1}B\sum_{j=1}^{N}\overline{L}_j(z)f_j(z))\right)$$

$$= \begin{pmatrix} (zI - A_1)\pi_-(zI - A_1)^{-1}B_1\sum_{j=1}^{N}\overline{L}_j(z)f_j(z) \\ \cdot \\ \cdot \\ \cdot \\ (zI - A_N)\pi_-(zI - A_N)^{-1}B_N\sum_{j=1}^{N}\overline{L}_j(z)f_j(z) \end{pmatrix}.$$

Using the coprime factorization (10.34), and noting that $\overline{L}_j = l.c.r.m.\{D_k | k \neq j\}$, and hence that, for $k \neq j$, $D_k(z)^{-1}\overline{L}_j(z) \in \mathbb{F}[z]^{m \times m}$, it follows that

$$\pi_{(zI-A_i)}B_i\overline{L}_jf_j = (zI-A_i)\pi_-(zI-A_i)^{-1}B_i\overline{L}_j(z)f_j(z)$$

$$= (zI-A_i)\pi_-N_iD_i^{-1}\overline{L}_j(z)f_j(z) = \begin{cases} 0 & j \neq i, \\ \mathscr{R}_if_i & j = i. \end{cases}$$

3. From the doubly coprime factorizations (10.29) and (10.33) one obtains the intertwining relations $\overline{X}_i(z)(zI-A_i) = D_i(z)X_i$ and $\overline{F}_i(z)D_i(z) = \overline{D}_i(z)F_i(z)$, respectively. Computing

$$\overline{F}_i(z)\overline{X}_i(z)(zI-A_i) = \overline{F}_i(z)D_i(z)X_i = \overline{D}_i(z)F_i(z)X_i \qquad (10.45)$$

proves (10.43).

The doubly coprime factorizations (10.29) and (10.33) provide eight intertwining relations and eight Bezout equations. These are utilized in checking (10.44). Since (10.43) has been established, one computes

$$\overline{D}_i(\overline{E}_i + F_iY_i\overline{L}_i) + \overline{F}_i\overline{X}_iB_i\overline{L}_i = \overline{D}_i\overline{E}_i + \overline{D}_iF_iY_i\overline{L}_i + \overline{F}_i\overline{X}_iB_i\overline{L}_i$$

$$= (I - \overline{F}_i\overline{L}_i) + \overline{F}_iD_iY_i\overline{L}_i + \overline{F}_i\overline{X}_iB_i\overline{L}_i$$

$$= (I - \overline{F}_i\overline{L}_i) + \overline{F}_i(D_iY_i + \overline{X}_iB_i)\overline{L}_i = I.$$

Next we check

$$N_iL_i(\overline{E}_i + F_iY_i\overline{L}_i) - (\overline{Y}_i + N_iE_i\overline{X}_i)B_i\overline{L}_i = N_iE_i\overline{L}_i + N_iL_iF_iY_i\overline{L}_i - \overline{Y}_iB_i\overline{L}_i - N_iE_i\overline{X}_iB_i\overline{L}_i$$

$$= N_iE_i\overline{L}_i + N_iL_iF_iY_i\overline{L}_i - N_iY_i\overline{L}_i - N_iE_i\overline{X}_iB_i\overline{L}_i$$

$$= N_iE_i(I - \overline{X}_iB_i)\overline{L}_i - N_i(I - L_iF_i)Y_i\overline{L}_i$$

$$= N_iE_iD_iY_i\overline{L}_i - N_iE_iD_iY_i\overline{L}_i = 0.$$

Finally, we compute

$$N_iL_iF_iX_i + (\overline{Y}_i + N_iE_i\overline{X}_i)(zI - A_i) = N_iL_iF_iX_i + \overline{Y}_i(zI - A_i) + N_iE_i\overline{X}_i(zI - A_i)$$

$$= N_iL_iF_iX_i + I - N_iX_i + N_iE_iD_iX_i$$

$$= I - N_i(I - L_iF_i - E_iD_i)X_i = I.$$

$\blacksquare$

The next result states the desired formula for open-loop control of parallel connections.

Theorem 10.21. *Here we use the same notation as in Theorem 10.20.*

1. A doubly coprime embedding of the intertwining relation (10.36) is

$$
\begin{pmatrix}
D(z) & -\bar{\Xi}_1(z) & \dots & -\bar{\Xi}_N(z) \\
\hline
H_1(z) & \Theta_{11}(z) & \dots & \Theta_{1N}(z) \\
\cdot & \cdot & \cdot & \cdot \\
\cdot & \cdot & \cdot & \cdot \\
\cdot & \cdot & \cdot & \cdot \\
H_N(z) & \bar{\Theta}_{N1}(z) & \dots & \bar{\Theta}_{NN}(z)
\end{pmatrix}
\begin{pmatrix}
\Theta(z) & \Xi_1(z) & \dots & \Xi_N(z) \\
\hline
-B_1 & zI - A_1 & & \\
\cdot & & \cdot & \\
\cdot & & & \cdot \\
-B_N & & & zI - A_N
\end{pmatrix}
=
\begin{pmatrix}
I & 0 & \dots & 0 \\
\hline
0 & I & & \\
\cdot & & \cdot & \\
\cdot & & & \cdot \\
0 & & & I
\end{pmatrix},
$$

$$
\begin{pmatrix}
\Theta(z) & \Xi_1(z) & \dots & \Xi_N(z) \\
\hline
-B_1 & zI - A_1 & & \\
\cdot & & \cdot & \\
\cdot & & & \cdot \\
-B_N & & & zI - A_N
\end{pmatrix}
\begin{pmatrix}
D(z) & -\bar{\Xi}_1(z) & \dots & -\bar{\Xi}_N(z) \\
\hline
H_1(z) & \Theta_{11}(z) & \dots & \Theta_{1N}(z) \\
\cdot & \cdot & \cdot & \cdot \\
\cdot & \cdot & \cdot & \cdot \\
\cdot & \cdot & \cdot & \cdot \\
H_N(z) & \bar{\Theta}_{N1}(z) & \dots & \bar{\Theta}_{NN}(z)
\end{pmatrix}
=
\begin{pmatrix}
I & 0 & \dots & 0 \\
\hline
0 & I & & \\
\cdot & & \cdot & \\
\cdot & & & \cdot \\
0 & & & I
\end{pmatrix},
$$

$$(10.46)$$

where

$$
\begin{aligned}
\bar{\Xi}_i &= \bar{L}_i \bar{F}_i \bar{X}_i, \\
\bar{\Theta}_{ii} &= \bar{Y}_i + N_i E_i \bar{X}_i, \\
\bar{\Theta}_{ij} &= -N_i D_i^{-1} \bar{L}_j \bar{F}_j \bar{X}_j, \\
\Xi_j &= F_j X_j.
\end{aligned}
\tag{10.47}
$$

2. For $\xi = \mathrm{col}\,(\xi_1, \dots, \xi_N) \in X_{zI-A}$, the map $\mathscr{R}^{-1} : X_{zI-A} \longrightarrow X_D$ is

$$
u(z) = \mathscr{R}^{-1} \xi = \sum_{j=1}^{N} \bar{L}_j(z) \pi_{\bar{D}_j} \bar{F}_j u_j = \sum_{i=1}^{N} \bar{L}_i(z) \pi_{\bar{D}_i} \bar{F}_i \pi_{D_j} \bar{X}_j \xi_j.
\tag{10.48}
$$

Thus $u(z)$, defined by (10.48), is the minimal steering controller to the state ξ.

Proof. 1. Our proof is computational. Let P_{ij} and Q_{ij}, $i, j = 0, \dots, N$, denote the i, j block element of the first product in (10.46) and of the second product in (10.46). Since the first double coprime factorization in (10.46) follows from the second one, only the formulas for Q_{ij} need to be checked. Our strategy is to derive the necessary conditions for the unknown parameters in the doubly coprime factorization (10.46) and then show sufficiency. We shall make use of the doubly coprime factorization (10.29), which is equivalent to the following equations ($i = 1, \dots, N$):

$$Y_i(z)D_i(z) + X_i(z)N_i(z) = I, \quad Y_i(z)\overline{X}_i(z) = X_i(z)\overline{Y}_i(z),$$

$$B_i\overline{X}_i(z) + (zI - A_i)\overline{Y}_i(z) = I, \quad B_iD_i(z) = (zI - A_i)N_i(z),$$

and

$$D_i(z)Y_i(z) + \overline{X}_i(z)B_i = I, \quad D_i(z)X_i(z) = \overline{X}_i(z)(zI - A_i),$$

$$\overline{Y}_i(z)(zI - A_i) + N_i(z)X_i(z) = I, \quad N_i(z)Y_i(z) = \overline{Y}_i(z)B_i.$$

Derivation of Formulas in (10.47).

1. We need to solve equations Q_{ii} for $\overline{\Xi}_i$ and $\overline{\Theta}_{ii}$ in $I = B_i\overline{\Xi}_i + (zI - A_i)\overline{\Theta}_{ii}$. From the doubly coprime factorization (10.29) we have that $I = B_i\overline{X}_i + (zI - A_i)\overline{Y}_i$. Subtracting the two equations one obtains $B_i(\overline{\Xi}_i - \overline{X}_i) + (zI - A_i)(\overline{\Theta}_{ii} - \overline{Y}_i) = 0$. From the doubly coprime factorization (10.29) we have that

$$\overline{\Xi}_i(z) = \overline{X}_i(z) - D_i(z)S(z),$$
$$\overline{\Theta}_{ii}(z) = \overline{Y}_i(z) + N_i(z)S(z) \tag{10.49}$$

for some polynomial matrix $S(z)$.

2. Inspection of equations Q_{ij}, $j \neq i$. From (10.46) we obtain

$$0 = B_i\overline{\Xi}_j(z) + (zI - A_i)\overline{\Theta}_{ij}(z).$$

This implies that, for some polynomial matrix $T(z)$,

$$\overline{\Xi}_j(z) = -D_i(z)T(z), \quad \overline{\Theta}_{ij}(z) = N_i(z)T(z).$$

Thus $\overline{\Xi}_j(z)$ is left divisible by all $D_i(z)$, $i \neq j$, and, hence, also by their least common right multiple $L_j(z)$, i.e., there is the factorization

$$\overline{\Xi}_j(z) = -\overline{L}_j(z)R(z). \tag{10.50}$$

Comparing (10.49) with (10.50) shows $\overline{X}_i - D_iS = -\overline{L}_iR$ or, equivalently, $\overline{X}_i = D_iS - \overline{L}_iR$. Now, the doubly coprime factorization (10.33) implies the Bezout identity $D_iE_i + \overline{L}_i\overline{F}_i = I$ and in turn the equality

$$\overline{X}_i(z) = D_i(z)E_i(z)\overline{X}_i(z) + \overline{L}_i(z)\overline{F}_i(z)\overline{X}_i(z).$$

Comparing the two representations of $\overline{X}_i(z)$ one infers $R(z) = -\overline{F}_i(z)\overline{X}_i(z)$ and $S(z) = E_i(z)\overline{X}_i(z)$. Substituting back into (10.49),

$$\overline{\Xi}_i = \overline{X}_i - D_iS = \overline{X}_i - D_iE_i\overline{X}_i = (I - D_iE_i)\overline{X}_i = \overline{L}_i\overline{F}_i\overline{X}_i.$$

Similarly,

$$\overline{\Theta}_{ii}(z) = \overline{Y}_i(z) + N_i(z)S(z) = \overline{Y}_i(z) + N_i(z)E_i(z)\overline{X}_i(z).$$

From the equality $D_iT = L_jR$ one obtains $T = D_i^{-1}L_jR = -D_i^{-1}L_j\overline{F}_j\overline{X}_j$, and this implies

$$\overline{\Theta}_{ij}(z) = -N_i(z)D_i(z)^{-1}\overline{L}_j(z)\overline{F}_j(z)\overline{X}_j(z).$$

Verification of Double Coprime Embedding (10.46).

1. Q_{i0}: We use the doubly coprime factorization (10.27), multiplied on the right by $L_i(z)$, to get

$$0 = (-B_iD_i(z) + (zI - A_i)N_i(z))L_i(z) = -B_iD(z) + (zI - A_i)H_i(z).$$

2. Q_{ii}: With $\overline{\Xi}_i = \overline{L}_i\overline{F}_i\overline{X}_i$ and $\overline{\Theta}_{ii} = \overline{Y}_i + N_iE_i\overline{X}_i$, one computes

$$
\begin{aligned}
(-B_i)&(-\overline{L}_i\overline{F}_i\overline{X}_i) + (zI - A_i)(\overline{Y}_i + N_iE_i\overline{X}_i) \\
&= B_i(I - D_iE_i)\overline{X}_i + (zI - A_i)\overline{Y}_i + (zI - A_i)N_iE_i\overline{X}_i \\
&= B_i\overline{X}_i - B_iD_iE_i\overline{X}_i + (I - B_i\overline{X}_i) + (zI - A_i)N_iE_i\overline{X}_i \\
&= -(zI - A_i)N_iE_i\overline{X}_i + I + (zI - A_i)N_iE_i\overline{X}_i = I.
\end{aligned}
$$

3. Q_{ij}, $j \neq i$: By (10.30), $\overline{L}_j(z) = l.c.r.m.\{D_k(z)|k \neq j\}$, which implies that $K_{ij} := D_i^{-1}\overline{L}_j = l.c.r.m.\{D_k(z)|k \neq i, j\}$ is polynomial. With $\overline{\Theta}_{ij} = -N_iK_{ij}\overline{F}_j\overline{X}_j$, then

$$
\begin{aligned}
(-B_i)(-\overline{\Xi}_j) + (zI - A_i)\overline{\Theta}_{ij} &= (-B_i)(-\overline{L}_j\overline{F}_j\overline{X}_j) + (zI - A_i)(-N_iK_{ij}\overline{F}_j\overline{X}_j) \\
&= B_i\overline{L}_j\overline{F}_j\overline{X}_j - B_iD_iD_i^{-1}\overline{L}_j\overline{F}_j\overline{X}_j = 0.
\end{aligned}
$$

Q_{00}: As previously, one computes

$$
\begin{aligned}
I = \Theta D + \sum_{i=1}^{N} \Xi_iH_i &= \Theta D + \sum_{i=1}^{N} F_iX_iN_iL_i \\
&= \Theta D + \sum_{i=1}^{N} F_i(I - Y_iD_i)L_i = \Theta D + \sum_{i=1}^{N} F_iL_i - \sum_{i=1}^{N} F_iY_iD \\
&= (\Theta - \sum_{i=1}^{N} F_iY_i)D + \sum_{i=1}^{N} F_iL_i.
\end{aligned}
$$

Q_{0j}: In the computation the following identity is used:

$$(\Theta - \sum_{i=1}^{N} F_i Y_i) = (I - \sum_{i=1}^{N} F_i L_i) D^{-1} = D^{-1} (I - \sum_{i=1}^{N} \overline{L}_i \overline{F}_i).$$

Note that from the Bezout identity $D_j E_j + \overline{L}_j \overline{F}_j = I$ one obtains $E_j = D_j^{-1}(I - \overline{L}_j \overline{F}_j)$. In turn, this implies $F_j D_j^{-1} \overline{L}_j \overline{F}_j \overline{X}_j = -F_j E_j \overline{X}_j + F_j D_j^{-1} X_j$. We also use the identity $\overline{F}_j(z) D_j(z) = \overline{D}_j(z) F_j(z)$ and the factorizations $D(z) = D_j(z) L_j(z) = \overline{L}_j(z) \overline{D}_j(z)$. With this information, a lengthy computation yields

$$-\Theta \overline{\Xi}_j + \sum_{i=1}^{N} \Xi_i \overline{\Theta}_{ij} = -\Theta \overline{\Xi}_j + \sum_{i \neq j} \Xi_i \overline{\Theta}_{ij} + \Xi_j \overline{\Theta}_{jj}$$

$$= -\Theta \overline{L}_j \overline{F}_j \overline{X}_j + \sum_{i \neq j} (F_i X_i)(-N_i D_i^{-1} \overline{L}_j \overline{F}_j \overline{X}_j) + F_j X_j (\overline{Y}_j + N_j E_j \overline{X}_j)$$

$$= -\Theta \overline{L}_j \overline{F}_j \overline{X}_j - \sum_{i \neq j} F_i (I - Y_i D_i) D_i^{-1} \overline{L}_j \overline{F}_j \overline{X}_j + F_j X_j \overline{Y}_j + F_j X_j N_j E_j \overline{X}_j$$

$$= -(\Theta - \sum_{i \neq j} F_i Y_i) \overline{L}_j \overline{F}_j \overline{X}_j - (\sum_{i \neq j} F_i D_i^{-1}) \overline{L}_j \overline{F}_j \overline{X}_j + F_j X_j \overline{Y}_j + F_j X_j N_j E_j \overline{X}_j$$

$$= -(\Theta - \sum_{i \neq j} F_i Y_i) \overline{L}_j \overline{F}_j \overline{X}_j - (\sum_{i \neq j} F_i D_i^{-1}) \overline{L}_j \overline{F}_j \overline{X}_j + F_j X_j \overline{Y}_j + F_j (I - Y_j D_j) E_j \overline{X}_j$$

$$= -(\Theta - \sum_{i \neq j} F_i Y_i) \overline{L}_j \overline{F}_j \overline{X}_j - (\sum_{i \neq j} F_i D_i^{-1}) \overline{L}_j \overline{F}_j \overline{X}_j + F_j X_j \overline{Y}_j + F_j E_j \overline{X}_j$$

$$- F_j Y_j (I - \overline{L}_j \overline{F}_j) \overline{X}_j$$

$$= -(\Theta - \sum_{i \neq j} F_i Y_i) \overline{L}_j \overline{F}_j \overline{X}_j - (\sum_{i \neq j} F_i D_i^{-1}) \overline{L}_j \overline{F}_j \overline{X}_j + F_j E_j \overline{X}_j + F_j Y_j \overline{L}_j \overline{F}_j \overline{X}_j$$

$$= -(\Theta - \sum_{i=1}^{N} F_i Y_i) \overline{L}_j \overline{F}_j \overline{X}_j - (\sum_{i \neq j} F_i D_i^{-1}) \overline{L}_j \overline{F}_j \overline{X}_j + F_j E_j \overline{X}_j$$

$$= -(I - \sum_{i=1}^{N} F_i L_i) D^{-1} \overline{L}_j \overline{F}_j \overline{X}_j - (\sum_{i \neq j} F_i D_i^{-1}) \overline{L}_j \overline{F}_j \overline{X}_j + F_j E_j \overline{X}_j$$

$$= -(I - \sum_{i=1}^{N} F_i L_i) D^{-1} \overline{L}_j \overline{F}_j \overline{X}_j - (\sum_{i=1}^{N} F_i D_i^{-1}) \overline{L}_j \overline{F}_j \overline{X}_j + F_j D_j^{-1} \overline{X}_j$$

$$= -D^{-1} \overline{L}_j \overline{F}_j \overline{X}_j + \sum_{i=1}^{N} F_i (\overline{L}_i D^{-1} - D_i^{-1}) \overline{L}_j \overline{F}_j \overline{X}_j + F_j D_j^{-1} \overline{X}_j$$

$$= -\overline{D}_j^{-1} \overline{F}_j \overline{X}_j + F_j D_j^{-1} \overline{X}_j = 0.$$

2. The direct sum representation (10.40) indicates that one can embed the spaces X_{D_i} in X_D to get a global steering controller. To this end, assume that $u_i(z) \in X_{D_i}$ is the minimal control that steers (A_i, B_i) to the state ξ_i at time zero. By Theorem 4.47, $u_i(z) = \pi_{D_i} \overline{X}_i \xi_i$. In view of the direct sum (10.40), the naive

embedding $u_i \mapsto \overline{L}_i u_i$ is not the right one. A correction proceeds by using the map $\mathscr{U}_i^{-1}$, defined in (10.41), to construct the embedding $\mathscr{E} : X_{D_1} \oplus \cdots \oplus X_{D_N} \longrightarrow X_D$ by

$$u(z) = \mathscr{E} \mathrm{col}\,(u_1(z), \ldots, u_N(z)) = \sum_{j=1}^{N} \overline{L}_j \pi_{\overline{D}_j} \overline{F}_j u_j.$$

Here $\overline{F}_i(z)$ arises from the doubly coprime factorization (10.33). Clearly,

$$u(z) = \sum_{j=1}^{N} \overline{L}_j \pi_{\overline{D}_j} \overline{F}_j u_j = \sum_{j=1}^{N} \overline{L}_j \pi_{\overline{D}_j} \overline{F}_j \pi_{D_i} \overline{X}_i \xi_i$$

$$= \sum_{j=1}^{N} \overline{L}_j \pi_{\overline{D}_j} \overline{F}_j \overline{X}_i \xi_i.$$

We prove next that $\mathscr{R}_i u = \xi_i$ for all i. To show this, we compute

$$\mathscr{R}_i u = \pi_{zI-A_i} B_i \sum_{j=1}^{N} \overline{L}_j \pi_{D_j} \overline{F}_j \overline{X}_j \xi_j = \sum_{j=1}^{N} (zI - A_i) \pi_- (zI - A_i)^{-1} B_i \overline{L}_j \pi_{\overline{D}_j} \overline{F}_j \overline{X}_j \xi_j$$

$$= \sum_{j=1}^{N} (zI - A_i) \pi_- (zI - A_i)^{-1} B_i \overline{L}_j D_j \pi_- \overline{D}_j^{-1} \overline{F}_j \overline{X}_j \xi_j$$

$$= \sum_{j=1}^{N} (zI - A_i) \pi_- N_i D_i^{-1} D \pi_- \overline{D}_j^{-1} \overline{F}_j \overline{X}_j \xi_j$$

$$= \sum_{j=1}^{N} (zI - A_i) \pi_- N_i L_i \pi_- \overline{D}_j^{-1} \overline{F}_j \overline{X}_j \xi_j$$

$$= \sum_{j=1}^{N} (zI - A_i) \pi_- N_i L_i \overline{D}_j^{-1} \overline{F}_j \overline{X}_j \xi_j = (zI - A_i) \pi_- N_i L_i \overline{D}_i^{-1} \overline{F}_i \overline{X}_i \xi_i$$

$$= (zI - A_i) \pi_- N_i D_i^{-1} \overline{L}_i \overline{F}_i \overline{X}_i \xi_i = (zI - A_i) \pi_- N_i D_i^{-1} (I - D_i E_i) \overline{X}_i \xi_i$$

$$= (zI - A_i) \pi_- N_i D_i^{-1} \overline{X}_i \xi_i = (zI - A_i) \pi_- (zI - A_i)^{-1} B_i \overline{X}_i \xi_i$$

$$= \pi_{zI-A_i} B_i \pi_{D_i} \overline{X}_i \xi_i = \mathscr{R}_i \mathscr{R}_i^{-1} \xi_i = \xi_i.$$

In this computation we used the representations of the projections π_{zI-A_i} and $\pi_{\overline{D}_j}$, the factorizations $D(z) = D_i(z) L_i(z) = \overline{L}_i(z) \overline{D}_i(z)$, the coprime factorizations $(zI - A_i)^{-1} B_i = N_i(z) D_i(z)^{-1}$, the Bezout equation $\overline{L}_i(z) \overline{F}_i(z) + D_i(z) E_i(z) = I$, and the fact that, for $i \neq j$, $L_i(z) D_j(z)^{-1}$ is a polynomial matrix. This completes the proof of (10.48).   ∎

We add a few remarks on the duality information encoded in the doubly coprime factorization (10.46). The factorizations $\overline{D}(z) = \overline{L}_i(z) \overline{D}_i(z) = D_i(z) L_i(z)$ lead to two direct sum decompositions, namely, $X_D = \overline{L}_1 X_{\overline{D}_1} \oplus \cdots \oplus \overline{L}_N X_{\overline{D}_N}$ and $X_{D^\top} = L_1^\top X_{D_1^\top}$ $\oplus \cdots \oplus L_N^\top X_{D_N^\top}$. These two direct sum decompositions are dual in the sense of the duality pairing

$$< f,g > = [D^{-1}f, g], \quad f(z) \in X_D, \ g(z) \in X_{D^\top}$$

and therefore imply $X_D^* = X_{D^\top}$. Also, the orthogonality relations $L_i^\top X_{D_i^\top} \perp \overline{L}_j X_{\overline{D}_j}$ are satisfied for $i \neq j$. To see this, one computes, with $f_i(z) \in X_{D_i^\top}$ and $g_j(z) \in X_{\overline{D}_j}$,

$$\langle L_i^\top f_i, L_j g_j \rangle = [(D^{-1})^\top L_i^\top f_i, L_j g_j] = [f_i, L_i D^{-1} \overline{L}_j g_j] = [f_i, D_i^{-1} \overline{L}_j g_j] = 0.$$

This computation depends on the fact that, because of (10.30), for $i \neq j$, $D_i(z)^{-1} \overline{L}_j(z)$ is polynomial. The doubly coprime factorization (10.46) yields basis matrices that correspond to the foregoing direct sums. Thus $\overline{\Xi}(z) = \left(\overline{\Xi}_1(z) \ldots \overline{\Xi}_N(z) \right)$ is a basis matrix for X_D, whereas $H^\top(z) = \left(H_1^\top(z) \ldots H_N^\top(z) \right)$ is a basis matrix for $X_{D^\top}$.

The previous theorem focused on constructing a minimal global steering controller from the minimal local controllers. However, the same methodology can be applied if one drops the minimality constraints.

Proposition 10.22. *The same notation and assumptions apply here as for Theorem 10.18. Assume further that $u_i(z) \in \mathbb{F}[z]^m$ are controllers that steer the nodes to prescribed states at time zero, i.e., $\mathscr{R}_i u_i = \pi_{zI-A_i} B_i u_i = \xi_i$, $i = 1, \ldots, N$. Then the input polynomial*

$$u = \sum_{j=1}^{N} \overline{L}_j \overline{F}_j u_j$$

is a global steering controller for the parallel connection.

Proof. The proof is by a computation analogous to the proof of Theorem 10.20. With (A,B) defined by (10.28), one obtains $\pi_{(zI-A)} Bu =$

$$\pi_{(zI-A)} B \sum_{j=1}^{N} \overline{L}_j \overline{F}_j u_j = \begin{pmatrix} (zI - A_1)\pi_-(zI - A_1)^{-1} B_1 \sum_{j=1}^{N} \overline{L}_j(z) \overline{F}_j(z) u_j(z) \\ \vdots \\ (zI - A_N)\pi_-(zI - A_N)^{-1} B_N \sum_{j=1}^{N} \overline{L}_j(z) \overline{F}_j(z) u_j(z) \end{pmatrix}.$$

Now, using the fact that, for $i \neq j$, $D_i(z)^{-1} \overline{L}_j(z)$ is polynomial, one can check that

$$\pi_{(zI-A_i)} B_i \sum_{j=1}^{N} \overline{L}_j \overline{F}_j u_j = \sum_{j=1}^{N} (zI - A_i)\pi_-(zI - A_i)^{-1} B_i \overline{L}_j \overline{F}_j u_j$$

$$= \sum_{j=1}^{N} (zI - A_i)\pi_- N_i D_i^{-1} \overline{L}_j \overline{F}_j u_j = (zI - A_i)\pi_- N_i L_i \overline{D}_i^{-1} \overline{F}_i u_i$$

$$= (zI - A_i)\pi_- N_i L_i F_i D_i^{-1} u_i = (zI - A_i)\pi_- N_i (I - E_i D_i) D_i^{-1} u_i$$
$$= (zI - A_i)\pi_- N_i D_i^{-1} u_i = (zI - A_i)\pi_- (zI - A_i)^{-1} B_i u_i = \xi_i.$$

∎

Next, the problem of recursive computation of (minimal) steering controllers is addressed. To this end, we introduce some notation. Assume all pairs (A_j, B_j) are reachable, and so, for all i, is the parallel connection of the first i nodes. Let

$$(zI - A_i)^{-1} B_i = N_i(z) D_i(z)^{-1}$$

be coprime factorizations. Let

$$\sum_{j=1}^{i} (zI - A_j)^{-1} B_j = \sum_{j=1}^{i} N_j(z) D_j(z)^{-1} = \overline{N}_i(z) \overline{D}_i(z)^{-1},$$

with $\overline{N}_i(z), \overline{D}_i(z)$ right coprime. Since $\overline{D}_i(z) = l.c.r.m.\{D_1, \ldots, D_i\}$, it follows that $\overline{D}_{i+1} = l.c.r.m.\{\overline{D}_i, D_{i+1}\}$. Because of our assumption of the reachability of the parallel connection, the polynomial matrices $\overline{D}_i(z), D_{i+1}(z)$ are left coprime. This implies the existence of right coprime polynomial matrices $L_i(z), \overline{L}_{i+1}(z)$ that satisfy the factorizations

$$\overline{D}_{i+1}(z) = \overline{D}_i(z) L_i(z) = D_{i+1}(z) \overline{L}_{i+1}(z).$$

Define

$$\overline{L}_i(z) = D_{i+1}(z), \quad \mathbf{D}_i(z) = \overline{L}_{i+1}(z),$$
$$\mathbf{L}_{i+1}(z) = \overline{D}_i(z), \quad \overline{D}_{i+1}(z) = L_i(z).$$

Using these, one obtains the factorizations

$$\overline{D}_{i+1}(z) = \overline{L}_i(z) \mathbf{D}_i(z) = \mathbf{L}_{i+1}(z) \overline{D}_{i+1}(z).$$

This implies the direct sum decomposition

$$X_{\overline{D}_{i+1}} = \overline{L}_i X_{\mathbf{D}_i} \oplus \mathbf{L}_{i+1} X_{\overline{D}_{i+1}}.$$

10.3   Open-Loop Control and Interpolation

The proof of Theorem 10.20 suggests that interesting connections exist between open-loop control and interpolation theory for rational functions. We illustrate this point by examples.

Example 10.23 (Scalar Lagrange interpolation).   The proof of the following lemma is by a straightforward computation.

Lemma 10.24. *Assume $f(z) \in \mathbb{F}[z]$ and $\lambda_i \in \mathbb{F}$ with $f(\lambda_i) \neq 0$. Then $z - \lambda_i$ and $f(z)$ are coprime, and the associated doubly coprime factorization is*

$$\begin{pmatrix} -\frac{1}{f(\lambda_i)} \frac{f(z)-f(\lambda_i)}{z-\lambda_i} & \frac{1}{f(\lambda_i)} \\ -f(z) & (z-\lambda_i) \end{pmatrix} \begin{pmatrix} z-\lambda_i & -\frac{1}{f(\lambda_i)} \\ f(z) & -\frac{1}{f(\lambda_i)} \frac{f(z)-f(\lambda_i)}{z-\lambda_i} \end{pmatrix} = \begin{pmatrix} 1 & 0 \\ 0 & 1 \end{pmatrix}.$$

Assume next that $d(z) = \prod_{i=1}^{N}(z - \lambda_j)$, with the λ_i distinct. Let the node pairs be $(A_i, b_i) = (\lambda_i, 1)$. The parallel connection of the nodes has the representation

$$A = \begin{pmatrix} \lambda_1 & & \\ & \ddots & \\ & & \lambda_n \end{pmatrix}, \quad b = \begin{pmatrix} 1 \\ \vdots \\ 1 \end{pmatrix}.$$

Let the **Lagrange interpolation polynomials** $l_i(z)$, $i = 1, \ldots, N$, be defined by

$$l_i(z) = \prod_{j \neq i} \frac{z - \lambda_j}{\lambda_i - \lambda_j}.$$

To apply Theorem 3.19, we compute $\overline{D}_i(z) = D_i(z) = z - \lambda_i$, $\overline{L}_i(z) = L_i(z) = \prod_{j \neq i}(z - \lambda_j)$, $D(z) = \prod_{j=1}^{N}(z - \lambda_j)$. The doubly coprime factorization (10.33) takes the form

$$\begin{pmatrix} -\frac{1}{L_i(\lambda_i)} \frac{L_i(z)-L_i(\lambda_i)}{z-\lambda_i} & \frac{1}{L_i(\lambda_i)} \\ -L_i(z) & (z-\lambda_i) \end{pmatrix} \begin{pmatrix} z-\lambda_i & -\frac{1}{L_i(\lambda_i)} \\ L_i(z) & -\frac{1}{L_i(\lambda_i)} \frac{L_i(z)-L_i(\lambda_i)}{z-\lambda_i} \end{pmatrix} = \begin{pmatrix} 1 & 0 \\ 0 & 1 \end{pmatrix},$$

which implies $E_i(z) = \overline{E}_i(z) = -\frac{1}{L_i(\lambda_i)} \frac{L_i(z)-L_i(\lambda_i)}{z-\lambda_i}$ and $F_i(z) = \overline{F}_i(z) = -\frac{1}{L_i(\lambda_i)}$. Similarly, the doubly coprime factorization (10.29) takes the form

$$\left(\begin{array}{c|c} 0 & 1 \\ \hline -1 & z-\lambda_i \end{array} \right) \left(\begin{array}{c|c} z-\lambda_i & -1 \\ \hline 1 & 0 \end{array} \right) = \left(\begin{array}{c|c} I & 0 \\ \hline 0 & I \end{array} \right),$$

which implies $H_i(z) = 1$, $X_i(z) = \overline{X}_i(z) = -1$, and $Y_i(z) = \overline{Y}_i(z) = 0$. With these identifications, one computes

$$\overline{\Xi}_i = -L_i \overline{F}_i \overline{X}_i = -L_i(z) \frac{1}{L_i(\lambda_i)} = -l_i(z),$$

$$\overline{\Theta}_{ii} = \overline{Y}_i - H_i E_i \overline{X}_i = -\frac{1}{L_i(\lambda_i)} \frac{L_i(z)-L_i(\lambda_i)}{z-\lambda_i} = \frac{1-l_i(z)}{z-\lambda_i},$$

$$\overline{\Theta}_{ij} = -H_i D_i^{-1} L_j \overline{F}_j \overline{X}_j = -\frac{1}{L_i(\lambda_i)} \prod_{k \neq i,j}(z - \lambda_k),$$

$$\Xi_j = -F_j X_j = -\frac{1}{L_j(\lambda_j)}.$$

The embedding in the doubly coprime factorization (10.46) is

$$
\begin{pmatrix}
\frac{\prod_{k=1}^{N}(z-\lambda_k)}{L_1(z)} & \frac{-l_1(z)}{z-\lambda_1} & \cdots & \frac{-l_N(z)}{z-\lambda_N} \\
 & \frac{1-l_1(z)}{z-\lambda_1} & \cdots & \frac{-l_1(z)}{z-\lambda_N} \\
\cdot & \cdot & & \cdot \\
\cdot & \cdot & & \cdot \\
\cdot & \cdot & & \cdot \\
L_N(z) & \frac{-l_n(z)}{z-\lambda_1} & \cdots & \frac{1-l_N(z)}{z-\lambda_N}
\end{pmatrix}
\begin{pmatrix}
0 & L_1(\lambda_1)^{-1} & \cdots & L_N(\lambda_N)^{-1} \\
-1 & z-\lambda_1 & & \\
\cdot & & & \\
\cdot & & & \\
\cdot & & & \\
-1 & & & z-\lambda_N
\end{pmatrix}
= I_{N+1}.
$$

Note that

$$
u(z) = \mathscr{R}^{-1}\xi = \pi_d\left(l_1(z) \ \ldots \ l_N(z)\right)\xi = \sum_{i=1}^{N}\xi_i l_i(z)
$$

is the required minimal-degree control sequence. So the minimal-degree control sequence is equivalent to the solution to the Lagrange interpolation problem of finding the minimal-degree polynomial $u(z) \in \mathbb{F}[z]$ satisfying the interpolation constraints $u(\lambda_i) = \xi_i$, $i = 1,\ldots,N$.

Example 10.25 (Scalar Newton interpolation). Again, assume that $d(z) = \prod_{j=0}^{n-1}(z-\lambda_j)$, with λ_i distinct. Define, for $i = 1,\ldots,n$,

$$
d_i(z) = \prod_{j=0}^{i-1}(z-\lambda_j), \quad p_i(z) = \prod_{j=i+1}^{n-1}(z-\lambda_j).
$$

We refer to $d_0,\ldots,d_{n-1}$ and to $p_0,\ldots,p_{n-1}$ as the **Newton interpolation polynomials** and **dual Newton interpolation polynomials**, respectively. Relative to the duality pairing

$$
< f,g >= [d^{-1}f,g], \qquad f(z),g(z) \in X_d,
$$

one has $X_d^* = X_d$. The Newton interpolation polynomials form a basis for X_d and the dual Newton interpolation polynomials form the dual basis, i.e., $< d_i, p_j >= \delta_{ij}$. Consider the input pair (A,b):

$$
A = \begin{pmatrix}
\lambda_0 & & & \\
1 & \lambda_1 & & \\
 & \ddots & \ddots & \\
 & & 1 & \lambda_{n-1}
\end{pmatrix}, \quad b = \begin{pmatrix} 1 \\ 0 \\ \vdots \\ 0 \end{pmatrix}.
$$

It is easily verified that (A,b) arises from the series connection of the node systems $\left(\begin{array}{c|c}\lambda_i & 1 \\ \hline 1 & 0\end{array}\right)$, $i = 0,\ldots,n-1$. Also, (A,b) is simply the matrix representation of (S_d,π_d)

in the Newton interpolation basis. In this case, the embedding of the intertwining relation in a doubly coprime factorization is

$$
\begin{pmatrix}
\Pi_{j=0}^{n-1}(z-\lambda_j) & -1 & -(z-\lambda_0) & \cdots & -\Pi_{j=0}^{n-2}(z-\lambda_j) \\
\Pi_{j=1}^{n-1}(z-\lambda_j) & 0 & -1 & \cdots & -\Pi_{j=1}^{n-2}(z-\lambda_j) \\
\cdot & \cdot & \cdot & \cdot & \cdot \\
\cdot & \cdot & & \cdot & \cdot \\
\cdot & \cdot & & \cdot & -1 \\
\Pi_{j=n}^{n-1}(z-\lambda_j) & 0 & & \cdots & 0
\end{pmatrix}
\begin{pmatrix}
0 & 0 & \cdots & & 1 \\
-1 & z-\lambda_0 & & & \\
0 & -1 & \cdot & & \\
\cdot & & & \cdot & \\
\cdot & & & & \cdot \\
0 & & & -1 & z-\lambda_{n-1}
\end{pmatrix}
$$

$$
=
\begin{pmatrix}
1 & 0 & \cdots & 0 \\
0 & 1 & & \\
\cdot & & \cdot & \\
\cdot & & & \cdot \\
\cdot & & & \cdot \\
0 & & & 1
\end{pmatrix}.
$$

Note that $\mathscr{R}^{-1}\xi = \pi_d\left(d_0(z) \cdots d_{n-1}(z)\right)\xi = \sum_{i=0}^{n-1}\xi_i d_i(z)$ is the required minimal-degree control sequence. Thus the minimal-degree control sequence is equivalent to the solution to the Newton interpolation problem of finding the minimal-degree polynomial $u(z) \in \mathbb{F}[z]$ with interpolation constraints $u(\lambda_i) = \xi_i$, $i = 0,\ldots,n-1$. We note that solving the scalar Newton interpolation problem is equivalent to expanding $u(z)$ with respect to the direct sum decomposition $X_d = \bigoplus_{i=0}^{n-1} d_i X_{z-\lambda_{i+1}}$ using the preceding interpolation constraints. In this form, the method can be extended to the multivariable, high-degree case.

Next, we proceed to explore the connection between the reachability of the parallel connection of N reachable nodes of McMillan degree one and tangential Lagrange interpolation. For a comprehensive treatment of tangential interpolation theory, in the spirit of this book, see Fuhrmann (2010).

Definition 10.26. Let $\lambda_i \in \mathbb{F}$, $\xi_i \in \mathbb{F}^{m\times1}$, $\eta_i \in \mathbb{F}^{p\times1}$ be such that the vectors $\{\frac{\xi_i}{z-\lambda_i} | i = 1,\ldots,N\}$ are linearly independent. The **right tangential Lagrange interpolation problem** asks one to find find a polynomial matrix $\Lambda(z) \in \mathbb{F}[z]^{p\times m}$ satisfying

$$
\Lambda(\lambda_i)\xi_i = \eta_i, \qquad i = 1,\ldots,N.
$$

It is worth pointing out that, contrary to the scalar case, the interpolation points λ_i need not be distinct. In fact, one can impose up to m linearly independent interpolation constraints at each interpolation point. Our subsequent treatment is based on Theorem 10.18, and we will use the same notation. Thus, assume the nodes are (A_i,B_i) with coprime factorizations $(zI-A_i)^{-1}B_i = N_i(z)D_i(z)^{-1}$. We restrict ourselves to the special case where all spaces X_{D_i} are one-dimensional, i.e.,

$\dim X_{D_i} = \deg \det D_i = 1$. Using the isomorphism $X_{D_i} \simeq X^{D_i}$ and the shift invariance of X^{D_i}, a nonzero $h_i(z) \in X^{D_i}$ is necessarily of the form $h_i(z) = \frac{\zeta_i}{z-\lambda_i}$, with $\zeta_i \in \mathbb{F}^m$ and $D_i(\lambda_i)\zeta_i = 0$. Define the polynomial matrix $D(z)$ by (10.30). The assumption that the $D_i(z)$ are mutually left coprime translates to the linear independence of the $\frac{\zeta_i}{z-\lambda_i}$. The assumption that $\dim X_{D_i} = 1$ means that A_i is scalar and necessarily $A_i = \lambda_i$ for some $\lambda_i \in \mathbb{F}$. Let now $(zI - A)^{-1}\xi = \sum_{i=1}^{N} \xi_i(zI - A)^{-1}e_i$ denote an element of X^D. Then

$$\mathscr{R}^{-1}(zI-A)^{-1}\xi = -\Xi(\sigma)\sum_{i=1}^{N}\xi_i(zI-A)^{-1}e_i = -\pi_-\Xi\sum_{i=1}^{N}\xi_i\frac{e_i}{z-\lambda_i}$$

$$= -\sum_{i=1}^{N}\xi_i\frac{\Xi(\lambda_i)e_i}{z-\lambda_i} = -\sum_{i=1}^{N}\xi_i\frac{\Xi_i(\lambda_i)}{z-\lambda_i}$$

$$= -\sum_{i=1}^{N}\xi_iF_i(\lambda_i)X_i.$$

As in Theorem 10.18, one has the direct sum representations $X_D = \oplus_{i=1}^{N}L_iX_{D_i}$ and, with $\overline{D}_i(z) := \mathrm{diag}\,(1,..,z-\lambda_i,1,..,1)$, $X_{zI-A} = \oplus_{i=1}^{N}X_{\overline{D}_i}$.
We can now analyze the open-loop control task for the parallel connection of N reachable nodes in terms of tangential interpolation. For simplicity, we restrict ourselves to the case of McMillan degree 1 nodes.

Theorem 10.27. *Let the pair (A,B), defined by (10.28), be reachable, and assume $(\lambda_i, B_i) = (A_i, B_i) \in \mathbb{F}^{1\times1} \times \mathbb{F}^{1\times m}$, $i = 1,\ldots,N$. Assume that $N_i(z)D_i(z)^{-1}$ is a right coprime factorization of $(zI - \lambda_i)^{-1}B_i$. Also, let $\{e_1,\ldots,e_N\}$ be the standard basis in $X_{zI-A} = \mathbb{F}^N$.*

1. For each i

$$H(\lambda_i)F_i(\lambda_i)X_i = e_i. \tag{10.51}$$

2. There exists a unique basis $\mathscr{B}$ in X_D that is the image under $\mathscr{R}^{-1}$, the inverse of the reduced reachability map, of the standard basis in $\mathbb{F}^n$. This basis is $\{\overline{L}_1(z)f_1(z),\ldots,\overline{L}_N(z)f_N(z)\}$, where the $f_i(z)$ are defined by

$$f_i(z) = \overline{D}_i(z)\frac{F_i(\lambda_i)X_i}{z-\lambda_i}.$$

3. The reduced reachability map $\mathscr{R}$ maps the basis $\{\overline{L}_1(z)f_1(z),\ldots,\overline{L}_N(z)f_N(z)\}$ onto the standard basis of X_{zI-A}.
4. Let $\mathscr{R}$ be defined by (4.55). Define the map $\hat{\mathscr{R}} : X^D \longrightarrow X^{zI-A}$ via the commutativity of the following diagram:

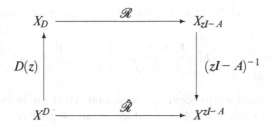

Then $\hat{\mathscr{R}}$ is

$$\hat{\mathscr{R}} = H(\sigma),$$ (10.52)

and its inverse is

$$\hat{\mathscr{R}}^{-1} = \Xi(\sigma).$$ (10.53)

5. Let $\mathscr{R} : X_D \longrightarrow X_{zI-A}$ be defined by (4.55), and let $\hat{\mathscr{R}}^{-1} : X_{zI-A} \longrightarrow X_D$ be defined through the commutativity of the following diagram:

$$
\begin{array}{ccc}
X_{zI-A} & \xrightarrow{\mathscr{R}^{-1}} & X_D \\
\downarrow{\scriptstyle zI-A} & & \downarrow{\scriptstyle D(z)^{-1}} \\
X^{zI-A} & \xrightarrow{\hat{\mathscr{R}}^{-1}} & X^D
\end{array}
$$

Then

$$\hat{\mathscr{R}}_i^{-1} \frac{e_i}{z - \lambda_i} = \frac{F_i(\lambda_i)X_i}{z - \lambda_i}.$$

6. The polynomial matrix $\Xi(z)$ satisfies the following right tangential interpolation conditions:

$$\Xi(\lambda_i)e_i = \Xi_i(\lambda_i) = F_i(\lambda_i)X_i, \qquad i = 1,\dots,N.$$ (10.54)

7. The minimal-degree control sequence that steers the system from the zero state to the state $\xi = \sum_{i=1}^{N} \xi_i e_i$ coincides with the coefficients of the polynomial vector $u(z) \in X_D$

$$u(z) = (\mathscr{R}^{-1}\xi)(z) = -\sum_{i=1}^{N} \xi_i \bar{L}_i(z)\left(\bar{D}_i(z)\frac{F_i(\lambda_i)X_i}{z - \lambda_i}\right),$$

i.e., of the polynomial vector $u(z) \in X_D$ *that satisfies the interpolation conditions*

$$\pi_{D_j}\overline{L}_i(z)\left(\overline{D}_i(z)\frac{F_i(\lambda_i)X_i}{z-\lambda_i}\right) = \begin{cases} 0 & j \neq i, \\ D_i(z)\frac{L_i(\lambda_i)F_i(\lambda_i)X_i}{z-\lambda_i} & j = i. \end{cases}$$

Proof. 1. From the doubly coprime factorization (10.46) follows the Bezout identity $H(z)\Xi(z) + \overline{\Theta}(z)(zI - A) = I$, which in turn implies $H(z)\Xi_i(z) + \overline{\Theta}(z)(z - \lambda_i)e_i = e_i$. Using (10.47) and evaluating it at $z = \lambda_i$ implies (10.51).

2. Let $\{e_1, \ldots, e_N\}$ be the standard basis in $X_{zI-A} = \mathbb{F}^N$. The map $\mathscr{R}^{-1}$, the inverse of the reduced reachability map, is given by (4.60). For $\xi = \sum_{i=1}^N \xi_i e_i \in X_{zI-A}$, one computes, using the identity $\overline{\Xi}(z)(zI - A) = D(z)\Xi(z)$ [obtained from (4.58) and (10.31)],

$$\mathscr{R}^{-1}\xi = \pi_D\overline{\Xi}\xi = \pi_D\overline{\Xi}\sum_{i=1}^N \xi_i e_i = \sum_{i=1}^N \xi_i \pi_D\overline{\Xi}e_i$$

$$= \sum_{i=1}^N \xi_i D\pi_- D^{-1}\overline{\Xi}e_i = \sum_{i=1}^N \xi_i D\pi_-\Xi(zI - A)^{-1}e_i = \sum_{i=1}^N \xi_i D\pi_-\frac{\Xi e_i}{z - \lambda_i}$$

$$= \sum_{i=1}^N \xi_i D\pi_-\frac{\Xi_i}{z - \lambda_i} = \sum_{i=1}^N \xi_i D\frac{\Xi_i(\lambda_i)}{z - \lambda_i} = \sum_{i=1}^N \xi_i\overline{L}_i\left(\overline{D}_i\frac{F_i(\lambda_i)X_i}{z - \lambda_i}\right) = \sum_{i=1}^N \xi_i\overline{L}_i f_i.$$

One can apply identity (10.43), which in our case takes the form $(\overline{F}_i(z)\overline{X}_i(z))(zI - \lambda_i) = \overline{D}_i(z)(F_i(z)X_i)$, to obtain

$$\overline{D}_i(\lambda_i)F_i(\lambda_i)X_i = 0. \qquad (10.55)$$

This implies $f_i(z) \in X_{D_i}$, and hence is a basis for it. That $\{\overline{L}_1(z)f_1(z), \ldots, \overline{L}_N(z)f_N(z)\}$ is a basis for X_D follows from the direct sum decomposition (10.40).

3. This is clear, since $L_i f_i = \mathscr{R}^{-1}e_i$. A direct proof is also possible using the identity (10.51) and the computation

$$\mathscr{R}(\overline{L}_i f_i) = \pi_{zI-A}B\overline{L}_i f_i = (zI - A)\pi_-(zI - A)^{-1}B\overline{L}_i f_i$$

$$= (zI - A)\pi_- HD^{-1}\overline{L}_i f_i = (zI - A)\pi_- HD^{-1}\overline{L}_i D_i\frac{F_i(\lambda_i)X_i}{z - \lambda_i}$$

$$= (zI - A)\pi_- H\frac{F_i(\lambda_i)X_i}{z - \lambda_i} = (zI - A)\frac{H(\lambda_i)F_i(\lambda_i)X_i}{z - \lambda_i}$$

$$= H(\lambda_i)F_i(\lambda_i)X_i = e_i.$$

4. Since $\mathscr{R}$ is an $\mathbb{F}[z]$-homomorphism, so is $\hat{\mathscr{R}}$. From the intertwining relation (4.57) one obtains, for $h(z) \in X^D$,

$$\mathcal{R}h = (zI-A)^{-1}\pi_{zI-A}BDh = (zI-A)^{-1}(zI-A)\pi_-(zI-A)^{-1}BDh = \pi_- Hh$$
$$= H(\sigma)h.$$

Thus (10.52) follows.

5. From the intertwining relation (4.61) one obtains, for $h(z) \in X^{zI-A}$,

$$\mathcal{R}^{-1}h = D^{-1}\pi_D \overline{\Xi}(zI-A)h = D^{-1}D\pi_- D^{-1}\overline{\Xi}(zI-A)h = \pi_- \Xi h = \Xi(\sigma)h.$$

6. We use representation (10.53) to compute

$$\mathcal{R}^{-1}\frac{e_i}{z-\lambda_i} = \Xi(\sigma)\frac{e_i}{z-\lambda_i} = \pi_-\frac{\Xi_i}{z-\lambda_i} = \frac{\Xi_i(\lambda_i)}{z-\lambda_i} = \frac{F_i(\lambda_i)X_i}{z-\lambda_i},$$

which proves (10.54).

7. Recall from Theorem 3.19 that $\overline{L}_i(z) = l.c.r.m.\{D_j(z)|j \neq i\}$. Noting that, for $j \neq i$, both $D_j(z)^{-1}\overline{L}_i(z)$ and $\overline{D}_i(z)\frac{F_i(\lambda_i)X_i}{z-\lambda_i}$ are polynomial, one obtains

$$\pi_{D_j}\overline{L}_i(z)\left(\overline{D}_i(z)\frac{F_i(\lambda_i)X_i}{z-\lambda_i}\right) = D_j\pi_-(D_j^{-1}\overline{L}_i)\left(\overline{D}_i\frac{F_i(\lambda_i)X_i}{z-\lambda_i}\right) = 0.$$

On the other hand, for $j = i$,

$$\pi_{D_i}\overline{L}_i(z)\left(\overline{D}_i(z)\frac{F_i(\lambda_i)X_i}{z-\lambda_i}\right) = D_i\pi_-D_i^{-1}\overline{L}_i\overline{D}_i\frac{F_i(\lambda_i)X_i}{z-\lambda_i} = D_i(z)\pi_-\frac{L_iF_i(\lambda_i)X_i}{z-\lambda_i}$$
$$= D_i(z)\left(\frac{L_i(\lambda_i)F_i(\lambda_i)X_i}{z-\lambda_i}\right).$$

Note that $D_i(z)\left(\frac{L_i(\lambda_i)F_i(\lambda_i)X_i}{z-\lambda_i}\right) \in X_{D_i}$. This follows from (10.55) since

$$D_i(\lambda_i)L_i(\lambda_i)F_i(\lambda_i)X_i = \overline{L}_i(\lambda_i)\overline{D}_i(\lambda_i)F_i(\lambda_i)X_i = 0.$$

10.4 Exercises

1. Show that $W(z) = \begin{pmatrix} (\frac{z+1}{z})^2 & -\frac{z+1}{z^2} \\ 0 & \frac{z+1}{z} \end{pmatrix}$ admits a minimal factorization $W(z) = W_1(z)W_2(z)$, with $\delta(W_1) = 2$ and $\delta(W_2) = 1$.

2. Show that $W(z) = \begin{pmatrix} z & 0 \\ 1 & z \end{pmatrix}^{-1}\begin{pmatrix} z-1 & 0 \\ 1 & z-1 \end{pmatrix}$ admits no nontrivial, minimal factorization $W(z) = W_1(z)W_2(z)$.

3. Let $(A,b) \in \mathbb{R}^{n \times n} \times \mathbb{R}^n$ be reachable, $q(z) = \det(zI - A)$.

   a. Prove that there exists a unique row vector $c \in \mathbb{R}^{1 \times n}$ with $cb = \cdots = cA^{n-2}b = 0$ and $cA^{n-1}b = 1$. Moreover, the observability matrix $\mathscr{O}(c,A)$ is invertible.
   b. Prove that for every trajectory $(x(\cdot), u(\cdot)) \in C^1([0,T], \mathbb{R}^n) \times C([0,T])$ of $\dot{x} = Ax + bu$ there exists a unique C^n-function $y \in C^n([0,T])$ with

$$\mathscr{O}(c,A)x = \begin{pmatrix} y \\ \vdots \\ y^{(n-1)} \end{pmatrix}, \quad u = q(\frac{d}{dt})y. \qquad (10.56)$$

   Conversely, for $y \in C^n([0,T])$, each pair of functions (x,u) satisfying (10.56) solves $\dot{x} = Ax + bu$.

4. Characterize when the circulant system $x(t+1) = \mathscr{A}x(t) + \beta u(t)$, with

$$\mathscr{A} = \begin{pmatrix} a_0 & a_1 & \cdots & a_{N-1} \\ a_{N-1} & \ddots & \ddots & \vdots \\ \vdots & \ddots & \ddots & a_1 \\ a_1 & \cdots & a_{N-1} & a_0 \end{pmatrix}, \beta = \begin{pmatrix} \beta_1 \\ \beta_2 \\ \vdots \\ \beta_N \end{pmatrix} \in \mathbb{R}^N,$$

   is reachable. For $N = 3$, $a_0 = 2, a_1 = 0, a_2 = 1$ and $\beta_1 = \beta_2 = 1, \beta_3 = -1$ construct a control that steers $(0,0,0)^\top$ to $(1,-1,1)^\top$.

5. Let $\omega_k > 0$, with $\omega_i^2 \neq \omega_j^2$ for $i \neq j$. Find an open-loop control that steers the zero state of the parallel connection of harmonic oscillators

$$\ddot{z}_k + \omega_k^2 z_k = u, \quad k = 1, \dots, N$$

   to $(L, 0, L, 0, \dots, L, 0) \in \mathbb{R}^{2N}$ in finite time $T > 0$.

6. Let $r \geq 0, \omega > 0$.

   a. Compute an open-loop control $u_*(t)$ that steers the systems

$$\frac{d}{dt}\begin{pmatrix} z \\ \dot{z} \end{pmatrix} = \begin{pmatrix} 0 & 1 \\ -\omega^2 & -r \pm 1 \end{pmatrix}\begin{pmatrix} z \\ \dot{z} \end{pmatrix} \pm \frac{1}{\sqrt{2}}bu_*(t)$$

   from the zero state to the state $(1,0)^\top$ in finite time $T > 0$.
   b. Compute a control $u(t)$ that steers the coupled system

$$\frac{d}{dt}\begin{pmatrix} x_1 \\ \dot{x}_1 \\ x_2 \\ \dot{x}_2 \end{pmatrix} = \begin{pmatrix} 0 & 1 & 0 & 0 \\ -\omega^2 & -r & 0 & 1 \\ 0 & 0 & 0 & 1 \\ 0 & 1 & -\omega^2 & -r \end{pmatrix}\begin{pmatrix} x_1 \\ \dot{x}_1 \\ x_2 \\ \dot{x}_2 \end{pmatrix} + \begin{pmatrix} 0 \\ 1 \\ 0 \\ 0 \end{pmatrix}u(t)$$

   from the zero state to the state $(1,0,1,0)^\top$ in finite time $T > 0$.

10.5 Notes and References

The case of two systems connected in series was treated in Callier and Nahum (1975), where reachability and observability characterizations were obtained. In a short note by Fuhrmann (1975), this was extended to parallel connections. The extension to more than two interconnected systems is done here for the first time. It leads in particular to new resultant-type conditions (10.7) and (10.11) for the mutual coprimeness of matrix polynomials. One can extend the analysis of open-loop control for parallel connections to the series connection case. This involves a generalized, tangential Newton interpolation; however, because of the computational complexity, the details are omitted. We observe that one might also be able to handle other structures, such as tree structures or series-parallel connections, in the same way. Clearly, the present results are in no way final from a computational point of view. There is a potential for deriving minimal-length, open-loop controls for more general structures. In a similar direction, one would also like to extend this circle of ideas to the computation of stabilizing, or optimal, feedback controls for networks using knowledge of stabilizing, or optimal, feedback controls for the individual node systems.

In this chapter, our focus has been on the analysis of systems arising from interconnecting simple systems in some structured way. Of significant interest is also the inverse problem, namely, to realize a desired system transfer function via standard interconnections of simpler systems. Representing a system whose transfer function has the coprime factorizations $G(z) = D_\ell(z)^{-1}N_\ell(z) = N_r(z)D_r(z)^{-1}$ as a parallel connection is tantamount to finding partial fraction decompositions of the form $G(z) = \sum_{i=1}^s D_{\ell,i}(z)^{-1}N_{\ell,i}(z) = \sum_{i=1}^s N_{r,i}(z)D_{r,i}(z)^{-1}$. If one wants minimal representations, i.e., satisfying $\delta(D) = \sum_{i=1}^s \delta(D_{\ell,i})$, then this is equivalent, to the mutual right coprimeness of the $D_{\ell,i}(z)$. Thus, partial fraction decompositions are the key to determining whether a transfer arises from a parallel interconnection. One way to obtain partial fraction decompositions is by applying Theorem 3.15. Consider a spectral factorization $d(z) = \det D(z) = d_1(z) \cdots d_s(z)$ with the $d_i(z)$ pairwise coprime. By Theorem 3.14, this leads to factorizations $D(z) = \overline{D}_i(z)E_i(z)$, with the $\overline{D}_i(z)$ mutually left coprime and the $E_i(z)$ mutually right coprime. These imply corresponding partial fraction decompositions. Of course, finer partial fraction decompositions may exist.

The inverse question for the series connection has been partially answered, in the Hilbert space context, by Sakhnovich (1976). An algebraic version of this result was derived in Bart, Gohberg and Kaashoek (1979), which is stated as follows.

Theorem 10.28. *Let $G(z)$ be a biproper rational matrix function with minimal realization (A,B,C,I) in the state space $\mathcal{X}$. There exists a minimal rational matrix factorization*

$$G(z) = G_1(z)G_2(z) \tag{10.57}$$

if and only if there exists a direct sum decomposition $\mathcal{X} = \mathcal{X}_1 \oplus \mathcal{X}_2$ for which $\mathcal{X}_1$ is A-invariant and $\mathcal{X}_2$ is $(A - BC)$-invariant.

Here a factorization (10.57) is called a **minimal factorization** if $\delta(G) = \delta(G_1) +$ $\delta(G_2)$. This was extended, using polynomial methods, in Shamir and Fuhrmann (1985).

Theorem 10.29. *Let $G(z)$ be a biproper rational matrix function with coprime factorization $G(z) = T(z)^{-1}D(z)$. There exists a minimal factorization (10.57) if and only if there exist factorizations $T(z) = T_1(z)T_2(z)$ and $D(z) = D_1(z)D_2(z)$ such that $X_T = T_1 X_{T_2} \oplus D_1 X_{D_2}$. In this case, $G_1(z) = T_2(z)^{-1}D_1(z)$ and $G_2(z) = \overline{T}_1(z)^{-1}D_2(z)$, where $E(z) = T_1(z)D_1(z) = D_1(z)\overline{T}_1(z)$ is a l.c.r.m. of $T_1(z)$ and $D_1(z)$ such that $T(z)^{-1}E(z)$ is biproper.*

For proofs of these theorems we refer the reader to the cited sources.

Inverse problems for feedback connections, or more complicated system architectures, are an open research problem. This indicates that the available results on inverse problems for networks are meager and hardly scratch the surface of this important research area. These problems have a close affinity to classical network synthesis and, after close to a century of research, remain mostly unsolved; see, however, Kalman (2010). It seems that finding a minimal realization, be it a state space or polynomial, is not sufficient in practice. One must take into account other considerations, such as, for example, hierarchy, efficiency, internal communication delays, and robustness against failures. It is our conviction that this is an important area for future research. One of the authors recalls a conversation with the late George Zames (in 1978 in Montreal), in which Zames emphasized the importance of this line of research. It seems to be an idea whose time has come.

Chapter 11
Synchronization and Consensus

Synchronization is one of the fundamental aspects of self-organization in networks of systems. More generally, the emergence of macroscopic states is frequently encountered in dynamical systems when one starts to study coupling effects. Well-known examples include synchronization of oscillators, the emergence of consensus states in models that describe the opinion dynamics of social networks or multiagent systems, or flocking phenomena in biological networks such as swarms of birds or a school of fish. In all these different network models the dynamics of the individual states may "cluster" together or "synchronize" toward a common state that exhibits the system with a unique characteristic identity. The analysis and control of such synchronized states thus becomes an interesting new task for the design of networks. The phenomenon of synchrony was apparently noticed first by Huygens, who was, alongside his scientific activity, also a clock maker. Huygens noticed that two pendulum clocks hanging on a wall tend to synchronize. With time, a multitude of synchronization phenomena were observed in different fields, including, for example, ciliary beating in biology, laser physics, and the firing of neurons as in Parkinson disease. Although most realistic models are of course nonlinear, it appears to be of fundamental interest to explore these issues first in the simplified context of linear systems theory.

The synchronization of interconnected dynamical systems refers to the task of studying and designing coupling laws that force the state vectors of the node systems to converge to each other. In a classical model by Vicsek et. al. (1995), a simple local averaging rule was introduced as a basis for studying the cooperative behavior of animals. The closely related concept of consensus has emerged through studies of multiagent systems and social and computer networks. Obviously, the process of achieving a consensus is a fundamental step that leads to coherent social network structures. A fundamental model for opinion dynamics is due to Krause (1997) and exhibits interesting clustering and consensus effects. The Hegselmann–Krause model is a simplification of the Vicsek model that makes it more attractive for engineering implementations.

© Springer International Publishing Switzerland 2015
P.A. Fuhrmann, U. Helmke, *The Mathematics of Networks*
of Linear Systems, Universitext, DOI 10.1007/978-3-319-16646-9_11

The coupling structure of a multiagent network is defined by a weighted graph whose vertices correspond to the agent dynamics, while the weights of the edges relate to the coupling strengths. Thus the underlying graph structure plays an important role in investigating synchronization phenomena. In the systems engineering literature, diffusive coupling models have been popular for studying synchronization, but more general coupling topologies are of obvious interest, too. Nondiffusive couplings arise, for instance, in systems biology; an example is a model of gene-regulatory networks with a cyclic graph structure, as proposed in Hori, Kim and Hara (2011). In the sequel, we will allow for more general interconnection structures than diffusive coupling, with the consequence that the interconnection matrix is no longer a Laplacian matrix. We present a fairly general approach to problems of synchronization and clustering for networks of identical linear systems. Since the mathematical problems of consensus are somewhat easier to describe, we first present an overview of the main models used in this field. We describe simple autonomous models for mean value consensus and then examine synchronization, for both first-order and higher-order system dynamics. For the latter system class, synchronization is often referred to as both partial state and output synchronization.

11.1 Consensus and Clustering in Opinion Dynamics

Dynamical systems that achieve consensus or clustering may serve as models for opinion dynamics in social networks. Consensus in a network occurs when the participants agree upon their opinions, while clustering refers to the formation of different opinion spectra. We describe simple models for social dynamics using systems of differential or difference equations that exhibit cooperative dynamics. Consider a finite number of agents that are allowed to communicate their opinion on a certain topic. We identify the opinion of the $i-th$ agent with a real variable $x_i \in \mathbb{R}$, while communication among the agents is modeled by an interconnection graph. The topology of the graph is determined by the pattern describing the extent to which the opinions of each agent are influenced by the opinions of the other agents. Thus the opinions of N agents in the network are described by the evolution of a vector $x \in \mathbb{R}^N$, while the communication pattern is specified by a real $N \times N$ matrix $A(x)$. The simplest class of models for consensus that have been described in the literature is of the form

$$\dot{x} = A(x)x, \tag{11.1}$$

where the equilibrium states of (11.1) correspond to the consensus states with equal components $x_1 = \cdots = x_N$. The task is to specify realistic interconnection matrices $A(x)$ such that all solutions of (11.1) converge to a consensus state. Thus consensus, or synchronization, is essentially a stabilization problem in a subspace of desired states. Monotone maps $x \mapsto A(x)x$ are one method of choice here. Another phenomenon of interest is clustering, i.e., the effect that only certain classes of agents become asymptotically equal to each other. Further related topics of interest are synchronization of higher-order systems, distributed computing, formation control and Markov chains on a graph.

In this section we analyze a linear model for mean value consensus and then turn to a brief description of nonlinear models for consensus. In the subsequent sections we will then completely focus on the synchronization of identical node systems.

1. Linear Models for Consensus. We describe a simple linear model for consensus that is already useful for a number of applications. Let $\Gamma = (V, E)$ denote a weighted digraph without loops on $V = \{1, \dots, N\}$ with adjacency matrix $A = (a_{ij})$. Every vertex i is thought of as representing a dynamic agent $x_i \in \mathbb{R}^n$ described by a linear system

$$
\begin{aligned}
\dot{x}_i &= \alpha x_i + \beta v_i, \\
w_i &= \gamma x_i,
\end{aligned}
\tag{11.2}
$$

with input v_i. A dynamical network with protocol v is then a feedback system

$$
\begin{aligned}
\dot{x}_i &= \alpha x_i + \beta v_i, \\
v_i &= \sum_{j=1}^{N} a_{ij} w_j.
\end{aligned}
\tag{11.3}
$$

Definition 11.1. System (11.3) solves the **consensus problem** for a function $y = k(x)$ if there exists an interconnection matrix A and an asymptotically stable equilibrium point $x^* \in \mathbb{R}^{nN}$ of

$$
\dot{x} = \Big(I_N \otimes \alpha + (I_N \otimes \beta) A (I_N \otimes \gamma) \Big) x
$$

such that $x_1^* = \cdots = x_N^* = k(x(0))$. If k is chosen as the mean value $k(x) = \frac{1}{N} \sum_{i=1}^{N} x_i$, then this is called the solution to the **mean value consensus (MVC) problem.**

In its simplest possible form, the mean value consensus system for a system of integrators (where $\alpha = 0$ and $\beta = \gamma = I_n$)

$$
\dot{x}_i = v_i, \quad i = 1, \dots, N
$$

is of the form

$$
\dot{x}_i = \sum_{j=1}^{N} A_{ij}(x_j - x_i), \quad i = 1, \dots, N,
$$

where $A \in \mathbb{R}^{N \times N}$ is the adjacency matrix of a digraph Γ. Let

$$L = \mathrm{diag}\,(A\mathbf{e}) - A$$

denote the associated Laplacian matrix and $\mathbf{e} := (1,\ldots,1)^\top \in \mathbb{R}^N$. Thus, for interconnection matrices of the form $A = L \otimes I_n$, one obtains the **standard MVC system** as

$$\dot{x} = -(L \otimes I_n)x. \tag{11.4}$$

For the general control system (11.2) and interconnection matrix $A = L \otimes K$ one obtains the **dynamic MVC system**

$$\dot{x} = \big(I_N \otimes \alpha - L \otimes \beta K \gamma\big)x.$$

This model will be studied in more detail in Section 11.2 on synchronization.

We begin with a discussion of (11.4). A directed weighted graph Γ with adjacency matrix A and Laplacian L is called **balanced** if

$$\sum_{j=1}^{N} a_{ij} = \sum_{j=1}^{N} a_{ji}$$

for all i. This is equivalent to $\mathbf{e}^\top L = 0$. Note that $L\mathbf{e} = 0$ is always satisfied for Laplacian matrices. Obviously, an undirected graph is balanced. The following characterization of systems that achieve mean value consensus is due to Olfati-Saber, Fax and Murray (2004).

Theorem 11.2. *Let Γ be a strongly connected weighted digraph and L the associated Laplacian. Then $L\mathbf{e} = 0$, and there exists a unique row vector $\mathbf{c} \in \mathbb{R}^{1 \times N}$, with $\mathbf{c}L = 0$ and $\mathbf{c}\mathbf{e} = 1$. Moreover, the following properties are satisfied:*

1. *Every solution of (11.4) converges to a scalar multiple of $\mathbf{e}$, i.e., for each initial condition $x(0)$,*

$$\lim_{t \to \infty} x(t) = (\mathbf{e}\mathbf{c} \otimes I_n)x(0). \tag{11.5}$$

   In particular,

$$\lim_{t \to \infty} e^{-tL \otimes I_n} = \mathbf{e}\mathbf{c} \otimes I_n. \tag{11.6}$$

2. *System (11.4) achieves mean value consensus if and only if Γ is balanced, i.e., if and only if $\mathbf{e}^\top L = 0$. In that case,*

$$\lim_{t \to \infty} e^{-tL \otimes I_n} = \frac{1}{N}\mathbf{e}\mathbf{e}^\top \otimes I_n.$$

Proof. Since Γ is strongly connected, Theorem 8.36 shows that the Laplacian matrix L has a simple eigenvalue $\lambda_1 = 0$, with eigenvector $\mathbf{e}$ and $N - 1$ eigenvalues $\lambda_2, \ldots, \lambda_N$ with positive real part. In particular, the left kernel of L is one-dimensional. After a similarity transformation S, one can assume that

$$-SLS^{-1} = \begin{pmatrix} 0 & L_{12} \\ 0 & L_{22} \end{pmatrix},$$

with L_{22} Hurwitz. Using the block-diagonal structure of SLS^{-1}, the matrix exponential of SLS^{-1} is readily computed. This shows that

$$\lim_{t \to \infty} e^{-tSLS^{-1}} = \begin{pmatrix} 1 & -L_{12}L_{22}^{-1} \\ 0 & 0 \end{pmatrix},$$

and the row vector $\lambda^{\top} = (1, -L_{12}L_{22}^{-1})$ satisfies $\lambda^{\top} SLS^{-1} = 0$ and $\lambda^{\top} e_1 = 1$. Thus $\mathbf{c} := \lambda^{\top} S$ satisfies $\mathbf{c}L = 0$ and $\mathbf{c}\mathbf{e} = \mathbf{c}S^{-1}e_1 = \lambda^{\top} e_1 = 1$. Since the kernel of L is one-dimensional, $\mathbf{c}$ is uniquely determined.

One concludes that $\lim_{t \to \infty} e^{-tL}$ exists and is of the form $\lim_{t \to \infty} e^{-tL} = \mathbf{ec}$. The identity

$$e^{-tL \otimes I_n} = e^{-tL} \otimes I_n$$

then implies

$$\lim_{t \to \infty} e^{-tL \otimes I_n} = \mathbf{ec} \otimes I_n.$$

This proves (11.6), which in turn implies (11.5). The stronger mean value consensus property is satisfied if and only if

$$\lim_{t \to \infty} e^{-tL \otimes I_n} x(0) = \frac{1}{N} (\mathbf{ee}^{\top} \otimes I_n) x(0)$$

for all initial conditions $x(0)$. By (11.6), this is equivalent to $\mathbf{ec} \otimes I_n = \frac{1}{N} \mathbf{ee}^{\top} \otimes I_n$, i.e., to $c = \frac{1}{N} \mathbf{e}^{\top}$. This completes the proof.   ∎

This result is easily extended as follows to graphs with time-varying interconnections. We present here only one such extension. Let $\mathscr{L}$ denote a compact set of real $N \times N$ matrices such that every element $L \in \mathscr{L}$ has nonnegative nondiagonal entries, satisfies $L\mathbf{e} = 0$, $\mathbf{e}^{\top}L = 0$, and is irreducible.

Theorem 11.3. *Let $L : [0, \infty) \longrightarrow \mathscr{L}$ be piecewise continuous. The solutions $x(t) \in \mathbb{R}^{nN}$ of*

$$\dot{x}(t) = -(L(t) \otimes I_n)x(t) \tag{11.7}$$

satisfy

$$\lim_{t\to\infty} x(t) = \frac{1}{N}(ee^\top \otimes I_n)x(0),$$

with error bounds

$$\left\|x(t) - \frac{1}{N}(ee^\top \otimes I_n)x(0)\right\|^2 \le e^{-2t\kappa}\left\|(I_{nN} - \frac{1}{N}ee^\top \otimes I_n)x(0)\right\|^2 \le e^{-2t\kappa}\|x(0)\|^2,$$

where $\kappa := min_{L\in\mathscr{L}}\lambda_2(L+L^\top) > 0.$

Proof. Since $\mathscr{L}$ is a compact set of irreducible Laplacians, 0 is a simple eigenvalue of each matrix $L \in \mathscr{L}$. Moreover, by Theorem 8.36, the assumption $e^\top L = 0$ implies that the symmetric matrix $P = L + L^\top$ is positive semidefinite with simple eigenvalue 0. Let $\lambda_1(P) = 0 \le \lambda_2(P) \le \cdots \le \lambda_N(P)$ denote the eigenvalues of P. Thus $\lambda_k(L + L^\top) > 0$ for all $L \in \mathscr{L}$ and $2 \le k \le N$. Since $\mathscr{L}$ is compact, κ exists and is strictly positive. Since $\lambda_2(L+L^\top)$ is the smallest eigenvalue of the restriction of $L+L^\top$ on the invariant subspace $(e \otimes \mathbb{R}^n)^\perp$, we see that

$$2\delta^\top L \otimes I_n \delta = \delta^\top \left((L+L^\top)\otimes I_n\right)\delta \ge \lambda_2(L+L^\top)\|\delta\|^2 \ge \kappa\|\delta\|^2$$

for all $\delta \in (e \otimes \mathbb{R}^n)^\perp$. Then every solution $x(t)$ of (11.7) satisfies $ee^\top x(t) = ee^\top x(0)$ since $(ee^\top \otimes I_n)(L \otimes I_n) = 0$. Thus

$$\delta(t) := (I_{nn} - \frac{1}{N}ee^\top \otimes I_n)x(t) = x(t) - \frac{1}{N}(ee^\top \otimes I_n)x(0)$$

is a solution of (11.7). This implies

$$\frac{d\delta^\top\delta}{dt} = 2\delta^\top\frac{d\delta}{dt} = -2\delta^\top(L \otimes I_n)\delta \le -2\kappa\|\delta\|^2.$$

Therefore,

$$\|\delta(t)\| \le e^{-t\kappa}\|\delta(0)\|,$$

and the result follows.   ∎

Similar results are valid in the discrete-time case when the Laplacian matrix L is replaced by the **normalized Laplacian** matrix

$$F = D^{-1}A.$$

Here $D = \text{diag}(A\mathbf{e})$ and A denotes the adjacency matrix of a weighted graph. If A is nonnegative, then the normalized Laplacian is a stochastic matrix. The following result gives a simple sufficient condition for consensus, which can be generalized in several directions (see Section 11.7, "Notes and References").

Theorem 11.4. *Let $x(0) \in \mathbb{R}^n$ and F be a primitive stochastic matrix. Then there exists a real number x_* such that the solution*

$$x(t+1) = Fx(t)$$

satisfies

$$\lim_{t \to \infty} x(t) = x_* \mathbf{e}.$$

In particular, this is satisfied for normalized Laplacians of connected graphs Γ with at least one loop around a vertex.

Proof. The primitivity of F implies (Theorem 8.18) that $\lim_{t \to \infty} F^t = \mathbf{e}c$ for some $c \in \mathbb{R}^{1 \times n}$. This implies the result. ∎

2. Distributed Algorithm for Solving Linear Equations. We illustrate the preceding analysis by showing how consensus algorithms can be used to design distributed methods for solving linear systems of equations. Distributed computing is of course a vast topic, and we leave it to the reader to explore further applications in this field.

Let us start with a classical problem from linear algebra, i.e., the computation of intersection points of a finite number of affine linear subspaces $L_i = a_i + V_i, i = 1, \ldots, N$, of a d-dimensional Hilbert space $\mathbb{H}$. We are interested in computing points $x^* \in E$ in the intersection $L_1 \cap \cdots \cap L_N$. For simplicity, let us assume that the vector spaces V_i are in general position in the sense that the direct sum decomposition

$$V_1^\perp \oplus \cdots \oplus V_N^\perp = \mathbb{H}$$

is satisfied. This implies both $V_1 \cap \cdots \cap V_N = \{0\}$ and $\sum_{i=1}^N \dim V_i^\perp = d$. In particular, whenever $L_1, \ldots, L_N$ intersect, they intersect at a unique point

$$L_1 \cap \cdots \cap L_N = \{x^*\},$$

and our goal is to compute the unique intersection point x^*. Of course, there are several methods available to do this, but our focus is on demonstrating how consensus algorithms can help in computing x^*. To this end, we introduce the selfadjoint projection operators $P_i : \mathbb{H} \longrightarrow V_i \subset \mathbb{H}$ onto V_i, with kernel $V_i^\perp$. Then the direct sum $P = \text{diag}(P_1, \ldots, P_N) : \mathbb{H}^N \longrightarrow \mathbb{H}^N$ is a selfadjoint projection operator onto $V_1 \oplus \cdots \oplus V_N \subset \mathbb{H}^N$.

We next present graph-theoretic ideas relevant for distributed computing. Fix an undirected and connected graph Γ whose set of vertices $\mathcal{V} = \{1, \ldots, N\}$ is labeled by the N linear subspaces V_i, together with a set of M edges $\mathscr{E}$ that define which vertices can interact with each other during the course of running the algorithm. Let $\mathfrak{A}$ denotes the associated $0,1$-adjacency matrix of the graph. For technical reasons we assign to each vertex a single self-loop of the graph, so that the diagonal entries of the adjacency matrix are all equal to 1. Let $d_i \geq 2$ denote the degree of the ith vertex, and set $D := \operatorname{diag}(d_1, \ldots, d_N)$. Let $B = (b_{ij}) \in \mathbb{R}^{N \times M}$ denote the oriented incidence matrix of the graph defined by (8.13). Since Γ is connected, the incidence matrix B has full row rank $N - 1$, and therefore the kernel of B has dimension $M - N + 1$. The normalized graph Laplacian is the stochastic matrix

$$\mathscr{L} = D^{-1}\mathfrak{A} = I_N - D^{-1}BB^\top.$$

We need the following lemma.

Lemma 11.5. *Assume that* $V_1^\perp \oplus \cdots \oplus V_N^\perp = \mathbb{H}$. *Then*

$$\operatorname{Ker} P(B \otimes I_d) = \operatorname{Ker}(B \otimes I_d).$$

Proof. We first prove

$$\operatorname{Ker}(B^\top \otimes I_d) \cap \operatorname{Im} P = \{0\}. \tag{11.8}$$

In fact, $x \in \operatorname{Ker}(B^\top \otimes I_d)$ implies that $x = \mathbf{e} \otimes v$ for a suitable element $v \in \mathbb{H}$. Therefore, $x \in \operatorname{Ker}(B^\top \otimes I_d) \cap \operatorname{Im} P$ if and only if $x = \mathbf{e} \otimes v$, and there exist elements $\xi_1, \ldots, \xi_N$ with $v = P_1\xi_1 = \cdots = P_N\xi_N$. Equivalently,

$$v \in \bigcap_{j=1}^N \operatorname{Im} P_j = \bigcap_{j=1}^N V_j = \{0\}.$$

This proves (11.8). By taking orthogonal complements,

$$\operatorname{Im}(B \otimes I_d) + \operatorname{Ker} P = \mathbb{H}^N.$$

Since $\operatorname{rk}(B \otimes I_d) = d \operatorname{rk} B = d(N - 1)$ and $\dim \operatorname{Ker} P = \sum_{i=1}^N \dim V_i^\perp = \dim \mathbb{H} = d$, one concludes that $\operatorname{Im}(B \otimes I_d) \cap \operatorname{Ker} P = \{0\}$. This implies the result. ∎

Proposition 11.6. *Assume that* $V_1^\perp \oplus \cdots \oplus V_N^\perp = \mathbb{H}$. *Then each eigenvalue* λ *of* $P(\mathscr{L} \otimes I_d)P$ *is real and satisfies* $-1 < \lambda < 1$.

Proof. The normalized Laplacian $D^{-1}\mathfrak{A}$ has nonzero entries on the diagonal. Thus Theorem 8.42 applies and implies that all eigenvalues of $\mathscr{L} \otimes I_d$ are real and are contained in the interval $(-1, 1]$. Applying a similarity transformation, we see that the same property is true for the symmetric matrix $D^{-\frac{1}{2}}\mathfrak{A}D^{-\frac{1}{2}} \otimes I_d$. Since P is a projection operator, the spectrum of $P(D^{-\frac{1}{2}}\mathfrak{A}D^{-\frac{1}{2}} \otimes I_d)P$ is contained in the convex hull of the spectrum of $D^{-\frac{1}{2}}\mathfrak{A}D^{-\frac{1}{2}} \otimes I_d$, i.e., it is contained in $(-1, 1]$. Using

$D^{\frac{1}{2}}P(\mathscr{L} \otimes I_d)PD^{-\frac{1}{2}} = P(D^{-\frac{1}{2}}\mathfrak{A}D^{-\frac{1}{2}} \otimes I_d)P$, one concludes that $P(\mathscr{L} \otimes I_d)P$ has all its eigenvalues in $(-1, 1]$. It remains to show that 1 is not an eigenvalue of $P(\mathscr{L} \otimes I_d)P$. In fact, otherwise there exists $v \in \operatorname{Im} P$ with $P(\mathscr{L} \otimes I_d)v = v$. Since $D^{-1}\mathfrak{A} = I - D^{-1}L = I - D^{-1}BB^{\top}$, this is equivalent to $P(D^{-1}BB^{\top} \otimes I_d)v = 0$. Since P and $D \otimes I_d$ commute, this is equivalent to $P(BB^{\top} \otimes I_d)v = 0$. Thus, using Lemma 11.5, $(BB^{\top} \otimes I_d)v = 0$ or, equivalently, $(B^{\top} \otimes I_d)v = 0$. This shows that $v \in \operatorname{Ker}(B^{\top} \otimes I_d) \cap \operatorname{Im} P$. By (11.8), thus $v = 0$, and we are done. ∎

After these preparatory remarks, we are now ready to introduce and study the distributed algorithm for subspace intersections. The key idea is very simple to describe. Suppose one has computed for each $t \in \mathbb{N}$ and $i = 1, \dots, N$ an element $x_i(t) \in L_i$. Then for each $u_i(t) \in \mathbb{H}$ the linear control system

$$x_i(t+1) = x_i(t) + P_i u_i(t)$$

evolves in the affine subspace L_i. In fact, the right-hand side describes all elements of L_i. Choose the input vector $u_i(t)$ such that the difference

$$\left\| x_i(t+1) - \frac{1}{d_i} \sum_{j \in \mathscr{N}_i} x_j(t) \right\|^2$$

is minimized. By a straightforward computation, this leads to the recursion

$$x_i(t+1) = (I_d - P_i)x_i(t) + P_i \left(\frac{1}{d_i} \sum_{j \in \mathscr{N}_i} x_j(t) \right), \quad x_i(0) \in L_i, \tag{11.9}$$

which is clearly in distributed form. Using the vector notation $x(t) = \operatorname{col}(x_1(t), \dots, x_N(t))$, this is equivalent to

$$x(t+1) = (I_{dN} - P)x(t) + P(\mathscr{L} \otimes I_d)x(t), \quad x(0) \in L_1 \times \cdots \times L_N.$$

Theorem 11.7. *Assume that $V_1^{\perp} \oplus \cdots \oplus V_N^{\perp} = \mathbb{H}$. Then the distributed algorithm (11.9) converges exponentially fast from every initial point $x(0) \in L_1 \times \cdots \times L_N$ to $e \otimes x^*$, where $x^* \in \mathbb{H}$ denotes the unique intersection point of $L_1 \cap \cdots \cap L_N$.*

Proof. Let $z(t) := x(t) - e \otimes x^*$. Then $z(t)$ satisfies the recursion

$$z(t+1) = P(\mathscr{L} \otimes I_d)z(t).$$

By Lemma 11.6, the eigenvalues of $P(\mathscr{L} \otimes I_d)$ are in the open unit interval $(-1, 1)$. Thus $z(t)$ converges exponentially fast to 0. The result follows. ∎

Of course, it is trivial to apply the preceding ideas to solving linear equations $Ax = b$. Let $A \in \mathbb{R}^{n \times n}$, with nonzero row vectors $A_1, \dots, A_n \in \mathbb{R}^{1 \times n}$ and $b = \operatorname{col}(b_1, \dots, b_n) \in \mathbb{R}^n$. Defining the affine hyperplanes

$$L_i = \{x_i \in \mathbb{R}^n \mid A_i x_i = b_i\}$$

with subspaces $V_i = \operatorname{Ker} A_i$ we see that the solutions of $Ax = b$ are just the intersection points in $L_1 \cap \cdots \cap L_n$. Moreover, A is invertible if and only if $V_1^\perp \oplus \cdots \oplus V_n^\perp = \mathbb{R}^n$. The projection operators are

$$P_i = I - \frac{A_i^\top A_i}{\|A_i\|^2}.$$

The next consequence of Theorem 11.7 is obvious.

Theorem 11.8. *Let A be invertible. Then the distributed algorithm*

$$x_i(t+1) = \frac{A_i^\top b_i}{\|A_i\|^2} + (I - \frac{A_i^\top A_i}{\|A_i\|^2}) \left(\frac{1}{d_i} \sum_{j \in \mathcal{N}_i} x_j(t) \right), \quad A_i x_i(0) = b_i,$$

converges from each solution of $A_i x_i(0) = b_i, i = 1, \ldots, n$, exponentially fast to $\operatorname{col}(A^{-1}b, \ldots, A^{-1}b)$.

3. Nonlinear Models for Consensus. Of course, the preceding, rather brief, discussion of linear consensus models can be extended in several directions, including to a study of the effects of dynamic or stochastic interconnections, allowing for nonlinear models and analyzing robustness issues. Here we focus on nonlinear models. We begin with a rather straightforward extension of Theorem 11.3 to nonlinear coupling models of the form

$$\dot{x} = -(L(x) \otimes I_n)x. \tag{11.10}$$

Of course, discrete-time models can be considered as well. Here $x \mapsto L(x)$ denotes a smooth function of Laplacian $N \times N-$ matrices

$$L(x) = D(x) - A(x),$$

where $D(x) = \operatorname{diag} A(x)\mathbf{e}$. To define $A(x)$, we introduce an **influence function** as a smooth strictly positive function $\chi : \mathbb{R} \longrightarrow [0, \infty)$ that is monotonically decreasing on $[0, \infty)$. The function χ is regarded as a measure of how strongly mutual agents influence each other. Thus, in applications to opinion dynamics, two agents, x_i and x_j, are thought of as influencing each other's opinions if the value of the influence function $\chi(x_i - x_j)$ is large, and small otherwise. Possible choices for such influence functions are, for example,

(a) Constant functions;
(b) The indicator function $\chi_r = \chi_{[-r,r]}$ on a compact interval $[-r, r]$;
(c) The potential $\chi(x) = k(1 + x^2)^{-\beta}$ for $\beta > 0$;
(d) The Gaussian distribution $\chi(x) = e^{-x^2}$.

Let $M = (m_{ij})$ denote a nonnegative matrix, for example, the weighted adjacency matrix of a graph. Thus M defines the relevant interconnections that are allowed between the various agents. Consider the adjacency matrix of a time-varying neighborhood graph defined as

$$A(x) = (m_{ij}\chi(\|x_i - x_j\|)) \in \mathbb{R}^{N \times N}.$$

Note that $A(x)$ is symmetric for all x whenever the scaling matrix M is symmetric.

Theorem 11.9. *Assume that M is irreducible, symmetric, and nonnegative. For each initial condition $x(0)$ the unique solution $x(t)$ of (11.10) exists for all $t \geq 0$. Moreover, $\|x_i(t) - x_j(t)\|$ converges to 0 as $t \to \infty$.*

Proof. Let $\Delta = \mathbf{e} \otimes \mathbb{R}^n$ denote the diagonal in $\mathbb{R}^{nN} = \mathbb{R}^n \otimes \cdots \otimes \mathbb{R}^n$, and let

$$\phi(x) = \min_{x \in \Delta^\perp} \frac{x^\top (L(x) \otimes I_n) x}{\|x\|^2}$$

denote the Fiedler number, i.e., $\phi(x) = n\lambda_2(x)$, where $\lambda_2(x)$ denotes the second smallest eigenvalue of $L(x)$. Thus

$$x^\top (L(x) \otimes I_n) x = \frac{1}{2} \sum_{i,j=1}^{N} a_{ij} \|x_i - x_j\|^2 \geq \phi(x) \|x\|^2.$$

For each solution of (11.10),

$$\frac{d}{dt} \|x\|^2 = -2x^\top (L(x) \otimes I_n) x \leq -\phi(x) \|x\|^2 \leq 0.$$

Thus the norm $\|x(t)\|$ decreases monotonically, and therefore $x(t)$ is positively bounded. This shows the existence of solutions for all $t \geq 0$. To proceed with the analysis, we need a lower bound on the Fiedler number. This is achieved as follows. Let $L_M = D_M - M$ denote the associated Laplacian of M. Since M is assumed to be irreducible, the Fiedler number μ of L_M is strictly positive. Moreover, by the monotonicity of χ,

$$\chi(\|x_i - x_j\|^2) \geq \chi(\sum_{i,j=1}^{N} \|x_i - x_j\|^2) = \chi((2N-1)\|x\|^2).$$

By the symmetry of L, we have that $(\mathbf{e}^\top \otimes x^\top)(L \otimes I_n) = 0$. Therefore, the orthogonal complement $\Delta^\perp$ is invariant under the flow of (11.10). Thus, for all $0 \neq x \in \Delta^\perp$,

$$\frac{d}{dt}\|x\|^2 = -\sum_{i,j=1}^{N} a_{ij}\|x_i - x_j\|^2 = -\sum_{i,j=1}^{N} m_{ij}\chi(\|x_i - x_j\|^2)\|x_i - x_j\|^2$$

$$\leq -\chi((2N-1)\|x\|^2)\sum_{i,j=1}^{N} m_{ij}\|x_i - x_j\|^2$$

$$= -2x^{\top}(L_M \otimes I_n)\,x\chi((2N-1)\|x\|^2)$$

$$\leq -2\mu\chi((2N-1)\|x\|^2)\|x\|^2 < 0.$$

This shows that each solution $x(t)$ of (11.10) satisfies $\lim_{t\to\infty} \mathrm{dist}(x(t),\Delta) = 0$. The result follows. ∎

The **Vicsek model** is one of the first nonlinear models for swarm formation in a multiagent dynamical system. The system studied by Vicsek et. al. (1995) is described as follows. Consider N agents $x_1,\dots,x_N \in \mathbb{R}^2$ in Euclidean space, modeled as simple integrators

$$\dot{x}_i = u_i, \quad i = 1,\dots N.$$

For positive real numbers $r > 0$, define as follows a time-varying undirected neighborhood graph $\Gamma(t) = (V,E(t))$, with the set of vertices $V = \{1,\dots,N\}$. The edges are characterized by the property that $\{i,j\} \in E(t)$ if and only if $\|x_i(t) - x_j(t)\| \leq r$. Let $\mathcal{N}_i(t) = \{j \mid \|x_i(t) - x_j(t)\| \leq r\}$ denote the set of $n_i(t)$ neighbors of agent i at time t. The agents are allowed to move with constant velocity but varying directions in the plane. The goal is to create a distributed feedback law that enables agents to synchronize the directions in which they move.

For discrete-time systems, a simple idea about how to achieve this is that each agent averages over the directions of all other agents within his/her neighborhood region. Following Jadbabaie, Lin and Morse (2003), this can be formalized in the following simple mean value consensus model. Fix a desired speed as $v > 0$ and

$$\theta_i(t+1) = \frac{1}{n_i(t)} \sum_{j \in \mathcal{N}_i(t)} \theta_j(t)$$

$$x_i(t+1) = x_i(t) + v e^{\sqrt{-1}\theta_i(t+1)}. \tag{11.11}$$

Here $0 \leq \theta_i < 2\pi$ denotes the polar angle of x_i. This can be conveniently rewritten as follows. Let $\theta = (\theta_1,\dots,\theta_N)^{\top} \in [0,2\pi]^N$, and let

$$\chi_r(t) = \begin{cases} 1 & 0 \leq t \leq r, \\ 0 & t > r \end{cases}$$

denote the characteristic function. The graph adjacency matrix of $\Gamma(t)$ is

$$A(t) := (\chi_r(\|x_i(t) - x_j(t)\|)).$$

Let $D(t) = \text{diag}(A(t)\mathbf{e})$, and let

$$F(t) = D(t)^{-1}A(t)$$

denote the normalized Laplacian. $F(t)$ is also called a **flocking matrix**. The discrete-time Vicsek model is then

$$\theta(t+1) = F(t)\theta(t). \qquad (11.12)$$

Consensus in the Vicsek system occurs, provided for each initial condition $\theta(0)$ there exists a constant $\theta_* \in [0, 2\pi]$ such that the solution of (11.12) satisfies

$$\lim_{t \to \infty} \theta_i(t) = \theta_*$$

for $i = 1, \ldots, N$. Sufficient conditions for consensus depend on certain connectivity properties of the time-varying graphs $\Gamma(t)$, such as, for example, that the flocking matrix is primitive for all t. Note that the flocking matrix $F(t)$ is a nonnegative matrix that is congruent to the adjacency matrix $A(t)$. Therefore, $F(t)$ is irreducible if and only if $A(t)$ is irreducible, or, equivalently, if and only if the graph $\Gamma(t)$ is strongly connected. Moreover, $F(t)\mathbf{e} = \mathbf{e}$. Therefore, $F(t)$ is a stochastic matrix with positive entries on the diagonal. Thus, Theorem 8.23 implies that $F(t)$ is primitive if and only if $\Gamma(t)$ is strongly connected. Note further that $F(t)$ being a stochastic matrix for all t implies that the solutions $\theta(t)$ of the Vicsek model leave $[0, 2\pi]^N$ invariant.

Although the results on linear consensus provide explicit sufficient conditions for a consensus of general time-varying systems (11.12), they cannot be directly applied to the Vicsek model (11.11). In fact, in the Vicsek model (11.11) the entries of the flocking matrix $F(t)$ depend on the relative distances $\|x_i(t) - x_j(t)\|$ of the agents and thus depend in turn on $\theta(t)$. This shows that connectivity assumptions on the time-varying neighborhood graph, such as, for example, the primitivity of $F(t)$, cannot be assumed a priori. What is needed are assumptions on, for example, the initial positions $x_1(0), \ldots, x_N(0)$ of the agents that guarantee that $F(t)$ will remain primitive for all t. This is difficult to verify for the Vicsek model because of the hard constraints defined by the characteristic function $\chi_r(t)$.

A simplification of the Vicsek model is due to Krause (1997). While the Vicsek swarm model is a second-order model that describes the evolution of points in the plane, the **Hegselmann–Krause model** Hegselmann and Krause (2002) works in the real line. Its main motivation is drawn from understanding the dynamics of social networks, in particular the formation and development of opinions in such networks. We regard an opinion of an agent as a real variable x. Thus the opinion dynamics in a network of N agents is described by the evolution of N real variables $x_1, \ldots, x_N \in \mathbb{R}$. In its simplest form, the Hegselmann–Krause model is

$$x_i(t+1) = \frac{\sum_{j:|x_i(t)-x_j(t)|\leq r} x_j(t)}{|\{j: |x_i(t) - x_j(t)| \leq r\}|}, \quad i = 1,\ldots,N. \tag{11.13}$$

Here agents i and j are thought of as influencing each other if and only if the distance between their opinions is small, i.e., $|x_i - x_j| \leq r$. This defines a state-dependent weighted adjacency matrix $A(x) = (a_{ij}(x))$, with

$$a_{ij}(x) = \begin{cases} 1 & \text{if } \ |x_i - x_j| \leq r \\ 0 & \text{else.} \end{cases}$$

More generally, using the notion of influence functions, one can define the adjacency matrix of a time-varying neighborhood graph of opinions as

$$A(x) = (\chi(\|x_i - x_j\|) \in \mathbb{R}^{N \times N}$$

and the flocking matrix as the normalized Laplacian

$$F(x) = D(x)^{-1}A(x), \quad D(x) = \text{diag}\,(A(x)\mathbf{e}).$$

Thus the ij-entry of $A(x)$ is small whenever agents x_i and x_j are not influencing each other's opinions. The generalized Hegselmann–Krause model for opinion dynamics is then

$$x(t+1) = F(x(t))x(t), \quad t \geq 0.$$

Thus, while the Hegselmann–Krause model (11.13) looks similar to the Vicsek model (11.11), the dynamics of (11.13) are simpler than that of (11.11). Without going into details, we mention that there exists an elegant convergence theory for the Krause model that is based on the theory of monotone operators.

11.2 Synchronization of Linear Networks

Using the theory of interconnected systems developed in Chapter 9, we now proceed to a general synchronization analysis of networks of linear systems. We consider networks of N identical interconnected linear systems, where the dynamics of each node $i = 1, 2, \ldots, N$ are described in state-space form as

$$\begin{aligned} \dot{x}_i(t) &= \alpha x_i(t) + \beta v_i(t), \\ w_i(t) &= \gamma x_i(t). \end{aligned} \tag{11.14}$$

Here the node system $\alpha \in \mathbb{R}^{n \times n}$, $\beta \in \mathbb{R}^{n \times m}$, $\gamma \in \mathbb{R}^{p \times n}$ is assumed to be reachable and observable. To define a network of such identical linear systems, we fix a **state interconnection structure**, defined by a matrix $A \in \mathbb{R}^{mN \times pN}$, and **input/output interconnection matrices** $B = (B_1^\top, \ldots, B_N^\top)^\top \in \mathbb{R}^{mN \times q}$ and $C = (C_1, \ldots, C_N) \in \mathbb{R}^{l \times pN}$, with $B_i \in \mathbb{R}^{m \times q}$ and $C_i \in \mathbb{R}^{l \times p}$, respectively. In the sequel, we will consider A as an $N \times N$ block matrix $A = (A_{ij})$, with blocks $A_{ij} \in \mathbb{R}^{m \times p}$. In particular, A is a square matrix if and only if $p = m$. Let $u = \mathrm{col}\,(u_1, \ldots, u_q) \in \mathbb{R}^q$ denote the external control input applied to the whole network. The input to node i is then

$$v_i(t) = \sum_{j=1}^N A_{ij} w_j(t) + B_i u(t). \tag{11.15}$$

Like the external input to the network, the output of the network is a linear combination of the individual node outputs w_i as $y(t) = Cw(t)$, with $w = \mathrm{col}\,(w_1, \ldots, w_N)$ and $y \in \mathbb{R}^l$. Let $x = \mathrm{col}\,(x_1, \ldots, x_N) \in \mathbb{R}^{nN}$ denote the global state of the network.

A directed weighted **state interconnection graph** $\Gamma = (V, E)$ is associated with the state interconnection structure of the system as follows. The set of vertices $V = \{1, \ldots, N\}$ corresponds to node systems (11.14). An edge $(i, j) \in E$ from system i to system j is defined if and only if $A_{ij} \neq 0$. We emphasize that the weights A_{ij} of the graph are matrices, unless $p = m = 1$. Similarly, graphs are defined for the input/output interconnection, respectively. So-called **diffusive coupling** refers to the special situation where $p = m = 1$ and A is the Laplacian matrix of an undirected weighted graph. Thus $a_{ij} < 0$ if and only if nodes $i \neq j$ are connected. Otherwise, for $i \neq j$ we define $a_{ij} = 0$. The diagonal elements of the Laplacian matrix A are defined by $a_{ii} = -\sum_{j \neq i} a_{ij}$. Then the interconnection law (11.15), with $B_i = 0$ and diffusive coupling, becomes

$$v_i(t) = \sum_{j \neq i} a_{ij}(w_j(t) - w_i(t)).$$

In the sequel, unless stated otherwise, we will not make restrictive assumptions on the structure of A. In particular, we will not assume that A is a Laplacian matrix or assume that the off-diagonal entries have a specific sign.

Using the interconnection matrices A, B, C and node dynamics α, β, γ, the resulting linear network has the form

$$\begin{aligned} \dot{x}(t) &= \mathscr{A} x(t) + \mathscr{B} u(t), \\ y(t) &= \mathscr{C} x(t), \end{aligned} \tag{11.16}$$

where

$$\mathscr{A} = I_N \otimes \alpha + (I_N \otimes \beta)A(I_N \otimes \gamma) \in \mathbb{R}^{nN \times nN} \quad \mathscr{B} = (I_N \otimes \beta)B \in \mathbb{R}^{nN \times q},$$

$$\mathscr{C} = C(I_N \otimes \gamma) \in \mathbb{R}^{l \times nN}.$$

Stated in terms of transfer functions, one obtains the node transfer function

$$G(z) = \gamma(zI_n - \alpha)^{-1}\beta$$

and associated left and right coprime factorizations

$$G(z) = D_\ell(z)^{-1}N_\ell(z) = N_r(z)D_r(z)^{-1} = V(z)T(z)^{-1}U(z).$$

Note that in this special case of a homogeneous network, our notation differs slightly from the preceding one. The network transfer function is

$$\mathcal{N}_G(z) = \mathscr{C}(zI_{nN} - \mathscr{A})^{-1}\mathscr{B}$$

and

$$\mathcal{N}_G(z) = C\Big(I_N \otimes D_\ell(z) - (I_N \otimes N_\ell(z))A\Big)^{-1}(I_N \otimes N_\ell(z))B$$

$$= C(I_N \otimes V(z))\Big(I_N \otimes T(z) - (I_N \otimes U(z))A(I_N \otimes V(z))\Big)^{-1}(I_N \otimes U(z))B.$$

In principle, there exist two different approaches to the design of such networks. The first one, on which we will mainly focus in the sequel, is to consider the interconnection terms A, B, C as free design parameters. A natural question in this direction then concerns the design of networks, i.e., how one can change the system dynamics of the network (11.16) by a suitable choice of the coupling parameters A, B, C. This is closely related to feedback control problems, such as stabilization or self-organization. A second approach would consist in assuming the interconnection structure to be fixed and designing local controllers for the node system to change the dynamics of the network.

To treat synchronization issues more broadly, we recall some basic terminology from geometric control theory. Consider an invariant subspace $\mathscr{V} \subset \mathscr{X}$ of a linear operator $A : \mathscr{X} \longrightarrow \mathscr{X}$. Then there are two induced linear maps, the restriction operator $A|_{\mathscr{V}} : \mathscr{V} \longrightarrow \mathscr{V}$ and the corestriction $A|_{\mathscr{X}/\mathscr{V}} : \mathscr{X}/\mathscr{V} \longrightarrow \mathscr{X}/\mathscr{V}$. The invariant subspace $\mathscr{V}$ is called **outer stable** if all eigenvalues of the corestriction $A|_{\mathscr{X}/\mathscr{V}}$ have negative real part. Consider a linear system with m inputs and p outputs,

$$\dot{x}(t) = \mathscr{A}\,x(t) + \mathscr{B}\,u(t),$$

$$y(t) = \mathscr{C}\,x(t),$$

on an n-dimensional state space $\mathscr{X} \simeq \mathbb{R}^n$. A linear subspace $\mathscr{V} \subset \mathscr{X}$ is called **controlled invariant**, or $(\mathscr{A}, \mathscr{B})-$ invariant, if

$$\mathscr{A}\mathscr{V} \subset \mathscr{V} + \mathrm{Im}\,\mathscr{B}.$$

Equivalently, $\mathcal{V}$ is controlled invariant if and only if there exists a state feedback matrix $F \in \mathbb{R}^{m \times n}$, with

$$(\mathcal{A} + \mathcal{B}F)\mathcal{V} \subset \mathcal{V}.$$

$\mathcal{V}$ is called an **outer stabilizable** controlled invariant subspace if $\mathcal{V}$ is an outer stable invariant subspace for $\mathcal{A} + \mathcal{B}F$. Similarly, $\mathcal{V}$ is called **conditioned invariant**, or $(\mathcal{C}, \mathcal{A})-$ invariant, if

$$\mathcal{A}(\mathcal{V} \cap \mathrm{Ker}\, \mathcal{C}) \subset \mathcal{V}$$

or, equivalently, if there exists an output injection transformation $J \in \mathbb{R}^{n \times p}$, with

$$(\mathcal{A} + J\mathcal{C})\mathcal{V} \subset \mathcal{V}.$$

If $\mathcal{V}$ is outer stable for $\mathcal{A} + J\mathcal{C}$, then $\mathcal{V}$ is called an **outer detectable** conditioned invariant subspace. A linear subspace $\mathcal{V}$ is called $(\mathcal{A}, \mathcal{B}, \mathcal{C})$ invariant if it is simultaneously controlled invariant and conditioned invariant.

The term **synchronization** is usually linked to a concept of stability requiring that the state trajectories of the coupled node systems converge asymptotically to each other. Thus, for the interconnected system (11.16) with input $u = 0$, we require

$$\lim_{t \to \infty} \|x_i(t) - x_j(t)\| = 0 \qquad (11.17)$$

for all $i, j = 1, \ldots, N$. Here $\| \cdot \|$ denotes the Euclidean norm on $\mathbb{R}^n$. Let $\mathbf{e} = (1, \ldots, 1)^\top \in \mathbb{R}^N$, and let

$$\Delta_n = \mathbf{e} \otimes \mathbb{R}^n = \{\mathrm{col}(\xi, \ldots, \xi) \mid \xi \in \mathbb{R}^n\} \subset \mathbb{R}^{nN}$$

be the linear subspace spanned by the column vectors of the matrix $\mathbf{e} \otimes I_n$. Thus Δ_n defines the diagonal in the N-fold direct sum space $\mathbb{R}^n \oplus \ldots \oplus \mathbb{R}^n$. We refer to Δ_n as the **synchronization subspace**. Let

$$\mathrm{dist}(x, \Delta) = \min_{v \in \Delta} \|x - v\|$$

denote the distance of a point $x \in \mathbb{R}^{nN}$ to Δ_n. Then, for the global state of the network $x(t)$, the convergence property (11.17) is equivalent to

$$\lim_{t \to \infty} \mathrm{dist}(x(t), \Delta) = 0.$$

For our purposes this property is a bit too weak because it does not imply an invariance of Δ under the flow. We therefore give the following stricter definition. The spectrum of a matrix M, that is, the set of eigenvalues, is denoted by $\sigma(M)$. The set $\mathbb{C}_- = \{z \in \mathbb{C} \mid \mathrm{Re}(z) < 0\}$ denotes the left half-plane in the field of complex numbers.

Definition 11.10. The state interconnection matrix A **synchronizes** (11.16), or (11.14), is **synchronizable** by A provided the following conditions are satisfied:

(i) **Synchronization Preservation:** The subspace $\Delta = \mathbf{e} \otimes \mathbb{R}^n$ is invariant under $\mathscr{A} = I_N \otimes \alpha + (I_N \otimes \beta)A(I_N \otimes \gamma)$.

(ii) **Outer Stability:** For all initial values $x(0) \in \mathbb{R}^{nN}$ and input $u(t) = 0$, the solutions of (11.16) satisfy

$$\lim_{t \to \infty} \|x_i(t) - x_j(t)\| = 0 \quad \text{for all } i, j = 1, \dots, N.$$

Similarly, system (11.14) is called synchronizable by state feedback, or output injection, provided conditions (i) and (ii) are true for $\gamma = I_n$ and $A \in \mathbb{R}^{nN \times pN}$ or $\beta = I_n$ and $A \in \mathbb{R}^{mN \times nN}$, respectively. In either case, the restriction of $I_N \otimes \alpha + (I_N \otimes \beta)A(I_N \otimes \gamma)$ on the invariant subspace Δ_n is called the **synchronization dynamics**.

Obviously, the attractivity condition (ii) for synchronization is equivalent to the condition that the corestriction $\mathscr{A}|_{\mathbb{R}^{nN}/\Delta}$ is Hurwitz. Thus one obtains the following proposition.

Proposition 11.11. *The state interconnection matrix A synchronizes (11.16) if and only if Δ_n is an **outer stable** invariant subspace for $\mathscr{A} = I_N \otimes \alpha + (I_N \otimes \beta)A(I_N \otimes \gamma)$.*

Note that $\mathscr{A}$ is, in output feedback form,

$$\mathscr{A} = I_N \otimes \alpha + (I_N \otimes \beta)A(I_N \otimes \gamma) \tag{11.18}$$

for the decoupled system $(I_N \otimes \alpha, I_N \otimes \beta, I_N \otimes \gamma)$. Moreover, for $\gamma = I_n$, (11.18) is in state feedback form, while for $\beta = I_n$ one obtains output injection transformations. Thus, in view of Exercise 1, the synchronizability of (11.14) implies that Δ is both an outer detectable conditioned invariant and outer stabilizable controlled invariant subspace of the decoupled system $(I_N \otimes \alpha, I_N \otimes \beta, I_N \otimes \gamma)$. However, this reformulation in terms of geometric control theory does not take into consideration the underlying graph structure that defines the network. The problem is that no characterization of conditioned invariant subspaces is known such that the associated output injection transformations have a prescribed pattern of unconstrained entries and zero entries.

Definition 11.10 imposes no restrictions on the synchronization dynamics. In particular, it is very well possible and allowable that the synchronization dynamics are asymptotically stable, which sounds counterintuitive. In many studies of the synchronization phenomena, therefore, additional assumptions, such as marginal stability of the synchronization dynamics, are imposed. In the sequel we will not require such additional assumptions because they are often easily handled in a second design step.

The outer stability condition can be replaced by the equivalent condition that there exists a solution trajectory of the form $\mathbf{e} \otimes \xi(t) \in \Delta$ of (11.16) such that

$$\lim_{t\to\infty} \|x_i(t) - \xi(t)\| = 0 \quad \text{for all} \quad i = 1,\ldots,N.$$

The existence of $\xi(t)$ is easily established; see, for example, the proof of Theorem 11.13.

We now aim at characterizing networks that are synchronizable. For simplicity, we focus on a special class of interconnection matrices that have been treated mainly in the literature on synchronization and consensus.

Definition 11.12. A state interconnection matrix A is called **decomposable** if there exist real matrices $L \in \mathbb{R}^{N\times N}, K \in \mathbb{R}^{m\times p}$, with

$$A = L \otimes K.$$

The matrix L then carries the main information about the underlying graph structure of the network, while K presents a uniform connection structure between the individual inputs and outputs of the nodes.

Theorem 11.13. *Assume that* $(\alpha,\beta,\gamma) \in \mathbb{R}^{n\times n} \times \mathbb{R}^{n\times m} \times \mathbb{R}^{p\times n}$ *satisfies* rk $\beta = m$, rk $\gamma = p$. *Let* $e = (1,\ldots,1)^\top \in \mathbb{R}^N$. *Then (11.16) is synchronized by* $A = L \otimes K$ *if and only if the following properties are satisfied:*

1. $Le = \lambda e$ *for some* $\lambda \in \mathbb{R}$. *Either* λ *is a simple eigenvalue of* L *or* $\alpha + \lambda\beta K\gamma$ *is Hurwitz.*
2. $\alpha + \mu\beta K\gamma$ *is Hurwitz for all other eigenvalues* $\mu \neq \lambda$ *of* L.

Proof. We first prove the sufficiency part. Let $SLS^{-1} = J$ be in Jordan canonical form, with eigenvalues $\lambda_1,\ldots,\lambda_n$ (counted with multiplicities). Then

$$(S \otimes I)\mathscr{A}(S^{-1} \otimes I) = I_N \otimes \alpha + J \otimes \beta K\gamma.$$

Thus, without loss of generality, one can assume that L is in Jordan canonical form with a block upper triangular matrix

$$I_N \otimes \alpha + J \otimes \beta K\gamma = \begin{pmatrix} \alpha + \lambda_1\beta K\gamma & * & \cdots & * \\ 0 & \alpha + \lambda_2\beta K\gamma & * & \vdots \\ \vdots & & \ddots & * \\ 0 & \cdots & 0 & \alpha + \lambda_N\beta K\gamma \end{pmatrix}.$$

Without loss of generality, assume that $\lambda = \lambda_1$. Clearly, e is an eigenvector of L if and only if $L\mathbb{R}e = \mathbb{R}e$. Equivalently, for all $v \in \mathbb{R}^n$, we have that $\mathscr{A}(e \otimes v) = (I_N \otimes \alpha + L \otimes \beta K\gamma)(e \otimes v) = e \otimes (\alpha v + \lambda\beta K\gamma v) \in \Delta$, i.e., $\mathscr{A}\Delta \subset \Delta$. Thus e is an eigenvector of L if and only if Δ is $\mathscr{A}$-invariant. If $\alpha + \lambda\beta K\gamma$ is Hurwitz, then, by condition 2, all block matrices $\alpha + \lambda_i\beta K\gamma$ are Hurwitz. Thus $\mathscr{A}$ is Hurwitz. Moreover, Δ is $\mathscr{A}$-invariant. Thus $A = L \otimes K$ synchronizes. If $\lambda = \lambda_1$ is a simple eigenvalue of L, then $(S^{-1} \otimes I)\Delta$ coincides with the subspace spanned by the first n basis vectors for the matrix $\alpha + \lambda \beta K \gamma$. By condition 2, all solutions of the transformed system $(I_N \otimes \alpha + J \otimes \beta K \gamma)$ converge to $(S^{-1} \otimes I)\Delta$. This again proves synchronization.

For the converse direction, note that the $\mathscr{A}$-invariance of Δ implies that $\mathbf{e}$ is an eigenvector of L. Assume that λ is not a simple eigenvalue, and $\alpha + \lambda \beta K \gamma$ is not Hurwitz. For simplicity assume, for example, that L contains a Jordan block of the form

$$\begin{pmatrix} \lambda & 1 \\ 0 & \lambda \end{pmatrix}.$$

The other cases are treated similarly. Then $I \otimes \alpha + J \otimes \beta K \gamma$ is upper triangular and contains a diagonal block of the form

$$\begin{pmatrix} \alpha + \lambda \beta K \gamma & \beta K \gamma \\ 0 & \alpha + \lambda \beta K \gamma \end{pmatrix}.$$

But the system

$$\dot{x}_1 = (\alpha + \lambda \beta K \gamma)x_1 + \beta K \gamma x_2,$$

$$\dot{x}_2 = (\alpha + \lambda \beta K \gamma)x_2$$

is not synchronizing, as can be seen by choosing $x_2 = 0$ and x_1 in the unstable eigenspace of $\alpha + \lambda \beta K \gamma$. This proves the first condition. The second condition follows by similar reasoning as above. ∎

Theorem 11.13 shows that the synchronization task for an interconnection matrix $L \otimes K$ is equivalent to a robust output feedback stabilization task. Such problems are in general hard to solve. In the next section, we derive a sufficient condition in the SISO case. The problem becomes simpler if we restrict ourselves to state feedback transformations, i.e., for $\gamma = I_n$ and $A = L \otimes K$, with $K \in \mathbb{R}^{m \times n}$. In that case, it is possible to solve the synchronization task by a simple constructive procedure based on algebraic Riccati equations. This is well known if the network is defined by a weighted Laplacian; see, for instance, Tuna (2009), and is still a viable approach for more general interconnection structures. The next result is a simple improvement of standard linear regulator theory.

Lemma 11.14. *Let (α, β) be stabilizable and $P = P^{\top} > 0, P \in \mathbb{R}^{n \times n}$ be the unique symmetric and positive definite solution of the algebraic Riccati equation*

$$\alpha^{\top} P + P\alpha - P\beta\beta^{\top} P + I_n = 0. \tag{11.19}$$

Then for all $\lambda \in \mathbb{C}$, with $\mathrm{Re}(\lambda) \geq \frac{1}{2}$, one obtains

$$\sigma(\alpha - \lambda \beta \beta^{\top} P) \subset \mathbb{C}_-.$$

Proof. Since $(\alpha, \beta, \gamma = I_n)$ is stabilizable and detectable, there exists a unique symmetric positive definite solution P of the algebraic Riccati equation (11.19). Thus, for complex numbers λ, one obtains

$$P(\alpha - \lambda \beta \beta^\top P) + (\alpha^\top - \bar{\lambda} P \beta \beta^\top)P$$
$$= P\alpha + \alpha^\top P - 2\operatorname{Re}(\lambda)P\beta\beta^\top P$$
$$= -I_n + (1 - 2\operatorname{Re}(\lambda))P\beta\beta^\top P.$$

Since by assumption $1 - 2\operatorname{Re}(\lambda) \le 0$, the matrix on the right-hand side of the equality is negative definite. The assertion then follows from Theorem 5.44 provided the pair

$$(F, G) := \left(\alpha^\top - \bar{\lambda} P\beta\beta^\top, I_n + (2\operatorname{Re}(\lambda) - 1)P\beta\beta^\top P \right)$$

is reachable. But this is obvious, because $G = I_n + (2\operatorname{Re}(\lambda) - 1)P\beta\beta^\top P$ is invertible. The result follows. ∎

Theorem 11.15. *Assume that (α, β) is stabilizable and $\gamma := I_n$. Let $P = P^\top$ be the unique positive definite symmetric solution of the algebraic Riccati equation*

$$\alpha^\top P + P\alpha - P\beta\beta^\top P + I_n = 0.$$

Let $L \in \mathbb{R}^{N \times N}$ be a matrix with $Le = \lambda e$ and simple eigenvalue λ. Assume that all other eigenvalues $\mu \ne \lambda$ of L satisfy $\operatorname{Re}(\mu) > 0$ [or satisfy $\operatorname{Re}(\mu) < 0$ for all eigenvalues $\mu \ne \lambda$]. Choose $\tau \in \mathbb{R}$ such that for all eigenvalues $\mu \ne \lambda$ the inequality $\tau\operatorname{Re}(\mu) \ge \frac{1}{2}$ is fulfilled, and set $K := \tau\beta^\top P \in \mathbb{R}^{m \times n}$. Then the state interconnection matrix $A = L \otimes K$ synchronizes the network (11.16).

Proof. $\operatorname{Re}(\tau\mu) \ge \frac{1}{2}$, and thus, by Lemma 11.14, we obtain

$$\sigma(\alpha - \mu\beta K) = \sigma(\alpha - \mu\tau\beta\beta^\top P) \subseteq \mathbb{C}_-.$$

The result follows from Theorem 11.13. ∎

The preceding result leads to an explicit sufficient condition for synchronization.

Corollary 11.16. *Assume that (α, β) is stabilizable and $\gamma := I_n$. Let $P = P^\top$ be the unique positive definite symmetric solution of the algebraic Riccati equation*

$$\alpha^\top P + P\alpha - P\beta\beta^\top P + I_n = 0.$$

Let L be the graph Laplacian of a weighted strongly connected digraph Γ, and let λ_2 denote the eigenvalue of L with smallest positive real part. Then $A = \tau L \otimes \beta^\top P$ synchronizes (11.16) for

$$\tau \geq \frac{1}{2\,\mathrm{Re}\,\lambda_2} > 0.$$

Proof. By Theorem 8.36, $\lambda_1 = 1$ is simple and all other eigenvalues $\lambda_i \neq 1$ have positive real part. Thus the result follows from Theorem 11.15. ∎

Stronger results can be obtained for restricted classes of node transfer functions. Recall that a square $m \times m$ strictly proper rational transfer function $G(z) = \gamma(zI_n - \alpha)^{-1}\beta$ is **positive real** if G has only poles in the closed left half-plane and

$$G(z) + G(\bar{z})^\top \succeq 0 \qquad\qquad (11.20)$$

is positive semidefinite for all complex numbers z, with $\mathrm{Re}\,z > 0$. Let (α, β, γ) be reachable and observable. The **Positive Real Lemma** then asserts that $G(z)$ is positive real if and only if there exists a positive definite symmetric matrix P such that

$$-P\alpha - \alpha^\top P \succeq 0,$$
$$\gamma = \beta^\top P. \qquad\qquad (11.21)$$

Moreover, $G(z)$ is strictly positive real, i.e., (11.20) is valid for all z, with $\mathrm{Re}\,z \geq 0$, if and only if (11.21) is satisfied with $-P\alpha - \alpha^\top P \succeq 0$ being replaced by $-P\alpha - \alpha^\top P \succ 0$.

The following lemma is proven next.

Lemma 11.17. *Assume that (α, β, γ) is reachable and observable, with $m = p$ such that $G(z) = \gamma(zI_n - \alpha)^{-1}\beta$ is positive real. Then for all complex numbers λ with $\mathrm{Re}\,\lambda > 0$, the spectrum of $\alpha - \lambda\beta\gamma$ satisfies*

$$\sigma(\alpha - \lambda\beta\gamma) \subset \mathbb{C}_-.$$

Proof. By the Positive Real Lemma, a positive definite symmetric matrix P exists, with

$$-(\alpha - \lambda\beta\gamma)^* P - P(\alpha - \lambda\beta\gamma) = -\alpha^\top P - P\alpha + \bar{\lambda}\gamma^\top \beta^\top P + \lambda P\beta\gamma$$
$$= -\alpha^\top P - P\alpha + 2\,\mathrm{Re}\,\lambda\gamma^\top \gamma$$
$$\succeq 2\,\mathrm{Re}\,\lambda\gamma^\top \gamma.$$

The pair $(\alpha - \lambda\beta\gamma, \gamma)$ is observable. Thus every complex eigenvector v of $\alpha - \lambda\beta\gamma$ with eigenvalue w satisfies $\gamma v \neq 0$ and

$$-2\,\mathrm{Re}(w)v^* P v \geq 2\,\mathrm{Re}\,\lambda\,\|\gamma v\|^2 > 0. \qquad ∎$$

This leads to the following simple sufficient condition for synchronization.

Theorem 11.18. *Let (α,β,γ) be reachable and observable, with $m = p$ such that $G(z) = \gamma(zI_n - \alpha)^{-1}\beta$ is positive real. Let L be the graph Laplacian of a weighted strongly connected digraph Γ. Then the state interconnection matrix $A = L \otimes I_m$ synchronizes the network (11.16).*

Proof. The eigenvalues of L are $\lambda_1 = 0, \lambda_2, \ldots, \lambda_N$, with $0 < \mathrm{Re}(\lambda_2) \leq \ldots \leq \mathrm{Re}(\lambda_N)$. Applying Lemma 11.17 one concludes that the eigenvalues of $\alpha - \lambda_i\beta\gamma$ have negative real part for $i = 2, \ldots, N$. Thus the result follows from Theorem 11.13. ∎

11.3 Synchronization of Homogeneous Networks

In this section, we consider networks of linear systems that are SISO, that is, the node systems (11.14) are defined by reachable and observable systems $\alpha \in \mathbb{R}^{n \times n}$, $\beta \in \mathbb{R}^n$, and $\gamma \in \mathbb{R}^{1 \times n}$. Let $g(z) := \gamma(zI_n - \alpha)^{-1}\beta$ denote the scalar strictly proper transfer function of the node system. Let

$$h(z) = \frac{1}{g(z)}$$

be the reciprocal of the transfer function. We allow for arbitrary multivariable interconnection matrices $(A,B,C) \in \mathbb{R}^{N \times N} \times \mathbb{R}^{N \times m} \times \mathbb{R}^{p \times N}$, with interconnection transfer function $\mathcal{N}(z) = C(zI_N - A)^{-1}B$. Thus the network transfer function is

$$\mathcal{N}_g(z) = C(h(z)I_N - A)^{-1}B,$$

i.e., can be represented as the composition

$$\mathcal{N}_g(z) = \mathcal{N}(h(z))$$

of the interconnection transfer function $\mathcal{N}(z)$ with $h(z)$. It is shown in Theorem 9.15 that a homogeneous network $(\mathcal{A},\mathcal{B},\mathcal{C})$ is reachable and observable if and only if (A,B,C) is reachable and observable. In the sequel we will always assume this. We next prove a simple frequency-domain characterization of the synchronizability of SISO node systems.

Theorem 11.19. *Assume that A satisfies $Ae = \lambda e$, with λ a simple eigenvalue. Then A synchronizes the homogeneous network (11.16) if and only if*

$$h(\overline{\mathbb{C}_+}) \cap \sigma(A) \setminus \{\lambda\} = \emptyset. \tag{11.22}$$

Proof. Let $g(z) = \frac{p(z)}{q(z)}$ be a coprime factorization. Suppose A synchronizes (11.16). Let $\lambda_1 := \lambda, \lambda_2, \ldots, \lambda_N$ denote the eigenvalues of A. By Theorem 11.13, the characteristic polynomials

$$\det(zI_n - (\alpha + \lambda_i \beta \gamma)) = q(z) - \lambda_i p(z)$$

are Hurwitz for $i = 2, \ldots, N$. This shows condition (11.22). Conversely, assume that (11.22) is satisfied. Then, for all $z \in \overline{\mathbb{C}_+}$ and all $2 \leq i \leq N$, one obtains $h(z) \neq \lambda_i$, i.e., $q(z) - \lambda_i p(z) \neq 0$. Thus $q(z) - \lambda_i p(z)$ is a Hurwitz polynomial for all $2 \leq i \leq N$. This completes the proof. ∎

This leads to the following explicit description of synchronizing homogeneous networks.

Theorem 11.20. *N identical SISO minimal systems*

$$\dot{x}_i(t) = \alpha x_i(t) + \beta u_i(t),$$
$$y_i(t) = \gamma x_i(t) \tag{11.23}$$

are synchronizable if and only if there exists a scalar proper rational real transfer function $f(z) \in \mathbb{R}(z)$ of McMillan degree $N - 1$ with

$$f^{-1}(\infty) \cap h(\overline{\mathbb{C}_+}) = \emptyset.$$

Proof. Suppose (11.23) is synchronizable through an interconnection matrix A, where λ is a simple eigenvalue of A and $A\mathbf{e} = \lambda \mathbf{e}$. Thus A is similar to an upper triangular matrix

$$\begin{pmatrix} \lambda & A_{12} \\ 0 & A_{22} \end{pmatrix}$$

such that $I_{N-1} \otimes \alpha + A_{22} \otimes \beta \gamma$ is Hurwitz. The set of such matrices L_{22} is open, and therefore one can assume, without loss of generality, that A_{22} has $N - 1$ distinct eigenvalues $\lambda_2, \ldots, \lambda_N$ that are distinct from λ. Thus we constructed a self-conjugate set of complex numbers $\lambda_2, \ldots, \lambda_r$ such that $q(z) - \lambda_i p(z)$ is Hurwitz for $i = 2, \ldots, r$. Choose a real transfer function $f(z)$ of McMillan degree $N - 1$ that has its poles exactly at $\lambda_2, \ldots, \lambda_N$. Then $f^{-1}(\infty) \cap h(\overline{\mathbb{C}_+}) = \emptyset$. Conversely, assume that $f(z)$ is a real rational transfer function whose poles are disjoint from $h(\overline{\mathbb{C}_+})$. Choose a minimal realization $f(z) = c(zI_{N-1} - M)^{-1}b$, and let λ be a real number disjoint from the poles of f. Let $S \in GL_N(\mathbb{R})$ be such that $S\mathbf{e}_1 = \mathbf{e}$. Then

$$A = S \begin{pmatrix} \lambda & 0 \\ 0 & M \end{pmatrix} S^{-1}$$

is an interconnection matrix that synchronizes. ∎

The proof shows that the existence of synchronizing interconnection matrices is equivalent to an inverse eigenvalue problem with spectral constraints. For instance, to construct an interconnection matrix A with nonnegative entries requires solving the inverse eigenvalue problem of finding a nonnegative matrix with spectrum in $h(\overline{\mathbb{C}_+})$.

11.4 Polynomial Model Approach to Synchronization

Synchronization is usually defined and studied in the state-space context. However, our example of coupled oscillators from the introduction shows that it might be preferable to perform the study in a functional context and use the concept of state maps to see the connection with the state-space analysis. In this section we recall the preceding chapters to aim at a polynomial approach to synchronization. Our starting point is a polynomial matrix representation of the node system by taking N identical higher-order systems of differential equations

$$D_\ell(\sigma)\xi_i = N_\ell(\sigma)v_i, \quad i = 1,\ldots,N. \tag{11.24}$$

Here $\sigma = \frac{d}{dt}$ denotes the differentiation operator, and $N_\ell(z) \in \mathbb{R}[z]^{p\times m}$ and $D_\ell(z) \in \mathbb{R}[z]^{p\times p}$ are polynomial matrices, with $D_\ell(z)$ nonsingular. We assume that the associated transfer function

$$G(z) = D_\ell(z)^{-1}N_\ell(z)$$

is strictly proper. More generally, one could incorporate outputs by considering the general class of Rosenbrock systems

$$T(\sigma)\xi_i = U(\sigma)v_i,$$
$$w = V(\sigma)\xi_i + W(\sigma)v_i.$$

We will consider these systems later on. Of course, one can always replace the differentiation operator σ with the backward shift, thereby studying higher-order systems of difference equations. We freely use the terminology and results from Chapter 4.

To simplify our discussion, we will initially assume that the couplings between these systems are defined by a decomposable state interconnection matrix $A = L \otimes K \in \mathbb{R}^{mN \times pN}$, leading to a state feedback transformation

$$v = (L \otimes K)\xi + Bu.$$

As previously, matrix L incorporates the underlying interconnection structure defined by the graph of the network. The resulting interconnected system is then of the form

$$\left(I_N \otimes D_\ell(\sigma) - L \otimes N_\ell(\sigma)K\right)\xi = \left(I_N \otimes N_\ell(\sigma)\right)Bu. \tag{11.25}$$

Defining

$$\mathscr{T}(z) = I_N \otimes D_\ell(z) - L \otimes N_\ell(z)K,$$
$$\mathscr{U}(z) = \big(I_N \otimes N_\ell(z)\big)B,$$

one obtains the associated transfer function of the network (11.25) as

$$\Phi(z) = \mathscr{T}(z)^{-1}\mathscr{U}(z) = \Big(I_N \otimes D_\ell(z) - L \otimes N_\ell(z)K\Big)^{-1}\big(I_N \otimes N_\ell(z)\big)B.$$

Note that $\mathscr{T}(z)$ is factored as

$$\mathscr{T}(z) = (I_N \otimes D_\ell(z))(I_{pN} - L \otimes G(z)K),$$

and $I_{pN} - L \otimes G(z)K$ is biproper. Thus $\mathscr{T}(z)$ is nonsingular and $\Phi(z) = \mathscr{T}(z)^{-1}U(z) = (I_{pN} - L \otimes G(z)K)(I_N \otimes G(z))B)$ is strictly proper.

To study the effects of couplings, it is crucial to compare the relevant polynomial model spaces. Let X_{D_ℓ} denotes the polynomial model of the individual node system (11.24), so that

$$X_{I_N \otimes D_\ell} = X_{D_\ell} \oplus \cdots \oplus X_{D_\ell}$$

denote the state space of the decoupled system. Similarly,

$$X_{\mathscr{T}} = X_{I_N \otimes D_\ell - L \otimes N_\ell K}$$

denotes the state space of the interconnected system. The connection between these two spaces is clarified by the following lemma.

Lemma 11.21. *1. Let $V(z) \in \mathbb{R}[z]^{r \times r}, U(z) \in \mathbb{R}[z]^{r \times m}$ be polynomial matrices, with $V(z)$ nonsingular, such that the transfer function $V(z)^{-1}U(z)$ is strictly proper. Then, for a matrix $A \in \mathbb{R}^{m \times r}$, the polynomial model spaces X_V and X_{V-UA} are equal as sets and the map*

$$\pi_{V-UA} : X_V \longrightarrow X_{V-UA},$$

$$\pi_{V-UA}(f) = (V - UA)\pi_-((V - UA)^{-1}f)$$

defines an isomorphism of vector spaces (but in general not of $\mathbb{R}[z]$-modules).
2. In particular, $X_{I_N \otimes D_\ell} = X_{\mathscr{T}}$ as sets and the map

$$\pi_{\mathscr{T}} : X_{I_N \otimes D_\ell} \longrightarrow X_{\mathscr{T}}, \quad \pi_{\mathscr{T}}(f) = \mathscr{T}(z)\pi_-(\mathscr{T}(z)^{-1}f(z))$$

is an isomorphism of vector spaces. Moreover, each $f \in X_{D_\ell}$ satisfies

$$\pi_{\mathscr{T}}(e \otimes f) = e \otimes f. \tag{11.26}$$

Proof. By the strict properness of $V(z)^{-1}U(z)$, we see that

$$(V(z) - U(z)A)^{-1} = (I_r - V(z)^{-1}U(z)A)^{-1}V(z)^{-1}$$

is fulfilled, with $I_r - V(z)^{-1}U(z)A$ biproper. Therefore, if f is a vector of polynomials, then $V(z)^{-1}f(z)$ is strictly proper if and only if $(V(z) - U(z)A)^{-1}f(z)$ is strictly proper. This shows the equality of the polynomial models $X_V = X_{V-UA}$ as sets. The linearity of π_{V-UA} is obvious. Suppose $f \in X_V$ satisfies $(V - UA)\pi_-((V - UA)^{-1}f) = 0$. Then $g(z) = (V(z) - U(z)A)^{-1}f(z) = (I_r - V(z)^{-1}U(z)A)^{-1}V(z)^{-1}f(z)$ is a polynomial. Since $V(z)^{-1}f(z)$ is strictly proper and $I_r - V(z)^{-1}U(z)A$ is biproper, this implies that the polynomial $g(z) = (V(z) - U(z)A)^{-1}f(z)$ is strictly proper. Therefore, $f = 0$. This shows the injectivity of π_{V-UA}. For surjectivity, let f be an element of $X_{V-UA} = X_V$. Since π_{V-UA} is a projection operator, $\pi_{V-UA}(f) = f$. This shows the surjectivity of π_{V-UA} and completes the proof of the first claim.

The second claim is a simple consequence of the first claim by setting $V = I_N \otimes D_\ell, U = I_N \otimes N_\ell, A = L \otimes K$. Finally, (11.26) follows from a simple calculation

$$\pi_{\mathscr{T}}(e \otimes f) = \mathscr{T}\pi_-(\mathscr{T}^{-1}e \otimes f) = \mathscr{T}\pi_-\left((I_{pN} - L \otimes G(z)K)^{-1}(e \otimes D_\ell^{-1}f)\right)$$

$$= \mathscr{T}((I_{pN} - L \otimes G(z)K)^{-1}e \otimes D_\ell^{-1}f)$$

$$= (I_N \otimes D_\ell)e \otimes D_\ell^{-1}f$$

$$= e \otimes f.$$

Here the third equation is true since $D_\ell^{-1}f$ is strictly proper and $(I_{pN} - L \otimes G(z)K)^{-1}$ is biproper. ∎

One can identify X_{D_ℓ} with the diagonal in the direct sum $X_{I_N \otimes D_\ell} = X_{D_\ell} \oplus \ldots \oplus X_{D_\ell}$, i.e.,

$$X_{D_\ell} \simeq \{e \otimes f(z) \mid f \in X_{D_\ell}\} \subset X_{I_N \otimes D_\ell}.$$

Similarly, using the identity (11.26), we define the diagonal in $X_{\mathscr{T}}$ as

$$\Delta := \{e \otimes f(z) \in X_{\mathscr{T}} \mid f \in X_{D_\ell}\}.$$

Δ is called the **synchronization space** of the network (11.25). Let $S_{I_N \otimes D_\ell}$ and $S_{\mathscr{T}}$ denote the shift operators on the polynomial models $X_{I_N \otimes D_\ell}$ and $X_{\mathscr{T}}$, respectively. By Lemma 11.21, we know that the vector space isomorphism $\pi_{\mathscr{T}}$ is not a module homomorphism, i.e., $\pi_{\mathscr{T}}$ is not commuting with the shifts. We now show that by restricting $\pi_{\mathscr{T}}$ to the respective diagonal spaces a module isomorphism is obtained.

Lemma 11.22. *Assume that e is an eigenvector of L with eigenvalue λ.*

1. The restriction of $\pi_{\mathscr{T}}$ on the diagonal X_{D_ℓ} of $X_{I_N \otimes D_\ell}$

$$\tau : X_{D_\ell} \longrightarrow \Delta, \ \tau(f) = e \otimes f$$

is an isomorphism of $\mathbb{R}$-vector spaces.
2. Δ is an invariant subspace of the shift, i.e., $S_{\mathscr{T}}(\Delta) \subset \Delta$.
3. There is a conjugacy of the shift operators

$$S_{\mathscr{T}}|\Delta \simeq S_{D_\ell - \lambda N_\ell K}.$$

In particular,

$$\det(zI - S_{\mathscr{T}}|\Delta) = \det(D_\ell(z) - \lambda N_\ell(z)).$$

Proof. That τ defines an isomorphism of vector spaces is an obvious consequence of Lemma 11.21. Each vector of polynomials $g(z)$ satisfies the identity $\mathscr{T}(z)e \otimes g(z) = e \otimes D_\ell(z)g(z) - Le \otimes N_\ell(z)Kg(z) = e \otimes (D_\ell(z) - \lambda N_\ell(z)K)g(z)$. Thus each vector of polynomials f satisfies $\mathscr{T}(z)e \otimes (D_\ell - \lambda N_\ell K)^{-1}f = e \otimes f$, and therefore

$$\mathscr{T}(z)^{-1}(e \otimes f(z)) = e \otimes (D_\ell(z) - \lambda N_\ell(z)K)^{-1}f(z).$$

For vector polynomials f, Lemma 11.21 implies that $f \in X_{D_\ell}$ if and only if $f \in X_{D_\ell - \lambda N_\ell K}$. Thus $f \in X_{D_\ell}$ satisfies

$$S_{\mathscr{T}}(e \otimes f) = \mathscr{T}\pi_-(\mathscr{T}^{-1}(e \otimes zf)$$

$$= (I_N \otimes D_\ell - L \otimes N_\ell K)\pi_-(e \otimes (D_\ell - \lambda N_\ell K)^{-1}zf)$$

$$= e \otimes D_\ell \pi_-((D_\ell - \lambda N_\ell K)^{-1}zf) - e \otimes \lambda N_\ell K\pi_-((D_\ell - \lambda N_\ell K)^{-1}zf)$$

$$= e \otimes \pi_{D_\ell - \lambda N_\ell K}(zf) \in \Delta.$$

Here the second equation follows from identity (11.4). This shows the invariance of Δ under the shift $S_{\mathscr{T}}$ as well as the conjugacy of $S_{\mathscr{T}}|\Delta$, with SD_ℓ. This completes the proof.   ∎

With all these facts at hand, we proceed to define synchronization for a higher-order system as follows.

Definition 11.23. Let $G(z) = D_\ell(z)^{-1}N_\ell(z)$ be a left coprime factorization of the strictly proper transfer function $G(z)$, and let $\mathscr{T}(z) = I_N \otimes D_\ell(z) - L \otimes N_\ell(z)K$. The network (11.25) of higher-order systems synchronizes with the interconnection matrix $L \otimes K$ if it synchronizes for the shift realization on $X_{\mathscr{T}}$. Equivalently, then $\Delta \subset X_{\mathscr{T}}$ is an outer stable invariant subspace for the shift operator $S_{\mathscr{T}}$ on $X_{\mathscr{T}}$.

We thus have defined the synchronization of polynomial models (11.25) by the synchronization of the associated shift realizations. For further clarification, let us consider the polynomial models and associated shift operators in more detail. Let (α, β, γ) denote the shift realization of the node identical transfer function $G(z) = D_\ell(z)^{-1} N_\ell(z)$. Thus the shift realization of the decoupled transfer function $I_N \otimes D_\ell(z)^{-1} N_\ell(z)$ on the state space $X_{I_N \otimes D_\ell}$ is

$$(I_N \otimes \alpha) f = S_{I_N \otimes D_\ell} f = \pi_{I_N \otimes D_\ell}(zf),$$

$$(I_N \otimes \beta)\xi = \pi_{I_N \otimes D_\ell}(I_N \otimes N_\ell(z)\xi) = (I_N \otimes N_\ell(z))\xi,$$

$$(I_N \otimes \gamma) f = ((I_N \otimes D_\ell)^{-1} f)_{-1}.$$

Similarly, the shift operator of the transfer function $\mathscr{T}(z)^{-1}\mathscr{U}(z) = (I_N \otimes D_\ell(z) - L \otimes N_\ell(z)K)^{-1} N_\ell(z) B$ on the state space $X_{\mathscr{T}}$ is

$$S_{\mathscr{T}} f = \pi_{\mathscr{T}}(zf).$$

Note that for $f \in X_{I_N \otimes D_\ell}$ one has $(\mathscr{T}^{-1} f)_{-1} = ((I_N \otimes D_\ell)^{-1} f)_{-1}$. Therefore,

$$
\begin{aligned}
S_{\mathscr{T}} f &= \mathscr{T} \pi_-(z\mathscr{T}^{-1} f) \\
&= zf - \mathscr{T} \pi_+(z\mathscr{T}^{-1} f) \\
&= zf - \mathscr{T}(\mathscr{T}^{-1} f)_{-1} \\
&= zf - \mathscr{T}((I_N \otimes D_\ell)^{-1} f)_{-1}.
\end{aligned}
$$

Similarly, for $f \in X_{I_N \otimes D_\ell}$,

$$S_{I_N \otimes D_\ell} f = I_N \otimes D_\ell \pi_-(z(I_N \otimes D_\ell)^{-1} f) = zf - (I_N \otimes D_\ell)(I_N \otimes D_\ell^{-1} f)_{-1}.$$

Using $\mathscr{T}(z) = I_N \otimes D_\ell(z) - L \otimes N_\ell(z)K$, this shows the identity of linear operators on $X_{I_N \otimes D_\ell} = X_{\mathscr{T}}$:

$$S_{\mathscr{T}} f = S_{I_N \otimes D_\ell} f + L \otimes N_\ell K(D_\ell^{-1} f)_{-1}.$$

Using the N-fold direct sum decomposition $X_{I_N \otimes D_\ell} = X_{D_\ell} \oplus \cdots \oplus X_{D_\ell}$, the shift operator $S_{\mathscr{T}}$ has the form

$$
\begin{aligned}
S_{\mathscr{T}} &= I_N \otimes \alpha + (I_N \otimes \beta)(L \otimes K)(I_N \otimes \gamma) \\
&= I_N \otimes \alpha + L \otimes \beta K \gamma.
\end{aligned}
$$

Therefore, Δ is an outer stable invariant subspace for the shift $S_{\mathscr{T}}$ if and only if $\Delta \subset X_{I_N \otimes D_\ell}$ is an outer stable invariant subspace for $I_N \otimes \alpha + L \otimes \beta K \gamma$. This leads us to the following characterization of synchronization in the polynomial model $X_{\mathscr{T}}$.

Theorem 11.24. *Let $G(z) = D_\ell(z)^{-1} N_\ell(z)$ be a left coprime factorization of the strictly proper transfer function, and let (α, β, γ) be a minimal realization of $G(z)$. Let $A = L \otimes K$ be a decomposable state interconnection matrix. Assume that $Le = \lambda e$ for a simple eigenvalue λ. The following statements are equivalent:*

(a) The network (11.16) of higher-order systems synchronizes.
(b) Δ is an outer stable subspace for $I_N \otimes \alpha + L \otimes \beta K \gamma$.
(c) The following two conditions are satisfied:

   (c1) $S_{\mathscr{T}}(\Delta) \subset \Delta$.
   (c2) For all eigenvalues $\mu \neq \lambda$ of L,

$$\det \left(D_\ell(z) - \mu N_\ell(z) K \right)$$

   is a Hurwitz polynomial.

Proof. The shift operator $S_{\mathscr{T}}$ is isomorphic to $\mathscr{A} = I_N \otimes \alpha + L \otimes \beta K \gamma$. Thus $\dot{x} = \mathscr{A} x$ synchronizes if and only if $S_{\mathscr{T}}(\Delta) \subset \Delta$ and the corestriction of $S_{\mathscr{T}}$ on $X_{\mathscr{T}}/\Delta$ has only eigenvalues with negative real part. Thus (a) and (b) are equivalent. It was shown in Lemma 11.22 that the restriction operator $S_{\mathscr{T}}|\Delta$ is conjugate to the shift operator $S_{D_\ell(z) - \lambda N_\ell(z) K}$. Therefore, $\det \mathscr{T}(z) = \det(zI - S_{\mathscr{T}})$ and

$$\det(zI - S_{\mathscr{T}}|\Delta) = \det S_{D_\ell(z) - \lambda N_\ell(z) K} = \det(D_\ell(z) - \lambda N_\ell(z) K).$$

Let $\lambda = \lambda_1, \ldots, \lambda_N$ denote the not necessarily distinct eigenvalues of L. Following the proof of Theorem 11.13 we see that

$$\det \mathscr{T}(z) = \prod_{i=1}^{N} \det(D_\ell(z) + \lambda_i N_\ell(z) K).$$

Since $\det(S_{\mathscr{T}}|\Delta) = \det(D_\ell(z) - \lambda N_\ell(z) K)$, we conclude from

$$\det \mathscr{T} = \det(S_{\mathscr{T}}|\Delta) \det(S_{\mathscr{T}}|(X_{\mathscr{T}}/\Delta))$$

that

$$\prod_{i=2}^{N} \det(D_\ell(z) - \lambda_i N_\ell(z) K)$$

equals the characteristic polynomial of the corestriction $S_{\mathscr{T}}|(X_{\mathscr{T}}/\Delta)$. This completes the proof. ∎

In the preceding approach, synchronization of a higher-order system (11.24) was defined via synchronization of the associated shift realization. One wonders if a direct approach is possible. We consider the synchronization task for Rosenbrock node systems of the general form

$$T(\sigma)\xi_i = U(\sigma)v_i,$$
$$w_i = V(\sigma)\xi_i + W(\sigma)v_i, \quad i = 1,\ldots,N. \tag{11.27}$$

We assume that $T(z)$ is a nonsingular $r \times r$ polynomial matrix and that the transfer function

$$G(z) = V(z)T(z)^{-1}U(z) + W(z)$$

is strictly proper. We use the following simplified notation for the decoupled system as

$$\mathscr{T} = I_N \otimes T, \quad \mathscr{U} = I_N \otimes U, \quad \mathscr{V} = I_N \otimes V.$$

We consider couplings among the partial state components defined by a coupling matrix $A \in \mathbb{R}^{mN \times pN}$. In the sequel, we will restrict ourselves to a special class of interconnection matrices.

Definition 11.25. An $N \times N$ block matrix $A = (A_{ij})$ with blocks $A_{ij} \in \mathbb{R}^{m \times p}$ is called **admissible** provided

$$< A >:= \sum_{j=1}^{N} A_{1j} = \cdots = \sum_{j=1}^{N} A_{Nj}.$$

This definition is general enough to cover all preceding cases of interest. Thus every decomposable matrix $A = L \otimes K$, with L a Laplacian matrix, is admissible, as is every finite sum $A = L_1 \otimes K_1 + \ldots \otimes + L_s \otimes K_s$, with $N \times N$ Laplacian matrices $L_1,\ldots,L_s$. Thus the interconnected closed-loop system on partial states is

$$\Big(\mathscr{T}(\sigma) - \mathscr{U}(\sigma)A\mathscr{V}(\sigma)\Big)\xi = 0. \tag{11.28}$$

Define the **partial state synchronization space** of (11.27) as

$$\Delta_{\mathrm{ps}} := \{(\xi,\ldots,\xi) \mid \xi \in \mathbb{R}^r\} \subset \mathbb{R}^{rN}.$$

Similarly, we define the **state synchronization space** as

$$\Delta := \{(f_1,\ldots,f_N) \in X_{\mathscr{T}-\mathscr{U}A\mathscr{V}} \mid f_1 = \cdots = f_N\}.$$

Here $X_{\mathscr{T}-\mathscr{U}A\mathscr{V}}$ denotes the polynomial model associated with the nonsingular polynomial matrix $\mathscr{T}(z) - \mathscr{U}(z)A\mathscr{V}(z)$. Note that, in general, the two spaces do not have the same dimension. For admissible interconnection matrics the polynomial matrix $\mathcal{T}(z) - \mathcal{U}(z)A\mathcal{V}(z)$ is nonsingular and maps polynomial vectors of the form $\mathbf{e} \otimes f(z)$ onto polynomial vectors $\mathbf{e} \otimes g(z)$. More precisely, for $f(z) \in \mathbb{R}[z]^r$, one has

$$\left(\mathcal{T}(z) - \mathcal{U}(z)A\mathcal{V}(z) \right) \mathbf{e} \otimes f(z) = \mathbf{e} \otimes (T - U < A > V)f. \tag{11.29}$$

We proceed by giving two definitions of synchronization, one for the partial states and the other one in the state space.

Definition 11.26. Let A be an admissible interconnection matrix.

(a) The **partial state** system (11.28) **synchronizes** provided all solutions $\xi(t) = (\xi_1(t), \dots, \xi_N(t))$ of (11.28) satisfy

$$\lim_{t \to \infty} \| \xi_i(t) - \xi_j(t) \| = 0. \tag{11.30}$$

(b) The **shift realization** of (11.28) **synchronizes** provided $\Delta \subset X_{\mathcal{T}-\mathcal{U}A\mathcal{V}}$ is an outer stable invariant subspace of the shift operator $S_{\mathcal{T}-\mathcal{U}A\mathcal{V}} : X_{\mathcal{T}-\mathcal{U}A\mathcal{V}} \longrightarrow X_{\mathcal{T}-\mathcal{U}A\mathcal{V}}$.

We next prove that these two definitions are actually equivalent and derive a polynomial matrix characterization. Our result is a natural extension of Theorem 11.24.

Theorem 11.27. *Partial state synchronization of the Rosenbrock system (11.28) is equivalent to the synchronization of the associated shift realization. In either case, synchronization is satisfied if and only if*

$$\frac{\det(\mathcal{T}(z) - \mathcal{U}(z)A\mathcal{V}(z))}{\det(T(z) - U(z) < A > V(z))}$$

is a Hurwitz polynomial.

Proof. By assumption on A, the higher-order system (11.28) induces a higher-order system on the quotient space $\mathbb{R}^{rN}/\Delta_{\mathrm{ps}}$. Thus the asymptotic stability condition (11.30) is equivalent to the asymptotic stability of the induced system on $\mathbb{R}^{rN}/\Delta_{\mathrm{ps}}$. In view of (11.29), this in turn is equivalent to the polynomial

$$\frac{\det(\mathcal{T}(z) - \mathcal{U}(z)A\mathcal{V}(z))}{\det(T(z) - U(z) < A > V(z))}$$

being Hurwitz. Similarly, (11.28) synchronizes for the shift realization if and only if Δ is an outer stable invariant subspace for the shift operator $S_{\mathcal{T}-\mathcal{U}A\mathcal{V}}$ on $X_{\mathcal{T}-\mathcal{U}A\mathcal{V}}$. The assumption on A implies that Δ is invariant under the shift. In fact, for $f = (f_1, \dots, f_N) = \mathbf{e} \otimes g \in \Delta$, and writing $\hat{T} := \mathcal{T} - \mathcal{U}A\mathcal{V}$ for short,

$$S_{\hat{T}}(f) = \hat{T}\pi_-(\hat{T}^{-1}zf)$$
$$= \hat{T}\pi_-(\mathbf{e} \otimes (T - U < A > V)^{-1}zg)$$
$$= \mathbf{e} \otimes (T - U < A > V)\pi_-(\mathbf{e} \otimes (T - U < A > V)^{-1}zg)$$
$$= \mathbf{e} \otimes S_{T-U<A>V}(f) \in \Delta.$$
(11.31)

This proves the invariance of Δ. For a nonsingular polynomial matrix $D_\ell(z)$, the characteristic polynomial of the shift operator S_{D_ℓ} on X_{D_ℓ} coincides with $\det D_\ell(z)$. By (11.31), the restriction of the shift $S_{\hat{T}}$ on Δ is conjugate to the shift operator $S_{T-U<A>V}$. Therefore, the characteristic polynomial of the $S_{\hat{T}}|\Delta$ is equal to $\det(T(z) - U(z) < A > V(z))$. Thus the characteristic polynomial of the corestriction $S_{\hat{T}}$ on $X_{\hat{T}}/\Delta$ is equal to

$$\frac{\det(S_{\hat{T}})}{\det(S_{\hat{T}}|\Delta)} = \frac{\det(S_{\hat{T}})}{\det(T(z) - U(z) < A > V(z))}.$$

This completes the proof. ∎

The following invariance principle states that partial state synchronization holds irrespective of the choice of coprime factorization for the node models.

Theorem 11.28. *Suppose that the Rosenbrock node systems $\Sigma_{VT^{-1}U+W}$ and $\Sigma_{\overline{VT^{-1}}\overline{U}+\overline{W}}$ are strictly system equivalent. Let A be an admissible interconnection matrix. Then partial state synchronization for $\mathscr{T}(z) - \mathscr{U}(z)A\mathscr{V}(z)$ is satisfied if and only if it is satisfied for $\overline{\mathscr{T}}(z) - \overline{\mathscr{U}}(z)A\overline{\mathscr{V}}(z)$.*

Proof. By Corollary 9.6, the two networks obtained from $\Sigma_{VT^{-1}U+W}$ and $\Sigma_{\overline{VT^{-1}}\overline{U}+\overline{W}}$ by coupling them with the same interconnection matrix A are strictly system equivalent. In fact, by a careful inspection of the proof of Corollary 9.6, the strict system equivalences can be seen to preserve the sets of synchronized states Δ and $\overline{\Delta}$, respectively. Thus

$$\det(\mathscr{T}(z) - \mathscr{U}(z)A\mathscr{V}(z)) = \det(\overline{\mathscr{T}}(z) - \overline{\mathscr{U}}(z)A\overline{\mathscr{V}}(z)).$$

Moreover, the same is true of the determinants of the corestrictions. The result follows. ∎

Output Synchronization The preceding results lead to a simple characterization of output synchronization. To simplify matters, we will work with discrete-time state-space systems. We begin with deriving a simple version of the internal model principle. Consider a linear discrete-time system

$$x(t+1) = Ax(t),$$
$$y(t) = Cx(t).$$

Then the set of possible output sequences $(y(t)|t \in \mathbb{N})$ is

$$\mathscr{Y} = \{(CA^t x_0) \mid x_0 \in \mathbb{R}^n\}$$

or, equivalently,

$$\mathscr{Y} = CX^{zI-A}.$$

We refer to $\mathscr{Y}$ as the **output behavior** of (C,A). If (C,A) is observable, then clearly this defines an autonomous behavior. In fact, for a left coprime factorization

$$C(zI - A)^{-1} = D_\ell(z)^{-1} N_\ell(z)$$

and (C,A) observable, Proposition 4.36 implies that

$$CX^{zI-A} = X^{D_\ell}.$$

Next we consider a second discrete-time system in first-order form as

$$x_1(t+1) = A_1 x_1(t),$$
$$y_1(t) = C_1 x_1(t).$$

Let $\mathscr{Y}_1 = C_1 X^{zI-A_1}$ denote the associated output behavior. We say that the output behavior $\mathscr{Y}$ is a subbehavior of $\mathscr{Y}_1$ whenever $\mathscr{Y} \subset \mathscr{Y}_1$ or, equivalently, whenever

$$CX^{zI-A} \subset C_1 X^{zI-A_1}.$$

We arrive at the following version of the **internal model principle**, i.e., the characterization of subbehaviors of an autonomous behavior.

Proposition 11.29. *Assume that the pairs (C,A) and (C_1,A_1) are observable. Then $\mathscr{Y}$ is a subbehavior of $\mathscr{Y}_1$ if and only if there exists an invertible transformation $T \in GL_{n_1}(\mathbb{R})$ such that*

$$TA_1 T^{-1} = \begin{pmatrix} A & A_1'' \\ 0 & A_1' \end{pmatrix}, \quad C_1 T^{-1} = \begin{pmatrix} C & C_1' \end{pmatrix}.$$

Proof. Using the observability of (C_1,A_1), one obtains the left coprime factorizations

$$C_1(zI - A_1)^{-1} = D_{\ell,1}(z)^{-1} N_{\ell,1}(z).$$

Moreover, (C_1, A_1) is isomorphic to the shift realization of $D_{\ell,1}^{-1} N_{\ell,1}$, and similarly for (C, A). Proposition 4.36 implies that $\mathcal{Y}_1 = X^{D_{\ell,1}}$. Therefore, $\mathcal{Y}$ is a subbehavior of $\mathcal{Y}_1$ if and only if the inclusion

$$X^{D_\ell} \subset X^{D_{\ell,1}}.$$

By Theorem 3.35, the subspace X^{D_ℓ} is a submodule of $X^{D_{\ell,1}}$, i.e., $S_{D_{\ell,1}} X^{D_\ell} \subset X^{D_\ell}$. In particular, from Theorem 4.26 one obtains the equivalence of shift realizations $A_1 | X^{D_\ell} \simeq S_{D_{\ell,1}} | X^{D_\ell} \simeq S_{D_\ell} \simeq A$, and $C_1 | X^{D_\ell} \simeq C$. This completes the proof. ∎

Now we apply these ideas to output synchronization. Consider an observable pair (γ, α) with left coprime factorization

$$\gamma(zI - \alpha)^{-1} = D_\ell(z)^{-1} N_\ell(z).$$

Then the interconnected system with coupling matrix $A = L \otimes K$ is

$$
\begin{aligned}
x(t+1) &= \mathscr{A} x(t), \\
y(t) &= \mathscr{C} x(t),
\end{aligned}
\tag{11.32}
$$

where $\mathscr{A} = I_N \otimes \alpha + L \otimes K\gamma$ and $\mathscr{C} = I \otimes \gamma$. Note that this system is output injection equivalent to the direct sum system $(I_N \otimes \gamma, I_N \otimes \alpha)$, and therefore $(\mathscr{C}, \mathscr{A})$ is observable. Let $\mathcal{Y}$ denote the output behavior of (11.32). From the left coprime factorization

$$\mathscr{C}(zI - \mathscr{A})^{-1} = \mathscr{T}(z)^{-1} \mathscr{U}(z),$$

with

$$\mathscr{T}(z) = I_N \otimes D_\ell(z) - L \otimes N_\ell(z)K, \quad \mathscr{U}(z) = I_N \otimes N_\ell(z),$$

we obtain

$$\mathcal{Y} = \mathscr{C} X^{zI - \mathscr{A}} = X^{\mathscr{T}(z)}.$$

Definition 11.30. The **synchronized output behavior** of (11.32) is defined as the intersection of the diagonal in $z^{-1}\mathbb{R}[[z^{-1}]]^{pN}$ with $\mathcal{Y}$, i.e.,

$$\mathcal{Y}_{\text{sync}} := \{(h_1(z), \dots, h_N(z)) \in X^{\mathscr{T}} \mid h_1(z) = \cdots = h_N(z)\}.$$

System (11.32) is called **output synchronized** if the following requirements are satisfied:

1. There exists an initial state x_0 with output $y(t) = \text{col}(y_1(t), \dots, y_N(t))$ satisfying $y_{\text{sync}}(t) := y_1(t) = \cdots = y_N(t)$ for all $t \geq 0$.

2. For all initial conditions,

$$\lim_{t\to\infty} \|y(t) - \mathbf{e} \otimes y_{\text{sync}}(t)\| = 0.$$

Arguing as in the proof of Theorem 11.27, it is easily seen that $\mathscr{Y}_{\text{sync}}$ is a sub-module of $\mathscr{Y}$. We obtain a very simple characterization of output synchronizability.

Theorem 11.31. *Assume $L\mathbf{e} = \lambda\mathbf{e}$. System (11.32) is output synchronizable if and only if*

1. The synchronized output behavior is nonempty, i.e., $\mathscr{Y}_{\text{sync}} \neq \emptyset$. Moreover,

$$X^{D_\ell - \lambda N_\ell K} \simeq \mathscr{Y}_{\text{sync}} \subset X^{\mathscr{T}}.$$

2. $\det(D_\ell(z) - \mu N_\ell(z)K)$ is a Hurwitz polynomial for all eigenvalues $\mu \neq \lambda$ of L.

Proof. Clearly, condition 1 is equivalent to the existence of an element in the output behavior $\mathscr{Y}$ of (11.32), with all components being equal. This proves the equivalence of condition 1 with $\mathscr{Y}_{\text{sync}} \neq \emptyset$. For the other points, note that output synchronization is equivalent to partial state synchronization of the system

$$\mathscr{T}(\sigma)\xi = 0.$$

Using $\operatorname{Ker} \mathscr{T}(\sigma) = X^{\mathscr{T}}$, Theorem 11.27 implies that output synchronization is equivalent to synchronization of the associated shift realization of $X^{\mathscr{T}}$, i.e., that

$$\frac{\det(I \otimes D_\ell(z) - L \otimes N_\ell(z)K)}{\det(D_\ell(z) - \lambda N_\ell(z)K)} = \prod_{\mu \in \sigma(L)\setminus\{\lambda\}} \det(D_\ell(z) - \mu N_\ell(z)K)$$

is a Hurwitz polynomial. This completes the proof. ∎

Clustering. Finally, let us briefly discuss the more difficult question of clustering partial state vectors. While synchronization deals with the issue of driving the states of all the node systems of a network to each other, clustering is concerned with the more general task of allowing for different states to which the system can be driven. For simplicity, we focus on just two clusters; the general case can be treated similarly at the expense of more involved notation. Thus, for $I = \{1, \ldots, r\} \subset \{1, \ldots, N\}, J := \{r+1, \ldots, N\}$, let

$$\Delta_{\text{ps}}(IJ) = \{(\xi_1, \ldots, \xi_N) \mid \xi_1 = \cdots = \xi_r, \ \xi_{r+1} = \cdots = \xi_N\}$$

denote the set of *IJ*-**clustered partial states**. Taking the union of all nontrivial subsets I one obtains the set of 2-**clustered partial states** as

$$\Delta_{\text{ps}}^{[2]} = \bigcup_{0 < |I| < N} \Delta_{\text{ps}}(IJ).$$

Similarly, we define the set of I-**clustered states** as

$$\Delta(IJ) = \{(f_1,\ldots,f_N) \mid f_1 = \cdots = f_r,\ f_{r+1} = \cdots = f_N\}$$

in the polynomial model $X_{\mathscr{T}-\mathscr{U}A\mathscr{V}}$. The same made be said of the subset of 2-**clustered states**. The counterpart to the set of admissible interconnection matrices is defined as follows (for IJ clustering only).

Definition 11.32. A block matrix $A = (A_{ij})$ with $m \times p$ blocks A_{ij} is called IJ-**admissible** if there are $m \times p$ matrices $\overline{A}_{II}, \overline{A}_{IJ}, \overline{A}_{JI}, \overline{A}_{JJ}$, with

$$\sum_{j=1}^{r} A_{1j} = \cdots = \sum_{j=1}^{r} A_{rj} := \overline{A}_{II}, \quad \sum_{j=r+1}^{N} A_{1j} = \cdots = \sum_{j=r+1}^{N} A_{rj} := \overline{A}_{IJ},$$

$$\sum_{j=1}^{r} A_{r+1,j} = \cdots = \sum_{j=1}^{r} A_{Nj} := \overline{A}_{JI}, \quad \sum_{j=r+1}^{N} A_{r+1,j} = \cdots - \sum_{j=r+1}^{N} A_{r+1,j} := \overline{A}_{JJ}.$$

Define

$$< A >_{IJ} = \begin{pmatrix} \overline{A}_{II} & \overline{A}_{IJ} \\ \overline{A}_{JI} & \overline{A}_{JJ} \end{pmatrix}.$$

The definition of IJ clustering then reads as follows.

Definition 11.33. Let A be an IJ-admissable interconnection matrix.

(a) The **partial state** system (11.28) IJ-**clusters** provided all solutions $\xi(t) = (\xi_1(t),\ldots,\xi_N(t))$ of (11.28) satisfy

$$\lim_{t\to\infty} \|\xi_i(t) - \xi_j(t)\| = 0 \text{ for all } i,j = 1,\ldots,r,$$

$$\lim_{t\to\infty} \|\xi_i(t) - \xi_j(t)\| = 0 \text{ for all } i,j = r+1,\ldots,N.$$

(b) The **shift realization** of (11.28) IJ-**clusters** provided $\Delta(IJ) \subseteq X_{\mathscr{T}-\mathscr{U}A\mathscr{V}}$ is an outer stable invariant subspace of the shift operator $S_{\mathscr{T}-\mathscr{U}A\mathscr{V}} : X_{\mathscr{T}-\mathscr{U}A\mathscr{V}} \longrightarrow X_{\mathscr{T}-\mathscr{U}A\mathscr{V}}$.

Following the preceding analysis, one can then easily prove the next result; we omit the straightforward details.

Theorem 11.34. *Let A be IJ-admissible. Partial state IJ-clustering of the Rosenbrock system (11.28) is equivalent to IJ-clustering of the associated shift realization. In either case, IJ-clustering occurs if and only if*

$$\frac{\det(\mathscr{T}(z) - \mathscr{U}(z)A\mathscr{V}(z))}{\det(T(z) - U(z) < A >_{IJ} V(z))}$$

is a Hurwitz polynomial. The eigenvalues of the interconnected system on the IJ-clustered states are the roots of the polynomial

$$\det(T(z) - U(z) < A >_{IJ} V(z)).$$

In the special case of a decomposable interconnection matrix $A = L \otimes K$, the result can be stated in a more convenient form as follows.

Theorem 11.35. *Let K be arbitrary, and assume there are real numbers $\lambda_{11}, \lambda_{12}, \lambda_{21}, \lambda_{22}$ such that*

$$L = \begin{pmatrix} L_{11} & L_{12}, \\ L_{21} & L_{22}, \end{pmatrix},$$

with $L_{11} \in \mathbb{R}^{r \times r}, L_{22} \in \mathbb{R}^{(N-r) \times (N-r)}$ and $L_{ij}E = \lambda_{ij}E$ for all ij. Here E denotes the matrix of appropriate size, with all entries equal to 1. Let

$$\Delta_{cl}(z) := \det \begin{pmatrix} T(z) - r\lambda_{11}U(z)KV(z) & -(N-r)\lambda_{12}U(z)KV(z) \\ -r\lambda_{21}U(z)KV(z) & T(z) - (N-r)\lambda_{22}U(z)KV(z) \end{pmatrix}.$$

Then IJ-clustering occurs in the Rosenbrock system (11.28) and $A = L \otimes K$ if and only if

$$\frac{\det(I_N \otimes T(z) - L \otimes U(z)KV(z))}{\Delta_{cl}(z)} \tag{11.33}$$

is a Hurwitz polynomial. The eigenvalues of the interconnected system on the IJ-clustered states are the roots of the polynomial

$$\det \begin{pmatrix} T(z) - r\lambda_{11}U(z)KV(z) & -(N-r)\lambda_{12}U(z)KV(z) \\ -r\lambda_{21}U(z)KV(z) & T(z) - (N-r)\lambda_{22}U(z)KV(z) \end{pmatrix}. \tag{11.34}$$

Example 11.36. We investigate clustering for the case of three symmetrically coupled oscillators ($\lambda \neq 0$):

$$\ddot{x}_1 + a\dot{x}_1 + bx_1 = \lambda \dot{x}_2,$$
$$\ddot{x}_2 + a\dot{x}_2 + bx_2 = \lambda(\dot{x}_1 + \dot{x}_3),$$
$$\ddot{x}_3 + a\dot{x}_3 + bx_3 = \lambda \dot{x}_2.$$

Thus $T(z) = z^2 + az + b, U(z) = 1, V(z) = z$. The interconnection matrix is $A = L \otimes K = L$, with $K = 1$ and

$$L = \lambda \begin{pmatrix} 0 & 1 & 0 \\ 1 & 0 & 1 \\ 0 & 1 & 0 \end{pmatrix}$$

having the characteristic polynomial $z(z^2 - 2\lambda^2)$. Note that e is not an eigenvector of $A = L$, and therefore the network does not synchronize independently of the parameter value. By permuting the second and third columns and rows of L, we see that L is permutation equivalent to

$$\begin{pmatrix} L_{11} & L_{12} \\ L_{21} & L_{22} \end{pmatrix} = \lambda \begin{pmatrix} 0 & 0 & 1 \\ 0 & 0 & 1 \\ 1 & 1 & 0 \end{pmatrix}.$$

Thus L is admissible for $I = \{1,3\}, J = \{2\}$ with $\lambda_{11} = 0, \lambda_{12} = \lambda, \lambda_{21} = \lambda, \lambda_{22} = 0$. The characteristic polynomial for the coupled system is

$$\det(T(z)I_3 - LU(z)V(z)) = T(z)(T(z)^2 - 2\lambda^2 z^2),$$

which is Hurwitz if and only if $a > \sqrt{2}|\lambda|$ and $b > 0$. Polynomial (11.34) is

$$T(z)^2 - 2\lambda^2 z^2.$$

Thus the quotient (11.33) is the polynomial $T(z)$. One concludes that asymptotic clustering occurs if and only if $T(z)$ is Hurwitz, i.e., $a > 0, b > 0$.

11.5 Examples: Arrays of Oscillators

Because synchrony is usually connected to periodic phenomena, the basic components for modeling are mostly taken to be nonlinear oscillators. However, simpler linear system models, such as the classical harmonic oscillator, can be used for the same purpose. Of course, because its failure is structurally stable, the harmonic oscillator is an unsuitable model for most periodic physical phenomena. Nevertheless, the analysis of synchronization phenomena for the harmonic oscillator provides important insights into the general theory of synchronization. Arrays of coupled damped oscillators are perhaps the simplest models for synchronization of linear systems. The theory of such networks can be developed quite generally using tools from spectral graph theory developed in Chapter 8. Here we focus on applying such an analysis to the case of a finite number of damped harmonic oscillators that are ordered either linearly or circularly. The exposition is largely based on Fuhrmann, Priel, Sussmann and Tsoi (1987).

I. Linear Arrays of Oscillators. We begin by discussing an extension of the example in the introduction to a linear chain of N coupled identical oscillators. Thus, consider the dynamic equations

$$\ddot{x}_1 + a\dot{x}_1 + bx_1 = 0,$$
$$\ddot{x}_2 + a\dot{x}_2 + bx_2 = \lambda\dot{x}_1$$

$$\vdots$$

$$\ddot{x}_N + a\dot{x}_N + bx_N = \lambda\dot{x}_{N-1}.$$

Here $\lambda \neq 0$ is assumed to be constant. Each SISO node system is in first-order form as

$$\alpha = \begin{pmatrix} 0 & 1 \\ -b & -a \end{pmatrix}, \quad \beta = \begin{pmatrix} 0 \\ 1 \end{pmatrix}, \quad \gamma = (0\ 1), \tag{11.35}$$

or, equivalently, via the strictly proper transfer function

$$g(z) = \gamma(zI_2 - \alpha)^{-1}\beta = \frac{z}{z^2 + az + b}.$$

Assuming $b \neq 0$, which we will assume from now on, ensures the coprimeness of the factors z and $z^2 + az + b$ (Figure 11.1). The state interconnection matrix for this system is decomposable as $A = L \otimes K = L$, with $K = 1$ and

$$L = \lambda \begin{pmatrix} 0 & & & \\ 1 & 0 & & \\ & \ddots & \ddots & \\ & & 1 & 0 \end{pmatrix}.$$

In particular, L is nilpotent and e is not an eigenvector of L. Thus synchronization for this chain of oscillators does not occur. We rewrite this vectorially as a first-order system $\dot{x} = \mathscr{A}x$, where $\mathscr{A} \in \mathbb{R}^{2n \times 2n}$ has the block lower triangular form

Fig. 11.1 Directed simple path

$$\begin{pmatrix}
0 & 1 \\
-b & -a \\
0 & 0 & 0 & 1 \\
0 & \lambda & -b & -a \\
& & & & \ddots & & \ddots \\
& & & & & \ddots & & \ddots \\
& & & & & & 0 & 0 & 0 & 1 \\
& & & & & & 0 & \lambda & -b & -a
\end{pmatrix}.$$

In this case, the spectral analysis becomes particularly simple because, by the lower triangular structure, one obtains

$$\det(zI - \mathscr{A}) = (z^2 + az + b)^N.$$

Clearly, this is also the minimal polynomial of our system. The system becomes a bit more interesting if we add inputs $u(t)$ to the system. For example, we might consider the case where only the first node is controlled, i.e., we consider

$$\ddot{x}_1 + a\dot{x}_1 + bx_1 = u(t),$$
$$\ddot{x}_2 + a\dot{x}_2 + bx_2 = \lambda \dot{x}_1$$

$$\vdots$$

(11.36)

$$\ddot{x}_N + a\dot{x}_N + bx_N = \lambda \dot{x}_{N-1}.$$

The network transfer function $\mathscr{N}(z)$ from the input to the states then becomes

$$\begin{pmatrix}
z^2 + az + b \\
-\lambda z & \ddots \\
& \ddots & \ddots \\
& & -\lambda z & z^2 + az + b
\end{pmatrix}^{-1}
\begin{pmatrix}
1 \\
0 \\
\vdots \\
0 \\
0
\end{pmatrix}
=
\begin{pmatrix}
(z^2 + az + b)^{-1} \\
\lambda z (z^2 + az + b)^{-2} \\
\vdots \\
(\lambda z)^{N-2} (z^2 + az + b)^{-N+1} \\
(\lambda z)^{N-1} (z^2 + az + b)^{-N}
\end{pmatrix}.$$

Obviously, for $\lambda \neq 0$, this is a left coprime factorization, and thus (11.36) is reachable.

Next we move on to symmetrically coupled arrays of oscillators (Figure 11.2).

Fig. 11.2 Undirected simple path

Specifically, we consider the system

$$\ddot{x}_1 + a\dot{x}_1 + bx_1 = \lambda \dot{x}_2,$$
$$\ddot{x}_2 + a\dot{x}_2 + bx_2 = \lambda (\dot{x}_1 + \dot{x}_2)$$

$$\vdots \qquad\qquad\qquad (11.37)$$

$$\ddot{x}_{N-1} + a\dot{x}_{N-1} + bx_{N-1} = \lambda (\dot{x}_{N-1} + \dot{x}_N),$$
$$\ddot{x}_N + a\dot{x}_N + bx_N = \lambda \dot{x}_{N-1}.$$

We assume that $\lambda > 0$. The matrix $\mathscr{A}$ has again the tensor product structure $\mathscr{A} := I_n \otimes \alpha + L \otimes \beta \gamma$, but L has the symmetric matrix representation

$$L = \lambda \begin{pmatrix} 0 & 1 & & & \\ 1 & 0 & 1 & & \\ & 1 & \ddots & \ddots & \\ & & \ddots & \ddots & 1 \\ & & & 1 & 0 \end{pmatrix}.$$

Matrix L has a tridiagonal symmetric Toeplitz structure. By Theorem 8.45, we know that the eigenvalues of L are distinct and equal to $2\cos\frac{k\pi}{N+1}$ for $k = 1,\ldots,N$, with an explicit formula for the eigenvectors, too. Thus the solutions of (11.37) can be written down in closed form. We will not do that here. Since $\mathbf{e}$ is not an eigenvalue of L, the system does not synchronize.

II. Circular arrays of oscillators. In the circularly oriented case, the interconnection structure is depicted as in Figure 11.3. Explicitly, we consider the system

Fig. 11.3 Directed cycle graph

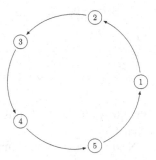

$$\ddot{x}_1 + a\dot{x}_1 + bx_1 = \lambda \dot{x}_N,$$
$$\ddot{x}_2 + a\dot{x}_2 + bx_2 = \lambda \dot{x}_1$$

(11.38)

$$\vdots$$

$$\ddot{x}_N + a\dot{x}_N + bx_N = \lambda \dot{x}_{N-1}.$$

Obviously, its state space is $2N$-dimensional and the dynamic equations can be written in first-order block circulant form, $\dot{x} = \mathscr{A}x$, as follows:

$$\frac{d}{dt}\begin{pmatrix} x_1 \\ \dot{x}_1 \\ \vdots \\ \vdots \\ \vdots \\ x_N \\ \dot{x}_N \end{pmatrix} = \begin{pmatrix} 0 & 1 & & & & 0 & 0 \\ -b & -a & & & & 0 & \lambda \\ 0 & 0 & 0 & 1 & & & \\ 0 & \lambda & -b & -a & & & \\ & & & \ddots & \ddots & & \\ & & & & \ddots & \ddots & \\ & & & & & 0 & 0 & 0 & 1 \\ & & & & & 0 & \lambda & -b & -a \end{pmatrix}\begin{pmatrix} x_1 \\ \dot{x}_1 \\ \vdots \\ \vdots \\ \vdots \\ x_N \\ \dot{x}_N \end{pmatrix}.$$

Using the special structure of matrix $\mathscr{A}$, its characteristic polynomial turns out to be

$$d_{\mathscr{A}}(z) = (z^2 + az + b)^N - \lambda^N z^N.$$

It is more convenient to analyze the system using the associated polynomial system matrices. We note that the coupling under consideration has a preferential direction. Later, we will also study more symmetric couplings. The interconnection matrix is $A = C$, where

$$C = \begin{pmatrix} 0 & & & 1 \\ 1 & \ddots & & \\ & \ddots & \ddots & \\ & & 1 & 0 \end{pmatrix}.$$

Note that C has the structure of a circulant matrix and thus can be diagonalized by the Fourier matrix. Its eigenvalues are exactly the $N - th$ roots of unity $1, \omega, \ldots, \omega^{N-1}$, where $\omega = e^{\frac{2\pi\sqrt{-1}}{N}}$ denotes the primitive root of unity. Note further that 1 is always an eigenvalue of C with associated eigenvector $\mathbf{e}$. To analyze the synchronization of (11.38), we consider the closed-loop polynomial system matrix $T(z) := (z^2 + az + b)I_N - z\lambda C$. The characteristic polynomial of $\mathscr{A}$ coincides with the determinant of $T(z)$. The zeros of $\det T(z)$ are equal to the roots of $z^2 + (a - \lambda\omega^k)z + b$ for $k = 0, \ldots, N - 1$.

Theorem 11.37. *Let $a, \lambda > 0$. System (11.38) synchronizes if and only if*

$$a > \lambda \cos \frac{2\pi}{N}, \quad b > 0.$$

Proof. Using the stability test for complex polynomials via the positivity of the Hermite–Fujiwara matrix (Theorem 5.52), a complex polynomial $p(z) = z^2 + uz + v$ is Hurwitz if and only if the Hermite–Fujiwara matrix

$$\mathbf{H}_2(p) = 2 \begin{pmatrix} \mathrm{Re}(u\bar{v}) & -i\,\mathrm{Im}\,v \\ i\,\mathrm{Im}\,v & \mathrm{Re}\,u \end{pmatrix}$$

is positive definite, i.e., if and only if

$$\mathrm{Re}(u) > 0, \quad \mathrm{Re}(u)\mathrm{Re}(u\bar{v}) > \mathrm{Im}^2(v).$$

In our situation, $u = a - \lambda \omega^k, v = b$, with $a, b, \lambda > 0$ real. Thus $z^2 + (a - \lambda \omega^k)z + b$ is Hurwitz for $k = 1, \ldots, N - 1$ if and only if $b > 0$ and $a - \lambda \cos \frac{2k\pi}{N} > 0$ for $k = 1, \ldots, N - 1$. The result follows. ∎

Note that, under our assumption that $a, b, \lambda > 0$, the inequality $|a - \lambda \omega^k| \leq a + \lambda$ is valid for all $1 \leq k \leq N$. Thus, if $(a + \lambda)^2 < 4b$, then there is oscillatory motion, with the different modes having damping terms of the form $e^{-\sigma_k t}$, with $\sigma_k = \mathrm{Re}\,\frac{a - \lambda \omega^k}{2}$. Obviously, the mode with the slowest rate of decay is the one where σ_k is (algebraically) the largest, and this occurs when $\omega^k = 1$, i.e., for $k = 0$. It is of interest to identify this mode. Indeed, if we look for eigenvectors of $\mathscr{A}$ of the form $(\xi, \ldots, \xi, \eta, \ldots, \eta)^\top$, then $\mathscr{A}x = \lambda x$ reduces to the pair of equations

$$\eta = \lambda \xi,$$
$$-b - a\eta + \lambda \eta = \lambda \eta,$$

which leads to the equation

$$(\lambda^2 + (a - \lambda)\lambda + b)\xi = 0,$$

whose roots are eigenvalues of $\mathscr{A}$. Thus we see that the slowest rate of decay is in the synchronized mode.

If we apply symmetric nearest-neighbor coupling, we obtain

$$\ddot{x}_k + a\dot{x}_k + bx_k = \lambda(\dot{x}_{k-1} + \dot{x}_{k+1}) \quad k = 1, \ldots, N, \tag{11.39}$$

with the understanding that $x_0 = x_N$ and $x_{N+1} = x_1$. The interconnection matrix A is the symmetric, circulant Toeplitz matrix

$$A = -\lambda \Gamma = -\lambda \begin{pmatrix} 0 & 1 & & & & 1 \\ 1 & 0 & 1 & & & \\ & 1 & \ddots & \ddots & & \\ & & \ddots & \ddots & 1 & \\ & & & 1 & 0 & 1 \\ 1 & & & & 1 & 0 \end{pmatrix},$$

which has **e** as an eigenvector. The associated eigenvalue -2λ is simple. By Theorem 8.48, the eigenvalues of A are $-2\lambda \cos(\frac{2k\pi}{N})$ for $k = 1,\dots,N$. The polynomial system matrix is $T(z) = (z^2 + az + b)I_N - \lambda z\Gamma$. System (11.39) can be written in state-space form as $\dot{x} = \mathscr{A} x$, where

$$\mathscr{A} = I_N \otimes \alpha - \lambda \Gamma \otimes \beta\gamma.$$

Here (α, β, γ) are as in (11.35). We observe that the eigenvalues of $\mathscr{A}$ are equal to the roots of the determinant $\det T(z)$, i.e., to the roots of

$$z^2 + (a - 2\lambda \cos(\frac{2k\pi}{N}))z + b = 0.$$

Note that $k = N$ corresponds to the synchronized mode **e**. Note further that $a - 2\lambda \cos \frac{2k\pi}{N} < a - 2\lambda$ for all $1 \le k < N$. We arrive at the following theorem.

Theorem 11.38. *The symmetrically connected cycle (11.39) synchronizes if and only if $z^2 + (a - 2\lambda \cos(\frac{2k\pi}{N}))z + b$ is Hurwitz for $k = 1,\dots,N-1$. Equivalently, synchronization occurs if and only if $b > 0$ and $a > 2\lambda \cos \frac{2\pi}{N}$. The eigenvalues for the synchronized dynamics are equal to the roots of $z^2 + (a - 2\lambda)z + b$. All other eigenvalues of the system matrix $\mathscr{A}$ have real part $< a - 2\lambda$.*

11.6 Exercises

1. Prove the following result from Trentelmann, Stoorvogel and Hautus (2001):
   A linear subspace $\mathscr{V} \subset \mathscr{X}$ is (A,B,C)-invariant if and only if there exists an output feedback transformation K such that $(A + BKC)\mathscr{V} \subset \mathscr{V}$.
2. (a) Show that a homogeneous synchronizing network for the real rational node transfer function

$$g(z) = \frac{\alpha z + \beta}{z^2 + az + b}$$

exists if and only if there is a complex number $\lambda \in \mathbb{C}$ satisfying the inequalities

$$a|\lambda|^2 - \alpha \operatorname{Re}\lambda > 0,$$

$$(a|\lambda|^2 - \alpha \operatorname{Re}\lambda)(ab|\lambda|^2 + \alpha\beta - (a\beta + b\alpha)\operatorname{Re}\lambda) > \beta^2(\operatorname{Im}\lambda)^2.$$

(b) Deduce that a homogeneous synchronizing network for

$$g(z) = \frac{\varepsilon z + 1}{z^2 + 1}$$

exists if and only if $\varepsilon \neq 0$.

3. Give necessary and sufficient conditions when the system of four coupled second-order systems

$$\ddot{x}_1 + a\dot{x}_1 + bx_1 = \lambda \dot{x}_2,$$

$$\ddot{x}_2 + a\dot{x}_2 + bx_2 = \lambda(\dot{x}_1 + \dot{x}_3),$$

$$\ddot{x}_3 + a\dot{x}_3 + bx_3 = \lambda(\dot{x}_2 + \dot{x}_4),$$

$$\ddot{x}_4 + a\dot{x}_4 + bx_4 = \lambda \dot{x}_3$$

clusters at x_1, x_4 and x_2, x_3, respectively.

11.7   Notes and References

There exists a huge literature from physics and systems engineering on synchronization, clustering, and consensus; we refer the reader to the survey paper by Doerfler and Bullo (2014) and the references therein. A new idea was recently proposed by R.W. Brockett in his 2014 Bernoulli lecture at the International Symposium on the Mathematical Theory of Networks and Systems (MTNS 2014) in Groningen, the Netherlands. Brockett asks a fundamental question concerning the potential mechanisms for synchronization: Given a symmetric matrix Q with distinct eigenvalues and a second-order system of the form

$$\ddot{x} + \eta(x,\dot{x}) + Qx = f(x,\dot{x},z),$$

$$\dot{z} = g(x,\dot{x},z),$$

what are the simplest, physically plausible choices of f and g that result in synchronization? He argues that the system

$$\ddot{x} + \eta(x,\dot{x}) + (Q+Z)x = 0,$$

$$\dot{Z} = -\alpha Z + x\dot{x}^\top - \dot{x}x^\top$$

should be an interesting candidate. See Brockett (2003) and Brockett (2013) for a study of closely related equations that underpin this belief.

Theorem 11.4 can be extended to normalized Laplacians of time-varying graphs $\Gamma(t)$ under weak connectivity assumptions. Let $\Gamma_i = (V, E_i), i = 1, \ldots, m$, denote finitely many weighted directed graphs on the same vertex set V with associated adjacency matrices $A_1, \ldots, A_m$. The union $\Gamma = \Gamma_1 \cup \cdots \cup \Gamma_m$ is the graph $\Gamma = (V, E)$ with edge set $E = E_1 \cup \cdots \cup E_m$. Consensus results for time-varying graphs were established by Jadbabaie, Lin and Morse (2003) under the assumption that there exists $T \in \mathbb{N}$ such that the union of graphs $\Gamma(kT) \cup \Gamma(kT+1) \cup \cdots \cup \Gamma(kT+T)$ is connected for all $k \in \mathbb{N}_0$. Cao, Morse and Anderson (2008) have derived more generally sufficient conditions for consensus if the time-varying graph of each flocking matrix $F(t)$ has a root for all t.

There exists by now a rich and rapidly growing literature on distributed control and distributed optimization. We refer the reader to Tsitsiklis, Bertsekas and Athans (1986) and the book by Bertsekas and Tsitsiklis (1989) for early contributions and further references. The distributed algorithm (11.9) for finding intersection points of affine subspaces has appeared several times in more general contexts; see, for example, Nedic, Ozdaglar and Parrilo (2010) for intersections of convex sets and for an explanation of the connection with the classical alternating projection method by von Neumann. Theorem 11.8 is due to Mou and Morse (2013).

Simple simulation experiments – see, for example, Blondel, Hendrickx and Tsitsiklis (2009) – show that trajectories in the Hegselmann–Krause model (11.13) do not converge to a common consensus state; instead, they **cluster** around certain limit points. Thus the Hegselmann–Krause model is really more a model for clustering rather than for consensus. The fine structure of the cluster states is quite interesting and requires further mathematical analysis. For example, it has been experimentally observed, and is conjectured to be true in general, that the solutions of (11.13) actually cluster in distances of $|x_i^* - x_j^*| \geq 2r$. Moreover, the distribution of the cluster points is not exactly evenly distributed, even for uniform distributions of the initial conditions. For extensions of the Krause model to continuous-time models, see Blondel, Hendrickx and Tsitsiklis (2009, 2010).

Consensus problems for second-order systems, including consensus among velocities, have been treated by, for example, Anderson, Lin and Deghat (2012) and Ren (2008). Cucker and Smale (2007) proposed a new consensus algorithm for the velocities of N second-order agents in $\mathbb{R}^3$ using the state-dependent graph adjacency matrix and Laplacian

$$A(x) = \left((1 + \|x_i - x_j\|^2)^{-\beta} \right)_{i,j} \quad \text{and} \quad L(x) = \operatorname{diag}(A(x)\mathbf{e}) - A(x),$$

respectively. They established asymptotic convergence results for the velocities in the network of second-order systems

$$\ddot{x} + (I_{3N} - L(x) \otimes I_3)\dot{x} = 0,$$

depending on whether $\beta < \frac{1}{2}$ or $\beta \geq \frac{1}{2}$. Extensions to general interconnection graphs are in Cucker and Smale (2007a).

From a systems engineering point of view, synchronization is a property that is desirable for the purpose of using feedback strategies. A well-known decentralized control approach to synchronization assumes a fixed diffusive coupling, together with N local feedback controllers around the node systems. The synchronization task then becomes to tune the local feedback controllers so that the network synchronizes. This design strategy is quite different from the approach taken here, where one aims to achieve synchronization via appropriate selections of the coupling terms. Whatever approach one prefers, there is plenty of room for further improvements. For example, one might replace the local feedback controllers by adaptive ones. Such an approach has been proposed by Helmke, Prätzel-Wolters and Schmidt (1991) and Ilchmann (2013), where synchronization is modeled as an adaptive tracking problem for networks of systems. This leads to synchronization results that are robust with respect to variations in both the interconnection and system parameters.

Synchronization problems for homogeneous networks of linear systems using state feedback transformations have been explored by Scardovi and Sepulchre (2009) and Tuna (2008), for example. Robust synchronization tasks using state feedback and output injection are studied in Trentelmann, Takaba and Monshizadeh (2013). Variants of Theorem 11.13 for diffusive coupling have been shown by several researchers, including Ma and Zhang (2010) and Lunze (2011). For networks with diffusive couplings and using state feedback with $\gamma = I_n$, Ma and Zhang (2010) have shown that synchronizability is equivalent to (α, β) being stabilizable and the graph being connected. Other versions of synchronizability via output injection were studied by Tuna (2009). Versions of the internal model principle for synchronization have been considered by Wieland, Sepulchre and Allgöwer (2011) and Lunze (2012), who proved a special case of Proposition 11.29. Using state-space methods, Lunze (2012) proved an extension of Theorem 11.31 for heterogenous networks, however under the strong additional assumption that the system matrices of the agents are diagonalizable. We believe that the results for higher-order systems introduced in Section 11.4 lead to a more natural approach to synchronization than standard state-space methods.

Chapter 12
Control of Ensembles

The purpose of this chapter is to provide an introduction to the emerging field of ensemble control for linear systems, i.e., the control of families of linear systems. Ensemble control refers to the task of controlling a large, potentially infinite, number of states, or systems, using a single-input function or a single-feedback controller. Thus, controlling ensembles is very much at the core of a robust theory of networks of systems. The study of ensembles is motivated from quite different applications.

- A classical example from physics concerns the conversion of heat into work by control of the heat flow and volume. Here the node systems are the gas molecules whose dynamics are described via the formalism of statistical mechanics. Of course, the sheer magnitude of the system, consisting roughly of $N \simeq 6 \times 10^{23}$ molecules per mole, calls for a thermodynamical or statistical mechanics approach rather than ab initio calculations.
- Quantum control of weakly coupled identical particles in nuclear magnetic resonance spectroscopy is another interesting area that has attracted much interest recently. Here the goal is to control a possibly large system of spins by applying short pulses of an electromagnetic field. The system is described by a controlled Liouville–von Neumann master equation evolving on density operators or Lie groups of unitary matrices. A closely related question concerns the control of parameterized families of systems using open-loop controls that are independent of the parameters. Open-loop control has recently acquired popularity in quantum control through the work of Li and Khaneja (2006), for example.
- The task of controlling probability distributions in the state space for a finite-dimensional control system leads to the control of Liouville transport equations or, more generally, of Fokker–Planck equations.

© Springer International Publishing Switzerland 2015

P.A. Fuhrmann, U. Helmke, *The Mathematics of Networks of Linear Systems*, Universitext, DOI 10.1007/978-3-319-16646-9_12

- Another area of interest is the control of parameter-varying systems using parameter-independent open-loop or closed-loop controllers. As a specific instance we mention the so-called **blending problem** from the late 1970s, which seeks to find parameter-independent feedback control laws that stabilize a family of linear systems.

In all these areas the question arises of how to approximately control a family of systems or of state variables. Thus the key ideas behind studying ensembles of linear systems are the same as those for studying large-scale systems. For complex dynamical systems the goals of controlling the entire collection of states is often asking for too much and needs to be replaced by more realistic goals, such as controlling, for example, the mean value or covariance of the state variables. This happens, for example, in the control of open quantum systems where the state variables are defined by density operators that describe an average of state variables. It also happens in daily life experience, for example, in cooking a meal in an oven. Here the interesting task is not so much to control the temperature distribution in the oven but rather to create an average temperature profile that yields the desired result. The same issue arises in motion control problems for infinite platoons of vehicles, as in the work by Rogge and Aeyels (2008), or for spatiotemporal systems described by partial differential equations (PDEs), on which see Bamieh, Paganini and Dahleh (2002). Using Fourier-transform techniques, spatially invariant control systems can be identified with parameter-dependent families of linear systems. This opens the door to applications of a variety of different approaches to distributed large-scale systems. Parameter-dependent systems can also be regarded as infinite-dimensional systems defined on suitable Banach or Hilbert spaces of functions. This brings about many opportunities for interactions between functional analysis and parametric systems. We also mention the theory of systems over rings as a systematic algebraic approach to analyzing parameter-dependent systems.

In this chapter we focus on the specific task of finding **open-loop controls** that steer a system from a family of initial states to another family of terminal states using a single open-loop control function that is independent of the parameters. We then illustrate how this question arises in a number of different applications, i.e., for controlling PDEs, controlling spatially-invariant systems such as platoons or to the robust control of networks of linear systems. We turn now to a more detailed description of such issues and begin to explain their mutual relationships.

12.1   Control of Parametric Families of Systems

The starting point for our analysis of ensembles of linear systems are parameter-dependent linear systems of the form

$$\frac{\partial}{\partial t} x(t, \theta) = A(\theta) x(t, \theta) + B(\theta) u(t), \quad x(0, \theta) = 0. \tag{12.1}$$

Of course, one could equally well consider families of discrete-time systems

$$x(t+1,\theta) = A(\theta)x(t,\theta) + B(\theta)u(t), \quad x(0,\theta) = 0,$$

and our results apply to this case, too. For simplicity we assume that the system matrices $A(\theta) \in \mathbb{R}^{n \times n}$ and $B(\theta) \in \mathbb{R}^{n \times m}$ vary continuously in a compact domain $\mathbf{P}$ of parameters θ in Euclidean space $\mathbb{R}^d$. The analysis of such families of linear systems can be carried out in several directions. A straightforward issue to begin with is the search for parameter-dependent controls that steer the systems from a family of initial states to a family of desired terminal states. A restriction here might be on the degree of continuity or smoothness in the parameters that is imposed on the controls. For instance, if the system matrices depend polynomially on a parameter, it may be desirable that the same be true for the feedback controllers and input functions. This leads to the control problems for systems over rings that have been discussed intensively in algebraic systems theory during the course of the past four decades. We refer the reader to the early work of, for example, Hazewinkel (1981), Sontag (1976), Tannenbaum (1981), and Conte and Perdon (2000) for further details. Another extreme case of studying families of systems – and this is the scenario that we will study in the remainder of this chapter – is to search for input functions or feedback laws that are independent of the parameters of systems and steer prescribed families of initial and terminal states arbitrarily close to each other. We refer to this as the **ensemble control** problem. Thus ensemble control refers to a specific class of robust control problems, and a priori it is not obvious whether or not such problems can be solved.

Let p and q be integers, with $1 \leq p, q \leq \infty$. The input to a state operator of (12.1) at time T is $\mathscr{R}_T : L^p([0,T], \mathbb{R}^m) \longrightarrow L^q(\mathbf{P}, \mathbb{R}^n)$

$$\mathscr{R}_T(u)(\theta) = \int_0^T e^{(T-s)A(\theta)} B(\theta) u(s) ds.$$

Note that $\mathscr{R}_T$ is an integral operator with continuous kernel $K : \mathbf{P} \times [0,T] \longrightarrow \mathbb{R}^n$,

$$K(\theta, s) = e^{(T-s)A(\theta)} B(\theta).$$

It is a well-known consequence of the assumed continuity of $A(\cdot), B(\cdot)$ on the compact parameter space $\mathbf{P}$ that $\mathscr{R}_T$ defines a bounded linear operator. In fact, $\mathscr{R}_T$ is a compact operator for $1 < p \leq \infty$ and $1 \leq q < \infty$. Moreover, by the continuity of $A(\cdot), B(\cdot)$, the operator $\mathscr{R}_T$ is compact, even as an operator from $L^p([0,T])$ to $C(\mathbf{P}, \mathbb{R}^n)$ and $1 \leq p \leq \infty$. From the compactness of $\mathscr{R}_T$ we deduce that $\mathscr{R}_T$ has a closed image if and only if $\mathscr{R}_T$ has finite rank. Clearly, $\mathscr{R}_T$ has finite rank for each linear operator of the form $K(\theta, t) = \sum_{j=1}^k \phi_j(t) \psi_j(\theta)$. In particular, this is true if A is parameter independent; however, for general parameter-dependent systems the finite-rank property cannot be expected.

After these functional analytic generalities, we proceed to introduce the notion of ensemble reachability that we are interested in.

Definition 12.1. Let $1 \leq p \leq \infty$. System (12.1) is **uniformly ensemble reachable** if for every continuous function $x^* : \mathbf{P} \longrightarrow \mathbb{R}^n$ and every $\varepsilon > 0$ there exists a control $u \in L^p([0,T],\mathbb{R}^m)$ such that the induced state trajectory satisfies

$$\sup_{\theta \in \mathbf{P}} \|x(T,\theta) - x^*(\theta)\| < \varepsilon. \tag{12.2}$$

Instead of trying to construct controls that achieve the uniform ensemble reachability condition (12.2), one can also search for controls $u(t)$ that minimize the L^q-norms for $1 \leq q \leq \infty$,

$$\left(\int_{\mathbf{P}} \|x(T,\theta) - x^*(\theta)\|^q d\theta \right)^{\frac{1}{q}} < \varepsilon. \tag{12.3}$$

We then say that the system is L^q**-ensemble reachable**. If the conditions in (12.2) or (12.3) are satisfied for $\varepsilon = 0$, then the system is called **exactly ensemble reachable**. Of course, the important point here is the ability to choose the input function independently of the parameter θ, and it is not at all obvious that systems of this kind do in fact exist.

This notion of ensemble reachability is intimately related to standard concepts from infinite-dimensional systems theory, such as approximate reachability and reachability. Let $\mathscr{A} : X \longrightarrow X$ and $\mathscr{B} : U \longrightarrow X$ be bounded linear operators on Banach spaces X and U, respectively. A linear system

$$\dot{x}(t) = \mathscr{A}x(t) + \mathscr{B}u(t) \tag{12.4}$$

is called **approximately reachable** if the reachable set of 0 is dense in X. See Fuhrmann (1972) and Curtain and Zwart (1995) for the (now classical) characterizations of approximate reachability in a Hilbert space via the condition that the reachability operator has a dense image. A result of Trigianni (1975) implies that parameter-dependent linear systems (12.1) are never exactly ensemble reachable. Thus the approximate notions of (uniform or L^q-) ensemble reachability are the only meaningful ones.

The mathematical connection between ensemble reachability and approximate reachability is easy to describe. Explicitly, for uniform ensemble control, let X denote the Banach space of $\mathbb{R}^n$-valued continuous functions on the compact parameter space $\mathbf{P}$, endowed with a supremum norm. Similarly, for L^q-ensemble reachability, choose $X = L^q(\mathbf{P}, \mathbb{R}^n)$. In either case, a continuous family of linear systems $(A(\theta), B(\theta))$ defines a linear system of the form (12.4) on a Banach space X with a **finite-dimensional** space of control values $U = \mathbb{R}^m$. Here

$$\mathscr{A} : X \longrightarrow X, \quad (\mathscr{A}x)(\theta) := A(\theta)x(\theta) \tag{12.5}$$

denotes the bounded linear multiplication operator, while the input operator

$$\mathscr{B} : \mathbb{R}^m \longrightarrow X, \quad (\mathscr{B}u)(\theta) := B(\theta)u \tag{12.6}$$

is defined via an m-tuple of Banach-space elements, i.e., by the columns of $B(\cdot)$.

Proposition 12.2. *Let bounded linear operators $\mathscr{A}$ and $\mathscr{B}$ be defined as in (12.5) and (12.6), respectively. The parameter-dependent system (12.1) is uniformly (or L^q-) ensemble reachable if and only if system (12.4) on the Banach space X is approximately reachable.*

Proof. We focus on uniform ensemble reachability and the Banach space $X = C(\mathbf{P}, \mathbb{R}^n)$; the proof for L^q-ensemble reachability goes mutatis mutandis. Let $t \mapsto x(t) \in X$ denote the unique solution to (12.4), with $x(0) = 0$. Then $t \mapsto x(t; \theta)$ is the unique solution to (12.1), with $x(0; \theta) = 0$ for $\theta \in \mathbb{P}$. The approximate reachability of (12.4) then says that for the continuous function $x^* : \mathbf{P} \longrightarrow \mathbb{R}^n$ and $\varepsilon > 0$ there exists $T > 0$ such that $\|x(T) - x^*\| = \sup_{\theta \in \mathbf{P}} \|x(T, \theta) - x^*(\theta)\|$. But this is simply the condition for uniform ensemble reachability. ∎

Thus the parameter-dependent system (12.1) is uniformly ensemble reachable if and only if the infinite-dimensional system (12.4) is approximately reachable. In the same way, by replacing the Banach space X with the Hilbert space $H = L^2(\mathbf{P}, \mathbb{R}^n)$, one concludes that the L^2-ensemble reachability of (12.1) becomes equivalent to the approximate reachability of the infinite-dimensional system (12.4). Unfortunately, the conditions for approximate reachability stated in Curtain and Zwart (1995); Jacob and Partington (2006) depend on an explicit knowledge of a Riesz basis of eigenvectors for the Hilbert-space operator $\mathscr{A}$. However, except for trivial cases where, for example, $A(\theta)$ has constant eigenvalues, the multiplication operator $\mathscr{A}$ defined by $A(\theta)$ does not have a point spectrum, and therefore the spectral conditions in Curtain and Zwart (1995) are not satisfied here. In the next section, we will explain how such difficulties can be avoided using tools from complex approximation theory.

12.2 Uniform Ensemble Reachability

Next we provide necessary, as well as sufficient, conditions for the uniform ensemble reachability of linear systems (12.1). These conditions are true, verbatim, for discrete-time systems as well. Let

$$(zI - A(\theta))^{-1}B(\theta) = N_\theta(z)D_\theta(z)^{-1}$$

be a right coprime factorization by a rectangular polynomial matrix $N_\theta(z) \in \mathbb{R}^{n \times m}[z]$ and a nonsingular polynomial matrix $D_\theta(z) \in \mathbb{R}^{m \times m}[z]$. We first state the necessary conditions for uniform ensemble reachability.

Proposition 12.3 (Necessary Conditions). *Let P be a subset of $\mathbb{R}^d$ such that the interior points of P are dense in P. Assume that the family of linear systems $(A(\theta), B(\theta))_{\theta \in P}$ is uniformly ensemble reachable. Then the following properties are satisfied:*

1. *For each $\theta \in P$ the system $(A(\theta), B(\theta))$ is reachable.*
2. *For finitely many parameters $\theta_1, \ldots, \theta_s \in P$, the $m \times m$ polynomial matrices $D_{\theta_1}(z), \ldots, D_{\theta_s}(z)$ are mutually left coprime.*
3. *For $m + 1$ distinct parameters $\theta_1, \ldots, \theta_{m+1} \in P$ the spectra of $A(\theta)$ satisfy*

$$\sigma(A(\theta_1)) \cap \cdots \cap \sigma(A(\theta_{m+1})) = \emptyset.$$

4. *Assume $m = 1$. The dimension of P satisfies $\dim P \leq 2$. If $A(\theta)$ has a simple real eigenvalue for some $\theta \in P$, then $\dim P \leq 1$.*

Proof. Consider a parameter value $\theta \in P$ and state vector $\xi \in \mathbb{R}^n$. Choose a continuous map $x^* : P \longrightarrow \mathbb{R}^n$, with $x^*(\theta) = \xi$. For $\varepsilon > 0$ there exists by assumption an input function $u : [0, T] \longrightarrow \mathbb{R}^m$ such that

$$\sup_{\theta \in P} \|x(T, \theta) - x_*(\theta)\| < \varepsilon.$$

In particular, we obtain $\|x(T, \theta) - \xi\| < \varepsilon$. Thus ξ is in the closure of the reachable set of 0; since the reachable sets of linear systems are closed in $\mathbb{R}^n$, this shows that $(A(\theta), B(\theta))$ is reachable. By the same reasoning, the ensemble reachability of the family $(A(\theta), B(\theta))_\theta$ implies reachability for the parallel interconnection

$$\bar{A} := \begin{pmatrix} A(\theta_1) & & 0 \\ & \ddots & \\ 0 & & A(\theta_s) \end{pmatrix}, \quad \bar{B} := \begin{pmatrix} B(\theta_1) \\ \vdots \\ B(\theta_s) \end{pmatrix} \qquad (12.7)$$

of finitely many linear systems $(A(\theta_i), B(\theta_i)), i = 1, \ldots, s$. By Theorem 10.2, the parallel interconnection (12.7) of reachable linear systems is reachable if and only if the $m \times m$ polynomial matrices $D_{\theta_1}(z), \ldots, D_{\theta_s}(z)$ are mutually left coprime. This completes the proof of the second claim.

The reachability of (12.7) implies that there are at most m Jordan blocks in $\bar{A}$ for each eigenvalue of $\bar{A}$. Thus $\sigma(A(\theta_1)) \cap \cdots \cap \sigma(A(\theta_s)) = \emptyset$ is satisfied for $s \geq m + 1$ distinct parameters $\theta_1, \ldots, \theta_s$, because otherwise there would exist an eigenvalue of $\bar{A}$ with at least $m + 1$ Jordan blocks. This proves the third claim.

The last claim follows from the third claim. In fact, let $\lambda(\theta)$ denote a branch of the eigenvalues of $A(\theta)$. Since the eigenvalues of a matrix depend continuously on the parameters θ, one concludes from the fourth claim that the functions $\theta \mapsto \lambda(\theta) \in \mathbb{C} = \mathbb{R}^2$ are continuous and injective. Therefore, since continuous injective functions do not increase dimensions, one concludes that $\dim P \leq \dim \mathbb{R}^2 = 2$. Moreover, if there exists a real branch of eigenvalues $\lambda(\theta)$ of $A(\theta)$, then $\dim P \leq 1$.

If $\lambda(\theta_0)$ is a simple eigenvalue of $A(\theta_0)$, then there exists an open neighborhood U of θ_0 in $\mathbf{P}$ such that for all $\theta \in U$ the eigenvalue $\lambda(\theta)$ is real. This completes the proof.   ∎

The preceding proof, using Theorem 10.2, shows an interesting connection between ensemble reachability for *finite* parameter sets $\mathbf{P}$ and reachability for parallel interconnection schemes of single-input systems.

Corollary 12.4. *Assume that $P = \{\theta_1, \ldots, \theta_s\} \subset \mathbb{R}^d$ is finite. Then a family of single-input systems $(A(\theta), b(\theta))_{\theta \in P}$ is uniformly ensemble reachable if and only if the following two conditions are satisfied:*

1. $(A(\theta_i), b(\theta_i))$ is reachable for $i = 1, \ldots, s$.
2. The characteristic polynomials $\det(zI - A(\theta_i))$ and $\det(zI - A(\theta_j))$ are coprime for all $i \neq j$.

Proof. This is an obvious consequence of Proposition 12.3.   ∎

In the discrete-time case, as is further explained in Chapter 10.1, one can strengthen this result by deriving explicit formulas for the inputs that steer to a desired state. In fact, the minimum-time ensemble control task for finite parameter sets becomes equivalent to the Chinese remainder theorem. We illustrate this approach for single-input systems. Let $\theta_1, \ldots, \theta_s \in \mathbf{P}$. The uniform ensemble reachability of the finite family $(A(\theta_i), b(\theta_i))$ is equivalent to the systems $(A_i, b_i) :=$ $(A(\theta_i), b(\theta_i)), i = 1, \ldots, s$ being reachable, with pairwise coprime characteristic polynomials $q_i(z) = \det(zI - A_i)$. Define $\hat{q}_i(z) := \prod_{j \neq i} q_j(z)$ and $q(z) := \prod_{j=1}^{n_s} q_j(z)$. Without loss of generality, we can assume that (A_i, b_i) are in controllability canonical form with local state spaces

$$X_{q_j} := \{p \in \mathbb{R}[z] \mid \deg p < \deg q_j = n_j\}.$$

Consider the parallel connection system

$$A = \begin{pmatrix} A_1 & & 0 \\ & \ddots & \\ 0 & & A_s \end{pmatrix}, \quad b = \begin{pmatrix} b_1 \\ \vdots \\ b_s \end{pmatrix}.$$

The state space of this global system is

$$X_q := \{p \in \mathbb{R}[z] \mid \deg p < \deg q = n\},$$

with direct sum decomposition

$$X_q := \hat{q}_1(z) X_{q_1} \oplus \cdots \oplus \hat{q}_s(z) X_{q_s}.$$

Assume that local target state vectors $x_j^* \in \mathbb{R}^{n_j}$ are chosen. These vectors uniquely define polynomial elements $r_j(z) \in X_{q_j}$ of degree $< n_j$ via

$$x_j^* = (x_{0j}^*, \ldots, x_{n_j-1,j}^*) \quad \text{and} \quad r_j(z) = \sum_{i=0}^{n_j-1} x_{ij}^* z^i.$$

Thus the components of x_j^* are simply the coefficients of the polynomial r_j. The ensemble control goal is then to find a polynomial $f(z) \in X_q$ such that its remainder modulo q_j is r_j. In fact, the coefficients $u_0, \ldots, u_{n-1}$ of the polynomial $f(z) = \sum_{i=0}^{n-1} u_{n-i-1} z^i$ are then simply the desired inputs that steer the system from zero to the local states x_j^*. In particular, the minimum length of such an ensemble control is $n = \deg q = \sum_{j=1}^{s} n_j$, as it should be. To compute $f(z)$, we apply the Bezout identity. Thus, by the coprimeness of $q_j, \hat{q}_j$, there exist unique polynomials $a_j(z)$ of degree $< n_j$ and $b_j(z)$ with

$$a_j(z)\hat{q}_j(z) + b_j(z)q_j(z) = 1. \tag{12.8}$$

Define

$$f(z) = \sum_{j=1}^{s} r_j(z)a_j(z)\hat{q}_j(z). \tag{12.9}$$

The Chinese remainder theorem then asserts that f is a unique polynomial of degree n that has r_j as remainder modulo q_j. The coefficients of f thus give the desired controls for (A, b).

Example 12.5. In the discrete-time case and for the parallel connection of s harmonic oscillators, it is very easy to carry out the calculations. For $\theta_1 < \cdots < \theta_s$, let

$$A_j := \begin{pmatrix} 0 & -\theta_j^2 \\ 1 & 0 \end{pmatrix}, \quad b_j := \begin{pmatrix} 1 \\ 0 \end{pmatrix}.$$

Thus (A_j, b_j) is in controllability canonical form, with $q_j(z) = z^2 + \theta_j^2, j = 1, \ldots, s$, pairwise coprime. In this example, the Bezout equation (12.8) is easily solved by the constant polynomial

$$a_k(z) = \prod_{j \neq k} (\theta_j^2 - \theta_k^2)^{-1}, \quad k = 1, \ldots, s.$$

For the local states $r_j(z) = \xi_1(\theta_j) + \xi_2(\theta_j)z$, formula (12.9) for the remainder polynomial is then

$$f(z) = \sum_{j=1}^{s} (\xi_1(\theta_j) + z\xi_2(\theta_j)) \prod_{k \neq j} \frac{z^2 + \theta_k^2}{\theta_k^2 - \theta_j^2}.$$

Note that this is exactly the degree $2s - 1$ **Lagrange interpolation polynomial** that satisfies

$$f(\pm\sqrt{-1}\theta_j) = r_j(\pm\sqrt{-1}\theta_j)$$

for $j = 1, \ldots, s$. However, for equidistant choices of the interpolation points, this solution suffers from the well-known Runge phenomenon. Thus, while $f(z)$ is a perfect match to the data at $z = \sqrt{-1}\theta_j$, the approximation error will blow up at the boundary points of the interval **P**. Better approximants can be obtained by interpolating at Chebyshev points.

It is considerably harder to establish sufficient conditions for uniform ensemble control, even if one restricts oneself to single-input systems depending on a scalar parameter θ. Li (2011) has proposed an operator-theoretic characterization of L^2-ensemble reachability for general time-varying linear multivariable systems. However, that characterization is stated in terms of the growth rates of singular values of the input-state operator and, thus, is difficult to verify, even for the time-invariant linear systems (12.1). We next state a result that leads to easily verifiable conditions.

Theorem 12.6 (Sufficient Condition). *Let $P - [\theta_-, \theta_+]$ be a compact interval. A continuous family $(A(\theta), b(\theta))$ of linear single-input systems is uniformly ensemble reachable (or, more generally, L^q-ensemble reachable for $1 \leq q \leq \infty$) provided the following conditions are satisfied:*

(a) $(A(\theta), b(\theta))$ is reachable for all $\theta \in P$.
(b) For pairs of distinct parameters $\theta, \theta' \in P, \theta \neq \theta'$, the spectra of $A(\theta)$ and $A(\theta')$ are disjoint:

$$\sigma(A(\theta)) \cap \sigma(A(\theta')) = \emptyset.$$

(c) For each $\theta \in P$ the eigenvalues of $A(\theta)$ have an algebraic multiplicity of one.

Conditions (a) and (b) are also necessary for uniform ensemble reachability.

Static output feedback control presents an interesting situation where all these assumptions fall easily into place. Thus, consider a fixed reachable and observable linear system (A, b, c). Let $P = [\theta_-, \theta_+]$ denote a compact interval of gain parameters. Then, for real values of θ, the closed-loop characteristic polynomial is $\det(A - \theta bc) = q(z) + \theta p(z)$, with p, q coprime and

$$c(zI - A)^{-1}b = \frac{p(z)}{q(z)}.$$

In particular, for distinct numbers $\theta \neq \theta'$ there exists no complex number z with $q(z) + \theta p(z) = 0 = q(z) + \theta' p(z)$. Theorem 12.6 therefore implies the following corollary.

Corollary 12.7. *Let* (A,b,c) *be a reachable and observable SISO system. The (discrete-time or continuous-time) output feedback system* $(A - \theta bc, b)$ *is uniformly ensemble reachable if the eigenvalues of* $A - \theta bc$ *are simple for all* $\theta \in [\theta_-, \theta_+]$.

∎

The proof of Theorem 12.6 is much easier for discrete-time systems, and therefore we first prove the result for discrete-time single-input systems. In this situation the uniform ensemble reachability condition can be restated in a more convenient form.

Lemma 12.8. *A family* $\{(A(\theta), b(\theta)), \theta \in P\}$ *of discrete-time single-input systems is uniformly ensemble reachable on* P *if, for all* $\varepsilon > 0$ *and all continuous functions* $x^* : P \to \mathbb{R}^n$, *there exists a real scalar polynomial* $p(z) \in \mathbb{R}[z]$ *such that*

$$\sup_{\theta \in P} \| p(A(\theta)) b(\theta) - x^*(\theta) \| < \varepsilon.$$

Proof. For an input sequence $u(0), \ldots, u(T-1)$ the solution is

$$x(T, \theta) = \sum_{k=0}^{T-1} A(\theta)^{kb}(\theta) u(T-1-k) = p(A(\theta)) b(\theta),$$

where $p(z) = \sum_{k=0}^{T-1} u_{T-k-1} z^k$. Thus the input sequence is parameter independent if and only if the polynomial $p(z)$ is parameter independent.   ∎

Using this basic observation we can characterize the uniform ensemble reachability property in explicit form as follows.

Proposition 12.9. *Assume that the discrete-time system* $(A(\theta), b(\theta))$ *is reachable for* $\theta \in P = [\theta_-, \theta_+]$. *The following statements are equivalent:*

(a) $(A(\theta), b(\theta))_\theta$ *is uniformly ensemble reachable.*
(b) *For continuous families of polynomials* $u_\theta(z) \in \mathbb{R}[z]$ *of degree* $< n$ *and* $\varepsilon > 0$, *there exists a polynomial* $p \in \mathbb{R}[z]$ *with*

$$\| p(A(\theta)) b(\theta) - u_\theta(A(\theta)) b(\theta) \| < \varepsilon.$$

(c) *For continuous families of polynomials* $u_\theta(z) \in \mathbb{R}[z]$ *of degree* $< n$ *and* $\varepsilon > 0$, *there exists a scalar polynomial* $p(z) \in \mathbb{R}[z]$ *with* $\| (p(A(\theta)) - u_\theta(A(\theta)) \| < \varepsilon$.

Assume that for each $\theta \in P$ *the eigenvalues of* $A(\theta)$ *are distinct. Let*

$$C := \{ (z, \theta) \in \mathbb{C} \times P \mid \det(zI - A(\theta)) = 0 \}.$$

Then each of the preceding conditions is equivalent to the following statements:

(d) *For continuous families of polynomials $u_\theta(z) \in \mathbb{R}[z]$ of degree $< n$ and a $\varepsilon > 0$, there exists a polynomial $p \in \mathbb{R}[z]$ with*

$$|p(z) - u_\theta(z)| < \varepsilon \quad \forall (z, \theta) \in C.$$

Proof. Let

$$R(A, b) = (b, Ab, \dots, A^{n-1}b)$$

denote the $n \times n$ reachability matrix. By reachability, the matrix $R(A(\theta), b(\theta))$ is invertible for each $\theta \in \mathbf{P}$. For $x^* : \mathbf{P} \longrightarrow \mathbb{R}^n$ continuous, define a polynomial $u_\theta(z) \in \mathbb{R}[z]$ of degree $< n$ as

$$u_\theta(z) = (1, z, \dots, z^{n-1}) R(A(\theta), b(\theta))^{-1} x^*(\theta).$$

Conversely, every continuous family of polynomials $u_\theta(z)$ of degree $< n$ can be written in this way. Since $u_\theta(A(\theta)) b(\theta) = x^*(\theta)$, Lemma 12.8 implies the equivalence of (a) with (b). Obviously, condition (c) implies (b). Assume that the estimate $\|(f - u_\theta)(A(\theta)) b(\theta)\| < \varepsilon$ holds. Then $\|(f - u_\theta)(A(\theta)) A(\theta)^{kb}(\theta)\| < \varepsilon \cdot \sup_{\theta \in \mathbf{P}} \|A(\theta)\|^k$. Therefore,

$$\|(f - u_\theta)(A(\theta))\| < c\varepsilon$$

for the constant $c = \sup_{\theta \in \mathbf{P}} \|R(A(\theta), b(\theta))^{-1}\| \max_{0 \le k \le n-1} \|A(\theta)\|^k$. Thus (b) implies (c). Now consider a matrix X with distinct eigenvalues $\lambda_1, \dots, \lambda_n$. Then, for real polynomials F and $\varepsilon > 0$, the operator norm bound $\|F(X)\| < \varepsilon$ is equivalent to $|F(\lambda_i)| < \delta(\varepsilon)$, $i = 1, \dots, n$. Here $\delta(\varepsilon)$ goes to zero if and only if ε goes to zero. This shows that condition (d) is equivalent to (c), and we are done. $\blacksquare$

1. Proof of Theorem 12.6 for Discrete-Time Systems. We now prove Theorem 12.6 for discrete-time systems. Our proof depends on Mergelyan's theorem, a rather deep theorem from complex approximation theory. It is stated here for convenience; see Chapter II in Gaier (1987) for a proof. Let $\hat{\mathbb{C}}$ denote the one-point compactification of $\mathbb{C}$, i.e., the complex plane which is extended by including the point ∞.

Theorem 12.10 (Mergelyan). *Suppose K is compact in $\hat{\mathbb{C}}$ such that the complement $\hat{\mathbb{C}} \setminus K$ is connected. Suppose that $f : K \longrightarrow \mathbb{C}$ is a continuous function that is holomorphic in the interior of K. Then for every $\varepsilon > 0$ there exists a complex polynomial $p(z) \in \mathbb{C}[x]$ such that, for all $z \in K$,*

$$|f(z) - p(z)| < \varepsilon.$$

We note that this theorem applies in particular in the special case where K has no interior points. In that case, the analyticity condition on f is superfluous.

Proof. The claims in Theorem 12.6 concerning necessity all follow from Lemma 12.3. Consider the compact set

$$C := \{(z,\theta) \in \mathbb{C} \times \mathbf{P} \mid \det(zI - A(\theta)) = 0\}$$

and the projection map $\pi : C \longrightarrow \mathbb{C}$ defined by $\pi(z,\theta) = z$. Since $\mathbf{P}$ is compact, it follows that C and, therefore, the image $K := \pi(C)$ are compact. Condition (b) of Theorem 12.6 is equivalent to π being injective on C, and therefore $\pi : C \longrightarrow K$ is a homeomorphism onto K. Thus, there exists a continuous map $\sigma : K \longrightarrow C, \sigma(z) = (z, \tau(z))$ that is a right inverse of π. The continuous map $\tau : K \longrightarrow \mathbf{P}$ has the property that $z \in K \iff (z, \tau(z)) \in C$. Since the eigenvalues of $A(\theta)$ define the continuous functions of θ and are assumed to be distinct for each $\theta \in \mathbf{P}$, one concludes that $K = K_1 \cup \cdots \cup K_n \subset \mathbb{C}$ consists of exactly n disjoint homeomorphic copies K_j of $\mathbf{P}$. Since $\mathbf{P}$ is simply connected, so is K, and thus the complement $\mathbb{C} \setminus K$ is connected. Moreover, K has an empty interior. Therefore, one can apply Mergelyan's Theorem 12.10 to uniformly approximate continuous functions F on K by polynomials. For a continuous family of desired states $x^*(\theta)$, define the polynomial u_θ in z as

$$u_\theta(z) = (1, z, \dots, z^{n-1}) R(A(\theta), b(\theta))^{-1} x^*(\theta).$$

Replacing the variable θ in u_θ with $\theta = \tau(z)$ we obtain a continuous function

$$F : K \longrightarrow \mathbb{C}, \quad F(z) := u_{\tau(z)}(z).$$

Thus the theorem by Mergelyan asserts that there exists a polynomial $p(z)$, with $|p(z) - F(z)| < \varepsilon$, uniformly on K. Equivalently, there exists a polynomial $p(z)$ such that

$$|p(z) - u_\theta(z)| < \varepsilon \quad \forall (z,\theta) \in C. \tag{12.10}$$

Note that $u_\theta(z)$ is a real polynomial. Thus, by a possible replacement of $p(z)$ by $\frac{1}{2}(p(z) + \overline{p}(z))$, we can assume that (12.10) is satisfied for a real polynomial $p(z)$. The result follows from Proposition 12.9. This completes the proof of Theorem 12.6 in the discrete-time case. ■

2. Proof of Theorem 12.6 for Continuous-Time Systems. For continuous-time systems

$$\dot{x} = A(\theta)x(t,\theta) + b(\theta)u(t), \quad x_0(\theta) = 0,$$

we apply a sampling argument. For each positive sampling period $\tau > 0$, consider the discrete-time system

$$x(t+1,\theta) = F(\theta)x(t,\theta) + g(\theta)u(t), \quad x_0(\theta) = 0, \tag{12.11}$$

where

$$F(\theta) := e^{\tau A(\theta)}, \quad g(\theta) = \left(\int_0^\tau e^{sA(\theta)} ds \right) b(\theta).$$

The result now follows from showing the next proposition.

Proposition 12.11. *Let $(A(\theta), b(\theta))$ satisfy the assumptions of Theorem 12.6. Then the sampled system (12.11) satisfies the assumptions, too.*

Proof. By the compactness of **P**, there exists $\tau_* > 0$ such that all conditions of Theorem 12.6 are satisfied for the discrete-time system $(F(\theta), G(\theta))$ and all $0 < \tau < \tau_*$. It is well known that the reachability of a continuous-time linear system (A, b) implies the reachability of the sampled discrete-time system (F, g) if the sampling period is sufficiently small. Thus the pointwise reachability condition (a) implies the same condition for $(F(\theta), g(\theta))$. The other conditions follow from the continuity of the eigenvalues and local injectivity of the matrix exponential function. ∎

Applying Proposition 12.11, the proof of Theorem 12.6 for the discrete-time case implies the uniform ensemble reachability of the discrete-time system (12.11). Note that, under sampling, the continuous- and discrete-time solutions coincide at the sampling points. Therefore, the finite-length input sequence u_k for the uniform ensemble reachability of $(F(\theta), g(\theta))$ induces a piecewise constant input function $u^\tau : [0, T] \longrightarrow \mathbb{R}$ that performs the uniform ensemble control task for the continuous-time system (12.1). This completes the proof of our main theorem. ∎

As mentioned earlier, conditions (a) and (b) in Theorem 12.6 are actually necessary conditions. The next result shows that the ensemble reachability of discrete-time systems can be shown under weaker assumptions than condition (c).

Proposition 12.12. *Let $P = [\theta_-, \theta_+]$ be a compact interval. A continuous family $(A(\theta), b(\theta))$ of linear discrete-time single-input systems is uniformly ensemble reachable provided the following conditions are satisfied:*

(a) $(A(\theta), b(\theta))$ is reachable for all $\theta \in P$.
(b) For pairs of distinct parameters $\theta, \theta' \in P, \theta \neq \theta'$, the spectra of $A(\theta)$ and $A(\theta')$ are disjoint:

$$\sigma(A(\theta)) \cap \sigma(A(\theta')) = \emptyset.$$

(c) The characteristic polynomial of $A(\theta)$ is of the form $z^n - a_{n-1}z^{n-1} - \cdots - a_1 z - a_0(\theta)$, with $a_1, \ldots, a_{n-1}$ constant real numbers.

Proof. Without loss of generality, we can assume that $A(\theta), b(\theta)$ is in controllability normal form for each θ. The polynomial $\pi(z) := z^n - a_{n-1}z^{n-1} - \cdots - a_1 z$ satisfies $\pi(A(\theta)) = a_0(\theta)I_n$. Moreover, $A(\theta)^{k-1}e_1 = e_k$ for $k = 1, \ldots, n$. By condition (b), we see that $a_0 : P \longrightarrow \mathbb{R}$ is injective, and hence the inverse function

$a_0^{-1} : a_0(\mathbf{P}) \longrightarrow \mathbf{P}$ exists and is continuous. Using the Weierstrass approximation theorem there exist polynomials $f_1(t), \ldots, f_n(t) \in \mathbb{R}[t]$ such that, for $k = 1, \ldots, n$,

$$\sup_{t \in a_0(\mathbf{P})} |f_k(t) - e_k^\top x_*(a_0^{-1}(t))| < \varepsilon$$

or, equivalently, $\sup_{\theta \in \mathbf{P}} |f_k(a_0(\theta)) - e_k^\top x_*(\theta)| < \varepsilon$. The real polynomial $f(z) := \sum_{k=1}^n f_k(\pi(z))z^{k-1}$ satisfies

$$f(A(\theta))e_1 = \sum_{k=1}^n f_k(\pi(A(\theta)))A(\theta)^{k-1}e_1 = \sum_{k=1}^n f_k(a_0(\theta))e_k.$$

This implies $\sup_{\theta \in \mathbf{P}} \|f(A(\theta))b(\theta) - x_*(\theta)\|_\infty < \varepsilon$, and the result follows. ∎

We illustrate the applicability of the preceding results by means of three examples.

Example 12.13 (L^2-Ensemble Observability). We briefly discuss the dual version of ensemble reachability, i.e., ensemble observability. Since duality theory is easier in a Hilbert-space context, we focus on the notions of L^2-ensemble observability.

Definition 12.14. Assume that $A(\theta) \in \mathbb{R}^{n \times n}, C(\theta) \in \mathbb{R}^{p \times n}$ vary continuously in a compact parameter domain $\mathbf{P} \subset \mathbb{R}^d$. The parameter-dependent system

$$\frac{\partial x(t, \theta)}{\partial t} = A(\theta)x(t, \theta), \quad x(0, \cdot) \in L^2(\mathbf{P}, \mathbb{R}^n),$$

$$y(t) = \int_{\mathbf{P}} C(\theta)x(t, \theta)d\theta \tag{12.12}$$

is called L^2-**ensemble observable** if there exists $T > 0$ such that $y(t) = 0$ on $[0, T]$ implies $x(0, \theta) = 0$ for all $\theta \in \mathbf{P}$.

Definition 12.14 implies that one can reconstruct the L^2-initial state $x(0, \cdot)$ of (12.12) from the average values

$$\int_{\mathbf{P}} C(\theta)x(t, \theta)d\theta, \quad 0 \le t \le T,$$

of the outputs $C(\theta)x(t, \theta)$. Thus ensemble observability is a rather strong property that is particularly useful in, for example, biological parameter identification tasks where often only an averaged type of output information is available.

System (12.12) is equivalent to the linear system

$$\dot{x}(t) = \mathscr{A}x(t), \quad x(0) \in L^2(\mathbf{P}, \mathbb{R}^n),$$

$$y(t) = \mathscr{C}x(t) \tag{12.13}$$

on the Hilbert space $X = L^2(\mathbf{P}, \mathbb{R}^n)$. Here $\mathscr{A} : X \longrightarrow X, \mathscr{C} : X \longrightarrow \mathbb{R}^p$ are bounded linear operators defined by

$$(\mathscr{A}x)(\theta) = A(\theta)x(\theta), \quad (\mathscr{C}x)(\theta) = \int_{\mathbf{P}} C(\theta)x(\theta)d\theta,$$

respectively. Thus, $\mathscr{A}$ is a multiplication operator while $\mathscr{C}$ is an integration operator. The preceding notion of ensemble observability is equivalent to the notion of the approximate observability of (12.13), as defined in Curtain and Zwart (1995), Definition 4.1.12. Moreover, Lemma 4.1.13 in Curtain and Zwart (1995) implies that (12.13) is approximate observable if and only if the dual system

$$\frac{\partial}{\partial t}x(t,\theta) = A(\theta)^\top x(t,\theta) + C(\theta)^\top u(t), \quad x(0,\theta) = 0, \tag{12.14}$$

is L^2-ensemble reachable. Therefore, Theorem 12.6 applies to (12.14) for $p = 1$. This shows that every continuous one-parameter family $(A(\theta), C(\theta)), \theta \in \mathbf{P} = [\theta_-, \theta_+]$, of single-output linear systems is L^2-ensemble observable provided the following three conditions are satisfied:

1. $(A(\theta), C(\theta))$ is observable for all $\theta \in \mathbf{P}$.
2. The spectra of $A(\cdot)$ are pairwise disjoint, i.e.,

$$\sigma(A(\theta)) \cap \sigma(A(\theta')) = \emptyset, \quad \forall \theta, \theta' \in \mathbf{P}, \theta \neq \theta'.$$

3. For each $\theta \in \mathbf{P}$ the eigenvalues of $A(\theta)$ have an algebraic multiplicity of one.

Example 12.15 (Robust Numerical Integration.). The simplest numerical integration method is certainly the Euler-step method. For a continuous-time linear control system $\dot{x} = Ax + Bu$, this yields the discrete-time system

$$x(t+1, h) = (I + hA)x(t, h) + hBu(t), \tag{12.15}$$

with a step-size parameter $h > 0$. One can then ask whether the family of discretized systems (12.15) can be robustly controlled using a control sequence $u(t), t \in \mathbb{N}$, that is independent of the step-size $h > 0$. Theorem 12.6 provides a simple answer. Assume that the pair (A, B) is reachable. Then for each parameter $h > 0$ the pairs $(I + hA, hB)$ are also reachable. Moreover, assume that A has only simple, distinct eigenvalues $\lambda_1, \ldots, \lambda_n$ that satisfy

$$\lambda_j \neq c\lambda_i \tag{12.16}$$

for all $c > 0$ and $i \neq j$. Then the eigenvalues $1 + h\lambda_1, \ldots, 1 + h\lambda_n$ of $I + hA$ are also simple and satisfy $1 + h\lambda_i \neq 1 + h'\lambda_j$ for $h \neq h'$. Thus the discretized system (12.15) is ensemble reachable for compact intervals $\mathbf{P} \subset (0, \infty)$ of step-size parameters provided (A, B) is reachable and A has distinct simple eigenvalues that satisfy (12.16).

This shows that the Euler-step approximation (12.15) of a continuous-time control system $\dot{x} = Ax + Bu$ inherits the reachability properties in a very strong sense, i.e., (12.15) can be controlled in a step-size independent way. It would be interesting to see whether this property carried over to more general, higher-order, Runge–Kutta methods.

Example 12.16 (Robust Open-Loop Synchronization.). We describe an application of Theorem 12.6 to the synchronization of N identical harmonic oscillators

$$\ddot{y}(t) + \omega^2 y(t) = v(t),$$

with state-space realization

$$A := \begin{pmatrix} 0 & -\omega \\ \omega & 0 \end{pmatrix}, \quad b := \begin{pmatrix} \frac{1}{\omega} \\ 0 \end{pmatrix}, \quad c := (0\ 1).$$

The identical frequency ω of the harmonic oscillators is assumed to be known. We assume that the oscillators are coupled in a ring with unknown coupling strength κ, which can vary over a compact interval $\mathbf{P} := [\kappa^-, \kappa^+] \subset (0, \infty)$ of positive numbers. Thus the network topology is described by a directed simple cycle graph with N nodes and weighted adjacency matrix κS, with the circulant matrix

$$S := \begin{pmatrix} 0 & 1 & & 0 \\ 0 & \ddots & \ddots & \\ & & \ddots & \ddots & 1 \\ 1 & & & 0 & 0 \end{pmatrix}.$$

We assume that the numbering of the harmonic oscillators is such that a single external input is applied to the first harmonic oscillator. Thus the input-to-state interconnection matrix is $e_1 = (1, 0, \ldots, 0)^\top$. The dynamics of the overall network is thus of the form

$$\frac{\partial}{\partial t} x(t, \kappa) = (I \otimes A + \kappa S \otimes bc)x(t, \kappa) + (e_1 \otimes b)u(t),$$
$$x(0, \kappa) = x^0, \tag{12.17}$$

where $x^0 \in \mathbb{R}^{2N}$ denotes the initial state of the network. Let $\mathbf{e} = (1, \ldots, 1)^\top$, and let $x^* \in \mathbb{R}^2$ denote the desired terminal state to which the harmonic oscillators are supposed to synchronize. The network of harmonic oscillators (12.17) is called **robustly synchronizable** from $x^0 \in \mathbb{R}^{2N}$ to $\mathbf{e} \otimes x^* \in \mathbb{R}^{2N}$ if for every $\varepsilon > 0$ there exists $T > 0$ and an input function $u \colon [0, T] \to \mathbb{R}$ such that the state $x(t)$ of (12.17) satisfies

$$\sup_{\kappa \in \mathbf{P}} \|x(T, \kappa) - \mathbf{e} \otimes x^*\| < \varepsilon.$$

We emphasize that the input u acts here as a universal input for the parameter-dependent network that steers x^0 to the synchronized state $1 \otimes x^*$ uniformly for all coupling strength parameters $\kappa \in \mathbf{P}$.

We next show, using Theorem 12.6, that the network (12.17) of harmonic oscillators robustly synchronizes from every initial state $x^0 \in \mathbb{R}^{2N}$ at a vector $1 \otimes x^* \in \mathbb{R}^{2N}$ of identical terminal states x^*. Let Φ denote the $N \times N$ Fourier matrix (9.39) and $\omega = e^{2\pi\sqrt{-1}/N}$. Thus, using (9.40), it follows that $\Phi^* S \Phi = \operatorname{diag}\left(1, \omega, \ldots, \omega^{N-1}\right)$ and $\Phi e_1 = \frac{1}{\sqrt{N}}\mathbf{e}$. Applying the state-space similarity transformation $z := (\Phi^* \otimes I_n)x$ using the unitary matrix $\Phi^* \otimes I_n$, we see that (12.17) is similar to the decoupled, parameter-dependent system

$$\frac{\partial}{\partial t} z_j(t, \kappa) = \left(A + \kappa e^{2\pi j\sqrt{-1}/N})bc\right) z_j(t, \kappa) + \sqrt{N}bu(t), \quad j = 1, \ldots, N$$

$$z(0, \kappa) = \Phi^* x^0.$$

This system is the parallel connection of reachable linear systems. Moreover, the eigenvalues of $I \otimes A + \kappa S \otimes bc$ are equal to

$$\bigcup_{j=1}^{N} \{z \in \mathbb{C} : z^2 + \omega^2 - \kappa \omega^j = 0\}.$$

These eigenvalues are distinct and simple if, for example, $1 \notin \mathbf{P} = [\kappa^-, \kappa^+]$. Moreover, under this condition, the eigenvalues for different $\kappa \neq \kappa'$ are distinct. This implies that $I \otimes A + \kappa S \otimes bc$ is reachable and conditions (a)–(c) of Theorem 12.6 are satisfied. One concludes that the network (12.17) is robustly synchronizable from every initial state provided $1 \notin [\kappa^-, \kappa^+]$. Similarly, robust synchronization can be established for more complicated network structures.

12.3 Control of Platoons

The Fourier transform provides an elegant way to utilize the preceding results on uniform ensemble control for the control of partial differential equations and platoons of systems. Although this point of view has appeared already in previous work by, for example, Green and Kamen (1985), Bamieh, Paganini and Dahleh (2002), and Curtain, Iftime and Zwart (2009), the control tasks that we consider here have not been addressed before. In fact, the previously cited works consider only scenarios where each subsystem is controlled by individual, independent input functions. Thus such approaches use an infinite number of control functions and are therefore severely limited in applicability. In contrast, we focus on the reachability of systems that employ a finite number of input functions that are distributed over the entire network. For controlling platoons or swarms of systems this approach appears to be more appropriate.

1. Finite Platoons Consider the task of controlling a finite platoon of N vehicles on a line. If each vehicle is controlled individually by independent input functions, then clearly the overall system will be reachable. A more interesting situation arises when one considers inputs that are broadcasted to all systems simultaneously. Thus all vehicles are controlled by the same input function (or by a small number of input functions). By assuming nearest-neighbor interactions, we obtain the control system

$$\dot{x}_1 = -x_2 + x_1 + u(t)$$
$$\dot{x}_2 = -x_3 + 2x_2 - x_1 + u(t)$$
$$\vdots$$
$$\dot{x}_{N-1} = -x_N + 2x_{N-1} - x_{N-2} + u(t)$$
$$\dot{x}_N = -x_{N-1} + x_N + u(t).$$

In matrix form the system is $\dot{x}(t) = Ax(t) + bu(t)$, where

$$A = \begin{pmatrix} 1 & -1 & & & & \\ -1 & 2 & -1 & & & \\ & -1 & \ddots & \ddots & & \\ & & \ddots & \ddots & -1 & \\ & & & -1 & 2 & -1 \\ & & & & -1 & 1 \end{pmatrix}, \quad b = \begin{pmatrix} 1 \\ 1 \\ \vdots \\ \vdots \\ 1 \end{pmatrix}. \tag{12.18}$$

To check for reachability, we apply the Hautus test. Recall from Theorem 8.46 that the eigenvalues of A are nonrepeated and are equal to $\lambda_k = 2 - 2\cos\frac{(k-1)\pi}{N}, k = 1,\ldots,N$. Moreover, $\xi^{(k)} = (\sin\frac{k\pi}{N+1},\ldots,\sin\frac{kN\pi}{N+1})^\top$ is an eigenvector for the eigenvalue λ_k. Define $\omega = e^{\frac{\sqrt{-1}\pi}{N+1}}$. Thus

$$b^\top \xi^{(k)} = \sum_{v=1}^{N} \sin\frac{kv\pi}{N+1} = \mathrm{Im} \sum_{v=1}^{N} \omega^{kv}$$

$$= \mathrm{Im}\left(\frac{1 - \omega^{k(N+1)}}{1 - \omega^k} - 1\right) = \mathrm{Im}\left(\frac{1 - (-1)^k}{1 - \omega^k} - 1\right)$$

$$= \begin{cases} 0 & \text{if } k \text{ is even,} \\ \frac{\sin\frac{k\pi}{N+1}}{1 - \cos\frac{k\pi}{N+1}} & \text{if } k \text{ is odd} \end{cases}$$

is zero if and only if $1 \leq k \leq N$ is even. This implies that the system is not reachable. In contrast, consider the case where $b = e_k$ for some $1 \leq k \leq N$. From the tridiagonal structure of A it is easily seen that (A, e_1) is reachable. The subsequent characterization of reachability shows an interesting connection to elementary number theory.

Theorem 12.17. *Let (A, b) be defined by (12.18).*

1. *The pair (A, b) is not reachable.*
2. *(A, e_k) is reachable if and only if k and $N + 1$ are coprime.*
3. *$(A, e_k + e_\ell)$ is reachable provided both pairs $k + \ell, N + 1$ and $k - \ell, N + 1$ are coprime.*
4. *The two-input system $(A, (e_k, e_\ell))$ is reachable if and only if there exists no $1 \leq r \leq N$ such that $N + 1$ divides both rk and $r\ell$.*

Proof. The first part has already been shown. For the second part note that $e_k^\top \xi^{(r)} = \sin \frac{kr\pi}{N+1} = 0$ if and only if $N + 1$ divides kr. Suppose that k and $N + 1$ are coprime. Then $N + 1$ divides kr if and only if $N + 1$ divides k. But this is impossible because $1 \leq k \leq N$. Thus the coprimeness of k and $N + 1$ implies $e_k^\top \xi^{(r)} \neq 0$ for all $1 \leq r \leq N$, i.e., it implies the reachability of (A, e_k). Conversely, assume that $d \geq 2$ is the greatest common divisor of $k, N + 1$. Then $k = k'd$ and $N + 1 = N'd$ for suitable integers $1 \leq k', N' \leq N$. Then $(N + 1)k' = kr$ for $r := N' \leq N$. Thus $e_k^\top \xi^{(r)} = 0$, and therefore (A, e_k) cannot be reachable.

By the Hautus criterion, the reachability of $(A, e_k + e_\ell)$ is violated if and only if there exists $1 \leq r \leq N$ such that $\sin \frac{kr\pi}{N+1} + \sin \frac{\ell r\pi}{N+1} = 0$. Recall that $\sin(\pi x) = \sin(\pi y)$ if and only if either $x + y$ is an odd integer or $x - y$ is an even integer. Thus, reachability holds if and only if $r(k - \ell) \notin (2\mathbb{Z} + 1)(N + 1)$ and $r'(k + \ell) \notin 2\mathbb{Z}(N + 1)$ is valid for all $1 \leq r, r' \leq N$. Certainly this is the case if $k + \ell, N + 1$ are coprime and $k - \ell, N + 1$ are coprime. This implies the result. Finally, the reachability of the two-input system $(A, (e_k, e_\ell))$ is equivalent to the condition that there exists no $1 \leq r \leq N$ with $\sin \frac{kr\pi}{N+1} = \sin \frac{\ell r\pi}{N+1} = 0$. This proves the result. ∎

Similarly, let us consider the situation where vehicles proceed on a circular domain. In this case, we obtain the linear control system $\dot{x}(t) = Fx(t) + gu(t)$ with system matrices

$$F = \begin{pmatrix} 2 & -1 & & & & -1 \\ -1 & 2 & -1 & & & \\ & -1 & \ddots & \ddots & & \\ & & \ddots & \ddots & -1 & \\ & & & -1 & 2 & -1 \\ -1 & & & & -1 & 2 \end{pmatrix}, \quad g = \begin{pmatrix} 1 \\ 1 \\ \vdots \\ \vdots \\ 1 \end{pmatrix}. \tag{12.19}$$

See Notarstefano and Parlangeli (2013) and Chapter 9 for reachability results on closely related linear systems evolving on circular arrays. The eigenvalues of the circulant matrix F are equal to $2 - 2\cos \frac{2k\pi}{N}, k = 1, \ldots, N$. Thus, for $N \geq 2$, the symmetric matrix F always has eigenvalues with a multiplicity of 2. Therefore, the circulant system (F, g) is not reachable. This also follows directly from the fact that

$Fg = 0$. However, contrary to the preceding case of platoons on a line, the occurrence of eigenvalues with multiplicities ≥ 2 in the Jordan canonical form of F shows that the system (F,g) is not reachable *for each choice* of g. Thus we need at least two independent inputs to control a platoon on a circle. The next result describes some of the possibilities for controlling with two inputs.

Theorem 12.18. *Let $1 \leq k < \ell \leq N$ and F be defined by (12.19). There exists no vector g such that (F,g) is reachable. The two-input system $(F,(e_k,e_\ell))$ is reachable if and only if N and $k - \ell$ are coprime.*

Proof. By Theorem 8.48, the eigenspaces of F are either one-dimensional and spanned by $u := e_1 + \cdots + e_N$ (for $r = N$) or by $u := e_1 - e_2 + e_3 - \cdots + e_N$ (for $N = 2m, r = m$) or are two-dimensional with a basis

$$
x^{(r)} = \begin{pmatrix} 1 \\ \cos(\frac{2r\pi}{N}) \\ \cos(\frac{4r\pi}{N}) \\ \vdots \\ \cos(\frac{2(N-1)r\pi}{N}) \end{pmatrix}, \quad
y^{(r)} = \begin{pmatrix} 1 \\ \sin(\frac{2r\pi}{N}) \\ \sin(\frac{4r\pi}{N}) \\ \vdots \\ \sin(\frac{2(N-1)r\pi}{N}) \end{pmatrix}.
$$

Thus reachability is satisfied whenever the inner products of e_k, e_ℓ with the basis of eigenvectors does not vanish. Computing the inner products we obtain

$$
< e_k, u > = < e_\ell, u > = 1, \quad < e_k, v > = (-1)^k, \quad < e_\ell, v > = (-1)^\ell,
$$

$$
< e_k, x^{(r)} > = \sin \frac{2(k-1)r\pi}{N}, \quad < e_\ell, x^{(r)} > = \sin \frac{2(\ell-1)r\pi}{N},
$$

$$
< e_k, y^{(r)} > = \cos \frac{2(k-1)r\pi}{N}, \quad < e_\ell, x^{(r)} > = \cos \frac{2(\ell-1)r\pi}{N}.
$$

Thus the inner products with the eigenvectors are nonzero for the simple eigenvalues λ_r, with $r = N$ or $r = m, N = 2m$. For the other cases assume that $v = \alpha x^{(r)} + \beta y^{(r)}$ is an eigenvector of F for λ_r, with $< v, e_k > = < v, e_\ell > = 0$. Then

$$
\alpha \sin \frac{2(k-1)r\pi}{N} + \beta \cos \frac{2(k-1)r\pi}{N} = 0,
$$

$$
\alpha \sin \frac{2(\ell-1)r\pi}{N} + \beta \cos \frac{2(\ell-1)r\pi}{N} = 0.
$$

This has a nonzero solution (α, β) if and only if $\sin \frac{2(k-\ell)r\pi}{N} = 0$. This is equivalent to the condition that N divides $(k - \ell)r$. Since $r < N$, this implies reachability if and only if N and $k - \ell$ are coprime.   ∎

Using the theory developed in Chapter 9, it is easy to extend the reachability analysis of platoons from first-order scalar systems to higher-order systems.

As an example, consider a homogeneous network of identical higher-order systems of the form

$$q(\frac{d}{dt})y_i(t) = p(\frac{d}{dt})v_i(t)$$

$$v_i(t) = y_{i+1}(t) - 2y_i(t) + y_{i-1}(t) + b_i u(t), \quad i = 1, \ldots, N. \tag{12.20}$$

Here $b = \mathrm{col}\,(b_1, \ldots, b_N) \in \mathbb{R}^N$, and $p(z)$ and $q(z)$ denote coprime real scalar polynomials with $\deg p < \deg q = n$. Defining $Q(z) = q(z)I_N$, $P(z) = p(z)I_N$, and A as in (12.18) we see that (12.20) is equivalent to the homogeneous network

$$\left(Q(\frac{d}{dt}) - P(\frac{d}{dt})A \right) y(t) = P(\frac{d}{dt})bu(t).$$

Applying Theorem 9.15, one concludes that the network (12.20) is reachable if and only if (A, b) is reachable. Thus, for $b = e_k$, the network (12.20) is reachable if and only if $N + 1$ and k are coprime, independently of the choice of coprime polynomials $p(z), q(z)$. Similarly, reachability results for platoons of higher-order systems with circulant interconnection matrices are obtained.

2. Infinite Platoons. We now turn to an analysis of infinite platoons and their reachability properties. Infinite platoons are infinite-dimensional control systems where the spatial variable is constrained to either $\mathbb{N}$ or $\mathbb{Z}$. The coordinates of infinite platoons are therefore defined by either one-sided infinite sequences $(x_k)_{k \in \mathbb{N}_0}$ or bi-infinite sequences $(x_k)_{k \in \mathbb{Z}}$ of elements $x_k \in \mathbb{R}^n$. In either case, we obtain a Hilbert space $\ell_+^2(\mathbb{R}^n)$ or $\ell^2(\mathbb{R}^n)$ of square summable sequences with norms

$$\|x\|^2 = \sum_{k=0}^{\infty} |x_k|^2, \quad \|x\|^2 = \sum_{k=-\infty}^{\infty} |x_k|^2,$$

respectively. The interpretation of, for example, the set of integers $\mathbb{Z}$ with the spatial domain of a platoon is due to the identification of curves $t \mapsto x(t)$ in $\ell^2(\mathbb{R}^n)$ with functions $x : \mathbb{Z} \times \mathbb{R} \longrightarrow \mathbb{R}^n, (k,t) \mapsto x(k,t) := x(t)_k$. In such an interpretation the space $\mathbb{Z}$ corresponds to the spatial variable, while t corresponds to time. We follow the work by Curtain, Iftime and Zwart (2009), who developed an operator-theoretic analysis of infinite platoons in the Hilbert space $\ell^2(\mathbb{R}^n)$ of bi-infinite sequences. We also refer to the more recent work by Feintuch and Francis (2012) for a deeper analysis of stability problems for platoons in the Banach space of ℓ^∞ sequences. Restricting the coordinates to the Hilbert-space context of ℓ^2 sequences enables us to employ Fourier-transform techniques. In a second step, we then connect the reachability analysis of infinite platoons to that for parameter-dependent linear systems.

Restricting the coordinates to the Hilbert-space context of ℓ^2 sequences enables us to employ Fourier-transform techniques. For bi-infinite sequences x of vectors $x_k \in \mathbb{C}^n, k \in \mathbb{Z}$, define the associated Fourier series

$$x(e^{\sqrt{-1}\theta}) := \sum_{k=-\infty}^{\infty} x_{ke}^{-\sqrt{-1}k\theta}.$$

Conversely, with functions $f \in L^2(S^1, \mathbb{C}^n)$ on the unit circle S^1 one associates the sequence of Fourier coefficients

$$x_k(f) = \frac{1}{2\pi} \int_0^{2\pi} f(e^{\sqrt{-1}\theta}) e^{\sqrt{-1}k\theta} d\theta, \quad k \in \mathbb{Z}.$$

The **discrete Fourier transform**

$$\mathscr{F} : \ell^2(\mathbb{C}^n) \longrightarrow L^2(S^1, \mathbb{C}^n), \quad \mathscr{F}(x) = x(e^{\sqrt{-1}\theta})$$

then defines a linear isomorphism whose inverse is defined by the Fourier coefficients via

$$\mathscr{F}^{-1} : L^2(S^1, \mathbb{C}^n) \longrightarrow \ell^2(\mathbb{C}^n), \quad \mathscr{F}^{-1}f = (x_k(f))_{k \in \mathbb{Z}}.$$

By the Plancherel formula, the discrete Fourier transform defines an isometry of Hilbert spaces, i.e., for all $x \in \ell^2(\mathbb{C}^n)$,

$$\|x\|_{\ell^2} = \|\mathscr{F}x\|_{L^2}.$$

Similarly, the discrete Fourier transform of a one-sided sequence $(x_k)_{k \geq 0} \in \ell_+^2(\mathbb{C}^n)$ is defined as

$$x(e^{\sqrt{-1}\theta}) := \sum_{k=0}^{\infty} x_{ke}^{-\sqrt{-1}k\theta},$$

which defines an element of the Hardy space $H^2(\mathbb{D}; \mathbb{C}^n)$ on the unit disc $\mathbb{D}$. The discrete Fourier transform

$$\mathscr{F} : \ell_+^2(\mathbb{C}^n) \longrightarrow H^2(\mathbb{D}; \mathbb{C}^n), \quad \mathscr{F}(x) = x(e^{\sqrt{-1}\theta})$$

then maps $\ell_+^2(\mathbb{C}^n)$ isometrically onto $H^2(\mathbb{D}; \mathbb{C}^n) = H^2(\mathbb{D})^n$, the Hardy space of n-tuples of holomorphic functions on the open unit disc that are Lebesgue square integrable on the unit circle.

Following these preliminaries, we now turn to a study of bi-infinite platoons on $\mathbb{Z}$. The simplest classes of such models are spatially invariant and have the form

$$\dot{x}_k(t) = \sum_{j=-\infty}^{\infty} A_j x_{k-j}(t) + \sum_{j=-\infty}^{\infty} B_j u_{k-j}(t), \quad k \in \mathbb{Z}, \tag{12.21}$$

where $A_k \in \mathbb{R}^{n \times n}$ and $B_k \in \mathbb{R}^{n \times m}$. Here we assume that the inputs and state sequences are in ℓ^2, i.e., $(x_i) \in X = \ell^2(\mathbb{C}^n)$ and $(u_i) \in U = \ell^2(\mathbb{C}^m)$, respectively.

To specify conditions that the convolution operators $\mathscr{A} : X \longrightarrow X$ and $\mathscr{B} : U \longrightarrow X$, with

$$(\mathscr{A}x)_k = \sum_{j=-\infty}^{\infty} A_j x_{k-j}, \quad (\mathscr{B}u)_k = \sum_{j=-\infty}^{\infty} B_j u_{k-j},$$

are well defined, we impose a condition on their Fourier symbols

$$A(e^{\sqrt{-1}\theta}) = \sum_{j=-\infty}^{\infty} A_j e^{-\sqrt{-1}j\theta}, \quad B(e^{\sqrt{-1}\theta}) = \sum_{j=-\infty}^{\infty} B_j e^{-\sqrt{-1}j\theta}.$$

Assume that both $A(\cdot)$ and $B(\cdot)$ are elements of $L_\infty(S^1, \mathbb{C}^{n \times n})$ and $L_\infty(S^1, \mathbb{C}^{n \times m})$, respectively. This guarantees that the associated multiplication operators

$$\mathscr{A} : L^2(S^1, \mathbb{C}^n) \longrightarrow L^2(S^1, \mathbb{C}^n), \quad x(e^{\sqrt{-1}\theta}) \mapsto A(e^{\sqrt{-1}\theta})x(e^{\sqrt{-1}\theta}),$$
$$\mathscr{B} : L^2(S^1, \mathbb{C}^m) \longrightarrow L^2(S^1, \mathbb{C}^n), \quad u(e^{\sqrt{-1}\theta}) \mapsto B(e^{\sqrt{-1}\theta})u(e^{\sqrt{-1}\theta})$$

are bounded linear operators with operator norms

$$\|\mathscr{A}\| = \|A(\cdot)\|_\infty = \text{ess sup}_{0 \le \theta \le 2\pi} \|A(e^{\sqrt{-1}\theta})\|,$$
$$\|\mathscr{B}\| = \text{ess sup}_{0 \le \theta \le 2\pi} \|B(e^{\sqrt{-1}\theta}).\|$$

It follows that the infinite-dimensional control system

$$\dot{x}(t) = \mathscr{A}x(t) + \mathscr{B}u(t) \tag{12.22}$$

that describes the platoon model (12.21) is well defined on the Hilbert space $L^2(S^1, \mathbb{C}^n)$. Note that the inputs for (12.22) assume values in the infinite-dimensional Hilbert space $L^2(S^1, \mathbb{C}^m)$. We present the following Hautus-type condition for approximate reachability.

Theorem 12.19. *The infinite platoon (12.21) is approximately reachable on the Hilbert space $L^2(S^1, \mathbb{C}^n)$ if and only if the reachability rank condition*

$$\text{rk}\left(zI_n - A(e^{\sqrt{-1}\theta}), B(e^{\sqrt{-1}\theta})\right) = n$$

is satisfied for all $z \in \mathbb{C}$ and almost all $\theta \in [0, 2\pi]$.

Proof. By the isometric properties of the Fourier transform, both systems (12.21) and (12.22) are similar and thus have identical reachability properties. Characterizations of the approximate reachability of linear systems on a Hilbert space are well known; see, for example, the textbook by Curtain and Zwart (1995). In fact, approximate reachability in finite time $T > 0$ is guaranteed for (12.22) if and only if the image of the reachability operator $\mathscr{R}_T : L^2([0,T],X) \longrightarrow L^2(S^1, \mathbb{C}^n)$,

$$\mathscr{R}_T u = \int_0^T e^{(T-s)\mathscr{A}} \mathscr{B}u(s)ds,$$

is dense in $L^2(S^1, \mathbb{C}^n)$. Since the closure $\overline{\mathrm{Im}\,\mathscr{R}_T}$ of the image of $\mathscr{R}_T$ coincides with the kernel of the dual operator, we conclude that the approximate reachability of (12.21) is equivalent to the condition

$$\int_0^{2\pi} f(e^{\sqrt{-1}\theta})^* \exp(tA(e^{\sqrt{-1}\theta}))B(e^{\sqrt{-1}\theta})d\theta = 0 \quad \text{for all } t \geq 0 \implies f(e^{\sqrt{-1}\theta}) = 0.$$

Of course, this is equivalent to the familiar Kalman rank condition

$$\mathrm{rk}\left(B(e^{\sqrt{-1}\theta}),\ldots,A(e^{\sqrt{-1}\theta})^n B(e^{\sqrt{-1}\theta})\right) = n$$

for almost all $\theta \in [0,2\pi]$. Thus the result follows by applying the standard Hautus conditions for linear systems.   ∎

A crucial implication for the convergence dynamics of platoons, when working in the Hilbert space ℓ^2, is that all trajectories $(x_n(t))_n$ converge to zero as $n \to \pm\infty$. Of course, this is a very restrictive assumption that is often not desirable in practice. For such reasons, Feintuch and Francis (2012) started an investigation of platoons in the Banach space ℓ^∞ of bounded bi-infinite sequences in $\mathbb{R}$. We endow ℓ^∞ with the norm

$$\|x\|_\infty = \sup_{n \in \mathbb{Z}} |x_n|,$$

which turns ℓ^∞ into a Banach space. Since methods from Fourier analysis cannot be applied, the analysis becomes more difficult. We do not go into details here but refer the reader to Feintuch and Francis (2012) for a discussion of several simple cases.

3. Finite-Dimensional Control of Platoons. A further drawback of the preceding analysis is that it assumes an infinite-dimensional Hilbert space of input values. We now extend the reachability analysis of platoons to the more difficult, and more interesting, case of finite-dimensional controls.

We start with a simple example of a one-sided infinite platoon over $\mathbb{N}_0$ that was first described decades ago by Fuhrmann (1972). Consider the discrete-time control system in $\ell_+^2(\mathbb{C})$

$$x(t+1) = Sx(t) + bu(t), \tag{12.23}$$

where $b \in \ell_+^2(\mathbb{C})$ and

$$S : \ell_+^2(\mathbb{C}) \longrightarrow \ell_+^2(\mathbb{C}), \quad S(x_0, x_1, x_2, \ldots) := (0, x_0, x_1, \ldots),$$
$$S^* : \ell_+^2(\mathbb{C}) \longrightarrow \ell_+^2(\mathbb{C}), \quad S(x_0, x_1, x_2, \ldots) := (x_1, x_2, x_3, \ldots)$$

denotes the right shift operator and left shift operator, respectively. Thus S and S^* are bounded linear operators on $\ell_+^2(\mathbb{C})$. S is an isometry while S^* is a contraction. The study of these shift operators is due to Beurling (1949). The spectral properties of S and S^* are well known; for example, S does not have eigenvalues. The spectrum of both S and S^* is equal to the closed unit disc $\overline{\mathbb{D}}$. The set of eigenvalues of S^* is $\mathbb{D}$, and the associated eigenvector of each eigenvalue $\lambda \in \mathbb{D}$ is $(1, \lambda, \lambda^2, \lambda^3, \ldots)$. Equivalently, expressed in the Hardy space $H^2(\mathbb{D})$, the associated eigenfunction is

$$e_\lambda(z) = \frac{\sqrt{1 - |\lambda|^2}}{1 - \lambda z}.$$

Spelled out in the coordinates of the sequence $x(t)$, system (12.23) is equivalent to the 2D system $[x_{-1}(t) := 0]$

$$x_k(t+1) = x_{k-1}(t) + b_k u(t), \quad t, k \in \mathbb{N}_0.$$

Let

$$b(e^{\sqrt{-1}\theta}) = \sum_{k=0}^{\infty} b_k e^{k\sqrt{-1}\theta}$$

denote the discrete Fourier transform of b. Since $b \in \ell_+^2(\mathbb{C})$, the Fourier transform $b(\cdot)$ extends to a holomorphic function in $\mathbb{D}$ and $b(z) \in H^2(\mathbb{D})$. Note that if $b(z)$ is a rational function with no poles and zeros in the closed unit disc, then b is an outer function. The following result by Beurling (1949) appears as Lemma 4.1 in Fuhrmann (1972). Since $b(z) = \frac{2}{2-z} = \sum_{k=0}^{\infty} 2^{-k} z^k$ is outer, the result implies that the platoon system $[x_{-1}(t) := 0]$

$$x_k(t+1) = x_{k-1}(t) + 2^{-k} u(t), \quad t, k \in \mathbb{N}_0.$$

is approximately reachable.

Theorem 12.20. *System (12.23) is approximately reachable in $\ell_+^2(\mathbb{C})$ if and only if b is an outer function.*

Proof. For the convenience of the reader we recall the main arguments from Fuhrmann (1972). System (12.23) is approximately reachable at zero if and only if the functions $e^{\sqrt{-1}k\theta} b(e^{\sqrt{-1}k\theta})$ span H^2. The span V is invariant under multiplication by $e^{\sqrt{-1}k\theta}$, and therefore Beurling's theorem implies that $V = qH^2$

for an inner function q. Thus $b \in qH^2$, i.e., q divides b. Thus $V = H^2$ if and only if b does not contain a nontrivial inner function. By the inner-outer factorization theorem, this is equivalent to b being outer.   ∎

The situation becomes quite different if we replace the forward shift operator S in (12.23) with the backward shift S^*. The approximate reachability of

$$x(t+1) = S^*x(t) + bu(t) \tag{12.24}$$

is equivalent to b being a cyclic vector for S^*. The question of characterizing cyclic vectors for the backward shift was first raised and answered by Douglas, Shapiro and Shields (1970) and extended to the multivariable case in Fuhrmann (1976b). Their characterization is, however, not as simple as that for S. A sufficient condition by Douglas, Shapiro and Shields (1970) for the cyclicity of b for the backward shift is that the Fourier transform $b(z) \in H^2$ can be analytically continued across all points of an arc in S^1, with the exception of an isolated branch point in the arc. An example of a cyclic vector is $b(z) = \exp(1/(z-2))$. No rational function $b(z) \in H^2$ is cyclic, and therefore rational stable functions lead to a nonreachable system (12.24). In fact, from Kronecker's theorem 4.18 one can deduce that the span of the orbit $(S^*)^{nb}$ is finite-dimensional.

It is possible to extend the analysis to broader classes of one-sided platoon models. Let $a(z) = \sum_{j=0}^{\infty} a_j z^j$ denote an analytic function in $H^2(\mathbb{D})$, with $\sup_{|z| \leq 1} |a(z)| < \infty$, and let $b(z) = \sum_{j=0}^{\infty} b_j z^j \in H^2(\mathbb{D})$. Then the discrete-time single-input linear control system on $\ell_+^2(\mathbb{C})$ is well defined as

$$x(t+1) = a(S)x(t) + bu(t) \tag{12.25}$$

or, equivalently, as

$$x_k(t+1) = \sum_{j=0}^{k} a_{k-j}x_j(t) + b_k u(t), \quad t, k \in \mathbb{N}_0.$$

Note that for $a(z) = z$ this specializes to (12.23). For the proof of the subsequent theorem, we apply methods from the theory of composition operators in Hardy spaces; see, for example, Douglas, Shapiro and Shields (1970) and Shapiro (1993) for further details. In particular, we make use of the following classical result by Walsh (1965) for polynomial approximations.

Theorem 12.21 (Walsh). *Let $a : \mathbb{D} \longrightarrow \mathbb{C}$ be an injective holomorphic function such that the boundary of $a(\mathbb{D})$ is a Jordan curve. Then the set $\{p \circ a \mid p \in \mathbb{C}[z]\}$ of polynomials in a is dense in $H^2(\mathbb{D})$.*

The discussion of the following two examples is taken from Bourdon and Shapiro (1990).

Example 12.22. The univalent function $a(z) = \frac{z}{2-z} \in H^\infty(\mathbb{D})$ maps the closed disc $\overline{\mathbb{D}}$ conformally into itself with fixed points $0, 1$. Note that $\sup_{|z| \leq 1} |a(z)| = 1$. The composition operator $C_a : H^2(\mathbb{D}) \longrightarrow H^2(\mathbb{D})$ is bounded but does not have cyclic vectors. Therefore, system (12.25)

$$x_k(t+1) = \sum_{j=0}^{k-1} 2^{j-k} x_j(t) + b_k u(t), \quad t, k \in \mathbb{N}_0,$$

is for no $b \in \ell_+^2(\mathbb{C})$ approximately reachable in $\ell_+^2(\mathbb{C})$. On the other hand, the univalent function $a(z) = \frac{1}{2-z} \in H^\infty(\mathbb{D})$ has $z = 1$ as its only fixed point and satisfies $\sup_{|z| \leq 1} |a(z)| = 1$. The composition operator $C_a : H^2(\mathbb{D}) \longrightarrow H^2(\mathbb{D})$ is bounded and cyclic. Thus there exists a generic set of elements $b \in \ell_+^2(\mathbb{C})$ such that

$$x_k(t+1) = \sum_{j=0}^{k} 2^{j-k-1} x_j(t) + b_k u(t), \quad t, k \in \mathbb{N}_0,$$

is approximately reachable in $\ell_+^2(\mathbb{C})$.

The following generalization of Theorem 12.20 is a simple consequence of the Walsh theorem.

Theorem 12.23. *Let $b \in \ell_+^2(\mathbb{C})$. Assume that $a \in H^2(\mathbb{D})$ defines an injective analytic function $a : \mathbb{D} \longrightarrow \mathbb{C}$ such that the boundary of $a(\mathbb{D})$ is a Jordan curve. Assume further that $\sup_{|z| \leq 1} |a(z)| < 1$. System (12.25) is approximately reachable in $\ell_+^2(\mathbb{C})$ if and only if b is an outer function.*

Proof. The reachable set V consists of all functions $(p \circ a)b$, where $p \in \mathbb{C}[z]$ is a polynomial. By Theorem 12.21 of Walsh, this implies that the closure $\overline{V}$ of the reachable set in H^2 is equal to bH^2. Let $b = fg$, with f inner and g outer, denote the inner-outer factorization of b. Then the closure of the reachable set $bH^2 = fH^2$ is equal to H^2 if and only if f is constant, i.e., if and only if b is outer. ∎

Actually, a stronger version of Walsh's theorem is valid asserting that the polynomials in a are uniformly dense in the space of uniformly continuous bounded analytic functions on $\mathbb{D}$. This implies approximate reachability in $\ell_+^2(\mathbb{C})$ with respect to the sup-norm on H^2. We next proceed to show how one can apply such uniform approximation techniques in a more general context. Consider bi-infinite platoons with spatial domain $\mathbb{Z}$, where m controls $u(t) = (u_1(t), \ldots, u_m(t))^\top \in \mathbb{R}^m$ are broadcasted to the platoon using

$$u_k(t) = \beta_k u(t),$$

where $\beta_k \in \mathbb{R}^{m \times m}$. This leads to the **broadcast control platoon system**

$$\dot{x}_k(t) = \sum_{j=-\infty}^{\infty} A_j x_{k-j}(t) + \sum_{j=-\infty}^{\infty} B_j \beta_{k-j} u(t), \quad k \in \mathbb{Z}.$$

We assume that the bi-infinite sequence of matrices $\beta = (\beta_k)$ defines an L^2 Fourier transform

$$\beta(e^{\sqrt{-1}\theta}) := \sum_{k=-\infty}^{\infty} \beta_{ke}^{-\sqrt{-1}k\theta} \in L^2(S^1; \mathbb{C}^{m \times m}).$$

The assumptions on A and B are as previously. Define matrix-valued functions as

$$A(e^{\sqrt{-1}\theta}) = \sum_{k=-\infty}^{\infty} A_{ke}^{-\sqrt{-1}k\theta}, \quad \hat{B}(e^{\sqrt{-1}\theta}) = B(e^{\sqrt{-1}\theta})\beta(e^{\sqrt{-1}\theta}).$$

The associated multiplication operators

$$\mathscr{A} : L^2(S^1, \mathbb{C}^n) \longrightarrow L^2(S^1, \mathbb{C}^n) \quad \text{and} \quad \hat{\mathscr{B}} : \mathbb{C}^m \longrightarrow L^2(S^1, \mathbb{C}^n)$$

are bounded linear operators, whereas the input-state operator $\hat{\mathscr{B}}$ is finite-dimensional and therefore defines a compact operator. We thus obtain the system on the Hilbert space $L^2(S^1; \mathbb{C}^n)$ with finite-dimensional controls as

$$\dot{x}(t) = \mathscr{A}x(t) + \hat{\mathscr{B}}u(t). \tag{12.26}$$

The associated finite-dimensional, parameter-dependent, linear system on $\mathbb{C}^n$ is

$$\dot{z}(t) = \mathscr{A}(e^{\sqrt{-1}\theta})z(t) + \mathscr{B}(e^{\sqrt{-1}\theta})\beta(e^{\sqrt{-1}\theta})u(t),$$

with parameter space S^1 being the unit circle. For a compact subset $\mathbf{P} \subset S^1$ consider the Hilbert spaces

$$L_{\mathbf{P}}^2(\mathbb{C}^n) = \{f \in L^2(S^1; \mathbb{C}^n) \mid f = 0 \text{ outside } \mathbf{P}\}, \quad \ell_{\mathbf{P}}^2 = \mathscr{F}^{-1}(L_{\mathbf{P}}^2),$$

with the isomorphism of Hilbert spaces $L_{\mathbf{P}}^2(\mathbb{C}^n) \simeq L^2(\mathbf{P}, \mathbb{C}^n)$.

By extending square-integrable matrix-valued functions $A(e^{\sqrt{-1}\theta})$, $B(e^{\sqrt{-1}\theta})$, and $\beta(e^{\sqrt{-1}\theta})$ on $\mathbf{P}$ by zero to functions on S^1 one can identify these matrix functions with unique elements of, for example, $L_{\mathbf{P}}^2(\mathbb{C}^{n \times n})$. Then the linear multiplication operators $\mathscr{A}$ and $\hat{\mathscr{B}}$ map $L_{\mathbf{P}}^2(\mathbb{C}^n)$ and $\mathbb{R}^m$ into $L_{\mathbf{P}}^2(\mathbb{C}^n)$. In particular, the linear system (12.26) is restricted to a control system on the Hilbert space $L_{\mathbf{P}}^2(\mathbb{C}^n)$. In the single-input case we obtain the following approximate reachability result.

Theorem 12.24. *Let $m = 1$, and let $P \neq [0, 2\pi]$ denote a nonempty compact interval contained in $[0, 2\pi]$. Assume that the matrix-valued functions $\theta \mapsto A(e^{\sqrt{-1}\theta})$, $B(e^{\sqrt{-1}\theta})$, and $\beta(e^{\sqrt{-1}\theta})$ are continuous on P. Assume further that the following conditions are satisfied:*

1. *The pair $(A(e^{\sqrt{-1}\theta}), B(e^{\sqrt{-1}\theta})\beta(e^{\sqrt{-1}\theta}))$ is reachable for all $\theta \in \mathbb{P}$.*
2. *The spectra of $A(e^{\sqrt{-1}\theta})$ and $A(e^{\sqrt{-1}\theta'})$ are disjoint for each $\theta, \theta' \in P, \theta \neq \theta'$.*
3. *The eigenvalues of $A(e^{\sqrt{-1}\theta})$ are simple for each $\theta \in P$.*

Then the restricted system (12.26) on the Hilbert subspace $L_P^2(\mathbb{C}^n)$ is approximately reachable in finite time $T > 0$.

Proof. Let $b(e^{\sqrt{-1}\theta}) := B(e^{\sqrt{-1}\theta})\beta(e^{\sqrt{-1}\theta})$. Theorem 12.6 implies that the parameter-dependent system

$$\frac{\partial x(t, \theta)}{\partial t} = A(e^{\sqrt{-1}\theta})x(t, \theta) + b(e^{\sqrt{-1}\theta})u(t), \quad \theta \in P,$$

is uniformly ensemble reachable in finite time. Note that

$$\int_0^T \exp((T - s)\mathscr{A})\hat{\mathscr{B}}u(s)ds$$

coincides with the L_P^2 function

$$\theta \mapsto \int_0^T \exp\left((T - s)A(e^{\sqrt{-1}\theta})\right) b(e^{\sqrt{-1}\theta})u(s)ds.$$

This implies that the image of the reachability operator $\mathscr{R}_T : L^2([0, T]; \mathbb{C}^m) \longrightarrow L^2(\mathbf{P}, \mathbb{C}^n) = L_P^2(\mathbb{C}^n)$,

$$\mathscr{R}_T(u) = \int_0^T \exp((T - s)\mathscr{A})\hat{\mathscr{B}}u(s)ds,$$

is dense in $L_P^2(\mathbb{C}^n)$. This completes the proof. ∎

12.4 Control of Partial Differential Equations

In this section we explore several instances where the control of PDEs interacts with networks of systems. This includes the realization of interconnected systems and platoons as discretizations of PDEs and showing how results from parametric linear systems can be used to gain further insight into PDEs using Fourier-transform techniques. Finally, we explain how one can control the state-space probability distributions of linear systems by solving the associated control task for the Liouville equation.

1. Networks as Discretizations of the Heat Equation. Let us consider one of the simplest PDEs, the classical **heat equation** in one spatial variable $z \in [0,1]$, with boundary controls

$$\frac{\partial \psi(t,z)}{\partial t} = \frac{\partial^2 \psi(t,z)}{\partial z^2},$$

$$\psi(0,x) = 0, \quad \psi(t,0) = u_0(t), \quad \psi(t,1) = u_1(t).$$

Here the boundary value functions $u_0(t)$ and $u_1(t)$ are regarded as control variables. For each nonnegative integer N and step size $h = \frac{1}{N}$ we subdivide the domain $[0,1]$ into $N+1$ equidistant points $z_i = ih, i = 0,\dots,N$. Consider a lumped discretization as $x(t) = (\psi(t,\frac{1}{N}),\dots,\psi(t,\frac{N-1}{N}))^\top \in \mathbb{R}^{N-1}$, and assume that the boundary value functions $u(t) = (u_0(t),u_1(t))^\top$ are known. Then the boundary value condition $\psi(0,x) = 0$ corresponds to the initial condition $x(0) = 0$. Using standard Taylor approximations we can replace the second-order differential operator $\frac{\partial^2 \psi(t,z)}{\partial z^2}$ with its associated difference operator as

$$\frac{\psi(t,z+h) - 2\psi(t,z) + \psi(t,z-h)}{h^2}.$$

By neglecting second-order error terms we end up with the discretized form of the heat equation as

$$\dot{x}(t) = \frac{1}{h^2}\left(Ax(t) + Bu(t)\right). \tag{12.27}$$

Of course, after rescaling time in x,u via $x(h^2 t), u(h^2 t)$ this system becomes equivalent to the linear system $\dot{x} = Ax + Bu$. Here $(A,B) \in \mathbb{R}^{(N-1)\times(N-1)} \times \mathbb{R}^{(N-1)\times 2}$ are the reachable pair

$$A = \begin{pmatrix} -2 & 1 & & & & \\ 1 & -2 & 1 & & & \\ & 1 & \ddots & \ddots & & \\ & & \ddots & \ddots & 1 & \\ & & & 1 & -2 & 1 \\ & & & & 1 & -2 \end{pmatrix}, \quad B = \begin{pmatrix} 1 & 0 \\ 0 & 0 \\ \vdots & \vdots \\ \vdots & \vdots \\ 0 & 1 \end{pmatrix}. \tag{12.28}$$

Note that (12.27) is an interconnected system for $N-1$ identical first-order integrators $\dot{x}_i = v_i$, coupled by interconnection matrices (12.28) for the states and inputs, respectively. Note further that A defines a Laplacian matrix for a tree, and its spectrum was analyzed in Chapter 7. It is easily seen that the discretized heat equation (12.27) is reachable.

As another example, consider the control of the heat equation on the unit circle, formulated as a periodic boundary value problem as

$$\frac{\partial \psi(t,z)}{\partial t} = \frac{\partial^2 \psi(t,z)}{\partial z^2} + u(t)g(z),$$

$$\psi(t,0) = \psi(t,1),$$

where $g(z) = g(z+1)$ is assumed to be periodic with period one. By discretizing this system using $x_N = x_0, x_{N+1} = x_1$ and setting $g(ih) = b_i$ we obtain the single-input control system $\dot{x} = \frac{1}{h^2}(Ax + bu)$ with

$$A = \begin{pmatrix} -2 & 1 & & & & \\ 1 & -2 & 1 & & & \\ & 1 & \ddots & \ddots & & \\ & & \ddots & \ddots & 1 & \\ & & & 1 & -2 & 1 \\ & & & & 1 & -2 \end{pmatrix}, \quad B = \begin{pmatrix} b_1 \\ b_2 \\ \vdots \\ \vdots \\ b_N \end{pmatrix}. \qquad (12.29)$$

Here, A is a circulant matrix, and the reachability of such systems was analyzed in Chapter 9 using the module structure on the set of circulant matrices. We refer to Brockett and Willems (1974) for further discussion of discretized PDEs from a module-theoretic point of view. Note that (12.29) is, up to sign, identical with the finite platoon system (12.19). In particular, Theorem 12.18 implies that (12.29) is not reachable.

The discretization process can be applied to more general boundary value problems. We illustrate this by the following variant of the controlled heat equation

$$\frac{\partial \psi(t,z)}{\partial t} = \frac{\partial^2 \psi(t,z)}{\partial z^2}, \quad (t,z) \in [0,T] \times [0,1],$$

$$\frac{\partial \psi(t,0)}{\partial z} = 0, \quad \frac{\partial \psi(t,1)}{\partial z} = u(t).$$

We refer the reader to Chapter 2.5.3 in the book by Coron (2007) for a detailed discussion of the reachability properties of this system, including an approximate reachability result Theorem 2.76 that is derived using flatness techniques. Using Taylor approximations (with step sizes identical to those for the second derivative term),

$$\frac{\psi(t,h) - \psi(t,0)}{h} \quad \text{and} \quad \frac{\psi(t,1) - \psi(t,1-h)}{h},$$

for $\frac{\partial \psi(t,0)}{\partial z}$ and $\frac{\partial \psi(t,1)}{\partial z}$, respectively, the boundary conditions become $\psi(t,h) = \psi(t,0)$ and $\psi(t,1) = \psi(t,1-h) + hu(t)$. Thus we arrive at the reachable single-input system

$$\dot{x}(t) = \frac{1}{h^2}\left(Ax(t) + hbu(t)\right),$$

with

$$A = \begin{pmatrix} -1 & 1 & & & & \\ 1 & -2 & 1 & & & \\ & 1 & \ddots & \ddots & & \\ & & \ddots & \ddots & 1 & \\ & & & 1 & -2 & 1 \\ & & & & 1 & -1 \end{pmatrix}, \qquad b = \begin{pmatrix} 0 \\ \vdots \\ \vdots \\ 1 \end{pmatrix}.$$

Again, this system is reachable and can be rescaled to standard form $\dot{x} = Ax + bu$. The spectral properties of matrix A were discussed in Section 8.7.

Instead of assuming a compact spatial domain, one can also consider the heat equation on an unbounded domain. This leads to infinite platoons of systems as their discretizations. Thus, consider, for example, the controlled heat equation on the nonnegative real line $[0, \infty)$:

$$\frac{\partial \psi(t,z)}{\partial t} = \frac{\partial^2 \psi(t,z)}{\partial z^2} + g(z)u(t),$$

$$\psi(0,z) = 0.$$

Consider a lumped approximation $x(t) = (x_k(t)) = (\psi(t,k))_{k \in \mathbb{N}_0} \in \ell_+^2$ of ψ, and $(b_k)_{k \in \mathbb{N}_0} := (g(k))_{k \in \mathbb{N}_0} \in \ell_+^2$ with step size $h = 1$. Using the standard discretization for the second-order derivative one obtains the infinite platoon

$$\dot{x}_k(t) = x_{k+1}(t) - 2x_k(t) + x_{k-1}(t)(t) + b_k u(t), \quad k \in \mathbb{N}_0.$$

This shows that the study of infinite platoons can be of use for the control of PDEs.

The preceding examples illustrate that interesting interconnection matrices arise as discretizations of PDEs. One can take this idea a step further by considering discretizations of parametric linear systems coupled by diffusive terms. This leads to networks of linear systems. Thus, consider, for example, the family of linear systems

$$\frac{\partial \psi(t,z)}{\partial t} = \alpha \psi(t,z) + \beta v(t,z),$$

$$y(t,z) = \gamma \psi(t,z),$$

$$v(t,z) = \frac{\partial^2 y(t,z)}{\partial^2 z} + Bu(t),$$

where the SISO system $(\alpha, \beta, \gamma) \in \mathbb{R}^{n \times n} \times \mathbb{R}^{n} \times \mathbb{R}^{1 \times n}$ is reachable and observable, $z \in [0, 1]$ and $\psi(t, z) \in \mathbb{R}^{n}$. By discretizing the second-order differentiation term as before we arrive at the interconnected linear system

$$\dot{x} = (I_N \otimes \alpha + A \otimes \beta \gamma)x + (B \otimes \beta)u(t),$$

where A is defined in (12.28) and B is arbitrary. By Theorem 9.15, this network is reachable whenever (A, B) is reachable. If the spatial domain $[0, 1]$ is replaced by the real line $\mathbb{R}$, this leads to infinite networks of linear systems.

2. Fourier-Transform Techniques. The preceding examples showed how large-scale interconnected control systems arise naturally as discretizations of boundary control problems for PDEs. Using Fourier-transform techniques, one can associate families of control systems to such PDEs on $\mathbb{R}^d$. To illustrate this idea, we consider the controlled heat equation in one spatial variable on the real line $\mathbb{R}$:

$$\frac{\partial \psi(t, z)}{\partial t} = a\psi(t, z) + \frac{\partial^2 \psi(t, z)}{\partial z^2} + u(t)\sqrt{\frac{2}{\pi}} \frac{\sin(Rz)}{z}, \tag{12.30}$$

$$\psi(0, z) = 0.$$

Our goal is to find a control function $u(t)$ that steers the initial temperature distribution $\psi(0, z) = 0$ to a final distribution $\psi(T, z) = \psi^*(z)$ in finite time $T > 0$. More specifically, for $\varepsilon > 0$ and a function ψ^* in the Sobolev space $W^2(\mathbb{R})$ we want to find $T > 0$ and a control $u : [0, T] \longrightarrow \mathbb{R}$ such that

$$\int_{-\infty}^{\infty} (\psi(T, z) - \psi^*(z))^2 \, dz < \varepsilon. \tag{12.31}$$

We refer to this as the **approximate reachability task** for ψ^* in (12.30). We emphasize that the control function in our problem is independent of the spatial variable z. Thus the freedom one has in controlling the system is quite limited. To approach such reachability questions, it is useful to note that (12.30) has a unique solution $\psi(t, z) \in C^1([0, \infty), W^2(\mathbb{R}) \cap W_0^1(\mathbb{R}))$ for piecewise smooth input functions $u(t)$; see Pazy (1983), Corollary 7.2.8. Thus the control problem is meaningful only by assuming $\psi^* \in W^2(\mathbb{R}) \cap W_0^1(\mathbb{R})$. Since the space $C_0^\infty(\mathbb{R})$ of smooth functions with compact support is dense in $W^2(\mathbb{R})$, one has $W_0^1(\mathbb{R}) = W^2(\mathbb{R})$. Thus the control problem for the heat equation (12.30) on the real axis takes place in $W^2(\mathbb{R})$. Our approach now is to replace the heat equation by an ordinary differential equation using the Fourier transform. This enables us to show that, under suitable assumptions on ψ^*, such approximate reachability tasks for PDEs are related to the ensemble control problem for parameter-dependent linear systems.

Recall that the **Fourier transform** on $\mathbb{R}$ defines the linear isometry $\mathscr{F}$: $L^2(\mathbb{R}) \longrightarrow L^2(\mathbb{R})$,

$$(\mathscr{F}\psi)(\theta) := \frac{1}{\sqrt{2\pi}} \int_{-\infty}^{\infty} e^{-\sqrt{-1}\theta z} \psi(z) dz,$$

with inverse transform

$$(\mathscr{F}^{-1}f)(z) := \frac{1}{\sqrt{2\pi}} \int_{-\infty}^{\infty} e^{\sqrt{-1}\theta z} f(\theta) d\theta.$$

It is well known that $\mathscr{F}$ maps the Sobolev space $W^2(\mathbb{R})$ exactly onto the space of all functions $\phi(\theta)$ such that $(1 + |\theta|)^2 \phi(\theta) \in L^2(\mathbb{R})$. Let

$$H(x) = \begin{cases} 1 & \text{for } x \geq 0, \\ 0 & \text{otherwise} \end{cases}$$

denote the Heaviside function. Note that for each $R > 0$ the Fourier transform of $\sqrt{\frac{2}{\pi}} \frac{\sin(Rz)}{z}$ is equal to $H(R - |z|)$. Thus, by Fourier-transforming equation (12.30), we obtain for $x(t, \theta) := (\mathscr{F}\psi)(t, \theta)$ the parameter-dependent control system

$$\dot{x} = (a - \theta^2)x(t,\theta) + H(R - |\theta|)u(t), \quad x(0,\theta) = 0. \tag{12.32}$$

By restricting ourselves to the compact parameter domain $\mathbf{P} = [0, R]$ we obtain the linear parameter-dependent system

$$\dot{x} = (a - \theta^2)x(t,\theta) + u(t), \quad x(0,\theta) = 0. \tag{12.33}$$

Since the Fourier transform defines an isometry on $L^2(\mathbb{R})$, the approximate reachability of the PDE (12.30) on the spatial domain $\mathbb{R}$ is equivalent to the existence of an input function $u(t)$ that is *independent of the parameter* $\theta \in [0, R]$ and has the following approximation property: For each function $\theta \mapsto x^*(\theta)$ of terminal states in $\mathscr{F}(W^2)$ and $\varepsilon > 0$ there exists $T > 0$ such that the solution of (12.30) satisfies

$$\int_0^R (x(T,\theta) - x^*(\theta))^2 d\theta < \varepsilon.$$

Thus we see that the approximate reachability task for the heat equation is equivalent to L^2-ensemble reachability for a simple family of first-order linear systems. Our previous results on uniform ensemble reachability apply and yield corresponding results on approximate reachability for PDEs. However, there is a "but" insofar as our results will only imply reachability within a certain frequency band. These observations are in harmony with stronger positive approximate reachability results established for the heat equation (12.30) on arbitrary bounded domains; see Theorem 2.76 in the book by Coron (2007).

Theorem 12.25. *Assume that $\psi^*(z) \in W^2$ is such that its Fourier transform $x^*(\theta) = \mathscr{F}(\psi^*)(\theta)$ is an even function of θ with support contained in $[-R, R]$. Then the heat equation (12.30) is approximately reachable to ψ^*.*

Proof. We first show that the approximate reachability of the heat equation to ψ^* is equivalent to the L^2-ensemble reachability of the linear parametric system (12.33) to x^*. By Theorem 12.6, the family (12.33) is uniform ensemble reachable on the parameter interval $[0, R]$. This then completes the proof. To prove the equivalence of the two notions of reachability, suppose that u is an input such that $\psi(T, \cdot) \in W^2$ satisfies (12.31). Then $x(t, \theta) = \mathscr{F}(\psi(t, \cdot))$ is a solution of (12.32). Since $\mathscr{F}$ is an isometry on L^2, we obtain for $x^* := \mathscr{F}(\psi^*)$

$$\int_{-\infty}^{\infty} (\psi(T, z) - \psi^*(z))^2\, dz = \int_{-\infty}^{\infty} (x(T, \theta) - x^*(\theta))^2\, d\theta = \int_{-R}^{R} (x(T, \theta) - x^*(\theta))^2\, d\theta$$

$$= 2\int_{0}^{R} (x(T, \theta) - x^*(\theta))^2\, d\theta < 2\varepsilon.$$

Here the second equation follows from the support property of x^*, while the third follows from the evenness of the functions $x(T, \theta)$ and $x^*(\theta)$. Conversely, for each solution $x(t, \theta)$ of (12.33) we note that $H(R - |\theta|)x(t, \theta)$ defines a solution of (12.32). Since $\theta \mapsto H(R - |\theta|)x(t, \theta)$ has compact support, $(1 + |\theta|)^2 H(R - |\theta|)x(t, \theta)$ also has compact support. Thus $H(R - |\theta|)x(t, \theta)$ is an element of $\mathscr{F}(W^2(\mathbb{R}))$ for all $t \geq 0$. This shows that the inverse Fourier transform $\mathscr{F}^{-1}(H(R - |\theta|)x(t, \theta))$ maps solutions to (12.32) bijectively to the solutions of the heat equation (12.30). ∎

3. Control of Liouville Equation. We now present a reachability result by Brockett on the Liouville equation, restricted to the space of Gaussian distributions. We begin by deriving the classical Liouville transport equation for time-varying vector fields. Let $f(x, t)$ be a time-varying complete C^k-vector field on a smooth orientable Riemannian manifold M. Let dx denote the canonical volume form on M and $\rho(0, \cdot)$ denote a smooth function with

$$\int_M \rho(0, x)\, dx - 1.$$

Let $\phi_{t,s}$ denote the flow semigroup of f. Thus $\phi_{t,s}$ is a diffeomorphism on M with $\phi_{t,s}\phi_{s,t} = id_M$ and

$$\frac{\partial \phi_{t,s}(x)}{\partial t} = f(\phi_{t,s}(x), t).$$

Define

$$\rho(t, x) = \rho(0, \phi_{0,t}(x)) \det D\phi_{0,t}(x),$$

i.e., $\rho(t,x)dx$ is the pullback of the volume form $\rho(0,x)dx$ by the diffeomorphism $\phi_{0,t}$. Applying the transformation theorem one concludes that, for all t,

$$\int_M \rho(t,x)dx = 1.$$

Consider a smooth function $\psi(x)$ on M with compact support. By a change of variables,

$$\int_M \psi(x)\rho(t,x)dx = \int_M \psi(x)\rho(0,\phi_{0,t}(x))\det D\phi_{0,t}(x)dx = \int_M \psi(\phi_{t,0}(x))\rho(0,x)dx,$$

and so, by differentiating both sides, we obtain

$$\int_M \psi(x)\frac{\partial \rho(t,x)}{\partial t}dx = \frac{d}{dt}\int_M \psi(x)\rho(t,x)dx = \int_M \frac{\partial \psi(\phi_{t,0}(x))}{\partial t}\rho(0,x)dx$$

$$= \int_M d\psi(\phi_{t,0}(x))f(\phi_{t,0}(x),t)\rho(0,x)dx$$

$$= \int_M d\psi(x)f(x,t)\rho(0,\phi_{0,t}(x))\det D\phi_{0,t}(x)dx$$

$$= \int_M d\psi(x)f(x,t)\rho(t,x)dx.$$

For time-varying vector fields F, the divergence on M satisfies the well-known identity $\mathrm{div}(\psi(x)F(x,t)) = d\psi(x)F(x,t) + \psi(x)\mathrm{div}\,F(x,t)$. Applying the divergence theorem, we obtain, for $F(x,t) = \rho(t,x)f(x,t)$,

$$\int_M \psi(x)\frac{\partial \rho(t,x)}{\partial t}dx = -\int_M \psi(x)\mathrm{div}(f(x,t)\rho(t,x))dx + \int_M \mathrm{div}(\psi(x)f(x,t)\rho(t,x))dx$$

$$= -\int_M \psi(x)\mathrm{div}(f(x,t)\rho(t,x))dx.$$

Thus we conclude that $\rho(t,x)$ satisfies the transport equation

$$\frac{\partial \rho(t,x)}{\partial t} = -\mathrm{div}(f(x,t)\rho(t,x)), \quad \rho(0,x) = \rho(x). \tag{12.34}$$

Conversely, if ρ is a C^k-function on $\mathbb{R}^n$, then $\rho(t,x) = \rho(\phi_{0,t}(x))\det D\phi_{0,t}(x)$ is the unique C^k-solution of the initial value problem (12.34). Applying this equation to a control affine vector field $f(x) + ug(x)$, a major distinction between open-loop and closed-loop control becomes manifest. In fact, for open-loop controls we obtain a control system on the space of density functions as

$$\frac{\partial \rho(t,x)}{\partial t} = -\mathrm{div}(f(x)\rho(t,x)) - u(t)\mathrm{div}(g(x)\rho(t,x)),$$

while smooth feedback control leads to

$$\frac{\partial \rho(t,x)}{\partial t} = -\operatorname{div}(f(x)\rho(t,x)) - u(x)\operatorname{div}(g(x)\rho(t,x)) - du(x)g(x)\rho(t,x).$$
(12.35)

The difference appears already for linear systems $f(x,u) = Ax + Bu$ in $\mathbb{R}^n$. We obtain the open-loop transport equation

$$\frac{\partial \rho(t,x)}{\partial t} = -\operatorname{tr}(A)\rho(t,x) - (Ax + Bu)^\top \nabla \rho(t,x),$$
(12.36)

while the closed-loop state feedback $u = Kx$ leads to

$$\frac{\partial \rho(t,x)}{\partial t} = -\operatorname{tr}(A + BK)\rho(t,x) - ((A + BK)x)^\top \nabla \rho(t,x).$$
(12.37)

The open-loop solution of (12.36) is

$$\rho(t,x) = e^{-t\operatorname{tr}A}\rho\left(0, e^{-tA}\left(x - \int_0^t e^{(t-s)A}Bu(s)ds\right)\right),$$

while the closed-loop solution of (12.37) under the state feedback $u = Kx$ is

$$\rho(t,x) = e^{-t\operatorname{tr}(A+BK)}\rho(0, e^{-t(A+BK)}x).$$

A **Gaussian distribution** function with positive definite covariance matrix $Q = Q^\top > 0$ and mean value $\mu \in \mathbb{R}^n$ is defined as

$$g_{Q,\mu}(x) = \frac{1}{\sqrt{(2\pi)^n \det Q}} \exp\left(-\frac{1}{2}(x-\mu)^\top Q^{-1}(x-\mu)\right).$$

Let $\mathscr{P}$ denote the convex set of positive definite real symmetric $n \times n$ matrices Q. Since the map $(Q,\mu) \mapsto g_{Q,\mu}$ is injective, we see that the set of Gaussian distributions forms a smooth manifold $\mathscr{G}$ that is diffeomorphic to $\mathscr{P} \times \mathbb{R}^n$ and is embedded into $C^\infty(\mathbb{R}^n)$. Moreover, the Gaussian distribution satisfies, for every invertible transformation $S \in GL_+(n,\mathbb{R})$ with positive determinant and $\mu \in \mathbb{R}^n$, the identity

$$g_{SQS^\top, S\mu}(x) = \frac{1}{\det S} g_{Q,\mu}(S^{-1}x).$$

We now consider the effect of transporting a Gaussian distribution by a linear control system. The solutions of the linear system $\dot{x} = Ax + Bu$ are

$$x(t) = \phi_{t,0}(x) = e^{tA}x + \int_0^t e^{(t-s)A}Bu(s)ds,$$

with

$$\phi_{0,t}(x) = e^{-tA}x - \int_0^t e^{-sA}Bu(s)ds.$$

Therefore,

$$g_{Q,\mu}(\phi_{0,t}(x)) = e^{t\,\mathrm{tr}(A)}g_{Q(t),\mu(t)}(x),$$

with

$$Q(t) = e^{tA}Qe^{tA^\top}, \quad \mu(t) = \phi_{t,0}(\mu) = e^{tA}\mu + \int_0^t e^{(t-s)A}Bu(s)ds.$$

This implies the explicit formula for the L^2-distance

$$\int_{\mathbb{R}^n}(g_{Q,\mu}(x) - g_{Q(t),\mu(t)}(x))^2 dx = \frac{1}{\sqrt{(4\pi)^n \det Q}} + \frac{e^{-t\,\mathrm{tr}A}}{\sqrt{(4\pi)^n \det Q}}$$

$$-\frac{2\exp\left(-\frac{1}{2}(\mu - \mu(t))^\top(Q + e^{tA}Qe^{tA^\top})^{-1}(\mu - \mu(t))\right)}{\sqrt{(2\pi)^n \det(Q + e^{tA}Qe^{tA^\top})}}.$$

Now consider the bilinear control system

$$\dot{x} = (A + BK(t))x + Bu(t),$$

where $K(t)$ and $u(t)$ act as independent control functions. Thus, by this process, we combine open-loop and closed-loop controls for the transport equation (12.39) on Gaussian density functions. This induces the control system on the parameter space for Gaussian densities as

$$\begin{aligned}\dot{Q}(t) &= (A + BK(t))Q(t) + Q(t)(A + BK(t))^\top, \\ \dot{\mu}(t) &= (A + BK(t))\mu(t) + Bu(t),\end{aligned} \tag{12.38}$$

which is equivalent to the restriction of (12.39) on $\mathscr{G}$.

Theorem 12.26 (Brockett (2012)). *Assume that (A, B) is reachable. Then the control system (12.38) on $\mathscr{P} \times \mathbb{R}^n$ is reachable. For pairs (Q_0, μ_0) and (Q_1, μ_1) in $\mathscr{P} \times \mathbb{R}^n$ there exists $T > 0$ and control functions $u(t)$ and $K(t)$ on $[0, T]$ that steer (12.38) from (Q_0, μ_0) to (Q_1, μ_1).*

Proof. The tangent space of $\mathscr{P}$ at Q consists of all matrices of the form $LQ + QL^\top$, with $L \in \mathbb{R}^{n \times n}$. Thus the solutions $Q(t)$ and $\mu(t)$ of (12.38) exist for all time and stay in $\mathscr{P} \times \mathbb{R}^n$. Every positive definite matrix Q can be factored as $Q = XX^\top$, with a factor $X \in GL_+(n, \mathbb{R})$ of positive determinant. Thus it suffices to prove reachability for the lifted system on $GL_+(n, \mathbb{R}) \times \mathbb{R}^n$:

$$\dot{X}(t) = (A + BK(t))X(t),$$
$$\dot{\mu}(t) = (A + BK(t))\mu(t) + Bu(t).$$

This in turn is reachable if and only if the decoupled system

$$\dot{X}(t) = (A + BK(t))X(t),$$
$$\dot{\mu}(t) = A\mu(t) + Bu(t)$$

is reachable. The second subsystem is linear and, thus, by assumption on (A, B), is reachable on $\mathbb{R}^n$ in time $T > 0$. The first equation is bilinear. The reachability of this system follows from a beautiful argument by Brockett (2012) that we now sketch. Recall that a bilinear control system on a Lie group is reachable provided the system is accessible and there exists a constant control such that the system is weakly Poisson stable; see Lian, Wang and Fu (1994). Since (A, B) is reachable, there exists a feedback matrix K such that $L := A + BK$ has distinct eigenvalues that are integer multiples of $2\pi\sqrt{-1}$. Thus there exists a (constant) control K such that e^{tL} is periodic. This shows that $\dot{X}(t) = (A + BK(t))X(t)$ is weakly Poisson stable for a suitable constant input. So it remains to prove the accessibility of the system. To this end, we compute the system Lie algebra $\mathfrak{g}$. Thus $\mathfrak{g}$ contains A together with every square matrix whose image is contained in that of B. The Lie bracket of A and BK is $[A, BK] = ABK - BKA$. Hence, $\mathfrak{g}$ contains every matrix whose image space is contained in that of AB. By iterating this argument, we see that $\mathfrak{g}$ contains all matrices whose image space is contained in the image space of $(B, AB, \ldots, A^{nB})$. By the reachability of (A, B), this implies that $\mathfrak{g}$ contains all real $n \times n$ matrices. Therefore, the Lie algebra rank condition shows the accessibility of the bilinear system. ∎

The preceding proof shows that open-loop control enables one only to control the mean value of a Gaussian state distribution. To control both the mean and variance of a Gaussian distribution, one needs to apply both open-loop and closed-loop controls in the bilinear affine form $u(t, x) = K(t)x + u(t)$. We conclude with the following straightforward consequence of Theorem 12.26.

Theorem 12.27. *Assume that (A,B) is reachable. The transport system*

$$\frac{\partial \rho(t,x)}{\partial t} = -tr\,(A+BK(t))\rho(t,x) - ((A+BK(t))x+Bu(t))^\top \nabla\rho(t,x), \quad (12.39)$$

with independent controls $u(t)$ and $K(t)$, leaves the manifold $\mathscr{G}$ of Gaussian distributions invariant. Using the controlled flow (12.39), one can steer in finite time $T>0$ two Gaussian distributions g_{Q_1,μ_1} and g_{Q_2,μ_2} into each other. Thus (12.39) is reachable on $\mathscr{G}$.

The preceding result can be generalized in several directions. First, one might consider replacing the set of positive definite matrices by the positive cone in a Euclidean Jordan algebra. Instead of using state feedback $A+BK$, one could study the effects of output feedback $A+BKC$. Finally, one might consider networks of systems and try to establish controllability results for the mean and covariance of state vectors in such interconnected systems. We leave these problems for future research.

12.5 Exercises

1. Show that the discrete-time system $x_{t+1}(\theta) = A(\theta)x_t(\theta) + b(\theta)u_t$, with

$$A(\theta) = \begin{pmatrix} \theta & 1 \\ 0 & \theta \end{pmatrix}, \quad b(\theta) = \begin{pmatrix} 0 \\ 1 \end{pmatrix},$$

   is not ensemble reachable over $\mathbf{P} = [0,1]$.

2. Show that the discrete-time system $x_{t+1}(\theta) = A(\theta)x_t(\theta) + b(\theta)u_t$, with

$$A(\theta) = \begin{pmatrix} 0 & 1 \\ 0 & \theta \end{pmatrix}, \quad b(\theta) = \begin{pmatrix} 0 \\ 1 \end{pmatrix},$$

   is ensemble reachable over $\mathbf{P} = [0,1]$.

3. Let $\mathbf{P} = [0,1]$. Consider the infinite sequence of systems

$$A_k(\theta) = \begin{pmatrix} 0 & -\theta^2 - 1/k^2 \\ 0 & 2\theta \end{pmatrix}, \quad b_k(\theta) = \begin{pmatrix} 1 \\ 0 \end{pmatrix}.$$

   a. Verify that $A_k(\theta)$ and $b_k(\theta)$ are uniformly ensemble reachable for each finite k, but the limiting system $A_\infty(\theta), b_\infty(\theta))$ is not uniformly ensemble reachable.
   b. Prove that the set of uniformly ensemble reachable SISO systems is neither open nor closed in the topology of uniform convergence.

4. Let $(A, b, c) \in \mathbb{R}^{n \times n} \times \mathbb{R}^n \times \mathbb{R}^{1 \times n}$ be reachable and observable, $R > 0$ fixed, and assume that $A - \theta^2 bc$ has distinct eigenvalues for $\theta \in \mathbf{P} = [0, R]$. Let $x_*(\theta)$ be a continuous function on $\mathbf{P}$. Prove that there exists $T > 0$ and a piecewise constant input function $u : [0, T] \longrightarrow \mathbb{R}$ such that the Fourier transform of the unique solution $\psi(t, z) \in \mathbb{R}^n$ of the coupled system of PDEs

$$\frac{\partial \psi}{\partial t}(t, z) = \left(A + \frac{\partial^2}{\partial z^2} bc \right) \psi(t, z) + bu(t) \sqrt{\frac{2}{\pi}} \frac{sin(Rz)}{z}, \quad \psi(0, z) = 0,$$

satisfies

$$\max_{\theta \in \mathbf{P}} \| (\mathscr{F} \psi)(T, \theta) - x_*(\theta) \| < \varepsilon.$$

5. Prove the following formula for the L^2-distance of two Gaussian distributions:

$$F(Q_1, Q_2, \mu_1, \mu_2) := \int_{\mathbb{R}^n} (g_{Q_1, \mu_1}(x) - g_{Q_2, \mu_2}(x))^2 dx = \frac{1}{\sqrt{(4\pi)^n \det Q_1}} +$$

$$\frac{1}{\sqrt{(4\pi)^n \det Q_2}} - \frac{2 \exp \left(-\frac{1}{2}(\mu_1 - \mu_2)^\top (Q_1 + Q_2)^{-1}(\mu_1 - \mu_2) \right)}{\sqrt{(2\pi)^n \det(Q_1 + Q_2)}}.$$

Show that for fixed values of Q_1 and μ_1 the function $(Q, \mu) \mapsto F(Q_1, Q, \mu_1, \mu)$ is strictly convex and assumes its minimum value at $Q = Q_1, \mu = \mu_1$.

12.6 Notes and References

Open-loop control issues for ensembles of systems have been studied in order to design robust compensating pulse sequences in quantum control and NMR spectroscopy; see, for example, Li and Khaneja (2006). The main motivation here comes from the known difficulties of quantum mechanical observations, which makes feedback strategies difficult to implement. Rigorous results for the ensemble control of infinite-dimensional bilinear systems have been obtained by Beauchard, Coron and Rouchon (2010). Open-loop control is also of interest for understanding biological formation control tasks for flocks of systems; see, for example, Brockett (2010). Perhaps the best control strategies are neither pure open-loop nor feedback control, and a mixture of the two seems more promising. This point of view has been stressed in a number of recent papers on minimum attention control, for example, by Brockett (1997), Brockett (2008), and Brockett (2012). These general control issues of how to best combine open- and closed-loop control strategies certainly deserve further study.

A well-studied issue in controlling parameter-dependent systems is that of robustness, where the goal is to find input functions that achieve a desired control objective, irrespective of parameter variations in the system. This can be done either in closed-loop, using feedback transformations, or open-loop control strategies. A classical robust feedback control problem is the so-called blending problem, namely, the task of finding a dynamic output feedback controller that simultaneously stabilizes a one-parameter family of scalar transfer functions; see, for example, Khargonekar and Tannenbaum (1985); Tannenbaum (1980) for a solution to special cases via Nevanlinna–Pick interpolation. To date, no general solution to the blending problem for families of MIMO systems is known. Proceeding in a different direction, we mention the pioneering work by Kharitonov on robust stability. In fact, the paper by Kharitonov (1978) has inspired several researchers to find switching controllers for the stabilization of polytopic families of linear systems; see, for example, Ghosh (1985).

The approximation theorem of Mergelyan (1952) is a very natural generalization of the Stone–Weierstrass theorem, but known proofs as in Gaier (1987) are not constructive. Runge's approximation theorem is a weaker version of Mergelyan's theorem, where the approximating polynomials get replaced by rational functions. Error results for the polynomial approximations in Mergelyan's theorem are obtained by Saff and Totik (1989). A potentially interesting method for computing input sequences for ensemble control is provided by the Faber polynomials (we are grateful to Christian Lubich for suggesting this to us). Faber polynomials $p_j(z), j \in \mathbb{N}$, allow one to approximate analytic functions $f(z)$ in a complex domain K by a convergent series of the form $c_0 + \sum_{j=1}^{\infty} c_j p_j(z)$, where only the coefficients c_j depend on $f(z)$. Such polynomials exist if the complement of K in the extended complex plane is simply connected.

Corollary 12.4 characterizes ensemble reachability for a finite set of parameters. The proof shows that this statement, i.e., the characterization of reachability for the parallel connection of finitely many SISO systems, is equivalent to the Chinese remainder theorem or to Lagrange interpolation. Thus corresponding reachability results should follow for parallel connections of countably many SISO systems via interpolation results for analytic functions such as the Mittag–Leffler theorem. We refer to Helmke and Schönlein (2014) for a proof of Theorem 12.6. Condition (c) in Theorem 12.6, stating that all eigenvalues of $A(\theta)$ are simple, cannot be removed easily. Exercise 1 gives a counterexample. Proposition 12.12 and Exercises 1 and 3 are due to Scherlein (2014).

It has been shown that, by restricting the set of parameters in (12.1) to a finite subset, a parametric family of systems is equivalent to the parallel connection of linear systems. This can be generalized as follows. Consider a mixed differential and integral Volterra equation of the form

$$\frac{\partial}{\partial t} x(t, \theta) = A(\theta)x(t, \theta) + \int_{\mathbf{P}} K(\theta, \theta')x(t, \theta')d\theta' + B(\theta)u(t). \tag{12.40}$$

Here we allow for rather general classes of kernel functions $K(x, y)$. Note that if $\mathbf{P}$ is a compact group and the integral defines a convolution operator with respect to the Haar measure, then the class of spatially invariant systems studied by Bamieh,

Paganini and Dahleh (2002) is obtained. Moreover, if one replaces the integral term with a Riemann sum and restricts oneself to the finite subset of N sampling points θ_i, then one obtains

$$\frac{\partial}{\partial t}x(t,\theta_i) = A(\theta_i)x(t,\theta_i) + \sum_{j=1}^{N} K(\theta_i,\theta_j)x(t,\theta_j) + B(\theta_i)u(t),$$

i.e., one obtains the general equations for a linear network of systems (9.5) studied in Chapter 9. This shows that spatiotemporal systems of the form (12.40) are the infinite-network counterparts of the finite system interconnections studied in Chapter 9. For general existence and uniqueness results for integral equations of the Volterra type we refer the reader to Väth (2000). We are not aware of systematic studies of the reachability or observability properties of systems of the form 12.40.

Theorem 12.19 and generalizations to exact reachability are due to Curtain, Iftime and Zwart (2009). The textbook by Curtain and Zwart (1995), Section 4.2, provides simple sufficient conditions for approximate reachability in a Hilbert space. These conditions require knowledge of a Riesz basis of eigenvectors and therefore do not apply to multiplication operators, which have a continuous spectrum. Multiplication operators on spaces of L^2-functions are not compact; the spectral approximation properties of such operators using finite-dimensional operators therefore become a nontrivial issue. We refer to Morrison (1995) for a nice presentation of results and examples in this direction.

The reachability properties of systems in Hardy spaces of analytic functions were derived by Fuhrmann (1972) and depend on Beurling's characterization of shift-invariant subspaces. We refer the reader to Fuhrmann (2012) for a discussion of SISO systems in a Hardy-space context that is close to the spirit of this book. A characterization of cyclic vectors for the backward shift is due to Douglas, Shapiro and Shields (1970). Composition operators on H^2 provide interesting examples of infinite-dimensional dynamical systems, and indeed of control systems. Littlewood's subordination principle, see, for example, Shapiro (1993), asserts that every composition operator $C_\phi(f) = f \circ \phi$ by an analytic function $\phi : \mathbb{D} \longrightarrow \mathbb{D}$, with $\phi(0) = 0$, takes the Hardy space H^2 into itself. This implies a generalization of the situation studied in Section 11.3 to infinite homogeneous networks, i.e., that the network transfer function $\mathcal{N}_g(z) = \mathcal{N}(h(z))$ of a homogeneous network with interconnection transfer function $\mathcal{N} \in H^2$ is always in H^2 provided $h(z) = 1/g(z)$ is an inner function with $h(0) = 0$. A nontrivial control system on spaces of univalent analytic functions is defined by Löwner's equation on the unit disc $\mathbb{D}$

$$\frac{\partial w(t,z)}{\partial t} = -\frac{e^{\sqrt{-1}u(t)} + w(t,z)}{e^{\sqrt{-1}u(t)} - w(t,z)}w(t,z), \quad w(0,z) = z.$$

Here the complex parameter z varies in the open unit disc. A generalization of this system on suitable matrix balls and its reachability properties would be interesting to study. For a study of the cyclicity and hyper cyclicity of composition operators, and further connections to universality in a function-theoretic context, we refer the reader to Grosse-Erdmann (1999). It seems that the connection to questions of reachability and observability has been overlooked within this circle of ideas.

A generalization of the Liouville equation (12.35) that has been frequently studied is the Fokker–Planck equation

$$\frac{\partial}{\partial t}\rho(x,t) = \frac{1}{2}\sum_{i,j=1}^{n}\frac{\partial^2}{\partial x_i \partial x_j}\left(a_{ij}(x,t,u)\rho(x,t)\right) - \sum_{i=1}^{n}\frac{\partial}{\partial x_i}\left(b_i(x,t,u)\rho(x,t)\right).$$

It is well known that the Fokker–Planck equation describes the evolution of probability density functions that are propagated by stochastic differential equations; see, for example, Hazewinkel and Willems (1981) for a collection of articles on the subject. Thus, similar to the Liouville equation, the Fokker–Planck equation is a natural object of study in ensemble control. We refer the reader to Jordan, Kinderlehrer and Otto (1998) for a demonstration of the connection with steepest descent flows on spaces of probability measures with respect to the Wasserstein metric and to Blaquiere (1992) and Poretta (2014) for reachability results on the Fokker–Planck equation.

References

W.A. Adkins, "The pointwise local-global principle for solutions of generic linear equations", *Linear Algebra Appl.*, vol. 66, (1985) 29–43.

H. Aling and J.M. Schumacher, "A nine-fold canonical decomposition for linear systems", *Int. J. Control*, vol. 39(4), (1984), 779–805.

B.D.O. Anderson and E.I. Jury, "Generalized Bezoutian and Sylvester matrices in multivariable linear control", *IEEE Trans. Autom. Control*, vol. AC-21, (1976), 551–556.

B.D.O. Anderson, Z. Lin, and M. Deghat, "Combining distance-based formation shape control with formation translation", in *New Trends in Control and Networked Dynamical Systems*, L. Qiu, J. Chen, T. Iwasaki, and H. Fujioka, Eds., pages 121–130, IET, 2012.

B.D.O. Anderson, C. Yu, S. Dasgupta and A.S. Morse, "Control of a three-coleader formation in the plane", *Syst. Control Lett.*, vol. 56, (2007), 573–578.

P.J. Antsaklis and A.N. Michel, *Linear Systems*, Corr. 2nd printing, Birkhäuser Publ., 2005.

L. Asimov and B. Roth, "The rigidity of graphs", *Trans. Am. Math. Soc.*, vol. 245, (1978), 279–289.

M.F. Atiyah and I.G. Macdonald, *Introduction to Commutative Algebra*, Addison-Wesley, Reading, MA, 1969.

B. Bamieh, F. Paganini, M. Dahleh, "Distributed control of spatially invariant systems", *IEEE Trans. Autom. Control*, vol. 47 (7), (2002), 1091–1107.

S. Barnett, "A note on the Bezoutian matrix", *SIAM J. Appl. Math.*, vol. 22, (1972), 84–86.

H. Bart, I. Gohberg and M.A. Kaashoek, *Minimal Factorization of Matrix and Operator Functions*, Birkhäuser, 1979.

G. Basile and G. Marro, *Controlled and Conditioned Invariants in Linear System Theory*, Prentice Hall, Englewood Cliffs, NJ, 1992.

K. Beauchard, J.-M. Coron and P. Rouchon, "Controllability issues for continuous-spectrum systems and ensemble controllability of Bloch equations", *Commun. Math. Phys.*, vol. 296, (2010), 525–557.

D.P. Bertsekas and J.N. Tsitsiklis, *Parallel and Distributed Computation: Numerical Methods*, Belmont, MA, Athena Scientific, 1989.

A. Beurling, "On two problems concerning linear transformations in Hilbert space", *Acta Math.*, vol. 81, (1949), 239–255.

G. Birkhoff, "Extensions of Jentzsch's Theorem", *Trans. Am. Math. Soc.*, vol. 85, (1957), 219–227.

M. Bisiacco, M.E. Valcher and J.C. Willems, "A behavioral approach to estimation and dead-beat observer design with applications to state-space models", *IEEE Trans. Autom. Control,* , vol. 51, (2006), 1787–1797.

R.R. Bitmead and B.D.O. Anderson, "The matrix Cauchy index: Properties and applications", *SIAM J. Control Optim.*, vol. 33(4), (1977), 655–672.

S. Bittanti and P. Colaneri, *Periodic Systems: Filtering and Control*, Communications and Control Engineering Series, Springer-Verlag, London, 2010.

A. Blaquiere, "Controllability of a Fokker–Planck equation, the Schrödinger system, and a related stochastic optimal control", *Dynam. Control*, vol. 2, (1992), 235–253.

V. D. Blondel, *Simultaneous Stabilization of Linear Systems*, Lecture Notes in Control and Inf. Sciences, Springer, Heidelberg, 1994.

V. D. Blondel, J.M. Hendrickx and J.N. Tsisiklis, "On Krause's multi-agent consensus model state-dependent connectivity", *IEEE Trans. Autom. Control*, vol. 54 (11), (2009), 2586–2597.

V. D. Blondel, J.M. Hendrickx and J.N. Tsisiklis, "Continuous-time average-preserving opinion dynamics with opinion-dependent communications", *SIAM J. Control Optim.*, vol. 48 (8), (2010), 5214–5240.

L.M. Blumenthal, *Theory and Applications of Distance Geometry*, Cambridge University Press, Cambridge, UK, 1953.

P.S. Bourdon and J.L. Shapiro, "Cyclic composition operators on H^2", *Proceedings of Symposia in Pure Mathematics* vol. 51, Am. Math. Soc., (1990), 43–53.

R. Bott and R.J. Duffin, "Impedance synthesis without the use of transformers", *J. Applied Physics*, vol. 20, (1940), 816.

S. Boyd, P. Diaconis and L. Xiao, "Fastest mixing Markov chain on a graph", *SIAM Review*, vol. 46 (4), (2004), 667–689.

R.W. Brockett, "The geometry of the set of controllable linear systems", *Research Reports of Autom. Control Lab., Faculty of Engineering, Nagoya University*, vol. 24, (1977), 1–7.

R.W. Brockett, "Linear feedback systems and the groups of Galois and Lie", *Linear Algebra Appl.*, vol. 50, (1983), 45–60.

R.W. Brockett, "Minimum attention control", *Proc. IEEE Conference on Decision and Control* (1997), 2628–2632.

R.W. Brockett, "Pattern generation and the control of nonlinear systems", *IEEE Trans. Autom. Control*, vol. 48(10), (2003), 1699–1712.

R.W. Brockett, "Control of stochastic ensembles", *Proc. of the 17th IFAC World Congress*, Seoul, 2008.

R.W. Brockett, "On the control of a flock by a leader", *Proc. Steklov Institute of Mathematics*, vol. 268 (1), (2010), 49–57.

R.W. Brockett, "Notes on the control of the Liouville equation", *Control of Partial Differential Equations*, Eds. F. Alabau-Boussouira et. al., Lecture Notes in Mathematics, vol. 2048, 2012.

R.W. Brockett, "Synchronization without periodicity", *in Mathematical Systems Theory, A Volume in Honor of U. Helmke*, (K. Hüper and J. Trumpf; Eds.), CreateSpace, 65–74, 2013.

R.W. Brockett and C.I. Byrnes, "Multivariable Nyquist criteria, root loci and pole placement: A geometric approach", *IEEE Trans. Autom. Control*, vol. 26, (1981), 271–284.

R.W. Brockett and J.L. Willems, "Discretized partial differential equations: examples of control systems defined on modules", *Automatica*, vol. 10, (1974), 507–515.

O. Brune, "Synthesis of a finite two-terminal network whose driving-point impedance is a prescribed function of frequency," *J. Math. Phys.*, vol. 10, (1931), 191–236.

P. Brunovsky, "A classification of linear controllable systems", *Kybernetika*, vol. 6.3, (1970), 173–188.

F. Bullo, J. Cortés and S. Martínez, *Distributed Control of Robotic Networks*, Princeton University Press, Applied Mathematics Series, 2009.

P.J. Bushell, "The Cayley–Hilbert metric and positive operators", *Linear Algebra Appl.*, vol. 84, (1986), 271–280.

C.I. Byrnes, "Pole assignment by output feedback", *Three Decades of Mathematical Systems Theory*, H. Nijmeijer and J.M. Schumacher, Eds., Springer, Heidelberg, 31–87, 1989.

C.I. Byrnes and T. Duncan, "On certain topological invariants arising in system theory", in *New Directions in Applied Mathematics*, P. Hilton and G. Young, Eds., Springer, Heidelberg, 29–71, 1981.

F.M. Callier, and C.D. Nahum, "Necessary and sufficient conditions for the complete controllability and observability of systems in series using the coprime decomposition of a rational matrix", *IEEE Trans. Circuits Syst.*, vol. 22, (1975), 90–95.

M. Cao, A.S. Morse and B.D.O. Anderson, "Reaching a consensus in a dynamically changing environment: A graphical approach", *SIAM J. Control Optim.*, vol. 47, (2008), 575–600.

R. Connelly, "Rigidity", Ch. 1.7 of *Handbook of Convex Geometry, Vol. A*, (Ed. P. M. Gruber and J. M. Wills), Amsterdam, Netherlands: North-Holland, pp. 223–271, 1993.

G. Conte, A.M. Perdon, "Systems over rings: geometric theory and applications", *Annual Reviews in Control*, vol. 24, (2000), 113–124.

J.-M. Coron, *Control and Nonlinearity*, Mathematical Surveys and Monographs, Vol. 136, American Mathematical Society, 2007.

G.M. Crippen and T.F. Havel, *Distance Geometry and Molecular Conformation*, J. Wiley and Sons, New York, 1988.

F. Cucker and S. Smale, "Emergent behavior in flocks", *IEEE Trans. Autom. Control*, vol. 52, (2007), 852–862.

F. Cucker and S. Smale, "On the mathematics of emergence", *Japanese J. Math.*, vol. 52, (2007), 197–227.

R.F. Curtain, H.J. Zwart, *An Introduction to Infinite-Dimensional Linear Systems Theory*, Texts in Applied Mathematics Vol. 21, Springer, New York, 1995.

R.F. Curtain, O.V. Iftime and H.J. Zwart, "System theoretic properties of a class of spatially invariant systems", *Automatica*, vol. 45, (2009), 1619–1627.

D. Cvetkovic, P. Rowlinson and S. Simic, *An Introduction to the Theory of Graph Spectra*, London Math. Soc. Student Texts 75, Cambridge University Press, Cambridge, UK, 2010.

P. J. Davis, *Circulant Matrices*, Wiley, New York, 1979.

L. de Branges, J. Rovnyak, *Square Summable Power Series*, Holt, Rinehart and Winston, New York, 1966.

R. Dedekind and H. Weber, "Theorie der algebraischen Funktionen einer Veränderlichen", *Crelle J. reine und angewandte Mathematik*, vol. 92, (1882), 181–290.

J.-M. Dion, C. Commault and J. van der Woude, "Generic properties and control of linear structured systems: A survey", *Automatica*, vol. 39, (2003), 1125–1144.

F. Doerfler and F. Bullo, "Kron reduction of graphs with applications to electrical networks", *IEEE Trans. Circuits Syst. I, Reg. Papers*, vol 60 (1), (2013), 150–163.

F. Doerfler and F. Bullo, "Synchronization in complex networks of phase oscillators: a survey", *Automatica*, vol. 50 (6), (2014), 1539–1564.

F. Doerfler and B. Francis, "Geometric analysis of the formation problem for autonomous robots", *IEEE Trans. Autom. Control*, vol. 55 (2010), 2379–2384.

R.G. Douglas, H.S. Shapiro and A.L. Shields, "Cyclic vectors and invariant subspaces for the backward shift operator", *Annales de l' institut Fourier*, vol. 20, (1970), 37–76.

A.W.M. Dress and T.F. Havel, "Distance geometry and geometric algebra", *Foundations of Physics*, vol. 23(10), (1993), 1357–1374.

E. Emre and M.L.J. Hautus, "A polynomial characterization of (A,B)-invariant and reachability subspaces", *SIAM J. Control Optim.*, vol. 18, (1980), 420–436.

P.L. Falb and W.A. Wolovich, "Decoupling in the design and synthesis of multivariable control systems", *IEEE Trans. Autom. Control*, vol. 12(6), (1967), 651–659.

A. Feintuch and B. Francis, "Infinite chains of kinematic points", *Automatica*, vol. 48(5), (2012), 901–908.

T.L. Fernando, L.S. Jennings and H.M. Trinh, "Generality of functional observer structures", *50th IEEE Conference on Decision and Control and European Control Conference (CDC-ECC)*, (2011), 4000–4004.

T.L. Fernando, H.M. Trinh, Hieu and L.S. Jennings, "Functional observability and the design of minimum order linear functional observers", *IEEE Trans. Autom. Control*, vol. 55, (2010), 1268–1273.

M. Fiedler, *Special Matrices and Their Applications in Numerical Mathematics*, Dover Publ. Inc., Mineola, NY, 2008.

M. Fliess, "Séries reconnaissables, rationelles et algébriques", *Bull. Sci. Math.*, vol. 94, (1970), 231–239.

M. Fliess, J. Levine, Ph. Martin and P. Rouchon, "Flatness and defect of nonlinear systems: introductory theory and applications", *Int. J. Control*, vol. 61, (1995), 1327–1361.

G.D. Forney, "Minimal bases of rational vector spaces, with applications to multivariable systems", *SIAM J. Optim. Control*, vol. 13, (1975), 493–520.

R.M. Foster, "A reactance theorem", *Bell Systems Technical J.*, vol. 3, (1924), 259–267.

B.A. Francis, "The linear multivariable regulator problem", *SIAM J. Control Optim.*, vol. 15(3), (1977), 486–505.

B.A. Francis and W.M. Wonham, "The internal model principle of control theory.", *Automatica*, vol. 12.5, (1976), 457–465.

P.A. Fuhrmann, "On the corona problem and its application to spectral problems in Hilbert space", *Trans. Am. Math. Soc.*, vol. 132, (1968), 55–68.

P.A. Fuhrmann, "A functional calculus in Hilbert space based on operator valued analytic functions", *Israel J. Math.*, vol. 6, (1968), 267–278.

P.A. Fuhrmann, "On weak and strong reachability and controllability of infinite-dimensional linear systems", *J. Optim. Theory Appl.*, vol. 9 (2), (1972), 77–89.

P.A. Fuhrmann, "On controllability and observability of systems connected in parallel", *IEEE Trans. Circuits Syst.*, vol. 22 (1975), 57.

P.A. Fuhrmann, "On series and parallel coupling of a class of discrete time infinite-dimensional systems", *SIAM J. Control Optim.*, vol. 14 (1976), 339–358.

P.A. Fuhrmann, "Algebraic system theory: An analyst's point of view", *J. Franklin Inst.*. 301, (1976), 521–540.

P.A. Fuhrmann, "On Hankel Operator ranges, meromorphic pseudocontinuations and factorizations of operator valued analytic functions", *J. London Math. Soc.*, Series II, 13 (1976), 323–327.

P.A. Fuhrmann, "On strict system equivalence and similarity", *Int. J. Control*, 25 (1977), 5–10.

P.A. Fuhrmann, "Linear feedback via polynomial models", *Int. J. Control* 30, (1979), 363–377.

P.A. Fuhrmann, "Duality in polynomial models with some applications to geometric control theory", *IEEE Trans. Autom. Control*, 26, (1981), 284–295.

P.A. Fuhrmann, "Polynomial models and algebraic stability criteria", *Feedback Control of Linear and Nonlinear Systems*, D. Hinrichsen and A. Isidori, Eds., Lecture Notes in Control and Inf. Sci., vol. 39, Springer, 78–90, 1982.

P.A. Fuhrmann, "On symmetric rational matrix functions", *Linear Algebra Appl.*, Special issue on Linear Control Theory, (1983),167–250.

P.A. Fuhrmann, "A duality theory for robust control and model reduction", *Linear Algebra Appl.*, vols. 203–204, 471–578.

P.A. Fuhrmann, "A study of behaviors", *Linear Algebra Appl.*, vol. 351–352, (2002), 303–380.

P.A. Fuhrmann, "Autonomous subbehaviors and output nulling subspaces", *Int. J. Control*, vol. 78, 1378–1411.

P.A. Fuhrmann, "On duality in some problems of geometric control", *Acta Applicandae Mathematicae*, vol. 91, (2006), 207–251.

P.A. Fuhrmann, "Observer theory", *Linear Algebra Appl.*, vol. 428, (2008), 44–136.

P.A. Fuhrmann, "A functional approach to the Stein equation", *Linear Algebra Appl.*, vol. 432, (2010), 3031–3071.

P.A. Fuhrmann, "On tangential matrix interpolation", *Linear Algebra Appl.*, vol. 433, (2010), 2018–2059.

P.A. Fuhrmann and U. Helmke, "Tensored polynomial models", *Linear Algebra Appl.*, vol 432, (2010), 678–721.

P.A. Fuhrmann, *A Polynomial Approach to Linear Algebra*, 2nd Edition, Springer, New York, 2012.

P.A. Fuhrmann and B.N. Datta, "On Bezoutians, Van der Monde matrices and the Lienard-Chipart stability criterion", *Linear Algebra Appl.*, 120, (1989), 23–37.

P.A. Fuhrmann and U. Helmke, "Parametrization of conditioned invariant subspaces", *Advances in Mathematical Systems Theory, A Volume in Honor of D. Hinrichsen*, (F. Colonius, U. Helmke, D. Prätzel-Wolters, and F. Wirth, Eds.), Birkhäuser, Boston, 103–134, 2001.

P.A. Fuhrmann and U. Helmke, "On the parametrization of conditioned invariant subspaces and observer theory", *Linear Algebra Appl.*, vol. 332–334, (2001), 265–353.

P.A. Fuhrmann and U. Helmke, "Strict equivalence, controllability and observability of networks of linear systems", *Math. Contr. Signals and Systems*, vol. 25, (2013), 437–471.

P.A. Fuhrmann and R. Ober, "On coprime factorizations", *T. Ando volume*, 1993, Birkhauser.

P.A. Fuhrmann, Z. Priel, H.J. Sussmann and A.C. Tsoi, "On self synchronization in ciliary beating", unpublished manuscript (1987).

P.A. Fuhrmann and J. Trumpf, "On observability subspaces", *Int. J. Control*, vol. 79, (2006), 1157–1195.

P.A. Fuhrmann and J. C. Willems, "Factorization indices at infinity for rational matrix functions", *Integral Equations and Operator Theory*, vol. 2/3, (1979), 287–300.

P.A. Fuhrmann and J.C. Willems, "A study of (A,B)-invariant subspaces via polynomial models", *Int. J. Control*, vol. 31, (1980), 467–494.

W. Fulton, "Eigenvalues, invariant factors, highest weights, and Schubert calculus", *Bull. Am. Math. Soc.*, vol. 37, (2000), 209–249.

D. Gaier, *Lectures on Complex Approximation*, Birkhäuser, Boston, 1987.

F.R. Gantmacher, *The Theory of Matrices*, 2 vols., Chelsea, 1959.

B.K. Ghosh and C.I. Byrnes, "Simultaneous stabilization and pole-placement by nonswitching dynamic compensation", *IEEE Trans. Autom. Control*, vol. 28, (1983), 735–741.

B.K. Ghosh, "Some new results on the simultaneous stabilizability of a family of single input, single output systems", *Syst. Control Lett.*, vol. 6, (1985), 39–45.

E.G. Gilbert, "Controllability and observability in multivariable control systems", *SIAM J. Control Optim.*, vol. 2, (1963), 128–151.

C.D. Godsil and G. Royle, *Algebraic Graph Theory*, Springer, New York, 2001.

I.C. Gohberg and I.A. Feldman, "Convolution Equations and Projection Methods for their Solution", English translation, *Am. Math. Soc.*, Providence, RI, 1974.

I.C. Gohberg and M.G. Krein, "Systems of integral equations on a half line with kernels depending on the difference of arguments", English translation, *AMS Translations*, vol. 2, (1960), 217–287.

I.C. Gohberg, L. Lerer and L. Rodman , "Factorization indices for matrix polynomials", *Bull. Am. Math. Soc.*, vol. 84, (1978), 275–277.

E. Gorla and J. Rosenthal, "Pole placement with fields of positive characteristic", *Three Decades of Progress in Control Sciences; X. Hu, U. Jonsson, B. Wahlberg and B. Ghosh, Eds.*, Springer, 215–231, 2010.

J.C. Gower, "Properties of Euclidean and non-Euclidean distance matrices", *Linear Algebra Appl.*, vol. 67, (1985), 81–97.

W.L. Green and E.W. Kamen, "Stabilizability of linear systems over a commutative normed linear algebra with applications to spatially-distributed and parameter-dependent systems", *SIAM J. Control Optim.*, vol. 23, (1985), 1–18.

U. Grenander and G. Szegö, *Toeplitz Forms and Their Applications*, University of California Press, Berkeley and Los Angeles, 1958.

K.-G. Grosse-Erdmann, "Universal families and hypercyclic operators", *Bull. of the Am. Math. Soc.*, vol. 36(3), (1999), 345–381.

S. Hara, T. Hayakawa and H. Sugata, "LTI systems with generalized frequency variables: A unified framework for homogeneous multi-agent dynamical systems", *SICE J. of Contr. Meas. and System Integration*, vol. 2, (2009), 299–306.

M.L.J. Hautus, "Controllability and observability conditions of linear autonomous systems", *Ned. Akad. Wetenshappen, Proc. Ser. A*, vol. 72, (1969), 443–448.

M.L.J. Hautus, "(A,B)-invariant and stabilizability subspaces, a frequency domain description", *Automatica*, vol. 16, (1980), 703–707.

M.L.J. Hautus and P.A. Fuhrmann, "On the zero module of rational matrix functions", *Proceedings CDC Conference, Albuquerque, 1980*.

M.L.J. Hautus and M. Heymann, "Linear feedback—an algebraic approach", *SIAM J. Control Optim.*, vol. 16, (1978), 83–105.

T.F. Havel and K. Wüthrich, "An evaluation of the combined use of nuclear magnetic resonance and distance geometry for the determination of protein conformations in solution", *J. Mol. Biol.*, vol. 182, (1985), 281–294.

M. Hazewinkel, "A partial survey of the uses of algebraic geometry in systems and control theory", *Istituto Nazionale di Alta Matematica Francesco Severi Symposia Methematica* , vol. XXIV, (1981), 245–292.

M. Hazewinkel and C.F. Martin, "Symmetric linear systems: an application of algebraic systems theory", *Int. J. of Control* , vol. 37 (1983), 1371–1384.

M. Hazewinkel and C.F. Martin, "Representations of the symmetric group, the specialization order, systems and Grassmann manifolds", *L'Enseignement Mathematique*, vol. 29, (1983), 53–87.

M. Hazewinkel and A.M. Perdon, "On families of systems: pointwise-local-global isomorphism problems", *Int. J. Control*, vol. 33, (1981), 713–726.

M. Hazewinkel and J.C. Willems, *Stochastic Systems: The Mathematics of Filtering and Identification and Applications*, Nato Advanced Study Institutes Series C78, D. Reidel Publ. Company, Dordrecht, 1981.

R. Hegselmann and U. Krause, "Opinion dynamics and bounded confidence models, analysis and simulations", *J. Artif. Societies Social Simul.*, vol. 5 (3), (2002).

G. Heinig and K. Rost, *Algebraic Methods for Toeplitz-like Matrices and Operators*, Akademie, Berlin, 1984.

U. Helmke, "The topology of a moduli space for linear dynamical systems", *Comment. Math. Helv.*, vol. 60, (1985), 630–655.

U. Helmke, "Rational functions and Bezout forms: a functorial correspondence", *Linear Algebra Appl.*, vol. 122–124, (1989), 623–640.

U. Helmke and P. A. Fuhrmann, "Bezoutians", *Linear Algebra Appl.*, vols. 122–124, (1989), 1039–1097 .

U. Helmke and P. A. Fuhrmann, "Tangent spaces of rational matrix functions", *Linear Algebra Appl.*, vol. 271, (1998), 1–40.

U. Helmke and J.B. Moore, *Optimization and Dynamical Systems*, Springer-Verlag, London, 1994.

U. Helmke, D. Prätzel-Wolters and S. Schmidt, "Adaptive synchronization of interconnected linear systems", *IMA J. Mathematical Control and Information*, vol. 8, (1991), 397–408.

U. Helmke and J. Rosenthal, "Eigenvalue inequalities and the Schubert calculus", *Math. Nachrichten*, vol. 171, (1995), 207–225.

U. Helmke and M. Schönlein, "Uniform ensemble controllability for one-parameter families of time-invariant linear systems", *Syst. Control Lett.*, vol. 71, (2014), 69–77.

H. Helson, *Lectures on Invariant Subspaces*, Academic Press, New York, 1964.

L. Henneberg, *Die graphische Statik der starren Systeme*, B.G. Teubner, Leipzig, 1911.

C. Hermite, "Sur le nombre des racines d'une equation algebrique comprise entre des limites donnes," *J. Reine Angew. Math.*, vol. 52, (1856), 39–51.

M. Heymann, *Structure and realization problems in the theory of dynamical systems*, CISM Courses and Lectures, vol. 204, Springer, 1975.

P. Hilton and Y. Wu, *A Course in Modern Algebra*, Wiley-Interscience, New York, 1974.

D. Hinrichsen and D. Prätzel-Wolters, "Solution modules and system equivalence", *Int. J. Control*, vol. 32, (1980), 777–802.

D. Hinrichsen and D. Prätzel-Wolters, "Generalized Hermite matrices and complete invariants for strict system equivalence", *SIAM J. Control and Optim.*, vol. 21, (1983), 289–305.

R.A. Horn and C.R. Johnson, *Matrix Analysis*, Cambridge University Press, Cambridge, UK, 1990.

R.A. Horn, N.H. Rhee and W. So, "Eigenvalue inequalities and equalities", *Linear Algebra Appl.*, vol. 270, (1998), 29–44.

Y. Hori, T.-H. Kim and S. Hara, "Existence criteria of periodic solutions in cyclic gene regulatory networks", *Automatica*, vol. 47(6), (2011), 1203–1209.

T.H. Hughes and M.C. Smith, "Algebraic criteria for circuit realizations", in *Mathematical Systems Theory, A Volume in Honor of U. Helmke*, (K. Hüper and J. Trumpf; Eds.), CreateSpace, 211–228, 2013.

T.W. Hungerford, *Algebra*, Graduate Texts in Mathematics Vol. 73, Springer, New York, 1974.

B. Huppert, *Angewandte Linear Algebra*, DeGruyter, Berlin, 1990.

A. Hurwitz, "Über die Bedingungen, unter welchen eine Gleichung nur Wurzeln mit negativen reellen Teilen besitzt", *Math. Annal.*, vol. 46, (1895), 273–284.

A. Ilchmann, "Decentralized tracking of interconnected systems", *in Mathematical Systems Theory, A Volume in Honor of U. Helmke*, (K. Hüper and J. Trumpf; Eds.), CreateSpace, 229–245, 2013.

B. Jacob, J.R. Partington, "On controllability of diagonal systems with one-dimensional input space", *Syst. Control Lett.*, vol. 55, (2006), 321–328.

A. Jadbabaie, J. Lin and A.S. Morse, "Coordination of groups of mobile autonomous agents using nearest neighbor rules", *IEEE Trans. Autom. Control*, vol. 48, (2003), 998–1001.

R. Jentzsch, "Über Integraloperatoren mit positivem Kern", *Crelle J. Math.*, vol. 141, (1912), 235–244.

J.Z. Jiang, and M.C. Smith, "Regular positive-real functions and five-element network synthesis for electrical and mechanical networks", *IEEE Trans. Autom. Control*, vol. 56(6), (2011), 1275–1290.

M.A. Jodeit, "Uniqueness in the division algorithm", *Am. Math. Monthly*, vol. 74, (1967), 835–836.

R. Jordan, D. Kinderlehrer and F. Otto, "The variational formulation of the Fokker–Planck equation", *SIAM J. Matrix Anal.*, vol. 29(1), (1998), 1–17.

T. Kailath, *Linear Systems*, Prentice-Hall, Englewood Cliffs, NJ, 1980.

R.E. Kalman, *Lectures on Controllability and Observability*, CIME Lecture Notes, Bologna, Italy, 1968.

R.E. Kalman, "Algebraic characterization of polynomials whose zeros lie in algebraic domains", *Proc. Nat. Acad. Sci.* vol. 64, (1969), 818–823.

R.E. Kalman, "New algebraic methods in stability theory", *Proceeding V. International Conference on Nonlinear Oscillations, Kiev*.

R.E. Kalman, "Old and new directions of research in system theory", *Perspectives in Mathematical System Theory, Control and Signal Processing*, Eds. J.C. Willems, S. Hara, Y. Ohta and H. Fujioka, Springer, vol. 398, 3–13, 2010.

R.E. Kalman, P. Falb and M. Arbib, *Topics in Mathematical System Theory*, McGraw-Hill, New York, 1969.

P.P. Khargonekar and E. Emre, "Further results on polynomial characterizations of (F, G)-invariant and reachability subspaces", *IEEE Trans. Autom. Control*, vol. 27 (10), (1982), 352–366.

P.P. Khargonekar, T.T. Georgiou and A.B. Özgüler, "Skew-prime polynomial matrices: the polynomial model approach", *Linear Algebra Appl.*, vol. 50, (1983), 403–435.

P.P. Khargonekar, A. Tannenbaum, "Non-Euclidean metrics and the robust stabilization of systems with parameter uncertainty", *IEEE Trans. Autom. Control*, vol. 30 (10), (1985), 1005–1013.

V.L. Kharitonov, "Asymptotic stability of an equilibrium position of a family of systems of differential equations", *Differencial'nye Uravnenija*, vol. 14, (1978), 2086–2088.

E. Kohlberg and J.W. Pratt, "The contraction mapping approach to the Perron–Frobenius theory: Why Hilbert's metric?", *Mathematics for Operations Research*, vol. 7 (2), (1982), 198–210.

S. Koshita, M. Abe, and M. Kawamata, "Analysis of second-order modes of linear discrete-time systems under bounded-real transformations", *IEICE Trans. Fundamentals*, vol. 90, (2007), 2510–2515.

I. Kra and S.R. Simanca, "On circulant matrices", *Notices of the Am. Math. Society*, vol. 59, (2012), 368–377.

U. Krause, "Soziale Dynamiken mit vielen Interakteuren. Eine Problemskizze", *Proc. Modellierung Simul. von Dynamiken mit vielen interagierenden Akteuren*, (1997), 37–51.

M.G. Krein and M.A. Naimark, "The method of symmetric and Hermitian forms in the theory of the separation of the roots of algebraic equations," English translation in *Linear and Multilinear Algebra*, vol. 10, (1981), 265–308.

M.G. Krein and M.A. Rutman, "Linear operators leaving invariant a cone in a Banach space" *Transl. Am. Math. Soc. Ser. 1*, vol. 10, (1950), 199–325.

M.A. Krasnoselskii, *Positive Solutions of Operator Equations*, Noordhoff, Groningen, 1964.

L. Kronecker, "Algebraische Reduktion der Scharen bilinearer Formen", *Sitzungsberichte Akad. Berlin*, (1890), 763–776.

V. Kucera, "A method to teach the parametrization of all stabilizing controllers", *Proc. 18th IFAC World Congress*, Milano, (2011), 6355–6360.

E. Kunz, *Einführung in die kommutative Algebra und algebraische Geometrie* , F. Vieweg & Sohn, Braunschweig, 1980.

E.L. Ladenheim, "A synthesis of biquadratic impedances", M.S. thesis, Polytechnic Institute of Brooklyn, 1948.

G. Laman, "On graphs and the rigidity of plane skeletal structures", *J. Engineering Mathematics*, vol. 4, (1970), 331–340.

S. Lang, *Algebra*, Addison-Wesley, Reading, MA, 1965.

P.D. Lax, "Translation invariant subspaces" *Acta Math.*, vol. 101, (1959), 163–178

P.D. Lax and R.S. Phillips (1967), *Scattering Theory*, Academic, New York, 1967.

N.E. Leonard and P.S. Krishnaprasad, "Control of switched electrical networks using averaging on Lie groups", *Proc. 33rd IEEE Conf. Decision and Control*, (1994), 1919–1924.

L. Lerer and M. Tismenetsky, "The Bezoutian and the eigenvalue-separation problem for matrix polynomials", *Int. Eq. Oper. Th.*, vol. 5, (1982), 386–445.

J. Levine and D.V. Nguyen "Flat output characterization for linear systems using polynomial matrices", *Syst. Control Lett.*, vol. 48, (2003), 69–75.

J.-S. Li, "Ensemble control of finite-dimensional time-varying linear systems", *IEEE Trans. Autom. Control*, vol. 56 (2), (2011), 345–357.

J.-S. Li, N. Khaneja, "Control of inhomogeneous quantum ensembles", *Phys. Rev. A*, vol. 73, (2006), pp.030302.

K.-Y. Lian, L.-S. Wang, and L.-C. Fu, "Controllability of spacecraft systems in a central gravitational field", *IEEE Trans. Autom. Control*, vol. 39(12), (1994), 2426–2441.

A.M. Liapunov, "Probleme general de la stabilite de mouvement", *Ann. Faculty Sci. Toulouse*, vol. 9, (1907), 203–474. (French translation of the Russian paper published in Comm. Soc. Math. Kharkow).

C.-T. Lin, "Structural controllability", *IEEE Trans. Autom. Control*, vol. 19 (3), (1974), 201–208.

D. Lind and B. Marcus, *An Introduction to Symbolic Dynamics and Coding*, Cambridge University Press, Cambridge, UK, 1995.

Y.Y. Liu, J.-J. Slotine, A.L. Barabasi, "Controllability of complex networks", *Nature*, vol. 473 (2011), 167–173.

M.S. Livsic, *Operators, Oscillations, Waves (Open Systems)*, Vol. 34. American Mathematical Society, 1973.

J. Lunze, "Dynamics of strongly coupled symmetric composite systems", *Int. J. Control*, vol. 44, (1986), 1617–1640.

J. Lunze, "Synchronizable modes in networked systems", *J. Phys. A, Math. Theor.*, vol. 44(4), (2011),045103.

J. Lunze, "Synchronization of heterogeneous agents", *IEEE Trans. Autom. Control*, vol. 47, (2012), 2885–2890.

D. G. Luenberger, "An introduction to observers", *IEEE Trans.*, vol. 8(2), (1964), 74–80.

D. G. Luenberger, "Observing the state of a linear system", *IEEE Trans. Autom. Control*, 16(6), (1971), 596–602.

C.C. MacDuffee, *The Theory of Matrices*, Chelsea, New York, 1933.

C.-Q. Ma and J.-F. Zhang, "Necessary and sufficient conditions for consensusability of linear multi-agent systems", *IEEE Trans. Autom. Control*, vol. 55, (2010), 1263–1268.

D.S. Mackey, "Minimal indices and minimal bases via filtrations", *MIMS Eprint*, Report 2012.82, (2012).

V. Mani and R.E. Hartwig, "Generalized polynomial bases and the Bezoutian", *Linear Algebra Appl.*, vol. 251, (1997), 293–320.

S.N. Mergelyan, "Uniform approximation to functions of a complex variable" *Transl. Am. Math. Soc.*, vol. 3, (1962), 294–391; *Uspekhi Mat. Nauk*, vol. 7(2), (1952), 31–122.

J. Mather, "Solutions of generic linear equations", in *Dynamical Systems* (M. Peixoto, Ed.), Academic, New York, 1973, 185–193.

J.C. Maxwell, "On governors", *Proc. Roy. Soc. Ser. A*, 16, (1868), 270–283.

M. Meshbahi and M. Egerstedt, *Graph Theoretic Methods for Multiagent Networks*, Princeton University Press, Princeton, NJ, 2010.

K.E. Morrison, "Spectral approximation of multiplication operators", *New York J. Math.*, vol. 1, (1995), 75–96.

A.S. Morse, "Structural invariants of multivariable linear systems", *SIAM J. Control Optim.*, vol. 11, (1973), 446–465.

A.S. Morse, "System invariants under feedback and cascade control", *Proceedings of the International Symposium, Udine, Italy, June 1975*, G. Marchesini and S.K. Mitter, eds. Springer, New York, 75–76, 1977.

A.S. Morse and W.M. Wonham, "Status of noninteracting control." *IEEE Trans. Autom. Control*, vol. 16(6), (1971): 568–581.

S. Mou and A.S. Morse, "A fixed-neighbor, distributed algorithm for solving a linear algebraic equation" *Proc. European Control Conference*, Zürich, Switzerland, (2013), 2269–2273.

P. Müller, "Primitive monodromy groups of polynomials", *Recent developments in the inverse Galois problem, M. Fried (Ed.)*, Contemp. Maths. vol. 186, Am. Math. Soc., 385–401, 1995.

C.T. Mullis and R.A. Roberts, "Roundoff noise in digital filters: frequency transformations and invariants", *IEEE Trans. Acoust. Speech Signal Processing*, vol. ASSP-24, (1976), 538–550.

H. F. Münzner and D. Prätzel-Wolters, "Minimal bases of polynomial modules, structural indices and Brunovsky transformations", *Int. J. Control*, vol. 30, (1979), 291–318.

A. Nedic, A. Ozdaglar and P.A. Parrilo, "Consensus and optimization in multi-agent systems," *IEEE Trans. Autom. Control*, vol. 55(4), (2010), 922–938.

M. Newman, "Matrix completion theorems", *Proc. Am. Math. Soc.*, vol. 94, (1985), 39–45.

N.K. Nikolski, *Treatise on the Shift Operator*, Springer, 1986.

G. Notarstefano and G. Parlangeli,"Controllability and observability of grid graphs via reduction and symmetries", *IEEE Trans. Autom. Control*, vol. 58 (2013), 1719–1731.

U. Oberst, "Multidimensional constant linear systems", *Acta Applicandae Mathematicae*, vol. 20, (1990), 1–175.

R. Olfati-Saber, J.A. Fax and R.M. Murray, "Consensus problems in networks of agents with switching topologies and time-delays", *IEEE Trans. Autom. Control*, vol. 49, (2004), 1520–1533.

R. Olfati-Saber, J.A. Fax and R.M. Murray, "Consensus and cooperation in networked multi-agent systems", *Proc. IEEE*, vol. 95, (2007), 215–233.

A. Ostrowski and H. Schneider, "Some theorems on the inertia of general matrices", *J. Math. Anal. Appl.* 4, 72–84.

G. Ottaviani and B. Sturmfels, "Matrices with eigenvectors in a given subspace", *Proc. Am. Math. Soc.* 141(4), (2013), 1219–1232,

A.B. Özgüler, *Linear Multichannel Control*, Prentice Hall, 1994.

F. Pakovich, "Algebraic curves $P(x) - Q(y) = 0$ and functional equations", *Complex Var. Elliptic Equ.*, vol. 56, (2011), 199–213.

G. Parlangeli and G. Notarstefano, "On the reachability and observability of path and cycle graphs", *IEEE Trans. Autom. Control*, vol. 57 (2012), 743–748.

B.N. Parlett and W.G. Poole, "A geometric theory for the QR, LU and power iterations", *SIAM J. Numer. Analysis*, vol. 10, (1973), 389–412.

P.C. Parks, "A new proof of the Routh-Hurwitz stability criterion using the second method of Lyapunov", *Proc. Cambridge Philos. Soc.*, vol. 58, (1962), 694–702.

A. Pazy, *Semigroups of linear operators and applications to partial differential equations*, Springer, New York, 1983.

J.W. Polderman and J.C. Willems, *Introduction to Mathematical System Theory*, Springer, New York, 1997.

M. Pollicot and M. Yuri, *Dynamical Systems and Ergodic Theory*, London Math. Soc. Student Texts 40, Cambridge University press, Cambridge, U.K., 1998.

V.M. Popov, "Invariant description of linear time-invariant controllable systems", *SIAM J. Control Optim.*, vol. 10, (1972), 252–264.

V.M. Popov (1973). *Hyperstability of Control Systems*, Springer-Verlag, Berlin, 1973.

A. Poretta, "On the planning problem for the mean field games system", *Dyn. Games Appl.*, vol. 4, (2014), 231–256.

W. Ren, "On consensus algorithms for double-integrator dynamics", *IEEE Trans. Autom. Control*, vol. 53, (2008), 1503–1509.

J. F. Ritt, "Prime and composite polynomials", *Trans. Am. Math. Soc.*, vol. 23(1), (1922), 51–66.

J.A. Rogge and D. Aeyels, "Vehicle platoons through ring coupling", *IEEE Trans. Autom. Control*, vol. 53 (6), (2008), 1370–1377.

H.H. Rosenbrock, *State-Space and Multivariable Theory*, Wiley, New York, 1970.

H.H. Rosenbrock and A.C. Pugh, "Contributions to a hierarchical theory of systems", *Int. J. Control*, vol. 19, (1974), 845–867.

G.C. Rota, "On models for linear operators" *Comm. Pure Appl. Math.*, vol. 13 (1960), 469–472.

P. Rouchon, "Flatness based control of oscillators", *ZAMM J. Appl. Math. Mechan.*, vol. 85 (2005), 411–421.

E.J. Routh, *A Treatise on the Stability of a Given State of Motion*, Macmillan, London, 1877.

E.B. Saff and V. Totik, "Behavior of polynomials of best uniform approximation", *Trans. Am. Math. Soc.*, vol. 316(2), (1989), 567–593.

L.A. Sakhnovich, "On the factorization of an operator-valued transfer function", *Soviet Math. Dokl.*, vol. 17, (1976), 203–207.

D. Sarason, "Generalized interpolation in H^∞", *Trans. Am. Math. Soc.*, vol. 127, 179–203.

L. Scardovi and R. Sepulchre, "Synchronization in networks of identical linear systems", *Automatica*, vol. 45(11), (2009), 2557–2562.

B. Scherlein, *Uniform Ensemble Controllability of Parameter-Dependent Linear Systems,* M.Sc. Thesis, University of Würzburg, 2014.

I.J. Schoenberg, "Remarks to Maurice Frechet's article: Sur la definition axiomatique d'une classe d'espaces vectoriels distancies applicables vectoriellement sur l'espace de Hilbert" *Ann. Math.*, vol. 36, (1935), 724–732.

I.J. Schoenberg, "The Chinese remainder problem and polynomial interpolation," *College Math. J.*, vol. 18, (1987), 320–322.

E. Seneta, *Non-negative matrices and Markov chains.* 2nd rev. ed., XVI, Softcover Springer Series in Statistics, 1981.

T. Shamir and P.A. Fuhrmann, "Minimal factorizations of rational matrix functions in terms of polynomial models", *Linear Algebra Appl.*, vol. 68, (1985), 67–91.

J. Shapiro, *Composition operators and classical function theory*, Universitext: Tracts in Mathematics. Springer-Verlag, New York, 1993.

K. Shoda, "Über die mit einer Matrix vertauschbaren Matrizen", *Math. Z.*, vol. 29, (1929), 696–712.

D.D. Siljak, *Decentralized Control of Complex Systems*, Mathematics in Science and Engineering Vol. 184, Academic Press, San Diego, 1991.

S. Smale, "Differentiable dynamical systems", *Bull. Am. Math. Soc.*, vol. 73, (1967), 747–817.

M.C. Smith, "Matrix fractions and strict system equivalence", *Int. J. Control*, vol. 34, (1981), 869–883.

M.C. Smith, "Synthesis of mechanical networks: the inerter", *IEEE Trans. Autom. Contr.* vol. 47, (2002), 1648–1662.

J. Snyders and M. Zakai, "On nonnegative solutions of the equation $AD + DA' = -C$", *SIAM J. Appl. Math.*, vol. 18(3), (1970), 704–714.

E.D. Sontag, "Linear systems over commutative rings: A survey", *Ricerche di Automatica*, vol. 7, (1976), 1–34.

E.D. Sontag, "On finitary linear systems", *Kybernetika*, vol. 15, (1979), 349–358.

E. de Souza and S.P. Bhattacharyya, "Controllability, observability and the solution of $AX - XB = C$", *Linear Algebra Appl.*, vol. 39, (1981), 167–181.

S. Sternberg, *Dynamical Systems*, Dover Publications, Mineola, NY, 2010.

S. Sundaram and C.N. Hadjicostis, "Control and estimation in finite state multi-agent systems: A finite field approach", *IEEE Trans. Autom. Control*, vol. 58, (2013), 60–73.

B. Sz.-Nagy and C. Foias, "Dilatation des commutants d'operateurs", *C.R. Acad. Sci. Paris*, A 266, (1968), 493–495.

B. Sz.-Nagy and C. Foias, *Harmonic Analysis of Operators on Hilbert Space*, North-Holland, Amsterdam, 1970.

H.G. Tanner, "On the controllability of nearest neighbor interconnections", *Proc. 43rd IEEE Conference on Decision and Control*, Atlantis, Bahamas (2004), 2467–2472.

A. Tannenbaum, "Feedback stabilization of linear dynamical plants with uncertainties in the gain factor", *Int. J. Control*, vol. 32, (1980), 1–16.

A. Tannenbaum, *Invariance and System Theory: Algebraic and Geometric Aspects*, Lecture Notes in Mathematics 845, Springer, Berlin, 1981.

H. L. Trentelmann, "On flat behaviors and observable image representations", *Syst. Control Lett.*, vol. 51, (2004), 51–55.

H.L. Trentelmann, A.A. Stoorvogel and M.L.J. Hautus, *Control Theory for Linear Systems*, Communications and Control Engineering Series. London: Springer, 2001.

H. L. Trentelmann, K. Takaba and N. Monshizadeh, "Robust synchronization of uncertain linear multi-agent systems", *IEEE Trans. Autom. Control*, vol. 58, (2013), 1511–1523.

R. Trigianni, "On the lack of exact controllability for mild solutions in Banach spaces", *J. Math. Analysis and Appl.*, vol. 50, (1975), 438–446.

J. Trumpf, *On the geometry and parametrization of almost invariant subspaces and observer theory.* Doctoral thesis, Institute of Mathematics, University of Würzburg, 2002.

J. Trumpf, "On state observers", in *Mathematical System Theory (Festschrift in Honor of U. Helmke), K. Hüper and J. Trumpf (Eds.)* , Amazon, CreateSpace, 421–435, 2013.

J. Trumpf, H.L. Trentelmann and J.C. Willems, "Internal model principles for observers", *IEEE Trans. Autom. Control* , vol. 59(7), (2014), 1737–1749.

J.N. Tsitsiklis, D.P. Bertsekas and M. Athans, "Distributed asynchronous deterministic and stochastic gradient optimization", *IEEE Trans. Autom. Control* , vol. 31(9), (1986), 803–812.

S. E. Tuna, "Synchronizing linear systems via partial-state coupling", *Automatica*, vol. 44, (2008), 2179–2184.

S. E. Tuna, "Conditions for synchronizability in arrays of coupled linear systems", *IEEE Trans. Autom. Control*, vol. 54(10), (2009), 2416–2420.

M.E. Valcher and J.C. Willems, "Observer synthesis in the behavioral framework", *IEEE Trans. Autom. Control*, vol. 44, (1999), 2297–2307.

A.I.G. Vardulakis, *Linear Multivariable Control*, Wiley, New York, 1991.

L.N. Vaserstein "An answer to a question of M. Newman on matrix completion", *Proc. Am. Math. Soc.*, vol. 97, (1986), 189–196.

M. Väth, *Volterra and Integral Equations of Vector Functions*, Marcel Dekker, New York, 2000.

T. Vicsek, A. Czirok, I. Ben Jacob, I. Cohen and O. Schochet, "Novel type of phase transitions in a system of self-driven particles", *Phys. Rev. Letters*, vol. 57, (1995), 1226–1229.

M. Vidyasagar, *Control Systems Synthesis – A Factorization Approach*, MIT Press, Cambridge, MA, second printing, 1987.

B.L. van der Waerden, *Modern Algebra*, New York, N. Y.: Frederick Ungar Publishing Co., 1949.

J.L. Walsh, *Interpolation and Approximation by Rational Functions in the Complex Domain*, American Mathematical Society, Providence, 1965.

S.H. Wang and E.J. Davison, "Canonical forms of linear multivariable systems", *SIAM J. Control Optim.*, vol. 14, (1976), 236–250.

X. Wang, "Pole placement by static output feedback", *J. Math. Systems, Estimation, and Control*, vol. 2, (1992), 205–218.

P. Wieland, R. Sepulchre and F. Allgöwer, "An internal model principle is necessary and sufficient for linear output synchronization", *Automatica*, vol. 47(5), (2011), 1068–1074.

E.P. Wigner, "The unreasonable effectiveness of mathematics in the natural sciences", *Comm. Pure Appl. Math.*, vol. 13, (1960), 1–14.

J.C. Willems, "From time series to linear systems. Part I: Finite-dimensional linear time invariant systems", *Automatica*, vol. 22, (1986), 561–580.

J.C. Willems, "Paradigms and puzzles in the theory of dynamical systems", *IEEE Trans. Autom. Control*, vol. 36, (1991), 259–294.

J.C. Willems and C. Commault, "Disturbance decoupling by measurement feedback with stability or pole placement", *SIAM J. Matrix Anal. Appl.*, vol. 19, (1981), 490–504.

J.C. Willems and P.A. Fuhrmann , "Stability theory for high order systems", *Linear Algebra Appl.*, vol. 167, (1992), 131–149.

A.R. Willms, "Analytic results for the eigenvalues of certain tridiagonal matrices", *SIAM J. Matrix Anal. Appl.*, vol. 30, (2008), 639–656.

H. Wimmer, "Inertia theorems for matrices, controllability, and linear vibrations", *Linear Algebra Appl.*, vol. 8, (1974), 337–343.

H. Wimmer, "Existenzsätze in der Theorie der Matrizen und Lineare Kontrolltheorie", *Monatsh. Math.*, vol. 78, (1974), 256–263.

W.A. Wolovich, *Linear Multivariable Systems*, Springer, New York, 1974.

W.A. Wolovich, "Skew prime polynomial matrices", *IEEE Trans. Autom. Control*, vol. 23, (1978), 880–887.

W.M. Wonham, *Linear Multivariable Control: A Geometric Approach*, Springer, 1979.

B.F. Wyman and M.K. Sain, "The zero module and essential inverse systems", *IEEE Trans. Circuits Syst.*, vol. 28, (1981), 112–126.

B.F. Wyman, M.K. Sain, G. Conte and A.M. Perdon, "On the zeros and poles of a transfer function", *Linear Algebra Appl.*, vol. 122–124, (1989), 123–144.

L. Xiao and S. Boyd, "Fast linear iterations for distributed averaging", *Syst. Control Lett.*, vol. 53, (2004), 65–78.

W.-C. Yueh, "Eigenvalues of several tridiagonal matrices", *Applied Mathematics E-Notes*, vol. 5, (2005), 66–74.

Index

© Springer International Publishing Switzerland 2015
P.A. Fuhrmann, U. Helmke, *The Mathematics of Networks*
of Linear Systems, Universitext, DOI 10.1007/978-3-319-16646-9

Printed in the United States
By Bookmasters